# BEYOND PATRONAG

# BEYOND PATRONAGE AND EXPLOITATION

*Changing Agrarian Relations in South Gujarat*

Jan Breman

DELHI
OXFORD UNIVERSITY PRESS
BOMBAY CALCUTTA MADRAS
1993

*Oxford University Press, Walton Street, Oxford OX2 6DP*
New York Toronto
Delhi Bombay Calcutta Madras Karachi
Kuala Lumpur Singapore Hong Kong Tokyo
Nairobi Dar es Salaam
Melbourne Auckland
and associates in
Berlin Ibadan

Part of this book, *Patronage and Exploitation : Changing Agrarian Relations in South Gujarat* was originally published by the Regents of the University of California Press 1974.



Oxford India Paperbacks 1993

ISBN 0 19 563087 4



Printed at Rajkamal Electric Press, Delhi 110033
and published by Neil O'Brien, Oxford University Press
YMCA Library Building, Jai Singh Road, New Delhi 110001

# Preface to the new edition

At the end of my studies at the University of Amsterdam in the summer of 1962, I left for India immediately to begin the research for which I had received a stipend from the Netherlands Organization for Scientific Research (ZWO/WOTRO). India was a late choice, since as a student of Wertheim, I had concentrated my academic interest on Southeast Asia, Indonesia in particular. In the beginning of the 1960s, however, Indonesia was out of bounds for Dutch scholars and I had to shift my basis for fieldwork to another country in Asia. Institutional contacts which already existed made India a natural choice. Dr Cora Vreede-de Stuers, Associate Professor at the University of Amsterdam had visited India on various occasions and she was well acquainted with the rather small community of social scientists at work in different parts of the country. M.N. Srinivas, at that time still with the M.S. University of Baroda, pointed out to her that Gujarat might be an attractive region for young Dutch students to get their research going. His successor, I.P. Desai, was willing to accept my affiliation to his department and to introduce me to the university library and to the field.

Both were necessary. The literature available in the Netherlands on India after Independence was extremely scanty and on arrival in Baroda, I knew next to nothing about the social fabric and dynamics of the society on which I was supposed to write a Ph.D. thesis. My immediate reaction was to indulge in reading and to continue doing so, but this was rudely disturbed by I.P. who exhorted me to go to the field.

But what was the field? In those days most certainly a village. To write another monograph on a local community, which was very much the thing to do for a young anthropologist who wanted to enter the profession, did not appeal to me much. To me the charm of a micro-study was that I could retreat from the wider social setting of India on which I then had no grip at all.

By escaping to a village, I was able to detach myself from an extremely complex setting, an arena without boundaries, in which I felt utterly lost. "The field" defined as a locality became not only more real but also seemed to be sufficiently manageable. In the good company of my friend and colleague Chris Baks, I started my novitiate as a fieldworker in a south Gujarat village hoping to improve our command over the language and, more generally, to come to terms with life in the countryside. After a month or so we went our separate ways—Chris to study the impact of tenancy abolition in an area more suitable for his subject, while I took up residence first in Chikhligam and later in Gandevigam, the fictitious names of villages situated in the two *talukas* to which they belong. We met in Baroda periodically, where we discussed our progress not only with I.P. but also with M.B. Desai, Professor of Agricultural Economics in the M.S. University. One of the few things that these eminent scholars had in common was that they were both Anavil Brahmins from south Gujarat. They were amused by the stories and gossip we brought back from the field where their roots lay and, although often in contradictory ways, coached us as only insiders can.

Coming from a working class background myself, I knew right from the beginning that my study would concentrate on the segment at the bottom of the caste hierarchy and the rural economy. On my initial travels in the central plain I could hardly have ignored the existence of a vast agrarian proletariat of tribal origin. The relationship between the dominated landless and the dominant landowners became a logical focus of my investigations.

In those years most rural studies tended to neglect agricultural labor. The important contribution made by Daniel and Alice Thorner was a major source of inspiration and so was the work of S. J. Patel. The monograph written by Dharma Kumar on land, caste and labor in south India's past, another milestone, had not yet been published. At the end of my fieldwork I felt a strong need to embed my contemporary data in a longer timespan. The same organization which had sponsored my stay in India from 1962 to 1964 enabled me in 1966 to continue my research along historical lines in the archives and other source material avaliable at the India Office Library in London.

After defending my thesis, which in the Dutch version had the title *Meester en Knechts* (Master and Servant), at Amsterdam University, in 1970, I took part in a follow-up project which started in the same year. Members of the research team who had participated in the first round invited several colleagues, both Dutch and Indian, to join them. The urban centre of Valsad district in south Gujarat became our headquarters for a period of two years. Two members of the Indo-Dutch group, S. D. Pillai and Chris Baks, edited the volume in which the collective results were published (*Winners and Losers: Styles of Development and Change in an Indian Region*. Popular Prakashan, Bombay 1979). I took time off from this project to go back to the villages of my earlier fieldwork. The report on what had happened since I had left ten years earlier became a new chapter, added as a postscript to the English translation of my thesis published in 1974. A major concern in the now rapidly increasing documentation on the plight of rural landless in different parts of the country was to deal with efforts to organize them as a social class. Stimulated by this on-going debate, I wrote a piece on the basis of data pertaining to the *Halpati Seva Sangh*, a caste association run for, but not by the Halpatis who constitute the main part of the proletariat in the countryside of south Gujarat ("Mobilisation of Landless Labourers", in *Economic and Political Weekly*, vol. IX, 1974: 489–96).

In later years I maintained contact with the village people who had initiated me into fieldwork. When I conducted local-level research on labor migration in the adjacent district of Surat, between 1977 and 1982, I made occasional visits to Gandevigam and Chikhligam. Within the framework of the Indo-Dutch Programme on Alternatives in Development, a new opportunity arose to study rural dynamics. This time I decided to concentrate on the increasing proportion of labor employed outside agriculture and the village household. Chikhligam and Gandevigam once again became the points of departure in my analysis of the trend towards economic diversification and the concomitant restructuring of the regional market for unskilled labor. These empirical investigations kept me busy over a five-year period, from 1986 to 1991, not continuously but intermittently. The following pages record what has happened to the people who make up the bottom and top of the hierarchy in the two villages with which I have remained in close touch over a thirty-year period. Patterns of work and employment, however, can no longer be understood in terms of a localized economy. The mobility of informal sector labor which I encountered in the field, going out as well as coming in, made it imperative to expand the horizon of my research to the whole of south Gujarat and even beyond. I have elaborated on this theme in another monograph, *A Footloose Proletariat*, which will go to press soon.

It is my hope to be allowed to return to the area of my interest in years to come. I am privileged to have had the rare opportunity to follow from nearby the slow and painful struggle for emancipation of those who are still down and out in south Gujarat. I would not know how to begin and where to stop were I to express my gratitude for all the help and support received. The exception that easily comes to mind is the director and staff of the Centre for Social Studies in Surat. I am indebted to them for their hospitality and support. Founded by the late I. P. Desai, this institute has become a second home for me rather than a temporary base.

Jan Breman
Amsterdam, December 1992
Surat, January 1993

# Foreword to the new edition

Thirty years ago I visited the state of Gujarat in India where, upon my proposal, Jan Breman as a young academic trained in sociology and cultural anthropology, had started his field research on relations between landowners and landless laborers. Together we visited the village Pipalgabhan in Chikhli *taluka*, where we were able to attend on 30 January 1963, a *panchayat* meeting. The *panchayat* consisted of representatives of the different *jatis* (caste groups) into which the village society was divided. Behind the table sat a member of the dominant *jati*, an Anavil Brahmin, who was the chairman; next to him came a Harijan who was acting as a secretary. Among the other members of the local council present were two women, belonging to the rather prosperous Patidar community.

But on the threshold of the entrance to the *panchayat* office a man was sitting, who by his dark complexion and simple clothes evidently could be identified as belonging to the bottom layer of the local rural society. I was told that he was a Dubla, a chosen representative of his tribal community who did not dare to seat himself at the table where he belonged. Like the two women, but unlike the Harijan secretary, he kept silent during the whole session. My wife was especially taken aback by the complete silence of the women, whereas I was particularly struck by the fact that the social position of the Dubla within the village society as a member of a tribal community was clearly inferior to that of a Harijan, and that his behavior appeared to be far more appropriate in qualifying him as an "untouchable" man than the Harijan, who evidently had enjoyed school education.

The following day, we could glance into a hut belonging to a Dubla family. It consisted of one small quadrangular room with a mud floor and a thatched roof, without any furniture and with only a few primitive household utensils.

In his description of the two villages where he had started his research 30 years ago, as he found them during his visit in 1991, Breman arrives at a picture which ostensibly differs from the image with which we came home in 1963: "Housing conditions have improved and inside the huts there are now a few more durable goods than there used to be" (pp. 328–29).

Breman was also impressed by "their growing social consciousness .... This change in mentality is particularly noticeable among the younger generation and is expressed in the ever louder demand that the quality of their life and work should be improved."

For this tribal community, Breman in Part II of his study consistently uses the term Halpati, coined by Mahatma Gandhi, since the former appellation Dubla has now acquired a derisive connotation. This again is a sign of the "growing social consciousness" within the community.

It is very important that Breman has decided to publish his study of the two villages, Chikhligam and Gandevigam as he found them in recent years, together with the original text of the English translation of this doctoral dissertation. In this way the combined studies form an exemplary specimen of social history, or rather historical sociology, as applied to two rural localities in the Third World.

Returning to the place of one's original research area is always a somewhat dubious undertaking. I have often warned my Ph. D. students when they announced that they were planning to return to their favorite spot in order to check their findings once more. Of course not in the sense that a criminal is warned not to return to the scene of his crime! But I have known some of my students to get confused about their original findings after visiting their research area several years later. Social change during the interval period had been so dramatic, that they felt too discouraged to present their dissertation. So my advice was: first get your degree and only then go back, in order to assess both the correctness of your analysis and possible changes that might have occurred in the meantime in the total picture. You might also like to discuss your findings with the people concerned.

This is precisely what Jan Breman did when he revisited the two villages, though only for short visits, in the early 1970s after having acquired his doctoral degree at Amsterdam University. As a consequence he could add an important "Postscript" to the English language translation of his dissertation, published in 1974. But now he has done more. He has returned to his two villages for a thorough re-study after the lapse of a full generation. It is actually quite a bold undertaking, since one may arrive at the conclusion of having been completely wrong about the predictions made earlier. This is what happened to Margaret Mead when she returned after a quarter of a century to Manus society where she had done research in 1928–29, resulting in the well-known study *Growing up in New Guinea*. She had then found a remarkable discrepancy between the values cherished by the elders and those adhered to by youngsters, but was convinced that in the end these children would develop into humans "like those same elders". But in her re-study of the same tribe in 1953, resulting in her book *New Lives for Old* (1956), she had to confess that the discrepancy between the values accepted in childhood and in adulthood had precisely been one of the potent factors bringing about social change. It is therefore very important to assess to what extent Breman's predictions in his 1974 study have come true.

I feel much honored that Jan Breman, who has developed into an accomplished sociologist, generally acknowledged as one of the foremost experts of Indian rural society, has invited me again to write a preface for his re-study. And I think that the way Breman has checked his earlier findings forms one of the most interesting aspects in the totality of the present volume.

In gauging the validity of the author's earlier predictions, we have to take account of the fact that he, in addition to his present re-study on a micro scale, again based on penetrating personal observations led by empathy, in the course of the past period has published several important studies which are similar in nature. Among them the most famous is his study of labour migration, titled *Of Peasants, Migrants and Paupers* (1985), dealing with the situation in another district of south Gujarat.

Breman himself admits (p. 282) that, whereas he had expected on the basis of his observations in 1971 that rice production in Gandevigam "would further increase", the rice crop by the late 1980s had "lost much of its significance".

A basic issue is, to what extent Breman's predictions regarding the future for the Dubla community have come true. In his doctoral dissertation Breman wrote: "The landless laborers have no access to the centers of power, and the isolation in which they find themselves is part of their economic and social pauperism. If we take their rapidly growing numbers into account as well, a prognosis cannot be other than pessimistic. It seems unlikely that the Dublas will soon know better times.... Their lives offer them no prospects of improvement and they show themselves indifferent to what the future may bring. It can never be better than it is today–and that is not enough" (p. 230).

After his return to his villages in the early 1970s, at a time when a "Green Revolution" was deemed to promise prosperity for those "wretched of the earth who have nothing to lose but their chains", his assessment of the future was darker still:

"The economic situation of the Dublas, miserable as it was, has deteriorated even more. As landless laborers for whom work has become a godsend even when it does not bring in enough to make ends meet, they have been forced further into a corner" (pp. 255–56). And he ends with the prediction "the situation does not offer much hope for the near future" (p. 260).

Breman, in his study of rural dynamics in Surat district (*Of Peasants, Migrants and Paupers*), even concluded with the sentence: "Thus, when I state that there is little reason to anticipate the near future with optimism, I do so in the consciousness that, for the pauperized sections among the landless, the situation can hardly become any worse than that it already is" (p. 445).

Should we assume, in view of the somewhat differential assessment by Breman of the present condition of the Halpatis in the two villages, that his conclusions were too pessimistic? I do not think so, Breman starts with the observation that "the impression of immense poverty . . . was the most striking feature" when he returned to the locality of his earlier fieldwork. It remains clear that for the majority of the Halpatis the significance of the visual improvement remains marginal (p. 325). Even for the 25 per cent who seem to have achieved some real advance, "the quality of life is still extremely poor" (p. 328).

In his final chapter titled "The Long, Long March Towards Emancipation", Breman states that "at least poverty conditions at the bottom of the village economy have not deteriorated". However, "the number of people who are unable to lead a life that is socially acceptable is greater than ever before". Breman thinks that "the significance of this observation is all the greater since it does not apply to a region with a stagnating economy" (p. 352).

Breman's analysis attributes the mounting discrepancy in welfare between the rich and the poorest of the poor mainly to the functioning of capitalist relations which have become increasingly dominant in the plains of south Gujarat. Moreover, he shows that in a rural society dominated by capitalist forces state protection extended to these "poorest of the poor" remains all but ineffective, owing to the fact that such powerless groups, lacking even solidarity organizations, can hardly benefit from protective measures administered from the center.

I consider Breman's new publication as highly significant because it is a convincing addition to the growing literature which proves that in this part of the world the so-called "free market economy" condemns a large section of the population to live in eternal misery.

W. F. Wertheim.

# Foreword

Jan Breman's present study originated in a research project carried out by a group of young Dutch sociologists and social anthropologists in southern Gujarat. Among the subjects were the effect of the abolition of tenancy (C. Baks), the impact of governmental planning activities at the local level (E. W. Hommes), and the phenomenon of hypergamy among the Anavil Brahmans, the dominant caste in a large part of the area (K. W. van der Veen). Although the researchers had their individual responsibility and did not operate as a team, yet their studies supplement each other in such a way as to present a more or less coherent picture of south Gujarat rural society.

All of them are highly indebted to the kind cooperation on the part of the Indian people and authorities, received throughout the period during which the researchers were in the field. Professors I. P. Desai and M. B. Desai, at that time both of them attached to the Maharaja Sayajirao University at Baroda, have earned our special gratitude through their never relenting interest in our project and their expert advice liberally given on many occasions. The studies resulting from the project were all accepted as doctoral dissertations at the University of Amsterdam, and published in Dutch in a limited number by the Department of South and Southeast Asia of the Anthropological-Sociological Center of that university. Efforts are being undertaken to make the other studies also available in English translation.

When, in January 1963, I visited the group at work in the area at that time, I once called upon one of the authorities, together with Dr. Breman. Asked about his project, the latter replied that he was concerned with the *hali* system. The official retorted that it was no use investigating that issue, since there were no longer any *halis*—the *hali* system had been abolished long ago. Not taken aback at all, Breman pursued his research project and found many data which are not generally known among Indian officials, and maybe not even among most of the Indian sociologists. I feel that Dr. Breman's study may substantially contribute to our insight into the dynamics of Gujarati society, and perhaps even of Indian society in general.

The "technique" applied by Dr. Breman looks very simple. He did not start with devising a questionnaire and administering it from a nearby city, but settled for several months among the villagers whose labour relationships he wanted to investigate. He followed, therefore, the observation "technique" usual among anthropologists. But in a highly stratified society like the Indian one, investigating a group like the Dubla land labourers presents an additional obstacle. Automatically, a foreign researcher finds his "natural" environment first and foremost among the members of the dominant caste, the patrons in the labor relationships. It is difficult for a researcher not to let himself be lured into looking at the relationships and at the opposite group mostly through their eyes. On the other hand, a researcher may be identified, by the servant group, with the people among whom he was mostly seen during the first weeks, and with whom he has several traits in common.

It seems to me that the way in which Breman approached the Dublas, and his serious attempts to complete his picture of Gujarati society by adding their perspective to the one representing the patrons' view, form the core of his "research technique." It is this aspect of his research which lends this research the character of a truly modern study.

I am convinced that as a pioneering approach to a society that has never been thoroughly studied, participant observation is a much more promising "technique" than administering questionnaires in order to collect quantitative data, which in such a situation mostly amount to a false pretense of exactitude.

Another aspect of the present study which deserves special mention is Dr. Breman's attempt to place his findings within a theoretical frame. Although it is not the first time that Indian land labor and *jajmani* relationships have been analysed in terms of patronage and clientele, it seems to me that Breman's sociohistorical analysis has added a new dimension to the way social change within Indian rural society should be interpreted in order to obtain a diachronic perspective.

A final point I would like to mention is the relevance of the crop pattern and of the general attitude of dominant caste members to employment alternatives among their land laborers. It is a highly interesting finding that despite the nearness of a small town the Dublas in one of the two villages investigated were not able to find the way towards urban employment, simply because the Anavils did not want them to leave their village, since they needed cheap labor for their mango gardens.

Finally, I would like to express my thanks to the Netherlands Organization for the Advancement of Pure Research for the liberal way in which, from the very first beginning, they sponsored and financed the south Gujarat research project. I also want to thank my colleagues Dr. A.J.F. Köbben and Dr. S.C.L. Vreede-de Stuers for their assistance in setting up and guiding the whole project.

W. F. Wertheim
Department of South and Southeast Asia
Anthropological-Sociological Center
University of Amsterdam

# Contents

# II

## DYNAMICS OF POVERTY AND PROSPERITY
## Rural South Gujarat Towards the End of the Twentieth Century

# Introduction

India has retained its character of a peasant society despite the accelerated process of urbanization and industrialization which it has undergone in the last few decades. As the census of 1961 showed, 82 percent of the population lives in rural areas, and roughly 70 percent is engaged in agriculture. These figures in themselves more than justify the continued interest of sociologists and anthropologists in India's rural structure.

Nevertheless, until recently only a limited part of the abundant literature on village India was concerned with agrarian sociology. A major preoccupation has been to study the caste society as a clear-cut system of social organization, while the network of social relationships involved in the organization of agrarian production has long remained of secondary interest.

Although it was generally recognized that India is not one of those peasant societies with little differentiation as to ownership and utilization of land on the village level, information on the division of the local community into landowners, small farmers, and agricultural laborers often had to give way before detailed description of the variations among the artisan and serving castes. What surprises us is that, in the sometimes very detailed listings of the activities of the various castes, agricultural labor — surely the principal activity in a peasant economy — is scarcely mentioned, except as work which is not very specific and can be performed by members of all castes.

Moreover, in many publications based on field work the propertied high castes in a village received proportionately more attention than the others. Concerning the members of the lower castes, often landless, data are much more scanty. As a result, we know comparatively little about the conditions in which the poorest people in India work and earn a living. They are the agricultural laborers, who are estimated to constitute approximately 25 percent of the agrarian population. It is true that two successive government reports — the records of the *Agricultural Labour Enquiry* of 1950–51 and 1956–57 — have made available a great deal of information in regard to this category, but the material they contain has not been interpreted in a wider framework.

My study is concerned with landowners and agricultural laborers in the southern part of the State of Gujarat. It is based on a combination of source studies and field-work reports. The relations between Anavil Brahmans, the principal local landowners, and Dublas, a tribal caste of landless laborers, still contain many elements of the form of service which existed in earlier times and which is generally known as the *hali* system.

For an understanding of the present situation an insight into the position of the agricultural laborer in the past is indeed of essential importance. After my field-work period, this conclusion led me to a study of the historical sources, which would enable me to analyze the master-servant relationship in the traditional society. Thus, the results of my investigation into the existing situation formed for me the starting-point for an interpretation of the past relationships between the two groups, and not the other way round. This made my book difficult to plan, but in the end I decided to arrange the chapters in chronological order, so that the material leads from the general to the particular. By employing data that go back to the beginning of the nineteenth century I have tried to record and explain the successive changes in the relationships between landowners and landless laborers, two important categories in the rural structure.

In order to place my argument against a more general background, I have begun this study with a brief outline of the development of agrarian labor relationships in India. This introduces a theoretical discussion of the relationships between landowners and agricultural laborers in the light of the *jajmani* system, the framework in which the intercaste relationships in the traditional society were institutionalized (part I). A brief description of the country and people of south Gujarat is then followed by a summary of the *hali* system as it functioned during the last century in the region under discussion, and of the changes which have occurred since then (part II). The conclusions from this part of my account are at the same time a starting-point for the report of the investigation into the existing relationships between landowners and agricultural laborers which I carried out in two south Gujarat villages during 1962 and 1963 (part III).

While my aim during field work was to gather data on the present situation, I supplemented my study by examining historical material after my return from India. A subsidy from the Netherlands Organization for the Advancement of Pure Research in 1966 enabled me to consult published and unpublished documents in a number of London libraries, and the same organization made possible the present English translation by Mrs. Wil van Gulik from the original Dutch version published in 1970.

The continuous exchange of ideas with Professor Wertheim—before my departure, during his visit in the field, and after my return—has clarified for me the concept of *guru*.

The friendly cooperation in India with Chris Baks, Enno Hommes, Igle Ronner, and Klaas van der Veen — my fellow-participants in the Gujarat project—has been of great importance to me, and a source of inspiration.

Miss H. L. Lambert, M.A., who was at the time attached to the London School of Oriental and African Studies and taught me Gujarati, made it easier for me to conduct my investigation and enabled me to be a little less of an outsider in the community with which I associated for a time.

Without wishing to understate the help of all those who assisted me in India when I was carrying out the investigation, I feel that special thanks are due to Professor I. P. Desai and Professor M. B. Desai. Their professional qualifications (sociology and agricultural economy, respectively) and their personal familiarity with the region under research, resulted in advice and suggestions that were very valuable to me. I am, moreover, indebted to Professor and Mrs. M. B. Desai for the hospitality and assistance they gave my wife during her stay in Baroda.

Professor K. de Vreese of the University of Amsterdam was so kind as to check the Glossary for imperfections. I would also like to express my appreciation to Professor Köbben for his careful reading of the manuscript. The final result bears to some extent the stamp of his many critical notes.

Finally, I gratefully acknowledge the aid I received from my wife. She corrected the final text, but, much more important, in the earlier stages of gathering and writing up materials she put me right many times.

"It is my duty . . ." In this way the Anavil Brahmans of south Gujarat deprecate expressions of gratitude, even if they wish to leave no doubt of the obligations under which their interlocutor has just come. It gives me pleasure that I am able at last to honor these obligations. Dutiful acknowledgement is, however, the last of the reasons why I have mentioned a number of people by name who helped me in the course of my study. Nevertheless I take this opportunity to express my most profound thanks to all those who enabled me in the field to collect the data of which this book is the result.

After an interval of seven years, I returned to south Gujarat in the summer of 1971 to carry out a new research project: an analysis of the institutional framework of unskilled labor and of the social mobilization of rural and urban laborers in the district. I used this opportunity to revisit the scenes of my earlier field work and to renew my association with the people I had known before. I was especially interested in finding out how the "green revolution" had manifested itself in the area.

In the Postscript, written after my return to the Netherlands in 1972, I comment on continuity and change in the relations between landlords and agricultural laborers in south Gujarat.

Jan Breman
Faculty of the Social Sciences
Erasmus University
Rotterdam

# I

# PATRONAGE AND EXPLOITATION

*Changing Agrarian Relations in South Gujarat, India*

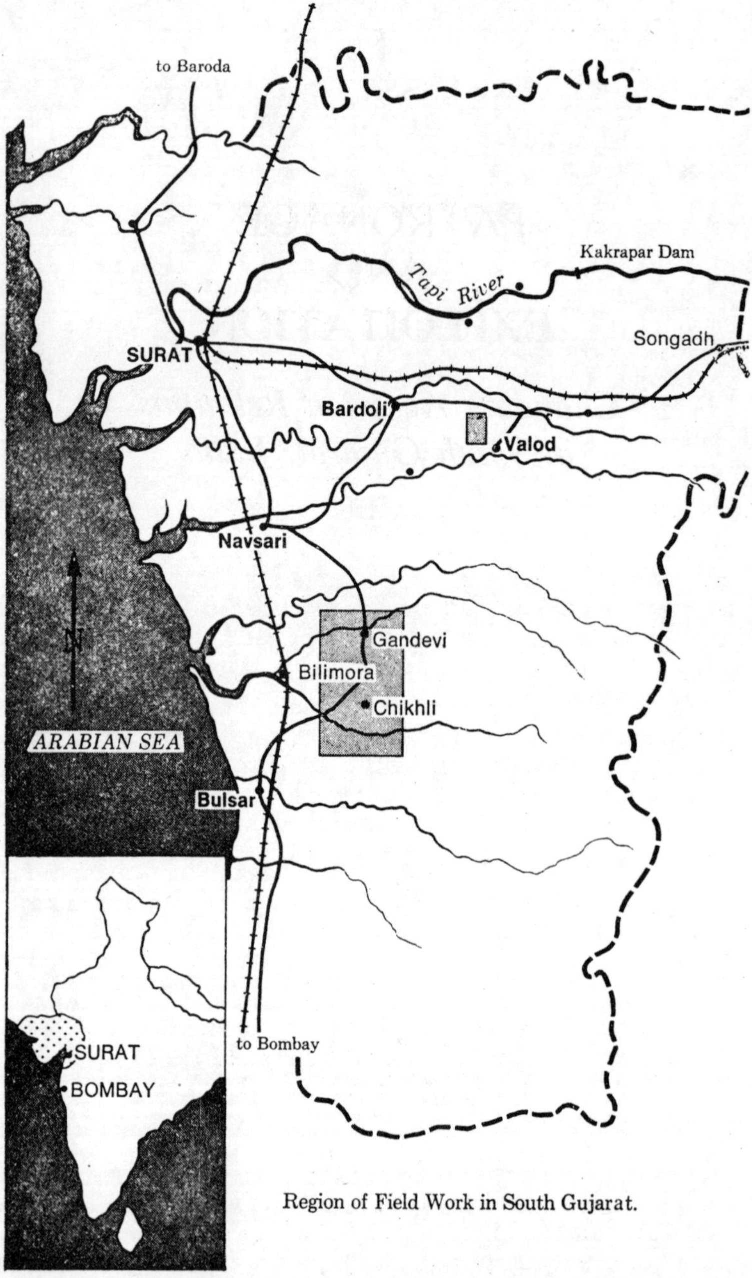


Region of Field Work in South Gujarat.

*Chapter 1*

# Agricultural Laborers in India

## *1.1 The Development of Agrarian Labor Relationships*

Recent data show that nearly a quarter of the agrarian population of India consists of agricultural laborers, a category long supposed to have been practically nonexistent at the beginning of the nineteenth century. In the literature of economic history, the precolonial society was usually depicted as a politically autonomous, economically self-sufficient, and more or less static community. Peasant castes on the one hand, and artisan and serving castes on the other, collaborated on a footing of equivalence and reciprocity. The peasant castes tilled the soil and exchanged part of their crops for the goods and services of the nonagrarian specialists. In this conception of the homogeneous village community, agricultural laborers did not figure, or at most in negligible numbers.[1]

The rise of a predominantly landless category of laborers was attributed until recently to changes in the economic structure of the local community under colonial rule. A complex of factors was thought to have caused a large-scale decline down the agrarian ladder, a process for which colonial rule was held responsible: peasant landholders became tenants and finally agricultural laborers.

British rule and the flooding of India with foreign manufactures destroyed domestic industries, and so drove the artisan on to the land. The British introduced in certain areas a system under which land revenue was assessed at high rates and was payable in cash, and which held individuals responsible for payments. This led to the destruction of the old village communities. The British brought about changes in the law which made

1. For many years this was the prevailing view. S. J. Patel summarized it as follows: "In pre-nineteenth century India . . . there were domestic and menial servants; but their numbers were small and they did not form a definite group . . . The large class of agricultural labourers represents a new form of social relationships that emerged during the late nineteenth and early twentieth centuries in India." (Patel, 1952, 32.)

it possible to sell land; the monetization of the economy, the stagnation of agriculture, and the pressure of population on resources increased the propensity to do so. The peasant was forced to sell land; it went either to the State for non-payment of taxes, or to the money-lender for non-payment of debt. This turned the peasant into a landless labourer.[2]

In these terms, D. K. Kumar has summarized the ideas that were long current in regard to the changes in the social and economic order since the beginning of colonial rule. On the basis of her research into the sources on the situation in south India in the early nineteenth century, she was the first to challenge that view in a fundamental way.[3]

Of course, the political, economic, and demographic changes of the nineteenth century did influence the relations of power and property in the rural areas. It is also quite possible that the percentage of agricultural laborers increased during that period, although the obscure and varying classifications of this category of labor in the successive censuses make it difficult to reconstruct the progress of this increase.[4]

During the colonial era there was undoubtedly a process of growing indebtedness and, in densely populated areas, increasing subdivision of the land. Did this, however, really lead to an agrarian regression which finally resulted in a proportion of agricultural laborers that has been rising ever since the previous century? It is impossible to analyze changes in the agrarian structure as a purely economic process. In peasant societies, changing relationships in landholding are reflected in changing positions in the social structure. If a substantial part of the rural population had drifted into a life of landless agricultural labor, a large-scale process of social descent would have resulted. There are, however, no signs to show that earlier generations of those who are now agricultural laborers, usually belonging to the lowest castes, occupied higher positions in the caste hierarchy at the beginning of colonial rule. Various non-agrarian castes worked on the land as well, and often it was their chief occupation, while they performed their caste functions — tanning, sweeping, weaving — incidentally. Moreover, in many areas

2. Kumar, 1965, 188.

3. In many earlier publications the existence of categories of agricultural laborers in precolonial India is indeed pointed out. See, for instance, Moreland, 1920, 111–5; R. Mukerjee, n.d., 62; Kosambi, 1956, 353; Neale, 1962, 22f.

4. See, for instance, Thorner, 1962, ch. 10; Kumar, *op. cit.*, ch. 10.

members of the low castes were excluded from landownership in order that their working power could be mobilized.[5]

Kumar calculated that, already in the early nineteenth century, agricultural laborers in south India formed at least 17 to 25 percent of the agrarian population,[6] and the situation in that area was no exception. An analysis of the first settlement reports drawn up by British district officials in the last century leads to the same conclusion for the area of south Gujarat.

That nevertheless so little attention was paid to the landless categories among the agrarian population is due to various factors. In the early colonial historiography, discussion of the Indian agrarian structure was largely restricted to the questions of who the landowners were and how they should be taxed. Nor does the more recent literature of economic history mention much differentiation as to landownership among the various castes in the precolonial period.

No doubt the picture of a closed, more or less harmoniously functioning "community" of peasants and nonagrarian specialists has been greatly influenced by the anticolonial and anticapitalist viewpoints that were brought to the fore in the struggle for national independence.[7] The image of the idyllic village community, characterized by reciprocity instead of competition, by solidarity instead of exploitation, compared favorably with the structure of the rural community as it was said to have come into being under British rule. This stereotype of a better past is still to be found in the ideology of the community-development organization, in which a plea is made for a return to the original village democracy based on harmonious cooperation and egalitarian relationships.[8] Although indeed in recent years this version has receded into the background, the ideological bias has not, a case in point being the decision to establish *panchayat* councils as the lowest politico-social units.

There is sufficient reason to suppose that, even at the beginning of the nineteenth century, a substantial part of the population con-

5. Pandya, 1959, 155; Kumar, *op. cit.*, 31.

6. Kumar, *ibid.*, 191.

7. Dumont, 1966, 67–89, and Neale, *op. cit.*, 45–47, were among those who conclusively refuted this representation of the precolonial village community, characterized in a more general sense by Wittfogel as a "beggars' democracy" (Wittfogel, 1963, 123). Such distortion of the past occurs both in the nationalist historiography on the traditional agrarian structure and in the Indian Marxist literature on the subject. For the latter, see, for instance, A. R. Desai, 1961, and B. Sen, 1962.

8. The writings of the former Minister of Community Development S. K. Dey are an example.

sisted of agricultural laborers. The differentiation among the agrarian population long antedates the colonial era. Investigators have come to realize that rights to the land were unevenly distributed among the various castes at a very early time. Agricultural laborers and members of the nonagrarian castes, who by way of compensation for services rendered could temporarily cultivate plots of land, should not be termed landowners. The complex character of the agrarian labor relationships and their wide regional variety make any categorization difficult. But to state, as is often done, that in the past agriculture as a sole occupation was open to all castes is to say that the members of the lower castes depended on the landowning castes for their living. They carried out all kinds of defiling and otherwise less-favored work and depended, whether as tenants, share croppers or agricultural laborers, on the higher castes, in which the bulk of ownership had always been concentrated.[9]

In this study, the total pattern of dependence resulting from an unequal distribution of landownership has not been considered, for I am concerned exclusively with the relations between landowners and agricultural laborers as they have developed since the last century.

## *1.2 Agricultural Laborers in the Traditional Society*

So far as any information on this subject can be found in the literature, it stresses the personal servitude of the members of the agricultural-laborer castes in the past. Many of them were permanently bound by debt to members of the landowning higher castes, by whom they were employed indefinitely.

Bondage existed in many gradations and was not restricted to agricultural laborers. Small farmers who entered into an agreement of tenancy or crop-sharing undertook to hand over part of the crop, but also to perform field services (*begar*), hence they could not till their own plots at the most suitable moment. In various ways and in varying degrees the landowning castes managed to establish complete control over the socially and economically weaker groups of the village and thus to monopolize their labor permanently or periodically. In general it can be said that the condition of dependence, on which the agrarian relationships were usually founded because

9. Leach, 1967, 12.

of the concentration of landownership, led to more or less permanent or even hereditary attachment. Nevertheless, bondage and permanence were preeminently characteristic of the category of farm servants, for whom they bore a far more personal character. It was in this category that the dependence of the landless laborers on the landowners was most pronounced.

Were there other categories of agricultural laborers aside from these bonded servants? Moreland thought not.[10] Kumar, in the beginning of her study, remarks incidentally that, in south India in the early nineteenth century, labor relationships varied from slavery to completely free labor. But the material she presents relates predominantly to relationships of bondage. Other publications also show that in many regions of India hereditary servitude, institutionalized in bondage, was the condition of a very large proportion of the caste of agricultural laborers.

The bondsmen were known as *halis* in Gujarat; *padiyals, pannaiyals* and *charmas* in Madras; *adimas* in Kerala; *huttalus, mannalus,* and *jeethas* in Mysore; *busaliyas* and *shalkaris* in Madhya Pradesh; *bhagelas* and *gassi-galus* in Andra Pradesh; *muliyas, gothis, chakars,* and *haliyas* in Orissa; *kuthias, krishans,* and *chakars* in West Bengal; *kamias, baramasiyas,* and *janouris* in Bihar; *harwahas, hariyas,* and *sewaks* in Uttar Pradesh; and *halis* and *sepis* in Punjab.[11] They were often referred to by their caste names, an indication that bondage was indeed the predominant condition of agricultural laborers. They belonged without exception to the untouchable or tribal castes, the lowest in the hierarchy.

> Agrestic serfdom is most prevalent in those parts of India where the lower and depressed orders are most numerous. Bombay, Madras, Malabar, Cochin, the Central Provinces, Berar, Central India and Chota Nagpur show the largest aboriginal population, and it is in these areas that the status of agricultural labourers verges most nearly on slavery.[12]

As for a large part of the tribal castes, it is doubtful whether in the Hindu social constellation they ever belonged to the class of landowning peasants. Their entry at a low level into the caste system inevitably led to their being taken up in the agrarian production process in a relationship of dependence, that is, as agricultural la-

10. Moreland, *op. cit.*, 115.
11. D. Desai, 1942; Sivaswamy, 1948; Lorenzo, n.d.; Singh, 1947; and Beidelman, 1959, 10.
12. R. Mukerjee, 1933, 225f.

borers or tenants. If their ancestors had had claims to the areas in which they wandered as "shifting cultivators," these claims were lost when their descendants were integrated into the caste system. This process of incorporation went on slowly, taking place in various parts of India at different times. In some regions it ended long ago, whereas in others it did not begin before the colonial era.

As agricultural laborers they were everywhere employed by members of high castes, who owned a disproportionately large part of the land. Landlords is a suitable term for these propertied peasants, who often owed it to their influential position and high ritual status to perform little or no agricultural labor themselves.[13]

Landlords and agricultural laborers, in other words, represented the extremes of the economic structure and of social stratification in the rural areas. A greater difference than that between the seignorial style of life of the richest and most powerful people in the village and the situation of poverty and bondage of the farm servants is hardly conceivable.

The servitude of the agricultural laborers was institutionalized in different ways from region to region, but the multiplicity of forms cannot conceal that the type of relationship was essentially the same everywhere. The dependent relationship usually came into being when a member of a caste of agricultural laborers accepted bondage in exchange for a loan in kind or in cash, most often to meet the expenses of his marriage. He undertook to work as a farm servant for a landowner until he had paid off his debt. But because of the low compensation he received for his services, his chances of discharging his debt were extremely slight. The debt tended to increase in the course of time, and the farm servant, with few exceptions, thus remained in bondage for the rest of his life. The attachment was continued by the next generation when the son of a farm servant also married at the expense of his father's master. In this sense the agreement between landlord and agricultural laborer was hereditary as well as permanent.

The origin of bondage has been explained in different ways. Adhering to the idea of a process of agrarian proletarization under colonial rule, S. J. Patel suggested that the members of the artisan and serving castes were driven by economic necessity to accept bondage as the traditional village community disintegrated. He ar-

13. Kumar, *op. cit.*, 189f; Lorenzo, *op. cit.*, 75; Singh, *op. cit.*, 23f; and Ramakrishnan, 1948, 18.

gued that this individual bondage took the place of the group bondage of the precolonial period.[14]

His assumption that, in the past, members of some castes were collectively bound in service to groups of landowners is nowhere confirmed by fact. Bondage has a long history; it existed on a wide scale in the Moghul period. When a region was opened up, the tribal population presumably lost their control of the land and were compelled to accept bondage to the incurrent caste Hindus. In some regions this process was still going on in the recent past, hence this seems a feasible reconstruction.

An analysis of the factors determining the relationships between landowners and agricultural laborers offers more scope for insight than do considerations of a more or less speculative nature on their causes. In other words, it is more to the point to know the societal background against which the servitude of the agricultural laborers functions than to reflect on the question of when and how it arose.

14. S. J. Patel, *op. cit.*, 87f.

*Chapter 2*

# Landlords and Agricultural Laborers in the Past: Parties in a System of Bondage and Patronage

## *2.1 Agrestic Serfdom*

Servitude as it existed in the past has been generally regarded as a form of unfree labor. The farm servants are invariably referred to as "landless serfs," "bonded servants," "agrestic slaves," "debt servants," and the like. The relationship of coercion derived mainly from the debt that originally led to the attachment, from the servant's inability to sever the relationship, and from the more or less automatic continuance of the bondage from father to son.

It has often been argued that servitude is associated with a need for regular laborers. In this view, it was the largest landowners, doing little or no work on the land themselves, who imposed servitude. Only in this way could they be sure of a steady supply of labor. When new regions were opened up, the local tribal population was subordinated by the incurrent Hindu groups.

> Where agriculture is practised with crude implements and without the aid of domestic animals, where the working population is scarce, where the land must be reclaimed from the wilds and marshes, and where soil and climate act as limiting factors for the employment of imported labour — it is not capital that is wanted, but native labour to reclaim the land and cultivate it under difficult environmental conditions. Under these circumstances bond-labour of the native population is introduced and it is pinned to the soil in conditions akin to slavery and serfdom.[1]

The theory formulated by Nieboer on the occurrence of slavery, extended by Kloosterboer to other forms of unfree labor, is completely

1. Lorenzo, 1945, 132.

in keeping with this interpretation.[2] For where it is possible for the cultivator to be self-supporting—that is to say, where land is not in short supply—no voluntary labor will be forthcoming, and forced labor in one form or another is inevitable.

Srinivas adduces arguments in support of that theory. He observes that in precolonial India there was a scarcity of agricultural laborers, and that dependent low castes could take refuge in as yet unreclaimed regions.[3] In such circumstances, forced labor flourishes. Conversely, the disappearance of servitude in its original form during this century is consistent with the theory. The increase of the population of agricultural laborers, it is suggested, decreased the need for landowners to apply labor compulsion.[4]

This is a decidedly one-sided presentation, and in direct contrast to the view that bondage was not associated with a lack of voluntary labor, but with a lack of employment, to which agricultural laborers reacted by accepting voluntary servitude.[5] In that case, the initiative was taken by the economically weak, who tried to find security by putting themselves in the hands of a landlord. For them, too, the sharply fluctuating demand for labor was an incentive toward accepting permanent bondage. A bondsman's son was entitled as well as obliged to apply to his father's master for a loan, and in this way the existing tie was prolonged into the next generation.

> It was not unknown for the descendants of labourers to claim work as a right from the descendants of former employers. After all, there were certain advantages about servitude; from one point of view, an obligation to work could be construed as a right to employment.[6]

To this it should be added that, as the rights in land were monopolized by a colonizing group, the local tribal population was practically forced to comply with terms of service which drastically restricted their freedom.

In this conception, it is true, the bondage of the agricultural la-

2. Nieboer, 1910, 2nd ed.; Kloosterboer, 1954. For a reinterpretation of Nieboer's theory, see Baks, Breman, and Nooy, 1966.

3. Srinivas, 1966, 42f.

4. Dovring, in an article on the economy of unfree labor, writes, "Landlords had an interest in bonded peasantry as long as labor was scarce in some sense they could experience. Liberation came above all when there was no longer scarcity; in the latter situation, economic pressure would replace the previous lack of civil freedom." (Dovring, 1965, 321.)

5. This is the essence of Patel's argument concerning the rise of categories of agricultural laborers under colonial rule (S. J. Patel, 1952, 88).

6. Kumar, 1965, 45.

borers resulted from their inability to earn an independent living, but it rests on the assumption that the dependent category was unable to withdraw from the authority of the dominant landowners by geographical mobility. The openness of the agrarian system, which plays a decisive part in Nieboer's theory, is in this view more apparent than real. In the regions that had already been reclaimed, at any rate, bondage was not simply imposed. While as a rule it is the way in which the relationship has come into being that is the criterium of unfree labor, for the attached servant the unfree element rested rather in his inability to break the bond once it had been contracted. For this reason traditional bondage was a special form of unfree labor, best described as voluntary servitude.

The last interpretation is one that does not fit the assumption of a lack of labor supply in the past. Only when employment outside agriculture became possible was the agricultural laborer's need of security lessened and, with it, his willingness to accept permanent bondage. This explanation throws a different light on both the incidence and the disappearance of bondage among the agricultural labor population.

In representing bondage as a form of unfree labor, investigators have heretofore paid attention only to factors arising either from a need of laborers, leading to forced attachment, or from a need of security, leading to voluntary servitude. These factors have in common that they explain bondage as it existed in the past on exclusively economic grounds and purely as a relation of labor.[7]

In my discussion of the *hali* system in south Gujarat I shall demonstrate in more detail that neither interpretation covers all the facts. The relationship between landlords and agricultural laborers as it developed in the course of time cannot be dissociated from the total context of the traditional Indian village or from the changes that have occurred within it. That is to say, the relations between landowners and farm servants must be judged in the light of a much more general pattern of relationships, which prevailed among the various castes in the traditional society and which has come to be called the *jajmani* system.

7. "Part of the reasons why bondage existed or did not exist are outside of the economic condition of the given moment. . . . [It is an] arrangement needed to maintain the sacral, hierarchical, or political institutions which, as we know, often are ends in themselves and not altogether subordinated to considerations of economic welfare." (Dovring, *op. cit.*)

## 2.2 *Agricultural Labor and the* Jajmani *System*

In studies of the *jajmani* system, emphasis is usually placed on the caste structure as a system of social stratification and of division of labor founded on the distinction between purity and impurity. In the past, every caste performed specific tasks, and in principle it enjoyed the monopoly in the region of carrying out the economic function allotted to it. The resulting interdependence among the castes was expressed in a local network of exchange relationships, which W. H. Wiser termed the *jajmani* system.[8]

The usual definition of this system as a mechanism of exchange of goods and services among the various castes on the local level would, at first sight, comprise the relation between landlords and agricultural laborers as well. Yet many authors do not regard the latter as partners in the *jajmani* system,[9] because, for one thing, the differentiation within the agrarian castes was not sufficiently recognized. This point of view was based on the conception of the traditional Indian village as a community of peasants tilling the soil along with a large number of specialist castes that were nonagrarian. Under these conditions, the exchange of goods and services must necessarily have been restricted to landowners on the one hand and artisan and serving castes on the other.[10]

There was, however, a more fundamental argument to the effect that the *jajmani* system did not include agricultural laborers. It was formulated concisely by D. F. Pocock:

> I would suggest that the heterogeneity of service justifies us in distinguishing one category of specialists to which, by virtue of its religious associations, the term *jajmani* may be applied. These are the specialists whose specialization derives from the exigencies of the caste system and not from economic needs or from the intricacy of a craft.[11]

He reasoned that the relationship between landlords and agricultural laborers was purely economic, on the assumption that this kind of relationship was outside the ideology of the caste system of which the *jajmani* network formed the structural sediment. Even though the members of some of the low castes undoubtedly worked pre-

8. Wiser, 1936.
9. For instance, Cohn, 1955, 56; Gough, in her review of Beidelman's study (see below), 1960; Gould, 1964; and Harper, 1959.
10. The *jajmani* system is explained in these words in Wolf's excellent exposition of the social and economic structure of the peasant society (Wolf, 1966, 38f.).
11. Pocock, 1962, 85.

dominantly as agricultural laborers, their function for other castes has been interpreted primarily in the light of their more specific caste occupations, such as leatherworking or waterbearing, that were exclusively reserved for them.

Although Pocock was right in arguing that the *jajmani* system was not simply an economic structure based on the principle of division of labor, his proposal to speak of *jajmani* relations only in the case of a few—the so-called religious—specialists is in essence equally restrictive. It excludes not only the agricultural laborers, but also the members of the artisan castes, who surely, as a rule, are regarded as having formed part of the system.[12]

To controvert this argument, it can be said that, in a peasant society with a limited functional autonomy of the various spheres, it is impossible simply to separate the performing of religious services from the network of social relationships. It is difficult to distinguish, as Pocock suggests, between the ritual services performed by some castes in support of the norms and values of the ritual hierarchy, and the work carried out by artisan and agricultural castes.

This is notably the case as regards agricultural labor. Pocock is mistaken in thinking that a *hali* in south Gujarat was taken into service only because the landlord needed an extra laborer.[13] No doubt such a need was an important consideration, but it was one that was also determined by the circumstance that agricultural labor—and especially handling the plow—was thought to be defiling. By leaving this kind of work to someone else, the landowner could enhance his prestige. Moreover, in many regions the agricultural laborer traditionally fulfilled some functions of a ritual nature in the house of the landlord.[14] In these cases there was question of a *jajmani* relation even by Pocock's rigid criteria.

As T. S. Epstein observed, it was only in regions with an outstandingly differentiated caste structure that the *jajmani* system comprised a broad diversity of castes.[15] In many parts of India, agricultural laborers fulfilled on behalf of the landlords widely varying tasks which elsewhere were carried out by members of different castes.

The nature of the services, and therefore the contacts among the castes, varied. There were great differences in the extent of speciali-

12. This is also the view of Vreede-de Stuers, 1965, 314–22.
13. Pocock, *op. cit.*, 89.
14. Srinivas, 1965, reprint, 41; Epstein, 1967, 233; Van der Veen, 1972.
15. Epstein, *op. cit.*, 232.

zation, in the exclusivity of the work, and in the frequency with which relations were maintained, but no clear line divided purely secular activities from ritual services. On the contrary, they were interwoven and embedded in a pattern of relationships, involving a number of castes which varied according to local differentiation.

Instead of a limited interaction among a small number of castes, the *jajmani* system had much wider implications, as Gould and others pointed out.[16] High castes tried to avoid low castes because the latter performed defiling or otherwise less favored work, but for that very reason they could not do without them. This dilemma was solved by the *jajmani* system. As the contacts were formalized the distance between the castes was accentuated and bridged at the same time. In this conception, the *jajmani* system was the institutionalization of the intercaste relationships, the mechanism regulating the intricate network of rights and obligations among members of different castes.

By refraining from doing defiling work and, more generally, from engaging in less favored activities, and on the other hand by calling upon the services of highly esteemed specialists, the members of a caste can both enhance their ritual status and emphasize their power and prosperity. Various dimensions of social position are involved and, though differing in weight for each caste, they are closely related. Agricultural labor, for instance, is regarded in many regions as an inferior activity, but only a small group is enabled by its wealth and predominance to live in accordance with this norm. As will appear in greater detail below, when landowners cease to perform agricultural work this is not only taken to be an indication of increased prosperity but also expresses a claim to higher ritual status and authority and, more generally, an attempt to acquire more esteem. Briefly, the *jajmani* relationships expressed differences in purity, material well-being, and power. It is these elements which, sometimes strengthening each other and sometimes mutually opposed, together determined the social positions of the members of a caste in the hierarchy.

This explanation leads to an interpretation of the *jajmani* system as a network of social exchange among the members of the various castes which had religious and economic as well as political facets.[17] However, knowledge of the exact nature of the *jajmani* system and

16. Gould, 1958, 431.
17. Beidelman, 1959, 30; Epstein, *op. cit.*, 15–16; Ishwaran, 1966, 39, and others.

the parties in it does not imply insight into the ways of exchange. It is this aspect that is essential for an understanding of the situation of the agricultural laborers in the past.

### *2.3 Landlords and Agricultural Laborers: Partners in the* Jajmani *System*

Opinions are divided as to the structure of the exchange relationships and the norms and values on which it was based. Wiser emphasized the complementary character of the occupational tasks, thus creating an impression of great role flexibility. The members of the castes were each in turn *jajmani*, receiving performances in the shape of goods and services, and *kamin*, providing such performances. In short, Wiser described a combination of exchange relationships among more or less equal partners. He claimed that the exchange was based on equality and reciprocity, although his factual description of the relations showed that he was not unaware of their asymmetrical character.[18]

Diametrically opposed to Wiser's view is that of Beidelman, who was convinced that inequality of power, based on landownership, was characteristic of the *jajmani* system and decisive of the way in which the exchange came into being.[19] He describes a relationship pattern with a component of exploitation, which obtained between landowning castes on the one hand and landless castes on the other. In general, the members of the ritually high castes were landowners and those of the ritually low castes were tenants or agricultural laborers. The roles of *jajman* and *kamin* were not flexible, but fixed. In the first instance, it was political and economic power, not ritual status, that determined what role was fulfilled by whom. The members of the locally dominant caste were the *jajmans par excellence*, the main landowners and at the same time the political power elite in the village or the region. They did not necessarily belong to the ritually highest or numerically most important castes.[20]

18. An often cited passage from Wiser's study is, "In turn each of these castes has a form of service to perform for the others. In this manner the various castes of a Hindu village in North India are interrelated in a service capacity. Each serves the others. Each in turn is master. Each in turn is servant." His remarks on the services provided by a number of individual castes, for instance the Chamars, testify to a more discriminating judgment. (Wiser, *op. cit.*, 10, 51–53.)

19. Gould, in a publication of 1958, arrived at a similar conclusion, which he retracted in an article of 1964.

20. For the concept of dominant castes see, for instance, Srinivas, 1955; Mayer, 1958; Béteille, 1965.

The members of the other castes depended as *kamins* on this caste of landowners, a dependence which varied with the economic importance and ritual status of the castes concerned. By virtue of their control of the soil, and thus of agrarian production, the members of the dominant castes could claim a disproportionately large part of the total package of goods and services for themselves. The inequality of power that was inherent in the *jajmani* system implied compulsory exchange on unequal terms.

The representation of servitude as a form of unfree labor fits completely into this interpretation. Beidelman regarded bondage and subservience as characteristic of the relation between *jajman* and *kamin* and wrote, not incorrectly, that there was little to distinguish a *jajmani* relation from servitude.[21] For him, the figure of the *jajman* coincided with that of the landlord, and he tended to identify the *kamins* with tenants or agricultural laborers.

Beidelman's study gave rise to a spirited discussion. Many authors agreed that members of the dominant castes occupied central positions in the network of exchange relationships.[22] But not all the members of the local elite were equally prominent, and in fact differed widely in wealth and power. The poorest and least influential among them could not afford to employ a large number of *kamins*, whereas, conversely, there were members of the nondominant castes who, on the strength of their position, managed to evade their obligations as *kamins* and themselves attempted to achieve *jajman* status with all the attendant privileges and advantages.[23]

It has not generally been recognized that the exchange was founded on exploitation. Several authors pointed out the paternalistic attitude of the *jajmans* towards their *kamins*. The *jajman* was expected to behave as a father does towards his son. It was incumbent on him to aid his *kamins*, and this obligation was religiously sanctioned by such values as liberality and magnanimity. The compensations which the *kamins* received were not seen as payment for the labor they had supplied, but as results of the responsibility which rested on every *jajman* for their welfare. This explains why the proportion of the produce distributed among them by the *jajman* at harvest time remained equal, although the sum total of the goods

21. Beidelman, *op. cit.*, 10f.
22. For instance, Pocock, 1957, 79; Gould, 1958, 431.
23. Beidelman, *op. cit.*, 17f.

and services he had received fluctuated. There was no question of a *quid pro quo*.[24]

The *kamins* had not only the duty but also the explicit hereditary right to serve their *jajman*, which guaranteed their basis of existence in an economy of scarcity. Since a paternalistic orientation, on the other hand, implies filial recognition of the father's authority and loyal behavior towards him, the *kamin* owed his *jajman* obedience, deference, and support.

In this description of protégé and protector, the element of patronage is not far to seek. By patronage I mean a pattern of relationships in which members of hierarchically arranged groups possess mutually recognized, not explicitly stipulated rights and obligations involving mutual aid and preferential treatment. The bond between patron and client is personal, and is contracted and continued by mutual agreement for an indeterminate time.[25]

A typification in this terminology makes sense only if by the *jajmani* system, as it functioned in the traditional society, we understand a mechanism not only for the distribution of goods and the exchange of labor but also for the allocation of power and prestige. It is interesting to note that Max Weber already mentioned the *jajmani* principle as a mechanism of patronage.[26]

In most regions of India, a lack of external outlets made local distribution of agricultural products largely inevitable. Any view of patronage in the traditional Indian village should take this fact into account. Maximalization of income remained subject to the increase of esteem and influence of the patron, and this he could not achieve without the support of his clients. As members of the higher castes, the dominant *jajmans* used the services of *kamins* to maintain or enhance their prestige, for instance by dissociating themselves explicitly from certain activities or, conversely, by laying claim to the help of highly favored specialists. As P. M. Kolenda wrote,

> Basic to the *jajmani* system is commitment to a royal or lordly way of life as an ideal.[27]

The *jajmans* tried to live up to this ideal by surrounding themselves with as many followers as possible. Srinivas speaks of an "invest-

24. For instance, Wiser, *op. cit.*, 63, 95; Epstein.
25. Breman, 1969, 397.
26. Weber, English translation 1958, 105.
27. Kolenda, 1963, 21.

ment in people," an endeavor to put others under an obligation and thus maneuver them into a relation of dependence.[28]

Besides prestige, the *jajmans* also drew power from the number of their followers, and it was important to have power because of the rivalry existing among the patrons. The dominant caste was divided into competing factions, alliances of families, each with a following which could be mobilized owing to the claim of the *jajmans* to their *kamins*. Conversely, the fate of the protégés was closely bound up with the well-being of their protectors. A member of a low caste who was a follower of a prominant patron was assured of security and safety in times of scarcity and tension.

Wiser was right to the extent that members of the dominant caste were committed to render counterservices; but these were services in which their patronship was manifest. The clients were proud of the esteem enjoyed by their lord and participated in it. Lewis and Barnouw cite the following passage from the autobiography of a member of an impure caste:

> This is the only means of livelihood open to us, and the richer the landlord we serve, the more prestige and honor we have.[29]

The clients profited by the prominence of their patron to improve their own position as much as they could.

As Gould and others wrote, the *jajmani* system was an expression of the extremely hierarchical character of the caste structure. The element of patronage, on the other hand, cut through the horizontal stratification of the village community. If the dominant caste was internally divided and did not form a closed front, it was equally difficult for the *kamins*, even if they belonged to the same caste, to take a common stand. The hierarchical social structure was split up into blocs of patrons with their clients.

Patronage was an essential aspect of the *jajmani* system. But we should not associate the patron-client relationship exclusively with this system of exchange on the local level. Village patrons often were themselves clients of regional "overlords," for whom they performed various services and to whom they owed loyalty and support.[30]

28. Srinivas, 1955, 30; see also Beidelman, *op. cit.*, 27.
29. Lewis and Barnouw, 1967, 113.
30. The importance of a regional elite as a link between the societal suprastructure and the village community was pointed out by Frykenberg and others (Frykenberg, 1965).

A representation of *jajmani* relationships in terms of patronage implies that servitude in the past cannot be simply described as a system of unfree labor. The servant was not only a laborer but also a client, and as such he was entitled to affection, generosity, and intercession on the part of his master, who, as a patron, had to guard and promote the interests of his subordinate. Conversely, the servant owed his master respect and loyalty. His obligations were not restricted to labor, but included serving the interests of his patron and standing by him in all contingencies. Total accountability on the one hand and unconditional loyalty on the other exceeded the obligations of the delivery of an amount of labor and a suitable remuneration.

Does this description of the *jajmani* system as a relation of patronage imply, then, that Wiser's view was correct? It is true that the group contrasts ran vertically as well as horizontally through the caste structure. But it would be premature to conclude that intercaste relations were harmonious and afforded maximum satisfaction and security to all concerned. In the relationship of patronage, the unequal control of scarce goods was at the same time accentuated and corrected.

That the *jajman* as a patron strove to increase his power and prestige, and not primarily his income, made the relationship pattern no less unequal. The patron's own interest was all important, but to reach his ends he had to make sure, to some extent at least, of the affection and loyalty of his *kamins*, his clientele. The *jajmani* system as a form of patronage implicity mitigated exploitation, while at the same time it contributed to its continued existence.

It should be borne in mind that not all *kamins* were in the same position. The distance between them and their *jajmans*, as well as the character of the services rendered, differed widely. The relationship certainly tended to contain more elements of exploitation as a *kamin* fulfilled less favored, lower skilled, and less monopolized tasks, for instance, those in agricultural labor. An oversupply of precisely this hardly specialized category made the agricultural laborer easy to replace and thus increased his dependence.

It is, finally, of importance that the members of the artisan and serving castes worked for various *jajmans* at the same time, while the farm servants by the nature of their work were mostly bound to only one *jajman*. Though not essential, this difference did have some influence on the relationship of patronage for, unlike the farm servants, the other *kamins* had to distribute their loyalty among different *jajmans*.[31] Even when the latter belonged to the same faction within the dominant caste, the less attached *kamins* formed an element of uncertainty for the patrons mutually competing for power and prestige. The great ritual distance, the laborers' small economic importance as measured both by indispensability and by exchangeability, their continuous bondage—all these factors operated against the farm servant and detracted from the principle of reciprocity in his relation to his master. But even in such cases, complete exploitation and excessively arbitrary treatment were contrary to the interests of the landlord, who had to be sure of the support of his followers.

To conclude, servitude in the traditional community should be interpreted as a form of unfree labor, complicated and mitigated by a relationship of patronage. For a detailed documentation of this view, the reader is referred to chapter 4, which includes a description of the *hali* system in south Gujarat.

The changes in servitude should therefore be judged against the background of changes in the relations between *jajmans* and *kamins*. I have consistently spoken of the *jajmani* system in the past tense, because the intercaste relationships have changed so radically that, from our point of view, the use of this term in the present village situation is no longer justified.

In the past, fluctuations of any kind did not essentially affect the foundation of the system. We do not overestimate the stability of the *jajmani* system if we say that, for instance, political and administrative changes, economic developments and demographic shifts did not assume anything like the proportions of those that took place in the twentieth century. Nor did a completely closed village economy ever exist. It remains true, however, that the relationships lost their local flavor in a process of enlargement of scale. Commercialization of agriculture and continuously increasing government intervention diminished the importance of local autarky and autonomy. The drift of members of the artisan and serving castes to the urban centers, from which they began to serve the surrounding countryside, contributed to the rise of an ever-increasing number of different contacts which went beyond the village.[32]

31. Wiser, *op. cit.*, 164; Srinivas, 1955, 30f.; Beidelman, *op. cit.*, 46; Gould, 1964, 24.

32. Wiser, *op. cit.*, sect. III; Beidelman, *op. cit.*, ch. IV.

Depending on the accessibility of the region, this development began early or late. In the countryside of south Gujarat, outside influences had begun to increase sharply in all respects by the end of the last century. Because goods and services were supplied from urban centers — these were in part new goods and services, and paid for in money — the relationships which had existed on the local level among the members of the various castes no longer fitted the new situation and became partly or completely superfluous.

The changing structure of exchange was accompanied by changes in the norms and values on which the intercaste relationships had been founded. The element of patronage within the framework of the village gradually receded into the background. In other words, a process of instrumentalization contributed to a decline of the traditional rights and obligations on either side. Transactions were increasingly concluded on a money basis. *Jajmans* and *kamins* came to regard each other as employers and employees rather than as patrons and clients, and felt themselves bound more by contract than by status. While the mutual separateness more clearly manifested itself as the vertical attachment disappeared, the powerlessness and subjection of the lower castes received a sharper emphasis.

These very general and brief remarks will be elaborated and elucidated in my analysis of the change in the relationships between landlords and agricultural laborers. Here I only wish to add that this is the situation to which, in fact, Beidelman's view applies. The disintegration of the *jajmani* system was due, I believe, above all to the depatronization of the intercaste relationships as they had been institutionalized in a more or less closed local order. In Beidelman's study the time perspective is vague, but the data with which he elucidates his arguments are relatively recent. On the other hand, what Wiser depicted was, as he himself stated, the traditional village community — and in an ideal-typical sense — as it no longer existed at the time of his stay in Karimpur. A more dynamic interpretation of the intercaste relationships is therefore needed, that is, an interpretation which takes into account the shift in these relationships and the disappearance of the *jajmani* system.

The *jajmani* system as such will not be discussed further. The remainder of my study is concerned exclusively with landlords and agricultural laborers in south Gujarat, the Anavil Brahmans and the Dublas. The data which I collected on the relations among the other castes and on those between these castes and the Anavil Brahmans and Dublas were incidental.

In my view, the developments in the relations between these two agrarian categories constitute an important aspect of the larger changes that have taken place in the rural social and economic structure of this region since the nineteenth century. Hence my study at the same time throws some light on the nature of the changes in the relations among the other castes in this period.

# PART II

*Chapter 3*

# South Gujarat: Region and Population

### *3.1 Geographic Aspects and Historical Development*

South Gujarat, in this study, is the name of the Indian region which administratively comprises two districts: Surat and Bulsar. Geographically a part of the western coastal plain of the Deccan plateau, it is an alluvial zone that lies more than 100 miles to the north of Bombay. This region can be divided into two parts. In the west is a densely populated and fertile plain composed of black clay and separated from the Arabian Sea by a narrow and infertile strip of saline soil. Toward the east, a barren area, with a comparatively sparse population and marked by an eroded and sloping landscape, gradually becomes hilly and wooded. In the south, on the boundary between Gujarat and Maharashtra, the black clay belt is almost wholly absent and the hills extend practically to the coast, whereas to the north, the plain broadens increasingly. The fertility of the clay region is surpassed only by that of the loamy soil on both sides of the rivers running from east to west and emptying into the Arabian Sea.

Despite some large irrigation projects completed during the last decade, agriculture is still largely dependent on the rains, which fall only during the monsoon, from the beginning of July to the end of September. Precipitation varies from about 40 inches in the west to 90 inches in the east. After the rainy season, winter extends from October to March, with temperatures averaging about 60° F. During the hot summer, from March to the end of June, temperatures rise to an average of nearly 100° F., although local differences are great.[1]

1. The *Surat District Gazetteer* of 1962, compiled on the model of the colonial *Gazetteers*, gives a detailed survey of the geography, history, and economy of the region, as well as an outline of the composition of the population. Various items of information in this chapter were taken from this work of reference.

The greater part of the population lives in the central plain and carries on intensive agriculture. The system of roads is largely concentrated in this area, and the most important urban settlements are near the coast along the railroad that connects Bombay with northern India.

In 1961, south Gujarat had almost 2,500,000 inhabitants. Eighty-five percent of these lived in the country, largely in villages of 500 to 1,500 inhabitants. The remaining 15 percent were in towns, that is, urban settlements of more than 5,000 inhabitants. In the central zone, the population is particularly dense, numbering upwards of 600 inhabitants per square mile—a figure exceeded in few other regions of the state.[2]

South Gujarat, an ancient culture area, is mentioned as a center of legendary kingdoms in the distant past. The more recent history of the region is closely bound up with the fortunes of the city of Surat, situated near the mouth of the Tapi River. In 1573, Akbar conquered the city, which under Moghul rule in the seventeenth century developed into the most important seaport and trading center on the west coast of India. During that period of prosperity, trading stations for resident factors were established in Surat by various European nations: there were Portuguese, English, Dutch, and French factories. Travel accounts of those days written by Europeans contain many details of daily life in the city and on the trade route to the north, but the region inland from the coast of this harbor principality is rarely mentioned. The rural part of south Gujarat belonged to the Moghul empire, but owing to the excentric situation of the region, central authority was always weak there. Reclamation of fresh areas by the founding of new villages was encouraged, and those who had already built up a local power domain were confirmed in their authority by Moghul overlords. As government functionaries—called *desais*—they were entitled to part of the revenues from the land tax. These frontier-region notables on the local and higher level successfully combined their intermediary function with a great deal of independence and recognized their subordination in name only.[3] The collapse of the Moghul empire brought with it a period of great unrest for the whole of south Gujarat, and a labile balance of power which lasted until the nine-

2. *Surat District Census Handbook*, 1962, 16.

3. The literature on the system of administration in south Gujarat during the era prior to colonial rule is extensive. Some of the main sources are Pedder, 1865; Watson, 1886; and Rogers, 1892.

teenth century. Historiographers have paid attention almost exclusively to events in the city of Surat, but the few data we have on the surrounding countryside show that there the administrative apparatus created by the Moghuls crumbled away fairly soon. Already in the second half of the seventeenth century the whole province suffered from raids by the Marathas, who wandered about ransacking the countryside. In the first half of the eighteenth century the Marathas conquered the whole region except Surat, which the Moghuls had leased to the English. Although the city still numbered over half a million inhabitants at the time, prosperity had ended, partly because the East India Company had moved its seat to Bombay.

The Polish traveler Hove, who made a journey southward from the city of Surat at the end of the eighteenth century, described a wooded and fairly inaccessible region. The population was concentrated in the coastal plain, and only there were they engaged in agriculture.[4] The power vacuum in this chaotic period was utilized by the local elite to enlarge its influence. In the absence of a regular administration its members acted as revenue farmers under Maratha rule. This country nobility came increasingly to regard the region in which they collected the taxes as their private domain, which they could dispose of at will. When they arbitrarily raised the tributes they levied from the population, the only form of protest available to the village inhabitants was migration, and this was widespread. One of the military leaders of the Marathas made himself independent of the ruling dynasty in Poona in the second half of the eighteenth century and founded the principality of Baroda. He then entered into an alliance with the East India Company, which enabled him to keep his territory in south Gujarat.

Then, after some campaigns at the beginning of the last century, the larger part of this province fell into the hands of the English. With this conquest, a long period of unrest ended. In the area under British rule the *desais* were displaced and, unlike their fellow caste members in northern India, they were not recognized as *zamindars*.[5] The system of revenue farming was replaced in 1817 by the so-called *ryatvari* system, under which each farmer was taxed individually. Settlement surveys in the second half of the nineteenth

4. Hove, n.d., 93–94.

5. In effect the situation did not change very much. At the end of the last century, Rogers wrote about the former power elite: "So thorough was the influence they originally obtained that up to the present day there are few villages in the southern districts of the Collectorate in which the . . . Patels are not descendants of the old Desais." (Rogers, 172.)

century served as a basis for the collection of land tax. The agricultural area of the plain gradually increased. Thanks to the *pax britannica*, the less fertile eastern part of south Gujarat became more accessible. The founding of villages in that region by members of high castes led to sedentarization of the local tribal population, which until then had led a migratory existence and had been engaged in primitive agriculture.

The construction of the railroad through the plain to northern India—the section between Bombay and Surat was finished in 1864—liberated the region from the relative isolation in which it had found itself after the fall of the seaport. With improved transportation, the production of cash crops was stimulated. In the north, cotton was introduced, a crop which thrived on the black soil. In the south, the cultivation of sugar cane was expanded. There was, however, no question of a sudden shift, no rapid transition from a subsistence economy to a market economy. Part of the produce had always been sold outside the village or transported to Surat. Moreover, although the construction of the railroad was an important advance, transportation of agrarian produce remained difficult as there was no system of roads connecting the villages with each other and with the railroad. Construction of such a network was not begun before the end of the last century, hence for a long time marketing continued along traditional lines. Until a few decades ago, urban traders bought up the crops at the farms and took care of the transport themselves. Early in this century the people of the villages still grew many different crops to provide as much as possible for their own needs. No rapid increase in the percentage of money income of the farmers occurred until the First World War.[6]

The same was true of industry, which after the decline of Surat had languished on in the small towns of the province, and which has grown rapidly only within the last few decades. After Independence, a number of new industries and workshops were started in urban centers along the railroad.

Finally, improved transportation has brought Bombay within easy reach, and a yearly migration to the brickyards and salt-pans near that city has begun. The migrants are small farmers and agricultural laborers for whom there is not enough work in agriculture during a large part of the year. As this rapid survey shows, a basis was laid during the nineteenth century for shifts in the economic

6. *Baroda Economic Development Committee Report*, 1918–19, 1920, 56.

structure which did not begin to take effect until the twentieth century.

Politically and administratively, too, colonial rule caused much less of a breach with the past than might be expected. In the part of south Gujarat that belonged to the principality of Baroda, the system of revenue farming continued to exist until the end of the nineteenth century, but the influence of the former authorities in the area under British rule also went on unimpaired. The lower functionaries were mainly recruited from their midst, and on the local level they could continue to exert their influence owing to the circumstance that, in the initial stages of colonial rule, decisions were largely made on the district level or below it. The far-reaching freedom of action enjoyed by the lower functionaries was curtailed only when a generally operative system of rules and procedures had been worked out and the powers of the various officials had been centralized and standardized.

In the principality of Baroda, government administration was increasingly adjusted to that of the neighboring English territory. Uniformity was completed when, a few years after Independence, the principality was incorporated into the State of Bombay. In 1964 the southern part of the Surat district, at that moment the largest in the State of Gujarat which had been formed in 1960, was split off as a new district with Bulsar as capital. Throughout this study the term of Surat district relates to this district in its old form.

### *3.2 Population*

In the social structure, the most striking aspect is the high proportion of tribal elements in the total population. More important than the classification according to caste is the division of the population into *ujliparaj* and *kaliparaj* (light-colored and dark-colored), that is to say, into caste Hindus and tribal communities. The *adivasis* or "scheduled tribes"—as the tribal groups are officially called instead of the denigrating *kaliparaj*—constitute a little over 10 percent of the inhabitants in the whole of south Gujarat, but in the southern part of the state they account for half the total population.[7] But even this is an arbitrary classification, an incidental statement of facts. "The entire course of Indian history shows tribal elements being fused into a general society," wrote the historian Kosambi, and this

7. The latest *Surat District Census Handbook* (1962, 22) gives a percentage of 49.96.

completely applies to developments in south Gujarat.[8] In the eastern area, which until recently remained difficult of access, tribal groups have long been able to retain their independence. The most important among them are Chaudharis in the north and the Dhodias in the south. At the beginning of the last century they were shifting cultivators, who migrated when the roughly tilled grounds were exhausted and who returned to them only after many years. Pacification by the British made possible an accelerated penetration of caste Hindus from the plain; their coming contributed much to the opening up of this region and to sedentary agriculture.[9] When the colonial authorities introduced the *ryatvari* system they took account of the weak economic condition of the autochthonous population; they determined the amount of land tax to be paid not only according to the fertility of the soil, but also in the light of the social position of the cultivator: a lower tax was imposed on plots tilled by tribal cultivators. However, this well-intentioned measure soon led to transfer of property rights, especially those of the best plots, to newcomers who were attracted by the low tax. Nevertheless, caste Hindus did not pour into this barren, hilly region in great numbers, and in these eastern areas more than 80 percent of the population still belong to tribal groups. Most of them have come to depend as tenants on, for instance, Banias and Parsis, who settled in small market towns as money lenders and licensees for distilling and selling spirits.[10] In the course of the years a great many religiously oriented movements occurred here; they opposed the consumption of meat and alcohol, and in general propagated Hindu values. This is now done in a more institutionalized way by education in *ashrams*, centers of social action along Gandhian lines that have been founded in recent decades with the aim of emancipating the tribal population.

Whereas in the eastern regions the incorporation of tribal cultures into the economy and the social and religious system of the Hindu society is still going on, this process has come to an end in the plain of south Gujarat. Here a hierarchy with many different agrarian, servant, and artisan castes crystallized long ago. Several groups in this region, however, appear to be of tribal origin. The Kolis, for in-

8. Kosambi, 1956, 25.

9. With reference to the autochthonous population, the British district officer Bellasis wrote in the mid-nineteenth century, "They rudely till the soil, but are very migratory and unsettled cultivators. . . . They are experts in the use of the bow." (Bellasis, 1854, 3.)

10. For a more extensive survey of the social history of this region, see Baks, 1969, ch. 3.

stance, who are now included among the *ujliparaj*, were regarded as "aboriginals of the plains" in the early days of colonial rule.[11] The Dublas, on the other hand, who inhabit the plain, are still classified as a "scheduled tribe." All such groups, both those in the plain and those in the eastern regions, firmly consider themselves Hindus. In 1911, the Dubla pupils of one school sent a request to the government for permission to state their religion as Hindu in the census of that year. Up to that time, they had been classified as animists, even though it was already known in the nineteenth century that they kept the fasts and festivals of the Hindu calendar. During the campaign led by Gandhi for the abolition of untouchability, some Dublas refused any contact with those caste Hindus who had admitted the impure Dheds to their houses.[12] A Brahman priest establishes for Dublas the most favorable day for marriage and performs the required ceremonies. Their dead are now usually cremated, not buried, although the survivors can hardly afford the expense. In the mutual contact among *adivasis*, status is apparent from their willingness to accept boiled food and water from each other. Dublas, for instance, are placed lower in the scale than Dhodias.[13]

As a rule, the groups with a tribal background regard themselves as *rajputs* and try to substantiate their claim to this status by genealogical myths.[14] Members of high castes put much stress on the deviating norms and customs of *adivasis*, for instance in family life and religion, without nowadays meaning to exclude them from Hindu society. The term of *adivasi* is here synonymous with low social status; it implies a pattern of behavior which is inferior to that of the higher castes. "Tribal caste" seems the most suitable name for these groups, which are placed below the artisans and above the impure castes in the hierarchy.

In the plain of south Gujarat, the caste system prevails in the form in which it is usually described for village India. Artisan and serving castes—such as Goldsmiths, Carpenters, Potters, Tailors, Washermen, Barbers, and Shoemakers—live in the larger villages, almost always represented by more than one household. This also applies to the impure castes, such as Tanners and Scavengers. Mediator groups are especially Banias and Parsis, a growing number of

11. *Census of India*, 1941, vol. XVII, pt. 1, 62.
12. *Census of India*, 1921, vol. XVII, pt. 1, 134. Much ethnographic material on the Dublas is to be found in the study of Shah, 1958.
13. Solanky, 1955. See also *Census of India*, 1901, vol. XVIII, pt. 1, 502.
14. *Census of India*, 1931, vol. XIX, pt. 1, 460.

whom have left the countryside and settled in towns. The Parsis are not caste Hindus. The Moslems have in most cases always lived in the large and small towns of south Gujarat, where they earn their living as shopkeepers and artisans. By far the larger part of the population in the plain belongs to the agrarian castes, among whom Anavil Brahmans, Kanbis, Kolis, and Dublas are the most important.

The Anavils of south Gujarat do not fulfill any priestly functions but are peasant Brahmans.[15] Although they are not very numerous, their position as the chief landowners and local power elite leaves no doubt that their dominance in this region is traditional. They are mentioned as those who developed south Gujarat in a remote past by using tribal labor. The remark by Baines, which I shall cite in full in the next chapter, refers in part to this colonization by Brahmans:

> Wherever they have settled in large masses . . . they have taken to cultivation on the same lines as the ordinary peasantry, except that they but very rarely put their hands to the plough, though they go as far as standing upon the crossbar of the harrow to lend their weight to that operation. Owing to this caste-imposed restriction . . . he is the centre of a well-defined system of predial servitude, his land being cultivated for him by hereditary serfs.[16]

It is not clear whether there were any Anavils among the Brahman groups who came to south Gujarat. It is not impossible that they were already present as notables in the area, and went through a process of brahmanization to legitimize their secular supremacy religiously. The Anavil Brahmans are subdivided into two status groups, the Desais and the Bhathelas, the latter regarded as being lower. In the precolonial period the Anavils were included in the village and regional administration as persons of authority, and later as revenue farmers. Indeed, the administrative and political functions were so closely connected with this caste that the professional title of such a functionary became the name of a group within the caste. Transformed into a landed gentry, the Desais increasingly removed themselves from their fellow caste members who continued to lead a peasant existence in the village. The traditional distinction resulted in a relation of hypergamy between the two sections.[17] Un-

15. Enthoven, 1920, pt. 1, 223.
16. Baines, 1912, 28. See also *Census of India*, 1911, vol. XVI, pt. 1, 282.
17. This subject is treated in Van der Veen, 1972.

der colonial rule, however, the Desais were not recognized as persons in authority, and when revenue farming was abolished they lost their chief source of influence and prosperity. The Bhathelas, on the other hand, managed to improve their economic condition. The social decline of the Desais and the economic rise of the Bhathelas ushered in an equalizing process that began in the nineteenth century and is still continuing. The Anavil caste as a whole has remained the leading group in the province as a caste of landowners, and its members have utilized their dominance to build up a lead of the other castes outside the village community in the new urban professions as well.

In the northern plain of south Gujarat, outside the territory of the Anavils, the Kanbi Patels are the dominant element. From simple farmers with a presumably tribal background they gradually developed into a local elite, but they always remained village notables who were not charged with administration on the local level. Only since the beginning of the twentieth century has their influence increased, partly in consequence of their participation in the independence movement. In the center of their territory, Bardoli *taluka*, Gandhi and Sardar Patel led a successful campaign in 1928 for reduction of the land tax. Since Independence, the Kanbis have increasingly outrivaled the privileged Anavil Brahmans, both in the villages where they live together and in the urban economy.

Kolis form the largest caste of south Gujarat. Their territory roughly coincides with that of the Anavils, i.e., they live in the western *talukas*. The Kolis have always been chiefly small farmers, but also sharecroppers, tenants, and agricultural laborers for higher castes who own more land. Especially among the younger generation there are many who have found employment, in the last few decades, as factory hands and lower technicians within and outside the area. Economically they have progressed, and as the political importance of numerical ratios in the village community is growing, the Kolis may be regarded as an ascending caste.[18]

The numerical strength of the castes in south Gujarat is not known, but the proportion of Dublas in the total population was established when they were counted separately as a "scheduled tribe." They form about 10 percent of the total population of the region; in the center of their territory, the central plain, this per-

18. For data concerning the Kolis, see Hommes, 1970.

centage is much higher.[19] Gandhi called them *halpatis*, and this name has found increasing acceptance as a caste name. The earliest colonial reports already refer to them as the agricultural laborers of the high landowning castes, who employed them as bound servants. Most members of this tribal caste are still agricultural laborers in the rural areas; in 1961 this was the occupation of 78 percent of the members of the caste.[20] There are a few other castes of agricultural laborers in the plain, for instance the likewise tribal Naikas, but it is especially the much more numerous Dublas who are looked upon as such. The population of agricultural laborers is strongly represented also in *talukas* with a high Dubla concentration. Outside the territory of the Dublas, notably in the eastern *talukas*, the percentage of tenants among the agrarian population rises sharply, as was shown in the 1951 census returns.[21] This is to be understood primarily as a different organization of agrarian production. It is doubtful whether Dublas in the fertile plain ever have owned any land on a large scale as independent farmers. We cannot even guess when, but it was certainly long ago that they became dependent as agricultural laborers. In the eastern area of south Gujarat, on the other hand, a relatively small number of incurrent members of high castes managed not very long ago to gain control of a large part of the arable land, which they subsequently leased to the original owners. In this area, too, a relation of dependence arose, but one of a different nature.

The low placement of the Dublas in the caste hierarchy parallels their weak economic position. Despite age-old contacts with landlords of higher caste, numerous tribal elements can be discerned in the social organization and culture of the Dublas. Their low living standard has probably prevented them from being sanskritized more rapidly. Recently, however, there has developed a tendency among the Dublas to abstain from meat and alcohol, to take morning baths, and to observe Hindu festivals.[22] Such attempts to conform to the behavior of higher social strata partly express the Dublas' protest against their dependent and less favored position in the social system.

Formerly, the members of the various agrarian, serving, and artisan castes in the plain of south Gujarat mutually maintained *jaj-*

19. *Surat District Census Handbook*, 1962, 200.
20. *Ibid.*
21. *Surat District Census Handbook*, 1952, 12–15.
22. Joshi, 1966, 91–93.

*mani* relations. In the network of exchange relationships, locally called *avat*, the Anavil Brahmans occupied a central position as the most important landowners and local power elite. "The social economy of the area is thus organized around the Anavil land owners, almost all the other castes being in functional dependence on them," as Naik wrote. Joshi arrived at the same conclusion: "It may be said that it was the group of Anavils on whom, more or less, depended the other caste-groups of the village for their earnings." [23]

Mukhtyar describes the founding of a village as it took place more than 200 years ago. The Anavils attracted members of agricultural, artisan and serving castes to work for them in exchange for a grain allowance or a piece of land.[24] In his monograph on a village in south Gujarat, Joshi makes it clear that agricultural laborers formed part of this system of exchange:

> As the Anavil masters supplied the necessary clothing to their Dubla-Naika Halis, the Darji's patrons were the village Anavils. Similarly the Mochi supplied shoes to the village people. As in the case of the Darji, the Mochi's patrons were the village Anavils because the latter provided shoes also to their Dubla-Naika Halis.[25]

In the remainder of this study I shall confine myself to the relations between landlords and agricultural laborers past and present. To conclude the above brief sketch of the area and the population, these two categories are not difficult to localize socially and geographically. They are primarily the Anavil Brahmans and the Dublas in the plain of south Gujarat.

23. Naik, 1958, 389; Joshi, *op. cit.*, 23. See also I. P. Desai, 1964, 83, 89, 160.
24. Mukhtyar, 1930, 44–45.
25. Joshi, *op. cit.*, 23.

## Chapter 4

# The *Hali* System in South Gujarat

### 4.1 Masters and Servants

*The Permanence of the Service Relationship.* In the earliest reports of British administrators, during the first decades of the nineteenth century, there was mention of a system of attachment—*hali-pratha*—which later became known as a form of bondage.[1]

*Hali* was the term applied to a farm servant who, with his family, was in the permanent employ of a landlord, a *dhaniamo*. The service was not contracted for a definite period; as a rule it continued indefinitely. It usually began when an agricultural laborer wished to marry and found a master who would pay for the marriage. The debt thus incurred attached the servant to the master for life. It increased in the course of the years, so that repayment was practically impossible. The service relationship ended only when the *hali* was taken over by another master. Although in such a case the original master received compensation, the transfer was not a business transaction; in several reports it was stated emphatically that *halis* were never sold.

Not only was entering into bondage the beginning of a lifelong service relationship, but the unfree elements in it were reinforced by some hereditary features. It is true that the debt incurred by the *hali* during his lifetime could not be passed on to the next generation, but the master had a claim to the son of his servant, as it was thanks to the master's beneficence that the boy had grown up, living on the *dhaniamo*'s food.[2] When the son married, usually at the age of sixteen, it was therefore natural that he should enter the employ of his father's master. Only when the master did not need another

1. *Correspondence between the Directors of the East India Company and the Company's Government in India on the subject of slavery. Account and Papers*, vol. 16, session 15–11–1837—16–8–1838, pt. LI, 433.

2. *Report of the Congress Agrarian Reforms Committee*, 1951, 129.

servant was the boy free to look for another master. These hereditary features were so essential to the relationship that, according to some reports, only those agricultural laborers who had been attached to the same family from father to son were called *halis*.[3]

*The Social Background of Masters and Servants.* As elsewhere in India, the servants came from castes of tribal origin. Most of the *halis* in the plain of south Gujarat were Dublas. In some regions the category of farm servants included Kolis, the poorest members of this caste of small farmers and sharecroppers, who, it was said, had fallen so low because of excessive drinking. They formed a separate group (*gulam* Kolis, i.e., slave Kolis), who had little contact with their fellow caste members. It was said that they were obliged to practice endogamy or even found partners among the Dublas. Members of other tribal groups were also mentioned as *halis*, but bondage has been identified with the Dublas to such an extent that for the members of this caste of agricultural laborers the term *halpatis* has become general. It is a euphemism for *hali*, introduced by Gandhi when he campaigned for the abolishment of *halipratha* in the nineteen-thirties.

Most *halis* were Dublas, but not all Dublas were permanently attached to a master. Lack of data makes it impossible to determine proportions; we can only say with certainty that bondage was the prevailing service relationship. In 1825, the Collector wrote that in Surat there were more "slaves" than anywhere else in India, and in one of the settlement reports of the end of the last century it was stated that the majority of agricultural laborers consisted of dependent farm servants.[4] There was, however, no clear-cut difference between bound and free agricultural laborers. The *chuta halis* (literally: free *halis*) belonged to the latter category. They, too, were indebted to a landowner, though not as deeply as the *bhandela halis*, the bonded servants. Unlike the latter, they did not work for the same farmer all the year round, but only when he needed them, and at other times they had to provide for themselves. The difference is not always very clear, but what it amounts to is that the *chuta halis* were less bound and, on the other hand, had less security. In

3. *Gazetteer of the Bombay Presidency*, 1877, pt. 2, 198.

4. "Some of these agricultural labourers are free to work wherever they like, but the majority of them are Hallee, either Bandhella or Chhuta," says the first survey and settlement report from 1893 of the subdistrict of Palsana, which is included in the *Revision Settlement Report Palsana Taluka*, 1911.

the sources I consulted there is no mention of a category of completely free labor.

The servants were invariably in the employ of members of prominent castes. These included Rajputs and Borahs, who were not very numerous but in some villages were the most important landlords. In the northern part of the plain, *halis* worked for Kanbi landowners, but it was preeminently the Anavil Brahmans who employed *halis*. They were the dominant caste in south Gujarat and, as such, the owners of the best and largest plots of land. The *hali* system, in other words, is especially identified with these two agrarian castes.

> The Dublas are the most degraded of all the agricultural classes. They were, for the most part, the "Halis" or slaves of the Anawala brahmins by whom they were employed in cultivation of their lands. Almost all families of the Dublas had been slaves for many generations.[5]

It would not be correct to represent the landlords as large landowners. Only a small portion of the Anavil Brahmans possessed so much land that they could rightly be called *zamindars*. These *zamindars*, the most powerful and prominent of the Anavil Brahmans, had many dozens of servants in their employ, and the same was true of some large landlords in villages where sugar cane, a labor-intensive crop, was grown. In contrast, most of the members of the dominant caste had only one or two servants. The least prosperous among them even had to do without any help whatsoever.

*The Origin of the Hali System.* "The Desaees Buttela Brahmins in Parchol, Soopa and some other pergunnahs, possess as large a portion of slaves as may be found perhaps in any part of India," wrote the Collector in reply to some questions on the incidence of slavery in his district shortly after Surat had come into the hands of the British.[6] Such clear evidence of the existence of the *hali* system even in the early nineteenth century refutes the view of Patel that this form of bondage arose under and in consequence of colonial rule.[7] As he mentions the *hali* system explicitly, not much is left of his argument as tested by the situation in south Gujarat.

Kosambi, and others, connected the rise of bondage with periods of famine. It was thought that in consequence of such a crisis, tribal groups became enmeshed in a debt relationship from which neither

5. Choksey, 1968, 55.

6. Letter from W. J. Lumsden, Collector and Magistrate of Surat, dated 8-8-1825 and included in the correspondence referred to in fn. 1 above.

7. Cf. my rejection of his argumentation in chapter 1.

they nor the following generations could escape.[8] As an explanation, however, famine seems too incidental an occurrence to give rise to an institution of such importance, which governed the relations between landlords and agricultural laborers in south Gujarat for so long a period. Instead of attributing decisive importance to events that took place in the historic past, it is more fruitful to retrace the general circumstances that favored the *hali* system.

In this connection, it is significant that the Anavil Brahmans are regarded as those who opened up the plain of south Gujarat. These Anavils are a caste of "*grahaste*" Brahmans, that is, those who are not allowed to perform any priestly functions. Concerning these landowning Brahmans, who also live elsewhere in India, Baine writes,

> Wherever they have settled in large masses, as in the Gangetic Doab or Oudh, or in compact local colonies, which probably preceded their advance as a sacerdotal body, they have taken to cultivation on the same lines as the ordinary peasantry, except that they but very rarely put their hands to the plough, though they go as far as standing upon the crossbar of the harrow to lend their weight to that operation. Owing to this caste-imposed restriction, probably, it may be noted that where the Brahman has settled otherwise than as a part of a large general community, he is the centre of a well-defined system of predial servitude, his land being cultivated for him by hereditary serfs of undoubtedly Dasyu descent. This is the case of the Masthan of Orissa and Gujarat [i.e., the Anavils], and with the Haiga or Halvika of Kanara, and the Nambutiri of the Malabar Coast, all of whom have settled in very fertile country.[9]

Various local legends, also in south Gujarat, recount how *Rama* distributed the land among colonizing Brahmans and provided them with tribal labor to reclaim and till this land. As in the case of the other peasant Brahmans, it is not certain whether the Anavils came from elsewhere. At any rate, there is no doubt that agricultural groups of this kind managed to lay claim to the land in their present territory at an early stage. Whether in doing so they dispossessed any tribal communities, particularly Dublas, is not clear. Presumably the latter, who were hunters and shifting cultivators, gradually—and especially in periods of scarcity—came to depend upon a class of substantial peasants, whether or not of alien origin, who had been hinduized at an earlier period, and who ultimately found recogni-

8. Kosambi, 1956, 353.
9. Baines, 1912, 28.

tion as Brahmans. The economic and social subjection of the tribal population in the plain may have laid the foundation for the Anavils' brahmanization.

This process cannot be dated with any certainty, but it probably took place long before the time of the Moghuls. As there are no historical data to confirm such a development as is sketched above, only the plausibility of the explanation can be demonstrated.

LABOR CONDITIONS. *Hali* means literally, "he who handles the plow (*hal*)." It is not an arbitrary designation, but indicates an activity which members of high castes in many regions of India try to avoid because of its impure character. The *hali* bound not only himself, but also his wife and children. The farm servant worked chiefly on the land, but he was also the personal servant of the landlord—an odd-job man who attended him, ran errands for him, in short, assisted his master in everything and did what could reasonably be expected from him.[10] Neither the work nor the working-hours of the *hali* were clearly defined. At all times of the day, and if necessary of the night, he was at his master's beck and call, always ready to carry out orders.

His wife, the *harekwali*, served as maid in the house of the master, usually only in the mornings. Her tasks ranged from grinding grain, fetching water, sweeping the floor, washing up after meals, and cleaning the stable to emptying the chamber pot on the dunghill. She did the rough household work which elsewhere was done by members of the impure castes. Moreover, she helped out on the land in the busy season.

The servant's daughter assisted her mother in the household of the master. His son—as *govalio* (cowherd)—tended the cattle and did various light jobs. In this period the master could test him to see whether he qualified both in ability and in personal behavior for later employment as a servant.

10. At the beginning of the nineteenth century the duties of a bound farm servant in Bihar were described as follows: "The master might employ slaves in baking, cooking, dyeing and washing clothes; as agents in mercantile transactions; in attending cattle, tillage, or cultivation; as carpenters, iron mongers, goldsmiths, weavers, shoe-makers, boatmen, twisters of silk, water-drawers, ferriers, bricklayers and the like. He may hire them out on service in any of the above capacities; he may also employ them himself, or for the use of his family in other duties of a domestic nature, such as in fetching water for washing, anointing his body with oil, rubbing his feet, or attending his person while dressing, and in guarding the door of his house, etc. He may also have connection with his legal female slave, provided she is arrived at the age of maturity and the master has not previously given her in marriage to another." Cited in Lorenzo, 1943, 61.

Attachment for an indeterminate period, severance of the relationship only in exceptional cases and often its prolongation into following generations, work obligations for the servant's whole family, and finally the nonspecific and exchangeable nature of the service were the chief elements of servitude in the past.

The counterservice expected from the master could be summarized in the obligation to provide for his servants. As one of the first reports on the *hali* system commented, "rearing and feeding in years of scarcity and the charges of settling them in marriage (about 30 rupees) are all borne by the master."[11] On the days that the *hali* worked, he was entitled to an allowance of grain—the *bhata*—which varied according to the size of his family. Instead of this grain he sometimes received a piece of land to till with his master's land; he was allowed to keep the produce for himself. In addition, he was given a slight breakfast in the master's house, and in the busy season also a midday meal on the land.

The master allotted him a site where he could build his hut, and supplied the materials for it. Each year the *hali* received some clothing: a headcloth, a jacket, a loincloth, a scarf for the winter, and a pair of shoes. His wife was entitled to a *sari* and some brass ornaments every year, and a meal daily after work. Further, the *hali* was allowed to gather firewood on the master's land. When the day's labor was over he received some tobacco and sometimes some toddy (fermented coconut milk). When he was ill, the master provided medicines, and if he lived to old age, the master gave him food every day when he could no longer work.[12] As is evidenced by the extent of the benefits enjoyed by the *hali*, he received total care, that is to say, he depended for his every need on what the master put into his hands. Practically everything he received was in kind. Only on special occasions, decided by the master and granted as a favor, did the servant get a small amount of money instead of his grain allowance. He received some coins as drink money (*pivanum*) especially at the high religious festivals of *divali* and *holi*, and on festive occasions in the master's house. By the end of the nineteenth century, money had become a somewhat more important component of the total package of the master's counterservices, but for the year as a whole it remained a trifling amount. Most of it was spent on spirits bought from the Bania or Parsi in the village.

11. Lumsden, *op. cit.*
12. *Gazetteer of the Bombay Presidency*, 1877, pt. 2, 198–200.

In this connection it should be noted that the debt of the *hali*, though recorded in money terms, in fact largely represented gifts in kind. For the marriage feast of his servant the master supplied great quantities of food and drink, and, in addition, some pieces of clothing for the *hali* and his wife. As for ready money, the servant was given the amount that he had to hand over to the bride's father, rarely more than 10 rupees. The money value of the total expenses with which bondage began in the last century did not exceed 100 rupees, and usually it was far less. This, however, was only the initial debt, which increased as the years went by.

The master did not have employment for his servants throughout the year, certainly not for all of them. When there was little or no work for a *hali* he was allowed to work for someone else and keep his earnings. Yet it was precisely in the slack period that other landowners needed little extra labor. At first the *hali* lived on the credit he had saved up, withdrawing only so much grain from his daily allowance as he really needed for food. The rest remained to his account, but he could never save enough to bridge the whole of the slack period. When his reserve was exhausted—and it did not last long, the master being the only one who kept book—he was allotted a quantity of grain, an advance payment which was now called *khavati* (literally, to eat). At the end of the rainy season the account was made up and the balance he owed was added to his total debt. Illness and other accidental circumstances caused the servant to be ever more deeply indebted to his master, and repayment was obviously out of the question.

### *4.2 Servitude as a Labor Relationship*

In the literature, the bondage of the Dublas in the past has been explained on purely economic grounds, and stress was laid on the fact that a number of landowners needed to have an adequate amount of labor permanently at their disposal. As stated above, it was the large landowners (chiefly Anavil Brahmans who did little or no work on the land themselves) who had one or more *halis* in their employ, as they were the only ones who could afford such assistance. The nature of the crops was another factor. In regions with an intensive crop system, especially where sugar cane was grown, there appear to have been more *halis* than elsewhere. They, however, had no year-round work either.[13] A motive at least equally important for

13. M. B. Desai, 1948, ch. 6.

the landlord was his desire to assure his having enough help at the peaks of the agrarian cycle. As agriculture depends on the rains, it is essential that certain activities, for instance sowing, transplanting, and harvesting rice, and plowing the land before a new crop is sown, should be finished within a few days. The normally abundant supply of labor very soon turns into a severe shortage. So as to have enough assistance at the most decisive moments, it was necessary for the landlords to take on permanently at least one farm servant, and if possible more, thereby monopolizing their labor and that of their families.[14] It should be added that to employ servants was not very expensive at the time.

The motivation of the landowning Anavil Brahman was as described above. Yet if it was advantageous for him to appoint Dublas in permanent service, what made the Dublas offer themselves as farm servants on such unfavorable terms?

That question is easy to answer when it is assumed that bondage was simply imposed. This is how the *hali* system has usually been represented. Great emphasis has been laid on the coercion applied by the masters, both in contracting and in maintaining the relationship. As already mentioned, it was not possible for the servant to abrogate the agreement, and his son was exposed to great pressure to follow in his father's footsteps. The servant was not allowed to leave his master until the debt had been paid off. It is therefore not surprising that the first British government officials in the early nineteenth century spoke of the *halis* as agrarian slaves, bound servants, praedial laborers, etc., a view which still has not essentially changed.

Unfree elements were certainly inherent in the *hali* system. But as pointed out in chapter 2, an interpretation which takes nothing else into account is too one-sided. For it is difficult to maintain that the servitude of the Dubla agricultural laborers was forced upon them against their will. On the contrary, for lack of continuous employment in the traditional agricultural economy, those who had managed to find someone to provide for them had every reason to consider themselves lucky.

It follows that servitude was in fact preferred to free labor by

14. "At first sight, this arrangement scarcely appears an economical one to the masters, but they have assured me that, everything considered, it is financially better than keeping no farm servants, and engaging labourers whenever they are wanted. For, when any of the principal operations, such as ploughing, reaping, etc., are in full swing, the demand for labourers is often far greater than the supply and not to have sufficient hands at such seasons must mean a heavy loss." *Papers Relating to the Revision Survey Settlement of the Bardoli Taluka*, 1897, 7.

landowners and agricultural laborers alike, and paradoxically enough for the same reason: the unequal distribution of the work on the land over the different seasons. Economically, the *hali* system was an attractive proposition for both parties. The Anavils wished to offset the risk of temporary labor shortage as much as they could, while the Dublas were well aware that there was work for them only during a limited period. As farm servants they did not receive any remuneration, but they were entitled to credit. Under the circumstances it was hardly relevant that their debt increased in this way. More important was the fact that their subsistence was assured in an economy of scarcity.

> The reason why he [the *hali*] sticks to the Dhaniama is that he helps him on all occasions of death, marriage, etc. In a way the Dhaniama is his sahukar [moneylender]. I called a few Hallees of each kind and questioned them closely and to my great surprise found that they were very contented with their lot.[15]

More telling than these testimonials of satisfaction before a British official are the statements in various reports from the last century that the *halis* were better dressed and had more and better food than those Dublas who were not attached.[16]

*Dhaniamo* means "he who gives riches."[17] At first glance such an appellation may seem somewhat exaggerated. The master's obligations did not go beyond providing the most indispensable means of existence for his *halis*. In the traditional agricultural economy, however, characterized as it was by widely varying crop yields and a strongly fluctuating labor demand, to be attached or unattached meant, in the final issue, either to have secured one's existence—however minimal—or to be so impoverished that a period of scarcity might easily bring starvation.

It is therefore more than doubtful that the *hali* strove to end his attachment. His being coerced to work is usually inferred from the condition that the servant was not allowed to leave his master as long as he was indebted to him. But the debt was rather fictitious in character and, if only for this reason, the term of debt slavery ap-

15. *Revision Settlement Report Palsana Taluka*, 1911, 5.

16. "I believe the slaves to be more comfortable than the free portion of their respective castes," wrote Lumsden in his letter of 1825. Half a century later this statement is unequivocally repeated: "*Halis* are still, as a rule, better off than those of their clan who are nominally free labourers." (*Gazetteer of the Bombay Presidency*, 1877, pt. 2, 201.)

17. B. H. Mehta, 1934, 339.

plied to this form of servitude is not very felicitous. Not only was repayment merely theoretical on account of the *hali's* minimal remuneration, but it was not envisaged by either of the parties.

For the master the expense involved in the beginning of the service relationship was an investment he had to make in order to obtain the services of a farm servant. He did, however, try to keep this debt within reasonable limits, as he well knew he could not recall it. The *hali,* on the other hand, did his best to maximalize it, and tried to get something out of his master as often as he could. That his dependence thus increased did not trouble him in the least. In view of his slender chances of finding work as an unattached laborer, repayment of his debt in order to end his bondage was the last thing he wanted. In short, servitude was sought rather than avoided by the Dublas.

So much was labor compulsion regarded as fundamental to bondage that some authors found it surprising that the *halis* were in no way inferior to the other Dublas.[18] Not only were they not held in contempt, a Dubla even was gratified if his daughter married one. Only then could he be sure she was provided for.

The option held by a master on the son of his farm servant also applied the other way around. Continuance of the tie from generation to generation was contingent upon the approval of the *hali* and his descendants. For them it was as much a right as a duty to succeed their fathers as farm servants.[19] The initiative to begin the relation was taken by both parties, and it was in the interest of both to maintain it.

Such an interpretation points in the direction of voluntary servitude. The motive of the master aside, the Dubla himself preferred permanent attachment, but it should be added that in general agricultural laborers chose dependence only to hold their own and to secure a livelihood. Nevertheless, it did not in any way alter their lack of freedom.

In both of the explanations of the *hali* system discussed above, the labor motive was the only one taken into account. In interpreting

18. "No social degradation attaches to the position of a Hali. Men of this class intermarry with the independent labourers of their own tribe." (*Gazetteer of the Bombay Presidency,* 1877, pt. 2, 198.)

19. Cf. also Harper, who describes a similar relation between landlords and agricultural laborers in Mysore: "Indentureship was a method of binding an employer and an employee into a stable alliance which ideally persisted from generation to generation. The servant was the lifelong employee of the master, and the servant's sons and the master's sons should, if possible, continue the alliance." (Harper, 1967, 40.)

this system, however, we should consider both the economic use of farm servants and also their function in the value system obtaining for landlords of the past.

To put it more concretely, *dhaniamo* and *hali* were not simply employer and employee. As explained in chapter 2, this was one aspect of the relationship; some elements of patronage could, however, also be distinguished. I shall examine this in the following paragraphs, beginning with the position of the dominant Anavil caste during the last century.

### *4.3 The Position of the Anavil Brahmans in the Rural Structure*

It is the life style of a landowner that determines, in part, whether he needs any laborers and how he employs them. That is to say, labor demand is determined by considerations of a social nature which entail economic consequences.

A landlord's need of the services of a hali was largely inspired by his desire to work on the land as little as he could, to be exempt from activities which he thought disagreeable and in any case undignified. In other words, a *hali* was taken on primarily to replace the labor of the master and the members of his household, not to enlarge the total effort by his contribution.

As landlords (which is the point) it is the Anavil Brahmans who have become widely known, especially the category of the Desais, of whom the most prominent, the Pedivalas, resided in the small towns of south Gujarat and enjoyed great esteem as local representatives of Moghul rulers. The term of landed nobility applies to them insofar as their income was derived from land and from land taxes.[20] They formed a regional elite and were no longer physically concerned in agriculture. Illustrative of the impressive style in which these most prominent members of the dominant caste lived is a report of 1863 on the outcome of a conflict between the prince of Baroda and the *desai* of the subdistrict of Navsari. Thanks to the intercession of the British administration, the *desai*, deprived of his rights as a revenue farmer, received in compensation almost 200,000 rupees and—a telling detail—annual payments of 300 and 1,000 rupees to allow him to keep a carriage and a palanquin.[21]

20. "The failure to define satisfactorily just what is rural or urban has led some social scientists erroneously to assume that because many in the upper class are large landowners, the elite are therefore largely rural. Others, however, have recognized the fundamentally urban nature of this stratum." (Sjoberg, 1965, 112.)

21. R. Wallace, 1863, 468. T. B. Naik, 1957, 178, also underlines the prom-

Less prominent, but still clearly landlords, were the Desais who acted as revenue farmers on a lower level—in one or two villages—and who had acquired the administrative powers conferred by this appointment.[22] For them, the regional elite was a reference group, with whom they were also allied through marriage ties. The landholdings of the village Desais never attained the size one would expect in a landed gentry, not the least reason for which was the subdivision of their property among all the sons. In their quality of landlords these Desais remained active in agriculture, even if they confined themselves to supervision. They left the physical labor to their *halis*.

The great majority of the Anavil Brahmans belonged to the lower Bhathela category. These, too, were above the average as landholders, but they lacked the opportunities open to the Desais to augment their property legally or illegally as members of the local power elite. In every way these peasant Anavils were inferior to their Desai fellow caste members who controlled the villages and in the precolonial period acted as representatives and confidential advisers to the regional elite. Wheras the indolent Desais, in their style of life, reminded the British government officials of the *zamindars* in the north of India, the Bhathelas were praised by them as industrious and conscientious farmers.[23]

In the early nineteenth century a great distance separated the two categories, but they were not strictly held apart. It was always possible to pass from the Bhathela group into the Desai group and, similarly, members of the latter could rise to greater eminence within their own status group.

The rights and emoluments accompanying the quality of revenue

inence of the Desais in the past: "Apart from the monetary and village grants some Desais had their own insignia: a cannon, a *phalki*, a canopy and a *chamar*, and had their own torch-bearers and attendants. Moreover they had the rights of being gifted in kind or cash by various other communities under customary rules."

22. This appears, for instance, from *Morrison's Report* of 13–11–1812, in which the system of tax collecting is outlined. "The zemindars (Desais and Patels) generally in the end agree to two or more persons and authorize them to settle the *juma* payable by the whole purgunnah and afterwards to distribute the amount on the different villages."

The district officers, as they were called by the British government, stood surety for each other. Each subdistrict had a number of principal *desais*, but the rights were subdivided among many others. In Bulsar, for instance, there were five "principal" *desais*, but 140 "subordinate sharers." For other subdistricts, such as Chorasi, Chikhli, Parchol, Supa, similar data are given. See also fn. 24, below.

23. *Gazetteer of the Bombay Presidency*, 1877, pt. 2, 192. Pedder states that, at the beginning of the nineteenth century, the Desais were regarded as *zamindars* and also called themselves that. (Pedder, 1865, 316.)

farmer were, in south Gujarat, traditionally divisible and transmissible. The members of the family all participated equally in the influence and prosperity which the office of *desai* brought with it.[24] Moreover, that office could be transferred from the family by sale or sublease. In practice, the delegation or sale of rights was almost always restricted to fellow caste members—in other words, the office remained especially a prerogative of the dominant Anavils.[25] In the villages under their jurisdiction the high Desais appointed fellow caste members as deputies who had to collect the taxes for them, to keep order and to dispense justice. It was in this way that the more prosperous and influential Bhathelas, utilizing the coupling of function to position within the caste, graduated to the Desai category.

The introduction of a number of Anavils into the local administration, therefore, did not lead to final partition. Power and riches were not concentrated within a limited and closed group, but the benefits of *desai*-ship always remained open to individual Bhathelas who, when they succeeded, ultimately found recognition as Desais.

Apart from this, mobility was institutionalized in another way, i.e., in the relation of hypergamy between Desais and Bhathelas. A Bhathela was prepared to pay a large bride-price to marry his daughter to a member of the higher status group.[26] In this way the girl could look forward to a life of comfort. No longer would she be obliged to do peasant work, and in the household she would be assisted by servants. These considerations were not primarily inspired by fatherly care and affection: marriage was a family affair rather than a personal matter, and could place the Bhathela on a good footing with his prominent and influential fellow caste members. Such contact with Desais in itself already gave prestige and

24. "Among the Desais of the Surat Athavisi there seems to have long existed, as there still exists, a jealousy of each other which would not permit any individuals to assume the position of principals or representatives of the family in watan duties; all sharers were considered equal, and as the watan families increased, the members of them divided among themselves the Desais' rights in different villages of their Purguna, and indeed the Desaigari of the same village was often divided into the 'khatas' (i.e., part) of two or more Desais, brothers or cousins." (Pedder, 1865, 316.)

25. Pedder states that Bulsar, for instance, was divided into four Desai watans, but the attendant emoluments benefited 350 separate shareholders belonging to seventeen different families, "whose rights have mostly originated in purchase or mortgage." (Pedder, 1865, 313.) In the principality of Baroda the office of *desai* was still sold until the end of the last century. (*Baroda Enquiry Commission*, 1873–75, pt. 1, 519.)

26. Concerning this marriage system, see Van der Veen, 1972, 75–82.

enabled the Bhathela to claim, if necessary, intercession and aid from the family to which he had married his daughter.[27] The connections of his new relatives, finally, improved the Bhathela's chances of acquiring a *desai*-ship for himself.

It is not impossible that stabilization of the Desais' position under colonial rule would have led ultimately to complete prohibition of intermarriage between the two categories. There was no progressive differentiation, however, and in the last century a leveling process took place instead. It was precisely the great power and autonomy of the Desais that induced the colonial administrators not to make use of them as intermediaries, so that they lost much of their influence and wealth. Only with great difficulty did they adjust to the altered circumstances, and in many cases they went into debt.[28] Under the circumstances, hypergamy was actually a boon. Gifts from their Bhathela relatives by marriage made continuation of their often extravagant way of life possible.

The following moralizing quotation expresses very well the behavior of the Desais and their attitude towards the Bhathelas:

> The Bhathella Desais, owing to their influence and cunning have succeeded hitherto in keeping the Government demand to a disproportionately low figure, which enables them to take less personal interest in the cultivation of the land. Hence all the fine masonry wells in Eroo are out of order, and the Dab, a weed which may be aptly called the Nemesis of improvident tillage, holds up its head to reprove the Desai, who thinks it beneath him to be a farmer or a landlord in the sense the Kaira Desai is. His pride takes another form, the pride of caste and privilege to marry 3 and 4 wives from the cultivating Bhatella class, who give him from 2000 to 4000 Rs to effect a settlement for their daughters in his home.[29]

After much squabbling, the British administration finally made a new arrangement for the Desais. On the advice of an enquiry committee, which made its report in 1865, the descendants of the former power elite received an annuity in compensation for the rights of

27. See *Census of India*, 1911, pt. 16, vol. 1, 282, and Naik, 1957, 178.

28. See the first cadastral report of Bulsar *taluka* from 1870, included as appendix R in *Papers Relating to the Revision Survey Settlement of the Bulsar Taluka of the Surat Colectorate*, 1900, 70. Cf. also the correspondence between the Collector and other government functionaries, which is included in *Papers Relating . . . Jalalpor Taluka*, 1900, 84, 99. See also the autobiography of G. G. Desai, an Anavil who in the colonial bureaucracy rose to be a magistrate but was discharged with ignominy. (G. G. Desai, 1906, 3.)

29. Cited from the first cadastral report of Jalalpor *taluka* in 1868 and included in *Papers Relating . . . Jalalpor Taluka*, 1900, appendix R, 11.

which they had been deprived. These annuities, called *desaigiri* after their former remuneration, were paid until 1953. The amounts were small, the more so as they had to be distributed among the ever-growing families. More important than the money, however, was the fact that the Desais could thereby legitimize and strengthen their claim to former importance.[30] Documents to show the rights formerly granted are still preserved with care. The solemnity and ostentation with which until recently the descendants of the *desais* went to town on the annual payday to receive their annuity had no bearing on the few rupees to which they were entitled, but was meant as a pointed reminder to the other village inhabitants of the illustrious past of their families.

A background which made it clear that a family had politically influential ancestors remained a prestigious factor until this century, as did residence in a village traditionally known as a domain of the higher status group. But from the early nineteenth century onwards—fairly soon in the British part, later also in the principality of Baroda—the Desais were compelled to live exclusively on their property in land. They did this in accordance with their aristocratic ideas, that is, as landlords, and looked down contemptuously on those of their fellow caste members who, even though they employed a servant, had to work with him in the field.

### 4.4 *The Anavil Brahman as a Patron*

A landlord's style of life, not the function of *desai*, had become the most important standard of evaluation for the Anavils. Having one or more servants at one's disposal indicated a certain degree of prosperity and enabled the landowners to behave in a manner which rated high in the scale of values and which, though on a more modest footing and carrying less power, in fact guaranteed for this dominant group continuity with the past.

30. "With regard to the future entry of names, the Collector may guarantee to the *Watandars* that a record shall be kept in a *Watan* book to show the shares of each sharer whose name is now entered, and that this book shall be corrected from time to time by the insertion of the names of heirs in application and legal proof of heirship. But no guarantee of payment of separate small sums in the form now paid should be given, because whether a settlement be accepted or not, this system will be superseded by one involving less labour and inconvenience in the public offices." (Pedder, *op. cit.*, 426.) The entry and periodical addition of names was for the Desais of the greatest importance. The above quotation illustrates very clearly the failure of the British government officials to understand this passion for registration where, after all, the yearly allowances were minimal. Entry in the register, however, was a confirmation of, or a claim to, high origin.

It was in this respect that the Bhathelas could more easily compete with the Desais than would have been possible in an earlier period. In contrast to the decline of the latter was the economic rise of the peasant Anavils in the nineteenth century. British reports of the period portray the Bhathelas as industrious and thrifty cultivators, but their efforts were aimed above all at imitating the Desais. They tried to ally themselves to their more prominent fellow caste members by marriage. The gifts to the bridegroom's family and the wedding feast itself cost large sums of money. To organize the most sumptuous of feasts, lasting several days, was expensive but created an ineradicable remembrance and, in the eyes of the village and especially of all Anavils, served as a measure of the family's position.[31] Not only did the well-to-do Bhathelas use this device to improve their own position and create relations with the higher status group, but they also tried in general to behave like Desais. They used their earnings from the cultivation of sugar cane to enlarge their property holdings, not to increase production in order to maximize their income. A Bhathela who bought land did not do so with an eye to the market: he merely wished to be the equal of the Desais, for whom land was a prestige good rather than a capital good. A landlord's behavior determined the perspective of the Anavils in this period, and the axiom they generally lived by was: the more land, the higher the status. T. B. Naik wrote:

It should be emphasized at this point that it is not monetary income but landed and other immovable and movable property that gives status among the Anavils. A man getting Rs 5.000/–per annum but having no land is much lower socially than another who has thirty acres of land giving an annual return of Rs 4.000/–. . . . these classes [the Bhathelas] try to get into the higher ones by getting more money and turning this money into landed property.[32]

Purchase of land alone was not enough. To conform to the life style of the Desais it was also necessary for the Anavils to limit their own

31. "Those who want to rise in social status can do so, especially through marriage into a higher family by paying a higher dowry and goods. Secondly, the classes as they rise in social ascendance spend more and more on marriage and other social ceremonies. Though conspicuous exhibitionism of one's status, *motia* or *mobhbho* (through a colorful decoration of the marriage booth, through invitations to singers and dancers, and through giving and displaying the costly gifts to the bridegroom), is the mark of all Anavil social classes, the tendency for showing off and spending increases as one rises (or wants to rise) in the social hierarchy. (The uppermost families could afford to go in for a much less elaborate affair, not requiring much money)." (T. B. Naik, 1957, 179.)

32. T. B. Naik, *ibid.*

physical work, to put an end to their wife's work in the field and to relieve her of all household duties except the preparation of meals and the care of the house altar.

This could only be done when the family took on at least one agricultural laborer in permanent service and delegated to him and the *harekwali* the most strenuous, unpleasant, and in part also impure tasks in the field and in the household. For the Bhathelas, too, it became increasingly true that esteem was accorded to him who led a comfortable life, who gave orders without having to work himself, whose wife had help in the household, who, in short, was a landlord in the same way as a Desai was. As the reports show, *halis* were recruited in times of prosperity, when a growing number of Bhathelas could afford to withdraw wholly or in part from work on the land and leave the toil to their servants.[33] When the economic tide turned, however, the number of *halis* diminished, and the masters were compelled once more to put their hand to the plow.[34] It is this alternation of extension and contraction of employment which suggests that the farm servant was to a large extent engaged for social reasons.

Although the process outlined above did much to equalize the two Anavil categories in the nineteenth century, it was chiefly the Desais who could afford a life of luxury. Their authority and wealth were undeniably affected by the government measures, but their lead was so great that only the more well-to-do Bhathelas could compete with them. Moreover, in the principality of Baroda the Desais kept their prerogatives until the end of the century. The administrative reports contain frequent complaints of their continuing influence, especially on the village level.

While the Desais clung to their privileged position, the majority of the Bhathelas was still condemned to live and work as peasants. Many of them even had to make do without the services of a *hali*. How much they were weighed down by peasant work, and its consequences, too, for their social position, is reflected in the mockery of the following Anavil song:

With an axe on his shoulder
The husband looks a Negro —

33. *Gazetteer of the Bombay Presidency*, 1877, pt. 2, 197; Shukla, 1937, 124.

34. "Halis are no longer maintained as the Anaolas find it hard enough to feed themselves, let alone their Halis." (*Papers Relating . . . Chikhli Taluka.*) The Census of 1901 states that during the famine of the preceding years many *halis* had been dismissed. (*Census of India*, 1901, pt. 18, vol. 1, 72–73.)

I wonder why
Father gave him his daughter.[35]

What was the cause of the Anavils' inclination to withdraw from all physical agricultural work? It seems reasonable to connect this attitude with their position in the ritual hierarchy as Brahmans, for whom agricultural labor, and physical work in general, are despised activities.

> The Anavil Brahmans, though intelligent and brave, are looked down upon by other Brahmans. The reason in this case is that the Anavils (as they are called by other castes) are agriculturalists and their women put on shoes while carrying the men's meals to the fields.[36]

Handling the plow, especially, is looked upon as impure, and wherever in India landowning Brahmans employ farm servants, this rule is usually given in explanation. At first view a Brahman's wish to meet the demands of the Brahman ideal seems to be the chief, if not the only reason for withdrawing from agricultural work. With a farm servant at his disposal the Anavil could lead a life suitable to a member of a high caste.

As mentioned earlier in this chapter, the farm servant and his wife, both on the field and in the master's household, carried out those tasks which elsewhere are left to the members of impure castes. Speaking of functional exchangeability as an attribute of the *jajmani* system, Gould remarks:

> A "clean" household cannot, by definition, retain its "clean" status unless certain kinds of work and ritual procedures with impure implications are transferred to others. If the traditional absorber of a specific form of impurity is not available, then the power and persuasion of the "clean" household must be employed to uncover someone else who will function as a substitute.[37]

In the absence of impure castes — a very low percentage of the rural population in south Gujarat — Dublas are employed to take over various less favored activities from those Anavils who have managed to improve their position.

To explain this as a desire for a Brahman way of life is only partly justified. In religious matters the Anavils are quite lax, and are not much concerned about the many rules to which as Brahmans

35. T. B. Naik, 1950, 33.
36. S. Mehta, 1930, 131.
37. Gould, 1964, 21.

they should subject themselves. In contrast to their respect for the Great Tradition as phrased in the holy scriptures they show little deference for the Brahman as village priest in their daily life. "Though Brahmans, the whole body of Anavils are laymen who never carry out priestly duties or accept alms from others. . . . The priesthood is unthinkable for them and considered below their dignity," wrote T. B. Naik.[38] Instead of aspiring to a secluded and puritanical life and recognizing obligations towards others by accepting gifts, the Anavils lord it over the village community. Their word is law for the other castes. In view of the central position they have always occupied as prominent landowners and as the local power elite, it is in fact more plausible to regard Kshatriyas as the reference group of the Anavils. In the seventeenth and eighteenth centuries they had the example of the Maratha rulers, while in the principality of Baroda the Kshatriya model was preserved in the life style of the upper stratum. As members of a dominant caste, the Anavils are comparable to rulers who have total authority over their subjects and who do not in the least detach themselves from their environment.[39]

It would, however, be going too far to explain the way of life of the Anavils one-sidedly in terms of a specific model, coupled to a specific caste which they try to imitate. The behavior of every dominant caste is unmistakably close to the Kshatriya ideal, but it does not necessarily have any reference to it. With equal justification it might be said that the ideal behavior pattern of the Anavil Brahmans was influenced by the example traditionally given by the urban elite in their territory, an elite which included not only Moslem governors representing the Moghul rulers but also members of the highest section of their own caste, the Pedivalas.[40]

If we apply this line of reasoning to the traditional village situation we may assume that the Anavil landowner's need of the services of a farm servant and his wife was in part ritually and in part socially determined. As soon as he could afford it he took a *hali* into his employ so that, as a member of the dominant caste, he could be free of physical work.

38. T. B. Naik, 1958, 389.
39. Cf. Srinivas, 1966, ch. 1, par. 3.
40. As to the style of life of the elite in preindustrial towns, Sjoberg writes, "Not only does the upper class function without engaging in physical labor, but it arrogates to itself 'luxury' items that enable it to achieve a life style that dramatically sets it apart from the lower class; in turn this ostentation reinforces its power and authority." (Sjoberg, 1965, 118–19.)

The Anavil with one or more farm servants no longer needed to work on the land and had acquired a position in which he could lead a life of conspicuous idleness, thus proving his superiority in the village community and in his own caste. *Halis* were for their master a symbol of the leisure which he enjoyed, but the master was not literally idle. He was continually busy enlarging his influence and reinforcing the commitment of his subordinates, while on the other hand he tried to damage, whether or not in coalition with others, the prestige and range of power of his peers.

In discussing the *jajmani* system I characterized the relationship between the members of the dominant caste and those of the other castes as one of patronage. The esteem enjoyed by the *jajman* was apparent from the size of his following. Landholding, chiefly concentrated in the dominant caste, was of decisive importance because, in view of the unequal distribution of property rights, it implied control of people. The clients shared in the crop, and some of them were also concerned in its production as tenants or agricultural laborers. To bind others and be assured of their dependence gratified the patron's self-esteem, but at the same time he could not dispense with it if he wished to be recognized as a prominent member of his caste and enjoy the attendant privileges.

In this light, it is not surprising that the social position of the Anavil Brahman was partly measured by the number of servants in his employ. By taking on more farm servants than he needed he reduced the proceeds of his holding. On the other hand, as his prestige was determined by the size of his property, not by the income he derived from it, he was evaluated by the number of his servants, not by the standard of whether, as a good cultivator, he knew how to use them in the most effective way in agrarian production.[41] *The question facing the Anavil landowner of the past was not how many servants he needed, but how many he could provide for.*

Extravagance in this respect was censured only when it caused the master's impoverishment. Living beyond his status, that is, taking on more farm servants than his economic capacity allowed, was

41. Writing about the production relations in Uttar Pradesh, Neale remarks, "The superior feels that he must support those who are loyal to him, whether as family head, patron, or faction leader. If he must provide for those under him, then he has no incentive to rationalize his economic operations. In effect, the followers must be paid in any case; to spread the load makes life easier and appears equitable." (Neale, 1962, 169.) In the context of this quotation, however, the author creates the erroneous impression that in such a situation there is question of a nonrational orientation.

bound to lead to loss of land, dismissal of *halis*, and in general, reduction of his group of clients. Economic decline was feared above all because it damaged the position of the landowner as patron.[42]

A massive following not only was a standard for conferring esteem upon the patron but also increased his power. This was of great importance especially in the precolonial period. Because of the unstable political situation of the seventeenth and eighteenth centuries, the revenue farmer sometimes had to extort payment by force, and only he who was able to mobilize a following if necessary could act as *desai*.

Connections with important people outside the village denoted a prominent position within it, and conversely. These Anavils, themselves influential patrons in the local community, were in their turn the clients of regional overlords to whom they owed assistance in case of conflicts. In their capacity of local power elite they profited by the weakening of central power. They began as mediators but became more and more autonomous; in the end they managed to build up a practically unassailable position and to exemplify as patrons the dominance of their caste.

In the course of nineteenth-century colonial rule, their supralocal power was curtailed. Within their village community, however, they remained completely sovereign. There, the members of the dominant caste vied with each other for the largest following in the village. In this situation of permanent rivalry, to have many followers at one's disposal was to be able to settle a feud by force if need be.[43] Under the circumstances, having several farm servants in the permanent nucleus of the client group bore witness to prudent foresight and was not at all an excessive luxury.

The power of the master rested on his ability to bind clients. It is not surprising that particularly the loyalty of the farm servants was beyond doubt. Placed low in the social and ritual hierarchy, little specialized and held in low esteem in their work, selected from an abundant supply, continually employed on the land or in the house-

42. A statement by Lorenzo concerning the *kamias* in Chota Nagpur, comparable to the *halis* in south Gujarat, also points in this direction: "The sale of the *Kamia* is supposed to be derogatory. It is often the result of the landlord's uncertain economic position, which makes it difficult for him to maintain his Kamias. The Babhan mahajans [the caste of the masters] take much pride in having a greater number of Kamias, and often as a result of this ráce for Kamia ownership, their sale is postponed as long as possible." (Lorenzo, 1943, 80–81.)

43. See also Cohn, 1955, 65.

hold of the master and therefore always at his disposal, these people were bound hand and foot.

In contrast to the members of the artisan and serving castes the Dubla servants had only one patron, of whom they were both the least favored and, perforce, the most loyal clients. The least favored since they carried out strenuous physical work and since, in a community in which obligations to others are to be avoided, their attachment to a master implied an extent of dependence that was looked upon as very demeaning.[44] If in chapter 3 the members of the dominant caste were called the patrons *par excellence*, the dependence and subordination of clients is preeminently represented in the category of farm servants. In the society of the time, no greater distance existed than that between the dominant Anavils and the lowly Dublas. The comprehensive inferiority of the latter lent even more luster to the independence and superiority of the Anavil masters.

A patron-like behavior was especially embodied in the way of life of the Desais. The system of prestige and power, in which the patronship of the members of the dominant caste was expressed, largely functioned in relation to the village community.[45] Because, by the end of the last century, landholding within the dominant caste became more equalized, an increasing number of Bhathelas successfully competed with the Desais. "Lordly behavior" and a "kingly style," or whatever such an orientation is called, has traditionally been the behavior that set the tone and was a worthy aim of Anavil Brahmans. In this century, however, it has become attainable by a growing number of the Desais' fellow caste members.

Administrative reports of the nineteenth century described the Anavils as rulers over the village population. Again and again we find complaints of the pressure exerted by them, and they are referred to as impertinent and quarrelsome. What the administrators regarded as stirring up trouble against the British and simple griev-

44. "Occupations involving heavy physical labor are generally esteemed to be very low and direct dependence on others for one's livelihood suggests lack of freedom, which is also not a sign for a high status occupation in the village." (Joshi, 1966, 28.)

45. Cohn represents the village as a small principality with a closed prestige and power system in the traditional community. "Before the end of the nineteenth century, the little kingdom was an almost closed prestige system; prestige depended on the amount of land one inherited from his ancestors, the status of his family in relation to the founding ancestors, and the low caste followers that he could muster." (Cohn, 1959–60, 92.) It is a characterization of local dominance in north India which also applied to the Anavils of south Gujarat.

ance-airing was really a symptom of the dominant position of the Anavils and of their mutual rivalry. Though intolerant of criticism directed against themselves, they are extremely outspoken and emphatic in their pronouncements on any subject, whether it concerns them or not. Their behavior is in marked contrast with that of lower castes, which as a rule is characterized by humble wariness and the fear of unnecessarily committing oneself and irritating one's superiors. The bossiness of the patrons, reflecting their arrogance and self-importance, contrasts sharply with the sometimes abject servility of the clients. In the presence of British officials, the patrons in their turn had to feign submissiveness, which evidently they did not find easy.[46] In the village system, however, the Anavils are accustomed to command, not to obey, and when they speak, the other villagers are expected to listen submissively. The Anavil's behavior displays irascibility and superiority. He is wont to dwell emphatically on the stupidity and backwardness of other castes, whether members of those castes are present or not. It is a consequence of their behavior

46. A small anthology of statements concerning the Anavils: "They are notoriously a quarrelsome and intriguing race, full of duplicity and chicanery, adepts in controversy, most obstinate in purpose, and extremely fond of litigation. They will dispute an usurped right, or the doubtful possession of a field or tree with a pertinacity unequalled even among the natives of India. Neither reason nor argument can convince them, and after petitioning every tribunal of appeal, I believe death alone puts an end to many of their inveterate disputes about lands." (Bellasis, 1854, 2.)

"I would have passed over any allusion to the character of the Bhatellas, were they not a power in the community, exercising great influence over the minds of the people who are numerically stronger than themselves." Captain Newport thus portrays them in 1822: "My intercourse with the Bhatella Brahmins has been extensive and intimate, and as individuals I found them obsequious and servile when they expected benefit; would seldom or never answer a question direct until its object had been guessed at, and then guided by interested motives, would either evade it by pretended ignorance or by direct falsehood. Indeed experience has proved to Government that some of the most respectable are not trustworthy."

"It rejoices me to think that both you [the Collector of Surat] and the revenue commissioner are not unacquainted with the turbulent and never-to-be-satisfied class of *ryots* in this district. Villagers who had not petitioned were threatened with excommunication, and in this manner it has been the Bhatella policy from the commencement of our rule to show as much general dissatisfaction as is possible, in the hope that per adventure their efforts may bring forth a little good." (*Papers Relating . . . Jalalpor Taluka*, appendix R [The first survey and settlement report, of 1868–69], p. 30.)

"These Bhatelas are notorious as the greatest intriguers in the country. This character, assisted by the hereditary influence I have described them in [sic] possessing, has enabled them to worm themselves into positions of power as patels of villages, which they have used to the injury of the revenues by exacting labour from the lower castes of cultivators, and thus preventing them to improve their own condition." (*Papers Relating . . . Chikhli Taluka*, 63.) Where the authors of these reports speak of Bhathelas it appears from the context that by this term they refer to all Anavils.

on the local level as *Rajas*, a term which they still acknowledge with a condescending smile.

### *4.5 The Dubla Farm Servant as Client*

Earlier in this chapter it was pointed out that the *hali* could claim total though extremely sparing support from his master. As a *kamin* received his grain allowance irrespective of the service he had performed, so the *hali* was not remunerated on the basis of his merits but on that of his needs, which were rated low indeed. There being no contractual agreement related to a market situation, the allowance of the farm servant was not very flexible. This allowance—a more accurate term than "wage"—was paid largely in kind and remained at the same minimal level throughout the last century. The patronship of the master was founded on the bondage of the *hali.* What he paid his Dubla farm servants always exceeded what they set against it in the form of labor, at any rate in his own calculation. The debt element, however, has been given undue importance in the literature. Writing about a similar service relationship in Uttar Pradesh, Singh remarks,

> The debt of a *harwaha* [comparable to a *hali*] is a fictitious asset of the money-lender. These people never take their debt seriously, for they know that their incomes being grossly inadequate redemption of debt is out of the question. They look upon the money that they are able to borrow from the Zamindar as their "income." And the Zamindar too does not consider the money advanced to a *harwaha* a loan. He does not want the debt to be redeemed.[47]

As this quotation shows, instead of wishing to terminate the relation as soon as they could, both parties aimed at continuing it as long as possible. As already mentioned, the master did not want the debt repaid. He did not make a "loan" to press for payment afterwards, but rather to be able subsequently to assert himself as a patron. He made it very clear who was the giving and who the receiving party, and in this way impressed both gratitude and loyalty upon the farm servant. The size of a servant's debt was a measure of his commitment, of the control which could be exerted on him by the Anavil. The master repeatedly, and often on the basis of false figures, told the Dubla that he got much more than he was entitled to. In this

47. Singh, 1947, 134.

way he managed at once to demonstrate his liberality and to emphasize his servant's dependence, in short, to behave like a patron.

Nor did the *hali* work to pay off his debt, as has often been suggested, for to be an unattached agricultural laborer was the last thing he wanted. Instead of keeping his debt low, the farm servant tried to exploit his privileged position and move his master to do him as many favors as possible. He appealed constantly to the Anavil's beneficence. By working on his sense of honor as a patron, by badgering him for extra compensation and pressing for all kinds of remuneration, he stimulated the master, as it were, to be liberal and to make a show of behaving like a good master. The farm servant was undoubtedly aware of the fact that by so acting he increased his dependence, but he accepted it as inevitable and considered himself entitled to protection.[48]

On the other hand, providing for the material needs of his farm servant was not the only obligation of the master. The protection offered by the Anavil to his *hali* reached far beyond giving him a living: he had to guarantee his existence in a much wider sense. This meant that the master should defend the social interests of his servant and that he was held responsible to some extent for all the actions of his subordinate.[49] Consequently, the *hali* of a prominent Anavil patron was assured of a reasonable amount of security and protection. For his part, he had to behave loyally, that is, do nothing that might provoke the displeasure of his master. Should conflicts arise, he was expected to take his master's side, even if it meant opposing his fellow caste members, the Dubla clients of the other Anavils.

As has been explained, the client enhanced the prominence of his patron. Apart from loyalty, the Dubla owed his Anavil master respect and submissiveness. His humbleness in word and gesture alone were sufficient to accentuate the superiority of the master. Old Anavil informants spoke to me of the great deference and timidity with which they used to be approached by the Dublas, while they censured severely the "impertinence" and "unmannerliness" of the agricultural laborers of today.[50]

48. "From close enquiries I have made, I can say with certainty that these poor people [*halis*] are treated with kindness and consideration, and are looked upon as humble dependents who have great claims on their masters and their families" is a comment from *Papers Relating . . . Chikhli Taluka*, 51. See also *Gazetteer of the Bombay Presidency*, 1883, pt. 7, 118.
49. See also Srinivas, 1955, 28.
50. Cf. Cohn, 1955, 61–62.

The Anavil's superiority was reinforced in a religious context when, for instance, at a marriage or a death in the house of his patron, the *hali* participated in the ceremonies in a clearly subordinate and servile way. This behavior was both a solemn confirmation of the tie between master and servant and an indication of the distance which separated them.

The dependence of the Dubla servants far exceeded that of the patron's clients who belonged to other castes. The Dublas had to rely on the benevolence of their masters for everything. The farm servant had no one else to turn to for favors; his Anavil master was the only one who was prepared to help him if he was in difficulties, but at the same time no one else could thwart and harass him as the master could.[51] Naturally this far-reaching bondage in the purely economic sense helped to inculcate loyalty and discipline in the farm servant, but it would be incorrect to suppose that only pressure could have induced him to go to such lengths.

Considering the weak position in which the *hali* found himself in all respects, it was in his own interest to side with his Anavil patron. By leaving no doubt of his loyalty, at least in his behavior, the Dubla hoped in his turn to compel his master to protect him. If in his attitude and behavior the farm servant also honored his Anavil as a benefactor in an exaggerated way, he did it partly to remind him of his obligations as a patron and to spur him to magnanimity.

Moreover, and this is characteristic of a relationship of patronage, the Dubla strengthened his own position in relation to third parties by identifying himself emphatically with his Anavil. The esteem and influence enjoyed by the master was shared by the servant, for a little of the farmer's importance was reflected upon his subordinate.

Earlier in this chapter it was maintained that bondage was attractive to the Dublas on account of the economic security it offered. Because of their contacts with the rich and powerful villagers and the privileges and protection that these contacts afforded them, the Dubla farm servants also rated higher in social esteem than their fellow caste members. The unattached laborers among the Dublas, some sources state, were even despised.

51. "Fear of eviction from the house site, fear of recall of loans, fear of loss of the additional income from the piece of land gifted to him and lastly, the security taken from a farm servant in the form of deferred wages to be paid only at the harvest, are more powerful and terrible sanctions than what the ordinary law may provide." (Sivaswamy, 1948, 30.) See also Beidelman, 1959, 36–37.

It has become a matter of prestige in the Dubla community to be a Hali, to work for a Dhaniamo. As a woman without having married a husband has no prestige in society, similarly a Halpati without a Dhaniamo as his master has no prestige in his community.[52]

The loyalty of the farm servant was said to be a byword. In the stereotype so stubbornly maintained by Anavils, the Dubla would go through fire and water for his master and identified his master's interest with his own. He could be counted upon to carry out as well as he could any instruction the master chose to give him. Attachment to a prominent Anavil gave the Dubla security and prestige, and the Dubla was devoted to him, if only for this reason. According to some authors, he was filled with contempt for those who deserted their *dhaniamo*.

Loyalty to the master figured as an important item in their moral code. They looked down upon those who were not and very often would not have anything to do with such deserters. The master on the other hand always appreciated this loyalty and put a premium on it by raising the status of his faithful dubla. There was not much of material reward but the labourer was admitted more and more into the confidence of the master, given a freer hand in his work, allowed a sort of familiarity and intimacy with the master's family and was trusted with work of a more delicate nature than that of a purely agricultural labourer.[53]

Owing to the frequent and intensive contacts between master and servant their relation lost much of its rigidity and stiffness, and came to bear a markedly personal stamp. The *hali* was the master's confidant. Being charged with the routine work he bore more responsibility than the casual laborers. Regarding all important agricultural questions the servant was consulted by his master, and the trust the latter put in him was apparent from his independence in carrying out orders. The *hali* knew about all kind of intimate family details of his Anavil which were hidden from outsiders and had to remain so. Even now every Anavil family can illustrate with incidents from the past the affection and loyalty of its former Dubla servants, which incidentally also enhance the glory of its own patronship. Such moving stories about the geniality governing the service relationship can usually be traced to an Anavil source and are not very convincing, if only for this reason.

52. J. Dave, translation from the Gujarati, 1946, 18. For a remark of similar purport, see *Report of the Hali Labour Enquiry Committee*, 1950, 48.
53. Kishore, 1924, 427.

### *4.6* Halipratha: *A System of Unfree Labor Mitigated by Patronage*

It is not difficult to recognize that servitude cannot have been so harmonious a relationship as is sometimes suggested in retrospect. The stress was on the "good" master and the "good" servant, that is, on the Anavil who carried out his obligations without stint or chicanery, and the Dubla who served his master loyally. Rather than the actual behavior it was the ideal relationship of *dhaniamo* and *hali* that was described, presumably on the strength of Anavil sources. Obviously the Anavils wished to create the impression that they treated their servants excellently. As patrons, moreover, they wanted to leave no doubt of their clients' loyalty.

Nor did the Dublas themselves, on being interviewed, give any evidence of dissatisfaction, sometimes to the rather naive-sounding astonishment of British government officials. As I explained, bondage was accompanied by many advantages, which partly compensated for the lack of liberty. But the servant had to avoid any semblance of impertinent action and behave in all respects as a subordinate and dependent. Open doubt of the master's good intentions, denial of his pretensions, was highly improper because it detracted from the Anavil's dominance. The *hali* should show himself satisfied and obedient. However, if the master refused to meet the obligations which had been established by custom, the *hali* could remind him of them by a show of indifference and indolence, or by remonstrating with his master in front of others. On the other hand, if the Dubla exceeded the limits of self-defense it might cost him dear.

Many reports of the nineteenth century asserted that the servants rarely ran away, but this, too, should not be adduced as proof of cordial relations between them and their master. To have a master meant for the Dubla to belong to the privileged category among the Dublas. The *hali* who deserted his master relinquished his social security, which he would do only in extremity. And where else could he have found work? In the traditional society, employment outside agriculture was extremely scarce. Moreover, the local isolation, a consequence of the poorly developed system of communications, made it difficult for him to judge the possible demand for labor elsewhere. But if alternative employment suddenly materialized nearby, he seized his chance with alacrity. In the eighteen-sixties, for instance, many farm servants left their masters when labor was needed for the construction of the railroad through the plain of south Gujarat.[54]

Such unexpected and temporary possibilities of escape aside, a *hali* decamped only if feelings had run too high. Practically the only way in which he could defend himself was by trying to withdraw completely from his master's authority. Precisely because this meant a total breach it was a step that the farm servant certainly would not take lightly, and moreover, it little profited the Dubla. At best he might succeed in finding a better master, but this prospect was too uncertain to warrant his severing the relation hastily.

Sometimes, however, the *hali* ran away on the instigation of another Anavil. As already mentioned, the relation between the most prominent members of the dominant caste as patrons was one of mutual rivalry. The master who could not keep his servants lost prestige. This was the more painful for him if a *hali* had been induced to leave his master to serve another Anavil in his own or a neighboring village. The code of the Anavils forbade such illegal transfer of a servant, and their caste *panch* saw to it that the rule was observed. The Anavil who engaged a *hali* who had left another master without the latter's permission was fined, and the Dubla himself was punished by caning. Yet it is difficult in retrospect to judge the effectiveness of these sanctions.

When a breach was not due to the struggle for power between competing patrons with the *hali* as prize, the *hali* could do little but fly from the village. Usually he went to the village of his mother or his wife, but he was not safe there. When informal collaboration between the Anavil Brahmans did not lead to his being brought back, the local authorities helped to trace runaway *halis*. In the first half of the nineteenth century, running away even rendered a *hali* liable to punishment by law.[55] Later, complaints about disloyal Dublas were no longer admitted, and officially the master could not do much against a runaway servant. The actual situation, however, had not changed a great deal. The Anavils remained the most important power elite in the principality of Baroda, and in the British

54. Enthoven, 1920, 347.

55. J. Vibart, Principal Collector of Surat, in a letter dated 16–12–1835, wrote, "By a letter from Government, dated 19th April 1822, the Magistrates are authorized to apprehend and return to his master any Halee who may abscond." G. Grant, Acting Judge and Session Judge in Surat, expressed the same view in a letter dated Feb. 22, 1836. This correspondence was included as Appendix to Report on Slavery in *Account and Papers*, vol. 16, session 15-11-1837 — 16–8–1838, pt. LI; see fn. 1 to this chapter.

part of south Gujarat, police functionaries and other local government servants were recruited from their midst. This meant that the masters could still informally appeal to the local authorities to compel a runaway Dubla to return.

Thus the Anavil landlords seem to have had the *halis* completely in their power. The members of the tribal and landless Dubla caste had no choice but to put themselves in the keeping of a master. This explains at the same time why the relationship began and continued with the approval of the servant. The tie attaching him to the master became increasingly close, and only on the master's initiative could it be severed. It is true that in the course of time the amount of support to which the *hali* was entitled had become standardized, but if necessary he had to accept less. The exceedingly unequal division of mutual rights and obligations was due to the difference in position between *dhaniamo* and *hali*. In theory, the right of the servant was the obligation of the master, but the former was not guaranteed in any way. The great economic, political, and social power of the Anavil made it possible for him to dictate, and in any case interpret, the service conditions one-sidedly. His obligations were rather in the nature of favors to be granted as he thought fit. The paternalistic attitude of the master could barely conceal the elements of enforcement that were inherent in the relationship. The servant was at his master's beck and call. Together with his family he had to comply with the latter's every wish, which might include sexual intercourse with the *hali*'s wife.

On the other hand, the element of patronage prevented any pronounced tyranny and complete exploitation, which would have detracted from the esteem in which the Anavil was held. Then, too, he had to rely on the support of his following. The patron's position stood or fell by his willingness to provide for and aid a group of followers, to let them participate in his goods, his power, and his esteem. The master must carry out a minimum of the obligations he owed to his servant. In the end, it was on the latter's labor that the success of the crop depended. In the traditional agrarian system, patronage not only served as a corrective of the lopsided distribution of scarce goods in traditional village India, but also accentuated inequality in all spheres of life. The presence and the symbols of subordination of the clients enabled the patron to demonstrate his superiority. As the economic and ritual distance between them widened, the relationship came to contain an element of exploitation, which in the case of the low-classed agricultural laborers was mitigated only by the mutual intimacy and sympathy inherent in the personal, usually hereditary tie with the landlord's family. In the concept of *patrimoniale Herrschaft* (patrimonial rule), Weber, perhaps harking back to his investigation into the conditions of agricultural laborers in the German provinces east of the Elbe River at the end of the last century, characterized the relationship in the following words:

> The master "owes" the subject something as well, not juridically, but morally. Above all — if only in his own interest — he must protect him against the outside world, and help him in need. He also must treat him "humanely," and especially he must restrict the exploitation of his performance to what is "customary." On the ground of a domination whose aim is not material enrichment but the fulfilment of the master's own needs, he can do so without prejudicing his own interest because, as his needs cannot expand qualitatively and, on principle, unlimitedly, his demands differ only quantitatively from those of his subjects. And such restriction is positively useful to the master, as not only the security of his domination, but also its results greatly depend on the disposition and mood of the subordinates. The subordinate morally owes the master assistance by all the means available to him.[56]

In south Gujarat it was the Anavil Brahmans who, as the dominant landowners, regulated the distribution of agrarian produce. It would be wrong to conclude that instead of production based on profit for a few, fulfilment of the material needs of all was aimed at. The Anavils were not altruistic to that extent. A more plausible explanation is rather that, for the landlords, maximalization of income took second place to maximalization of prestige and power, which in the nineteenth century amounted to behavior as a patron in the village. This objective required a minimal fulfilment of the need of clients for aid and protection. At that time, the Dubla *halis* were objects in the endeavor of the members of the dominant caste to acquire more power and esteem, not subjects in a surplus-oriented

56. Weber, 1962, 682. Concerning the past situation of agricultural laborers in the eastern regions of Germany, Weber remarks in another publication: "Except in times of extreme political turmoil, a class consciousness among the rural proletariat directed against the masters could only develop purely individually with relation to one master alone insofar as he failed to show the average combination of naive [sic] brutishness and personal kindness. That, on the other hand, the agricultural laborers were not normally exposed to the pressure of a purely commercial exploitation was in complete accordance with this. For the man opposite them was not an 'entrepreneur' but a territorial lord in miniature." (Weber, 1924, 474.)

market economy. An important factor in this process was the effort of the Bhathelas to attain the same position as the Desais partly by taking *halis* into their employ, which enabled them to imitate the life style of the higher status group.

On the basis of all these considerations it can be concluded that servitude in south Gujarat during the nineteenth century was essentially a form of unfree labor that was complicated and mitigated by a relationship of patronage.

# Chapter 5

# The Disintegration of the *Hali* System

## *5.1 Changes in the Relationships between Landlords and Agricultural Laborers*

During the past few decades, drastic changes have taken place in the relations between landlords and agricultural laborers in south Gujarat—changes which, in fact, amount to the distintegration of the *hali* system. We read in numerous reports that Dublas offered themselves as farm servants with growing reluctance or, if they were employed as such, that they ran away from their masters. For their part, the Anavils began to be less inclined to take on farm servants, and came to prefer casual labor, which they could engage and dismiss according to their needs. Similar developments are reported for other regions in India, and here, too, the greater freedom of the agricultural laborers is stressed and contrasted with their attachment in an earlier period.[1]

The decline of the *hali* system was gradual. There was certainly no sudden turn in the relationship between Anavils and Dublas. Nor is it readily conceivable that such an important institution in the social and economic life of a large part of the agrarian population should come to an end within a short span of time. Although complaints about runaway *halis* have never been scarce, it was from the beginning of the twentieth century onwards that mention was made of increasing friction between masters and servants. Time and again we read in such later reports that servitude has fallen into disuse, or at any rate will very soon be a thing of the past.[2] Nevertheless, in publications from the nineteen-thirties and forties, the Dublas were still called unfree laborers, in the employ of the large landowners.

1. See, for instance, the report of the eighth conference of agricultural economists, which was devoted to systems of wage payment in agriculture. Published in *I.J.A.E.*, III, 1948, no. 1. Cf. also Thorner, 1962, ch. 3.

2. *Census of India*, 1921, vol. III, pt. 1, 222; *Gazetteer of the Baroda State*, 1923, vol. I, 352; *Census of India*, 1931, vol. XIX, pt. 1, 254–55.

Occasionally these accounts contain the prediction that there was no reason to expect that the situation will soon change.[3] It appears that the latter view was the correct one, as was evidenced by the fact that in 1947, after India had become independent, the government of the then State of Bombay appointed a committee to inquire into the situation of the *halis* in south Gujarat.[4]

If anything becomes clear from all this source material it is that the change observed in the relationship between Anavils and Dublas was a gradual one. It is therefore impossible to say precisely when there was no longer question of a *hali* system. There are indications, however, that Anavil-Dubla relations changed most drastically after the First World War. According to Joshi, the *hali* system disintegrated between 1930 and 1950 in his village of field work. In another village situation, I. P. Desai believes this change took place between 1940 and 1960.[5] Their conclusion could be considered to apply to the whole of south Gujarat.

Although in the literature we find many more or less elaborate discussions as to how the disappearance of the *hali* system should be explained, it is striking that the authors dealt very briefly with the labor relationship that replaced it. Most of them confined themselves to the statement that day-laborers came to form an increasing percentage of the total labor force. As was pointed out earlier, that category was never absent in the traditional society, but the point is that their proportion rose sharply in recent decades — a development which also took place elsewhere in India. The authors refer only incidentally to the recruitment of these casual laborers and their labor conditions, and more generally to the nature of their contacts with employers. They create the impression that servitude was replaced by a casual and intermittent service relationship which was entered into only for limited periods varying from one day to a season. In this view, the close and permanent personal tie that had existed between employer and employee was replaced by a superficial and ephemeral contractual relationship.[6]

In part III of this study will be a found a detailed description, based on materials gathered during field work, of the pattern of re-

3. Shukla, 1937, 132–3; *Census of India*, 1941, vol. XVII, 74.

4. *Report of the Hali Labour Enquiry Committee*, delivered by M. L. Dantwala and M. B. Desai. Not published.

5. I. P. Desai, 1964, 162.

6. "During the fourth and fifth decades of this century, the method of permanent 'employment' of Halis gave way to casual labourers on a seasonal wage basis. This made the village labour market more or less 'free'" (Joshi, 1966, 57.)

lationships that has come to prevail between Anavils and Dublas in south Gujarat. At this point I intend only to sketch the circumstances under which the *hali* system disappeared. After a discussion of the factors which, in the literature, are regarded as responsible for this disappearance I shall briefly outline my own interpretation of the change that has taken place in the relationship between landlords and agricultural laborers.

### *5.2 Current Views on the Disappearance of the* Hali *System*

To explain the disappearance of a system of unfree labor—and the *hali* system has always been described in these terms—the obvious theory to apply is that of Nieboer, which was commented upon in chapter 2. His thesis can be summarized in the statement that an abundant supply of laborers obviates the necessity of their bondage. In such circumstances the landowners no longer need to protect themselves against shortage by assuring the permanent bondage of one or more agricultural laborers. This explanation of the disappearance of the *hali* system conforms with the widespread assumption that, owing to a process of agrarian pauperization under colonial rule, the percentage of agricultural laborers has risen sharply. As I said before, this assumption needs some modification. The returns of the successive censuses—its chief foundation—are too unreliable to be compared without correction, one of the reasons

TABLE 5.2.1

SHIFTS IN THE NUMBER OF AGRICULTURAL LABORERS, 1881–1921

| *Year of census* | *Agricultural laborers* | *Remaining agrarian population* | *—(percent of total agrarian population) Agricultural laborers* |
|---|---|---|---|
| 1881 | 259,205 | — | — |
| 1891 | 190,896 | 1,196,580 | 14 |
| 1901 | 372,964 | 634,688 | 37 |
| 1911 | 313,479 | 970,675 | 24 |
| 1921 | 295,815 | 1,058,182 | 22 |

being a varying definition of the concept of agricultural laborer. The same is true of the data concerning south Gujarat, as is evident, for example, from the figures for the state of Baroda.[7] The size of the

7. Data taken from *Census of the Baroda Territories*, 1881, and *Census of India*, 1921, vol. XVII, pt. 1, 364.

category of agricultural laborers, as indicated in the table, shows improbable fluctuations. Depending on the point of view chosen by the interpreter, their percentage of the total agrarian population can be said to have either fallen or risen in the course of time. A number of socioeconomic village monographs from the first half of this century contain more detailed information to go by.[8] However, although it can be concluded from them that land was increasingly fragmented, there is no evidence to show that in the end landholding groups became landless on any great scale. There are no indications of important shifts in land ownership during that period. A few smaller castes aside, Anavil Brahmans remained the most important landowners.[9] Their economic dominance in the plain of south Gujarat continued to be uncontested. Nor did there occur any drastic change in the position of the Kolis, most of whom continued to be peasants and tenants. It is true that they occasionally worked for the larger landowners, but such an additional source of income for them had already been noted in mid-nineteenth century reports. As before, it was especially the Dublas who worked as agricultural laborers. In short, there is no reason to assume that landowners or their descendants have become landless in any considerable number since the beginning of this century in south Gujarat.[10]

Growth of the proportion of agricultural laborers could, however, also be due to a sharp increase of the social category from which they were traditionally recruited. To test this assumption I have compared, in the following table, the rise of the number of Dublas in the state of Baroda with the population increase of their most important employers, the Anavil Brahmans.[11]

The two population categories show a clearly different demographic pattern. The numerical strength of the locally dominant caste has changed very little in the course of five decades. Birth control to lessen the risk of many daughters,[12] the prohibition for

8. Especially Mukhtyar, 1930, and Shukla, 1937.

9. Joshi (1966, 67) calculated that in 1901 the Anavils in his village of field work possessed 90.4 percent of the land, and in 1957, 93.1 percent.

10. Some reports state that, at the beginning of this century, landless laborers became owners of the unreclaimed, usually infertile village land. (*Land Revenue Administration Report*, 1923–24, 37; *Census of India*, 1921, vol. XVII, pt. 1, 360–367.)

11. The data have been taken from the 1881 *Census of the Baroda Territories*, 1883, 143; *Census of India*, 1921, vol. XVII, pt. 1, 344; *Census of India*, 1931, vol. XIX, pt. 1, 431.

12. *Census of India*, 1901, vol. XVIII, pt. 1, 234; see also *Census of India*, 1911, vol. XVI, pt. 1, 278, and Mukhtyar, *op. cit.*, 56.

TABLE 5.2.2

NUMERICAL INCREASE OF DUBLAS AND ANAVILS IN THE STATE OF BARODA, 1881–1931

| *Year of census* | *Dublas* | *Anavils* | *Ratio of Dublas to Anavils* |
|---|---|---|---|
| 1881 | 42,197 | 10,335 | 4.0 |
| 1891 | 48,886 | 11,148 | 4.4 |
| 1901 | 41,043 | 10,862 | 3.8 |
| 1911 | 50,623 | 9,916 | 5.1 |
| 1921 | 51,834 | 10,751 | 4.8 |
| 1931 | 65,459 | 11,818 | 5.9 |
| 1941 | 76,479 | — | — |

widows of this high caste to remarry, and, especially, migration are the chief factors that explain the stagnation of the Anavils in their native region. The landless Dublas, on the other hand, increased numerically by leaps and bounds. Although it was they and other economically weak castes who were most affected by crises, such as the famine at the turn of the century and the influenza epidemic of 1917,[13] the relatively higher birth rate of the lower castes combined with a gradual long-term decrease of mortality led in the end to an accelerated growth of this least-favored group. As the census returns of 1921 showed, the number of children per 1,000 married women between fifteen and forty years of age was much higher in the tribal castes than in the castes of Brahmans, being respectively 240 and 188.[14]

It could be concluded from this outline that, in consequence of a demographic shift, the supply of agricultural laborers gradually increased to such an extent that the Anavil landlords no longer needed to bind Dublas as farm servants (see table 5.2.2). Did the *hali* system, in other words, therefore disappear as a result of the accelerated expansion of the population of agricultural laborers?

A first objection to this line of argument is that the rise of a surplus in the first decades of the twentieth century would imply a

13. *Report of the Famine Operations in the Baroda State for the Year 1899–1900*, 1900, 3; *Census of India*, 1901, vol. XVIII, pt. 1, 231; *Census of India*, 1921, vol. XVII, pt. 1, 37, 42.

14. *Census of India*, 1921, vol. XVII, pt. 1, 212; also *Census of India*, 1931, vol. XIX, pt. 1, 115. To my knowledge, the demography of the caste system has not been much researched, although the social consequences of unequal numerical increase have great weight.

previous shortage of agricultural laborers. Now this I refuted emphatically in the foregoing chapter. At most we can speak of periodic scarcity of labor, but in this respect the situation did not change. Complaints on the part of landowners that there were not enough hands at the peak of the season to cope with all the work that had to be done were a recurrent theme.[15] If it was true that, in the traditional society, the *hali* system did not function in circumstances in which there was a shortage of agricultural laborers, it may be too hazardous to explain its disappearance as being the result of a surplus.

To summarize, there are not enough indications of a process of agrarian impoverishment leading to a rise in the percentage of landless laborers. We do know that, owing to a differential demographic pattern, the proportion of agricultural laborers within the agrarian population has increased, while on the other hand the dominant caste has lagged far behind in population growth. It is undeniable that this change in numerical ratios has affected the relations between Anavils and Dublas, but in my opinion the suggestion of a direct connection between the growth of the category from which the agricultural laborers were mainly recruited and the disappearance of the *hali* system does not take into account all the facts.

It is therefore not surprising that the Anavils' tendency to stop employing attached laborers is usually attributed to completely different reasons. Of these, the cost argument is the most important. Because of a sharp rise in wages — they are reported to have doubled within a span of ten years, from 3.6 annas in 1907 to 7 annas in 1917 — few landlords would have been able to provide for one or more laborers throughout the year.[16] The amount to be paid by the master when the *hali* entered his service also rose considerably, while owing to the pull of urban centers the risk of the *hali* running away was much greater than before. Finally it is suggested that subdivision of the land had affected the landlord's financial capacity.[17] Especially in the crisis years after 1930, many of them were obliged to dismiss their farm servants.

Yet this line of argument, which presupposes a deterioration in the landlords' economic standing, is equally unconvincing. In the

15. Mukhtyar, *op. cit.*, 159–61; M. B. Desai and C. H. Shah, 1951, 49.

16. *Census of India*, 1921, vol. XVII, pt. 1, 42; see also *Census of India*, 1931, vol. XIX, pt. 1, 255.

17. *Papers Relating . . . Jalalpor Taluka*, 1900, 8. This argument is difficult to reconcile with the slight numerical growth of the largest group of landlords, the Anavil Brahmans.

first place, the statements about the large wage increases should not be taken too seriously. It was the money value of the allowance that increased, for the allowance itself continued to be paid largely in kind for a long time and certainly did not increase.[18] This was only slightly less true of the "loan" contracted by the farm servant at the beginning of his service. This loan was usually meant to cover the expenses of his marriage, and as a rule the master himself continued to supply the necessary food and other requirements for this purpose.[19]

As production for the market expanded — this was especially the case after the First World War — the proportion of the total wage that was paid in money did increase a little. But owing to the rise of prices of articles which the farm servant formerly was given but now had to buy — salt, tobacco, cooking fuel, etc. — this partial changeover to a money wage was not detrimental to the employer. Again and again we read that the rise in the wages of agricultural laborers lagged behind that of the cost of living.[20] The same is true of the allowance received by the servant on the occasion of his marriage.[21]

The economic crisis of the nineteen-thirties may have accelerated the disintegration of the *hali* system. Whereas formerly *halis* are said to have been dismissed in years of adversity but taken on again when times were better, the farm servants were now no longer taken on at all. Nevertheless, the alleged weakening of the economic position of the landlords does not, in my view, offer a satisfactory explanation of the fact that bondage fell into disuse.

Much more plausible is the finding that, because of a drastic change in the crop system, the landlords no longer needed to use permanent labor. In a village study of 1927, Mukhtyar posited a connection between the cultivation of sugar cane and the existing system of labor relations.

> In this village, and, in point of fact, in the whole of Southern Gujarat, there still obtains a system of labour called the Hali system in which a

18. J. M. Mehta, 1930, 127; Shukla, *op. cit.*, 123–28.

19. "The loans in kind form a substantial proportion of the total advance, as almost the entire quantity of foodgrains, pulses and vegetables required for feasting on the occasion are advanced by the Dhaniamo. The other significant factor about the marriage loans is that as compared with the rise in the prices and the cost of living, the increase in the total computed expenses on the Hali's marriage has been much lower." *Report of the Hali Labour Enquiry Committee*, 26.

20. *Census of India*, 1931, vol. XIX, pt. 1, 269; *Report of the Hali Labour Enquiry Committee*, 30; C. H. Shah, 1952, 440–61; Shirras, 1922.

21. See fn. 18 above.

labourer mortgages his labour to the farmer for a loan he takes for celebrating his marriage. A capitalistic cultivator keeps one or two Halis for performing field operations. He is bound to maintain them whether he exacts work from them or not. He, therefore, deems it wise and profitable to occupy them in sugarcane-cultivation.[22]

The decline in the cultivation of this highly labor-intensive crop after the First World War led to vastly diminished demand for permanent labor. Joshi, in the history of another south Gujarat village, also sketches a contraction of agricultural activities after the nineteen-thirties, when Anavils first began to cultivate mango trees. Whereas laborers formerly were needed throughout the year for sugar cane and garden crops—such as ginger, spices, root crops, and bananas—most of the work could now be performed by temporary hired laborers in the picking season. The new crop was introduced only gradually, and this prevented a sudden end to bondage. New *halis* were no longer engaged. Dublas who were employed as farm servants remained in service.[23] On the basis of these data it is plausible to say that the cause of the disappearance of the traditional service relation should not primarily be sought in the impoverishment of the Anavils, but rather in the changeover to a crop system not entailing very intensive labor.

However, it would be incorrect to create the impression that the landlords took upon themselves the initiative of abolishing the *hali* system. A number of reports hold the opposite view, that is, that the change in the labor pattern was forced upon them. In the preceding chapter I explained that bondage was certainly not one-sidedly imposed on the Dublas. On the contrary, by becoming *halis* they were assured of at least a minimal existence in the more or less closed village economy. With increasing opportunities for employment outside the village, servitude lost much of its attraction for these landless laborers. Fewer and fewer Dublas were prepared to enter into an agreement which, in practice, bound them for life.

As time went on, the possibilities of finding a livelihood outside agriculture for themselves in the region gradually increased. After a period of stagnation in the last century, the cities of Surat and Navsari began to prosper again, and smaller towns along the railroad began to develop into industrial centers of some importance.[24] Moreover, seasonal migration over longer distances contributed

22. Mukhtyar, *op. cit.*, 97.
23. Joshi, *op. cit.*, 57.
24. See, for instance, Kapadia, 1966.

considerably to the expansion in employment. On a small scale at first but in ever-increasing numbers, especially since the Second World War, agricultural laborers and small peasants migrated during part of the year from the south Gujarat villages to the environs of Bombay. They worked there in the dry season—the period between November and the end of May—in salt-pans and brickyards in the open air. It might therefore be concluded that the Dublas felt less and less attracted to agricultural servitude. The same line of thought led to the observation that they no longer relied on the support of the landlords who enabled them to marry; it should be borne in mind that precisely the essence of the *hali* system has been deemed to be the necessity for the *hali* to take a "loan" for that purpose. The Dublas began to feel that complete dependence on the master, which they formerly accepted as security, was a heavy burden. This is the tenor of reports in which the disappearance of the *hali* system is attributed to increased mobility of the Dublas. Conversely, of course, the risk of losing money lessened the landlord's inclination to take on new *halis*. Another indication in this direction is that at one time the Anavils advocated the introduction of work-cards in order to prevent a farm servant from leaving his master without permission.[25]

The disintegration of the *hali* system can undoubtedly be interpreted in various ways. Some authors are inclined to put the stress on the inability or unwillingness of the Anavils to take farm servants into their employ. Others think it was the Dublas who, against the will of the landlords, withdrew from the traditional service relation. In both cases, however, the explanation of the disappearance of bondage is a purely economic one: a change in the situation which determined the demand for labor or, conversely, the need for work.

The available data show unmistakably that, economic changes aside, factors of a politico-social nature played a role. The traditional relationship between master and servant was increasingly undermined by external pressure. In the first place, the elements of coercion inherent in the *hali* system were no longer thought to conform with principles of good colonial government. Although nineteenth-century reports never glossed over the lack of freedom of the *halis*, the judgment of administrators in the present century was more harsh. They compared the *halis* to pre-Civil War plantation

25. Memorandum by Rao Saheb B. M. Desai to the *Royal Commission on Agriculture in India*, 1927, vol. II, pt. 2, 577, 601.

slaves in the southern United States.[26] In the state of Baroda in 1923, it was announced that to force laborers who were indebted to a landowner to perform labor in compensation was prohibited.[27] As for the British part of south Gujarat, the *hali* system was never prohibited there officially, but local officials were forbidden long ago to assist in tracing runaway farm servants. We should of course not delude ourselves that this regulation was carried out. In practice the Anavils could always count on the help of the lower government officials. It is therefore not surprising that an official prohibition such as that issued by the state of Baroda was not very effective either.

A man refusing to give his work, for which he has taken a loan is sometimes improperly prosecuted for cheating; and I regret to say that magistrates have sometimes taken a strangely perverted view of such prosecutions.[28]

Yet the fact that the landlord could no longer legitimately appeal to the local authorities when a farm servant had deserted him is of some importance. As I explained in chapter 3, the bureaucratic apparatus gradually became a little less susceptible to the interests of the Anavils. Thanks to the recent centralization of administrative decision-making on the district level or higher, the arbitrary activity of lower government officials — themselves often Anavil Brahmans — was increasingly weakened. While, formerly, to favor their own relatives and fellow caste members was a perquisite of their authorized position, Anavils who now wished to take advantage of their official standing for this purpose were guilty of unlawful actions.[29] As landlords the members of this caste remained dominant in the region, but the increasing bureaucratization of local authority inevitably led to a lessening of their authority. Those who had acquired a position in the government apparatus now had fewer opportunities to give priority to the interests of their own group than they had fifty years earlier.

The diminishing influence of the Anavil Brahmans is, for instance, apparent from their vain attempts to persuade the government to take measures against runaway *halis*. Their spokesmen ar-

26. *Census of India*, 1921, vol. III, pt. 1, 220.
27. *Census of India*, 1931, vol. XIX, pt. 1, 255.
28. Symington, n.d., 50.
29. "Several cases have been recorded in which the Dhaniamo employed the agency of the Government under the old regime for the purpose. It is only recently that the farmer's position has weakened in this regard." *Report of the Hali Labour Enquiry Committee*, 24.

gued that legal sanctions were needed in order to safeguard agrarian production. This attempt to curtail the mobility of the agricultural laborers was no more successful than the landlords' proposal that requests by agricultural laborers for a wage increase should be made an indictable offense.[30]

The attitude of the colonial government towards the *hali* system was also influenced to some extent by the pressure exerted by the independence movement. Gandhi, who was exceedingly familiar with the situation in south Gujarat, was one of the first national leaders to condemn *halipratha* in no uncertain terms. During his stay in the region he repeatedly declared that a struggle against colonial suppression was justified only if it put an end to lack of freedom in indigenous society, such as was suffered by the Dubla *halis*. The denunciation of the traditional service relationship, repeated explicitly in a report of the land-reform committee of the Congress Party in 1951, was a setback for the landlords.[31]

Partly in consequence of the Gandhian social movement, *halis* no longer acquiesced in bad treatment, but left their masters for good.

> The Dublas are undoubtedly becoming more independent, and more capable of combining in defence of their rights or interests. If they think they are stinted of their food ration or not paid a fair wage, they will desert the village in a body, as happened this year in Afva, Vankaner, Timberva, and other places.[32]

It should be added that increased contacts with the towns and the possibility of finding work outside agriculture if desired has made for a more independent attitude among agricultural workers.

Such reports show that those laborers rebelled against their bondage, and suggest that more and more Dublas preferred the freedom enjoyed by day-laborers.

### *5.3 A Supplementary Explanation*

Whether the changes in relationships are primarily attributed to landlords or to agricultural laborers, or whether they are connected with economic or politico-social factors, the common denominator of most of these views is that they portray the disintegration of the *hali* system in terms of a changing *labor* relationship: a changeover from the bondage of servitude to the freedom of the day-laborers'

30. *Baroda Economic Development Committee, 1918–1919*, 1920, 131–32; Dave, 1946, 33.

31. *Report of the Congress Agrarian Reforms Committee*, 1951, 128–30.

32. *Report of the Special Enquiry . . . Bardoli and Chorasi Talukas*, 1929, 15.

existence. Although this representation is not wholly incorrect in itself, it is far from complete.

In the preceding chapter I distinguished another dimension of the *hali* system. Servitude is more than a labor relationship characterized by lack of freedom. The relationship between the *hali* and his master has been explained by me partly in terms of patronage. There is reason to assume that the patron-client dimension in the relationship between landlord and agricultural laborer has lost much of its importance since the beginning of this century.

This background should figure largely in any explanation of the disappearance of the *hali* system, and it is an aspect that emerges clearly from the information on changes in the relations between landowners and agricultural laborers elsewhere in India in recent decades. Srinivas and Bailey even say explicitly that the elements of patronage ceased to exist when, in the villages, traditional servitude fell into disfavor with the rise of a market economy.[33] Similar observations are made by other sociologists who have remarked on the loss of the close tie that had bound landless families to landowning families for generations.[34] The vertical pattern of relationships that was characteristic of the traditional situation was replaced by mutually antagonistic attitudes. Landowners on the one hand, and agricultural laborers on the other, have become parties with conflicting interests, and contacts between them increasingly reflect that situation. This process of change was closely related to the disruption of the stationary village economy and a break-through to an expanding economy on capitalist principles. Improved communications with the outside world, along with increasing production for the market, resulted in an enlargement of scale that put an end to the more or less closed order of the traditional rural community. The relationships between the various categories which it comprised were organized on new lines. In the various regions of India this process of course took place at different times, but in general the turn of the century is considered the point of beginning, and the First World War and the period after Independence as phases of acceleration. On the basis of my own investigation I shall set forth in the following chapters the course taken by this process in south Gujarat, as reflected by the changes in relationships between landlords and agricutural laborers.

33. Srinivas, 1955, 27; Bailey, 1957, 143–44.

34. See, for instance, Gough, 1955 and 1960; Cohn, 1955; Neale, 1962, ch. 9; Sivertsen, 1963; Béteille, 1965, ch. 5; Orenstein, 1965; and Harper, 1968.

# *PART III*

*Chapter 6*

# Present Relationships between Landlords and Agricultural Laborers in South Gujarat

## *6.1 Plan of the Field Work*

An analysis of the relationships between landowners and agricultural laborers after the disappearance of the *hali* system was, in brief, the starting-point of my field work in 1962–63. There seemed no doubt that the *hali* system no longer existed. "There are no *halis* any more, they belong to the past," I was told at the highest administrative level when I paid a courtesy call there and explained the purpose of my stay.

This remark by an official was intended as more than a statement of fact. The word *hali* has ominous overtones in official ears, for its connotations with bondage—for many people even with a kind of slavery—are inconsistent with the image of India after Independence. It seemed not only obvious but also reasonable that officials preferred to speak of the *hali* system in the past tense. In various publications during the past few decades its imminent dissolution was indeed announced. The reliable study by V. H. Joshi on the process of social and economic change in a south Gujarat village, moreover, confirms this assumption. Joshi arrives at the conclusion that the *hali* system no longer exists. He describes a situation in which the agricultural laborers dispose freely of their labor; as a rule they work as day-laborers for one Anavil landowner or another.[1]

In short, these conclusions seem to confirm that the repeated predictions have at last come true. Formerly bound farm servants, the Dublas have now become free day-laborers.

Yet when I began orienting myself in the area of research in preparation for field work, I soon learned that this view is not en-

1. V. H. Joshi, 1957, 53–64.

tirely balanced. It is true that some of the agricultural laborers do not work for the same landowner day after day. Aside from these casual laborers, however, there is still a category, differing widely in numerical strength and depending on the locality, that consists of farm servants. On inquiry they turned out to have been employed for years, sometimes all their lives, by the same landowner. And I could not escape the impression that their labor agreement paralleled strikingly that of the *halis* of the past.

Should I conclude that the traditional service relation still existed, though to a differing extent in different places? This was indeed the opinion of a number of local political leaders and social workers whom I interviewed about it. Nevertheless their view may well lack discrimination. As I soon discovered, the nature of the farm servant's relation to the landlord has changed so drastically that it would be erroneous nowadays to call him a *hali.*

To begin with, I sought the changes especially in the disappearance of the unfree elements that had been inherent in the service relationship. In other words, certain Dublas are still servants but are no longer bound. But I was quickly compelled to abandon this point of view. The bondage of the agricultural laborers has not wholly ended. To suggest a transition from bondage to freedom is to give an incomplete picture of the process of change that has taken place. In the course of my field work I arrived at the conviction that this approach does not cover all the aspects and is even somewhat incorrect. Gradually I shifted the emphasis to the elimination of the element of patronage in the relationship between landlords and agricultural laborers in south Gujarat. Particularly in this sense, I believe, the term of *halipratha* no longer applies.

Thus, at the beginning of the field work proper I was burdened with a confused mass of contradictory information, and, being unfamiliar at the time with the past history of the situation, I opted for an open statement of the problem, which I formulated as follows:

—How is the disappearance of the *hali* system to be explained?

—What relationship between landlords and agricultural laborers has replaced it?

Without knowing clearly what I would find, I could in this way express my conviction that a good understanding of the pattern of relationships that has come into being required the gathering of data on both the traditional service relationship and the manner in which it changed during the past few decades. As I have provided a detailed description of the past situation in part II, I shall concentrate on present conditions in the report of my village study.

The way in which I stated the problem further implied that I should give close attention to the change in those contacts which are determined by the work situation: the recruitment of labor past and present, the varying conditions of employment of agricultural laborers by landowners, the duration of the service relationship, etc. But any shifts in this pattern of economic interaction, in the labor system, reflect a process of change which reaches much farther: a change in the relationship of Anavil Brahmans and Dublas, the wealthiest and poorest groups in the rural society.

### *6.2 Method*

Before venturing to prepare a plan of research, I spent a month in a centrally situated, randomly chosen village in the plain of south Gujarat to get my bearings and to improve my command of the language. During this period I could familiarize myself with the field-work situation and accustom myself to local conditions. The advantages were many—for instance, I need not fear any too drastic consequences of my investigation proper if I should trespass in some way, and in particular if I should infringe any rules, an error which every stranger in a caste society seems to commit sooner or later. That initial stage was especially valuabe because it enabled me to obtain a better insight into the subject of inquiry which I had formulated in a distant country with no knowledge of the local situation. This first and as yet sketchy orientation was much more effective and profitable than a study of the literature during the same short period could have been. Traveling throughout the area, I was hospitably received in a great many villages; I talked to numerous informants, such as civil servants, "social workers," landowners, and agricultural laborers. During this phase of reconnoitering I decided on the definitive plan of my investigation and formulated the basic questions with which I began my field work. This first orientation, finally, prepared the way for the work and enabled me to select on less arbitrary grounds the two villages in which I was going to stay longer to collect data.

My subject of study was the changes in the relationships between landlords and agricultural laborers. From the outset I intended to select villages where part of the population consists of Anavil Brahmans and Dublas. For it was principally the members of these castes who were looked upon as the elements in the *hali* system, and it is they who still form the most important groups of landowners and agricultural laborers, respectively, in south Gujarat. But the two castes are not represented everywhere in equal strength. The Anavil Brahmans live chiefly in the *talukas* farthest to the south, hence the northern part of the plain did not primarily qualify for field work. Although there, as elsewhere, the majority of the agricultural laborers are Dublas, the landowners mostly belong to the caste of the Kanbi Patidars. To be quite clear, in the region to which the field work was largely confined there were other landowners besides Anavil Brahmans, whereas the agricultural laborers were not all Dublas. I concentrated my attention especially on those two castes, however, because the aim of the study was not to analyze the relationship between all employers and employees in agriculture but principally to trace changes in the relations between Anavil landlords and Dubla agricultural laborers.

From conversations with informants during the period of orientation, the fact emerged time after time that, aside from the social position of the Anavils as landlords, their relationship to the Dublas depends to a great extent on the following circumstances:

- —The nature of the crop system: When agricultural operations are restricted to a few months of the years, the landlords prefer to employ casual labor. The use of permanent farm servants, on the other hand, is more feasible in areas of perennial agriculture.
- —The extent of geographic isolation: If urban centers and the railroad in the west are nearby and easy to reach, the agricultural laborer can more easily find work outside the village. The farther from the city and the less convenient the transportation, the smaller his chance of alternative employment. Geographic isolation thus retards the disruption of the traditional relationship between master and servant, as I was told several times.
- —The political background: It was suggested that in the British part of south Gujarat the *hali* system fell into disuse earlier because of the activity of the Congress Party and especially through the influence of Gandhi and his followers. In the territory of the principality of Baroda the process of change was

much slower. There the Anavils had full scope and obstructed the emancipation of the Dublas.

I made the study a comparative one by dividing the field work between two villages that differed in the above respects. These factors, which predicted different patterns of relationship between Anavil landlords and Dubla agricultural laborers, were not used as a hypothesis, that is, they were not carefully formulated in a preliminary stage and verified by field work afterwards. They simply served as a lead for me to make a selection for research on the village level.

The choice finally fell on Chikhligam and Gandevigam, two villages which I named after the *talukas* in which they are situated and which are comprised in the present Bulsar district. To arrive at a decision I proceeded as follows. On the basis of data given in the district handbook, I listed a number of villages which at first sight seemed suitable for field work.[2] I visited some of these villages myself and made inquiries as to the social and economic background of others from local authorities. In continuous consultation with Professors I. P. Desai and M. B. Desai in Baroda I finally made my choice, in which an important consideration was also the possibility of separate accommodation for me.

With regard to the three factors outlined above, the two villages that were selected present contrasting aspects, which enabled me to carry out my plan of establishing, in the course of the investigation, the weight of each of those factors separately and explaining any difference in the results in this way.

Labor demand in Chikhligam is greatest in the periods before and after the rainy season. The winter months are the slack period in agriculture. Gandevigam, situated in a region called the garden of south Gujarat, has irrigation agriculture. The crop system there requires continuous employment of labor. The yield of agricultural production in this village is much higher than in Chikhligam and the produce is very largely sold on the market. Chikhligam is difficult to reach, and alternative nearby employment is completely lacking.

2. *Surat District Census Handbook*, 1951. At the time of my stay, the district handbook based on the 1961 census was not yet available. For the social-research worker the handbook is an excellent guide, providing various facts on every village: total number of houses and households, number of inhabitants divided according to men and women, number of illiterates, working population in and outside agriculture, the nearest railroad station and market place, and finally the facilities available in the village, such as school, cooperative, etc. As others discovered before me, however, the data given are not wholly reliable.

Gandevigam is on an asphalt road within easy reach of some towns containing various industries. Finally, in former times Chikhligam was under British rule, whereas Gandevigam belonged to the principality of Baroda.

In both villages — and this was the first criterion of selection — Anavils and Dublas are the leading categories of landowners and agricultural laborers. Labor demand varies not only with the nature of the crop system but with the social position of the landowners. The category of Anavil Brahmans which is regarded as higher, the Desais, have always been landlords. Some of the Bhathelas, on the other hand, still work on the land themselves, although the number of active cultivators among them has sharply declined in recent years. The greater differentiation within the latter group makes it possible to trace the influence of the landowner's status, which is determined by the amount of physical work he does on the land, and on the nature of his relationships with agricultural laborers. It is for this reason, not primarily because they form the majority of Anavil Brahmans, that I concentrated my research especially on the Bhathela group. Yet in this respect, too, there is a difference between Gandevigam and Chikhligam. Like the other garden villages in this region, the former is typically a Bhathela village, but Chikhligam is situated in a markedly Desai area, and one of the Anavil families in the village belongs to this category. The Desai style of life, in other words, has always been an example which could be observed at close range in Chikhligam.

Both Chikhligam and Gandevigam have in common with the other villages, which I considered in the first instance for field work, that the population of each was said to be less than 1,500. The division of the field work between two villages precluded my staying much longer than three months in each. By selecting two villages that were not very large, I hoped to avoid as much as possible the difficulties incident to the short duration of each stay. I am well aware of the limitations imposed on the investigator by a brief sojourn. I had no time, for instance, to interview systematically any village inhabitants other than Anavils and Dublas on the subject of the relationship pattern between the members of these castes of landlords and agricultural laborers. And I am aware that other such gaps exist.

Finally, I used my time as follows: The first phase was spent, between October 1962 and January 1963, in the city of Baroda and in south Gujarat, where I stayed in a pilot village for an experimental period. Early in February I installed my family, who had arrived from the Netherlands, in Baroda. The greater part of the field work proper took place between February and July. During the rainy season, from July to September, I was occupied mainly with literature study in Baroda and Bombay. I resumed the field work in September and concluded local research on the village level at the end of November.

### *6.3 Procedure*

"May I see your questionnaire?" I was often asked when I explained the aim of my investigation in a university or the office of a government body. This question is in accordance with the marked preference in India for numerical data which are collected in great quantities during fleeting contacts with respondents. As mentioned in the preceding paragraph, I used the research technique of participant observation in attempting to gain an insight into the relationship between landlords and agricultural laborers in two villages in south Gujarat. At the end of my stay I verified my findings by a brief investigation in a third village on which I shall report in the final chapter.

Three months' field work in each of the two villages is scarcely adequate for participant observation. It could be said that the shorter the period spent in the village by the investigator, the more emphasis is laid on interviewing rather than on observing. It is true that I confined my attention in the main to the Anavils and Dublas, but the presence of other castes partly determined their relationship with each other. A minimum of knowledge of the whole village situation was therefore indispensable.

The central question was: What type of relationship has replaced the *hali* system? I tried to find the answer by collecting data concerning contacts between them from all Anavils and Dublas in Chikhligam and Gandevigam. I interviewed the members of the Anavil caste as to the size and composition of their property in land, the nature of their crops, the number of laborers in permanent service, their need of day-laborers, the wages they paid, working hours, and the like; insofar as they themselves or any members of their households were employed outside agriculture I questioned them concerning their work and education. The Dublas, on the other hand, provided me with information on the nature of their service status; for whom and for how long they had worked as farm servants or whether they were day-laborers, their wage and indebtedness, the daily routine of an agricultural laborer, which members of their households also worked on the land or in the master's house, and possibly what work they did, temporarily or permanently, other than agricultural. I have not wholly succeeded in quantifying my data on the economic situation and style of life of the two castes, owing to the short duration of my stay as well as to my impression that I had not been able to put into figures all the relevant facts needed for a really reliable report.

Thus I tried in vain to establish the average indebtedness—a wage component, as I will explain—of the agricultural laborers. The farm servants sometimes could not (would not?) tell me even approximately how much they owed and by how much the debt had increased in the course of time, while the amounts stated by the landlords were certainly too high.

A budget survey had to be omitted, because I could obtain only an incomplete picture of some important components of receipts and expenditure and their distribution throughout a year. The distortion caused by the sharp seasonal fluctuations can be corrected only by continuous registration. It is even more difficult to calculate the income from theft of crops or the expenditure on illegally distilled spirits, concerning which, understandably, the Dublas are not very communicative.

Nor have I been able to state with precision the frequency of running away among farm servants. The landlords' statements on this point turned out to be exceedingly unreliabe, yet I could not ask the Dublas who had left the village the reason of their departure. The Dublas who had come to live in the village were sometimes apt to conceal their motive for coming, but I could by no means assume that in all cases they had run away from a master in another village.

The condition of pauperism in which the Dublas live does not easily lend itself to the collection of numerical data. Many of them reacted vaguely and evasively to concrete and exact questions, because of ignorance, indifference, and suspicion.

Finally, too, the lack of stability of the unit of co-residence, another phenomenon of poverty, was responsible for the respondents' inability to give exact replies to questions. They sometimes had to remain silent when I asked after a husband, children, or parents because they did not know their whereabouts or present circumstances. Then, too, facts are concealed for completely different reasons. I did not succeed, for instance, in proving numerically that only a very small proportion of the Dubla children attend school. I was readily given access to the list of registered children, which contained a great many Dubla names. In conversation, the teacher did not hesitate to admit that most of them permanenty stayed away. But my attempts to take samples were resented and had to be abandoned.

Instead of suggesting an exactness that would be unfounded I have used the material I collected to give a broad outline and sketch some tendencies. A further quantification is certainly possible but requires a different method from the one I used.

My field work was appreciably influenced by the antagonism that exists between Anavil Brahmans and Dublas. It was not difficult to arrive at a good understanding with the members of the dominant caste. I was introduced into the villages by them and in their houses. They regarded it as evident that I would stay in their village quarter and that they were my obvious, even my only informants. I also found it easy to associate socially with them, not least because the English of some of them was better than my Gujarati. Especially at the beginning of my stay I spent most of the day in their company, and in fact this was expected of me. During that period I was able to collect material relating to the whole village system and to the Anavils themselves. The information they gave concerning other castes was generally very superficial and, moreover, not very reliable. The strikingly unfavorable picture, for instance, that the Anavils gave me of the behavior and way of life of Dublas was of interest as evidence of their dissatisfaction with the attitude of these subordinates but turned out to be completely incorrect in various respects.

Whereas the Anavil Brahmans were relatively easy to approach, it was much more difficult for me to build up a confidential relationship with the Dublas. I had to use some tact to overcome the resistance of the landlords of the village, who were not very pleased with my association with agricultural laborers. "You should ask us what you want to know," they gave me to understand. I made as many allowances as I could for their susceptibility, but ultimately I went my own way.

A much more pressing problem was the suspicion exhibited by the agricultural laborers as to my intentions. They identified me with the dominant caste, which was understandable enough.[3] I stayed in the midst of the Anavil Brahmans, which made a close connection with them inevitable. In their quarter I had been accommodated in an empty house. The Dublas met me for the first time in the houses of their masters, where I had been invited as a guest, or on the fields in the company of their landlords, who were accustomed to speak of the Dublas in the rudest and most offensive terms even if a servant or maid was present, with me as a silent observer.

The much greater distance that separated me from the Dublas may explain the reticence and suspicion with which they reacted to my first attempt at approaching them. For me this was the tensest period of my stay in the village. Fortunately, like the rest of the inhabitants the Dublas knew I was an outsider after all, as by my attitude and behavior I did not correspond with the picture of a high-caste Hindu: for instance, I had no objection to drinking illegally distilled spirits with them and tried to associate with them on a footing of equality. The friendship I had formed with a Mochi boy in Chikhligam, and with a Moslem boy from a village adjoining Gandevigam, stood me in good stead in this phase of the field work. Thanks to them the communication barrier with the Dublas was broken through for me—my Gujarati not being up to the dialect of this tribal caste—because, being members of the less favored groups in the villages, the boys were acceptable to the agricultural laborers as my interpreters.

Contacts with the Dublas multiplied as my stay in the village continued. The evident vexation this aroused in some of the Anavils, who regarded my behavior as improper, enhanced my reliability in the eyes of their subordinates. Answering my questions, however, continued to make many Dublas uneasy. After all, to give information about one's master was to commit oneself. To allay their suspicion I therefore rarely made notes on the spot of the conversations I had with them.

Usually I managed to keep out of the quarrels between Anavils and Dublas. If I did become involved for some reason or other and my role might be in doubt, I never hesitated to take the side of the latter. The following occurrence is illustrative. One evening, while I was sitting with a Dubla in front of his hut, his master came to call him in angry tones because he had not turned up for

3. See Baks, and others, 1965, 171–73.

work that day. A dispute followed, and when the landlord, obviously expecting approval, drew me into it I said something to the effect that anyone might like to take a day off occasionally and told him straight out that I did not care for his remarks about the laziness and mendacity of Dublas in general. As a result, my relations with this Anavil lost much of their cordiality, but this was something I had to accept. Emotionally, any such rather noncommittal expression of sympathy with the agricultural laborers came easy for me; moreover, it was of the utmost importance for my investigation that I should remain on good terms with this not easily approachable group.

For the same reason I addressed myself directly to the agricultural laborers in the investigation I carried out in a third south Gujarat village for checking purposes. Considering the short time I could spend there, it was impossible for me to win the confidence of both the landlords and the laborers, and the greater usefulness of good contacts with the Dublas weighed more. Since I did not live in the village itself but at a short distance in the small town of the subdistrict, there was, moreover, no need to take much notice of the landlords' feelings. Instead of going to them first I went, from the first day onwards, straight to the Dubla village quarter. Thanks to an introduction which I had likewise acquired without resort to the landlords, I stood well with the Dublas from the beginning. But the local power elite were greatly offended. After a few days the village headman—seated on a horse, the only one in the village, to emphasize his dignity—came to the Dubla quarter and without giving me so much as a glance interrogated the Dublas in a surly tone about "him there": "who is he?" and "what is he after?" This was wholly unnecessary as the landlords certainly knew all about my conversations with their subordinates. Partly to prevent them from venting their annoyance with my actions on their servants, I tried towards the end of my stay to get into touch with the landlords after all. I was accorded an icy reception. Some did not wish to receive me at all, and sent word from the inner recesses of the house that they were not at home.

The atmosphere of friction and conflict prevailing between landlords and agricutural laborers unquestionably made my field work more difficult. By including both these groups in the investigation, however, I could arm myself against partiality and avoid becoming too easily accustomed to caste cohesion and group judgment. A further advantage was the possibility of checking the reliability and completeness of information. Many of the discrepancies between the statements of Anavils and Dublas turned out, on further inquiry, to be due to a difference in emphasis. Widely disparate statements on the agricultural laborers' working day, for instance, were explained by the fact that Dublas measured it by the working hours at the peak of the season, whereas Anavils thought in terms of a normal day in the slack period. In this way I was able to correct the distortion that inevitably arises when one gathers information from a single caste.

To the extent that a subject of inquiry is socially more delicate, and certainly in a situation of conflict, the chance increases, especially when participant observation is the method of investigation, that the fieldworker has to take sides. It is a choice he may feel obliged to make or to which he may be compelled by his surroundings. In the course of my own field work I never made a secret of my own stand when I was explicitly asked about it. This was one reason why I endeavored to proceed as objectively as possible in collecting the data comprised in the following chapters.

*Chapter 7*

# Chikhligam

*My Subodbhai is a vakil,*
*And Vinodbhai is a doctor,*
*Ramanbhai begins his studies*
*With the English A, B, C.*
T. B. Naik, *Songs of the Anavils of Gujarat*

*O mother, I can't dehusk the grain,*
*I can't grind the corn,*
*I want to go away to Bombay.*
*To hell with your money,*
*To hell with your bride-price,*
*My married one lives in an old tottering hut.*
*Mother, I want to go away to Bombay,*
*I don't want your drinks,*
*Nor do I want your food,*
*Mother, I want to go away to Bombay.*
*O mother, to Bombay*
*I would have gone by train.*
*O mother, in Bombay*
*I would have sat at* [*a*] *window and seen through it.*
*O mother, in Bombay*
*I would have eaten* rotli *with tea*
*And fine rice too.*
*In Bombay, O mother,*
*I would have walked on a footpath.*
*I want to go away to Bombay.*
T. B. Naik, *Dubla Songs*

### *7.1 Region, Village, and Population*

Chikhli *taluka* is situated in the transitional area between the densely populated and fertile western plain and the relatively inhospitable and infertile eastern part of south Gujarat. The local town, with over 5,000 inhabitants, is a trade and civic center for the villages in the subdistrict. Here are the local government offices, some doctors and barristers, and the traders and shopkeepers on whom the villagers of the region depend for the sale of their agrarian products and the purchase of consumer goods. Industrial activity is limited to a number of workshops, each with a few employees. As there are bus connections with the urban nuclei of the surrounding subdistricts and with the railway line in the west, the town is a link in the contacts of the rural population with the outside world. The national highway from Bombay to Ahmedabad also runs through this regional center; it is mainly important for through traffic.

Chikhligam is about four miles from the *taluka* town. Rough country roads, running through terrain that is dry and arid both in summer and in winter, connect the village with those surrounding it and with the highway at two miles' distance. From that point, the inhabitants of Chikhligam can go to town by bus four times a day, but they rarely use this means of transport. In the wet season, when the country roads become impassable and the village is practically isolated from the outside world, the villagers can get to town only along the highway, which is normally not used except by those who own a bicycle. Two inhabitants subscribe to a regional newspaper, and there is a single radio set in the village, which was out of order during my stay. There is no running water, electricity, or any other modern convenience.

On the way to the village, the visitor passes through undulating terrain. The high, dry ground (*jarayat*) is stony and uneven, and dotted with *babul* trees (*Acacia arabica*). This *jarayat*, which in Chikhligam constitutes two-thirds of the total village land, is of inferior quality throughout the subdistrict and not very suitable for agriculture. The depressions in the terrain, on the other hand, have been leveled and are surrounded by embankments. This is the *kyari* soil on which in the wet season the rain water is collected for rice cultivation. Along a small river, which forms the eastern boundary of the village area, there is, finally, the very fertile *bagayat*, consisting of irrigated garden plots.

The returns of the 1951 census—on which my selection of field villages was based—had made me expect a village population of a little more than 1,000, which was well within the limits I had set. But whereas in 1951 only 900 inhabitants were registered, their number had risen to over 1,800 at the 1961 census. The explanation of this apparent doubling is the seasonal migration of a large proportion of the inhabitants of Chikhligam during about seven months of the year. In 1951, Chikhligam probably already had about 1,500 inhabitants, but the census of that year was taken in a period when the migratory workers had not yet returned to the village (see table 7.1.1).

TABLE 7.1.1

DISTRIBUTION OF THE INHABITANTS OF CHIKHLIGAM, ACCORDING TO CASTE*

| | Number |
|---|---|
| *High castes* | |
| Anavil Brahman (Landlord) | 125 |
| Bania (Merchant) | 5 |
| *Intermediate castes* | |
| Darji (Tailor); Ghanchi (Oil presser); Valand (Barber); Mochi (Shoemaker); Vairagi (*saddhu*) | 75 |
| *Low castes* | |
| Dhodia (Petty cultivator) | 1,360 |
| Dubla (Agricultural laborer) | 200 |
| *Untouchable castes* | |
| Garuda (Priest); Dhed (Weaver) | 90 |
| *Moslems* | |
| Shopkeeper; Barber | 5 |

* These figures, which I obtained from the village clerk, are only approximately reliable. I have been able to verify only the figures for Anavils and Dublas.

Despite the large number of inhabitants I persisted in my choice of Chikhligam. The members of the tribal Dhodia caste, who form three-quarters of the total village population, live outside the village nucleus in ten hamlets scattered over the whole village area. Each of these *phalias* comprises ten to twenty houses, whose inhabitants are mutually related by patrilocal residence. The hamlets, which are connected by narrow paths, are surrounded by the fields belonging to the inhabitants. Not only are the Dhodias spatially separated from the rest of the village population, but apparently they do not depend much on other castes for their needs. Their economic and social life is set within their own *phalia.* Each of the two largest *phalias* even has its own shop, with a modest stock of goods. Nor do the Dhodias mix with the other inhabitants at village festivals, as I observed when the bonfire was lighted on the occasion of *holi.*

Under those circumstances, I could easily restrict my field work to the nuclear village, in which the members of the other castes live, with Anavils and Dublas as the largest groupings. A pond, a large public well, a small school building, a decayed *dharmasala* (guest house), and a shabby little mud temple constitute the village center. As a center, however, it is especially recognizable because the Anavils live here. The members of this dominant caste tolerate no other neighbors but a Bania family in their street. Most Anavils live in large brick houses containing a series of three spacious rooms with a veranda in front, and the newly built houses in the street have an upper storey. The stable is separated from the living quarters by a large yard. Some Anavils have a private well behind the house.

A second street is the domain of the Mochi caste, at present still represented in the village by six households. The houses in the Mochi street are also of brick, but much smaller and lower, and a number of them are empty. Two are used as shops, one of which belongs to a Ghanchi family and the other to a Moslem shopkeeper, who is in the village only during the daytime. Most of the scattered houses of the three Darji households and the Valand have tiled roofs but mud walls.

The Moslem barber lives among the Dublas, whose block has a cluttered appearance. The huts in which they live are not built in rows, but in a rambling order. The Dublas' huts, the worst in the village, generally have only one room and are built of mud brick; their thatched roofs hang so low that one has to stoop to enter. There is no ventilation except by the open entrance. The total area of the hut is often so small that there is not enough sleeping space for all the members of the household. About six years before my stay, nearly half of the Dublas moved to a piece of ground that belongs to the neighboring village. Administratively they no longer live in Chikhligam, but because they still work temporarily or permanently for the Anavils of the village as agricultural laborers, I included this group in my investigation. Their dwellings in the new quarter, built with financial support from the government, are larger than their former huts and have tiled roofs. Finally, on the fringe of the village nucleus are to be found the houses of the untouchable Garudas and Dheds. Their surroundings look very well kept.

Little can be said with certainty about the age of the village. The Anavil Brahmans assert that their ancestors of several generations ago came here from villages farther to the west. As I noted in chapter 3, the *pax britannica* put an end to the original way of life of the tribal Dhodias who, as primitive cultivators, used to shift periodically to other grounds, returning after some years to their point of departure. It appears from the old administrative records that they were still very mobile at the beginning of the colonial era. Their sedentarization was promoted by the founding of villages in this area on the initiative of high-caste Hindus from the plains. The Anavils of previous generations, in particular, enjoyed a reputation as founders of villages. The origin of Chikhligam is probably similar to that of a nearby village described by Mukhtyar.[1]

This author states that immigrant Anavil Brahmans persuaded members of artisan and serving castes to settle in the newly reclaimed area. To this it should be added that, in this preponderantly tribal region, no fully formalized caste structure arose. Many local-level studies are reports of villages which the investigators selected precisely because of the presence of a large number of castes. In comparison with these, the caste composition of Chikhligam is relatively simple. No doubt this is connected with the relatively undeveloped functional interdependence of the inhabitants in the region. The Dhodias, who form a large part of the population of the subdistrict, have remained more or less autonomous as members of a caste of tribal origin.[2] Only on a limited scale do they use goods and services supplied by other castes, and even now they maintain few contacts with the other village inhabitants.

For the Anavil Brahmans, highly placed in the hierarchy, it seems to be much less easy to dispense with a large number of castes. As I said earlier, however, their Dubla farm servants are accustomed to perform numerous services for them — from water-bearing to

1. Mukhtyar, 1930, 44–45.
2. For an extensive description of the Dhodias, the reader is referred to the ethnographic study of this tribal caste by A. N. Solanky.

washing garments, cleaning the house and stable, etc.—which elsewhere in India are carried out by diverse castes to whom these tasks have been specifically allocated.

### *7.2 Economic Position and Activities of the Inhabitants*

In an agrarian community the possession of land is naturally an important indication of position in the economic structure. Anavil Brahmans, less than 10 percent of the inhabitants, own almost as much land as the Dhodias, who form nearly three-quarters of the population. The land owned by an Anavil household is many times larger than that of other households, averaging 19.4 acres as against an average of, for instance, 2.2 acres per Dhodia household. Another indication of the ascendancy of the Anavils is that much the greater part of the coveted *bagayat* is in their hands. This concentration of their tenure in the arable land is a measure of their dominance.

It may be assumed that the Anavil Brahmans have laid claim to a disproportionately large part of the total holdings ever since their arrival. Informants from various castes expressed the belief that at the beginning of this century proportions were no different. If the members of the dominant caste deprived Dhodia farmers of their property rights, this must have happened in a more distant past; it is not due to a development which began only after the turn of the century. As early as 1865, in the first cadastral registration of Chikhli *taluka*, it is reported that

> The Desais have managed through the influence in the revenue administration of the pargana they have retained from the time of our predecessors in the country, and by pandering to the passion of these degraded creatures [the *adivasis*] for drink, treating them at the same time with a rude sort of kindness, to get possession of the richest lands.[3]

This quotation answers the question of how the titles came to change hands. Excessive drinking caused the tribal population to fall into debt to the newcomers and the Anavils, who in the villages represented the government, exploited their politically dominant position. They probably compelled the Dhodias to become sedentary by laying claim to part of the area in which they wandered as shifting cultivators. It was penetration from the west that made land a scarce

3. Cited from the original report in *Papers Relating to the Revision Survey Settlement of the Chikhli Taluka of the Surat Collectorate*, 1897, 62.

commodity, and the subjection of the autochthonous population to the incoming group was reinforced by taxation.

TABLE 7.2.1

CASTE MEMBERSHIP AND LANDHOLDING IN CHIKHLIGAM*

| Caste | Percent of village population | Property Acres | Property Percent | Extent of gardens Acres | Extent of gardens Percent |
|---|---|---|---|---|---|
| Dhodias | 73 | 350 | 44 | 7 | 7.5 |
| Dublas | 10 | 33 | 4 | 4 | 4 |
| Anavils | 7 | 297 | 37 | 77 | 81.5 |
| Harijans (Dheds and Garudas) | 5 | 49 | 6 | 3 | 3 |
| Mochis | 3 | 47 | 6 | — | — |
| Banias ) | ) 2 | 14 | 2 | 3.5 | 4 |
| Outside ownership ) | | 10 | 1 | — | — |
| Total | 100 | 800 | 100 | 94.5 | 100 |

* Figures provided by the village accountant.

The Dhodias are the small peasants of the village. Two processes led to the existing pattern of small landownership in this caste: first, the accumulation, as described above, of the most valuable soil in the hands of the Anavil Brahmans during the previous century, and secondly, fragmentation into scattered plots because of demographic growth in the more recent past. With a few exceptions, all Dhodia households own a 2 to 3 acre plot of land which mainly consists of not very fertile *jarayat*. The rapid increase of their numbers in the past few decades owing to the falling death rate has resulted in increasing pressure on the land. The labor of the members of each individual household usually suffices for their own plot. Apart from that, the Dhodias of the same hamlet occasionally work for each other reciprocally at harvest time. Formerly this exchange of labor (*sandhal*) was more widespread. Only a few of the larger farmers among them, who do not wish to supply labor on a footing of reciprocity, employ paid casual labor from time to time. Owing to resources other than those acquired through agriculture, some differentiation in economic position has gradually come about among the Dhodias. Income obtained from outside the village has enabled some of them to extend their holdings and to raise themselves above the level of their less fortunate fellow caste members. In former times many Dhodias were sharecroppers of Anavil Brahmans, and in that capacity they were obliged to assist the landlord on his land at the peak of the season. With the termination of nearly all those crop-sharing agreements by the Anavils some years ago, such field services provided by the Dhodias also came to an end.

Thanks to seasonal migration, the small cultivators are now less dependent on the dominant caste of the village. Some of them still occasionally perform agricultural labor for Anavils against payment, but they do it grudgingly and only if they need the money badly. With their own team of oxen, some Dhodias plow the land of those members of the dominant caste who keep no servant and, sometimes, no cattle.

Most of the land owned by Mochis and Harijans was bought some decades ago. With money earned in town, some members of this caste purchased plots in order to provide for the relatives they left behind. A few Dheds, whose ancestors had been entitled to the produce of certain plots in exchange for services rendered to members of the dominant castes, such as weaving garments and disposing of dead cattle, have now been given ownership of this land. Some Mochis who emigrated to East Africa bought land as an investment and leased it out to Anavils, Dhodias, or Dublas on a crop-sharing basis. In most cases they lost this land because of the tenancy reforms; only those who returned to the village kept their property. Like those of the Dhodias, the holdings of the families of these non-agrarian castes are so small that they do not need the regular labor of other workers, and it is beneath their dignity to work on the land of others.

The agricultural laborers of Chikhligam are mostly Dublas. They are, in large majority, completely landless and do not even own the piece of land on which their hut stands. That there are a few landholders among them does not signify that this caste originally owned more land and was gradually dispossessed. When asked about it, my informants told me that their plots, mainly consisting of infertile *jarayat*, had come into their possession at the beginning of this century. According to the Dublas, the land then belonged to no one, and the government had parceled it out to those interested—Mochis and Dublas. Informants claimed that the majority of the land now owned by Dublas had originally been the property of the Anavil caste, but had been relinquished about the turn of the century by the Anavil generation of the time. Some of the plots had belonged to an Anavil widow, who died the last of her family and left the land to her Dubla farm servants. This explains why one Dubla even owns a plot of the highly prized *bagayat*. Finally, after having worked for years in Bombay, two Dublas managed to acquire a plot of land with their savings. Actually the landed property of this tribal caste is even lower than appears from table 7.2.1. The Dublas who removed to the new houses are no longer registered in Chikhligam, although they still work there. All the landholders are among those covered by the table, i.e., those who remained behind. The Dubla agricultural laborers work almost exclusively for the Anavils of the village, in permanent service or on daily wages.

The economic positions of the members of the dominant caste also show considerable differences. In certain cases, Anavils have less land than the largest Dhodia cultivators. Even so, however, they are more favorably placed, for their land is of far better quality, and moreover they can count on the support of richer relatives. They clearly profit by their membership of the dominant group in the village. They can, for instance, borrow more money, and on more favorable terms, than the Dhodia farmers.

The Dublas, on the other hand, are not all landless, as was explained above. Six of those who call themselves cultivators own more land than the poorest Dhodias. In comparison with the other members of their caste, they are undoubtedly better off. They have one, sometimes two oxen, a cow, a couple of goats, and some simple agricultural implements. Their houses are larger than those of the landless Dublas, they own a greater variety of household goods, and they consume more, if not better, food. Nevertheless they are the poorest of the cultivators, and all of them carry out additional agricultural labor for the Anavils of the village. There is, in other words, a clear correlation between caste membership, property in land, and economic position. With a few exceptions, the Anavil Brahmans can be characterized as landlords and the Dublas as agricultural laborers.

Nevertheless, only for a small part of the village population is agriculture the sole means of existence. Despite the geographic isolation of their village, the majority of the inhabitants depend to some extent on income acquired elsewhere. Living in Chikhligam and employed in the village in other than agricultural work are: three shopkeepers (one Ghanchi, two Dhodias), three tailors (Darjis), two barbers (one Valand, one Moslem), one temple servant (Vairagi), one shoemaker (Mochi), two smiths-wainwrights (Dubla father and son), and one *panchayat* servant (Dubla). A few of the inhabitants work in surrounding villages or in the *taluka* town as schoolteachers (two Dhodias), postmen (two Dublas), counter-clerk (Anavil), shoemaker (Mochi), and *panchayat* servant (Dhed). Finally, seven Dhodias cycle each day to Bilimora, on the west-bound railway line, where they work as factory hands. The Garudas in the village are priests and genealogists; they visit their clients, the untouchable Dheds, who live scattered everywhere in south Gujarat.

As these data show, nonagrarian employment in Chikhligam and its immediate surroundings is of little significance, less than 10 percent of the male working population being fully or partly employed in this way. However, the income from agriculture is supplemented in another manner. A great majority of the inhabitants of Chikhligam leave the village for shorter or longer periods, and some even stay away and return at most for a brief visit. Where they go and what they do is determined mainly by the caste to which they belong. The chances of finding work outside the village depend on migration traditions, on the nature of the work, and on the education of the worker, factors that differ for each caste.

The Anavil Brahmans are employed in the highly skilled and best-paid professions. Among the members of this caste who have left the village are a doctor, a college teacher, a stationmaster, an engineer, a manager in the state life insurance company, a money-lender, two garage owners, a mechanic, and a worker in a chemical factory. Five other Anavils are employed in government services. All live in towns, and all but one in south Gujarat. Many still own land in the village or are entitled to part of the produce of the family holding. Their agricultural interests are taken care of by relatives or have been entrusted to fellow caste members.

Most of the Mochis who have emigrated in the past have gone to East Africa, where they are engaged in their caste occupation or as shopkeepers. Others work in these trades in Bombay or Bilimora. The Dheds used to work as house servants for Europeans in Bombay and are now employed as such by well-to-do and westernized Indian families in that city. The Garudas were at no time economically attached to Chikhligam. In recent decades a number of them have gone to Bombay, where they practice their occupation of priests or are otherwise employed, for instance as drivers.

Relatively few Dhodias and Dublas have permanently left the village, apart from those who have gone to live in their wives' villages after marriage (*ghar jamai*). Usually, those who did go away are employed as unskilled laborers in Bombay, Surat, or the smaller railway towns in south Gujarat. The parents of many of them have not heard from their sons in years, and do not know where they are or what they are doing. Upon the death of relatives left behind, those who have gone away are sure to be forgotten. Their exact number is therefore hard to determine. According to my data, nineteen Dubla men from twelve huts are permanently employed outside the village.

Contacts of the tribal castes, small cultivators, and agricultural laborers with the city environment are as yet not very intensive. They have no specific knowledge that would equip them for nonagrarian employment, and being unskilled they qualify for only the least-favored and lowest-paid jobs.

For both Dhodias and Dublas, migratory labor is of far greater importance. Seven or eight months a year are spent by more than half of them in the brickyards near Bombay. The exodus begins after the wet season, but most of the migrants do not depart until the grass has been cut in mid-November. They stay away until the end of May or the beginning of June. The brickyards are small enterprises operated in the open air, and activities are concentrated around one or more kilns. The laborers are accommodated on the site in improvised huts of loosely piled bricks with roofs of thatch, branches, or leaves. Clay is brought in from the vicinity and made into bricks by teamwork. The *patla*, a gang of six laborers, molds the clay, forms the bricks and puts them out to dry. Because work is done in the open, it must be suspended during the monsoon. During the winter and summer months, only the aged and infirm, and occasionally women and children, remain in the village. Especially in the Dubla quarter, many houses are closed up for these periods. Complete families leave Chikhligam, and all—men, women, and children—work in the brickyards.

This form of migratory labor has increased considerably since the Second World War. Although a number of men of the older generation worked in the brickyards before the war, the demand for laborers in this branch was low, and many brickyards were functioning only part of the dry season. This occasionally occurs even now, so that the laborers must already return to Chikhligam in March. During and after the war, however, building activities in

Bombay and its surroundings expanded enormously. The number of brickyards increased accordingly, and their production was sharply accelerated.

The laborers are engaged by a labor-recruiter (*mukadam*), who is connected with one or more brickyard owners directly or through a contractor in Bombay. The recruiter often lives in the village, belongs to the same caste and has been a laborer himself. Insofar as the migrants are not picked up by trucks, the recruiter provides the money for the journey, and usually they all travel together to Bombay by train. The labor-recruiter is the link between village and brickyard. He goes with the laborers to Bombay, keeps accounts if he is literate, and pays the weekly allowances. Male laborers receive 6 rupees a week for living expenses, and women are given 4 rupees a week, but wages are not paid until the end of the season. By this system the laborers are attached to their place of work, and the brickyard owner does not run the risk of losing laborers halfway through the season. Accounts are settled upon their return to the village at the beginning of July, when the laborers receive what is owing to them. The amounts they save in this way vary widely, but even when the whole family works, they rarely have more than 100 rupees left, usually less. During the monsoon, when the laborers are in the village, they can get advances on their wages of the coming season through the labor-recruiter. In this way the brickyard owner is assured that he will have enough labor the following year, whereas the laborers, for whom there is not enough work on the land in the rainy season, can pay their daily expenses in the village.

This system makes for a lasting contact between the parties concerned. The laborers find employment waiting for them at a large distance from the village, and the brickyard owner constantly has all the labor he needs. On the other hand, whereas the party offering employment (the brickyard owner) has insured himself against nonfulfilment, the laborers have no redress against unjust treatment. The labor-recruiter receives 150 rupees for each gang delivered by him. If the gang does not appear at the beginning of the season he loses the premium and is, moreover, answerable for the advances that were paid through him to the members of the gang during the last wet season. His interests therefore coincide with those of the brickyard owner. Inasmuch as the wages are not settled until the end of the season, the laborers cannot leave the brickyard ahead of time. If they do, and if at the moment of leaving they are still indebted to the brickyard owner, as they often are at the beginning of the season, the other members of the gang are jointly answerable for their debts. Usually the laborers, nearly all illiterate, do not know exactly what amount is owing to them at the end of the season, so that the owner can give them less than they are entitled to. At the instance of the brickyard owners the labor-recruiters nowadays keep in touch with each other. Laborers who have run away, or who have taken up advances in the wet season without reporting for duty afterwards, are excluded, and occasionally civil actions are brought against them. Although it is a foregone conclusion that they cannot repay the money, such actions are used as means of intimidation. To organize the laborers into unions is difficult because of the seasonal character of the work. Moreover, the trade unions in Bombay regard these migrants as unwelcome intruders who depress the local wage level.

In an underdeveloped economy, the mechanism described above ought to balance wide local differences of demand and supply. But the laborers are bound hand and foot to the labor-recruiter, on whose collaboration they depend if they want to be employed again the following year. On the other hand, the power of the *mukadam* should not be overestimated. He lives in the same village and belongs to the same caste as the laborers. As a young man he usually has been away under the same conditions as those who now depend on him and who are sometimes related to him. Social pressure prevents him from totally identifying himself with the brickyard owner's interests or from systematically exploiting his contract laborers for his own ends.

Dhodias and Dublas participate in equal proportions in the yearly migration to Bombay. Yet there is an important difference in the economic position of the two castes. Their income from the brickyards has enabled the Dhodias to keep their family property, if only by progressive fragmentation, and to hold their own as petty cultivators. This also explains why more members of this caste remain in the village. It is the young people who go away, and only the women who have not been married long accompany them for a few years. Then they, too, remain behind to look after the cattle and harvest the winter crop, and their children go to school. The Dublas, on the other hand, having no land to provide for any relatives remaining in the village, must take their families with them. A Dubla can go to the brickyards only if the other members of his household work too, for earnings are so low that he cannot send money home. Migratory labor becomes impossible in cases of poor health or pregnancy of the wife.

The Dhodias' landownership, unremunerative in itself, puts them on a higher level than the landless Dublas. On the other hand, the Dublas continue to depend on agricultural labor because their income from seasonal migration is temporary; they thus remain at the disposal of the Anavil landlords of the village.

### *7.3 Caste Hierarchy and Political Structure*

The settlement pattern and the villager's economic condition and activities indicate the caste to which he belongs. His position within the caste system appears from the pattern of social interaction. Norms regarding the accepting of food and water are indications of ritual status in the caste hierarchy. Thus, members of a high caste will in general not accept food from or eat together with those of lower castes. To some extent there is a relationship between a caste's economic and ritual status. This is true especially of the mutual allocation of position among the Anavil Brahmans, Dhodias and Dublas, the largest castes. However, the untouchable Dheds and Garudas, lowest in the hierarchy, are in better economic circumstances than the members of the tribal castes.

As elsewhere, actual behavior in Chikhligam deviates from the norm. In contradiction to caste endogamy, the Vairagi lives with the daughter of a Darji in the village. Although Anavils do not accept water from the untouchable Dheds, the village headman, a member of the dominant caste and a politically active member of the local section of the Congress Party, says he does accept water from Dheds in other villages. Again, the behavior pattern on which the place of the Anavils in the caste hierarchy is based includes vegetarianism and abstinence from alcohol, but some Anavils barely succeed in hiding the fact that they indulge in drinking illegally distilled spirits at the Ghanchi's, or worse, buying it from Dublas. It is an open secret that there are Anavils who commit the even greater sin of going to the dwelling of one of the Mochis to eat meat. In the eyes of their fellow caste members these Anavils have fallen very low because they associate with Mochis and Dublas, especially because they do it in the village itself.

These are individual deviations from the prevailing norm. Of more importance is the fact that many people of some castes leave

Chikhligam for years at a stretch and return only occasionally for short visits to relatives. Whereas interdependence to some extent implies the necessity of recognizing the relative positions of the castes, long absence may lessen commitment in the village community. Little is left to bind Mochis, Dheds, and Garudas to Chikhligam. Their children grow up in the village, but nearly all go away as soon as they are old enough, mainly to Bombay.[4] The Garudas occupied a marginal position in the village community even in the past when they all practiced their priesthood. The relatives they have left behind depend on the money sent from outside and have practically no contacts in the village.

A change in the pattern of a caste's economic activities may mean that its members are striving after a higher position in the traditional status hierarchy. When, however, many members of a caste leave the village for long periods to earn their living elsewhere they may also be withdrawing from the caste relationships in their village. The way in which the Mochis who emigrated to East Africa behave will differ profoundly from that of their fellow caste members who are still employed in the village economy. As for Dheds who work for well-to-do Indian or European families in Bombay, and who return to their native village for only a few days a year, dressed in trousers and nylon shirts and showing off their knowledge of English to the investigator, other norms and values obtain than those that were traditional for their parents, who were the weavers of the village and carried out a number of tasks regarded as defiling.

Seasonal migration, too, is bound to influence the relations between the villagers. For the Dublas this will be discussed below. In the Dhodias' case, the yearly migration has accentuated the economic and social isolation already inherent in their residential separation, and lessened their influence on the organization of village matters.

For the pattern of interaction, not only the extent of incorporation of the various castes in the village economy is of importance, but also their numerical proportions, which greatly influence intercaste relationships. Although it may be useful to know that, in the social hierarchy, a Ghanchi is higher than a Dhodia, this knowledge

4. I could not ascertain when members of these castes began leaving the village. Dheds from the nearby village of Atgam worked for Europeans in Bombay as early as the nineteen-twenties (Mukhtyar, 57). At the turn of the century, emigration from the *taluka* was still negligible. Presumably it has gathered momentum since then. Especially after the First World War, the population of Greater Bombay increased rapidly.

is not very relevant if three-quarters of the village inhabitants belong to the Dhodia caste, and some other castes are represented by only a single or a few households each. The arrangement of castes by ritual standards, therefore, does not reflect the actual distribution of power in the village, which is determined by the dominance of the Anavil Brahmans and the numerical preponderance of Dhodias.

The Anavil Brahmans have always been regarded as the people in authority in the subdistrict,[5] and their ascendancy continues today. Although this caste comprises less than 10 percent of the inhabitants, it has as many representatives in the village council as do the Dhodias, i.e., three. This *panchayat* meets only a few times a year. The daily course of events is wholly governed by the village headman, who has always come from the ranks of the Anavils. The police *patel*, who has to report all trespasses and crimes to the police in the *taluka* headquarters, is an Anavil. The local credit cooperative presents the same picture—most of its ninety-one members are Dhodias, but five of the seven members of the board, including the secretary, are Anavil Brahmans.

Anavil ascendancy is apparent in various ways aside from the fact that they occupy the formal positions of power. They act as arbiters in quarrels between members of other castes, and the leading Anavil faction has put a number of influential Dhodias under obligation, so that its power basis is safe.

The influence of the dominant caste is apparent, for instance, in the effect of tenancy reform in Chikhligam. The purpose of the Tenancy Act was to have the land revert to the actual cultivator, i.e., the tenant. In the past, landholding villagers who had left Chikhligam—mostly Anavils and Mochis, but also a few Banias and a Brahman from the *taluka* town who owned land in the village—used to lease out their property. When the new act became operative, the property rights of non-Anavil absentee owners shifted to their former tenants, mostly Anavils and Dhodias. The Mochis from Chikhligam who were in East Africa lost their *jarayat* to Dhodias and Dublas, and their *bagayat* to Anavils, who rented only the most valuable land. Most of the land leased in the village, however, belonged to Anavils. The income of one Desai family derived

5. Compare the report cited in fn. 3 above, pp. 62–63: "These Bhathelas are notorious as the greatest intriguers in the country. This character, assisted by the hereditary influence I have described them as possessing, has enabled them to worm themselves into positions of power as patels of villages, which they have used to the injury of the revenues, by exacting labour from the lower castes of cultivators, and thus preventing them to improve their own conditions."

almost wholly from land they had leased out, chiefly to Dhodias. As did the other Anavils, this family lost very little property in the tenancy reform. A good example is the case of a Dhodia who grew millet on *jarayat* he had rented from the village headman. He wished to have the land registered in his name, as he was entitled to do under the Tenancy Act. The village headman had been able to prevent this, as he himself said, by exerting some pressure. Although for him, in his function, this sort of thing is very easy, the same is true to some extent for all his fellow caste members. Some of them still lease out land illegally, but without anxiety, to Dhodia cultivators on a crop-sharing basis. The Anavils think this land reform unfair. It is not right to take away a man's property, they say. In regard to an Anavil who had told me indignantly how a Dhodia had cheated a Mochi owner by "stealing" land from him in this way, I was told later that he had "received" a *bagayat* lot from the same Mochi under the Tenancy Act. Not only did the members of the dominant caste not lose any property, they added to it. Some Dhodias and Dublas managed to become the owners of the land they had rented from Mochis living in East Africa and from a number of outsiders, but not from Anavils living in or outside the village.

The exercise of power by the Anavil Brahmans might in theory be counterbalanced by the numerical strength of the Dhodias. But, as has become clear, the latter's weak economic position makes them extremely vulnerable. It is therefore not surprising that Anavils speak kindly of Dhodias — a mild-tempered caste who do as they are told. The landlords can harass them in various ways, for instance by denying credit to "troublesome" and "unwilling" Dhodias, either directly by not lending them money individually, or indirectly by using the Anavils' majority vote in the board of the credit cooperative. Contacts with the government often run via the influential members of the dominant caste, and Dhodias who know how to behave can be sure that their business with the government will receive proper attention. A frequently used device in the "dry" state of Gujarat is to prosecute unruly Dhodias for illegally distilling and drinking alcohol, that favorite pastime of all castes of tribal origin in the region.

The following occurrence is typical of the behavior of the Anavils. One night during my stay in Chikhigam, police from the *taluka* town raided some Dhodia hamlets. They seized three men, whom the police had already tried to apprehend by day several times but who had always managed to evade them. They were wanted for running away from a brickyard while owing money to the owner. Ostensibly, however, they were taken into custody for distilling and drinking alcohol, an offense for which practically all members of the caste could have been arrested. Distilling equipment was found in the houses of two of the prisoners. The Anavil police *patel* had been informed of the time of the raid, and had chosen some fellow caste members, one of whom was a notorious drinker, to assist the police. Early in the morning the raiding party returned to the Anavil street in high spirits, and spent the rest of the day exchanging stories about the successful expedition. Relatives of the arrested Dhodias came to ask the Anavils to put in a word for them with the police; it was harvest time, and the men could not be spared. They were told that, first of all, two Dhodias who had escaped the night's raid because they had been fishing on the river must give themselves up. Shortly afterwards the two did come to report to the police *patel*. As for the others, the Anavils said, it had been their own fault, and if they had done wrong they must pay for it. To my question why certain Dublas had not been arrested for the same offense that night, the *patel* answered indignantly that these were their own laborers and they could not possibly spare them.

Nevertheless the power structure is beginning to change. Because so many of them go to work at the brickyards, the Dhodias are no longer quite so poor as before, nor so completely bound economically to the Anavils. Apart from renting land from Anavils, many Dhodias used to perform field service for their landlords, but now only the poorest of them do this. More and more of their children attend school. The Anavils were shocked when I figured out for them that, in all, more Dhodia children from Chikhligam (ten) go to secondary school in a nearby larger village than Anavil children (six). The son of a labor-recruiter from the village is at college in Bilimora.

The strengthening of the position of the Dhodias has been apparent, too, in the siting of a new building for the lower classes of the village school. On the Dhodias' request, this building was not put near the Anavil street, but halfway between the nuclear village and some Dhodia hamlets, so that the children of both castes have to walk the same distance. The headman, who is a contractor and with whom the order to build it was placed, complied with their request against the wishes of several Anavils.

At the supralocal level, the growing emancipation of the Dhodias begins to have effect. In accordance with the government policy of giving the population more control of their own affairs, *taluka* councils were established in 1963. As chairman in this subdistrict a Dhodia was elected, one of the very few *adivasis* to be elected to such a post in Gujarat. Although he does belong to the Congress Party, locally the instrument of power of the Anavil Brahmans, and was put up as their candidate so as to ensure the support of the village headmen from the eastern part of the *taluka*, the Anavils in Chikhligam see his appointment as the writing on the wall. They are less than happy with present developments, which they regard as threats to their position. Despite the lack of unity among them, they usually take a common stand where their interests are concerned. They are particularly alarmed at the comparatively rapid population increase among the *adivasis*, and blame the government for supporting these tribal castes. In another dozen years, they fear, the Dhodias will have their own political leaders, who will manage their affairs at higher levels than that of the village. Differences of opinion among the Anavils are put aside for the sake of unity. Only mutual cooperation, they argue, will enable them to withstand the much more numerous Dhodias. As a prognosis for the future this may be dubious, but at present the dominance of the Anavils is still undisputed.

### 7.4 The Pattern of Agricultural Activities

Agriculture in Chikhligam largely depends on the rain, which falls only during the southwest monsoon, from the end of June until the beginning of October. During this period the *taluka* has an average rainfall of about seventy inches, with a possible variation of 50 percent either way. After the wet season, winter sets in, and lasts from the middle of October to the beginning of March, with an average temperature of 60° F. The third season is summer, the hot period in which daytime temperatures rise to over 100° F.

Of the total arable land in Chikhligam (about 800 acres), half consists of the rather infertile and stony *jarayat*. In other parts of south Gujarat, where this soil is more fertile and level, *jowar* (sorghum), cotton, or groundnuts are grown in the monsoon period. My informants affirmed that these crops grew in Chikhligam on the most level high village lands before the Second World War, together with some inferior varieties of grain (*nagli* and *kodra*) and castor seed, from which oil was pressed for use in house lighting. In the agricultural economy of Chikhligam these products probably have never been of great importance. As in other villages on this infertile *jarayat*, their cultivation was abandoned during the interwar depression.

At present the *jarayat* is used only for growing grass, which is cut some weeks after the end of the monsoon. The hay is intended for the owners' own cattle and is not sold outside the village. During the winter and summer months only the *babul* tree, so characteristic of the landscape in this region, grows on the dry and barren soil. Its wood is used as firewood, and its thorny branches serve to enclose the *kyara* and *bagayat* lots as protection against cattle and other wandering animals.

In the *kyari*, which lies scattered among the *jarayat*, only rice is grown. Formerly sugar cane was cultivated here, but this ended several decades ago. At the first rainfall the soil is plowed under and the rice plants are set out in seedbeds. Fertilization is often confined to these small pieces of land. About three weeks later, the topsoil is kneaded with water into mud, and the seedlings are transplanted. No thorough weeding is done. Immediately after the rice harvest, the plots that are still wet enough to plow are sown to *val* (a leguminous plant) as a winter crop. If, because of lack of rain, the rice is late in ripening, the soil is too hard to plow, and in many cases no *val* can be grown. The *kyari* crops are consumed mainly by the owners, and only a few small cultivators sell part of their produce either to other inhabitants of Chikhligam or to merchants in the *taluka* town, usually not because they have a surplus but because they need money. Later in the year they have to buy rice again at a higher price.

Only on the fertile *bagayat*, which constitutes less than 10 percent of the total arable land, is there some well irrigation. At present this garden land is completely under mango. Orchard cultivation, too, depends on rainfall, but in the dry winter and summer months irrigation is desirable. Formerly, twenty-three wells were used in the gardens, water being brought up from them by the labor of yokes of oxen. Only a limited area could be irrigated, and a great many laborers had to be employed as well. At present, four wells are in use, and for these, motor pumps have been installed. Their total capacity is such that all the gardens can be irrigated and a great deal of labor is saved.

The mangoes are picked from the middle of May to the middle of June. During the rest of the year the trees demand little attention. In winter, until the hot season begins in March, the leaves are regularly sprayed with insecticide, and in the dry period occasional irrigation is necessary. Especially when fertilizer has been added, the soil must be plowed; this is usually done at the end of the wet season and in summer. Some large landowners in Chikhligam grow mango seedlings, which they sell even outside the *taluka.* The cultivation of these young trees requires great expertise and care. Kernels of mango fruits are set in rows before the beginning of the monsoon. At the end of the rainy season of the same or the following year, the young trees are potted and put into pits. Daily watering is necessary until the rainy season starts, and the seedlings are then ready to be sold. They are watered every week until April, when summer has begun, and at that time the seedlings are grafted upon mango trees in order to produce the desired species.

The gardens have not always been under this monoculture of mangoes. The first cultivated trees were planted in Chikhligam only forty years ago. Prior to that time, crops in wide variety were cultivated, largely for home consumption, on the *bagayat*: sugar cane, elephant's foot (*suran*), sweet potatoes and other vegetables, bananas, and turmeric. Sugar cane was the only real money crop with which tax was paid, and with which large purchases — of cattle, for example — outside the village were financed.

Reduction of the range of crops has taken place not only in Chikhligam. In the course of a few decades, mango cultivation has superseded the traditional garden culture in the whole of south Gujarat, and the mango has replaced sugar cane as the most important commercial crop. It was especially during the nineteen-forties that mango-growing assumed large-scale proportions. Whereas in south Gujarat the mango orchards covered 4,095 acres in 1938–39, they had increased to 21,538 acres in 1953–54.[6] The owners of *bagayat*, who now depend on the produce of the orchards for their money income from agriculture, devote much more attention to them than to rice-growing. Special care is given to mango trees, as is indicated by the use of manure and other fertilizer,

6. B. A. Desai, 1–10.

thorough weeding, spraying with insecticides, and other measures. This greater concern is also apparent in the cultivation of mango plants, which requires expert knowledge. Fruit-growing has proved so profitable that recently tracts of moderately rough *jarayat*, insofar as they adjoined the *bagayat* area, were leveled and planted to mango trees. After six years the trees begin to bear fruit. So long as they cast little shadow, the traditional *bagayat* crops can be grown among them. Although this was done during the introduction and extension of mango-growing in the existing gardens, the new *jarayat* lots are now confined solely to young mango plants.

The government is attempting to introduce the so-called Japanese method of rice-growing, which guarantees the cultivator a higher yield but requires more labor and capital. The farmer has to level the *kyari*, use superior seed, sow in rows and on elevated beds, observe certain distances between plants, weed intensively, and use more fertilizer. Not a single cultivator in Chikhligam had adopted this method.

As indicated by this brief survey of present and past agricultural economy in Chikhligam, the present crop system is not characterized by highly intensive labor. More than half the arable soil is not cultivated, and agricultural activities are unevenly distributed over the year. They are concentrated mainly in the periods just before and after the rainy season, i.e., the months of April, May, and June, especially in the gardens, and October and November, especially on the *kyari* and *jarayat*.

Inasmuch as agriculture in the village is so extensive, it is impossible for most Dublas — and their number has rapidly increased in the past few decades — to subsist on it the year round. Their seasonal migration to the brickyards fits well into the agrarian cycle; most of the activities fall in the period directly before and after the wet season, when the laborers are still in the village. Nor are their employers, the Anavil Brahmans, much interested in agriculture; they say the profits are too low. More and more Anavils, especially the young, find employment in urban centers.

In the circumstances, the fact that both landlords and agricultural laborers prefer to work outside the village is understandable. To say that it offers them a way out of the restricted possibilities is to state the present position, not to explain it. On inquiry I found that in the past agriculture figured much more largely in the lives of both Anavils and Dublas. How has it come to pass that a large

TABLE 7.4.1
AGRICULTURAL CALENDAR IN CHIKHLIGAM

| *Month* | *Jarayat (grass)* | *Kyari (rice and val)* | *Bagayat (mangoes)* |
|---|---|---|---|
| July through September | After the first rainfall, cutting of green grass | Mixing the soil, transplanting rice seedlings, cutting green grass on embankments, weeding | Cutting green grass, potting young plants |
| October through November | Making hay some weeks after the end of the monsoon | Reaping and threshing rice, plowing, sowing *val* | Mowing grass in orchards, plowing the ground |
| December through March | — | Picking *val* | Irrigating orchards, spraying trees with insecticide, watering young potplants |
| April through June | Cutting down *babul* trees if necessary | Repairing field embankments and fences, plowing ground and fertilizing seed-beds, setting rice seedlings at first rainfall | Grafting young plants on trees and daily watering, fertilizing, irrigating and plowing the ground, pruning trees, mango-picking |

proportion of the caste of agricultural laborers now migrates for a considerable period every year, and that more and more members of the dominant caste even leave the village definitively? The answer must be sought in the character of the changes that have taken place during the last few decades.

### *7.5 The* Hali *System in Chikhligam, and Its Disappearance*

"Formerly the Dublas were our devoted servants, they listened to what we said, were satisfied with what we gave them, and worked hard for us. Now everything has changed. The agricultural laborers are in the village only during the rainy season. They have abandoned us and no longer honor their superiors." This was the purport of what some old Anavils told me on the day of my arrival in

Chikhligam. Seated on their porch swings in front of their houses, they reminisced about the good old days and regretted their disappearance. With unending patience and many digressions, members of this caste were prepared to embroider on the theme that day and throughout my stay, for they had nothing else to do.

To inquire after past conditions is risky business. Even information about events of a few years earlier sometimes proved to be unreliable. To reconstruct the relationship that existed several decades ago between landlords and agricultural laborers is more difficult still, in part because young informants have to base their statements on hearsay. I tried to verify the data I was given on past events as carefully as possible, for instance by asking not only Anavils and Dublas but some members of other castes as well about these matters. Comparison of the accounts of agricultural laborers with those of landlords showed that the latter tended to exaggerate by overstating the number of *halis* employed by their fathers. As the discrepancies in the numbers stated by the two groups were considerable I could not quantify the data on this past period. What was important, however, was that all those concerned described the same type of relationship: they all dwelt at length on the *hali* system. The information given me by the villagers on this subject agrees with the picture that emerges from the government reports of the nineteenth and the beginning of the twentieth century which I cited earlier.

The relationship between *hali* and *dhaniamo* was rarely severed. One instance in which it did end during the participants' lifetime involved an Anavil family which is now the most prominent one in the village. At the beginning of this century the family's fortunes were at their lowest ebb. A contemporary official report describes the period as one of famine in the *taluka*, and *halis* were dismissed everywhere.[7] Moreover, three very aged brothers in this family told me their parents' house was destroyed by fire and a large part of the livestock was lost. Their father was obliged to dismiss all his *halis*, twenty of them as they claimed. Aside from such setbacks as this, however, the tie between the families of landlords and agricultural laborers remained intact for many generations.

Why did the Anavil Brahmans in Chikhligam take *halis* into their service? In explaining it to me, both landlords and farm

7. The same report (1897, 7–8): "*Halis* are no longer maintained as the Anaolas find it hard enough to feed themselves, let alone their *Halis*."

servants were inclined to ascribe it to economic motives. As I noted above, agriculture was much more intensive in the past. In the *bagayat* plots along the river, the landlords grew various garden crops, chiefly for home consumption. Most of these, and especially sugar cane, which was a commercial crop, required highly intensive labor. At that time the *jarayat* was also under cultivation. In winter, when work was scarce, the servants took oxcarts to the forest area east of the *taluka* to cut wood for their masters. The necessity of using their laborers throughout the year made it worthwhile for the landowners to take *halis* into their employ. Moreover, the Anavil Brahmans desired to refrain from agricultural labor as much as possible themselves. Not all members of the dominant caste could completely avoid working, but the largest landowners among them did, especially members of the Desai family. Many Bhathela families in Chikhligam, however, still worked with the servants on the land. They now say euphemistically that they occasionally "lent a hand," but they were not in a position even to avoid plowing, a demeaning activity for Brahmans.

The economic situation of the members of the dominant caste determined whether they employed servants, and how many. Most Anavils had one or two *halis*, only the most prominent among them having more. In this respect, too, the situation in Chikhligam conformed with that reported for other villages in the same region at that time. The Anavil Brahmans have always given high priority to their wives' inactivity as well. Where all the men of the caste tried to avoid working on the land, it is obvious that their wives did not assist in the fields. In the house the women could likewise be spared the necessity of doing less favored or defiling work by leaving it to the *hali*'s wife, the maid.

For a Brahman, taking a *hali* was evidence of his prosperity and enhanced the esteem in which he was held. His doing so conformed with what was considered fitting for a member of the caste of landlords. The local Desai family set the tone. As my informants assured me, this family had the greatest number of servants. The Bhathelas in the village tried to emulate this "lordly" example, but mostly in vain. To me they boasted of the many *halis* employed by their fathers, but as I said, on this point their information was unreliable.

To dismiss a *hali* was an action of which the master was ashamed. He did it only in extreme cases, for by dismissing a servant he admitted that he was no longer able to fulfil a patron's obligations of aid and protection. The ill luck of one Anavil in the village was explained to me in one sentence, "His father had five daughters and five *halis*," an allusion to the obligations of this family in the previous generation. To live above one's social position, that is, for a landlord to have more servants than he could afford, was finally to lose land and be compelled to discharge the servants. In this connection I was told about the decline of an Anavil from a neighboring village in the previous generation. As master of a *hali* with six sons, this man had been so conceited as to think it humiliating not to make use of his preferential rights. To avoid what he thought of as loss of face he took all six of them into his employ by giving them marriage money. My informants gave me graphic descriptions of these Dublas doing nothing and living at the expense of their master. Pride once more went before a fall: when the man died, his sons were left without resources. The greater part of the land was lost, and the servants went to another master.

Many Anavils had to work with their servants on the land, and some had no servant at all. Only occasionally did these poorer Anavils use the labor of the unattached Dublas in the village. For them the seignorial life style of the larger landowners remained an unattainable ideal.

Among the Dublas, the first consideration in becoming *halis* was the security they found in servitude. Not all members of this caste of agricultural laborers succeeded in finding a master, an Anavil who was prepared to provide them with a livelihood. Both Anavils and Dublas confirmed that there had always been a category of casual laborers in the village. It might be important to know the former proportions of unattached laborers and *halis* in the village, but the statements I have on this ratio are not exact and diverge widely. Most of the informants of both castes, however, stated that the number of farm servants was many times larger than it is now. They mentioned especially the smaller size of the Dubla caste at that time.

The casual laborers, too, mainly depended on the landlords in Chikhligam for their livelihood, but they added to their earnings by work in the surrounding villages. Anavils and Dublas were unanimous in their belief that before the Second World War the *halis* were better off than the casual laborers. Especially during the slack winter months, the unattached Dublas often had no income and led a miserable existence.

This is the background against which must be judged the accounts of the affection and loyalty felt by the servants for their masters, so highly praised in retrospect by the Anavils of the village. Evidently, to put his security at stake by risking a conflict with his *dhaniamo* was the last thing a *hali* would do. Moreover, such norms as obedience and fidelity were guiding principles in the Dubla mind. Running away was a sin for which a Dubla would be punished in another life by having to serve his master as an ox. Thus the behavior of a "good" servant was supported by religious norms.[8] A more immediately operative sanction was that a deserted master could count on the cooperation of the local authorities, themselves usually caste members, in hunting down the Dubla.

For another Anavil to take on a servant without the permission of his former master was also prohibited. Things were different when, about thirty years ago, a Mochi bought *bagayat* land in Chikhligam with money he had earned in East Africa and took a Dubla into his employ by giving him marriage money just as the Anavil landlords did. His venture failed, as the Anavils took pleasure in telling me. This Mochi *dhaniamo*, of low caste and incompetent as a landlord, did not compel respect. His servant's impudence and laziness were too much for him. After a time the Dubla entered the service of an Anavil. The new master, however, did not pay the Mochi for the debt incurred by this *hali*, as would have been usual. In other ways, too, the Mochi, the only known example in the village of a *dhaniamo* of low caste, came to grief. He leased out his land on a crop-sharing basis to an Anavil landlord, and when the latest tenancy reform occurred this Anavil had himself registered as the owner.

On the other hand one should not take an exaggerated view of the collaboration and harmony within the dominant caste in the past. The relationship between the various patrons, who were divided into factions, was one of competition. By luring a servant away from a fellow caste member they could cause the latter to lose prestige, for this proved that his servant was not loyal. Here was a

8. "The Dhaniamas, belonging to the higher castes, have sedulously developed a deep-rooted feeling among the Halis that running away from the master is a great sin, for which God will punish the Hali's descendants." In *Report of the Congress Agrarian Reforms Committee*, 1951, 129. See also Lorenzo, 1943, 66.

possibility, although a very small one, for the Dublas to manipulate their masters by reminding them of their obligations as patrons and addressing them as such in the presence of others. Nevertheless, the pressure which the *dhaniamo* could apply was considerable, and the chances for his subordinates to escape it were slight. "They did anything we asked of them or more; they were our slaves (*gulam*)," the Anavils of Chikhligam now say.

In a situation of complete powerlessness the only thing to do is to recognize one's dependence. By clearly showing subservience and docility the servant could try to gain extra advantages and to receive better treatment than his fellow servants, thus persuading himself that his position was higher than theirs. A Dubla prided himself on belonging to the household of a prominent Anavil. This attitude has not yet wholly disappeared. An old Dubla, the servant of a Desai in the village, whispered to me that his master was the only true Anavil in the village, the other landlords being only Bhathelas.

Despite the inferior position in which the members of the castes of agricultural laborers found themselves, servitude offered them opportunity to gain some influence and esteem, if only among their equals. Tales about the good old times, the bond and mutual sympathy that used to exist between master and servant, are told even now by the older Anavils and Dublas. Concerning the traditional situation in Tanjore, Gough writes:

> In recalling the past, it is the mutual love between master and servant, not the Brahmans' power to make unlimited exactions, in which they take pride. Powerful economic, ritual and emotional ties bound each Brahman household to its own hereditary labourers.[9]

The statements of landlords and agricultural laborers in Chikhligam alike confirm the conclusion drawn in chapter 4 about the situation that obtained some decades ago: they were partners in a system of bondage and patronage.

The lines on which the relationship between landlords and farm servants in Chikhligam are now organized are quite different. The change is recent and is closely bound up with the introduction of mango cultivation just before the Second World War. The mango was not a new fruit to the population of Chikhligam: they knew it as a country fruit for their own use, which was picked from

9. Gough, 1960, 50. Similar statements can be found in several other anthropological publications on village India.

trees growing on the enclosures or in their yards. The first man to cultivate mangoes was the late husband of an Anavil woman who now manages the family property for her sons who work in town. All fellow caste members have followed his example. It was a small-scale change at first, and in the initial stage other crops were still grown among the young trees. This interesting possibility of lowering the cost of investment has been wholly neglected in the recent extension of the orchards to favorably situated *jarayat* lots. When I asked why, the Anavils replied that it was too much trouble.

The changeover to the relatively nonintensive cultivation of fruit trees cannot be explained by an existing or imminent shortage of labor. Although, as I was told, some Dublas worked in the brick-yards near Bombay for a few months of the year, their number was small. Such employment as existed outside agriculture before the war became even scarcer during the depression years. Moreover, my informants emphasized, the Dublas were practically all attached at the time. The landlords would never have permitted their laborers to leave the village in large numbers. Only when there was less to do in agriculture, when, in other words, the Anavil Brahmans no longer needed them, did the yearly migration to the brickyards get into its swing. Eight older Dublas told me that they had begun work as *halis*, but that they had been dismissed as redundant. This was confirmed by their former masters. It was not the increase of alternative employment in town that was responsible for the seasonal migration of the members of the caste of agricultural laborers, but rather their redundancy in the village economy.

Anavil landlords themselves mentioned the following reasons for the change of crops: plant diseases in some garden crops, the fall of cane-sugar prices, and the growing demand for mangoes in the towns. Although such considerations in the field of agronomy and market economy no doubt contributed to the replacement of the traditional *bagayat* crops, they do not explain the preference for mango cultivation in the localities inhabited by Anavil Brahmans. The role played by social factors in the introduction and diffusion of the new crop was, in my view, more important. From the beginning the members of the dominant caste had strived after a seignorial status, but their active participation in agriculture was unavoidable even if they did not till the land themselves. They had to be continually present in their fields or gardens to supervise and direct the work of their farm servants. Fruit-growing, however, enabled them to expend even less energy on managing their property. Especially for the Bhathelas among them, who could not afford much help and of necessity worked their own land, the cultivation of mangoes meant almost complete deliverance from personal activity. A changeover to this crop requiring little labor enabled them to lead the same kind of life as their more well-to-do fellow caste members. V. H. Joshi has correctly pointed out that the introduction of mango cultivation in south Gujarat was to some extent due to the Anavils' desire for a higher money income, therefore to a commercialization of agriculture.[10] It is, however, doubtful whether the landlords gave evidence of a capitalist mentality — in the sense of careful calculation of marketing possibilities for their agrarian output — when they so clearly were motivated by their wish to exert themselves as little as possible. An illuminating comment by G. Keatinge on the rapid expansion of cotton cultivation in central Gujarat in the first decades of the twentieth century points to the same conclusion:

> It is even asserted by careful observers that the keenness of cultivators to grow cotton is due not only to the fact that they can usually make good profits from it, but also that it is an easy crop to grow and leaves them plenty of leisure.[11]

So far the change in the crop system seems to fit in completely with the long-standing ideal of the members of the dominant caste to live in lordly style. At the same time, their wish for a less active role in managing their property also arose from a more general loss of interest in agriculture and in village life. Even in the previous generation, the Anavil Brahmans showed an increasing preference for living and working outside the village. This tendency has increased among the present generation, as I shall substantiate below. It is obvious that the lessening of the landlords' involvement in agriculture and village life has had far-reaching consequences for the position of the Dubla agricultural laborers.

Thus, the relationship of patronage between landlords and agricultural laborers ended. The time was past when the Anavils felt obliged to provide for Dublas by taking them into their employ as *halis* and thereby recognizing them as clients who were entitled to aid and protection. That the landlords were no longer able to take on a large proportion of the agricultural laborers, owing to the

10. Joshi, 1966, 49.
11. Keatinge, 1921, 146.

demographic shift between the two categories, was also a factor. In recent years the number of Dublas in Chikhligam has risen rapidly. I was told by various informants that there are now many more huts of this caste than there used to be. The number of Anavils, on the other hand, has decreased. As noted above, a relatively large number of them have left the village permanently, while the natural increase of the caste in the village has remained below that of other castes. The result has been that the ratio of agricultural laborers to Anavil households has changed. For the latter it was simply impossible to provide the same proportion of Dublas with a permanent living and, in general, to behave like patrons towards them. Nor were they sufficiently motivated to endure any privations in order to do so.

I have not found any indications that the disappearance of the *hali* system in Chikhligam could be ascribed to outside pressure. The British authorities in south Gujarat did not in any way intervene in favor of the agricultural laborers, and "social workers" who, influenced by the Gandhian movement, fought for the interests of the Dublas were not noticeably active in this *taluka*.[12] At most it can be said that some landlords were vulnerable to the criticism, especially emanating from Gandhian circles, that they practiced complete bondage of agricultural laborers in the period of *halipratha.* At the same time, however, they argue inconsistently that the greater freedom of the Dublas has simply led to more ingratitude. It annoys them that complaints about the past, however justified, are extended to the present.

A changing orientation of the landlords was, I think, of decisive importance, and the changeover to mango cultivation was an expression of it. The wide range of obligations to *halis* and their families was increasingly felt to be a burden. By means of detailed calculations, Joshi shows that in the span of a year a *hali* proved to be much more expensive than a casual laborer who was needed for only a limited number of days.[13] These views concur with those in some earlier reports in which the *hali* system was called uneconomic, a statement that was usually elucidated by a comparison of cost as between a day-laborer and a farm servant, which was always to the detriment of the latter.[14] It remains to be seen whether the

12. *Report of the Hali Labour Enquiry Committee*, 52.
13. Joshi, *op. cit.*, 56–57.
14. *Census of India*, 1921, vol. III, pt. 1, 222; Mukhtyar, *op. cit.*, 162–66; Shukla, *op. cit.*, 121–25.

return of a *hali*'s labor in the past was relatively much higher. At any rate, those who put so much stress on wage cost and labor productivity do not distinguish the element of patronage inherent in the *hali* system. Although the population of agricultural laborers has indeed proportionally increased, the "surplus of labor" is especially the result of a different definition of the laborer's economic use, this time solely in terms of work performance.

In their dealings with Dublas, the Anavils generally assumed the attitude of employers toward employees. The question they put to themselves was no longer "How many farm servants can I provide for?" but "How many hands do I need, and when?"

It was from sheer necessity that the Dublas, no longer assured of subsistence, reacted by partially withdrawing from the village economy, which they could do because of the opportunity of seasonal employment in the brickyards near Bombay. In my discussion of the conditions I found in the course of my field work I shall elucidate this further.

### *7.6 The Present Situation*

THE DUBLAS. My data on the Dublas relate to fifty-one huts with 327 occupants in all, that is, an average of 6.4 per hut. The village nucleus comprises twenty-six huts, and twenty-five have been built just outside the territory of Chikhligam. Of the male working population over sixteen years of age — ninety-five by my count — thirty-eight are employed in agriculture. Most of them work as farm servants (twenty-one) or as day-laborers (eleven), the others (six) cultivate their own land and add to their meager income by regularly performing field labor for the landlords of Chikhligam (see subsection 7.2). Five Dublas are employed in the village but outside agriculture, and at the time of my stay, fifty-two had gone to the brickyards.

When a man marries, a new household is usually formed and new accommodation is sought. Of the fifty-one huts, thirty-three contain a household of two generations, usually parents (or a parent) with unmarried children. The joint-family character of other households is found chiefly among those Dublas in the village who are permanently employed outside agriculture or are landowners or both (eleven in all). Compared with their landless fellow caste members, these Dublas are economically better off also because their position is augmented by a greater number of male earners per hut (averaging 2.8 as against 1.3).

In Chikhligam, as in other villages of south Gujarat where the gardens have been transformed into orchards within the past few decades, employment in agriculture has declined sharply. Attached labor, however, has not declined. A surprisingly large number of Dublas (twenty-one) are still farm servants; one is employed by the Bania of the village, the others by Anavil landlords. More than half of them work for the most active members of the dominant caste in the village — three Vashi families — who not only cultivate fruit trees but also grow mango plants, which require more, and more frequently applied, labor. Most of the other Anavil households also have one or more servants. They themselves do not work, and the size of their property is such that it is easier for them, despite the relatively nonintensive crop cultivation, to appoint a Dubla on a yearly basis than to employ day labor regularly at small intervals. The oldest farm servants in the village entered service as *halis* thirty or more years ago, that is they married at the expense of their fathers' masters after having worked for them as cowherds. They are exceptional, however, and most farm servants today entered service at a later age, and only recently. When younger, nearly all of them worked in the brickyards for a number of years. Some found the work too heavy, but others say they will certainly go again sometime in the near future. In contrast to its former state, the service relationship is now a temporary one. The servant status is no longer more or less automatically transferred from father to son. Neither party deliberately intends to maintain the relation for life. Landlords as well as farm servants reckon with the possibility — indeed, the probability — that their agreement will come to an end. The Dubla, who is no longer of the sort who worked for the master from childhood, knows that he can find work in the brickyards if necessary, and he has become less compliant. The Anavil aims to give his servant as little credit as possible, so as to limit possible monetary loss in advance. Neither puts trust in the other's intentions, partly because the personal tie between them is lost. Servitude, in fact, has been reduced to a businesslike agreement with a limited purpose between partners who mutually distrust each other.

This is also evident from the Dubla's work performance. At the peak of the season he works from 5 or 6 a.m. to 7 p.m. At other times he begins a little later and stops earlier, and his midday break is longer. Jobs in and around the master's house are not much to his liking. The servant no longer wishes to be at his landlord's disposal from morning till night, and tries to limit his obligations to agricultural labor if he can. Nor does he take it for granted that his wife should work as a maid in the master's household, and his son as a cowherd. The Dubla women working in Anavil houses in the morning are by no means all married to servants. *Govalia*, too, are now separately hired. The contract, therefore, is not only temporary but restricted, relating only to the labor of the servant and no longer including that of the members of his household as a matter of course. Nevertheless, the close relatives of the Dublas are still expected to hold themselves at the master's disposal as extra hands at harvest time.

The remuneration of Dublas in permanent service has likewise become more specific. In kind, the daily ration has remained at a level that has been fixed since the last century: two *seers* (a cubic measure, weighed out in a hollow wooden cylinder) of *jowar* or four *seers* of unhusked rice.[15] In money, their daily wage at the time of my field work was 8 annas. On the mornings when he begins early, the servant is entitled to a sort of breakfast (a *rotli* and some vegetables) and a cup of tea. For some activities he also receives a midday meal. In addition there are the time-honored perquisites: some garments, a pair of shoes, a shawl (also used as a blanket), reeds for roof-mending, permission to gather wood in the gardens, a few sheaves of rice or some *val* at harvest, *chas* (the skimmed milk left over from the preparation of *ghi*), etc. The Anavils manipulate these wage elements as means of pressure to make an unwilling and misbehaving servant see reason.

Some landlords attempt in every possible way to skimp the extras given the servants, and increasingly they prefer to replace the emoluments by money. For the maid and the cowherd this has become the rule. Although the fourteen Dubla women who work in the houses of the Anavils in the mornings still receive a midday meal (rice and *dal*) when they have to stay later than noon, their money wages are about 1 rupee a month. The cowherds, too (eleven in all), who earn 1 or 2 rupees a month, receive

15. Mukhtyar has calculated that on a daily ration of five *seers* of unhusked rice a Dubla family of five persons suffers from chronic undernourishment (Mukhtyar, appendix VIII).

no clothing or shoes, only their food. The Anavil women consider it inconvenient and beneath their dignity to prepare food for Dubla farm servants. The changeover to money payment is undoubtedly due in part to the rise of different habits of consumption. One Anavil, who wished to adopt the kind of agreement prevailing in another village, offered to pay his servants a daily wage of 1 rupee and stop all allowances in kind. Understandably enough, they did not accept his proposal. Transformation of allowances into money payments is to the servants' disadvantage. In the past, a *hali* was allowed to cultivate a plot himself, and the produce it yielded was an important supplement to his daily earnings. During the war years, a period of high grain prices, this *vavla* was replaced by a yearly payment, and in Chikhligam this bonus now amounts to between 25 and 50 rupees, depending on the skill and bargaining position of the farm servant. It is paid in small amounts spread over the whole year, and only when the servant asks for it and the master has no objection. Although originally intended to cover unforeseen expenses, the wage is now necessarily spent on articles that the servant used to receive from his master, on his daily ration of tobacco and oil (formerly pressed from the castor seeds he cultivated on the *jarayat*), and on clothing, shoes, and the like.

Around *divali,* when the accounts are settled, the servant rarely finds that there is money owing to him. Not without reason, the Dublas suspect their master of double-crossing them. On the other hand, the master is likely to deduct from the year's allowance the drink money he has occasionally handed over to persuade the servant to do something outside his routine activities, or else he writes it down as advance payment if the amount is exceeded. This seems reasonable to the Anavils. Formerly the Dubla was at his master's beck and call every hour of the day without receiving extra pay for such service.

The allowance received by the servant for days when he does not work — because of heavy rain in the monsoon period — is also debited to him. Although the master is expected in any case to give this *khavati* (literally, "in order to eat"), he deducts it from the year's allowance. The debt run up in this way, sometimes added to by marriage money, also becomes collectable at *divali*. The Anavils admit that a Dubla cannot possibily discharge the debt so long as he works as a servant, but when he runs away to the brickyards, as farm servants frequently do, the Anavil tries to recover the debt from him there or, as I witnessed, even holds his relatives responsible if they have remained in the village.

This increasingly businesslike character has also entered the relationships among the Anavils themselves. When a Dubla farm servant finds employment with a new master, the latter no longer considers himself answerable, as would have been true in the past, for the obligations incurred by his subordinate toward a previous employer. Even within his own village there is no longer any question of compensation on the part of the new employer. In contrast to former usage, the servant himself is now held responsible.

Conditions of employment for Dublas in permanent service display much that is reminiscent of the *hali* system, but some differences are fundamental. The relationship is no longer hereditary and remains essentially confined to a work performance by one party and an exactly calculated remuneration by the other. The Dubla does not feel bound to do more than his agreed portion of the work, and as for the Anavil, he does not consider himself obliged to pay more than the minimal amount of compensation. If the servant says he cannot live on it, that is regarded as his own problem. There is no longer any question of complete and lifelong support. Two servants in the village who are too old to work are still to some extent useful as *govalia*. But three others, who had worked for their master from childhood, were dismissed a year ago, in certain instances after a provoked quarrel, and are now, sparingly, supported by their children. Nor is continuance of the grain allowance during illness a self-evident right. Although the Dubla may leave his landlord more easily, it will be clear that it has been he who came off the worse when their relationship changed. This is largely the result of the increase of monetary transactions in an agrarian subsistence economy.

Aside from the Dublas in the permanent service of landlords, there is a category of casual laborers. When classifying the economic activities of Dubla men in the productive age group in Chikhligam, I arrived at the figure of eleven day-laborers. They, too, chiefly depend on the Anavils for their livelihood. It is true that casual laborers are free to work outside the village, but only some Kolis in the neighboring village employ them occasionally. These Kolis, like the landowners among the other castes in Chikhligam, are petty cultivators who can usually manage the work with their own families. Inasmuch as the Anavils do not continually need the services of these laborers, there is employment for them during only part of the year. The advantage of a slightly higher wage — they received 12 annas a day during my stay in the village — is outweighed by frequent long periods of unemployment. Their freedom is therefore rather debatable. On days without work they ask the large Anavil landowners of the village for *khavati*, and in compensation they bind themselves to work for the landowners on days selected by the latter. Those Anavils who can afford it make sure of a labor reserve at the peak of the season by giving advances in this manner. Moreover, they pay casual laborers only 8 annas a day during these periods, the same wage as the farm servants. The day-laborers much prefer to go to the Kolis who, they say, make them work harder but are more liberal and join in working with them in the field. They give the laborers 1 rupee and also spirits. These small farmers of a lower caste are, in short, less high-handed than the Anavil Brahmans.

The prospect of a day-laborer's life is not very inviting, and most Dublas in this category regard it as a temporary solution. Not, for instance, the three who were dismissed in old age and now try to earn a little here and there as day-laborers, but those who are prevented by circumstances (a pregnant wife, a baby newly born, or aged parents to be cared for) from going to the brick-yards. Work in the brickyard, not permanent service, is for them the alternative to the meager existence of the day-laborer. Their stay in Chikhligam is a breathing space; those who are young and healthy hope to go away again in the following year.

Considering all this, it is not surprising that the majority of Dublas is on the move every year. There is virtually no hut from which one or more occupants are not absent, working in the brick-yards. Of the men in the productive age-group, fifty-two were absent during the period of my field work. In most cases the whole family goes, so as to add the earning power of the woman and the children. Only the aged, women with very young children, and otherwise unproductive family members remain behind. They usually must provide for themselves, as weekly allowances in the brickyards are low and strictly for those who work there, so that little or nothing is left to support those at home.

The organization of this migratory labor has already been explained above. Seasonal employment in the brickyards has undoubtedly lessened the dependence of the Dublas on the landlords in the village. Assisted by relatives, some manage to save enough for their marriage money. This is exceptional, however, for once in the brickyards the Dublas adopt an easier and less frugal way of life. They eat more and better food — for instance, the meat and fish they cannot afford in the village — take tea with milk, and buy various cheap commodities ranging from clothes to household utensils. Moreover, a great deal of drinking goes on. The amount they finally save is therefore hardly ever enough to pay the costs of a marriage. Occasionally the labor-recruiter or the owner of the brickyard is prepared to help. A seasonal laborer who is known as diligent, honest, and well-behaved can obtain an advance if he promises to hold himself at the disposal of his creditor until the debt is paid off. In this way there results a relationship of dependence of the same type as that with which the Dubla was familiar in the village community.[16]

During the past decade, labor requirements in the brickyards of Bombay have risen sharply. This industry, however, is very susceptible to economic fluctuations and a regression will undoubtedly lead to large-scale shut-downs. Even in the present period of expansion, the migratory laborer's existence is precarious and uncertain. Most of the brickyards are small enterprises with no more than two or three kilns, badly managed, scantily financed, and operating in an unstable market. Marginal production frequently leads to an owner's bankruptcy. so that the laborers have to return to the village prematurely and without earnings. The labor-recruiter is another uncertain element. For placement, the gangs recruited by him depend on the contacts that he has established, often over a period of many years. If he drops out, or if he quarrels with the brickyard owners, it is the laborers who suffer for it. Thus the Dublas are not at all sure of finding work every year, whether in the same brickyard or in another. One Anavil told me that he was convinced that the majority of the migratory laborers would prefer to work for him if he guaranteed them a daily wage of a little over 1 rupee all the year round. But in my view he underestimates the strong desire that has arisen among the Dublas to escape, if only temporarily, from the yoke represented by landlords and by agricultural work. On the other hand, the Anavil's claim confirms my impression that seasonal migration does not hold any great advantages for the Dublas.

16. For a similar type of relationship among migratory labor in Japan, cf. Bennett and Ishino, 1963, ch. 8.

The temporary or permanent absence of Dubla men from the village may lead to termination of marriages. The women who remain behind without support return to their native villages with their children, or they marry again. Divorce occurs frequently (I counted sixteen cases), for instance, when one of the partners is in poor health and cannot work. Many children grow up in the households of relatives owing to the absence of one or both parents; in Chikhligam twenty-three children have been taken into households to which neither of their parents belongs. Similarly, a number of incomplete families have joined other households: I found nine cases of Dubla men or women who had been left behind and who, with their children, had moved in with their parents, a brother or a sister. The relatively low stability of Dubla households — which, I regret to say, struck me as significant only when I was writing up my data — is undoubtedly connected with the precariousness of their existence.

Only a small number of the Dublas from Chikhligam, nineteen in all, have succeeded in finding regular work in town. Nonagrarian employment is scarce and not very accessible to members of a caste of unskilled agricultural laborers unfamiliar with the industrial scene. Nor are there as yet many Dublas who are established in town and can help their fellow caste members in finding work and living quarters. Seasonal migration as a first step towards permanent settlement in town is therefore a solution available to no more than a few Dubas, and most of them return year after year to the village at the end of the dry spell. As they grow older, they usually are completely reabsorbed into the village economy. Work in the brickyards is arduous, the largest contingent there consists of young people, with their greater endurance. The farm hands in the village are mostly older men, who feel no longer capable of the exertions demanded by brickyard employment.

For most Dublas, the escape from agricultural labor by seasonal migration is temporary in yet another respect. The amount of money they bring home from the brickyard — varying with the size of the family, but usually not exceeding 50 to 75 rupees — is not sufficient to tide them over the rainy season. Even with advances borrowed from the labor-recruiter they are rarely able to make ends meet. During the months when they are in Chikhligam, most Dublas are therefore compelled to work in agriculture. At the time of the inquiry, about half of them — twenty-five of the fifty-two Dublas who had been away — had put themselves at a landlord's continuous disposal during the wet season. This category I have described as monsoon servants. Their tasks consist of cutting a head-load of green grass daily and of weeding occasionally. At the beginning and end of the rainy season, with much work to be done in the fields, activities multiply (see table 7.4.1). The workers receive the daily wages of servants on a yearly basis, but not the perquisites to which the latter are entitled. Unlike the Dublas in permanent service, the monsoon servants are not entitled to *khavati* on days when there is little or nothing for them to do, although they try to wheedle it out of their employers. As for the Anavil, he is always assured of cheap labor in this way.

One may ask what induces a Dubla to bind himself under such unrewarding conditions. Frequently an existing dependence on the landlord leaves him no choice. The "obligations" of a monsoon servant may be purely personal. In several cases, Dubla migratory laborers had been employed originally as farm servants, but they had run away to the brickyards without discharging their debts. Seemingly, the members of the dominant caste are resigned to this, for them, disagreeable habit, but on the laborer's return from Bombay they demand payment, and sometimes even before that.[17] So long as the Dubla is indebted, and this may be permanently, he is obliged to work for his former master in the rainy season. In other cases the pressure is more direct. The landlord sees to it that on their return he will be the only one to make use of the services of relatives and certainly of members of the household of his permanent servant. After a few days — to avoid giving an impression of wanting them badly — the Anavil goes to the hut in which they live, asks after their health, and tells them when he expects them. For members of the dominant caste, the possibility of obtaining an extra amount of labor and, in this way, protecting themselves in advance against shortage, is an important consideration in favor of taking a Dubla into permanent service during the rainy season. At that period there are usually enough agricultural laborers available, but at some critical moments the demand for them is so high that even if the supply is maximal, i.e., if all the Dublas are there, not all the landowners can procure the labor they need. In the competition for laborers it is the largest landowners who come out best, as I noticed at the beginning of the rice harvest. Through

17. By visiting a number of brickyards in Bombay I tried to gain an impression of the migratory labor. On one occasion I observed that an Anavil came and fetched a runaway servant home.

various "obligations," some landlords had been able to monopolize the labor of a large part of the Dubla work-force, and moreover, against low wages. While the Dublas worked with their families in the fields of their "benefactors" as servants, monsoon servants, and day-laborers entitled to *khavati,* the smaller landowners had to delay harvesting until the others had finished, pay higher wages to the Dublas then available, and in some cases even give up sowing the winter crop, as by that time their soil had become too dry to be plowed.

Seasonal migration has become a vital necessity for the majority of Dublas. Though such migration is partly due to the rapid growth of their numbers, the changeover to mango cultivation is chiefly responsible for the fact that so few of them find continuous employment in agriculture.

The yearly migration to the brickyards fits neatly into the agricultural calendar. The Dublas' presence in Chikhligam coincides with the period of greatest activity on the land. This is not as fortuitous as it seems, for although they could have found work in Bombay at an earlier date, the Dublas who work as monsoon servant are not allowed to leave until after the grass harvest (the middle of November), when the interests of the landlords are safe. The Anavils have discharged their obligations towards the Dublas, i.e., to assure their livelihood, but have managed to retain their traditional claim to the working-power of the members of this caste. Although the seasonal migration has lessened the Dublas' dependence, it has not put an end to it. Many of the Dublas are temporarily bound to work as monsoon servants, but in the long run most of them are reintegrated into the village economy as servants or day-laborers. As one of the Anavils succinctly put it, the Dublas work for them on their conditions, "under obligation of debt and for future credit."

THE ANAVIL BRAHMANS. The seasonal migration of the Dublas from Chikhligam was an effect, not the cause, of a shift towards a less varied and labor-intensive agrarian economy. In the preceding pages, I interpreted the introduction of mango cultivation as primarily a reflection of the lessening involvement of the Anavil Brahmans in agriculture. The cultivation of the traditional garden crops is undoubtedly much more profitable, but it would require of the landowners an amount of exertion and trouble which they are not prepared to devote to it. Already in the past generation several members of the dominant caste went to live in town, and recently this movement has gained impetus. Of the male Anavils in the productive age group, distributed over fourteen households in the village, sixteen live on their land in the village, thirteen are employed in nonagrarian occupations, and all but one in the latter category live outside Chikhligam (see subsection 7.2). Three Anavil landlords have additional incomes, respectively as a contractor, a postal agent, and an insurance representative.

As town amenities increased, village life became less attractive to the dominant caste. The young people in particular yearn to live and work in the town. Urban Anavils are preponderantly professional men, clerks with the government or in the private sector, and teachers in primary and secondary schools. It is the typical white-collar work in the administrative sector that draws the young Anavils. Education has become an imperative necessity for them, as they need an S.S.C. (secondary-school certificate) to qualify for even a simple administrative position. The poorer Anavils of the village also send their children to school and count on the help of relatives and fellow caste members in town to find them jobs. The household is prostrated if the son fails in his examination. Again and again the candidate tries to surmount this barrier to a better social position, mooning about the village between examinations. One landlord was overcome with joy when his unclever daughter, after having failed four times, managed to cross the finish line after all. Even without a high dowry she was now regarded as more eligible to marry an urban fellow caste member. Through schooling, the younger generation tries to escape the "backward" village life, and the older generation helps them.

Those who remain behind in Chikhligam feel that they have had bad luck and look upon themselves as misfits. They have not been able to leave the village, and long for the amenities and pleasures of town life. Without much zest, and under the pressure of circumstances, they manage the family property and complain that agriculture is not very profitable. This argument seems ill-founded, for in the first place it assumes as a norm the most successful town-dwelling Anavils, not the majority of their fellow caste members who lead simple clerks' lives. Moreover, a more active management and more investment in agriculture would easily better their situation. But the village Anavils endeavor as conspicuously as possible to be nonfarmers. Their dilemma is that they want to gain as high an income as they can while working as little as possible. Although they rarely visit the *jarayat* or *kyari* plots they possess, most of them take a walk to their gardens in the afternoons in order to inform themselves of the state of the trees. Wearing white *topis* and carrying impressive long canes, they demonstrate their position of landlords and inspect the work of their laborers.

The cultivation of mangoes is the main topic of conversation. "The king of fruits for the kings among men" was the slogan they repeated to me with a mixture of self-mockery and conceit. Seated on their porch swings and looking out over their front yards, they never tire of discussing the quality of the bloom, prospects of the crop, marketing possibilities, price fluctuations, and such matters.[18]

Although most of the village Anavils have been able to abstain from agricultural labor, two of them cannot afford the luxury of not working. These two do not possess much land and are not so comfortably housed as their fellow caste members. Yet they conform as much as possible to the way of life they consider fitting for an Anavil. For plowing, an activity despised by Brahmans, they hire Dhodias. Day-laborers help them with the rest of the work in the fields. One of them also employs a Dubla woman to do the dirty housework. Their children go to school, and one son is a post office worker in the *taluka* town. His father has managed to offer his children a better future, but it is doubtful whether the new generation of the other Anavil family will ever be in a better position. This family is, unfortunately, very large and, worse still, includes many daughters for whom marriage money must be set aside.[19] Unlike the other members of his caste this Anavil has not had his wife sterilized, and the consequences prevent him from improving his weak economic position. As working cultivators, these two Anavils strike a discordant note in their caste.

Eight Anavils have one or more farm servants in their employ (twenty-one in all). Supervising their Dublas is the only activity they engage in. Some of them lend a hand occasionally, but real work is left to their servants, supplemented by day-laborers if necessary. Most of the masters confine themselves to the shouting of instructions or abuse. Four Anavils do not have enough land to employ a farm

18. Cf. also Joshi, *op. cit.*, 53.

19. As to the amounts of money required for such marriages, see Van der Veen, *op. cit.*, 97–99 and 135–37.

servant, but feel that to work would be beneath their dignity. They have leased out their *kyari* land on a crop-sharing basis to fellow caste members who in this way can fully utilize the working power of their own servants. The produce of their orchards is usually sold in advance (when the trees are in bloom) to a factor outside the village who takes care of all the work involved. Their income might be higher if they exerted themselves, but some supplement it with part of the earnings of a son in town. They have a simple routine: morning tea and breakfast — bath — meal — afternoon nap — walk to the garden — tea — evening meal. In the evening as in the daytime they can be found on their porch swings in front of their houses. They are landlords so far as their life style is concerned, but certainly not farmers.

Those Anavils who employ farm servants are more active, for they supervise the work of the Dublas during mango-picking and the rice and *val* harvests, plan the day's work with their laborers, visit their gardens more frequently than the others and also grow young mango plants. Nevertheless, they are not overly busy. Illustrative of the indolence of this caste is the Anavil who has his servant carry an easy-chair to the orchard, so that he can watch the work of his underlings in comfort.

The reputation of the landlords of today — that they have forgotten how to drive an oxcart or to plow a straight furrow — is a dubious compliment in the eyes of members of other castes. Generally speaking, the Anavils themselves are in no way ashamed of their sometimes feigned ignorance in such respects, but the most competent of them regret that so many of their fellow caste members have thrown all responsibility in agriculture to the winds, even that of overseeing the work performance of their laborers.

Still another instance of their lack of interest in agriculture may be cited. In 1962 Chikhligam was connected up with the irrigation network of the large Kakrapar project in south Gujarat by the extension of a branch canal. At no time during my stay was there any indication that the landlords were going to seize the opportunities thus offered by, for example, leveling and irrigating the *jarayat*. The new potential was at best regarded as a reserve in the event that rainfall at the end of the monsoon should fall below expectations. Nor have Anavils in the surrounding villages, where irrigation has been available for a longer period, taken any steps towards intensification of agriculture.

Mango-growing has, above all, made possible a "more dignified" style of life for all Anavil Brahmans. The distance between members of the lower Bhathela section and the Desais, which must have been considerable, has lessened, and the difference in life style between them has faded. During the previous generation, the Desai family had already invested in education, for which those who remained in the village sold their land. Their former gardens are now largely in the hands of the Vashi family, who are, in their turn, spending their acquired riches on nonagrarian purposes. The young people all receive educations that are meant to have them end up in town. "Conspicuous leisure" has become possible for nearly all Anavils, who — with or without servants — can now lead a landlord's life, or at least pretend to. They no longer derive prestige from the number of their laborers, or in general from the number of clients of other castes retained by them. The dominant caste in Chikhligam is still divided into factions, but for exercising influence at the village level the support of a group of local clients is no longer the only relevant factor. As the economic, political, and administrative scale was enlarged, the display of power within the village community declined and became subject to external structures of decision-making. In this respect, the former patrons have become clients. The success of their activities outside the local framework is partly determined by the weight and esteem they have acquired on other grounds than the criteria of former times. On the other hand, to have a body of support in the village is to be able to exert influence legitimately at a higher level. In this form, patronage has come to have a more restricted purpose and usually remains limited to specifically political matters. Conversely it is true that connections with sources of power — particularly with the government bureaucracy and with the political parties located in town — is still a sovereign means of commanding respect and support in the village. Then, too, in their new scale of desiderata the Anavil Brahmans give the highest priority to education. The profits returned by this investment are considerable, because they enable the Anavils to orient themselves towards town life.

Thus the traditional patron-client relationship has come to an end. Obviously such an important institution as patronage — a whole network of personal relationships resting on very pronounced norms and values — did not disappear overnight. In Chikhligam there are still recognizably patronage-like elements in the behavior of Anavil Brahmans. That the personal character of the landlord concerned makes a difference is illustrated in the attitudes of the village headman and his cousin (the son of his father's brother). They belong to the prominent Vashi family and own an equal amount of land. But whereas three farm servants are enough for the village headman, his cousin requires no less than six. The latter also employs more Dubla women in his household, one of whom is solely charged with the care of the children. This cousin proudly enumerated for my benefit the names of all the persons dependent on him for their livelihood. He justified their large number by referring to the many guests he had to entertain and whom he had to receive in the proper fashion, that is to say, surround them with a great deal of service. He belongs to a newly rich Bhathela family and has married a Desai woman. I, for one, willingly fell a victim to the catering from his kitchen. The food he served was more varied and of better quality than that of any other Anavil in the village. The fact that he is the only one to possess a *savari galli*, a vehicle used exclusively by prominent Desais for the transport of persons, also testifies to his eminent position. It is in his lavishly furnished house, not in that of the village headman, that important receptions and meetings are held.

The headman is the opposite of his cousin in many respects. A fairly young man whose political ambitions go beyond Chikhligam, he is a member of the Congress Party, for which he performs many tasks in the *taluka*. Besides being a landlord, he is a contractor, and through his connections in the capital of the subdistrict he has received various lucrative commissions. His preoccupations often take him outside the village, while his cousin stays at home. He has arrived at the conclusion that Chikhligam as a base is so small and so far from the center of power that it lacks scope for full development of his leadership. At the time of my departure the village headman was thinking of following the example of a rich Anavil landlord from a neighboring village who moved to the *taluka* town. From there he could regularly come by car to Chikhligam to supervise his property and interests. He was the first to try to do away with the perquisites to which his servants are traditionally entitled, which he proposed to replace by money payments. As yet, however, his plan has not succeeded.

Individual differences aside, all Anavils who have one or more Dublas in their employ still present, more or less, the traits of a patron of times past, if only because they are sometimes addressed by their subordinates in this role. The older servants in particular endeavor to play upon their master's liberality and high position by a show of servility that young Dublas think revolting. Furthermore, the landlord is expected to maintain his dignity under all circumstances. Brutish behavior is condemned by all, including his fellow caste members. On the other hand, the Anavil wins admiration when he has a natural authority over his subordinates, when he is tolerant and yet not to be trifled with. The servants can also count on the intercession of their master when they get into difficulties, usually over drink. The Anavils, however, do not concern themselves unduly with their Dublas. Insofar as members of the dominant caste still admit to having "obligations" towards their subordinates, they do so to be assured of an adequate supply of labor rather than from a desire to aid and protect the servants who depend on them.

Aside from a few individual exceptions and some vague notions of duty and honor that remain from an earlier period, a businesslike attitude of the landlords towards their agricultural laborers has come to prevail.

THE RELATIONSHIP BETWEEN ANAVILS AND DUBLAS. The Dublas in Chikhligam feel that the Anavil Brahmans have abandoned them to their fate, and are embittered by the realization that the landlords no longer need them. *Khavati*, formerly resulting from the responsibility of the *dhaniamo* for his subordinates' welfare, has increasingly come to represent an advance payment on labor not yet performed and is considered a favor that can be arbitrarily withheld by the master. Prior agreements among the members of the dominant caste concerning the amount of wage they are prepared to pay have resulted in more uniform treatment of the agicultural laborers. The landlords have become impervious to flattery, and appeals on the part of the Dublas to their magnanimity are usually in vain. The agricultural laborers, on the other hand, when they are paid the minimal remuneration, know that they are being deceived into the bargain. According to a report on their situation, a check of the servants' daily grain allowances showed 10 to 25 percent short weight.[20]

Have the developments of the last few decades increased the poverty of the Dublas in Chikhligam? A clear-cut answer to this is difficult, for it depends partly on how their condition in the past is evaluated. The Dubla farm servants themselves believe they are worse off; they extol the liberality and justness of the masters of the

20. *Report of the Hali Labour Enquiry Committee*, 27.

past. It is true that nowadays they are treated in a more impersonal way, but the stereotypes about the loving care they enjoyed in the past are not very convincing. Patronage does not exclude a precariously minimal livelihood for the clients.

Nevertheless I believe that the agricultural laborers among the Dublas have indeed become poorer. In the first place, the farm servants' social insecurity has increased since the disappearance of *hali-pratha.* They are now in an unstable category and their attachment no longer entitles them to any support, however slight; on the contrary, in old age and in case of illness they can be dismissed without much ado. Apart from this, the changeover to money wages has contributed to the deterioration of their condition. Insofar as the daily allowance is given in kind, it has remained at the same level for more than a century. Paid in money, this living wage has risen from 2 to 8 annas during the past fifty years and will no doubt further increase. But the prices of the articles they can buy for it have risen much more sharply. Instead of the emoluments in kind that the Dublas used to receive from their masters' gardens, they now are given a yearly bonus which, like their wages, is not adjusted to the sharply rising prices. Transformation of a subsistence economy into a money economy has taken place at the cost of the economically weak Dublas. Their wages are so low that even a box of matches must be made to last, although it costs 1 anna. Now that the tinder-box has fallen into disuse, they postpone lighting their *bidis* (country cigarettes) until the master comes to the field for inspection and can give them a light.

In contrast to the extreme poverty of the Dublas is the wealth of most members of the dominant caste. Nearly all of its members live in brick houses; only a few still have old-style dwellings of mud brick. Their rooms are furnished more expensively than in former times, with a sofa, cupboards, a clock, sometimes a table and chairs, and kitchen utensils of the coveted stainless steel. There is a wide array of "novelties," which the proud owners are eager to show. Men, women, and children are better dressed now, and the food and other items they use are of a better quality, for instance, wheat instead of millet, refined sugar instead of *gur* (raw cane sugar), and cigarettes instead of the local *bidis*; and milk tea flavored with spices has found general acceptance among the Anavils. A wide gap between the standards of living of landlords and agricultural laborers has, of course, always existed, but in the traditional economy the difference was one of quantity rather than quality. Owing to the process of economic enlargement of scale, the consumption patterns of the two castes have sharply diverged.

In a more or less closed village community it is natural that the landowners should share the crop with members of other castes. Partly, however, because of the disappearance of local isolation, agricultural products have attained a higher market value. The material results of the changes in agricultural production were not used by the landlords of Chikhligam in the traditional way, i.e., in sharing the profit with their Dubla clients — a change of preferences to the disadvantage of the agricultural laborers, for the money earned in mango growing was invested by the Anavils in education for their children. Apart from this, maximalization of income has become an end in itself for them. Besides undiminished "conspicuous leisure" it is now "conspicuous consumption" that characterizes the life style of the Anavils of the village. They enjoy their newly acquired wealth and buy luxury goods previously unknown in the village community, thereby once more emphasizing their distance from the other inhabitants. The conclusion is that the changeover to a money economy should be understood and evaluated in terms of both production and consumption.

Economically the gap between landlords and agricultural laborers has become wider. Although the latter may not be much worse off than before in an absolute sense, they measure their poverty against the increased wealth of the former. The Anavils admit that the wages of an agricultural laborer are low, but claim that they cannot pay more, an argument which fits in with their complaints about the low productivity of agriculture. A second motive they adduce is that the Dublas do not need much. The kind of life that a member of the dominant caste leads, or is expected to lead, requires a high income. The Anavils have become more cash-minded during the last few decades and in the calculation of cost and benefit of a farm servant, the profitability of his labor is now the chief criterion.

> It is obvious that the average of wages paid to the Hali is higher than that paid to the free labourer, while the actual output of work per day by the former is only equal to, if not less than, that of the latter, as the former, unlike the latter, is an irresponsible fellow. Thus evidently the Hali system is uneconomical and inefficient.[21]

It is interesting to note that the changed interpretation of the use of Dublas led at the same time to a change in stereotyped concepts

21. Mukhtyar, *op. cit.*, 170. See also Shukla, *op. cit.*, 130.

about them. Whereas writers of the nineteenth century had termed them hard-working, honest, polite, and hospitable, they are now looked upon as lazy, dishonest, and impertinent.[22]

At the time of my stay in Chikhligam a landlord dismissed a farm servant who was "overwise"; he did not know his place and always replied insolently when his master browbeat him. The Anavils complain that nowadays the promises of a farm servant are not worth a thing. Their proverbial devotion to the master's interests has been replaced by indifference and unreliability as the relationship between them has become more impersonal. A farm servant sometimes stays away for a whole day without notice, and should there be a dispute, he is likely to go off to the brickyards. "Like a hare, a Dubla does not remain long in the same place," is a proverb which the landlords like to cite. The lack of understanding among laborers is annoying, say the Anavils, their responsibility is nil, and their idleness knows no bounds. They are good-for-nothings who do not want to work so long as there is still something to eat. It seems likely, however, that it is not the performance of the agricultural laborers that has changed, but rather the weight given it by the landlords. While minimizing their obligations they now demand a maximal work performance, an effort which the *halis* were never asked to make in the past.

The Anavils are especially indignant that nowadays they sometimes have to go to the huts of the farm servants to ask them to come to work. Such a forced journey to the Dubla's quarter particularly annoys them. It is improper and far beneath their dignity, they feel, and they consider that they should be approached by the landless, not the other way round. For people who have been accustomed to being begged for employment, this is indeed a distasteful symbol of changed times.

As for the Dublas, they discovered, often the hard way, that the Anavils no longer wished to behave like patrons. No longer could the Dublas claim total care and generous aid as clients. "If the master's work does not suffer in the event of the Hali's sickness, he may be allowed to rot in his cottage," Shukla already noted in about 1930.[23] Although we may well wonder whether in the traditional community the master treated his *halis* as well as he claimed to do, a completely different relationship has surely come into being when the value

22. *Gazetteer of the Bombay Presidency*, 1882, vol. XIII, pt. 1, 158, reports favorably on the Dublas. For a negative judgment, see, for instance, Mukhtyar, 1930, 167–68, and *Report of the Hali Labour Enquiry Committee*, 1950, 45.

23. Shukla, *op. cit.*, 123.

of a farm servant is measured by his output, not by his behavior as a client. The landless laborers feel unprotected and exploited. They have seen the prosperity of the landlords increase and are aware that this has been at their cost.

Formerly the son of the farm servant grew up in the house of the master where his mother worked. He played with the son of the Anavil, who later became his boss. This intimate relationship between master and servant has now ceased. Subservience and respect are still demanded from the servant, but the personal tie that mitigated the inequality is no longer there. The landlords do not trust us, say the Dublas in permanent service. They give directions even when it is not necessary, only to make their authority felt. Never does a master say he is satisfied; whatever the farm servant does, the landlord always has some fault to find. Running away under cover of night, for the landlord a proof of faithlessness and dishonesty, is for the farm servant the only way to escape the authority of his master. To ask for permission is to be refused and to give the master an opening to exert pressure so as to prevent his servant from carrying out his plan.

That they depend on their landlord in everything is drummed into the minds of the Dublas. *Khavati* on non-workdays must be literally begged by the servant, and he does not get it until he has done some odd jobs around the house of the Anavil. Sometimes these jobs are clearly unnecessary, and are only meant to show who is master. The servant tries to avoid this confrontation by sending his children to his master's house to get the daily allowance or perhaps ask for *khavati*.

As the relations between the families of master and servant became more impersonal, the humble character of the tasks performed by the Dublas received more emphasis. It is the most disagreeable jobs that fall to them, and this applies even more to their wives who work in the Anavil houses. The women are often on bad terms with the mistress, who leaves the unpleasant work to them and has assumed more and more airs as the gap between them has widened. Among the Dublas there is a growing awareness that the low esteem they enjoy in the village is closely connected with the kind of work they do. The weak social and economic position of this tribal caste is, however, perpetuated by their illiterateness. Whereas the Anavil boys and girls go to school, very few Dubla parents have their children educated, and then only in the lower forms, to have them out of the way in the daytime. At the age of eight or nine the boys work as *govalia* and the girls look after their smaller brothers and sisters in their mother's absence. When the family goes to Bombay, the children contribute to its income by working in the brickyards.

Without wishing to keep up the former relationship, both parties appeal to their traditional rights while rejecting the obligations that went with them. The Anavil Brahmans no longer treat their laborers as clients, but still, and as a matter of fact, expect gratitude and meekness. On the other hand, the Dublas, especially the older people, still occasionally try to approach their landlord in his quality of patron. Because they have always been accustomed to receive a living wage, they do not recognize the recoverable character of advance payments. Nowadays, however, they try to avoid situations in which they are committed to unconditional obedience and loyalty. A relationship of attachment, formerly much valued because of the social security it offered, is now rejected. This is partly because the servant can no longer count upon the support and protection of his master, but above all because the Dublas regard the bondage inherent in attached labor as unacceptable.

For a Dubla with a family, it is impossible to live on a day-laborer's income, and this is the sole reason why some of them accept permanent service. But in order to reduce their obligations and to be free to sever the tie at a moment's notice if necessary, they try to pay for their marriage without the master's help.

Among the members of their caste, the farm servants are of little account. In the present situation, the recognition of dependence offers few advantages, while the disadvantages are many. Nowadays no Dubla girl likes to marry an agricultural laborer who is in permanent service, for she probably will have to work in his landlord's house and perform all the humble and dirty tasks that the Anavil women look down on. The farm servants are no longer a privileged category who are proud of their relationship with an Anavil. It is the migratory laborers whom the Anavil Brahmans have in mind when they say that the Dublas have failed their former "benefactors." Labor discipline in Chikhligam is said to have disintegrated through the irresponsible behavior of the seasonal migrants. Particularly the young people are looked upon as insolent, that is to say, they no longer display any humility in the presence of Anavils and no longer allow themselves to be abused, let alone thrashed, a treatment that in the past many a Dubla had to endure. Those remaining at home are led astray by the drinking and extravagant spending of the migratory laborers, say the Anavils. In the brickyards the Dublas develop a craving for various luxuries — they drink milk tea, for instance, and smoke real cigarettes, and one of them even wore a nylon shirt when he returned. In short, there they acquire habits to which members of the dominant caste feel that they alone are entitled.

Because of general dissatisfaction with the present situation, both parties tend to idealize the past. On the other hand, it cannot be denied that relations between landlords and agricultural laborers have deteriorated in recent years. The mutual distrust and antagonism of which I sketched the sources still prevail. The Anavil Brahmans fear the rise of the lower castes. The influence of the Dhodias increases, as I described earlier in this chapter. The rapid demographic growth of the tribal castes heightens the anxiety of the members of the dominant caste. They feel deluged by the *kaliparaj*, whom they believe inferior. The traditional power elite of the villages in south Gujarat resent that the government promotes the interest of these backward castes, at least in theory. In their opinion it would be much more useful if the authorities directed the attention of the agricultural laborers to their obligations towards their employers and taught them to respect their betters. That the members of the younger Dubla generation fail to show subservience is a source of boundless irritation for the landlords. They feel hardly able to hold their own in the new situation, and moreover their number in the village is diminishing. At present their caste is still dominant, and quarrels between caste members are smoothed over for the sake of unity. As they have come to resemble each other in their life style, Desais and Bhathelas are no longer strictly separated. Especially in the presence of outsiders, the earlier rigid distinction between the high and the low section of Anavil Brahmans has become blurred. They all speak of "we Anavils." Their feeling of solidarity has intensified under the threat presented by the low castes.

From the Dublas the Anavils have little to fear as yet. Even when Dublas are present the landlords speak rudely and offensively about the failings they attribute to this caste. By removing to a new village quarter, about half of the Dublas have managed to extricate themselves to some extent from the power of the dominant caste. When the land settlements were in preparation, the Anavils feared that the agricultural laborers would claim title to the land near the village pond on which their huts stood. The Vashi owners allotted other places on uncultivated and rough *jarayat* to the Dublas concerned, but the laborers refused. The affair was brought before the government authorities, since Dublas who wish to buy land for their huts can obtain a subsidy from the government. In this manner an important means of pressure is taken away from the landlords, who, in case of conflict, can summarily evict a laborer and make him homeless.

When in this case the Dublas in Chikhligam appealed to the government, it was ruled on the district level that they had to relinquish the land on which they lived. After this victory the Vashis also refused to accept the Dublas as tenants. A Bania from the *taluka* town offered to sell them a piece of land just outside the village which he would have lost anyway under the Tenancy Act. For the land and the twenty-eight houses to be built on it, the authorities gave a subsidy of 21,000 rupees, 750 rupees per hut. The Dublas had to contribute 250 rupees each in the form of labor, as the subsidy had been granted on condition that they build the houses themselves under the supervision of a contractor. But even I could see that the value of the material supplied to the Dublas was far below the allotted amount. The houses have numerous defects. It is their own fault, say the Anavils, that "the Dublas now have been swindled by officials of the subsidizing body as well as by the contractor. They should have asked our advice." For the Dublas, however, it made little difference whether they were swindled by officials or deceived by the Anavils, who would have done so if they had had the opportunity.

Just as expected, the Anavils' manipulations in this affair to prevent the agricultural laborers from obtaining the land to which they were entitled had the support of the regional bureaucracy. As evidenced by this example, the laborers now find themselves in an isolated position. They must still communicate with the authorities through the landlords, by whom their claims are put forward unconvincingly, if not damaged. Nevertheless, the landlords have not been able to prevent these Dublas from gaining a small amount of independence when they settled on their own land in the new quarter.

The migratory labor of the agricultural laborers has affected the Anavils' influence still further. The landlords' complaints about the faithlessness of the Dublas spring from rancor about their own declining authority rather than from any anxiety about an imminent shortage of labor. For every year at the peak of the agrarian season, and after a number of years permanently, most migratory laborers, however discontented and "insolent," are compelled to put themselves in the landlords' hands again.

Apart from the new possibilities of escape the Dublas occasionally resort to time-honored means of resisting the pressure of the Anavils. They can play upon the Anvils' fear of witchcraft, which can be traced to the tribal background of the Dublas. Some of them are renowned for their magic incantations, and the Anavils are among those who fear their sinister power. If the worse comes to worst, they can still escape from the landlord's control by running away to the village of a relative, usually that of a wife or a mother's brother.

The means of resistance available to the Dublas are not impressive. They are the second-largest caste in Chikhligam, but their power is not in keeping with their number, and politically they have not as yet been activated. The functionary of the C.P.I. (Communist Party of India) in the *taluka* capital devotes his efforts exclusively to the small farmers in the region. The Congress Party is so much a stronghold of the dominant caste that the caste of landless laborers does not receive much attention.

Although their migration to the brickyards near Bombay has heightened the Dublas' awareness of their position, they lack the means of changing it. In the brickyards they are removed from the urban sphere, uninteresting for trade unions and political parties, and uncanvassed. They leave the brickyard areas only for brief visits to a tea-stall, a cinema, or another brickyard where fellow-villagers and relatives are employed. Thus far no Dubla leaders have come to the fore to give guidance to the protest of their fellow caste members against the low social position and economic distress in which they live. Moreover, the level of education in this caste is so low that there is no prospect of an early change.

All in all, the authority of the caste of landlords over that of the agricultural laborers in Chikhligam is still considerable. The only way of protest open to the Dublas is that of patient and unremitting resistance. Hate and enmity towards their exploiters remain below the surface, and in the presence of a landlord these feelings are rarely evidenced. On the whole, the laborers behave in a lethargic and indifferent fashion, avoid giving answers, know nothing and can do nothing, take no initiative, and accept no responsibility. The inertia of the Dublas has become proverbial. For the Anavil Brahmans this mechanism of passive resistance is merely another proof of the Dublas' inferiority.

# Chapter 8

# Gandevigam

*On whose head did I see a green basket?*
*I saw it on the head of Dolto, the son-in-law of the family.*
*In whose stable did I see him picking up cowdung?*
*I saw him in my brother's stable.*
*I took him to be a Khalpo and beat him with shoes*
*I took him to be a Dhed and pelted him with stones*
*I took him to be a Dubla and beat him with my sticks*
*He said, "For God's sake do not beat me,*
*I am Khandubhai's son-in-law."*

T. B. Naik, *Songs of the Anavils of Gujarat*

*Leaving your father's place*
*Why have you come here?*
*Your father-in-law has a ramshackle hut*
*Your father-in-law is a serf of the whole village*
*Why have you come here?*
*Your father is an honourable man in the village*
*What brings you here?*

T. B. Naik, *Dubla Songs*

## *8.1 Region, Village, and Population*

Gandevigam is one of the smallest subdistricts of south Gujarat. This Gandevi *mahal* forms part of the central plain, an ancient culture area with intensive agriculture. The subdistrict has always exceeded the others in soil fertility and population, as early government reports show.[1] The urban center of Gandevi has long served the surrounding countryside. Agricultural products from the villages are marketed here. The town's situation on a navigable river not far from the coast formerly facilitated shipment of goods

1. See the first survey and settlement report of Jalalpor *taluka* of 1868, included as appendix R in the revised report dated 1900: "Gundevee, a purguna belonging to the Guicowar, and famous for being the most fertile as well as the highest assessed in the Western Presidency," 1868, 1.

from and to important ports, especially Surat, the large seaport on the west coast of India. The smaller population centers for miles around availed themselves of the facilities of trade and the artisanal activities that flourished in this harbor principality in its heyday. Its sphere of influence, no doubt, also comprised the region of Gandevigam, and it is not impossible that the ancestors of the present village population contributed to the food supply of the city, which, by the end of the eighteenth century, had about 600,000 inhabitants.

The political turmoil that led to the decline of Surat ushered in a time of stagnation for the whole of south Gujarat. The vicissitudes of war between the Moghul forces and the Marathas brought insecurity and economic distress. Bands of robbers ransacked the countryside, and the population long remembered, in particular, the plundering raids of Shivaji, the Maratha chief. Even now, certain villagers can describe how and where their ancestors used to hide the crop and other possessions at his approach. Profiting by the political vacuum, men wielding local power levied such high taxes that a large part of the land remained uncultivated and whole villages were depopulated. This may explain why in a European travel record of the late eighteenth century the region round Gandevigam was described as unreclaimed.[2]

After many decades the struggle for power was decided in favor of the English and the prince of Baroda, a vassal who had managed to free himself of the Maratha rule. They divided the territory of south Gujarat between them, and Gandevi *mahal* fell to the principality. The town remained a center of local administration. It was the seat of a very prominent Desai family which had long ago received from a Moghul prince various seignorial rights over a number of villages in the vicinity and which now pledged loyalty to the new ruler. The influence of this family of landed nobles declined sharply in their traditional domain when, in the second half of the nineteenth century, the system of revenue farming was abolished, although they continued to hold a prominent position as large landowners.

The political and administrative stabilization that took place during the early nineteenth century did not lead at first to an economic revival. The river mouth silted up and Gandevi town became

2. "Although the soil is very favourable about Gaundevee, I saw but little culture: the greatest part of this district, as far as the eye could reach, is one field of high pasturage." (Hove, 1787–88, 93–94.)

inaccessible to shipping, and consequently artisanal activity deteriorated even further. Only after the north-south railroad was constructed around 1860 did commerce and the economy in general begin to expand again, though mainly in the new towns along the railroad. In population and economic importance Bilimora came to far outdistance Gandevi town. Whereas in 1881 it was an insignificant little railway village with fewer than 5,000 inhabitants, Bilimora has since grown into a town of over 30,000, with some textile mills, a paper factory, and a number of small industries in the metal sector. It is also a center of trade in timber, which is brought out from the woods in the east over a narrow-gauge railway.

Although Gandevi town was connected fairly soon with the railroad by the building of a highway to Bilimora, it remained at a disadvantage on account of its offside location in the new communication network. Its population long experienced no growth, and early in the twentieth century a plague epidemic reduced it by about one-fourth. Since then the population increased very gradually, and after Independence with greater rapidity, to 11,000 in 1961. The building of a cooperative sugar factory, the promotion of textile manufacture in small enterprises or as domestic industry, and the extension of local government services contributed to this growth. Gandevi has remained the administrative center of the subdistrict and has an economic function as distribution point for the surrounding villages, including Gandevigam.

After Independence, the principality of Baroda was incorporated into the State of Bombay, later Gujarat. The various parts of the former principality were merged into the local administrative structure. Gandevi *mahal* was included in the Surat district, which was divided into two parts in 1963. Like Chikhli *taluka,* Gandevi *mahal* is now part of the Bulsar district.

The subdistrict is known as a very prosperous agricultural area. It owes this reputation chiefly to the fertile soil found along the upper reaches of the Ambika River to the northeast of the town. This region is called the garden of south Gujarat, which is the more striking as the whole southern part is one of the most fertile areas of Gujarat. The great fertility of the soil along the river has led to the high population density of 916 per square mile in 1961 for the subdistrict as a whole, a figure far above the average of Gujarat and exceeded by none of the subdistricts in the southern part of the state. Within a radius of less than two miles from the village nucleus, Gandevigam is surrounded by five other villages, which is an indication of the extremely high concentration of population in the area. This is the first important contrast with Chikhligam. There is little to distinguish Gandevigam from the other garden villages in their rural surroundings and, as we shall see, they are closely interconnected economically and socially. Soil conditions are noticeably different from those in Chikhligam, where much of the surrounding land is, at present, considered uncultivable. In contrast to the arid and barren aspect of that landscape in the summer and winter months, the much more intensive agriculture in the garden region gives the land around Gandevigam a lively and pleasing appearance. Mango orchards, cane fields, rice fields, beds of vegetables, and banana plantations make an attractively diversified picture; compared with this variegation of color, the rural scenery in the vicinity of Chikhligam is monotonous after the monsoon ends. Finally, being situated in a more urban and economically more developed subdistrict than Chikhligam, Gandevigam is easy to reach. The village borders on the asphalt road that runs from Bilimora by way of Gandevi to Navsari, the chief town of an adjoining subdistrict. In the daytime there are buses in both directions every two hours from the village stop on the road. The distance to Navsari is eight miles, to Bilimora seven miles. Halfway between the village and Bilimora the road passes the town of Gandevi at three miles' distance.

I have not been able to learn the age of Gandevigam with any certainty. Its site in an ancient culture area suggests a much longer history than that of Chikhligam, but the village inhabitants could not verify this. They say that the village was founded long ago by a shepherd who had a garden in this spot. The first Anavil Brahmans to settle here were two brothers, to whom some families still trace their ancestry. Gandevigam was an *inamdar* village. The prince of Baroda had ceded land titles and a number of other prerogatives, not police rights, to an *inamdar* for services rendered. This *inamdar* lived in Baroda, where he was in the maharajah's employ, and visited Gandevigam once or twice a year. He was represented in the village by a steward. After the annexation of the principality in 1950, the rights of the *inamdars* were annulled. Since then, the land revenues no longer go to these large absentee landholders, whose main privilege it was, but the farmers pay taxes to the government. For the village population this change had few, if any, consequences. The steward still lives in the village and has received land from the *inamdar* to assure him a living.

In 1963 Gandevigam had slightly more than 1,000 inhabitants, according to information given me by the local authorities. With a few exceptions the villagers are members of three castes: Anavil Brahmans, Kolis, and Dublas. When I arrived at the village the first thing I saw was the huts of the Dublas on both sides of the asphalt road. A large part of the village territory lies between this road and the Ambika River. Where the two meet is the village nucleus with the houses of the other inhabitants. There are two Anavil streets and three Koli streets, spatially separated but connected by paths. The rows of houses are behind trees and invisible from the road. A little farther along, wedged in between the road and the river, are the gardens, nowadays mainly planted with fruit trees. The land across the road is given over to completely different crops: rice, bananas, vegetables, or sugar cane according to the season. The dense vegetation makes for a relatively temperate microclimate in the village territory even during the hot season, a circumstance which I, as a West European, highly appreciated, especially in the afternoon hours. In the shaded *bagayat* villages the temperature is always a few degrees lower than in the more open *jarayat* villages of the plain.

In the Anavil streets many houses are empty for the greater part of the year. It is characteristic of the villages in the *bagayat* region that during the dry winter and summer months the Anavil Brahmans live in their gardens, where they are surrounded by the crops they cultivate. They used to have small and simple wooden dwellings there, not much more than sheds. For in the narrow streets of the village nucleus, where the cattle usually remain much nearer the houses, insects are a veritable plague. To stay in the shaded gardens is far more agreeable, certainly in summer. During the rainy season, however, the narrow paths become all but impassable, and the Anavils were wont to return to their street before the monsoon broke. Increasingly, the wooden cottages in the gardens have been replaced by large two-storey brick houses, with spacious front and back yards and separate outhouses, and as these new abodes are much more comfortable, many Anavils remain there the year around, neglecting their permanently closed houses in the village street. I was lodged in one of these. When the family estate is divided, one brother sometimes gets the house in the village nucleus, the other the house in the garden. Several Anavils are too poor to have a house built in their gardens, especially as it means the loss of a parcel of the most fertile arable land. Nevertheless there is, generally speaking, a tendency within the dominant caste to take up permanent residence in the gardens.

The majority of the Kolis' houses are not built in rows but are scattered along the three streets they inhabit. Their houses are smaller than those of the Anavils, lack electricity, are built of wood and wattle instead of brick, and have roofs of tiles or corrugated iron. Each street has one well, whereas many houses in the Anavil quarter have their own well. The cattle are generally sheltered in the house.

The Dublas live along the asphalt road on the fringe of the village. Their small, low mud huts with overhanging roofs of straw or sugar-cane leaves are on rough and low-lying terrain. A few members of this caste are better housed: their dwellings are partly built of wood or brick and have tiled roofs. None of the houses, however, has more than one room. Although this Dubla quarter is of fairly recent date, and is the largest in the village, its inhabitants have only one well among them. Until the early nineteen-fifties their huts were scattered in the gardens of the Anavils who employed them. Concentration of the Dublas and a tendency towards deconcentration among the Anavils are therefore important changes in the residence pattern of the past few decades.

Facilities in Gandevigam are few. The stock of the two village shops, both run by Kolis, is confined to victuals and other daily consumer goods. One shop is situated on the asphalt road near the Dubla quarter and is exclusively patronized by members of this caste; the other is nearer to the village nucleus. Here, too, is the village school. For all other amenities the inhabitants depend on the neighboring larger villages or on the town.

As in Chikhligam, collecting data on Anavil Brahmans and Dublas was the main purpose of my stay in Gandevigam. In 1963 I counted in the village 185 Anavils distributed among twenty-six houses, and 381 Dublas in sixty-one huts. This figure is lower than that mentioned in the village records,[3] but the local government data are not wholly reliable. The steward who remained in the village, for instance, has not been registered as a Maratha Brahman but as an Anavil. A Dhodia and Naika family living among the

3. In these records, 225 Anavils and 429 Dublas are registered as inhabitants.

Dublas have been included in the count of their fellow inhabitants of the quarter.[4] I further learned that a number of persons who no longer live in the village are still registered there. Finally, it is uncertain whether some Dublas live in the territory of Gandevigam or in that of the neighboring village. They were probably incorrectly registered as inhabitants of Gandevigam in the latest censuses; when sugar was rationed during the Second World War, they had to get their ration cards in the other villages. How much lower my total count is I do not know, for I did not include the Kolis in my census. I have taken data from and about them only insofar as such facts could throw light on the relationship between landlords and agricultural laborers. As will become clear, they belong to neither category. Background information was given me by randomly selected Kolis, a sample of thirty-seven taken in three streets. On the other hand, I included in my investigation all Anavil Brahmans and Dublas who were present in the village. Where not expressly stated otherwise, the quantified data that will be discussed below have been taken from my own census.

## *8.2 Economic Position and Activities of the Inhabitants*

The inhabitants of Gandevigam differ widely in prosperity, by and large according to caste. The disparate quality of their accommodation was mentioned above. Two years before my stay the richest Anavil of the village had a house worth 30,000 rupees built in his garden, whereas the value of most Dubla huts does not exceed 100 rupees. The middle position of the Kolis, better than that of the Dublas but worse than that of the Anavils, is also apparent in other respects. For instance, the average amount of money spent on a marriage, according to my informants, is for Anavils more than 5,000 rupees, for Kolis 1,000 rupees, and for Dublas about 300 rupees. It is true that the most prosperous of the Kolis are better off than the poorest of the Anavils, and the way of life of the poorest Kolis is very similar to the penurious existence of the Dublas. Nevertheless, a connection exists between living standard and caste membership, which appears most clearly from differences in landholding among the various castes. The Anavil Brahmans, as in Chikhligam the smallest group, control a disproportionately large part of the land. The Dublas, numerically the largest caste, are with one exception completely landless. In the preceding generation

4. Like the Dublas, the Naikas are a tribal caste in the plain of south Gujarat.

one Dubla received a plot from the *inamdar* for whom he had worked as a farm servant. On this plot, far into the gardens, are to be found the huts of his two sons and three other relatives.

TABLE 8.2.1

CASTE MEMBERSHIP AND LANDHOLDING IN GANDEVIGAM*

| | *Property* | | *Population* | |
|---|---|---|---|---|
| *Caste* | *Acres* | *Percent of total* | *Number* | *Percent of total* |
| Anavil | 260 | 71 | 225 | 22 |
| Koli | 62 | 17 | 350 | 35 |
| Dubla | 0.25 | — | 429 | 43 |
| Others | 42.50 | 12 | — | — |

* Data obtained from the village records.

In the whole of south Gujarat the Kolis are petty cultivators, and those in Gandevigam are no exception. The members of other castes listed in table 8.2.1 are inhabitants of surrounding villages who possess land in Gandevigam.[5] Conversely, landowning villagers, especially Anavils, possess plots elsewhere, which illustrates the intermingling of the village economies in this *bagayat* region. The picture emerging from the record of land rights kept by the village clerk (*talati*) could be supplemented by the following survey, for which I myself gathered the data. These figures relate to holdings both within and outside Gandevigam. The inequality of the landowning castes in terms of separate households is even more pronounced when it is taken into account that the Anavils possess the great majority of gardens, i.e., the most valuable plots. Kolis own mainly less valuable *kyari* and *jarayat* land.

It is important to make a distinction between ownership and cultivation. Of the Kolis I interviewed, sixteen are sharecroppers of Anavil landlords. The land concerned is mostly *kyari*, on which they cultivate sugar cane, bananas, or vegetables. When they grow vegetables, the Kolis vacate the land by the end of the summer, so that the owner can grow rice on it during the monsoon. Landlords and sharecroppers each pay half of the cost of seedlings, fertilizer, and irrigation water, and share the produce equally. The crop-sharing system has enabled the Kolis to hold their own as

5. Some of these are also Anavil Brahmans, who together possess about thirty acres. The rest of the arable land in this category belongs to a Bania and some Moslems from a neighboring village.

smallholders. Tenancy, as well as sharecropping, however, is now officially prohibited, and the Anavils run the risk of losing their land. Now that crop-sharing arrangements no longer have a legal basis, it is not surprising that the number of sharecroppers among the Kolis has recently declined. Moreover, to limit their risks as much as possible, the Anavils now lease out the plots for one season at a time, which means great insecurity for the tenants.

TABLE 8.2.2

CONCENTRATION OF LANDHOLDINGS, BY CASTE, IN GANDEVIGAM

| *Landholding (acres)* | *Caste of owners* | |
|---|---|---|
| | *Anavils* | *Kolis* |
| — | 1 | 5 |
| Less than 3 | 3 | 20 |
| 3 to 6 | 6 | 4 |
| 7 to 16 | 7 | 2 |
| More than 16 | 9 | 1 |
| Total | 26 | 32[1] |

[1] The information given by five Kolis was too vague to be classified.

A large proportion of the Kolis (thirteen of my thirty-seven informants) have to carry out field labor for others occasionally to make ends meet. The extent to which they succeed in doing this depends on the quantity of land they cultivate themselves and on their income from other sources. Two Kolis in Gandevigam depend entirely on their earnings from agricultural labor. The father of one of them was a tenant of an Anavil in a neighboring village until he died, and the informant himself was until recently a sharecropper of a landlord in Gandevigam. He has lost his team of oxen since then, and now he possesses nothing. He can no longer afford the education of his son, who went to secondary school. In the socioeconomic scale he has clearly lost ground.

Yet there is no reason to speak of a general deterioration within this caste. Ten of the Kolis in my sample have managed to add to their property under the Tenancy Act, though almost none by more than 1 or 1.5 acre. In all these cases the landlord lived outside the village. Then, too, more Anavil landlords now leave the plowing of their land and the transporting of the crop to these small cultivators. An increasing number of Koli families, finally, supplement the family income by the earnings of a member working permanently or temporarily outside the village. Thus, while the middle position of this caste as a whole has not fundamentally changed, a process of differentiation in economic activities and prosperity level has begun.

By far the majority of Dublas are laborers on the land, but some have lately been able to rent some land for sharecropping from Anavil landlords who did not want to lease it out to Kolis. It is to be expected that this trend will continue. The conditions under which the Dublas obtain this land, on which they cultivate bananas on a crop-sharing basis, are less favorable than those allowed to Kolis. The latter are entitled to half the produce, whereas Dublas receive only one-third. Although the landowner in the Dublas' case is obliged to pay all expenses, he often demands repayment of half of them when the accounts are settled at the end of the season. For such an intensive crop as bananas, some Koli informants said, crop-sharing under these conditions is not economic. A Dubla, who measures his earnings by those of an agricultural laborer, thinks otherwise.

Is there any reason to assume that in earlier times the land was distributed differently among the inhabitants of the village? The Dublas, I believe, have always been in the same circumstances. There are no indications that members of this tribal caste have ever been engaged in agriculture as independent cultivators.

For the Kolis, too, large-scale loss of land seems improbable. They were mentioned as sharecroppers and agricultural laborers in the employ of Anavil Brahmans as long ago as the mid-nineteenth century. Although several Kolis were dispossessed and became tenants on their own *kyari* and *jarayat* land in the nineteen-thirties, the Anavils also had to give up land in that period of world crisis, mostly to Banias (traders and moneylenders). In the ensuing years, many Anavils and Kolis could buy land again, and the situation of both castes improved considerably. During the Second World War, a time of scarcity of food crops, the cultivation of sugar cane and bananas in Gandevigam was extremely profitable.

Paradoxically enough, the Tenancy Act, which was meant to help the Kolis, will probably weaken the economic position of many of them. In practice this act has confirmed the traditional ascendancy of the Anavils as landowners in the village community.

As a rule, the Kolis need little extra labor on their land, for the assistance of members of the family, including women and children, is usually sufficient. At the peak of the season, collaboration with other Koli families is one solution. Yet, despite the pattern of small landholding within this caste, about half the households about which I collected data employ from time to time the labor of others against payment. Although the normal daily activities can be carried out by the family, outsiders must be taken on at harvest time, especially when the family does not wish to exchange labor with other Koli families — often a source of annoyance. One of the reasons for this periodic need of labor from outside the family is that Kolis grow highly labor-intensive crops, such as sugar cane, vegetables, and bananas, on land which a number of them still cultivate under a crop-sharing agreement. I found that Kolis employ mainly their fellow caste members for casual labor. This occurs not so much because they are inspired by a magnanimous wish to support members of their own caste, as some of them claimed, but because the services of the agricultural laborers *par excellence*, the Dublas, have largely been monopolized by the Anavils.

Most of the Kolis who hire labor incidentally are themselves employed as day-laborers from time to time. Understandably, as small farmers they give priority to working on their own land, but this means that, when they are free, other landowners have little to do either.

Koli sharecroppers were formerly obliged to perform field labor for their landlords if necessary. As already described, however, the Anavils enter into fewer crop-sharing arrangements with Kolis, and the Kolis, for their part, are less disposed to work for Anavil Brahmans in the fields, for they no longer feel obliged to help their former landlords.

As the tenancy relationship diminished, a small proportion of the Kolis have come to depend increasingly on income from wage labor for others. They prefer to work for fellow caste members who, they say, treat them better and pay a little more. The category for whom agricultural labor is the main source of income is not large. Most Kolis have remained small farmers who try to supplement their earnings by means of various agrarian and nonagrarian side activities. Some distill and sell spirits illegally, breed and trade in cattle, or are small-scale mango buyers who put up the fruits for auction in towns along the railroad against prices that are slightly higher than the local ones. Koli women leave Gandevigam early in the morning with baskets of vegetables on their heads to hawk their produce around in the villages or to offer it for sale in the markets of neighboring towns. Many Kolis are cart-drivers for members of the dominant caste and also, with their oxen, plow the landlords' ground for 10 rupees a day. They regard agricultural work for others as beneath their dignity and try to avoid it as much as possible. This attitude is prompted both by the low remuneration of this work and by the slight prestige that it carries. Although the average landholdings of the Kolis are small, few — not more than five of the thirty-seven households in my sample — are completely landless. Of these, three are sharecroppers. Possibly, as population growth continues, a higher percentage of the members of this middle caste will retrogress to an existence of landless laborers, but this is not very probable. An increasing number of Kolis manage to find work outside agriculture in the neighboring towns.

On the basis of landholding, therefore, and leaving individual differences within each group out of consideration, three categories can be distinguished in Gandevigam:

1. Anavils, who, as in Chikhligam, own most of the land and the most fertile part of it, and who moreover possess relatively more equipment, such as motor pumps, agricultural implements, and cattle. Most of them employ one or more farm servants permanently, and rarely work on the land themselves. It is mainly for this reason that I have classified them as landlords, not farmers.

2. Kolis, who as small farmers mainly possess *kyari* and *jarayat* land, and who supplement their income by a series of other activities. It is true that members of this caste sometimes call upon the labor of others, and themselves occasionally work for others. As a rule, however, these roles of employer or employee are temporary, and often they are combined in one individual.

3. Dublas, who largely earn their livelihood as agricultural laborers. This made it possible for me to concentrate my investigation into the relations between landlords and agricultural laborers on the Anavil Brahmans and Dublas in Gandevigam, as I had done in Chikhligam.

Thus far I have confined my remarks to activities and positions connected with the agrarian economy of Gandevigam. As already mentioned, the total income of Chikhligam is earned in large part outside the village and agriculture. It might be expected that this proportion would be even higher in Gandevigam, for the village is situated in the immediate vicinity of some urban centers, with which it is connected by an excellent road. But, as the census returns of 1961 showed, 88 percent of the working population is engaged in agriculture. As compared with Chikhligam, the proportion of inhabitants of the garden village who have nonagrarian sources of income is relatively much smaller. In view of the more isolated location of my first field village, at a greater distance from the cities and less easy to reach, the pronouncedly agrarian basis of existence among the population of Gandevigam is noteworthy.[6]

TABLE 8.2.3

NUMBER OF INHABITANTS OF GANDEVIGAM EMPLOYED OUTSIDE AGRICULTURE

| *Caste* | *Living and working in Gandevigam* | *Living in Gandevigam and working elsewhere* | *Temporarily living and working elsewhere* | *Permanently living and working outside Gandevigam* |
|---|---|---|---|---|
| Anavils | — | 4 | — | 9 |
| Kolis[1] | 4 | 8 | 6 | 13 |
| Dublas | — | 25 | 5 | 24 |

[1] The data on the Kolis relate only to those included in my sample. I estimate the total number of Kolis employed in nonagrarian occupations to be higher than that of the two other castes, both absolutely and relatively.

Within the village itself only four inhabitants are engaged outside agriculture: two Kolis are shopkeepers, one is a carpenter, and the fourth is a tailor. Aside from these occupations, they own a plot of land and partly depend on its produce for their maintenance. The reverse is true of some of their fellow caste members, who are primarily small cultivators but add to their earnings by such activities as dealing in wood and illegally distilling spirits.

The proportion of inhabitants who live in Gandevigam but work elsewhere is larger. A small category, consisting entirely of Kolis and Dublas, leaves the village for regular periods and returns before the rainy season.

Many inhabitants may have gone to live in the neighboring towns, and their definitive departure would render the proportion

6. Cf. the continuing and even increasing stress on agriculture which T. S. Epstein found in Wangala, and her contrasting findings in Dalena, the other village comprised in her field work. (Epstein, 1962, especially ch. II.)

of the village population working outside agriculture much higher. I therefore inquired about those who had left the village in the past few years.[7] Once more, Chikhligam had the larger number of former inhabitants who had permanently departed. Thus, while in the vicinity of Gandevigam much more alternative employment is available than in Chikhligam, relatively few inhabitants make use of it.

My data show that, outside agriculture, Anavil Brahmans are engaged in the highly skilled and best-paid professions, and most often they are government employees. Dublas are largely unskilled and perform low-paid work. A job with the government or in a factory is much coveted by them, as it means a fixed income. When they succeed in obtaining such a job they get only the less-favored tasks, such as cleaning and maintenance work. While the Anavils from the village who work for the railway are employed as station-masters or ticket-clerks, Dublas from Gandevigam are luggage porters, watchmen, track walkers, or car cleaners. There are few artisans among them: two tinkers (father and son) and one brick-layer.

The Kolis who live in Gandevigam but work outside the village are, except for one teacher, carpenters or bricklayers. Six members of this caste leave the village periodically to work in an umbrella factory in a town north of Bombay. They leave their wives and children in the village, unlike the Dublas in this category, who migrate to the brickyards on the fringe of Bombay. Their number is much smaller than in Chikhligam. The figures show that Kolis, too, perform more skilled and better-paid labor than Dublas. The Dublas who work outside the village total twenty-five, of whom two are itinerant tinkers who travel about the surrounding country-side and three are timber cutters for a merchant in a nearby village, where two other Dublas work in a grain mill; and after the rainy season, one Dubla goes daily to a brickyard at a short distance from Gandevigam. Of the other seventeen, who work in the capital or in Bilimora, three are hands in a paper factory, three roadmen, six servants in shops or workshops, seven bricklayers, and four transport workers. Many of them walk the 3 miles to their work in Gandevi every day, and some even go 7 miles on foot to Bili-

7. As in Chikhligam, the statements are only approximations of reality. The longer the time since the departure from the village, the greater the chance that the inhabitants will forget those who departed, especially when no close relatives have remained behind.

mora. Anavils and Kolis who work in these towns all travel by bus or by bicycle. In what way the Dublas who permanently live and work outside Gandevigam are employed I could not find out, except in a few cases. Some work in factories or as railroad workers and sweepers, and some are vegetable growers in Bombay.

Obviously, Dublas who have had little or no schooling are not considered for skilled and well-paid jobs. Yet education is not the only factor in the selection of employees for nonagrarian occupations. As in Chikhligam, alternative employment has come to mean something different for each caste. The position in the village community and the economic and social content of caste membership determine to a large extent the kind of work that is considered suitable outside it. This confirms my findings with regard to Chikhligam.

### *8.3 Caste Hierarchy and Political Structure*

The small extent of caste differentiation is the most striking feature of the social structure of Gandevigam. Economically, politically, and ritually the Anavils, who all belong to the Bhathela section, are at the top. They are followed by the Kolis, and the Dublas are classified as lowest. That is to say, high, middle, and low are represented in this village by only one caste each.[8] Curiously, Chikhligam, situated in a frontier area, contains more castes than Gandevigam. For a village in the central plain one would expect a more varied constellation. Serving and artisan castes, whose presence would suggest a division of labor characteristic of the Indian village, are completely absent. Perhaps it is precisely its location in a more developed region, with urban centers nearby, that contributes to the simplicity of stratification in Gandevigam. The villagers all belong to castes that, originally, were decidedly agrarian. Nonagrarian castes may have been represented in the village at an earlier period. At the beginning of the twentieth century there was mention of migration of artisans in south Gujarat to the towns as a result of the disintegration of the closed economy,[9] but I found no

8. Aside from a few households, which have erroneously not been registered separately in the village administration and the census report.

9. "The complete equipment of artisans and menials with which the old type of village was furnished is being dissolved by the force of competitive tendencies. As villages become larger, the village barber, blacksmith, carpenter or potter seems to lose the indefiniteness of his circle of clientele. The influence of custom in fixing the remuneration for the hire of labour is also giving way gradually to the laws of supply and demand. In many important directions the village services are being

indications that this applied to Gandevigam. More probably, a number of requirements in this area have always been supplied by urban centers. For instance, the rough clothing of the tribal castes was traditionally woven in the nearby town, whereas the demand of the inhabitants of Gandevigam for services of all kinds is mainly filled from neighboring larger villages, which contain various artisan and serving castes.

Even so, however, there is no question of a profound division of labor. As in the whole of south Gujarat, the Dublas in Gandevigam carry out widely varied services for the Anavil Brahmans, whereas the Kolis are largely autonomous and do their own cleaning, scavenging, and the like. When more specialized tasks must be done, they call in outsiders; for such services as those of a priest, hairdresser, blacksmith, or potter, they seek members of those castes from neighboring villages. The higher their own position in the caste hierarchy and the more prosperous the household, the more likely they are to call upon specialists. This means that it is primarily the most prosperous Anavil Brahmans who act as typical *jajmans*.

The calling in of specialists is made possible by the short distances between the various settlements in this densely populated region. It follows that Gandevigam cannot be studied as a separate social system. Only at the supralocal level has caste hierarchy a real meaning.

That Gandevigam forms part of a regional community also finds political expression. To obtain an insight into the local power relationships it is, therefore, not primarily the village system that is relevant. Athough the Anavil Brahmans are less numerous than the two other castes, they have three representatives in the village council as against two each of the Kolis and Dublas. The overrepresentation of the highest caste in the *panchayat* is a reflection of their ascendancy. Moreover, both the village headman and the police *patel* are Anavils.

The supremacy of this caste in Gandevigam is undisputed but means little. In the first place, decisions are not made in the village council. The *sirpanch* — village headman and chairman of the council — has no influence at all in Gandevigam. He is an aged bachelor who lives on his pension as a teacher and on the income

depleted by the discontented village artisan or menial leaving for towns or large centres in the hope that with better wages and in newer surroundings his ambitions can be satisfied." (*Census of India*, 1921, vol. XVII, pt. 1, 358.)

from his small mango garden. Nor are the other Anavil councillors influential personalities. This is not due so much to factionalism with behind-the-scenes activity as it is to lack of involvement among the Anavil Brahmans in the village as a social unit. They regard themselves not as inhabitants of the village but as *bagayat* landlords who live in the region extending on both sides of the river. This whole area on the Ambika is the domain of the Anavil Brahmans, or more precisely, of the Bhathelas. The dominance of this group is not confined to their functions in the village system, but affects also the regional frameworks of integration, such as the secondary school and the agricultural cooperative.

Together with the other children of the neighborhood, the boys and girls of Gandevigam attend the secondary school that has been built amidst a number of villages. In the past few years an increasing number of pupils come from the middle castes, notably the Kolis. Yet it is Anavils who rule the roost in the school board, and a disproportionately large number of pupils still belongs to this caste. This is even more true of the teaching staff. Some years ago, a Koli from Gandevigam succeeded in becoming a teacher at this school. However, as his brother told me, owing to the opposition and jealousy of the Anavils his situation became untenable. The Anavils did not think it proper that a Koli should hold such a post, above all at their school. In the end, he emigrated to East Africa.

In the agricultural cooperative, the dominance of the Anavil Brahmans is even more evident. The storage shed and office of this institution are situated on the asphalt road just outside the territory of Gandevigam. This cooperative has a pronouncedly regional function, not only economically but also socially. The Anavils of the village do not meet each other in the village nucleus. In the afternoon they emerge from their houses and set out for the site of the cooperative, where they congregate and meet their fellow caste members from other villages. Koli members go there only for business, not to have a chat. Upon my arrival in Gandevigam, I was not taken to the village headman but directly to this meeting-place. When, after my introduction into the circle, the question of my lodgings came up, this at first appeared to present some difficulties. But any objections were put aside as irrelevant by all those present. Even if I were to persist in my plan of exclusively collecting information about farming as it was done in Gandevigam, they thought it did not matter where I stayed. Not only because distances were short: I could daily meet the landowners of Gandevigam (meaning the members of the dominant caste) at the cooperative building.

The more than 700 members forming the cooperative live in nine villages in the neighborhood and belong to twelve different castes. Nevertheless this institution is entirely an Anavil affair. All of the fifteen members of the board are Anavils. The Koli members from Gandevigam feel discriminated against. The Anavils are, as it were, preferential members who can obtain credit more rapidly, on easier terms, and in greater amounts. This is no doubt partly due to the smallness of the Kolis' holdings, and in general to their dubious solvency as small cultivators. As a rule, large farmers everywhere have more influence in most cooperatives. It is quite likely, however, that Kolis are systematically put at a disadvantage for noneconomic reasons, and especially on account of their low position in the local hierarchy. The less prosperous Anavils benefit by belonging to the dominant group. They enjoy numerous advantages because of their caste affiliation, whereas the richer Kolis have an inherent handicap. Their stronger economic position in comparison with that of the least well-to-do Anavils is offset by their lower social status.

Because of its density of population, the whole region forms an ecological unit whose inhabitants know that they are closely interconnected socially and economically. This impression is strengthened by the regional control exerted by the Anavil Brahmans. Twice within a short time, during my stay, the chairman of the cooperative gave an expensive wedding feast for a daughter, and on both occasions sumptuously entertained many hundreds of guests from the environs. By this exploit, informants assured me, he stood a fair chance of becoming one of the ten most prominent Anavils of the region.

The background of the leading role played by the Anavil Brahmans must be sought in their privileged position at the time of the principality of Baroda, when members of this caste had administrative rights as revenue farmers. The system of revenue farming was abolished in the principality by the end of the nineteenth century, later than in the British area. Although the abolishment caused their powers to be curtailed, the caste as a whole retained its proprietary rights to the largest part of the arable land. This fixed the basis for a continuance of power, which was exerted by the members of the Bhathela section as well, even when they had not fulfilled any administrative functions on a regional level. As everywhere, the Anavils of the *bagayat* region are accustomed to being obeyed. They act with assurance and are described by the other castes as hard, cunning, avaricious, and rude. Such pejorative stereotypes also figure in the proverbs that circulate about them and that contain warnings about their greed, untrustworthiness, and rude language.

During my stay there was a great commotion in the adjoining village, where a member of the caste of Blacksmiths was anonymously threatened with a nocturnal attack on his house. According to local gossip he had earned much money with his engineering workshop and by dealing in mangoes. This had apparently provoked the jealousy of the Anavils, who were even more indignant when on the occasion of his daughter's wedding he held an expensive feast. It was a public secret that the threatening letter came from some members of the dominant caste. Everyone knew that it was a warning to the Smith to keep to his place, on pain of meeting the same fate as a Bania who had been compelled to leave the village about six years previously. This Bania had been robbed of all his possessions in a nocturnal attack planned by Anavils but executed by their Koli clients. It was characteristic of the Anavils, my informants said, that they had been the first to go to the Bania's house the following morning to express their sympathy. I was told that the Smith, realizing what was in store for him, chose to resist. Three times an attack was announced by letter, and although everyone knew it would not be carried out, the Smith each time informed the police in the *taluka* town. Several village policemen kept watch over his house on the three evenings concerned and, besides food, demanded a kind of hazard money. In this way the Smith paid a penalty of 150 rupees; in addition he made it publicly known that he would stop dealing in mangoes.

Gandevigam has some Anavils, too, who buy up mangoes from the smaller garden owners and sell them independently of the cooperative; they feel threatened by the competition of some Kolis from the village. One Koli, who had bought up the crop of a Moslem grower from another village, complained to me that the Moslem had gone back on his decision; he had been advised by certain Anavils that the Koli was not solvent.

Finally, the ascendancy of the dominant caste is clearly shown by the ineffectiveness of the Tenancy Act of 1957. It is true that ten Kolis managed to obtain some land, but in all those cases the landlord lived outside the village and was not an Anavil Brahman. The Anavils have evidently been able to prevent the transfer of land to their former sharecroppers and tenants. Various Kolis told me that they had not wanted to lay claim to any land to which they had no rights. Their views on unlawful ownership are undoubtedly strengthened by the pressure of the Anavils for whom they are still sharecroppers in secret. A Koli who had himself registered as owner would of course thenceforth not be considered for a sharecropping partnership. It is therefore not surprising that some Koli farmers, most of whom had not enough land of their own to live on, spoke with abhorrence of some fellow caste members who had availed themselves of the opportunity to acquire a piece of land formerly belonging to their landlord.

Nevertheless, as in Chikhligam, the power relationships in the garden village are changing. As I shall explain below, the dominant caste has as yet little to fear from the landless Dublas. Among the Kolis, however, there is a tendency to rise in the social scale, as everywhere in the plain of south Gujarat. This is due to the bettering of the economic position of this middle caste in recent decades. A minority has declined, but most of the Kolis have managed to improve their income. Higher earnings in agriculture are only part of the reason for this improvement. Increasing numbers of Kolis have found work outside agriculture, in which the rise of the level of education in this caste seems to be a factor. Many more of the young people go to secondary school and even to a university. In the process of differentiation that is taking place among the Kolis, those who are most successful try to utilize their stronger economic position to gain political power, by making their weaker fellow caste members feel obligated to them. In several villages the Kolis have already taken over the leadership of the village council,[10] but such shifts of power remain confined to single villages. This largest caste of south Gujarat has not yet succeeded in ousting the Anavil Brahmans from their dominant position in the region, that is, in getting into a position themselves in which they can impose their will on others.

Gandevigam's Kolis are as yet no match at all for the Anavil

10. For extensive documentation on the improvement of the Kolis' position, see Hommes, 1970.

Brahmans of the village, but their more favorable social and economic circumstances make them less dependent than the landless Dublas. This explains why Kolis venture more openly to oppose the landlords of the village and speak bitterly of them. In my conversations with Anavils, their fear of the rise of the Kolis was a recurring theme. They have found that the Kolis no longer behave with docility and are showing a self-esteem which, in the Anavils' eyes, is misplaced. They cite local examples of "faithless" Kolis, and claim that the leasing of land for sharecropping to members of that caste is not without risk.

Yet the dominant caste of the Anavil Brahmans is not really in danger. They act very high-handedly and show uninhibitedly their indifference toward the lower castes. One of the many instances of this is the way in which they behaved over the inclusion of Gandevigam in the electricity network. Electricity was installed only in the streets and houses of the Anavils because, they said, the Kolis could not afford the luxury of electric lighting. This is true of most Kolis: the cost of putting in an electric outlet has to be paid by the occupant of the house. But public lighting, of which the construction and operation costs are financed from the village funds, was confined to the Anavil streets as well, despite a request by the Kolis that some lights be provided in their village quarter. The Kolis are indignant about this. The Dublas are not, because they had never expected to be considered for such facilities in any case.

Many Kolis still rent land from Anavils for sharecropping on an informal basis or are in their debt. For these reasons they owe their landlords support and obedience. The interests of the members of this middle caste are, in other words, divided, which thus far prevents them from combining their strength. They find the path to prosperity and influcnce barred by the caste of landlords. As one Koli said dejectedly, when he summarized our conversation on this subject, "Where there are Anavils, others cannot get on."

### *8.4 The Pattern of Agricultural Activities*

The distance between Gandevigam and Chikhligam is not, I think, more than 10 miles as the crow flies. The climate in the two villages is therefore very similar. However, owing to the dense vegetation of the garden region the day temperature in Gandevigam fluctuates less. The subdistrict has slightly more rainfall than the neighboring Chikhli *taluka*. Irrigation renders the farmers of the garden region more independent of the quantity and yearly distribution of rain. Gandevigam counts fifty-six wells, which chiefly irrigate the gardens. In 1939 the prince of Baroda had a pumping station built in an adjoining village which, like Gandevigam, is on the Ambika River. From that point the water is carried through irrigation canals to the surrounding villages — in Gandevigam to the *kyari* area. The greater part of the arable land can thus be permanently irrigated. As the soil is highly fertile, crops can be grown more intensively than in Chikhligam. To a far greater extent, therefore, the life of the inhabitants of Gandevigam remains concentrated round agriculture.

TABLE 8.4.1

THE CULTIVATION PATTERN OF GANDEVIGAM IN 1961*

| *Crop* | *Acres* |
|---|---|
| Fruit | 131 |
| Grass | 66 |
| Pulses | 57 |
| Rice | 54.5 |
| Tuberous plants, vegetables, and spices | 37.5 |
| Bananas | 34 |
| Sugar cane | 10 |
| Total | 390[1] |

* Data obtained from the village records.

[1] Of this total, 42 acres produced two crops in 1961. In addition, 9 acres lay fallow that year.

As in Chikhligam, the soil can be divided into *bagayat, kyari*, and *jarayat*. Of the roughly 360 acres of arable land, 145 belong to the first type, whereas the second and third types cover, respectively 105 and 110 acres. The *jarayat* is mainly under grass. In recent years the cattle holdings of the large farmers have sharply decreased. The water for irrigation used to be brought up from a well by animal traction, which required two teams of oxen working in turns, but now motor pumps are used. Despite the decrease of the cattle herds, the grass crop cannot feed the animals of all the farmers, so that many of them have to procure feed from elsewhere.

On the *kyari*, rice is grown during the monsoon, followed by vegetables (potatoes, tomatoes, onions, and cabbage), pulses (especially *val*), and red pepper. These crops alternate with sugar cane or bananas. The *bagayat*, the garden area of the village, is largely under orchards, especially mango trees, but to a lesser extent also *chiku* trees. The cultivation of *chiku* (*sapodilla*) was introduced in south Gujarat in the nineteen-fifties. The *chiku* has a moist tropical habitat and bears fruit three years after planting. Another of its advantages, compared with mango, is the higher yield per tree. In the yards and along the paths in the gardens, finally, there grow the traditional coconut palm, lemon shrub, and jackfruit (*phanas*).

Over the years the agricultural economy of Gandevigam has altered markedly as a result of the changeover to the cultivation of fruit trees. At the beginning of this century, sugar cane was the staple crop in the subdistrict, and until after the Second World War it was the chief source of income for the farmers of Gandevigam. Aside from sugar cane, ginger was grown as a market crop. A specialty of Gandevigam at that period was the cultivation of a spice called *pipar*, used in medicine, which was so profitable that it was sometimes called black gold. The agriculture of the area during the early decades of this century, however, was commercialized to only a slight extent. In the *bagayat* region many diverse crops were grown, as appears from the report of the revised settlement survey of the subdistrict in 1907.[11] That report does not give cultivation data for Gandevigam, but it does include two neighboring villages. There, sugar cane covered about 18 percent of the arable land. In the rest of the area some fifteen other products were grown, for instance cereals (including some species of millet, but especially rice), several pulses, vegetables of various kinds, spices, bananas, oil seeds, and tobacco. The Anavils of Gandevigam told me that these crops were grown primarily for home consumption.

Very gradually the proportion of products leaving the village increased, and this was accompanied by more and more crop specialization. At first the products were the traditional market crops, chiefly sugar cane, but in the past few decades the cane fields were transformed into orchards. The mango was not a new fruit in this region: the trees growing in the yards had long been familiar. The uncultivated mangoes were especially used for home consumption. From an administrative report of 1900 it seems clear that about that time some rich Parsis near Navsari began growing the fruit in orchards.

11. *Revision Settlement Report of the Gandevi Taluka of the Navsari Division*, Baroda, 1907, appendix VII. I tried in vain to obtain a copy of the first survey and settlement report on this subdistrict, dated 1891.

Some country mangoes are also to be sent to the North from the British and Gaekwari groups of bagayat villages round Amalsad. As the fruit trees are grown on the field boundaries, graft mangoes, on account of their stunted growth, are not countenanced in these villages; but the rich Parsis of Navsari have bought up land in Vijalpur and Jalalpor and have converted them into pleasant gardens full of superior mango trees.[12]

It was especially after the nineteen-forties that mango culture became widespread in Gandevigam, but agriculture has remained much more intensive in this garden village than in Chikhligam, as is shown by the agricultural calendar.

The following survey is intended only to convey an impression of the most important activities. It is based on information from a number of farmers and agricultural laborers, verified so far as possible by personal observation. I could not establish in detail the labor required for each of the crops and its distribution over the year, although I was aware that more detailed knowledge on this point would enable me to analyze the labor relationships more thoroughly. The duration of my stay both in Chikhligam and in Gandevigam was not long enough for such a detailed investigation, nor was I sufficiently schooled in agricultural economy.

On the *jarayat* the grass is cut some weeks after the end of the rainy season. During the rest of the year some work must be devoted to cutting down *babul* trees if necessary, removing thorns that have fallen from the trees and might be dangerous for cattle, and mending fences.

The *kyari* is largely under rice during the rainy season, and under pulses or vegetables afterwards, when, as a rule, the ground is plowed and harrowed. Rice and pulses are cultivated in the same way as in Chikhligam. Similarly, the cultivators in Gandevigam have not adopted the so-called Japanese method of rice-growing, which requires more labor and capital but results in a higher yield. For the growing of vegetables and red peppers, seed plots are prepared and fertilized, and small canals are dug to convey the water to the beds. After three or four weeks the young plants are set out in well-fertilized soil, which is irrigated every ten or fifteen days. Weeding is done very intensively both among the plants and along the rows. Three or four months later comes the harvest, after which the ground lies fallow until the rainy season. The culti-

12. *Papers Relating to the Revision Survey Settlement of the Jalalpor Taluka of the Surat Collectorate, Bombay*, 1900, 7.

vation of red pepper requires a great deal of labor. Picking can begin as early as eight weeks after transplantation and may go on until far into the summer.

Sugar cane, formerly the most important product, is now grown only in a small area. The ripening process requires a period of twelve to fourteen months, and of all crops it requires the greatest number of laborers per land unit. Depending on the species and the time of harvesting, sugar cane is usually planted on narrow ridges of earth between the end of October and February. Before this is done, the soil has been plowed, fertilized, and harrowed five or six times. After planting, the fields are irrigated every three or four weeks. Up to four weeks after planting fertilizer may still be added. When two months have passed, the ground between the plant rows is turned. Weeding is very thorough, and the ridges on which the cane has been planted are periodically raised to reinforce them and to keep the roots covered. To prevent impoverishment of the soil, sugar cane is alternated with rice as a monsoon crop and pulses as a winter crop.

The *bagayat* region is mainly devoted to mango orchards. There are not many differences in cultivation practices among the growers or between the two villages, although better farmers use more fertilizer, spray the trees more regularly, and irrigate and plow more intensively.

*Chiku* trees require more maintenance. The harvest period begins at the end of October and lasts through March. After the monsoon the ground is worked; under the trees, whose branches hang lower than those of mangoes, this is always done with a hoe. Then the ground is harrowed, and some time later it is irrigated and again plowed. During harvest, irrigating is done weekly, for the ground should be continually wet. At the end of the harvest time, and before the monsoon breaks, the orchard is again irrigated, plowed, and fertilized. In Chikhligam several garden owners heap up small mounds of earth around mango trees, but in Gandevigam this is chiefly done under *chiku* trees.

Under the young fruit trees which have as yet few leaves and do not give much shade, cultivation of other crops is possible: bananas, tuberous plants (especially *suran*), and ginger (*adhu* and *haldar*). Several Anavils told me that such cultivation of crops under trees is unprofitable on account of high labor costs. The members of a government experimental farm emphatically denied this and said that the reason for the decline in the cultivation of crops under fruit trees must be sought in the present-day habit of the garden owners *par excellence*, the Anavil Brahmans, of taking life easy.

Bananas, which are also grown on the *kyari*, thrive better on the well-drained loamy soil in the gardens on the *bagayat*. The banana plant needs much water, and to reduce the cost of irrigation it is planted before the rainy season, in the months of April through June. Until the rains come, weekly irrigation is necessary. After a period varying from fourteen to eighteen months the bananas are harvested. The plant, being an annual, must then be taken out. Like sugar cane it is cultivated on ridges that are maintained by raising. As with all wet crops, intensive weeding around the banana plant is necessary. Between the plant rows this is done by means of a harrow, but the weeds among the plants on the ridges are removed by hand. If growth is slow at the beginning of the monsoon, fertilizer is added, being mixed with the upper layer of the soil by harrowing.

The chief tuberous plant cultivated in Gandevigam is elephant's foot (*suran*). Shoots of the ripe tubers are stored in a dark place and are planted on ridges in the gardens about May. Planting in this period enables irrigation cost to be kept low. As with sugar cane and bananas, *suran* is a crop that requires highly intensive labor. Until the rainy season sets in, the fields must be irrigated every ten days. By the end of March or in April, at any rate before planting, the ground is green-manured, after which it is irrigated and plowed. The *suran* takes three years to ripen, by which time the tubers weigh five pounds or more. They are uprooted in January or February and are kept in a dark place until May. *Suran* that is not fully developed is sometimes taken out of the ground and sold in August, when prices for vegetables and tubers are high.

Ginger is cultivated in much the same way. To begin with, the soil is prepared by irrigation, fertilizing, plowing, and preparing ridges. After planting, irrigation is necessary every eight days until the rains come. During this period weeding takes place and the ridges are raised again. The ginger is ripe when the leaves of the plants become yellow and fall off; this is about seven months after planting. The tubers can be left in the ground for some time without spoiling, provided that irrigation is omitted. As is indicated by the foregoing brief outline of the agrarian cycle, in Gandevigam as elsewhere most care is devoted to the cultivation of the market crops. The production of rice, pulses, and grass could undoubtedly be increased and their quality improved, but these crops are grown mainly for home usage. Understandably, it is the market products that receive most attention.

TABLE 8.4.2

AGRICULTURAL CALENDAR OF GANDEVIGAM

| *Period* | *Jarayat* | *Kyari* | *Bagayat* |
|---|---|---|---|
| July through September | After rainfall, cutting green grass | Mixing the soil with water ("puddling"), transplanting rice seedlings; cutting green grass on field embankments; weeding | Cutting green grass in gardens and along paths; banana harvest (planted the previous year); weeding and raising ridges of bananas, *suran*, ginger, etc.; harvesting unripe *suran* and ginger; if necessary, fertilizing orchards |
| October and November | Making hay some weeks after end of monsoon | Harvesting and threshing rice; preparing soil for winter crops: fertilizing, plowing up, harrowing, digging irrigation canals; sowing *val* and planting vegetables and seedlings of red pepper in seed plots; harvest of sugar cane | Irrigating, plowing up and harrowing orchards; harvest of ginger; beginning of *chiku* harvest |
| December through March | Cutting down trees, if necessary | Transplanting vegetables and red pepper plants from seed plots; harvest of sugar cane continued, harvest of *val*; irrigating, harrowing, weeding, and raising ridges of sugar cane, vegetables, and red pepper | Harvest of *chiku* continued; irrigating, plowing up, and harrowing mango orchards at first bloom; weekly irrigation of *chiku* trees; harvest of *suran* |

TABLE 8.4.2—*Continued*

| *Period* | *Jarayat* | *Kyari* | *Bagayat* |
|---|---|---|---|
| April through June | — | Harvest of vegetables and red pepper continued; mending embankments and fences; plowing up soil and preparing seed plots; at first rain, planting rice seedlings | Mending fences; if necessary, green-manuring with hemp and sugar cane leaves in preparation for cultivating tubers; irrigating, plowing up, and harrowing gardens; planting *suran*, bananas, and ginger; harvest of mangoes, pruning trees |

The crop pattern of Gandevigam is more intensive than that of Chikhligam in all respects. Not only is most of the arable land in the garden village under cultivation all year, but the crops themselves require more labor and capital. During a certain period of the year, even more than one crop is grown on parts of the *bagayat*, although this area is rapidly diminishing. Owing to the varied crop pattern there is continuous employment in agriculture, but it declines in the months before and after the rainy season. In April, May, and June, the *bagayat* crops require many hands, while after the rainy season there begins a busy time especially on the *kyari*.

### *8.5 Landlords and Agricultural Laborers in the History of Gandevigam, and the Changes in the Crop System*

Until several decades ago, the relationships between Anavil Brahmans and Dublas in Gandevigam were institutionalized in the *hali* system. As in Chikhligam, I often had to rely on distorted accounts of the way it functioned. Distorted, because the description of a better past reflected the dissatisfaction of both landlords and agricultural laborers with the present. Essentially, however, their accounts were confirmed by what is recorded in colonial administrative reports and older monographs about villages of the plain of south Gujarat.

The Dublas of Gandevigam were employed as *halis* by the landlords of the village. According to my informants there was no category of casual laborers in those days. The Dublas, much less numerous than today, all lived in the gardens and, if only for that reason, all of them had a master. Most members of the caste of agricultural laborers worked throughout the year for their landlord, but when the Anavil was not in a position to maintain his subordinate continuously, the latter was allowed to work for someone else occasionally and keep the earnings. Yet he was regarded as a *hali*, not a casual laborer. In other words, all agricultural laborers were in principle attached; this is the chief difference from the traditional relationships in Chikhligam, where besides *halis* there were also casual laborers. This difference is no doubt due to the dissimilar character of the agrarian economy of the two villages in the past. Because of the very intensive garden culture of Gandevigam, it was advantageous to employ agricultural laborers in permanent service. The need for extra hands was not confined to a few months but was evenly distributed over the seasons. This was the main reason for the Anavils' preference for *halis* over casual laborers — as extra labor, it is true, for the Anavils of Gandevigam used to work the land themselves. Although a few of the dominant caste, the Desais, had acquired various seignorial rights as revenue farmers and occupied an unassailable position as the regional elite, the majority of Anavil Brahmans led a relatively simple existence on the local level.

Within this framework they were without a doubt the most prominent villagers, owners of the majority of the arable land. Yet they were primarily peasants, who distinguished themselves from the Desais by a more unpretentious way of life. Nevertheless the Bhathelas of Gandevigam also kept one or more farm servants if possible. The task of the housewife was then lightened by the maid.

Those who could not afford a farm servant depended in the first place on their own working ability. Of necessity they carried out the physically strenuous and less favored jobs, such as drawing water and looking after the cattle, and they were not even free from plowing. The Bhathelas who employed farm servants were themselves actively concerned in agriculture, and they worked with their servants on the land. At that time, the difference between master and servant must have been apparent chiefly in the kind of work they did.

The Anavils of the garden region certainly knew about the "lordly" way of life of their more illustrious fellow caste members. In the capital of the subdistrict there lived a Desai family regarded as one of the most prominent in the whole of south Gujarat. Undoubtedly individual Anavils in Gandevigam aspired to a life in such grand style. In the village nucleus is to be seen a large three-storey brick house, now in a state of decay. It bears witness to the former wealth of a Bhathela family, whose last son departed in relative poverty to Africa years ago. For a few generations, members of this family had lived extravagantly, like maharajas, some villagers told me with a mixture of admiration and derision. But they overshot their mark, and in the end all their possessions — land, cattle, and also the many *halis* in their employ — had to be given up.

Another Anavil, one of the richest men in the village, told me that his grandfather had become very prosperous in his time. He had owned 100 acres of land and twenty oxen, and employed twelve *halis*. The grandson still took pride in the fact that his father had sworn never to eat without guests. His fellow caste members did not contradict that the oath had been sworn, but they did deny that the father had kept it.

Stories of a glamorous personal past are, however, exceptions. The present generation of Anavils in Gandevigam still remembers the meager existence they led during the crisis years before the Second World War. Many lost their land and had to dismiss their farm servants temporarily or even permanently. But they depict the life of earlier generations in brighter colors, and this agrees with the description of the situation of the Bhathelas in administrative reports of the last century.

The position of the ordinary bhathela peasant has improved under British rule. Though less frugal than the Kaira Kanbis, they are successful cultivators.[13]

They in general possess the largest holdings of land. They are active and industrious, and, with the assistance of their Halees or hereditary servants, are enabled to bestow much labour and care on cultivation. They cultivate the major portion of the superior crops and to their exertion and labour the great increase in the growth of sugarcane during the last ten years is to be ascribed. The result of their industry and activity is, that they enjoy a degree of comfort and wealth unknown to the cultivator of the Deccan. The Battellah always lives in a well-built brick house. He is always well-clothed, has the best cattle in the village, and the command of a little money. He is looked upon as a person of influence and substance by others.[14]

13. *Gazetteer of the Bombay Presidency*, vol. IX, Baroda, 6.
14. Bellasis, 1854, 2.

In the eyes of the writer of these reports, the Bhathelas distinguished themselves favorably from the lazy and extravagant Desais by the care they devoted to agriculture.

These are doubtful compliments. It was because of their status and behavior as agriculturalists that the Bhathelas were looked down upon in their caste. The villages in the *bagayat* region, a domain of these peasant Anavils, were regarded as decidedly backward. The functioning of the *hali* system in Gandevigam should be seen in this light. For the majority of Bhathelas, employing servants had a predominantly economic basis, that is to say, it was dictated above all by the very varied and labor-intensive crop system. Extra labor was indispensable in addition to the personal efforts of the members of this landholding caste *par excellence*. In Gandevigam, *halis* were not primarily taken on for reasons of "conspicuous leisure." The patronage character of the *hali* system, although it was certainly not absent, was therefore much less pronounced here than in villages that contained Desai families.

The distance between the *halis* and their Bhathela masters was less marked than that between *halis* and the Desai landlords in other villages. The Bhathelas were less well housed, and they dressed as peasants. A few older Anavil women in the village still wear their *saris* bound up between their legs, a *kaccha* way of dressing which is found only among women of low castes who work on the land. When I asked whether the wives of Anavils had ever been actively concerned in agriculture, my informants always denied it emphatically and indignantly. True or not, this denial was hardly surprising in view of the derogatory proverb that states that the women of this caste used to take their husbands' midday meals to the fields. This in itself is an unheard of and improper activity for Brahman women, let alone agricultural labor. The less refined life of the Bhathelas was also indicated by their food, which differed from that of the other villagers in quantity rather than in quality. This is not to say that the Anavils of Gandevigam approached the subsistence level of the caste of agricultural laborers, or even that of the Kolis. As measured by external criteria of prosperity, however, there was a striking difference between the seignorial style of the Desais and the peasant way of life of the Bhathelas.

The small distance between them and the Dublas was apparent in their daily interaction. The Bhathelas worked on the land with their farm servants and shared their efforts. They were continuously in their vicinity. The Dublas of Gandevigam put great emphasis on the simplicity of the masters of the old days, with whom they worked together, as they claimed, and with whom they were on good terms. Daily contact, not only between master and servant, but also between the members of their families, tended to lead to considerable intimacy. References to a happier past should of course be regarded with caution. Owing to the present antagonistic relationship pattern, both parties are inclined to depict the past situation in rosy colors. Nevertheless, it seems that, for lack of a clear-cut conflict of interests, easy ways and mutual affection were not necessarily exceptional in the relations between the masters and their servants.

The relationships between landowners and agricultural laborers in Gandevigam were altered by the changes of the past few decades in the pattern of agricultural activity. To summarize these changes as the transition from a closed village subsistence economy to an open market economy would be an unwarranted simplification. It is true that in former times most of the produce remained in the village or was directly exchanged for other agrarian and non-agrarian consumer goods in local markets.[15] As has been mentioned, however, everywhere in south Gujarat the growers sold sugar cane and, in the *bagayat* region, other garden products outside the village. Nevertheless, the terms of money economy and market production apply only in part to the traditional agrarian structure of Gandevigam.

Money was used for only a few purposes, especially for paying land tax, for buying land and cattle, and, above all, for financing marriages. It was scarce, and was rarely used by the villagers in everyday life.[16]

It should be added that the closed nature of the village economy

15. "Nowsaree, Gundevee, Chikhlee, Balda, Pardi, and Damaun, and Surat, are all places at which the agricultural produce of these purgannas is largely consumed, and such surplus produce which is not consumed at these market towns is readily bought up for exportation, the facility of water-carriage to other parts being great. Besides these market towns, much business is carried out at the Atwaras, or weekly fairs, held in different villages of the Bhugwara and Parnera purgunnas. Banians from Bulsar, Gundevee, Chikhlee, Balda, Pardi, and Damaun regularly frequent these fairs, with a variety of price goods, cloths, cutlery, cooking utensils, beads, bangles, Native ornaments, pepper, ginger, tobacco, and such articles of general consumption and luxury. The neighbouring villagers attend, bringing with them garden produce, wood, salt, and grain. The mode of transacting business is generally by barter, and little money is used." (Bellasis, 1854, 6.)

16. Joshi, 1966, 38–40.

was breached to only a limited extent by the cultivation of the cash crops referred to above. The contacts of the landowner with the market remained confined to the individual trader from the regional town who visited the villages at harvest time to buy up the produce, either for his own account or on commission for merchants in Bombay or north Gujarat. The sugar cane was pressed by the farmers themselves and from its juice *gur* was made, a compact brown substance that was put into earthenware jars and transported by oxcart. Often the urban trader acted as moneylender, and the peasant became indebted to him when he obtained an advance for one of the spending purposes that have been described. Thus, money transactions were clearly separated from the traditional local form of exchange.

The modernization of agriculture in Gandevigam was gradually completed. As in Chikhligam, it was the result of two reciprocally strengthening processes. First, there was the contraction of the very varied crop system. The wide range of crops, intended to provide as much as possible for the needs of the village itself, was replaced by a much narrower range of produce intended for the market. The extension of the irrigated area in the village was of great importance. Just before the Second World War a distribution network was completed in the *bagayat* region, utilizing the water from the Ambika River. Labor demand was greatly influenced by the fact that the extension of the market production was accompanied by the gradual transition from sugar cane to mangoes as the main money crop.

Secondly, the marketing of this more commercial agriculture was organized along other lines. By the end of the nineteen-thirties a cooperative sugar factory began operating in the subdistrict to process the cane from the surrounding villages. The farmers in Gandevigam became more market-minded: formerly, the factor-moneylender had, as it were, screened them from the market mechanism. Improved road communications with the outside world changed all this. As no permanent crossing facility existed, the farmers used to ford the river early in the morning with their oxcarts and travel the country roads to the railroad town where their garden products were auctioned. On their return they often had to wait for hours on the river bank, because at that point on the river the tides are still perceptible. It was therefore an enormous step forward when a bridge was built just outside the village, and when the road between the two subdistrict towns, which passes Gandevigam, was paved.

The greater accessibility of the *bagayat* region facilitated the founding of an agricultural cooperative with its own selling apparatus. This cooperative, which was initiated by some Anavils from Gandevigam and other garden villages in 1944, has flourished. In 1960, a total turnover of 700,000 rupees was reached. The cooperative is situated on the asphalt road just outside Gandevigam. There is a large shed, in which the produce — especially mangoes, but also *chikus*, bananas, and *suran* — is stored. Operations are conducted by two fully salaried employees, who work in an administrative building equipped with office furniture and a typewriter, as well as a telephone, by local standards the last word in modernity. During the mango harvest, extra hands are hired to unload the oxcarts which bring in the produce, to weigh the baskets under supervision, or to load them directly on to the trucks of the urban merchants. These casual laborers work on piece-rates, and are mostly women from the caste of shepherds (*Ahirs*), who are selected because the Anavil board members say they work much harder than Dublas. The value of the lots is provisionally determined on the basis of the current day's price, which is passed on from the regional auctions by telephone. The amounts are entered on the members' accounts, which are settled at the end of the season.

Early in 1963 the cooperative bought a tractor, which is hired out to members, together with a wage laborer. During my stay the board was planning to install a small factory for canning mangoes. Various negotiations were being conducted, and some members of the board traveled to central Gujarat to see a canning factory in operation.

Improved communications not only have led to commercialization of the marketing organization, but have also increased the contacts of Gandevigam's inhabitants with the outside world, and this was further facilitated when bus services began operating. Owing to the contraction of the agricultural economy, local demand cannot wholly be met by home produce, on the other hand the increasing circulation of money has brought many new articles into the village. Commercialization, in other words, is not confined to production and distribution, but has entered the consumer sphere as well. Villagers regularly go to town to buy what they need.

Owing to these developments, contact with the urban environment has also been intensified.

The commercialization sketched above was partly due to the extension of contacts beyond the village system. Encouraged by the improved means of communication, landowners have allocated an ever larger proportion of their arable land to market production.

In explaining the changeover to mango cultivation, the Anavils, like their fellow caste members in Chikhligam, give as the most important reason the occurrence of plant diseases among the traditional garden crops. They also stress the expectation that the increasing demand for mangoes would render cultivation of these fruits very profitable. The transformation of the garden lands into orchards began before the Second World War, but did not cover large proportions until the end of the nineteen-forties. Low prices for the traditionally cultivated crops cannot have been the reason. Financing the new culture was made possible precisely by the high profits during the war years on such crops as sugar cane and bananas. Referring to that period, one Anavil said to me: "Money came like rain in the monsoon." The garden owners did not seize this golden opportunity to intensify agriculture by using more capital to bring about increased production. On the contrary, they preferred to invest the profits from the war years in such a way that they could adjust their behavior to what they regarded as representative of their caste. Mango cultivation freed them from manual work and enabled them to reduce their active participation in agriculture. This was their most important incentive for changing to the new crop, which does not require intensive labor. As in Chikhligam, the planting of orchards still goes on. Only the least prosperous Anavils have not yet changed their gardens into orchards, partly because they cannot as yet afford the large investment, amounting to about 2,000 rupees per acre. By sinking new wells, the larger landowners have recently reclaimed even the high *kyari* land for mango cultivation. Formerly they grew many different crops under the young trees to offset their temporary loss of income. Nowadays they do not trouble to do it themselves, but lease out the new orchards for this purpose to Kolis. Most of the Bhathelas from the *bagayat* region gradually realized their aim of improving their reputation within the Anavil caste. No longer can they be ridiculed as farmers — toiling drudges who behave like peasants and are no credit to their caste.

Improved communications with the outside world, so closely connected with increasing integration into the market, also contributed to the increased social acceptability of the Bhathelas. In former times an isolated location was looked upon as a measure of backwardness.

The Anavlas at this side of the khadi were formerly loth to give their daughters in marriage to the other side; but now that the danger and inconvenience of the crossing have been removed, the khadi villages do not pay a larger dowry than is customary in other villages.[17]

The Bhathelas from the remote garden villages were at a disadvantage in the marriage market. As they came to lead a more "civilized" life, the pattern of marriage relationships changed. Although I have no specific information — one of the lacunae in my research data — I think I am justified in asserting that phenomena of poverty such as occurred among the older Anavils of Gandevigam (marriage by exchange, bachelor households) are no longer found among the younger generation. Partner selection today is more in accordance with the Anavil ideal. Marriages are contracted with Anavil families from villages which in the past would not have dreamed of entering into any relationship with the Bhathelas of this garden village.

I have attributed the change in the crop system of Gandevigam during the last few decades primarily to the endeavors of the farmer Anavils to attain more esteem within the caste by drastically reducing their participation in physical work. As in Chikhligam, however, the rapid diffusion of mango cultivation brought with it a commercialization of the agricultural economy which had far-reaching consequences for the position of the Dublas in the village.

One of these consequences was the introduction of money wages. After the war ended, the Anavils ceased to pay the wages of their farm servants in kind, and also, as occurred elsewhere in south Gujarat, abolished the various extra allowances that used to be customary.[18] The increasing orientation of the Anavils towards the market brought with it a more businesslike management of their affairs. During that period of high prices for agricultural products it was more advantageous to pay money wages. At the same time, because of the reduced cultivation pattern it was no longer possible to set aside a substantial part of the total produce for redis-

17. *Papers . . . Jalalpor Taluka*, 1900, 5.
18. C. H. Shah, 1952, 448–50.

tribution within the village, for instance as emoluments to farm servants. An ever greater proportion of the agrarian produce was sold in the market. The growing trend toward a money economy in the village depressed the level of existence of the Dublas, for whom the extra allowances in kind had formed, particularly from the qualitative standpoint, an important wage component.

The more general use of money has, moreover, widened the differentiation in ranges of consumer goods used by landowners and agricultural laborers. Because of the enlargement of economic scale, numerous new articles were introduced in Gandevigam, which are exclusively in the possession of members of the dominant caste. They are the only villagers who can afford to buy a great many previously unknown articles — new foodstuffs and luxury goods, such as superior kinds of cereals, canned cooking oil, cigarettes, and tea. There are also the more or less durable consumer goods, varying from cigarette lighters to spectacles, fountain pens, rubber slippers (made in the Bata factories in Bombay), bicycles, large steel cupboards, sofas, stainless steel kitchen utensils, radios, table fans, tape recorders, and other "novelties." It would be incorrect to suggest that previously the difference in housing, clothing, and food between landowners and agricultural laborers in Gandevigam was small, but without a doubt the prosperity gap has widened enormously, both qualitatively and quantitatively. As I have said, the members of the dominant caste have become strongly urban-directed in their habits. More than the other villagers they go by bus or bicycle to the neighboring towns to shop, to arrange business matters in government offices, to go to the barber, to visit tea-stalls or the cinema, in short, to do all those things which notably the Dublas cannot possibly afford.

Production for the market and a money economy, more generally a process of enlargement of scale, have led to increasing differences in prosperity between the landowners and the agricultural laborers of Gandevigam. Gradually the relationship between Anavils and Dublas has changed from a personal into an impersonal one. An increasingly functional evaluation of the Dublas' profitability as labor has been accompanied by progressively smaller remuneration of their services.

Like those in Chikhligam, the garden owners of Gandevigam did not change over to mango cultivation because of fear of labor shortage. If the Anavils now say that at the time their preference for the new crop was partly inspired by their wish not to have to depend any longer on the agricultural laborers of the village, they mean that they simply no longer want to be responsible for them. Epstein has formulated the social background of the patron-client relationship as follows:

In a system in which employers are obliged to provide for the subsistence of their hereditary servants and the latter have no other income earning opportunities, masters will employ labourers as long as the total product does not decrease through the employment of an additional worker. Since the employer would only have to give in charity what he saved in wages, he incurs no net loss by engaging labourers whose employment does not add anything to the total product. The number of workers employed in India's non-market peasant societies is thus determined by the maximization of total product, rather than by the interaction of marginal productivity and wage rates.[19]

This principle, on which the traditional master-servant relationship had functioned, no longer obtained. The familial character of servitude was lost. The contrasts between the two parties sharpened, and neither felt obliged any longer to promote the other's interests. The distance between the two categories, formerly bridged by common aims, became wider, and the *hali* system was bound to disintegrate.

It seems likely that the enlargement of economic scale described above has played a more important role in this disintegration than political and administrative intervention from outside. In Gandevigam, as in Chikhligam, the disappearance of the *hali* system was not due to measures on the part of the authorities, although according to the letter of the law it should have been. Whereas in the part of south Gujarat that was under British rule the *hali* system was never officially prohibited, the principality of Baroda declared it to be against the law as a form of unfree labor in 1923. Nevertheless, it soon became evident that the prohibition had little effect.

In July 1923, the Government of the State by proclamation declared this whole system of forced indenture as illegal and allowed the Raniparaj serf to repudiate it if he chose to. But the intentions of the Government were not properly interpreted by subordinate revenue officials and the operation of the Government's order is therefore not effective.[20]

19. Epstein, n.d., 9; this mimeographed article was published in a slightly different version in 1967.

20. *Census of India*, 1931, vol. XIX, Baroda, pt. 1, 255.

These negative results are not surprising, considering that most of the lower authorities, who had to implement the law in the territory of the principality in south Gujarat, came from the dominant caste, that is, from among the landlords who would be vulnerable if lawsuits followed. As a rule, any conflict between *dhaniamo* and *hali* remained out of reach of the government. If local functionaries were called in, this was usually on the request of a landlord who wanted his rights as a master — not legally sanctioned — to be restored to him. In the existing relationship pattern it was simply unthinkable that a Dubla should address himself to the government with a request for aid. Yet the prohibition did not quite remain without consequences. Its enactment undoubtedly had the psychological effect of making it clear to the Anavils that the government no longer allowed itself to be guided by the Anavils' interests alone. It was an indication of the widening separation of the former power elite from the administrative apparatus in a process of bureaucratization.

Nor did Independence bring immediate improvement in the position of the agricultural laborers in the Gandevigam area. The political and social climate of the principality did not favor the struggle for national freedom, and the lower authorities in particular were unwilling to promote the interests of such underprivileged groups as agricultural laborers. Social workers did not take up the case of the Dublas of the Gandevigam area so long ago that the disintegration of the *hali* system there can be ascribed to their efforts.

With regard to Chikhligam I found that the migratory labor of the Dublas was a result, not the cause, of the changed relationship between landlords and agricultural laborers. In the case of Gandevigam, too, the rise of alternative employment was probably a secondary factor in the disappearance of the traditional service relationship, although after the event such things are difficult to prove. It is true that in the central plain of south Gujarat, notably in the towns along the railroad, a process of economic expansion began even before the Second World War, but agricultural laborers from the villages in the neighborhood were especially at a disadvantage when it came to profiting by the labor demand that accompanied it. As was apparent in Chikhligam, and as will be discussed in further detail below, the Dublas' mobility is subject both geographically and economically to the assent of their traditional employers, the Anavil Brahmans. That is to say, the labor needs of the village ultimately determine whether any Dublas succeed in getting work outside it. In this respect, however, there is a marked difference between the situation in Chikhligam and that in Gandevigam.

To conclude, in both villages it was primarily the landowners who started to change the nature of the service relationship with agricultural labor. In the framework of a process of economic enlargement of scale they began to manage their property in a more businesslike way. One aspect of this change was a pronouncedly impersonal evaluation of the use of those who traditionally served them. Under the influence of their increasing orientation to the market, the Anavil Brahmans of Gandevigam renounced their obligation to provide for the Dublas, and the chance offered to these landowners by the new crop to reduce their own physical activity widened the gap between the two categories.

### *8.6 The Present Situation*

THE DUBLAS. As to the Dublas who live in Gandevigam — 381 in all, distributed among sixty-one huts — my data show that 103 belong to the male working population over sixteen. Of this number, seventy-three work in agriculture and thirty in other occupations. The agrarian Dublas can be divided into farm servants (fifty-one), sharecroppers (seven), and day-laborers (fifteen), but they are only a part of the whole working population of this caste. Most of the female members of the Dubla households also work on the land if there is a demand for their labor, and in addition several of them serve as maids in the houses of the landlords. Secondly, I did not include working children, especially boys, in my census. As in the case of the Dubla women, to include them in the working population would be completely arbitrary. The Dubla boys, who usually work as farm helpers for the Anavils, are no problem, but the question of whether children between, say, eight and fifteen years of age belong to the working population in fact depends on how regularly they are called in. This I could not systematize. Younger children occasionally help out on the land, but they are not remunerated separately. They are not recognized or paid as independent hands until they have reached the age of sixteen.

Taking the foregoing facts into consideration, the involvement of the Dublas of Gandevigam in agriculture is even more marked than may appear from the above figures. The differences in the economic activities of this caste, as between Gandevigam and Chikhligam, are great. Whereas in the latter village 40 percent of the Dublas gain their livelihood by agricultural labor, in the garden village 70 percent of their fellow caste members depend upon agriculture.

FARM SERVANTS. The proportion of farm servants is higher in Gandevigam than in Chikhligam, being 70 percent as against 55 percent of all those engaged in agriculture. The disparity in the composition of the working population within this caste becomes even more apparent if it is taken into account that in Gandevigam almost half of all Dubla men in the productive age group are in permanent service with a landlord. With the exception of one Koli, the latter are all Anavil Brahmans, most of whom also live in Gandevigam.

Servitude is still a relationship that is entered into at an early age, but today it is rarely preceded by a period in which the Dubla boy works as *govalio*. The landlords' stock of cattle has considerably diminished, partly because of the mechanization of irrigation, so that most Anavils no longer employ cowherds. They have found a cheaper solution by jointly appointing a village *govalio*, an old and redundant farm servant. He collects the cattle from the various farms in the mornings and tends the whole herd on the fallow portion of the village *jarayat*.

Even more markedly than before, an advance of money to a Dubla is the beginning of an agreement that nearly always is contracted on both sides for an indeterminate time. Most often a Dubla's servitude is initiated by his marriage. Of the fifty-one farm servants I interviewed, thirty-nine had received marriage money — to the extent of 300 to 400 rupees — from their present master. Their marriage at the age of seventeen or eighteen had been wholly or partly financed by an Anavil landlord, at whose disposal they had put their labor from that moment onwards. The other servants, who had entered service at a later age, had obtained loans for other purposes, such as to pay the expenses of illness in the household. Sometimes the attachment had developed gradually by an accumulation of small advances. What should be emphasized is that, as a rule, servitude begins with a debt.[21]

21. The only cases known to me of servants who did not begin service with a debt are two unmarried Dublas boys who bound themselves to a Koli for two

Even when I did not ask about it, all those concerned told me emphatically that the agreement no longer is valid for a lifetime. Payment of the debt releases the servant from his obligations. Yet the fact that payment is a possibility, recognized by both parties, of terminating the service relationship does not mean that it ends this way in practice. Like all their fellow caste members, the Anavil Brahmans of Gandevigam complain of the faithlessness of the servants they take on. On the other hand, the impression created by the Anavils that running away is usual nowadays was not confirmed by my data: thirty-six of the fifty-one Dubla farm servants have worked for the same master since their marriage. In this respect there is a marked difference from the situation in Chikhligam. In Gandevigam a period of thirty or forty years' service is much less exceptional. It even appears that servitude in half the cases is still hereditary: twenty-seven Dublas are in the service of Anavil families who employed their fathers before them. Least constant are the Dublas from outside the village, for example, the category of sons-in-law (*ghar jamai*) who settled in the village after marriage. Nevertheless, the observer is struck rather by the continuity in the relationship patterns, once they are established, than by a large turnover.

The exaggerated accounts of the Anavils about the Dublas' faithlessness are undoubtedly inspired by their fear of losing their money. But if breach of agreement is the exception rather than the rule, why do the Dublas themselves think of servitude as a temporary matter? Above all because they attempt to deny lasting dependence even if they know better. This is wishful thinking, because repayment of the amount they have received usually turns out to be impossible, and few Dublas in Gandevigam have managed to do it. As a rule, their monetary obligations to the landlord tend to increase in the course of the years.

The doggedness with which the Dublas maintain that the relationship is temporary is due largely to their unwillingness to look upon themselves as permanently attached. For most of them, this is simply self-deception. The lack of employment both in agriculture and in other fields compels them to enter and continue the service which makes their dependence so apparent. Thus the relationship between master and servant has been reduced to a labor agreement based on a condition of debt.

years, in exchange for food, clothing and 60 rupees a year. They will not receive the money until the period has expired.

The nature of the work and the manner of remuneration make it clear that attachment is at present above all a labor relationship. On days when there is enough work, the servants are occupied in the fields or in the gardens of their masters from 7 in the morning until 7 or 8 o'clock at night, except for an interval at midday. In a peak period it may happen that work is prolonged into the evening. Now that the servants no longer live in the gardens, their effort is confined almost exclusively to agricultural labor. The Dublas are no longer in attendance to do all kinds of jobs in and around the Anavils' houses. The master is rarely seen in the village quarter of his subordinate. Only two of the wealthiest landords have a house servant, one of whom is a Dubla bachelor who lives in his master's house. The other is a Koli who, besides acting as major-domo, is charged with the supervision of the agricultural laborers. Some Anavils employ a Dubla boy who performs various jobs and services that were formerly done by the farm servant. On the whole, the work of the farm servant has become more specific: he is responsible for agricultural tasks in the hours appointed for them. Contacts with the master take place chiefly within this framework of time and place. Moreover, today the agreement is restricted to the servant personally, as it is no longer a matter of course that his wife, and in time his children, will be attached to the master's household. It is true that many a Dubla woman works as a maid, but quite often she does so in another household than that of her husband. The diminished demand for cowherds has also been partly responsible for a more individual relationship between master and servant, in which the latter's wife and children are not automatically included. Both the more specific content and the limited character of the servitude indicate that these Dublas are much less incorporated into the landlord's household than were the *halis*.

Accordingly, the remuneration of the servant is based on his work performance and not, as under *halipratha*, on the needs of himself and his family. For each working day the farm servant receives 8 annas, a morning and a midday meal, and tea twice daily. An increasing number of Anavils have now slightly raised the money wage in the harvest period but no longer provide any food. The farm servant is annually entitled to some pieces of clothing, a pair of shoes, and a shawl, and to remuneration for cutting grass and looking after the cattle during the monsoon. He should be allowed to gather wood in the master's garden, and at the beginning of the monsoon the landlord should provide him with thatch to mend his roof. A little tobacco should be a regular wage component, and in the event of illness the master is expected to pay for the medicines. At harvest time, finally, the Dubla should be able to count on a small share of each crop.

This is how the relationship should be if custom were adhered to. However, the Dublas complain that they are given worn clothes and old shoes, sometimes bought for the purpose in the neighboring town. Medicines either are not paid for by the landlord at all or their cost is added to the existing debt. Prolonged illness has become grounds for dismissal.

Aside from skimping, which became systematic as the money economy became more prevalent, the replacement of part of the wages in kind by payment in cash has been to the disadvantage of the servant. What the Dublas used to be paid in produce they now have to buy at the higher retail prices. The effect of the unprofitable terms of exchange is a curtailment of their consumption.

On days when there is no work the farm servant cannot claim *khavati*. The master expects him first to try to find work with another farmer. Only in a period of continuing and complete unemployment, notably during the monsoon, does the master feel obliged to yield to the Dubla's repeated pleading and pay him at least a minimal allowance. The reluctance of the Anavils to extend this form of credit is due, to no small extent, to the fact that advances, from marriage money to *khavati*, are nowadays given wholly in money. The amount that the Anavils are forced to spend in this way deprives them of it for a number of alternative spending purposes to which they accord higher priority. For the same reason the landlords are not prepared to comply with the requests of their servants to defray high incidental costs — for instance, expenses incurred in cases of illness, a marriage, or a death. The compensation paid to a servant is determined by the work he delivers or will deliver. The time is past when compensation was based on the master's moral obligation to maintain his *halis*.

A continuous need of labor explains in part why the landlords of Gandevigam still take Dublas into permanent service. The Anavils are guided in this not only by the nature of the crop system — which even now is labor-intensive during a large part of the year — but also by their desire to abstain from physical agricultural work. Employing one or two servants means having a cheap, fixed nucleus of labor and a claim to temporary extra hands, for although the agreement covers the servant alone, the latter is expected to mobilize his relatives for the benefit of his master when they are needed. Thus, while reducing his responsibility the landlord can enlarge and contract his stock of labor at short notice. He of course runs the risk of losing the amount of his investment if the Dubla leaves him without notice. He cannot appeal to the government, for the lending of money is now subject to a system of permits. There is even an explicit prohibition against the discharge of debts by means of work. Although a Dubla knows better than to adduce this argument even if he should be aware of its existence, it is of course not without consequences that the claim of the landlord is not recognized as legally valid.

The Anavils arm themselves against the risk of desertion by their laborers by not letting the debt become too high. Restriction of credit is not, however, their only guarantee. Since the Dublas have ceased to live in the gardens, various landlords have proceeded to farm out cattle for sharecropping, as it were. A newborn calf is brought to the hut of the servant, who undertakes to look after the animal, but he can do this only if he has received permission to cut grass in his master's garden. When the cow is in milk it goes back to the owner, and the servant is entitled to half the estimated value of the animal. A variant is that an agricultural laborer receives half the calves bred from a cow that he has looked after. In Chikhligam, where most of the Dublas are absent during a large part of the year, this "lending" of cattle does not occur, nor are all the landlords in Gandevigam prepared to do it. Sometimes the care given the animals by the Dublas leaves much to be desired. On the other hand, this is an excellent means of binding the farm servant who no longer lives on his master's land. It is especially desirable for the Anavils, because in this way they earn back part of their investment and can go on giving credit, while the debt of their subordinates remains within reasonable limits.

The immediate reason why a Dubla becomes a farm servant is usually that for most members of this caste this is the only way to finance their marriage. Yet the loan is concomitance rather than cause. For the great majority of Dublas who perform agricultural labor, servitude still offers the best guarantee of a living. Although their daily wage is extremely low, they are assured of employment on most days of the year. The relation between master and servant has become markedly businesslike. The Anavils have unilaterally fixed a standard wage, above which they need not go. The masters are generally not susceptible to any appeals by the servants for further assistance. Anavils told me that they give their Dublas something extra from time to time when they are in a good temper. The reaction of their subordinates indicated that a good-tempered landlord must indeed be a rare figure in Gandevigam.

On the other hand, there is no longer any question of many-sided service by the family of the Dubla. The Dubla's contribution is personal and as a rule confined to agricultural labor. The allowance he receives is not sufficient for him to be able to cope with eventualities such as divorce and remarriage, illness, continuous rain during the monsoon, or collapse of a wall of his hut. On such occasions the master cannot refuse too often to come to his assistance — at least if he wishes to keep his servant. But what formerly was regarded as part of the maintenance wage is now booked as debt. However, the farm servant has at least a possible source of credit, even though laboriously obtained. Aside from the security of a minimal income this is for the Dubla the compensatory side of servitude. The drawback, however, is profound dependence, for the master manipulates the debt as a means to compel obedience, and sometimes even as an excuse for not fulfilling his minimal obligations. He knows how to demonstrate his authority in numerous ways. Servants are not given the perquisites to which they are entitled, nor do they get *khavati* unless after prolonged insistence. If it does not suit him to do so, the Anavil refuses to pay the wage and tells his subordinate to return the next day. Should the servant be ill, the master comes to his hut only to ask impatiently when he will be fit for work again. An "insolent" Dubla is sometimes not given any work for days on end to teach him a lesson. It is in an attempt to escape from the tutelage and subservience imposed upon him that the farm servant tries to pay off the amount he received for his marriage. My data show that only eight Dublas in the village have succeeded in doing so, owing to a combination of advantages. One had extra earnings as a drum-player at marriages; another worked with his master's permission in a factory for two years and gave him his savings on his return. The others reared cattle for their masters for some years and paid off their debts in this way. All of them pointed to the absence of setbacks over longer periods in their career as servants, such as long illness, having to care for aged parents, death of calves reared, or many small children — that is to say, the absence of circumstances that are normal rather than exceptional for the situation in which the Dublas find themselves.

The example of the fortunate ones is what the Dublas have in view when they say that the relationship is temporary. Yet it would be incorrect to infer that their desire to pay back the marriage money indicates an effort to break the servant relationship. Without a doubt, the large majority wishes, for lack of a better alternative, to remain in servitude. For them, repayment of debt is first of all an attempt to render themselves less dependent on their masters, to lessen their control. Of the eight Dublas who paid back their marriage money, three are still employed as servants, though no longer bound by debt.[22] Four Dublas who were freed from debt servitude through no action of their own are in a similar situation. They became redundant when their former master sold or lost his land, left the village, or died. These unindebted servants now receive from their new masters a slightly higher daily wage, which has only heightened the desire of their less fortunate fellows for this form of service.

It is precisely for these reasons that the Anavils appear to be less than happy with this kind of relationship. They are therefore not at all sorry that so few farm servants manage to repay their debts in the end. For that matter, the landlords have it in their own power to prevent repayment, and they do not shrink from wrongfully adding all kinds of expenses to the debt. The Dublas are well aware of this trickery, but they are powerless. A striking instance is that of the master who went to the hut of a Dubla servant to lecture him about his absence. On the way there he hurt his foot on a thorn, with the result that the Dubla's account was debited with 5 rupees, the amount the master had spent on medicines. Another Dubla entered into marriage at the same time as his brother, and together they obtained 275 rupees from an Anavil. When, after three months, one of them died, the master charged the remaining brother with the total amount.

It is not to be wondered at that in most cases the debt does not decrease but becomes larger. It was my impression that most of the farm servants have long since abandoned hope of ever becoming

22. Of the others, one has found work outside agriculture, and four have become day-laborers.

free of debt. They keep up appearances for themselves and in the presence of others, but at the same time they try to persuade their masters to give them as much credit as possible. Once attached they have little to lose. It is therefore understandable that severance of the relationship is usually due to a conflict between master and servant which has flared up suddenly. This does not happen as often as the landlords claim it does, but no doubt more often than termination by mutual consent after repayment of the debt.[23]

To break with the landlord means in many cases to break with the village and to give up shelter on a piece of land of one's own, at least temporarily, without having any clear notion of how or where to find other employment. A Dubla will think twice before he deserts his master. Often this also means a break with his family. The deserted wife returns to her native village with the children and enters the household of her parents or a brother. Economic hardship of this nature is no doubt the chief explanation of the relatively high frequency of broken families in the quarter of the landless laborers. There are a considerable number of children who, for this reason, are taken in by their mother's brother or mother's sister. Several Dublas can tell me only one thing about their father — that he ran away — and they do not know where he went nor whether he is still alive.[24]

Running away is for a Dubla an act of desperation rather than an expression of indifference and unreliability, as the Anavils claim. The master's reluctance to give credit is, in other words, inspired by an abundance of labor supply, and not primarily by the Dublas' faithlessness. An illuminating remark was made by an Anavil who told me that a farm servant leaves only when his master does not fulfil his minimal obligations. A resolute and repeated refusal to give credit, for instance in the form of *khavati*, means in fact that the advantages of servitude are lost and the disadvantages of attachment remain. Both parties reckon with the possibility of termination of the agreement, but their views as to how it should end differ. The diminished stability of the relationship pattern is the result

23. Five Dublas in Gandevigam admit that they ran away from their masters after a quarrel. Actually there were more who did so, but their number can no longer be ascertained, as several left the village. Conversely, more than one of the numerous sons-in-law in the Dubla quarter came from elsewhere for this reason.

24. In eleven huts, one of the parents of the children living there is absent for another reason than employment of the father outside the village; nineteen children, born outside Gandevigam, grow up without their parents in households of relatives in the village. These are incomplete but telling data.

of the landlords' indifference and denial of total responsibility rather than of emancipation of the farm servants.

SHARECROPPERS. It is open to question whether the sharecroppers among the Dublas, whom I have placed in a separate category, should not rather be included among the farm servants. In giving a plot of land to Dublas for banana cultivation, the Anavils have found a new means of assuring themselves of dependable labor. Only reliable agricultural laborers are considered for sharecropping. Part of their time is at the landlord's disposal, and often a son of the sharecropper works as a farm servant. Apart from this it is profitable for the Anavils to have Dublas, because they are satisfied with a smaller proportion of the produce than Kolis. Finally, there is virtually no chance that the agricultural laborer will have himself registered as owner under the new law. For all these reasons the number of sharecropping farm servants will probably increase rapidly in Gandevigam.

Harper mentions such an agreement between landlords and agricultural laborers elsewhere in India:

> In more recent years a few Haviks have leased land to their servants to induce them to continue in service. . . . Had it not been for the Tenancy Act in 1947 this method of enforcing the system of indentureship might have succeeded. The Tenancy Act gave the servant permanent rights to the land and, in addition, limited the amount of rent that the landlord could ask.[25]

In view of the relations between Anavil Brahmans and Dublas in south Gujarat this last remark seems somewhat unrealistic.

DAY-LABORERS. During my stay in Gandevigam I counted fifteen day-laborers among the Dublas. This is a moderate number, but it should be borne in mind that, according to my informants, some decades ago no one at all in this category was to be found in the garden village.

Although they work mainly for the local landlords, the day-laborers are less bound to Gandevigam and the Anavils than are the farm servants. They look for casual work in the other villages as well, and are called in from time to time by Kolis in the neighborhood. As a rule they receive 12 annas and two meals a day from the farmer for whom they work. At harvest time they are not

25. Harper, 1968, 58.

usually given meals, but the daily wage is raised to 1 rupee or higher. They also occasionally undertake to perform a specific task for a prearranged lump sum of money. Sometimes they are bilked of part of the agreed remuneration for the work assigned to them. When, for instance, an amount of 10 rupees was arranged, the landlord may decide after completion of the job that 6 rupees is really more than enough. The powerlessness of the agricultural laborers in the face of such trickery does nothing to diminish their rancor.

Some Dublas in this category have always been day-laborers. A few are unmarried boys, and for that reason as yet unattached. Others are former farm servants who paid off their debts or were dismissed. Especially among the sons-in-law in the Dubla quarter there are several who have escaped from the control of their former masters and now live as free laborers in their wives' villages. The background of this category is therefore varied. The group is not in the least closed; in several cases I found farm servants and day-laborers in one household. It is difficult to ascertain which of the two is better, or rather less badly, off — the day-laborer or the farm servant. Some of the former much prefer to remain unattached. Being self-assured and resourceful, they succeed in scraping together a day's wage even in periods of continuing unemployment. They belong to the type that can always look after itself and does not wish to be subjected to a master. Others prefer the security of a minimal existence to the risks of an unattached life. The offer of a higher average daily wage is for them no compensation for the possibility of receiving no allowance for days on end. Their employment as day-laborers is often only temporary.

There is indeed no clear-cut difference in conditions of employment. In cases of continuing unemployment or more incidental reverses, the day-laborers try to procure an advance from a landlord. In exchange they undertake to work for their creditor on such days as he may need them and against the remuneration of a farm servant. They are then merely unattached in name. If the debt runs up they become in fact part-time servants, a status which either continues or finally changes into complete servitude according as the Anavil thinks fit.

EMPLOYMENT OUTSIDE AGRICULTURE. Thirty Dublas in Gandevigam work outside agriculture. This represents nearly 30 percent of the male working population of that caste. The rise of their proportion in the past few decades reflects the expansion of alternative employment in the neighboring towns and the improvement of communications. In subsection 8.2 I have given a survey of their activities.

I did not get the impression that these Dublas isolate themselves from the agricultural laborers in their caste and form a separate category, nor did I find any concentration within a limited number of households. Those who have managed to find employment outside the village, however, do try to assist their relatives to find such jobs as well. Owing to their greater familiarity with the urban environment and the means of gaining a livelihood there, they are in a somewhat better position to do this than those who work in the village, although they can rarely exert any influence in the matter. Recommendation by the landlord usually carries more weight. Several Dublas therefore owe it to the support of an Anavil — their father's master, for example — that they found employment in a workshop or obtained a job with some government agency. The Dublas recognize the influence of the landlords in this respect, but are convinced that it is mainly used negatively, that is, to prevent them from finding employment outside the village. They complain that "the Anavils speak badly of us and see to it that we are not accepted." Local government functionaries admitted to me that these charges contain a germ of truth. In their view the landlords are afraid that if alternative employment expands they will not have enough labor themselves, or at least that their authority will be impaired and the wage level will rise.

The twenty-five Dublas who leave the village early in the morning and return in the evening are without a doubt the most fortunate members of their caste. In the first place they have a fixed income, with a higher daily wage than that of the agricultural laborers. Secondly, they have escaped the control of the landlords in the village. Their good luck is relative, however. In the workshops and government offices they receive the lowest pay; most of them do not earn more than 1.50 rupees a day. Those who work in Bilimora, seven miles away, cannot afford to go there and back by bus, which costs 0.40 rupees each way. They, too, have to incur debts in order to live. One Dubla, who works in a paper factory, and two others, who are employed by a wood merchant, have been given advances by their employers to marry. In other words, their independence has become limited.

Of all the Dublas in Gandevigam, only a few seem to migrate to the brickyards every year. During my stay they numbered not more than five. Unfamiliarity with the channels leading to work in the environs of Bombay is only a small factor. Most of them know quite well to whom they should apply, but say they do not feel attracted to this kind of work. The difference in agrarian employment between Gandevigam and Chikhligam, where during the greater part of the year there is little or nothing to do for agricultural laborers, no doubt explains why so few from the garden village avail themselves of this opportunity. It may be that intervention by the landlords is also a factor. Nevertheless, in the existing situation migration is evidently not a desirable alternative to agricultural labor. The Dublas talk at great length about the disadvantages — the strenuous work, the insecurity of the new surroundings, the temporary character of the employment, the small difference in wages, and the uncertainty inherent in migratory labor. With some exaggeration it might be said that the push factor is lacking and the pull factor is not strong enough.

A GENERAL SURVEY OF THE CONDITION OF THE DUBLAS. From the foregoing analysis, it seems clear that the Dublas who work outside the village are least badly off. Some possess a few more

TABLE 8.6.1

NUMBER OF OCCUPANTS PER HUT AND THE DISTRIBUTION OF THE MALE WORKING POPULATIONS OF THE DUBLAS IN GANDEVIGAM

| *Number of Occupants per Hut* | *Number of Huts* | *Number of Occupants* | *Male Working Population* | | |
|---|---|---|---|---|---|
| | | | *Average per Hut* | *Totals* | *Agricultural Laborers* |
| 4 or less | 18 | 56 | 0.8 | 15 | 9 |
| 5 through 7 | 27 | 168 | 1.7 | 45 | 32 |
| 8 and 9 | 11 | 94 | 2.2 | 24 | 13 |
| 10 or more | 5 | 62 | 3.6 | 18 | 11 |
| Totals | 61 | 380 | 1.7 | 102 | 65 |

material goods and live in slightly better huts than those of their agricultural fellow caste members. In general, however, the various categories of Dublas do not live in separate households. This means that differences in the degree of poverty are determined primarily by the ratio between productive and unproductive members of each household and only secondarily by the nature of their work.

The Dubla quarter presents a cheerless sight. It is on rough and uneven terrain that is waterlogged during the monsoon. The huts, many of which are in a state of decay, were constructed in a rambling order. A dwelling consists of a single room which one must stoop to enter, so as to avoid the low, overhanging roof. The inhabitants crowd together in a space of a few square yards, in continuous twilight and, except for the doorway, without ventilation. To say that this interior is scantily furnished is to put it mildly. In a great many cases even the simplest household goods for daily use are lacking. I stopped trying to record the goods they did own, because my interpreter found it too painful to go on enumerating so many articles which obviously were not there. Most of the inhabitants sleep on jute sacks or home-made reed mats. Kitchen utensils are usually limited to some earthenware pots and a few copper pans and mugs. An official report on the condition of the *halis* prepared shortly after the Second World War stated that less than half the Dubla households owned a grinding stone. Further, one-third of the 285 families that were interviewed did not possess even the simplest ornaments, which are usually acquired at marriage; in other families their worth did not exceed 5 rupees.[26]

As regards Gandevigam, the condition of the Dublas has not improved in any way since the war. There is so little floor space that not all the members of the household can sleep inside. The furnishing is so meager that, in my eagerness to record at least something, I found myself counting the nails in the beams, on which here and there some clothes were hanging — rags to me, but to the owner a precious second garment. Only the men wear shoes, which are indispensable for work in the field. The children in particular are scantily dressed. As for food, cooking oil is rarely used. Salt and spices are bought in the village shop, sometimes in exchange for the eggs of the hens they keep. Only farm servants and maids regularly eat vegetables as a component of the meals they are given in the household of the master. During the monsoon, when the income of the agricultural laborer drops even further, they are reduced to living on a gruel of water and rice (*barka*), occasionally supplemented by a species of weed as a vegetable. The Koli shopkeeper in the Dubla quarter sells many articles in smaller quantities than the normal weights: a pinch of tea, a handful of chillies, a small bag of salt, etc. Most households are in his debt to an amount

26. *Report of the Hali Labour Enquiry Committee*, 40.

of 3 or 4 rupees, and the shopkeeper does not allow them credit beyond that amount. Accounts are settled when the wages rise at harvest time.

The miserable living conditions and the undernourishment result, of course, in bad health. More or less chronic illness is a familiar and accepted phenomenon in every household. A widespread high consumption of home-distilled alcohol completes the picture of pauperism. Equally typical of the situation is the view of the landlords that alcoholism is the cause of the Dublas' poverty instead of an escape from daily misery. In the 1950 report cited above, the term "subhuman standard of living" was used, a description as valid today as when it was written.

The Dublas have seen the Anavils' prosperity rise rapidly in the last few years and compare their own deplorable conditions with the landlords' comfortable way of life. Most of them feel that their situation has deteriorated within the past few decades. It is true that formerly they lived in the gardens of the Anavils and were attached for life, but the masters shared the effort and remunerated them more liberally in those days. They were part of the landlord's household and their living was secure. The daily allowance was paid in grain, and the prices of foodstuffs and other articles that had to be bought were much lower. The picture they gave me of the earlier situation is certainly too optimistic, for it took no account of arbitrariness and compulsion. On the other hand, a more miserable existence than that led today by, especially, the agricultural laborers among the Dublas is inconceivable. Aged farm servants who can no longer work are dismissed without ceremony. They can only fall back on the support of any children or relatives who are able or prepared to give it, but these find it difficult enough even to keep their own heads above water. In such a situation of extreme scarcity, mutual assistance even among members of the family is weakly developed, and the prevailing hardness between near relatives accentuates the hopelessness of existence.

The low standard of living of the Dublas in Gandevigam is closely connected with the disproportionate growth of their numbers. All informants pointed to the high increase in the number of Dubla huts since the previous generation. So far as I could ascertain, this demographic development is due chiefly to a migration surplus and falling death rates, the latter factor being undoubtedly the more important. Famines and epidemics, to which many formerly fell victim everywhere in India, especially among the castes of agricultural laborers, have become fairly exceptional. The expansion of agrarian employment in Gandevigam has not kept pace with the rapid rise in the number of landless laborers. On the contrary, the crop system has become less labor-intensive, and the changeover to pump irrigation in the gardens has reinforced that tendency. Nor has it been counteracted by the fact that the Anavils themselves no longer work with their servants, and that few Kolis are now prepared to hire themselves out as day-laborers for 8 or 12 annas.

Diminished demand and increased supply of labor has contributed greatly to the weakening of the position of the agricultural laborers. One of the consequences has been the rise of a category of day-laborers. This labor reserve has lessened the necessity for Anavils to safeguard themselves against shortage. The number of farm servants in the employ of landlords is no longer adjusted to the maximal demand of labor at a few peaks of the agrarian cycle. Now that the risk of temporary scarcity has lessened, the investment and maintenance costs of a farm servant, converted into money as they are, are weighed against other priorities.

Yet attachment is by no means disappearing. The distribution of the work throughout the year is such that the Anavils still find it desirable to maintain a permanent nucleus of labor, the more so since most of them no longer work the land themselves. Appointing one or more Dublas in permanent service is, in the long run, easier and cheaper than having to call in day-laborers every time.

Nor have the Dublas found employment outside agriculture for their growing numbers. As I said, the landlords oppose it, because a cheap and abundant labor reserve is more to their interest. Apart from that, alternative employment is hard to get, because of the low level of schooling among the members of this caste. They are considered only for the humblest and lowest-paid jobs, and this will be true for a long time to come. Only a handful of Dubla children attend the village school fairly regularly and for some consecutive years, even though education is compulsory. The school has registered a larger number, but checks and inquiries made it clear that most of the children are rarely if ever present. Whereas the Kolis avail themselves of education as an avenue for strengthening their social and economic position within and outside the village community, most of the Dublas are not much concerned about the schooling of their children. The indifference of the older generation, however, is not the chief reason for the prevailing illiteracy among the younger generation. In the first place, older children must stay home to look after their younger brothers and sisters when both parents are working. Secondly, a number of Dubla boys, who formerly tended the cattle of the Anavils, are now *govalia* for their own father. They are charged with the care of the cow or buffalo which the landlord has farmed out with their father. Finally, the social climate is scarcely conducive to regular attendance of a large number of Dubla children.

Settlement of the Dublas in a village quarter of their own has somewhat lessened their dependence on the landlord. This applies primarily to those who work outside Gandevigam. They are the ones who most strongly resist the Anavil yoke under which their fellow caste members are laboring and who say that they themselves would refuse to work for 8 annas. Yet they fear being driven back into agriculture and thus, finally, into their former subjugation. On account of their vulnerability they dare not risk an open conflict with the power elite. Nevertheless, behind the scenes it is they who fan the discontent that prevails in their caste, who express it, and whose voice is persuasive.

Among the Dublas, the awareness of their subjugation has grown. They are beginning to feel that they should shake it off and cease to concede the superiority of the landlords. Yet the weak economic position in which they find themselves perpetuates their dependence and perhaps aggravates the subjugation by its depersonalized character.

The situation of the other Dublas does not fundamentally differ from that of the farm servants. Although there is some differentiation among the caste members, it rests mainly on differing extents of poverty and attachment, on different forms of dependence.

THE ANAVIL BRAHMANS. Within the past few decades, the Anavil Brahmans of Gandevigam have become landlords. This term does not refer to the quantity of land they possess — which has probably not changed very much, taking the caste as a whole — but more particularly to the degree to which they themselves are physically active on the land. The drudges of former times have developed into gentleman farmers who rarely if ever work the land. It should be added that this does not apply to all the members of the dominant caste. Of the twenty-six Anavil households in the village, seventeen employ a total of forty-three servants, and the remaining nine households have to get along without laborers in permanent service. Three of these landowners do not work at all. They subsist on the proceeds of their neglected gardens or small orchards, and leave the annual picking to a factor who has bought the crop long in advance. Their meager income is supplemented by the incidental allowances from relatives outside the village.

The other six households own too little land to be able to afford the convenience of a servant. Daily work in the fields and gardens is done by the male members of these households. Only when they cannot cope do they call in Dublas or Kolis for the least-favored jobs, notably plowing. Usually such temporary servants are day-laborers, but sometimes they are farm servants who have been given a few days' leave by their masters. The Anavils of this group are hard-working farmers who still grow the traditional garden crops, partly because they are more profitable than fruit. They strive to free their sons from agricultural labor when the time comes. Their poverty is apparent not only from the absence of permanent help on their land, but also from such evidence as a shabby house or the bachelor life of one or more brothers. They remain in the background, and some of them live more or less hidden, deep in the gardens. Although the older Anavils made no secret of their presence they told me with some emphasis that the way of life of these simple farmers was not what might be expected of Anavils.

It is an existence that most of their fellow caste members have happily left behind them. They have attained prosperity and surrounded themselves with a comfort designed to demonstrate their living as landlords. For miles around, the Anavil Brahmans of the garden region enjoy the reputation of possessing two houses, an unbelievable luxury amidst poverty and scarcity. The largest and most recently built garden houses cost from 15,000 to 30,000 rupees, and the wealth of their owners is also apparent from the way the houses are furnished. In contrast to their sober dress of former years the most prominent Anavils now wear immaculate white *dhotis;* they no longer have stubbly beards but are clean-shaven, thanks to their electric shavers.

By avoiding agricultural labor themselves, the members of the dominant caste have accentuated the social function of servitude. Light work is allowed, such as sheaving or mango-sorting, but the

Anavils apply themselves chiefly to the organizational aspects of their undertaking. They distribute the labor tasks and supervise the progress of the various activities. Their old people tell "when I was young" stories and grumble about the boundless laziness of the young. They still have memories of the privations of the nineteen-thirties, and reproach their sons for playing cricket rather than helping in the garden. Nevertheless the older generation, too, accepts with visible pleasure both the newly acquired wealth and the convenient way of life it has brought. Their social need for servants has increased to such an extent that, in spite of the changeover to mango cultivation, which requires less intensive labor, it is doubtful whether the number of agricultural laborers in permanent service has really diminished. Perhaps it is only in a comparative sense that servitude has become less important. For the Anavils, the cultivation of fruit trees has satisfactorily reconciled their two divergent requirements: the highest possible income and the least possible effort. They look with disfavor upon some Kolis who have followed their example on a small scale, but when a Dubla planted a small tree in front of his own hut in the new quarter, this was regarded as intolerable, so some Anavil boys pulled it up.

Between the Anavils of Chikhligam and those of Gandevigam there is one great difference. In the former village the members of the dominant caste have removed themselves, as a rule, much farther from agriculture. Many of them are employed in urban professions, a development that began even in the previous generation. For the Anavils of the garden village, on the other hand, land property has remained their chief source of income, hence they are considerably more active in it. Such complete idleness as in the *jarayat* village is not often encountered here. The landlords of Gandevigam are much more frequently in their gardens or fields, where they supervise the work of their subordinates. Their greater commitment to agriculture is apparent, for instance, in the recent introduction of *chiku* cultivation. This fruit tree requires more care (irrigation and labor) than mango, which for their fellow caste members in Chikhligam is a reason for not taking it up despite the attractive market prices. The Anavils of the *bagayat* region try to avoid activities that would label them as working farmers, but they have no objection to directing a farming enterprise. This means, for instance, attending with great concentration to the daily schedule, watching the cost sharply, and continually checking the progress of the undertakings. In other words, they lay stress on their managerial role.

I have described in detail the consequences of this increasingly depersonalized way of management for the master's relationship to his agricultural laborers. It should be noted, too, that the relations among the Anavils themselves tend to become more business-like as well. I was told more than once that the old hospitality that marked their social intercourse has disappeared. They themselves explain their preference for living in their gardens by their wish to remain secluded as much as possible, or at any rate to limit contact to what is strictly necessary. The cordial and spontaneous aid that formerly would have characterized their mutual relations has been replaced by an unmistakable aloofness. The members of the dominent caste are always prepared to vilify each other, and with no hesitation they accuse their neighbors of greed or denounce them as liars. When still in Chikhligam I had been warned that the Anavils in Gandevigam "do not collaborate well with each other." Jealousy and spite aroused by newly acquired wealth or esteem lead to all kinds of personal and family quarrels — about land, about laborers, about taking unripe fruits to the cooperative, and so on. The pejorative stereotypes current among outsiders about the dominant caste are also used by its members, but only in relation to other households than their own.

Nearly all of the Anavils of Gandevigam have retained their interest in agriculture, even though most of them no longer work the land. Notwithstanding the vicinity of some urban centers and the excellent communications, only a few of them are employed outside the village. Their fellow caste members in more remote and isolated Chikhligam are much more urban-oriented. What is the cause of this difference?

In the first place, the Anavils of the *bagayat* region belong to the section of the Bhathelas who have always led a peasant existence. In villages where the dominant caste is represented by Desais there was an early tendency towards separation from agriculture. Education took root and developed much earlier in the *jarayat* villages precisely for this reason. A school education facilitated escape from the agrarian environment and was a condition for obtaining one of the coveted positions in the rapidly growing framework of local government.

Because they are still so much concerned with agriculture, the

Anavil Brahmans of the garden villages are held in lower esteem within the caste than are those of the *jarayat* villages. In Chikhligam I was given to understand that I should not go to Gandevigam if I wished to know how Anavils live. Gandevigam was portrayed as a real peasant village populated by Bhathelas.

The urban orientation of the dominant caste has been greatly influenced by the Desai behavior pattern, and the Anavils of the *bagayat* region showed, as it were, a delayed reaction. The time lag, however, is not the only reason why they are as yet so rarely employed in nonagrarian occupations. The profitability of agriculture during the past few decades has undoubtedly been another factor. On the basis of categories which I supplied, three Anavil informants classified their fellow caste members in the village as follows: eight rich; seven well-to-do; seven moderately well off; four poor. Only for the first category had I mentioned a demarcation line, an annual income exceeding 10,000 rupees. In the year preceding my stay the two largest garden owners had earned 20,000 rupees by the produce of their fruit trees alone. The prices of *bagayat* plots have soared, but land is rarely on the market. An Anavil who had been classed as moderately well-to-do told me that he would not sell his garden even for 50,000 rupees. The owners cling to their land because it enables them to have a high income, not simply through pride of possession.

Despite the Bhathelas' adjustment to the style of life of the Desais, their status within the caste has not greatly improved. As mentioned earlier, a number of the traditional standards of prestige have lost significance. A landlord's way of life was the model for the previous generation, but now the general preference, as in Chikhligam, is for residence in town and an existence outside agriculture. It is clear that the Anavils of Gandevigam do not conform to this standard. The high income they draw from agriculture can only partly modify the negative evaluation of their rural way of life. They are now endeavoring to follow the example of their more "advanced" fellow caste members in their orientation towards the town. It was therefore to be expected that I found among the younger generation the same contempt for agriculture and village life as elsewhere. "For farming you need no brains; it is dull and stupid work" was the way in which the educated son of a landowner summarized the attitude of his age group.

After secondary school, many Anavil boys, and increasing numbers of girls, go to college in the neighboring town to try for a B.A. or a B.Sc., the threshold to the middle cadre of government and industry. The richest landlord of the village plans to send his son to the U.S.A. for a year to pursue his studies there, which will cost him about 15,000 rupees. After his return, the son will be able to ask a higher dowry and be in line for a relatively well-paid job in town.

The growth of the number of Anavil Brahmans, though not nearly so rapid as that of the lower castes, stimulates departure from the village. By enabling one or more of his sons to take up employment outside agriculture, a father prevents weakening of the family's economic position in the next generation by subdivision of the land. The son who leaves agriculture thinks himself fortunate, whereas the one who stays behind feels injured.

A Bhathela background is undoubtedly still an important factor in the evaluation of members of the dominant caste in south Gujarat. A humble origin, however, may be partly or wholly redeemed by a number of more recent, chiefly individual, prestige criteria, such as income, level of education, and a nonagrarian occupation. Owing to the new scale of values the distinction between the two sections is blurred, certainly for outsiders, and the common features are stressed.[27] A common Anavil life style has gradually come into being, inspired, indeed, by the behavior pattern of the traditional notables, but also shaped by the adoption of various elements from the new urban and secular culture. It is abundantly clear that the Bhathelas of the garden villages are rapidly making up their disadvantage. The nine members of this caste who work outside the village all belong to the younger generation.

If the present evolution continues, will not all members of the dominant caste in the end have left the village community and taken up urban professions? That remains to be seen. Economic success has become a very important determinant for social status. The Anavils of Gandevigam, after all, attained prosperity through agriculture. As already stated, I am convinced that intensification of crop cultivation would yield even higher revenues. Yet even a rather inactive management provides the landlord with an income well above that earned by most of his fellow caste members in town. In other words, agriculture is so extremely profitable that it cannot simply be abandoned. By its revenues, moreover, it enables

27. Cf. Joshi, *op. cit.*, 108–10.

the landlords to realize their influence and prestige in both the "traditional" and the "modern" way — in buying, as it were, esteem for themselves by means of suitable marriage arrangements as well as in equaling their urban fellow caste members in education and material comfort.

A continuing transformation of the fields into mango orchards, and informal sharecropping arrangements with reliable Kolis and Dublas for the exploitation of the remaining land, will enable the *bagayat* Anavils to continue to draw income from their agrarian possessions. It is the perfect way out of the dilemma posed by their desire to concern themselves as little as possible with agriculture and at the same time to enjoy as high an income as possible. The most prosperous among them are considering whether to follow the example of one of their number and appoint a Koli as supervisor, which would accentuate their freedom from this humiliating labor.

Another factor of importance is that the garden villages have become much more agreeable to live in. Improved road connections have brought the towns in the vicinity within easy reach. More and more Anavils have their own means of transport; many have bicycles, and a few rich landlords of neighboring towns can afford an automobile. Housing in the villages has become much more comfortable and is, moreover, often cheaper than in town. Facilities, too have been improved; the installation of electricity and water has especially benefited the members of the dominant caste.

Those who combine a job in town with the management of their property have completely cast off the despised peasant existence. To this category in Gandevigam belong a number of Anavils who are primary or secondary schoolteachers and in their leisure hours keep a watchful eye on their agrarian interests. Their fellow caste members in the surrounding villages who daily go by bicycle or bus to town, where they have administrative jobs, are in the same position. To others they are shining examples. While profiting by the returns from agriculture they succeed in drastically limiting their dependence on it.

THE RELATIONSHIP BETWEEN ANAVILS AND DUBLAS. I could not have chosen a more desirable period for my stay in Gandevigam than the mango season. In the Anavil street, where I had been allocated an empty house, I passed many evenings in hospitable company. They were pleasant gatherings, usually on the veranda of the house of my neighbor, a large landowner. At a gesture from him, his servant, a Dubla boy, brought in a large basket of mangoes. These were the windfalls, the bruised or overripe fruits of the day. With some other neighbors, regular visitors like myself, we ate as much as we wanted. The servant stood in the background awaiting further orders. He would drag in another basket if the gentlemen had not had enough, and after we finished he would pick up the pits we had thrown down. Insofar as the servants are still allowed to take mangoes home at harvest time, the quantities are certainly smaller than they used to be. The servants, however, cannot afford to eat them, and sell them to a buyer from town who waits at the end of the day just outside the village.

The mango season largely coincides with the period in which most marriages are performed, and during the months I spent in Gandevigam I attended many weddings. In the most prosperous Anavil fashion these weddings are lavish occasions celebrated with a great deal of pomp and circumstance. In the marriage ceremony, which expresses all kinds of relations of reciprocity, superordination, and subordination, Dublas also take part. The contribution of servants and maids of the household consists, for instance in carrying the pot with earth from the bride's native village, and holding the jars from which water is sprinkled. These are symbolic of services that require physical effort, and the Anavils, having ensured the assistance of Dublas in their household, can demonstratively and to their greater glory abstain from them. Often many hundreds of guests partake of the marriage meal, but members of the caste of agricultural laborers are of course not included. The Dublas stood looking on from a corner until after the meal, when they could take the leftovers, chasing away the stray dogs of the village.

The foregoing paragraphs illustrate once again the contrast between the plentitude in which the landlords live and the want to which the landless laborers are reduced. Moreover, the two situations express the nature of the social relation between the two castes: the subordination of the Dublas to the Anavils. This applies in the first place to the farm servants, a numerous category in the garden village, but in a wider sense to all Dublas.

After the disappearance of the *hali* system the Dublas' subservience continued to exist, though no longer as a goal in itself.

That is to say, attachment served to maximize the income of the landlords, while the laborers were forced to work harder and they were used more effectively, at the same time receiving minimal remuneration. The Dublas' standard of living has fallen accordingly.

The Dublas never tire of discussing the egotism and greed of their employers. They are convinced, not incorrectly, that it is due to their effort that the members of the dominant caste can lead lives of luxury and idleness.

Conversely, the Anavils believe they are the ones who have reason to complain. In their eyes, the Dublas have changed from faithful, industrious, and obedient subordinates to recalcitrant good-for-nothings, dishonest and underhanded in their dealings with superiors. To the Anavils' displeasure, the Dublas stay away for days without notice, or work in secret for a Koli who pays them a little more. A recurrent scene in the village is that of an irate and scolding master on his way to fetch a servant for his daily job.

They frequently accuse the Dublas of theft, remarking wryly that the agricultural laborers work harder in their masters' fields during the night than they do by day. The neighboring towns contain some merchants who are known as receivers of stolen vegetables and fruit. By various means the landowners try to put an end to these practices. Some inspect the small bundles of the laborers when they leave the land in the evenings, many make nocturnal tours of their gardens. The large landowners recently hired a field watchman from another region during the harvest period of the chief crops, but this measure turned out to be rather ineffective. At one time during my stay there was a great commotion when it came to light that one of these Bayias was in league with a number of Dublas and an Anavil from the adjoining village.

The foregoing observations reflect clearly the fact that the judgment of the Anavils about the Dublas is in general negative. Agricultural laborers, they say, have the minds of animals — they take no thought for the future, are extremely uncivilized, have dirty habits, are stupid, and are unpredictable in their behavior.

On the other hand, several members of the dominant caste are aware that this judgment is severe and unfair. One of them remarked that, although the Dublas' alcoholism is always mentioned, the consumption of alcohol among the Kolis is certainly not less widespread. They are aware that they exploit the weak position of the agricultural laborers. These Anavils regret "the decline of the moral standard" during the last few decades. In their opinion, the result has been that the landlords think in terms of money alone, and that the agricultural laborers, for their part, do not wish to work. The decline began as early as the nineteen-forties. One Anavil told me that during the war period a loyal servant used to hand over his sugar-ration card of his own accord (the servant did not eat sugar anyway, he added). If the servant did not, however, and preferred to sell it in the black market, the master took it from him. That was wrong, just as, in his opinion, it would be wrong for the master not to give the servant any food, if it suited his convenience. The same applies to a landlord's unwillingness to give *khavati* and his refusal to help in case of illness.

Some Anavils frankly admit that they treat their laborers badly. "We look after our cattle better," one of them literally said. A landlord who conducts himself as a patron is praised by his fellow caste members. He should not belong too unmistakably to the type of working farmer, nor be too industrious and thrifty. A natural ascendancy over members of lower castes, combined with a gracious familiarity towards subordinates, is generally appreciated. Greed and brutishness are unworthy of an Anavil, as they are inconsistent with a landlord's pattern of behavior. Physical punishment belongs to the past, although a furious master may be excused if he kicks his servant now and then, and not all Dubla boys working in a household are spared an occasional beating. The landlord whose laborers do not run away wins admiration. A grasping, short-tempered master, on the other hand, makes himself ridiculous, for he has lost his ascendancy. "Good" Anavils therefore have no trouble at all in finding new servants. Their benevolent and accommodating behavior, however, is not due to compassion for the Dublas' predicament, but is practised as one of the virtues of a dominant caste to validate its supremacy. Lest I should be misunderstood, this is the recommended behavior. Reality is usually different. All Anavils wish to have cheap, hard-working labor, and most of them are indifferent towards the conditions under which the laborers live.

The landlords no longer even observe the unwritten rule of allotting the work as much as possible to Dublas from the village. At the peak of the season they hire labor from another subdistrict. These *biladia* are contracted as gangs for one season; they find their own food and build temporary huts in the fields and gardens. Another advantage in employing them is that because of their unfamiliarity with the local situation, they have no outlets for stolen fruits or vegetables. By making use of temporary labor in this way the Anavils show that they do not wish to accept any obligations over and above remuneration of the work performance. Employing casual labor on a daily basis relieves them of a great deal of "unnecessary trouble," but on the other hand weakens their grip on the Dublas.

The landless laborers have found, in their growing mutual alliance, a new means of defense against the Anavils. The Anavils themselves say that the Dublas, living together in the new quarter, have developed a "group instinct." The laborers left the gardens about ten years ago and bought pieces of land, just outside the village, on which to build their huts. Each of them paid the sum of 6 rupees and 3 annas, which went into the village exchequer. Because of the lapse of time, it was no longer possible for me to find out who had initiated the move. At any rate, it was not done against the will of the Anavils. Some of them feared they would lose authority over their subordinates, but most hoped that there would be fewer thefts of garden crops. The former expectation was justified, the latter was not. The Dublas have become a little less susceptible to the sanctions of the Anavils. The scattered location of the huts in the gardens had reinforced the Anavils' grip on their laborers. Now that the Dublas have settled together in the new quarter, their solidarity has increased. In the evenings after work they assemble in groups to talk, and inevitably the information they exchange fans the flame of existing feelings of dissatisfaction. The Anavils now not only have to reckon with the individual resistance of a subordinate, but in the last resort they may have to deal with the whole caste. At present, group control is primarily inward-directed. A man who has given in too easily to a superior's wishes is criticized by the other Dublas, sometimes even censured. It is, for instance, less easy now for the landlord to have sexual relations with the wife of an agricultural laborer. In the first place this is made difficult by the grouping of the Dublas' living quarters. Then, too, the farm servants themselves are under pressure from their own group to refuse to accept this behavior on the part of the master. Although I did hear of a few cases of heterosexual or homosexual relationship with subordinates, they appear to occur less frequently nowadays, and criticism of such relationships has become stronger.[28]

A quarrel between individual Anavils and Dublas may develop more easily than before into a larger conflict. When, a few years ago, some agricultural laborers were accused of attempted theft and were beaten by Anavils, a protest meeting was called in the Dubla quarter on the following day. A leader from the Praja Socialist Party in a neighboring town was invited to speak. It was not much more than a symbolic gesture of opposition to the unlimited authority of the landlords, and indeed was understood as such by the latter.

Another recent phenomenon is that the Dublas may go on strike. Some years ago the farm servants of Gandevigam stopped work for a few hours when they were denied the right to pick wood in the gardens and the landlords refused to raise their wages in compensation. This action failed. The Anavils fetched laborers from other villages and, to demonstrate their firmness, went to work themselves.

It may be concluded from the above-mentioned instances that protest has in some cases become an expression of group behavior. Does this mean that Anavil Brahmans as a group have become opposed to Dublas in general and that there is question of progressive intercaste antagonism? Joshi, in his study of a village of about the same composition in south Gujarat, does arrive at that conclusion. In my view, however, it should not immediately be concluded from the increasing antagonism between landlords and agricultural laborers that one or both parties present a solid front. I have described in great detail the continuing subservience of the Dublas in Gandevigam. Those who work outside the village are exceptions. Half the Dubla men are bound as farm servants. The dependence inherent in this relationship of service is utilized by them to obtain various benefits, amounting to credit in the broadest sense of the word: incidental aid, cattle to rear, land for sharecropping, a job for a son or brother in town, intercession with a third party. By granting favors of different kinds, or by holding out the prospect of favors, the members of the dominant caste maintain a system of individual alliances which at the same time keeps the Dublas divided among themselves. Day-laborers may be

28. Joshi arrives at a similar conclusion. (*Ibid.*, 107–08.)

absorbed into this network when they have taken up an advance and become indebted. The Dublas of the village compete with each other for the places and privileges which the Anavils have to distribute. Divergent, and partly conflicting, interests prevent the rise of a widely shared feeling of solidarity.

Unity no more exists among the members of the dominant caste than among the Dublas. The lack of concord among those in Gandevigam had been pointed out to me in advance by the Anavils of Chikhligam. This does not mean, as I thought it did, that the Anavil group in the garden village has broken up into opposing factions. Characteristic of them is rather a pronounced individualism. They oppose each other through indifference rather than because of willful contentiousness. Several Anavils no longer observe the unwritten rule of not taking on each others' agricultural laborers unless permission has been explicitly given. Some less prosperous masters have recently lost their farm servants to a larger landowner, who did not even compensate them for the Dublas' debts. "What business is it of mine?" they say. They are unmoved if their fellow caste members call them "thieves of other people's servants." No one, however, pushes matters to extremes. The garden owners look after their interests outside the village individually. Gandevigam is not an arena in which they fight out their differences of opinion. In general they pay little attention to each other's comings and goings, or rather, they reject undue efforts to influence their own acts and behavior.

The relations between Anavils and Dublas no longer have the patronage character of *halipratha.* On the other hand, the network of individual alliances controlled by the landlords has prevented the development of horizontal solidarity among the agricultural laborers. Although the latter are beginning to show some unity, no indissoluble solidarity or class consciousness has yet come into being.

The Dublas lack the power to improve their situation themselves. Resistance is limited to short-lived actions, and unity of purpose is confined to at most one village. As agricultural laborers they have hardly been touched by political activity. The parties of the left have thus far mainly directed their efforts towards the urban scene and, in the rural area of south Gujart, chiefly to the category of small farmers. At elections, the Dublas in many villages vote for the same political party as the landlords, usually the Congress

Party. Sometimes this reflects direct pressure exerted on subordinates, but often the mechanism works in a subtler way.

Key figures in the countryside are the so-called social workers, mostly members of the higher castes who, inspired by a mixture of social awareness and political ambition, stand up for the underprivileged and backward groups. Associating themselves with a social movement, usually of a Gandhian type, they are of interest because they control a bloc of votes in local political parties. Many of them are affiliated with the Congress Party, long the most important political entity in the area. Thus, although landlords and agricultural laborers often vote for the same party, they do so from dissimilar backgrounds and for different reasons. The "social workers" make themselves mediators in local disputes between the Anavils and Dublas. They urge the former to treat their subordinates better, and warn the latter against the dangers of a too militant attitude. When, despite their efforts, a conflict flares up, the mediators rarely take the side of those whose rights they claim to champion. Having learned by experience, the Dublas have little confidence in them. On the other hand, the "social workers" are among the few channels through which agricultural laborers can obtain a hearing in the higher realms of the state bureaucracy, while, in the Dublas' name, they maintain contacts with those government bodies that allocate favors. Four Dublas, who regard themselves as spokesmen for their caste in Gandevigam, have successfully used these channels and received subsidies through a "social worker" (an Anavil in a neighboring village) to build larger huts of better material, comparable to those in Chikhligam.

The government does not evince much interest in the condition of the agricutural laborers. Although a number of existing regulations bar arbitrary acts by landlords, this passive support rarely changes into active concern and aid for the Dublas. The government has not been able to guarantee the assistance and support which used to be institutionalized in the *hali* system. Employment is limited, and social security is lacking. The only progress that has been made in south Gujarat is in the housing of the Dublas on land of their own.

What little the government does, however, draws the reproaches of the Anavils, who would prefer the authorities to prohibit the seeking of work elsewhere or to forbid strikes. Either directly or through political channels the members of the dominant caste have access to the local government bureaucracy, and they know how to manage it. Officials visiting the village meet Kolis only rarely and Dublas in the area not at all. This makes it clear why "depatronization" has placed the Dublas in an even more isolated position.

The rapid numerical growth of the Dublas, their powerlessness, the contraction of the agricultural economy, the limited employment outside it, certainly for them — all these factors point to progressive pauperization. The agricultural laborers react by passive resistance, the chief means of protest they possess. Their behavior is characterized only in part by docility and servility. They do not openly refuse to carry out the landlords' orders, but sabotage them. They do not want to work hard for a minimal daily wage, but will not say so openly. Their servility is exaggerated, their obedience is too demonstrative not to arouse suspicion. It is the impertinence of the powerless. Their behavior has all the features of what has been called a culture of repression.[29]

The conviction that things were better in the past does not imply that the majority of the Dublas yearn for the shelter and security of former times. To have these meant at the same time to admit and accept a far-reaching inferiority. They are all too aware that their low position in the social structure is closely connected with the dependence in which they have always found themselves. This subjugation, which detracts from the feeling of self-respect that has been growing among them, not only affects the farm servants but includes all the members of this tribal caste. Their continuing servitude fills the Dublas with a veiled but profound hatred against their lords and masters, the Anavil Brahmans.

29. Holmberg, 1967.

# Chapter 9

# Summary and Conclusions

## 9.1 Chikhligam and Gandevigam Compared

These two villages differ greatly as to the share of the population engaged in agriculture. In Chikhligam, for some decades, increasing numbers of both Anavil Brahmans and Dublas have sought and found work outside the village. For the Anavils, nonagrarian employment meant departure for urban centers both within and outside south Gujarat. Most Dublas, on the other hand, leave the village only temporarily and return every year, from necessity, before the beginning of the monsoon. Agriculture has remained the main source of income for the large majority of Gandevigam's population. Most members of the castes of Anavils and Dublas here are still landlords and agricultural laborers, respectively.

In the garden village, Dublas are mostly employed as farm servants. This category is also represented in Chikhligam, though in smaller number, and the day-laborers of both villages, who work all year in agriculture, form a minority among the agricultural laborers. The villages therefore differ mainly in the number of laborers engaged in agriculture throughout the year, not primarily in the nature of the prevailing agrarian labor relationship. However, owing to the seasonal migration of the majority of Chikhligam's Dublas, attached labor in that village is characterized by a less lasting relationship, and the Anavil Brahmans' control of their farm servants is less effective than in Gandevigam.

An important factor in explanation of the difference in agrarian employment is soil fertility. Because of the poor soil quality and indifferent use of irrigation, much of the village land in Chikhligam is not cultivated intensively, and agricultural labor is in fact confined to a few months of the year. This has greatly reinforced the urban orientation of the members of the dominant caste in the village. Partly because of the contraction of the crop system, a process that is far more advanced in Chikhligam than in Gandevigam, agrarian production is much lower than it could be if available labor and capital were fully utilized. The difference in extent of involvement in agriculture of the Anavils of Chikhligam and those of Gandevigam — also traceable to the time lag in the efforts of the Bhathelas of the garden village to attain the life style of the leading section of their caste — is reflected in the labor intensity of the agrarian cycle and in the number of Dublas who succeed in finding a livelihood outside the village. Alternative employment is, apparently, not an independent variable, and the vicinity of towns is not decisive for its realization. Inasmuch as, owing to its very profitable character, agriculture has remained the chief preoccupation of the Anavils of Gandevigam, they use their influence on the Dublas at times to obstruct any efforts of their laborers to find work outside the village. In Chikhligam, on the other hand, where the members of the dominant caste have changed their way of life, prefer urban professions, and are no longer actively engaged in agriculture, there was an early lack of employment for agricultural laborers, and, consequently, many of the Dublas were pushed out of the local economy.

The changed attitude of the Anavils resulted in depersonalization of their relationships with the agricultural laborers, who nevertheless did not become any less dependent. The economic distance between the two castes as a whole increased greatly. Differences in the prosperity of Anavils are more conspicuous than differences in the poverty of Dublas. In this latter caste, differentiation has hardly begun, whereas it is progressive among Dhodias and Kolis. The rapid demographic growth of the Dublas has aggravated their social insecurity and thus contributed to the continuance of the landlords' control of their actions. Although they prefer not to be attached, more Dublas are day-laborers from necessity than from conviction. Only those Anavils who are known for exceptional rudeness, unreliability, and greed have trouble finding new farm servants. Solidarity among the Dublas increased after they had moved to a village quarter of their own, but as yet there is no question of a common front against the local power elite; nor is such a development systematically encouraged by outside parties. Thus far, resistance has been incidental and limited to individual cases. In these respects the difference between the two villages is only one of degree. Although less weight is attached to agriculture in Chikhligam than in Gandevigam, the relations in both villages between

Anavil landlords and Dubla agricultural laborers are organized more or less on the same lines.

### *9.2 Parties to a Process of Depatronization*

The disintegration of the *hali* system marked the disappearance of patronage. The possibility of selling an increasing proportion of the agrarian output on the market led to commercialization of the relationships between landlords and agricultural laborers. The allowance paid to Dublas, which in the subsistence economy had been based on various "obligations" towards the landlord, increasingly took on the character of a wage payment in exchange for a labor performance in the market economy, and it was paid in money. The amount was no longer related to the minimally defined needs of the agricultural laborers, but was based on the supply of labor and on the control exerted over them by the landlords.

The patron-client relationship has not disappeared in all respects, but the locus of patronage has shifted to a supralocal level and is framed in a more limited context, being chiefly economic or political in nature. The network of brokerage in the new situation, the parties in it, and its character are subjects which are outside the scope of this study. I have related "depatronization" to an enlargement of scale which put an end to the closed nature of the traditional social and economic order, but both the course of this process and its results were determined to a large extent by the reactions of the parties involved towards these outside influences.

THE LANDLORDS. There is no doubt that the landlords desired and promoted the disruption of the *hali* system. The change in the scale of preferences of the members of the dominant caste — high investments in education and in a consumption pattern adopted from the urban culture — was to the detriment of their subordinates. The changeover to market production made possible alternative ways of spending the crop profits and diminished the superiors' social responsibility towards their servants. Many Anavils seized the opportunity of cultivating fruit trees so as to free themselves from agricultural labor. Even without employing farm servants they can now lead a landlord's life as behooves a member of the dominant caste. The social gap between the Desai and Bhathela sections, which used to be extremely important, has narrowed as their economic positions have become comparable, with the consequence that their life styles are more alike. Urban orientation has enhanced the feeling of unity within the caste. Whereas, in 1921, 785 out of every 1,000 Anavil Brahmans were still landlords,[1] this ratio has drastically changed. Concerning the members of this caste, I. P. Desai remarks:

They are now convinced that not agriculture but salaried employment or occupations . . . are the means of livelihood. In their evaluation of occupation, "service" comes first and probably agriculture has no place. (Education is, for the Anavils, the suitable preparation for life in the city.) Now it is fixed from the day the boy goes to school that he has to be a salaried employee and they think of what he has to be and think of what they call the "line" for the boy. Education is becoming a normal necessity and so much so that they are sending their girls also to the secondary school. Some girls every day walk four miles to the school and back to the village. Even for a girl matriculation is becoming the minimum requirement.[2]

Their integration in the urban environment has given importance to other standards of values than the traditional ones, especially as to education, income, and a profession outside agriculture. The village community is now only in part the framework in which the position of the Anavils is determined. The dependence of members of other castes is no longer an aim in itself, but has become a means to enhance the masters' own prosperity and to finance preparation of their children for nonagrarian professions. Ostentatious leisure is now possible without a landlord's life, and the desire for conspicuous consumption is satisfied by the display of riches. Even where the landlords are still actively concerned in agriculture, the grip of the members of the dominant caste on the Dublas has become more depersonalized, being dictated by income maximization. The agricultural laborers' right to work is no longer recognized, their social security no longer guaranteed. In the endeavor of the members of the dominant castes to attain more esteem and influence within and outside the village, the Dublas have been changed from subjects into objects.

THE AGRICULTURAL LABORERS. Descriptions of the disappearance of the *hali* system often create the impression that it was the agricultural laborers who initiated it — that when alternative em-

1. *Census of India*, 1921, vol. XVII, Baroda, pt. 1, 363.
2. I. P. Desai, 1966, 81.

ployment arose, increasing numbers of them severed the relationship of servitude. My data confirm this assumption only to a limited extent. The need of the local landlords determines whether any Dublas succeed in finding work outside the village, and if so, what proportion of them. Once a migration tradition has arisen, as in Chikhligam, this changes. Contact with alternative sources of income in the cities then takes place through the Dublas' own mediators, the *mukadams*, and is no longer controlled by the Anavils. Even then, escape is often only temporary, because of the unskilled, low-paid, and unstable character of urban employment that is available to Dublas. The termination of the traditional service relationship may also be connected with the Dublas in another way. It might be argued that the rapid growth of their number makes it impossible for the landlords to provide for them all. This fact undoubtedly reinforced the effect of "depatronization," but it did not cause it. The agricultural economy of south Gujarat could have been much more labor-intensive if the crop system had been different. The abundant supply of labor, in other words, is not traceable simply to the rapid demographic growth of the population of agricultural laborers, but is also connected with the preference of the dominant landowners for pursuing a kind of agriculture that is not highly intensive.

The elements of bondage that still exist are especially apparent in attached labor. Weber, who himself studied a similar shift in the relationship between landlords and agricultural laborers in eastern Germany in the late nineteenth century, found that farm servants in their turn began to resist their continuing servitude.[3] The same applies to the Dublas in south Gujarat. A fundamental difference exists, however, in that the rise of alternative employment in India has been much slower, whereas the population is increasing much more rapidly than it did in Germany in the late nineteenth century.

Although indeed aspirations have been aroused among the Dublas that are incompatible with the bondage inherent in attachment, with dependence, and with subordination in general, it seems that the members of that caste were above all subjected to the process of depatronization. In more than one respect they were the losing party.

3. Weber, 1892, 797–98.

EXTERNAL INFLUENCES. Factors other than those mentioned above played a role as well. In many government reports the *hali* system was criticized. Steps were taken to abolish it, but they started from the assumption of debt slavery and, if only for that reason, were not very effective. These regulations, moreover, were too half-hearted to allow of serious implementation, even if the local bureaucracy had cooperated. The governmental sanctions were chiefly effective in a passive sense, the semblance of legitimacy having been removed from the arbitrary actions of landlords towards run-away or otherwise "disobedient" farm servants. Administrative intervention from outside, a very important component of the enlargement of scale, affected the traditional authority of the local power elite. Yet the fact that the government administration was made more autonomous — that rules and procedures were standardized and the staff of officials gradually became professionalized — could only offset in part the influence of the dominant caste. In the first place, members of that caste had been incorporated into the local administration as village headmen and police *patels* from the early nineteenth century onwards. Moreover, in their entry into the higher echelons — which was subject to education and connections — the traditional elite, in the course of time, had built up a lead which greatly enhanced their influence.

These, in brief, were the main reasons why the regional bureaucracy always accommodated itself to the interests of the dominant caste, and also why the local administration made few if any efforts to improve the condition of the agricultural laborers. The use of brute force by landlords is, of course, illegal, but it is tacitly sanctioned by the authorities.[4] Only such intermediate castes as the Kolis in Gandevigam have succeeded to some extent in shaking off the yoke of the dominant landowners, but in a struggle for power the former can rarely hold their own against the latter.

Political movements have had no more important influence than has the government. Since the nineteen-twenties the leaders of *ashrams* in the neighboring Surat district have striven for the abolishment of the *hali* system, to which end they called upon Gandhi, Sardar Patel, and others for help. An agreement to this effect between the landlords' representatives and agricultural laborers was ratified in 1939 by Gandhi, though not without demur, for he thought the conditions of the agreement favored the land-

4. Béteille, 1965, 182–84.

lords and were disadvantageous to the laborers. The main stipulations were that the Dublas would thenceforward receive their wages in money, and that their debts would be canceled. The agreement was confirmed some years after the Second World War. This well-meant intervention, however, resulted in a deterioration of labor conditions, for since then the amount of wage, rather low even at the time, has not increased correspondingly with the price rises of the articles that the Dublas used to receive as wages in kind. Reduced consumption was the immediate result. The leaders of the emancipation movement in the region have not unequivocally declared themselves against the dominant landowners; they try to bring about a change of mentality among this local elite by persuasion and collaboration. Moreover, the *ashrams,* whose leaders belong to high castes, are greatly influenced by a Hindu doctrine that rejects some tribal culture elements in Dubla life. In moralizing and even reprimanding terms, the members of tribal castes are lectured by social workers about all kinds of "deviations" in their social life that are looked upon as inferior, and are censured especially for their lack of hygiene and their consumption of alcohol. Rather unrealistically, the Dublas are advised to avoid contracting a debt, as they lose their freedom in this way. Under these circumstances it is not surprising that the ideas propagated by the *ashrams* scarcely take root, and that social workers attached to these institutions enjoy the confidence of the tribal castes only to a limited extent. They have not been able to bring about any change in the existing relationships of power and property by their efforts. It appears from the report written by the leader of this movement in the northern *talukas* that the landlords took little notice of the pressure exerted on them to treat their agricultural laborers more humanely.[5] From its beginning, the emancipation movement, which spread gradually to other parts of south Gujarat through *ashrams* and social workers, confined itself to persuasion and reconciliation and did not essentially affect the social system in which landlords and agricultural laborers function.

### *9.3 The Present Situation*

What relations between landlords and agricultural laborers have replaced the *hali* system? — this was the second question that I posed at the beginning of my field work. The labor system in a market

5. Dave, 1946.

economy is primarily determined by labor supply and demand. Whereas in some regions the demand has remained about equal, and in other parts of south Gujarat has considerably shrunken, the caste from which the laborers were traditionally recruited has grown rapidly. Opportunities for employment outside agriculture have not kept pace with these developments. The kind of work available to Dublas in the urban centers reduces them to a life of pauperism, and their condition is not much better than that of the rural Dublas, whose labor conditions are imposed by the landlords. The living standard of the agricultural laborers, low to begin with, deteriorated further when allowances in kind were replaced by money wages.[6] Most of them are continually indebted, and this is the main reason why, in budget calculations of households of agricultural laborers, expenditure always turns out to exceed income.[7] The debt binds the Dublas and provides the landlords with a means of pressure. In the long term it is cheaper and safer for the landlords to give a limited loan to a farm servant than to hire day-laborers. Moreover, they select the most industrious and obedient laborers for farm service.

The mutual rights and obligations have always been one-sidedly weighted. I pointed out in an earlier chapter the disguised exploitation that could be perceived in the *hali* system. On the other hand, patronage-like elements still occur in the present relations between landlords and agricultural laborers. I know of a number of Anavils who still feel obliged to aid and protect their subordinates, and these servants, in their turn, sometimes speak of their masters in terms of affection and devotion. As for the landlords' endeavor to acquire political support within the village for the purpose of obtaining advantages outside, it might be described as patronage for new ends.[8]

To the conclusion that exploitation was inherent in the *hali* system, and that patronage may occur in a situation of exploitation, should be added the comment that the difference between past and present is more than one of degree.

It would, therefore, be incorrect to interpret as a form of patron-

6. Concerning the Dubla agricultural laborers in his field village, I. P. Desai writes, "They live from day-to-day and even by any standard starvation among them is general, even in 1963, in spite of the five-year plans." I. P. Desai, 1966, 140.

7. By adding to their income the amount by which the debt of the agricultural laborers increases every year, this misunderstanding would be solved.

8. Cf. Béteille, 1965, 182–84.

age the network of obligations with which the Anavils have surrounded themselves. Although the landlords wish to preserve their traditional rights, they are not prepared to accept the attendant responsibility for the well-being of the Dublas who depend on them. On the contrary, they try to limit their obligations to the utmost, that is to say, to specify them and reduce them to a minimum. In other words, the Anavils do not behave as patrons, but still expect a maximal benefit of varied counterservices. Their demands increase as the dependence of their subordinates grows. All agricultural laborers among the Dublas depend on them, but dependence is particularly inescapable for farm servants. They are bound to about the same extent as the *halis* were, but they do not receive the aid and protection that their predecessors enjoyed. The farm servants in their turn therefore try to confine themselves as much as possible to a fixed amount of work, to render their obligation specific instead of diffuse. Whenever possible, they sabotage any claims by the master that exceed that amount. Both parties have exorbitant expectations towards each other, but they try to get away with fulfilling partial roles as employer and employee, respectively. In the end the Anavils, by virtue of their dominance, succeed better at this than do the Dublas. The risk of subsistence has been shifted onto the agricultural laborers. The distinction made by DeVos between expressive exploitation on the one hand (in which the mere fact of recognition of dependence by others creates satisfaction and is a purpose in itself) and instrumental exploitation on the other hand precisely illustrates the change which has taken place in the relationship between the landlords and agricultural laborers.[9] The widening social gap is even harder to accept for the Dublas than is the growing economic disparity. The Anavil who used to work alongside his subordinate now keeps aloof from agricultural labor. By freeing himself as much as possible from his undignified work the landlord has at the same time made the landless laborers more conscious of their humble position.

The Dublas are increasingly aware of the fact that their subjection is one of the main causes of their low social status. Economic disadvantages aside, weakness and dependence are looked upon as dishonarable and deplorable.[10] Their resistance takes various forms.

9. DeVos, 1966, 68–72.

10. Cf. Barth, 1960, 140, and Harper, 1968, 37. Also Weber, 1892, 797, where he says, "Dissociation from the patriarchal community of home and economy at any price, even at that of passing into the homeless proletariat, is the most marked

They no longer submit to beatings, nor do they tolerate landlords maintaining sexual relations with Dubla women. In the tense situation that now exists, incidents of this nature occasionally lead to sudden and violent explosions.

To sum up, however, the agricultural laborers cannot escape the control of the dominant landowners. Attachment as farm servants symbolizes the dependent condition in which more or less the great majority of Dublas lives. They still depend on the landlords for their livelihood. Those who have escaped and have found work outside the village and agriculture are highly esteemed by their fellow caste members but are looked upon with great irritation by the local power elite. When the members of the dominant caste oriented themselves towards the urban environment while maintaining their interests in the village community, their indifference to the fate of the Dublas became even more evident.[11] Dependence in an economic sense is used by the landlords to further political ends as well. This explains why the Anavils, though not very numerous, have largely been able to maintain their supremacy in the rural area of south Gujarat. They mobilize the support of the members of the lower castes to legitimize their dominance, and thus to perpetuate it in the new political structure.

The process of "depatronization" has left the agricultural laborers in a condition of isolation. The landlords are averse to being reminded of their former obligation to look after the interests of their subordinates, and sometimes flatly refuse to intercede for them with a third party. Nor are there any indications that the Dublas are trying to join forces with other weak categories, for instance the small farmers. The basis for such a collaboration is lacking.

The government has not taken over the functions of patronage, especially those of protection and social security. Industrial employment is expanding at a slow pace, and so far has not been sufficient to counteract the increasing pressure on the agrarian sources of existence. A proposal on the part of the central government to introduce a minimum wage for agricultural laborers was delayed for a long time; in the end, most states did not adopt it, and the

tendency in precisely the ablest elements among the laborers. . . . It is the tremendous and purely psychological magic of 'freedom' which finds expression in it. It is largely a question of a grandiose illusion, but after all, man, and thus the agricultural laborer, does not live 'by bread alone.'"

11. See also Béteille, 1965, 201–02.

remaining ones obstruct the implementation of provisions to this end. Nor do the states enforce compulsory education, which was introduced as an important medium for the improvement of the economic and social position of backward groups. Not only the older Dublas but a large segment of the younger generation are illiterate, and only thanks to the activities of an emancipation movement is this situation beginning to change locally. On a modest scale, the government has enabled the members of this tribal caste in south Gujarat to settle on land of their own in the villages, and also occasionally has subsidized the construction of huts on the land. The amount of money allocated to this project is low, however, and moreover it was not completely spent in the past few years. The government is perfectly cognizant of the miserable existence of the agricultural laborers. The commission appointed in 1947 to enquire into the extent of bondage in which the Dublas lived was instructed to draft proposals "enabling them to live a life consistent with human dignity and self-respect." The official reports concerning the backward regions and groups which have appeared since then, however, state that little has been done for the Dublas and that their circumstances have not changed.[12]

The Dublas have not succeeded in turning their numerical strength into political power. Nor has mobilization of their dissatisfaction by political parties yet been attempted. Especially in the Surat district, communist leaders have tried in recent years to win agricultural laborers over to their party, but in the state of Gujarat this party is too weak to have any influence. The votes of the Dublas in the villages are controlled by the landlords or by social workers. The emancipation movement that is operating in south Gujarat has founded a society to promote the welfare of Dublas (*Halpati seva sangh*), with a leadership consisting mainly of social workers and landlords who are affiliated to the Congress Party. At a conference that I attended, for instance, the Dublas were exhorted by the guest of honor, Morarji Desai, to break their bonds of dependence and not to submit to the bad treatment that falls to their lot, but at the same time they were urged to be patient. To control the increasing protest and rebelliousness seems to be the main function of this organization, but stabilization of the existing relationships amounts, for the Dublas, to stagnation, if not to deterioration of their condition.

12. See, for instance, *Report of the Scheduled Areas and Scheduled Tribes Commission, 1960–61*, New Delhi, 1961, 384.

The landless laborers have no access to the centers of power, and the isolation in which they find themselves is part of their economic and social pauperism. If we take their rapidly growing numbers into account as well, a prognosis cannot be other than pessimistic. It seems unlikely that the Dublas will soon know better times. Legislation which will guarantee a higher wage for the agricultural laborers is inadequate, as are other measures aimed at bettering their lot. Government intervention to this effect may aid in improving their material circumstances, but it must include economic and social reforms that will reinforce the position of the Dublas in the labor market. Industrialization, the change which in the late nineteenth century in Germany interrupted the process of proletarization of the agricultural laborers in the eastern areas, as described by Weber, progresses too slowly in India to provide a solution. The ratios between urban and rural population and between agrarian and industrial employment in south Gujarat have not appreciably changed in recent decades. This means that Dublas increasingly compete for the available labor places in agriculture — to the greater comfort and gain of the landlords.

Wider employment in agriculture appears to be a more realistic possibility. The much-discussed "green revolution," a program of measures in the field of agricultural economy and technology, which has recently led to a spectacular rise of agrarian production in several regions, also could enhance, and sometimes has enhanced, the labor-intensity per land unit. This does not, however, necessarily imply a rise in the real income of the agricultural laborer. There are, on the contrary, many indications that, because of the green revolution, the economic and social gap between landowners and landless laborers in the rural areas has widened further.[13] Improvement of the condition of the Dublas can be expected only if there is a fundamental change in the relationships of power and property, but the possibility of this is remote. The administrative decentraliza-

13. Cf. also the statement of Vyas in a report on the conditions of agricultural laborers in four villages of west India: "In most of the regions there are no organisations of agricultural labourers which can represent them adequately before the employers and secure for them higher wages and better conditions of work. The attitude of the State in this field is reflected more in pious hopes than in any active steps. The way the provisions of the Minimum Wages Act for Agricultural Labourers are flouted in almost all areas is a significant indicator. Because of all these factors it is quite possible that though the demand for agricultural labour may show an increase in relation to supply, the wage rates may remain depressed, as has been happening for the last decade or so. In such circumstances, even if there are no large-scale disturbances, a situation of accentuated tension may arise in many areas." Vyas, 1964, 23–24.

tion that has been introduced, as well as the consequent tendency of the local government bureaucracy to become politicized, have had, on the whole, advantageous consequences for the dominant landowners.

Among the Dublas, there is increasingly a collective awareness of their condition of subjection, a feeling of being wronged, and the beginning of an attempt at collective opposition against their exploiters. But it must be added immediately that the conditions that are prerequisite to implementing their protest organizationally in an effective and lasting way are lacking. Their rebelliousness is of short duration and is often directed against individual landlords. Like their companions in misery elsewhere, the Dubla agricultural laborers have become sunken in apathy and suspicion. Their lives offer them no prospects of improvement and they show themselves indifferent to what the future may bring. It can never be better than it is today — and that is not enough.

*Chapter 10*

# The Findings Reexamined

To what extent are the findings of my investigation generally valid? Chikhligam and Gandevigam were not chosen arbitrarily; they are typical of a great number of villages in south Gujarat having a similar social and economic composition. In generalizing my conclusions, however, I must make one reservation.

In most of the villages of the northern plain of south Gujarat, Kanbi Patels as well as Anavil Brahmans are the dominant landowners employing Dublas as agricultural laborers. An insight into the nature of agrarian labor relationships in this area, the Surat district in its present form, would therefore yield an answer to the question whether my findings are indicative of castes of landlords other than the Anavil Brahmans, that is, whether it would confirm my assumption that it is economic position, not caste membership, that determines the way in which agricultural labor is employed.

Moreover, such a comparison would make it possible to devote greater attention to the political background of changes in the relationships between landlords and agricultural laborers. In Gandevigam and Chikhligam, each with its own administrative past, politics were not a factor in the disappearance of the *hali* system. Informants assured me, however, that politics did play a role in the northern *talukas*, where there was an active emancipation movement along Gandhian lines. In this region, I was told, it was through political action that *halipratha*, denounced by Gandhi as immoral, fell into discredit.

To check my earlier conclusions, therefore, I spent a brief period in one of the villages in the Surat district.

THE AREA AND THE VILLAGE. The villages in those *talukas* that have formed the Surat district since 1964 are situated in a plain of black and fertile soil, the basin of the Tapi River. A railroad and a bus line serve to maintain contacts within the area and with the city of Surat, the center for the whole region. The village that I selected in the eastern part of the plain, and which I will call Valodgam after the subdistrict to which it belongs, has about 600 inhabitants, most of whom are engaged in agriculture. This applies particularly to the members of the two largest castes, Kanbi Patels and Dublas, respectively landlords and agricultural laborers. In addition, several artisan and serving castes are represented by one or more households, whose members carry out the traditional crafts and skills (as blacksmiths, carpenters, shoemakers, hairdressers, etc.) or are employed in new occupations in nearby towns.

In Valodgam, as in many other villages of this region, Kanbi Patels, being the most important landlords, are in a dominant position. Wherever members of this caste settled — tradition has it that they originated in central Gujarat — they succeeded in monopolizing most of the arable land. Dispossession of the tribal population living in the region had already begun before the inception of colonial rule. With few exceptions, the Dublas who live in Valodgam are completely landless. The adjoining village is inhabited exclusively by members of the tribal castes, especially Gamits, but also Dublas, most of whom have held their own as small landowners. Yet in the past the inhabitants of this tribal village also lost a large part of their land, which was taken by Kanbis of Valodgam. Moreover, many of the tribal cultivators came to depend as tenants on Banias, who for a long time have inhabited the small market towns of this region. Among the autochthonous tribal groups there is an awareness that they have been partly or wholy dispossessed by these incoming elements. During recent years their protest has been expressed in a local movement of unrest, socioeconomic in nature and with a religious undertone. In this study, however, the data I gathered concerning this *sati-pati* movement are left out of consideration.

The plain in which Valodgam is situated is an important agricultural area. Towards the end of the last century, cotton was introduced as a market crop. Besides food grains, the growing of groundnuts (peanuts) in particular has gained importance. All of these are cultivated in Valodgam. Only the dry and hot summer is a period of rest in the agricultural cycle. On the low-lying *kyari* soil, rice is grown during the monsoon, and pulses (especially *val*) afterwards as a second crop. On the higher fields, the farmers grow millet, cotton, and groundnuts.

Employment outside agriculture in the region itself, which has no large towns, is not very common. In some small towns along the railway, a few workshops have recently been established, though as yet their output is not large and they employ only a few hands. Larger industries are to be found exclusively around Surat, and owing to improved transportation the pull of this city on the inhabitants of the inland territory is increasing.

THE LANDLORDS. The time at my disposal was too short to permit collecting information on the situation of all the Kanbi Patels, and I confined my researches to ten households of this caste, approximately one-third of the total number. Of these households, two possessed less than ten acres of land, four had between ten and thirty-five acres, and the remaining four had more. The largest landowner, who was also the village headman, actually owned over 100 acres. Although in this sample the large landlords are overrepresented, it is a fact that the average land holdings of the other Kanbi Patels are also rather extensive. Most of them have much larger herds of cattle than do the Anavil Brahmans. The village headman even boasts a horse, the only one for miles around, on which he hurried to the Dubla quarter as soon as I had been seen there.

Together the ten Kanbi households employ fifty-five farm servants, an important indication of the intensiveness of agriculture and the large size of their properties. Of this number, nineteen are in the employ of the headman alone. The permanent servants comprise cowherds, house servants who at the same time are agricultural laborers (*chakars*), and field laborers. Besides these, most households employ one or more maids. Agricultural laborers are chiefly Dublas from the village itself, but some also come from the adjoining village, inhabited by tribal castes.

The large and spacious brick houses along the village street have tiled floors as well as beautifully decorated wooden front doors. The rich furnishings bear witness to the wealth of the occupants, the Kanbi Patels. The members of that caste in Valodgam all belong to the Leva section, a dominant group in the villages of this region. In the past they were peasants who tilled the land together with their laborers. During and after the Second World War they became prosperous, and now many of them no longer do any physical work on the land. The fact that their prosperity is of recent date, however, does not imply that the previous generation were very poor. As early as the turn of the century, the Collector of Surat wrote about the Kanbis:

The present men, as they live less laborious lives, so they have more expensive tastes than their forbears and to gratify them will resort to the *savkar* [moneylender] if there is no money in the house. Formerly the ordinary cultivators to a man wore country cloth; now they must have it of finer texture from Manchester. Cheap local rice, *dal* and *gul* [*gur*] were enough for the daily food; now vegetables, imported rice and refined sugar are in demand. A more luxurious generation seeks after *pan-supari* [betelnut], cheroots, hired servants, sweetmeats and American watches.[1]

A number of old district reports also mention the *hali* system. Servitude is certainly no recent phenomenon in the area, and the Kanbis of Valodgam were among those who at an early date employed *halis* and completely controlled them.

Now they feel hampered in the exercise of their power by the activities of social workers. At a short distance from the village there is an *ashram*, founded some decades ago, which since then has been a center from which attempts are made to improve the condition of Dublas and members of similar castes. This institution has been closely associated since it was founded with the Congress movement, the strongest political current in south Gujarat. Although the Kanbis of Valodgam were zealous supporters of Congress in the struggle for independence, they lost their confidence in the party because of the activities of the *ashram* workers among the tribal groups. They transferred their sympathy and political support to the Swatantra Party, which is popular among the large landowners in other parts of south Gujarat as well. The Kanbis are convinced that the Dublas are being set against them and are incited to leave their masters. They fear an increasing political awareness among their subordinates, but nevertheless still feel that they are lords and masters in the village, able to rebuff any "subversive" activities on the part of outsiders.

Although, like the Anavil Brahmans in the more southern and western *talukas* of south Gujarat, the Kanbi Patels are landlords, they have remained much more involved in agriculture, and are as yet only slightly urban-oriented. The younger generation attends

1. Broomfield and Maxwell, Bombay, 1929, 65.

schools and universities, and an increasing number of Kanbis are engaged in nonagrarian occupations in Surat or have left the country for East Africa or England, but this is especially to prevent landownership from being excessively fragmented among succeeding generations. In general the Kanbis have remained agriculturists much more than have the Anavils. They manage their property in a businesslike way and are solidly integrated in the village community.

THE AGRICULTURAL LABORERS. As mentioned earlier, the landlords employ large numbers of laborers, more than are available in Valodgam. My data, however, concern only the landless Dublas from the village itself. Of the men in the productive age group, thirty-six work as field laborers (here still called *halis*) and eleven as *chakars*, and six boys are *govalia.* Apart from these agricultural laborers in permanent service, twenty Dublas are day-laborers, and seven own a piece of land themselves or are illegal sharecroppers for Banias from a nearby town, in some cases both. Inasmuch as the tribal peasants from the adjoining village supplement their income from land by periodically working for Kanbis as well, these figures do not give an exact picture of the total employment pattern. Except for some scarcity at the peak of the season there is an ample supply of labor.

*Govalio, chakar,* and *hali* used to be successive phases in a Dubla's service relationship with a master. He began as a cowherd and after some time became a *chakar.* A few years after his marriage he usually entered the category of *halis.* Such careers still exist, but are no longer the rule. Moreover, a servant changes landlords much more often than in the past. In recent years the Kanbis have had some difficulty in finding new stableboys or house servants. Owing to the extreme dependence of such servants, who carry out part of their tasks in the master's house and have to be at his disposal from morning until night, most of the Dublas now become field servants without having passed through the preceding stages.

Stableboys (*govalia*) are as a rule young Dublas about fourteen years old, who tend the cattle and if necessary help out on the land. They live in the house of the Kanbi and receive, besides their meals and clothing, an annual wage of about 75 rupees. After some years they become *chakar.* They work on the land but also perform various jobs in and around the house. Their daily task begins at 4 o'clock in the morning and ends at about 9 in the evening. The unmarried house servants receive the same amount of food and clothing as the stableboys, but a higher annual wage, ranging from 100 to 125 rupees. Often this wage is not actually paid, but is partly or wholly entered in the books as owing to the servant until he marries. Once married, he begins to urge the master to relieve him of service in the house and restrict his task to agricultural labor, for which he will be paid a daily wage instead of an annual one — changes that are decided improvements for the Dubla. The Kanbi may try to delay this transfer to the category of field servant as long as possible, but if he does not wish to lose his servant he will finally have to give in.

Most of the attached laborers are farm servants who mainly work on the land. A number of them have been bound to the landlord since their youth; for others the attachment begins at the moment that they receive an advance from a Kanbi in order to marry. It has become very difficult for a Dubla to obtain a large initial loan, for instance some hundreds of rupees. This is not because the landlords fear that they will not be repaid — in fact, they know that repayment is practically impossible for the farm servants and are under no illusion in this respect. On the contrary, they are well aware that their claim will increase in the course of time. What worries them is the possibility that the local government will once again cancel the debt, so that the Dublas in permanent service will be released from their obligations towards them. The Kanbis must then make new advances of money so as to be able to exert pressure and ensure the labor of their subordinates, just as in the old days when most of the farm servants, after their debt had been remitted, soon put themselves in their masters' debt again.

The field servant's working hours vary according to the season, but they are always less than those of the house servant. As elsewhere in south Gujarat, the servants formerly received their wages in kind: daily meals and a grain allowance, and periodically clothing and other perquisites. After the agreement achieved some decades ago by the Gandhian workers, the customary distribution of small quantities of the produce at harvest time was ended. Buttermilk (*chas*), for instance, one of the few protein-containing components of their diet, used to be given free. Now the laborers have to buy it from their landlords, and most of them can rarely afford to do so. Meals are provided only on very busy days, when the farm servant works from 6 o'clock in the morning until 7 at night. At the time of my stay the daily wage was 1 rupee. On the request of their laborers, some Kanbis resumed distributing, on some days of the week, the traditional amount of grain instead of money.

The day-laborers in Valodgam form a residual category among the landless Dublas. They are those who have not (yet) managed to find a master, and their existence is precarious because they are not assured of work every day. The wage they receive is not higher than that of farm servants, and their chances of getting credit are even slighter.

In the region itself, alternative employment scarcely exists, and it is difficult for the Dublas to obtain such jobs farther away. The employment of a Dubla man from the village in a small town in Saurasthra set in motion the seasonal migration of some of his fellow caste members to the same town. The members of a few households migrate each year to the brickyards, and a small number of Dublas have succeeded in finding permanent employment in Surat and have left the village permanently. All in all, agricultural labor has remained the chief source of income for the members of this caste. That is to say, in the agricultural pattern of Valodgam the demand for Dubla labor is such that relatively few of them dare to take the risk of seeking work elsewhere.

A striking contrast between the Dublas of Chikhligam and Gandevigam and those of Valodgam is that many more of the younger members of this caste in Valodgam receive education. This is certainly due to the effort of the social workers here. Boarding-schools are attached to the *ashrams*, where boys of tribal castes, and increasingly girls, are schooled. Nevertheless, this improvement, however important socially, has not led to an improvement of the economic position of the Dubla agricultural laborers. The attempt to start a house industry by distributing spinning-wheels is completely in accordance with the Gandhian philosophy on which the *ashrams* are founded, but has not sufficiently taken hold among the tribal population to be of any real importance.

However, their activities in the educational field did have the result that the tribal groups have become aware of their dependence much earlier in Valodgam than elsewhere in south Gujarat. They gave a content of their own to their protest in the *sati-pati* movement mentioned above. The leaders of the movement were Gamits and Dublas who had been educated in the local *ashrams* but turned away from them in disappointment.

THE RELATIONSHIP BETWEEN LANDLORDS AND AGRICULTURAL LABORERS. The results of an investigation carried out by the agricultural economist J. M. Patel in 1961–62 into the condition of the agricultural laborers in a village near Valodgam largely agree with my own findings.[2] It appears from this report, which only recently came into my hands, that agricultural laborers, practically all Dublas, devote 86 percent of all their expenditure (an average of 593.15 rupees per household) to food, and not more than 3.5 percent to "social and miscellaneous expenses," whereas the Kanbi landlords allocate 46 percent of their expenditures (an average of 4,210.63 rupees) to food, and have close to 33 percent left for "social and miscellaneous expenses."[3]

The fundamental pattern of the relationship between Kanbi Patels and Dublas in Valodgam is identical with the situation in Chikhligam and Gandevigam which was described in preceding chapters. The agricultural laborers are conscious of servitude as a form of attachment that detracts from their own dignity. The rude treatment that is often their lot emphasizes their subservient condition. This applies especially to the *govalio* and *chakar*, who work in the master's house and are continually in his vicinity. For the Dubla, that security of regular work and the possibility of obtaining small loans outweigh the disadvantages of dependence. The Kanbis manipulate their subordinates' debt as means to exert pressure but, taught by experience, they try to keep the amount as low as possible. They grant a larger loan only when an agricultural laborer has worked on probation for a year, and then only if another Dubla, one who has proved to be reliable, will stand security for him.

In Valodgam, too, the changeover to the money wage was part of a more comprehensive process of commercialization. The Kanbi Patels conduct themselves like managers, running their land prop-

2. J. M. Patel, 1964, 92–119. A difference is that Patel defines the field servants among the Dublas as "casual labourers"; this is misleading, as his own description of this category shows (*ibid.*, 107). He only refers to the Dublas in the category of *chakars* as "attached labourers."

3. Patel does not mention how he collected these data, nor does he relate them to the incomes of the households. However, even if we do not attach too absolute a value to the amounts below and to their mutual ratios, his figures are useful indications of the differences between living standards of landlords and those of agricultural laborers.

erty in an efficient manner. They complain of the high labor costs and take on no more agricultural laborers than are strictly necessary, but if it suits the landlords better, they too take on laborers from elsewhere. They no longer feel obliged to give the Dublas of their own village an opportunity to work and provide for their existence. Without a doubt the rapid rise in the number of members of the Dubla caste has contributed to their indifference.

Relations became strained with the spread of commercialization. The personal tie of affection between master and servant is tenuous now because the relationship often begins at a later age, and severance occurs more frequently than it formerly did. Dublas

TABLE 10.1

PERCENT EXPENDITURE ON VARIOUS ITEMS BY ANNUAL HOUSEHOLD EXPENDITURE OF AGRICULTURAL LABORERS' HOUSEHOLDS, LANDOWNERS' HOUSEHOLDS, AND AVERAGE VILLAGE HOUSEHOLD

| *Item* | *Percent of total expenditure of agricultural laborers' households* | *Percent of total expenditure of households of cultivating group* | *Percent of total expenditure of all households in village* |
|---|---|---|---|
| Food | 85.94 | 45.59 | 56.85 |
| Clothing | 6.84 | 7.29 | 7.56 |
| Education | 0.16 | 1.89 | 1.44 |
| Fuel and light | 1.92 | 4.90 | 4.20 |
| House rent and repairs | .33 | 1.25 | 1.43 |
| Health and recreation | 1.14 | 3.62 | 3.30 |
| Traveling | .14 | 1.84 | 1.50 |
| Social and miscellaneous expenses | 3.53 | 32.62 | 23.72 |
| Total | 100.00 | 100.00 (read: 99.00) | 100.00 |
| | *Agricultural laborers* | *Landowners* | *All* |
| Total annual expenditure per household (rupees) | 593.15 | 4,210.63 | 1,624.28 |

complain of the greed, exactingness, and hardness of their masters, and in their turn appear to be marked by the same objectionable characteristics as those of their fellow caste members elsewhere. The Kanbis call their subordinates lazy ("they come to work only five days of the week"), dishonest, secretive, and impertinent. Their disobedience and rebelliousness taxes the patience of the masters, but ultimately it is the Dublas who come off worse.

As in Chikhligam and Gandevigam, the nostalgic way in which both parties in Valodgam speak of their former association does not ring very true and seems primarily to reflect the friction characterizing their present relationship. Nevertheless, the conflict of interests between them has undeniably increased with time. The Dublas are convinced that in the past they were better off, and this view is confirmed by the local social workers.

The pattern of relationships between landlords and agricultural laborers has changed in Valodgam in the same way as that described above for the two other villages. It is true that the emancipation movement in the region has given this process a particular content, but it has not much affected its direction and outcome. The booklet in which the movement's leader reports on its activities bears witness to the opposition and sabotage which his group encountered from the Kanbis, and also to its inability to change the condition of the Dubla agricultural laborers for the better.[4]

Local variations aside, the situation that exists in Chikhligam and Gandevigam is unquestionably not peculiar to these two villages. I conclude from this that my findings there are indicative of the pattern of relationships between landlords and agricultural laborers that has developed throughout south Gujarat.

4. Dave, 1946.

*Chapter 11*

# Postscript. Polarization: More of the "Same" for Everyone

### *11.1 The Villages Revisited*

In the summer of 1971, when I returned to south Gujarat, my primary concern was not the village situation, but I used the opportunity to make several trips to Chikhligam and Gandevigam. In each village I spent about two weeks in all — not very much, but enough to form an impression of what had happened in the meantime.

What struck me first was the absence of visible change. As before, Chikhligam could be reached only by a rough country road, and the bus services, though greatly extended in recent years, still do not reach there. Nor has the village been supplied with electricity. Gandevigam, which already had more public facilities seven years ago, has retained its lead. My impression of greater prosperity in Gandevigam was confirmed when I saw that this village in the garden region had already recovered from the consequences of a flood some years ago. There were few traces of the havoc caused in 1968, when the Ambika River inundated the riverside villages after heavy rainfall in the monsoon.

The population of the district has greatly increased, although growth shows a differential pattern. The birth rate among the higher castes was already relatively low before sterilization came into use as the chief means of reducing it. Among the members of the lower castes, however, restriction of the number of children was much less practiced. The government is pursuing in earnest its plans to curb the rapid population growth, and the organization charged with the implementation of these plans has taken several steps, including the enlistment of the local authorities. The first time I returned to Chikhligam the village headman kept away, because he thought he recognized the car of the district family-planning unit in the jeep that brought me. He was afraid he was once again going to be pestered about the number of applications for sterilization made by men in the reproductive age range. Propaganda to this end is conducted by a female staff member of the family-planning organization, a Dhodia girl recently stationed in the village, who directs her efforts almost exclusively to the members of the lower castes. It is the latter who are most susceptible to the promised bonus which, in specially conducted campaigns, amounts to 80 rupees, a sum that for an agricultural laborer is more than twice his monthly wage. It seems, however, that the possibility of greater results is not the only reason for singling out the tribal castes, such as the Dublas and Dhodias. Among members of the higher castes as well as in government circles the opinion is often voiced that the rapid numerical increase of the lower castes not only affects economic development adversely but also lowers the quality of the population — a dubious argument. However, it is clear that, as a result of population growth, the pressure on the agrarian basis of existence has exceeded the tolerance limit. For those categories that own disproportionately small amounts of land or none at all, intensification of production merely offers a short-term solution.

Because there are few possibilities of finding work outside agriculture and outside the village, the problem is aggravated. Economically, the small district towns in particular are characterized by stagnation rather than growth. Employment in the urban centers has increased very little, so that it can scarcely acommodate the existing local labor supply, let alone absorb the large surplus from the countryside.

### *11.2 Changes in the Pattern of Agricultural Activities*

No drastic changes have taken place in the crop system of either village. In Chikhligam, mangoes and rice have remained the principal market products. Several Dhodia farmers recently turned to chillie cultivation. For this purpose, new wells were sunk and, in some cases, oil pumps installed. In Gandevigam, rice cultivation has greatly increased. With a further extension of the orchard area in this village the shift from mango to *chiku* cultivation — even more profitable and less risky — has continued. The Koli landowners have concentrated more on the cultivation of vegetables and chillies. The increased yields of the crops are not due to any sudden breakthrough in a stagnating agricultural economy, nor does it proceed with such rapidity in south Gujarat that the use of the term "green revolution" is justified. Nevertheless, the New Agrarian Strategy launched by the government in the late nineteen-sixties has undoubtedly resulted in a changeover to new agricultural methods. The use of fertilizers and insecticides is becoming general, and in rice cultivation the introduction of higher-grade seeds increases steadily.

This evolution has, however, only widened the gap between the two villages. A large part of the fertile soil in Gandevigam is permanently irrigated by water from the canals and wells. The fields are scenes of great activity almost the year round. "Progressive" farmers in this garden region achieve two rice crops a year, some of them even three. The intensification of agriculture has increased employment, but mechanization has also progressed. Tractors have come into general use, and these vehicles are also employed for the transport of agricultural products. During the rush season the roads in this region are crowded with traffic.

In comparison, agriculture in Chikhligam is still labor-extensive. Although an irrigation canal crosses the village area, the land lies too high and is, moreover, to uneven to allow irrigation. At present the Land Development Bank of the district is preparing a program of leveling of land which, combined with the use of electricity, could facilitate irrigation in a larger area. It is not easy, however, for Chikhligam to make up the deficiency. The extension of employment has not kept pace with the growth of the population in general and of the agricultural population in particular, while the distribution of agricultural activities over the year is also much more irregular than in Gandevigam.

Differentiation has increased both regionally and socially. It is the large farmers that have profited by the support given by the government to promote production. As they control the cooperative movement and have great influence on the local bureaucracy they are generally the ones to benefit by the facilities offered in this way. The *gram sevak*, whose task as a local-level worker has been reduced from community development in a wide sense to agricultural extension work only, directs his efforts almost exclusively to members of the dominant caste in both villages. Kolis and Dhodias rarely qualify for any credit granted by the government, such as for land improvement. Insofar as they are able to buy better seed and fertilizer for rice cultivation most of them use the crop for household consumption. At any rate this makes it unnecessary for them to buy grain for their daily needs at steadily rising prices in the village shops, as the agricultural laborers have to do. The contribution of the small farmers to production for market consists chiefly of a few garden crops. These labor-intensive products form the principal source of their money income from agriculture and enable them to find employment within their own households. In this way they can avoid the dependence associated with the performance of wage labor for large farmers. In a situation of enlarged socioeconomic inequality, the small landowners strive more or less successfully to retain their independence. They are not systematically included in the development programs, and the advantages that some of them obtain from these programs are derivative and limited in nature.

### *11.3 The Political Framework*

In both Chikhligam and Gandevigam the hegemony of the Anavils is undiminished. In the village council as well as in the local cooperative they are still in control. Various facilities offered by the government are distributed through these institutions, and the mediating role of those in charge enables them to perpetuate their local power. Any request made by a Dhodia from Chikhligam for a loan towards buying an oil pump, for instance, depends on the intercession of the village headman and that of the secretary of the cooperative, both Anavils. The weak economic condition of the small farmers contributes to the political ascendancy of a small elite, and the New Agrarian Strategy has not altered this in any way. Fertilizer, for example, can be bought from the cooperative at prices fixed by the government, but only when paid for in cash. Several Dhodias who cannot do so are supplied with small quantities on credit by a few prominent Anavils at higher prices or in exchange for support in local affairs.

In consequence of the administrative and political decentralization in recent years, which has shifted the center of authority to district level, key positions within the village have become more important than before because of the weight they carry outside it. The village headman, for instance, occupies a seat in the regional council in that capacity. Because he is a member of one or more council committees that allocate government money for such purposes as public works, health care, and education, he is able to obtain considerable facilities, either for the village or for his own benefit. The two Anavils of Chikhligam and Gandevigam meet their fellow caste members or other locally influential people in the council, and the various benefits are distributed within this group on a basis of mutual favoritism. Aside from this they have easy access to the regional bureaucracy. The present village headman of Chikhligam prides himself on his familiarity with some prominent district officials and reinforces his leading position by occasionally also letting non-Anavils profit by his good contacts. Some years ago his cousin, who is now head of the village cooperative, gained his influential position of secretary of the *taluka* branch of the Congress Party, once a stronghold of the dominant caste in south Gujarat. After the split in the top ranks of the party, a large majority of the party supporters continued to follow Morarji Desai, a man of the region. Much the same thing happened in other parts of this conservative state. But their support rapidly crumbled away as the popularity of Indira Gandhi increased, reinforced by her slogans to ban poverty and the successful end of the war against Pakistan. In the months preceding the election of March 1972, many local cadre members joined the New Congress Party. It was not so much ideological motives that made them change their allegiance, but rather fear of losing access to the external sources of power. As already noted, local leadership is based to a large extent on the distribution of facilities provided by government bodies. By changing his allegiance in time, the district branch secretary from Chikhligam was able to avoid being excluded from benefits after the election.

It would not be correct to conclude from such occurrences that the old political system has remained completely unchanged. Notably, the ruling party can no longer count on the influence of the largest landowners in obtaining the votes of the small farmers and the agricultural laborers. This mechanism, based on local patronage, has been replaced by direct mobilization of the weak but numerically dominant groups by organizers from the regional power centers. Nevertheless the old elite has managed to accommodate itself anew to a governing party and therefore has lost little of its authority.

### *11.4 The Condition of the Intermediate Castes*

Whereas the top group has generally succeeded in consolidating its privileged position, most of the Dublas are just as badly off as they were some years ago, and perhaps even more so. Although this account is primarily intended to establish the present relationship between these two categories, I shall begin by giving a brief description of the condition of the Kolis and Dhodias who are between the two strata in Gandevigam and Chikhligam, respectively.

The generally accepted view is that, in the district as a whole, these castes have been able to improve their position. Those who have found work outside agriculture have been especially successful. The skilled employees of the factories and workshops in the small towns of the district are usually members of these castes. A greatly increased interest in education has played an important role in their social rise. With a college education, Kolis and now also Dhodias increasingly succeed in obtaining clerical jobs in government or industry formerly reserved for members of the higher castes. Participation in education is facilitated by the circumstance that many of the schoolteachers, particularly those in the countryside, belong to these intermediate castes. The Kolis and Dhodias have gained political influence as well. By appealing to collective loyalty, candidates of these numerically influential castes were elected to various important posts. Another dimension of the improvement in their position is expressed by urban-based sectarian movements — such as the Moksh Marg, which is active among the Dhodias — urging sanskritization of their style of living.

Does this mean that the Kolis and Dhodias have shaken off the dominance of the Anavils in village life? Any general statement concerning either of these groups would be premature, as a process of differentiation is taking place precisely in the intermediate castes.

As to the Kolis of Gandevigam, some of them have succeeded, partly owing to the Tenancy Acts, in rising to the position of medium-sized landowners. In contrast to the majority of their fellow caste members, who work exclusively with their own relatives, they call on Dublas as additional labor for longer periods. Whenever possible they are inclined to avoid manual labor, if not for themselves, at least for their wives and children, in imitation of the Anavils. These Kolis are better housed (in houses with stone walls and tiled roofs), own more capital goods (especially agricultural implements), and further distinguish themselves by a higher level of consumption. Together with the artisans they form an elite within their caste. Children from such successful Koli families are most likely to take secondary or higher education; after completion of their studies, jobs are found (sometimes being bought) for them in a factory or government office in one of the neighboring towns. On the other hand, the situation of most of the other members of this caste has changed little. Their holdings are small, and they are still typically thrifty and hardworking peasants, in whose households all the members have to take part in the work on the land. They have perceptibly remained behind in income and status. In families with many children, above all, the prospects for the new generation are extremely gloomy.

Among the Dhodias of Chikhligam, who, taken as a whole, are below the level of the Kolis in all respects, this differentiation is even more pronounced. Here, too, the bettered position of a number of households is indicated by, for instance, tiled roofs and the possession of bicycles and transistor radios. During my stay in the village, preparations were being made for the building of the first brick house in one of the Dhodia quarters. Even less than in Gandevigam do the widening economic gaps within this caste reflect disparate incomes from agriculture. Most successful are the labor-recruiters who locally enlist labor gangs for the brickyards. One of them — the most successful — contracts about 500 laborers a year. Together with a brother he owns two trucks, his younger brother is enrolled in the university at Surat, and he himself bought a mango orchard from an Anavil. In short, he is a man of substance. Through the influence of the recruiters the slightly better-paid jobs in the brickyards — supervising labor gangs, driving trucks — usually fall to their fellow caste members. The boys from these financially more secure Dhodia households no longer participate in the yearly migration to Bombay. Several have been trained for higher-qualified jobs and are now employed as schoolteachers, factory hands, or diamond cutters. This top category in particular profits by the special facilities for tribal caste members. The general impression of increased power and independence is above all derived from this minority.

In speaking of ascending castes in a general sense, therefore, insufficient account is taken of the growing differences among Kolis and Dhodias. Those that have lagged behind are not concerned with accentuating the widening distance between them and their successful fellow caste members, while the latter hope to improve their chances of competing with the high castes for economic advantages and political influence by appealing to group solidarity. In the intermediate castes, the process of socioeconomic differentiation remains veiled, so to speak, by the personal interests of both those ascending and those remaining behind.

The continuing polarization in the rural regions of south Gujarat may be most adequately demonstrated by a comparison of the situation in which Anavils and Dublas in Chikhligam and Gandevigam find themselves.

### *11.5 The Anavils*

The Anavils have not lost the stamp of a rural caste. For most members, agriculture is still the principal basis of existence. To be sure, secondary education has become a matter of course for the younger generation, but the possibilities of nonagrarian employment have not much increased, so that migration to towns in or outside the district is impeded. A long period of unemployment after completion of college is the rule rather than an exception, and many find themselves finally forced to accept jobs far below their level of training. There is sharp competition for the few places available in government offices or industry, and candidates from ascending Koli and Dhodia households also seek those jobs in increasing numbers. Although it is true that Anavils, through the influence of fellow caste members in established positions, have succeeded in maintaining their social lead, the remuneration of those urban jobs for which they are qualified is far below the income they can derive from farming. In addition, the influx of members of lower castes has diminished the prestige of the various occupations in which the members of the dominant castes used to be strongly represented — teacher, government employee, bank clerk, and the like.

Urban orientation among the higher castes has, finally, decreased because public facilities in the countryside have improved. Electricity and tap water supply have made life more comfortable in many villages. Bus services have been extended and their frequency increased, so that the villages are less isolated. It should be borne in mind, however, that these facilities benefit, in the main, the rural elite. Their way of life differs little from that of the upper urban class.

The Anavils generally have done very well for themselves. The extent to which they have improved their position is shown, for instance, in the successful development of the regional cooperative of Gandevigam and neighboring villages. Whereas this institution — still primarily an instrument of the dominant caste for their own interests — listed sales totaling about 750,000 rupees in 1962, they had risen to 2,500,000 rupees for the year 1971. Membership increased from 700 to over 900, but cash reserves trebled in the same period, and now amount to nearly 1,000,000 rupees. The price of *bagayat* land, mostly owned by Anavils, has risen to an average of 25,000 rupees an acre in Gandevigam, that of *kyari* to 10,000 rupees or higher. One informant has given up his teaching post. As a result of more careful management of his farm, he now has a substantial income: his medium-sized orchard alone provides him with a yearly revenue of over 9,000 rupees, more than twice the salary he earned as a teacher.

The Anavils of the garden region are not the only ones who have profited by this favorable trend. While the exceedingly fertile soil in Gandevigam yields more per acre, the Anavils in Chikhligam possess, on an average, more land. This does not mean, of course, that there are no differences in prosperity among them. The richest members of the caste had acquired new durable goods in addition to those I noted seven years earlier. Their owning a motorcycle or even an automobile is becoming less unusual. Their houses have been improved, for instance, by tiling on the walls and floors. One Anavil from Gandevigam told me about a holiday tour he had recently made through India which cost him a sum that would have provided a living for two agricultural laborers for a whole year. Following a more traditional spending pattern, another member of the caste recently invited some 1,500 guests on the occasion of the sacred-thread ceremony of his son's son. It is true that only a minority can afford such luxuries, but in general the situation of the Anavil households is one of increased prosperity, and I met with no examples of real deterioration among them. Practically all the Anavils have succeeded in realizing the aim of not performing agricultural labor themselves, and this has further lessened the social distance between Bhathelas and Desais. The sociocultural homogeneity within the dominant caste is reinforced by their strongly developed sense of being the backbone of the social system.

Awareness of their common interest does not, however, preclude clashes. In Chikhligam, for instance, a conflict has arisen between the former village headman and his cousin, the former secretary of the cooperative. They have now exchanged functions, but behind this change — not sensational in itself — there is a fierce struggle for power. Both men strive to be acknowledged as the village leader, a condition for influence outside the village. For this kind of political competition, it would not be correct to use the term factionalism, which would imply the existence of alliances within the local system that are kept up by means of patronage-like relationships. In this sense the Anavils are no longer split into different camps, each with its own superiors. There is, rather, progressive individualization, in which members of the dominant caste, as businesslike farmers, keep a sharp watch on each other and regularly exchange allegations of improper practices in using the canal water, hiring agricultural laborers, and the like. Among the various households, suspicion and jealousy prevail.

Sincere collaboration is therefore still rare, yet there is increasing awareness of the fact that mutual opposition at any rate damages the common interest, for the Anavils feel more threatened in their privileged position than ever before. A feeling for the political scene has, it is true, induced many leaders to pronounce for the N.C.P. at the right moment, but in this they have been followed by only a minority of their conservatively thinking fellow caste members. The rural elite has little use for the socialist course to which the new party has ideologically committed itself. Most Anavils feel shocked by the success of the populist strategy of the N.C.P. They explain the striking victory of Indira Gandhi in the spring of 1972 as a transition to a more left-wing rule — a shift that does not hold much promise for them. They are, however, somewhat reassured by their conviction that the *garibi hatao* program will fail once again. The fact that the regional leadership of the N.C.P. consists predominantly of Anavils and members of other high castes assures continuity and presents a possibility of more or less legally invalidating any too drastic proposals — for instance, the land reform proclaimed by the government.

In the latest election, most members of the dominant caste remained faithful to the Old Congress Party. The young people in particular, however, feel attracted to the Jan Sangh. This militant urban-based right-wing party has recently won many members among the sanskritizing castes in the countryside. A number of young Anavils and Kolis from Gandevigam — disappointed, like so many of their age group, in their expectations of getting good jobs after completing their studies — have joined a newly established branch, in the *taluka* town, of the R.S.S., the paramilitary organization affiliated with this orthodox Hindu party.

The members of the upper social stratum deeply regret the fact that, in the present political system, strength is measured in numbers. They see it as the writing on the wall that the N.C.P. owed its victory to massive support from the lower castes. In the election campaigns, they complain, these categories were incited to offer resistance against the rule of the higher castes and were lured with false promises or money. A widespread allegation against the politicians is that they bought votes of the population in poor neighborhoods through trusted agents.

From the height of their elite position the Anavils give vent to a deep aversion to the democratic system in which they feel outvoted by an illiterate, irresponsible, and inert mass. In their opinion, participation in the elections should be reserved for those who are qualified to judge because of their parentage and education. It is a view in which the superiority of the high castes is asserted and the inferiority of the low castes emphasized.

## *11.6 The Dublas*

My first walk through the Dubla quarter on the fringe of Chikhligam showed me that their material situation had not improved. Many huts were in a state of collapse and offered little or no protection against the winter cold and the monsoon rains. The average number of occupants per hut has greatly increased, and the available space makes it impossible for more and more households to shelter all their members at the same time. Although the hamlet is surrounded by a great deal of waste land belonging to the village, any applications for permission to build new huts there, transmitted to the district authorities with a negative recommendation by the village headman, have remained unanswered.

In Gandevigam it was the Dubla quarter, inundated even in normal rainfall, that suffered most in the flood of some years ago.

It is true that the total damages in this quarter were relatively low, but that was because of the low value of the huts and household effects. The little that the occupants possessed was mostly lost. The aid offered from outside was distributed by the local elite and, if only for that reason, passed the Dublas by, or benefited only the confidants of some of the prominent Anavils. Government money was used to build fifty-two new huts with tiled roofs, but not in the old place. Because the village authorities refused to put higher-lying ground at the Dublas' disposal, a large part of them had to move to a distant site on the territory of a neighboring village. One water well is the only public facility there. Electric power wires pass right over the rows of huts, but there is no connection to them anywhere. Fraudulent economizing in material led to jerry-building. The Anavil secretary of the regional cooperative had to resign when it became publicly known that he had sold wood intended for these huts and had pocketed the proceeds. Lastly, the new Dubla quarter is difficult of access. The men and women now have to traverse much greater distances to reach the houses and fields of the farmers, which creates much hardship especially during the rainy season.

The deplorable condition of the Dublas is, of course, closely connected with the lack of sufficient and continuous employment. From Chikhligam, the yearly migration to the brickyards goes on as before, because of need but not by choice. The continual alternation of work situations, which often entails the breaking up of households for long periods, makes any stable existence impossible. It also means complete subjection to a labor system in which the employer, owing to the temporary nature of the engagements, rejects any obligation to provide for the most elementary needs of his employees. This sometimes leads to their being treated inhumanely. During my stay, for instance, a grandfather, a father, and two daughters returned to the village in mid-season. They suffered from typhoid fever and had been sent away from the brickyard so as not to infect the other laborers. Gravely ill and penniless, they had found their way back, traveling surreptitiously by train and getting a lift in an oxcart. Once they were in Chikhligam, not one of the farmers for whom members of this household sometimes worked gave them any aid. I found these people in their hut, uncared for and without food, lying on jute bags. Two of them died of the fever the next day.

The Dublas of Chikhligam are dependent on their income from agriculture for at least part of the year, and most of their fellow caste members in Gandevigam depend on it all of the time. During recent years, the living and working conditions of agricultural laborers in India have aroused increasing interest, particularly as to how the New Agrarian Strategy has affected them. The information that is gradually coming to light shows that the situation of this numerous category of the population has not noticeably improved, and in many cases has deteriorated. This seems to be true of the Dublas as well. Although in both villages agricultural laborers' wages have practically doubled as compared with seven years ago — farm servants now earn 1 to 1.5 rupees and day-laborers 1.5 to 2 rupees — the prices of basic living needs have risen correspondingly or even more, so that a decent existence is still impossible. Employment in Chikhligam has not greatly increased, although in Gandevigam the demand for casual laborers to work for farmers with large and medium-size holdings has risen. Even in the labor-intensive crop pattern of Gandevigam, however, the labor supply shows a much greater rate of increase. This impression of the situation in the two villages is confirmed by the evolution of the wage level and of employment in the whole of south Gujarat. On the basis of 1962–63 equal 100, the input of labor in 1967–68 had risen to 114, but this was offset by a fall in the real wage level to 78 in the same period.[1]

These data, though illustrative in themselves, reflect a more fundamental transformation. They indicate an alteration in the nature of rural organization — a changeover to a capitalist mode of production. The commercialization of the relationship between farmers and agricultural laborers is reflected in a number of labor forms which have become increasingly conspicuous during the past decade. The percentage of servants has decreased but will probably remain at the present level in view of the need among the large farmers of a small fixed core of labor. For attached labor, employers prefer Dublas who have some skills, can handle the new tools, and show an insight into the new agricultural techniques. A submissive attitude is another requirement. Agricultural laborers who are known to be "troublesome" and "pugnacious" are rejected, insofar as these more militant Dublas have not already renounced

1. P. Bardan, "Green Revolution and Agricultural Labourers," in *Economic and Political Weekly*, vol. V, no. 29–31, July 1970.

servitude themselves. Casual laborers are employed on a contractual basis, and the landowners can usually draw on an abundant supply. For the harvest of some products, for instance chillies and cotton, piece-work has been introduced on a large scale, a form of labor in which children from agricultural laborers' households are employed more than before. Thirdly, contract labor has gained ground. Labor gangs undertake to carry out a specific amount of work — harvesting rice, grass, or mangoes — at a prearranged price per day or land unit.

The most drastic change in the labor system is undoubtedly the mobilization of agricultural labor across considerable distances. Although Dublas from many villages in south Gujarat have been accustomed for a generation to migrate to the brickyards and saltpans every year, agricultural work is now increasingly being performed by laborers from elsewhere. This migratory labor has assumed enormous proportions. For the harvest of sugar cane, for instance, the cooperative factories with which the bigger landowners are associated recruit thousands of laborers every year in the western districts of Maharashtra. The workers are small peasants and landless laborers, who have no means of existence during the dry season in their home area. With or without their oxcarts they come to south Gujarat for a period of five or six months each year, traveling from village to village cutting cane and transporting it to the factories. In the same way, individual owners of mango orchards and grass fields or dealers in these products contract laborers in distant villages for the harvest period through the agency of labor-recruiters. The labor supply is less and less determined by local conditions. Large-scale migratory labor has a wage-depressive effect and facilitates replacement in case of insufficient performance or too drastic demands. The abundant supply combines with the enlargement of scale to foster competition, and labor comes on the market at a price far below the minimum level of existence.

Yet the economic advantage gained by the farmers is not the only factor that has led to this state of things. The new forms of labor enable the Anavils to limit to a large extent their dependence on the performance of the Dublas, a source of boundless irritation to them. Accelerated changeover to mechanization of the agricultural production, even when clearly not profitable from the viewpoint of efficient management, reflects the desire of the farmers to do away as much as possible with their need for agricultural laborers. Another solution for them is to keep their relations with these subordinates as distant and businesslike as possible by leaving them to intermediaries. The large farmers appoint overseers, charge servants with the negotiations to be conducted with casual laborers, or deal exclusively with gang-foremen or recruiters. The latter receive advance money, arrange the distribution of wages on pay-day and hold themselves responsible for the completion of the arrangements. Conversely, these mediating roles provide an answer to the growing aversion among agricultural laborers to binding themselves individually to landowners. Extending credit to Dublas is increasingly done through *mukadams,* who in this way provide themselves with a pool of labor which they can utilize as it is needed.

The loose and businesslike ties with the farmers which manifest themselves in all these labor forms have led to a further hardening of the relationship on both sides. In the eyes of many Anavils, Dublas are good-for-nothing *kamchor,* labor thieves. Not all landlords, it is true, act with the same harshness, but as a Dubla from Gandevigam put it, "One Anavil kicks you from his field when you have gathered wood without permission, another tacitly allows it; but both of them refuse actual help." Fearing the destruction or theft of crops, more and more landowners have decided to fence in their property with barbed wire, as well as to have a closer watch kept on the fields. The polarization between, on the one hand, the larger farmers who do not personally work, and, on the other, the landless laborers rests on the fact that the former no longer recognize the right of the latter to share and participate in agrarian production. I pointed out these indications of a large-scale rural pauperization when I closed my previous investigation. What is, therefore, quite frequently regarded as a negative consequence of the green revolution is, in my view, actually the condition on which this process started. Partly as a result of government policy the contrasts in rural regions between the weakest and the strongest party have sharpened. The term "green revolution" still applies, but not, as I see it, primarily to a sudden spurt in farm production owing to a technological breakthrough. More important is the fundamental shift in the mode of production to capitalism as the base of agriculture.

The economic situation of the Dublas, miserable as it was, has deteriorated even more. As landless laborers for whom work has become a godsend even when it does not bring in enough to make ends meet, they have been forced further into a corner. Providing for a household requires continuous improvisation.

Besides their exploitation, another aspect of the poverty of the Dublas is their increasing isolation. Not only has the landlords' traditional help and protection disappeared, but they do their utmost to obstruct any direct contact of their former clients with external sources of power. By taking advantage of the ignorance and powerlessness of the Dublas, some Anavils use their influence and their familiarity with laws and regulations to prevent outside support for Dublas. The Dublas in both villages complain of the opposition they encounter in their attempts to qualify for the facilities which the government expressly makes available to the tribal castes. Anavils and members of other high castes, they say bitterly, bar the way for us everywhere. Those few Dublas who have heard of the *garibi hatao* program have no illusions whatsoever that its generous promises will ever be fulfilled. That the social isolation in which the Dublas find themselves has increased is strikingly demonstrated by their removal from the residential nucleus in both Gandevigam and Chikhligam. Regrouped on the fringes of these and many other villages they may rightly be called the new untouchables of south Gujarat.

From various parts of India there have recently been reports of agricultural laborers who are beginning to offer resistance against the exploitation to which they are exposed. It is rightly pointed out, however, that if growing economic inequality is a prerequisite for political unrest, it is nonetheless not sufficient in itself.[2] Large-scale protest movements in reaction against a deteriorated position in the production system need to be organized, but in a situation of pauperization the chances of joint action are drastically limited.

The horizontal ties among the Dublas have remained weak. Some are more oppressed than others. As sharecroppers, servants, casual laborers, and gang laborers they have no parallel interests. The limited favors that can be obtained in a situation of extreme and continuing scarcity lead to patronage-like relationships. An attitude of protest, for instance, is incompatible with the need of showing oneself worthy of consideration by behaving "well." Intra-class antagonism manifests itself in dissension and mutual watchful-

2. T. K. Oommen, "Agrarian Tension in a Kerala District: An Analysis," *I.J.I.R.* Reprint Series no. 15, n.d.

ness. If someone refuses to work for 1 rupee, another is always prepared to take his place, as I was told time and time again. The Dublas are aware of their isolation and attribute it to a lack of leaders of their own, "who can show the way." But who will venture to risk standing up for others against the powerful Anavils, and who, for that matter, expects to get a hearing from the arrogant and unwilling bureaucrats? In the situation of poverty itself there are undoubtedly elements that constitute obstacles to mutual solidarity. Incompetence and ignorance, often considered typical of pauperized groups, have also rendered the members of the landless castes helpless and passive in their dealings with outsiders. On the other hand it is the aggregate of inequality — economic and political — along with social polarization that forms the basis for a growing collective perception of injustice, even if fear of sanctions usually limits any expression of it to verbal avowals during encounters in their own hamlets. Indeed, their awareness of unity is further stimulated from outside because Anavils regard all the Dublas as being cut from the same cloth. The landlords are strikingly ill-informed about anything concerning their laborers, including the contrasts among them, and what is more, they are totally indifferent to them.

One organization that is active among the Dublas is the *halpati seva sangh*, stemming from the Surat district, which is still its main field of action (see chapter 10). In recent years it has extended its activities to the Bulsar district. With financial support from the government it has opened small offices in nearly all *taluka* towns as bases for one or two staff members who try to get through to the Dublas in the countryside. Social workers and local politicians form the nucleus of this voluntary movement, which aims at improving the social position of this landless caste. Among the efforts to that end, education occupies an important place. In the Dubla quarters of an increasing number of villages the organization's members have established kindergartens, with free meals for the children during school hours. Some six months ago the *halpati seva sangh* opened a primary boarding school in Chikhligam, but despite the efforts of the local staff, school attendance remains low. Pupils of the two institutions of secondary education that the organization has started elsewhere in the district belong in the majority to other castes, for lack of Dubla applicants.

As in the other activities undertaken by the *halpati seva sangh*, the education they offer aims at adjustment rather than awareness. Alcoholic excess is combated mainly on moral grounds. Groups may apply for musical instruments so as to hold *bhajan* gatherings, in which religious songs extolling Hindu values are sung. To keep the Dublas from consulting the *bhagat*, their own medicine man, chests containing medicines have been distributed among the villages. Most components of the program are on this sociocultural level. The movement has little to offer in the way of measures towards improvement of the Dublas' economic condition. By paying the expenses of artisan training, the organization tries to provide an alternative to the forlorn existence of the agricultural laborers, but only a limited number of young people can profit by it. Still awaiting implementation is a plan to establish some labor cooperatives whose members would work for farmers on a contractual basis and, in the slack period, would contract for the construction and maintenance of local roads and irrigation canals.

The *halpati seva sangh* can best be characterized as a movement aiming at adapting the Dublas to the existing social system on the most favorable conditions possible. The movement derives from Gandhian principles of gradualism and conciliation of contrasts. In this view, any schooling to promote class consciousness and class struggle is not so much tactically injudicious as sinful and therefore objectionable.

At present a more radical approach would, it appears, have few chances of success. In south Gujarat, the Land Grab Movement, launched in the autumn of 1970 by left-wing parties, ended in failure. The agitation conducted by a bourgeois leadership from the district towns only impelled the Dublas of a few hamlets to take action. The risks attached to occupation of the land owned by large farmers were too great. In some villages near Chikhligam, where the action was led by the C.P.I. — a party with only a handful of members in the whole Bulsar district — the laborers were taught their place. A strong police force dispersed the crowd of men, women, and children by brute force. Many participants were taken to distant prisons and the instigators received harsh sentences. The *halpati seva sangh* refused to help any of those that had joined. The landlords went round the laborers' hamlets to say that this had been a warning and that any further action would be put down with even greater severity. Although such experiences may intensify the existing awareness of oppression, they do not necessarily lead to more vehement confrontation.

The *halpati seva sangh* maintains close ties with the government party. Through its staff members the Dublas are urged to vote for the N.C.P. After the split in the Congress movement some years ago, the district authorities of Bulsar — who before the recent election supported the O.C.P. — promptly withheld the subsidy for the *halpati seva sangh*. The split in the formerly united party has brought some clarity, inasmuch as the political choice of the upper stratum at least no longer coincides with that of the lower classes. This is not to say that the interests of the landless laborers are now better protected. It seems probable that the Dubla hamlets have become the vote banks of the regional leaders of the N.C.P., most of whom derive from the higher social strata. Members of the landowning castes are amply represented among the leaders of the *halpati seva sangh* — indeed, they are headed by an Anavil, who in his attitude and actions corresponds perfectly to the traditional image of the *dhaniamo*. Small wonder, therefore, that the Dublas talk with cynicism about this organization. They have learned to distrust profoundly the promises made by anyone. They regard the *halpati seva sangh* members who function as brokers and mobilizers as agents of the upper castes, but this is not always true. The local cadres, mostly consisting of young and skilled Dublas, admit to being not much more than pawns and only dare to voice their criticism in whispers for fear of dismissal. In their thought, however, they show themselves much more radical, and a change of course might be possible if the organization were taken over by them.

In this analysis I have emphasized the obstacles in the path of those trying to create organized solidarity. Intraclass tensions play a role, and nonantagonistic mobilization of the agricutural laborers conceals fundamental contradictions. It would nonetheless be incorrect to suppose that awareness of unity still needs to be awakened. Closer contact shows that class feelings are certainly present, but under the existing power structure it would be imprudent and injudicious for the laborers to display them too openly. The exploitation is heavy and the laborers' situation is confused. Complete bondage to a single landlord has for most of them been replaced by partial dependence on farmers, shopkeepers, labor-recruiters, government functionaries, and politicians presenting themselves as social workers. But the different forms of oppression are often very sharply distinguished and brought under a common denominator by those who suffer from them.

The situation does not offer much hope for the near future. Escalation of violence seems almost unavoidable. Despite the striking moderation of its demands, the *halpati seva sangh* remains a thorn in the side of the farmers. Shortly after my departure a local leader of the organization — a Dubla whose father fled from Gandevigam years ago after a quarrel with his landlord — was murderously attacked by a field watchman under orders of an Anavil. The Dublas, on the other hand, are losing their fear precisely as a result of pauperization. When I was taking leave, one of them said that, before my next visit, I would have heard of large-scale open clashes between farmers and agricultural laborers.

# II

# DYNAMICS OF POVERTY AND PROSPERITY

*Rural South Gujarat Towards the End of the Twentieth Century*

*Chapter 12*

# Revisiting the Field over a Thirty-year Period

My anthropological fieldwork in the countryside of western India during 1961–62 focused on the interaction between the landowners and the landless, members of castes at the top and bottom of the local hierarchy. I stayed for over a year in two villages that differed from one another in some respects, but which showed similar dynamics in terms of agrarian relationships. What I described and analyzed was first and foremost the end of a system of bonded labor. In actual fact, the disintegration process had already occurred during the preceding few decades and what I noted during my fieldwork was its results and effects. Due mainly to economic changes, the landed elite no longer found the attachment of labor an attractive proposition; at the same time, bondage to the households of dominant landowners had become less and less acceptable to the male and female farmhands. State agencies had played only a marginal role in the drastic change that had come about in relations between employers and employees in the agrarian economy.

The slow erosion of the attached labor system that had been prevalent until the late-colonial era meant that men and women of the landless caste gradually reached a stage where they were able to exercise their own choice in deciding to whom to sell their labor power. At the same time, however, they lost the basic subsistence security which earlier generations had enjoyed: a daily pittance even when there was no work to be done. In other words, lacking any means of their own, they lost the support which was so essential if they were to survive the recurrent crises in their lives, and also the credit that had enabled them to bridge the gap in their means of existence. That deficiency was due to a wage far below the amount needed to remain independent. The possibility and even necessity for landless people to "live in freedom" in future brought an end to existing relations of patronage. Redemption from bondage, under the laws of the new mode of production, however, reduced agrarian workers to the status of a commodity for which the demand was limited. The disappearance of the elements of patronage which had helped to conceal the earlier abundant supply of rural labor put an end to the state of unfreedom in which this class used to live, but not to the stark poverty that accompanied it. That was how I described the reversal in the labor system which had come about at the time of my first fieldwork in western India.

Early in the 1970s I returned to south Gujarat, this time to investigate conditions of work and labor relations in what was soon to become known as the urban informal sector (see Breman 1976, 1977). The district town of Valsad was the location of that study, undertaken in 1971–72, and the base from where I made repeated visits to the villages of my earlier fieldwork. The result was the addition of a postscript to the English edition of my doctoral thesis, in which I reported the changes that had taken place in the preceding ten years.

In my subsequent academic life I continued my pre-occupation with the working class in that part of South Asia which I had selected, more or less by chance. In 1977, I started on a long-term project in the countryside of south Gujarat, in a district next to where I had started my anthropological fieldwork 16 years earlier. My point of departure was the acceleration that I had observed in the transition to a capitalist mode of production, the effects of which on the labor system seemed to me a fascinating theme of research (Breman 1978–79, 1985). The nature of the phenomenon precluded that I could suffice by staying merely in one or even more villages. The fieldwork consequently shifted to the camps in open terrain or along the roadside in which labor migrants put up their shelters, and to their widely-dispersed work sites. This extension in scale of the research and the inadequacy or even irrelevance of the village as a unit of study in charting the processes of rural transformation has been discussed elsewhere (Breman 1989).

In the mid-1980s, the social dynamics that I had continued to study at the local level caused me to re-phrase the problem, focusing now on the consequences of diversification of the rural economy for employment and labor relations. The increasing significance of work outside agriculture is closely tied to the fact that labor has become increasingly mobile over longer distances than ever before. This latest research was started by going back to my two initial fieldwork villages as bases of operation, partly because of my familiarity with the local situation but more particularly due to the fact that the inhabitants had become accustomed to my recurrent presence without experiencing any detrimental effects from it. Another advantage was that the return to this familiar milieu made it possible for me to gain better insight into the direction and pace of the social changes taking place in this part of India. Some time perspective was essential if I was to find an answer to the question of what impact the process of economic growth had made on the conditions of poverty that I had found earlier among the landless rural masses.

My attention continued to be directed primarily towards this class at the bottom of the rural hierarchy in the two villages. To obtain a quantitative basis for my fieldwork I held a census in 1986–87 in those localities inhabited by members of the agricultural laborer caste. A number of details were noted with regard to each Halpati household, particularly about employment and living conditions. Similar data were registered with regard to all Anavil households, i.e. the dominant caste of landowners by whom, a quarter-century earlier, the majority of Halpatis had been employed. Between these lower and upper strata are the castes of small farmers, Dhodias in Chikhligam and Kolis in Gandevigam, who form a considerable part of the total population. A more elementary survey was held among a random sample of these households, who belong to the broad middle sector of society. The totality of the material compiled in this way, together with the bench-mark data which I had initially collected, makes it possible to take stock of the dynamics of development on a micro scale and also of the differential impact of this process of change within the local order. The official census reports held at ten-yearly intervals provide additional information on shifts that have occurred in the composition of the working population at the village level. Classification of the inhabitants into various economic and social categories, further sub-divided according to age and sex, gives the impression of a very detailed stock of statistics. One problem immediately arises, however, which is that over the years these census reports have not always maintained the same definitions and that straight comparisons would therefore not be very meaningful. Moreover, it would be unwise to think that this statistical presentation is complete or entirely reliable, or to assume that the quantitative distortion of reality in successive censuses has always been biased in the same direction or to the same degree. My experiences during the first fieldwork had already brought me to this conclusion, and subsequent research at the local level has given me no cause to quell my skepticism of official data.

It should be understood that the latest fieldwork was intended to be a follow-up to, rather than a mere repetition of, the earlier study. The results of my research in the early 1960s and the late 1980s are thus not compatible, let alone directly comparable. In the first place, the main focus of research had changed to the significance of work for landless households outside agriculture and the village. Secondly, I did not originally intend to write a village monograph, nor can this sequel be described as such. I have no reason at all to change my earlier statement that the village in itself is not an obvious unit of research. When I first lived in Chikhligam I ascertained that roughly half the Halpatis had moved, a few years earlier, to a colony just outside the village where they had built their huts on a piece of waste land (Breman 1974: 96–97). Although they were thus no longer included in the village administration, they continued to work for local farmers and I, therefore, included them in my study. On returning to Gandevigam at the beginning of the 1970s I found that here too a section of the landless laborers had settled outside the village. Huts that had been destroyed in the 1968 floods had been replaced by a new colony along the roadside just outside the village boundary where houses with tiled roofs had been built with financial aid from government (ibid.: 252). From the economic point of view, the households that had found new accommodation here also remained dependent on their previous abode. My follow-up research was localized in the same old and new localities where Halpatis resided with whom I had been in close contact during my initial fieldwork. During my later stay in Gandevigam, however, I found out that a small and distant Halpati street consisting of eleven huts was actually sited on the territory of another village (ibid.: 154). More important than that was that the members of these households had always worked outside the village of my research. Since these households were not included in my latest fieldwork, I had to remove this street from my first census, for comparative purposes. After making allowance for this correction, I ascertained that the Halpati population in the two villages had increased between 1962–63 and 1986–87 from 632 to 889.

| *Village* | *1962–63* | *1986–87* |
|---|---|---|
| Chikhligam | 51 H. H. 327 members | 86 H. H. 493 members |
| Gandevigam | 50 H. H. 305 members | 79 H. H. 396 members |
| Total | 101 H. H. 632 members | 165 H. H. 889 members |

To put these figures into perspective they must be placed in a broader demographic framework. The total population of the district more than doubled between 1951 and 1991. The area lying within the present district boundaries of Valsad had 554,833 inhabitants in 1901, somewhat more than half the number that was registered by the first census held after Independence.

TABLE 1
POPULATION INCREASE IN VALSAD DISTRICT 1951–91

| *Year* | *Valsad district* | *Chikhli* taluka | *Gandevi* taluka |
|---|---|---|---|
| 1951 | 937,380 | 123,872 | 109,371 |
| 1961 | 1,137,801 | 139,986 | 130,304 |
| 1971 | 1,428,742 | 176,089 | 159,483 |
| 1981 | 1,774,136 | 214,128 | 190,049 |
| 1991 | 2,172,992 | 260,779 | 217,286 |

The figures in Table 1 thus point to a demographic acceleration in the second half of the 20th century. This trend has been caused in the first place by a fall in mortality rates. According to conventional wisdom, the birth rate remained at a high level initially, an assumption that is undoubtedly partly justified. However, more in-depth research than has been undertaken so far is needed to show how a persistently high birth rate is the principal cause of progressive population increase. In my opinion, that growth must have been influenced by migratory flows from across the district boundaries. My micro-studies have caused me to deduce that insufficient allowance is made for migration when drawing conclusions from census results, particularly for the temporary presence or absence of seasonal migrants. Only when allowance is made for this factor, for which the material available so far is still inadequate, will it be possible to trace the course of the birth rate with any certainty. My own data suggest that women in the fertile age brackets produce fewer children than a generation ago. According to the provisional results of the 1991 census, the crude birth rate in Gujarat has come down to 30 per thousand. Can this decrease be attributed to all classes into which the population is divided?

The Halpatis belong to the category of scheduled tribes who make up half the total population of south Gujarat. A popular notion. which also seems to pervade government thinking, is that the *adivasis* are distinguished from other social classes by exercising less restraint in their demographic behavior. This is due, it is said, to the fact that they have only recently become exposed to the civilization process. The results of a series of census reports, however, show no proof of this assumption. The share of scheduled tribes remained unchanged between 1961 (the first census in which they were separately classified) and 1991, i.e. 52.5 and 52.8 percent.

These figures do conceal important differences, however. In the first place, tribal castes are unequally spread over the district. In Gandevi *taluka* they form less than 30 percent of the population, against 70 percent in Chikhli, a disproportion which has shown little change in recent years. In the second place, 'scheduled tribes' is a collective term for a large diversity of social identities constituted by groups such as the Halpatis and Dhodias in south Gujarat. A majority of the latter own some land; the Halpatis are usually land less other than in exceptional cases. Such diverse economic backgrounds do not preclude a distinction in demographic profile. The lack of extensive and detailed demographic research makes it impossible to say anything about this with any certainty. According to the 1971 census, Halpatis represent about 20 percent of all *adivasis* in the district of Valsad, the majority living in Gandevi and in other *talukas* on the south Gujarat plain. Apart from a small minority the Halpatis are landless, and this is also the main feature of most households belonging to this tribal caste in my research villages. The report on my initial fieldwork mentions that in Chikhligam, where the soil is not very fertile and population pressure is less, a few Halpatis owned fields on which they grew their own crops (Breman 1974: 100–01). Their number has not increased in the meantime, and farming has become a sideline for almost all these households. Among Halpatis in Gandevigam landownership is even more exceptional. One of the very few owners is a newcomer to the village, a Halpati who has worked for some years in the Gulf region as contract laborer. He has utilized his earnings to buy a small plot of rice land which he then converted into a mango grove.

The fact that most Halpatis are landless does not mean that all landless are Halpatis. I have earlier described the lower middle castes, i.e. the Dhodias in Chikhligam and the Kolis in Gandevigam, as petty farmers. This still applies to the majority among them, although in this milieu the significance of work outside agriculture and usually away from the village has strongly increased. Income earned from these off-farm activities has actually enabled the Kolis to maintain their position as small landowners. Many Dhodias of Chikhligam, however, show signs of increasing landlessness. In 1986–87 almost a quarter of the households included in my survey, i.e. 17 of the 72 covered by my random sample, owned no land. This certainly did not make them equal in all respects to the Halpatis, a conclusion to which I shall return later. Nonetheless, it should be realized that, from the social point of view, the composition of the rural proletariat is becoming more diverse. In the following, the Halpati will still be presented as the prototype of the landless laborer, but this does not alter the fact that caste and class are not categories that can be regarded as completely synonymous.

Is it true that households living in poverty attempt to alleviate their economic vulnerability by producing more children? An affirmative answer would seem to reinforce the persistent stereotype of unbridled demographic behavior among the lowest social classes. That this is not due to supposedly foolhardy indifference on the part of these weaker segments but rather to rational consideration of costs and benefits makes little difference to the ultimately negative effect on the welfare level of society as a whole. On the basis of anthropological-demographic research carried out in the 1970s the thesis was put forward that a continual and acute shortage of employment opportunities in households dependent for their livelihood on the sale of labor power demands a high fertility rate, in order that the chances of earning an income may be spread as widely as possible and be as large as possible in peak periods (cf. Epstein & Jackson 1975).

It is no easy matter, however, to find empirical evidence of this alleged strategy. Analysis of the 1981 census data for Valsad shows that the size of landless families is smaller rather than larger as compared to the average for the total population of the district. Based on my quantitative data collected in 1986–87, I have further ascertained that Halpati households in the fieldwork locations average fewer members than do the landowning castes of Chikhligam and Gandevigam. This agrees with calculations in other macro-and micro-studies recently carried out, according to which laborers' households in the Indian countryside are the smallest among agrarian classes (Krishnaji 1989: 125). It must be remembered, on the other hand, that living together in a large family unit very rarely occurs in the landless milieu, while such a joint household pattern is far more customary among landowning classes.

Sweeping conclusions should not be drawn from the changes in numbers over time among Halpati populations of the two fieldwork locations. Without more accurate quantitative data, such as age distribution, number of children per woman in her productive years, in-migration and out-migration, little can be said with any pertinence about the demographic profile of Halpatis. A growth rate of 40 percent over a quarter-century is not noticeably high. Out migration from the villages of individual members and even of complete households has certainly occurred in the intervening years; however, during brief visits made over a number of years (1971, 1977–78, 1981–82, 1984, 1989, 1990, 1991) I have not observed that more Halpatis have left rather than entered these localities. That natural population growth seems to decelerate is corroborated by the fall in average numbers of members per household.

TABLE 2
CHANGING SIZE OF HALPATI HOUSEHOLD, 1962–87

| | *1962–63* | *1986–87* |
|---|---|---|
| Chikhligam | 6.4 | 5.7 |
| Gandevigam | 6.1 | 5.0 |

*Chapter 13*

# Work and Employment in the Landless Milieu

The census of Halpati caste members in the two villages made it possible to form an idea of the shift that had occurred over time in the work patterns of the rural proletariat. In the 1986–87 survey I enquired into the labor status of male Halpati household members of 16 years and over. The information thus obtained gave some indication of the size of the total working population in this landless milieu and of the changing pattern of employment since the beginning of the 1960s.

TABLE 3
WORKING HALPATI MALES, 1962–87

| | *1962–63* | *1986–87* | *Increase* |
|---|---|---|---|
| Chikhligam | 95 | 164 | 72% |
| Gandevigam | 85 | 124 | 46% |

In the two villages together, the number of working male Halpatis has leapt from 180 to 288, an average increase of 60 percent. Clearly, this figure does not reflect the total size of the working population. The poverty of landless households necessitates that as many of their members as possible, regardless of age and gender, should be economically active throughout the year. The problem, however, is that a single census is not sufficient to ascertain how much waged labor is performed outside the home by women and children. I realized only after completion of the survey in the Halpati colonies that it ought to be possible to quantify the differential participation in the labor process of both sexes and of various age cohorts. All my later visits were too brief, however, to enable me to carry out such a census of all households of this landless caste or, better still, to repeat it at various times of the year.

The absorption of children into the work force is a gradual process. Their economic life starts in the domestic sphere at a very early age. Unpaid work delegated by adults to the children includes getting water from the well, collecting wood and dung for fuel and grass for a cow or goat, caring for the younger children, and running errands. Girls are called upon more often than boys for such jobs. When they reach 8 to 10 years of age, the children follow their parents as unpaid helpers outside the home. Between the ages of 10 and 15 they are increasingly recognized as workers in their own right, though given less heavy tasks and therefore earning only half the adult wage. After reaching 15 or 16 years, boys are expected to do the same amount of work as the adult men, while girls handle jobs which equal those of fully-grown women. As we shall see, school attendance has only a marginal effect on this work pattern.

Apart from the rather limited category of maid servants, paid wage labor by landless women is very irregular. Need for their labor power arises only intermittently in agriculture or in the rural economy at large, although more in Gandevigam than in Chikhligam. Women who do not leave the village as working migrants, whether individually or with other household members, earn a wage for only part of the year; I estimate this to be from 60 to a maximum of 90 days in Chikhligam, and from 120 to 150 days in Gandevigam. Men's participation in the wage labor process in the village or its immediate neighborhood is considerably greater, ranging from a minimum of 180 days to 240 days per year.

Since women have less frequent employment outside the home they also add less to the total income. The different economic valuation in monetary terms between the sexes is strengthened by the fact that women's labor power is remunerated at a lower rate than that of men. This does not signify that only marginal use is made of female labor. Women are responsible for care of the household: preparing meals, repair and maintenance work, looking after the children and also cattle when available. Women are tied to the home far more than men, especially when children are small. A major part of the cost and energy spent on social reproduction from generation to generation is attributed to women, "We wait until we are called for" was their regular response when asked how they sought waged employment. As we have seen, however, the maximal number of household members, irrespective of gender, is mobilized whenever a chance of paid labor arises. In that sense, participation in economic production could not be higher. Insofar as the labor process is concerned, landless life is characterized by early entrance and late withdrawal by both males and females.

TABLE 4
WORKING MEMBERS AND DEPENDENTS OF HALPATI HOUSEHOLDS, 1986–87

| | *Chikhligam* | *Gandevigam* |
|---|---|---|
| Members | 5.7 | 5.0 |
| Dependents | 2.2 | 1.9 |
| Economically active, | 3.5 | 3.1 |
| Out of which: men | 1.9 | 1.6 |
| Women & children | 1.6 | 1.5 |

The greatly increased pressure on the meagre sources of subsistence is shown by the fact that the fall in average numbers of Halpati household members during the last quarter-century has not been coupled with a proportional reduction of the economically active. The figure for the latter has remained constantly high for the men of Chikhligam and Gandevigam, while the participation of women and children in the labor process has dropped only due to lack of demand and not to reduced supply.

One result of the changes in the rural economy is that female paid labor has become marginalized in the village economy, a trend that has weakened the economic position of women of landless households. It should be noted here that withdrawal from waged work for status reasons has had a similar effect on women in landowning households.

The employment of male Halpatis in and out of agriculture can be broken down as shown in Table 5.

TABLE 5
EMPLOYMENT OF HALPATI MEN, 1962–87

| | *Agriculture* | | *Non-Agriculture* | |
|---|---|---|---|---|
| | *1962–63* | *1986–87* | *1962–63* | *1986–87* |
| Chikhligam | 38 | 32 | 57 | 132 |
| Gandevigam | 62 | 72 | 23 | 52 |
| Total | 100 | 104 | 80 | 184 |

Notwithstanding the growth of the working population, total absorption in agriculture of labor hailing from the bottom class of the two villages has shown little change. On the other hand, there has been a considerable increase in numbers of landless who have found work in other economic sectors. The identification of Halpatis with agricultural labor is definitely a thing of the past. Households in this caste now depend for their livelihood on non-agrarian activities to a ratio of almost 2:1.

Within the villages of Chikhligam and Gandevigam the modest diversification in the pattern of economic production has shown little increase. Such local jobs that were available in the past to the landless were mainly semi-artisanal and used to be carried out by a handful of Halpati families. A few more occupations have emerged since the beginning of the 1960s: helper or owner of a neighborhood shop, proprietor of a roadside teastall or bicycle repair shop, a tailor, two laborers employed by the agricultural cooperative in Gandevigam. For the great majority of Halpatis, however, work outside agriculture entails that they must leave the village to seek a livelihood.

Population density in the district more than doubled between 1951 and 1991, while still maintaining the differences that had long existed between the *talukas* in this respect.

TABLE 6
POPULATION DENSITY PER KM² AND
URBAN GROWTH

| *Year* | *Valsad* *district* | *Chikhli* taluka | *Gandevi* taluka | *Urban* *population* | *Total %* |
|---|---|---|---|---|---|
| 1951 | 179 | 216 | 388 | 123,409 | 13.2 |
| 1961 | 217 | 244 | 462 | 181,655 | 16.0 |
| 1971 | 273 | 306 | 565 | 256,416 | 18.0 |
| 1981 | 338 | 373 | 674 | 388,862 | 21.9 |
| 1991 | 416 | 454 | 770 | 530,788 | 24.4 |

*Source*: Census reports, 1951–91.

The urban population more than quadrupled between 1951 and 1991, at a considerably higher growth rate than that of the rural population. These different dynamics are indicative of clear sectoral shifts in the economy. With one or two exceptions, urban growth poles are localized in a fairly narrow strip of the western plain. This zone runs from north to south, bordered on one side by the railway from Bombay via Surat and Baroda to Ahmedabad, and on the other side by the parallel highway. Here the industrialized belt is to be found that has developed in south Gujarat. In 1981 Gandevi *taluka*, situated in the heart of that zone, was already urbanized to an extent of 36.5 percent of the total population. In comparison, Chikhli *taluka*, situated to the east, has retained a more pronounced rural character. At the time of my first visit, its chief town had not yet attained the minimal size that determines an urban nucleus: 5000 inhabitants and a male working population of whom at least three-quarters are employed outside agriculture. In 1991 the urban residents of Chikhli amounted to 6.9 percent of the total population living in this sub-district.

It seems reasonable to assume that the position of two research locations in such different economic landscapes would be expressed in their patterns of employment. One would expect Chikhligam to be embedded in a rural-agrarian environment, while Gandevigam should similarly be characterized in terms of a more urban-industrial orientation. However, the figures given in Table 5 regarding the employment of male Halpatis in the villages of my fieldwork show no such contrast. In attempting to explain this apparent paradox, I first have to draw attention to the strong increase in labor mobility.

One of the most noticeable changes that have occurred in the thirty years since my first visit is that the countryside is much more accessible and, conversely, easier to leave. This observation applies in particular to Chikhligam, situated in the hinterland. In the past it used to take much time and trouble to reach this village from the nearest railway station. In the rainy season in particular, the rough country road between the village and the only town in the sub-district was almost if not totally impassable for weeks at a time. Nowadays, 21 buses run daily to Chikhligam and one is able to travel from there in various directions. Gandevigam is sited on a main road between two towns and has been in direct contact with the outside world since long ago. That external orientation has been increased in the last few decades by road improvements and a greater frequency of bus services. In addition to electricity, which was brought in far earlier than in Chikhligam, Gandevigam also has a few telephone connections. Of late it has even become possible to call overseas direct. The sense of isolation, which can still be felt in Chikhligam, has totally disappeared in Gandevigam.

Modernization of the physical infrastructure and of the communications system is only one of the indicators that characterize the rapid transformation of the regional economy. These dynamics particularly affect production and employment in sectors outside agriculture, in both town and countryside. In the 1970s and 1980s in particular, the southern part of Gujarat state underwent a vigorous process of industrialization. An annual increase of non-agrarian production of as much as 8 percent during this period shows an impressive rate of growth, but to draw general conclusions from such a leap forward would not be advisable. Similar emergent poles of industrial activity may be found elsewhere on the South Asian subcontinent. Conditions in other parts of India, however, comply far more with what development economists derogatorily used to call the Hindu model of economic growth, i.e. fluctuating between stagnation and slow expansion. Regional contrasts have therefore undoubtedly become more acute during the recent past. Nevertheless, I would dare to hypothesize that the societal dynamics observed in the area of my fieldwork indicate a trend that will probably become more widespread in the immediate future. More and more of the country will become involved in a process of change such as that manifested during the past few decades in south Gujarat.

At the start of my analysis of the impact this transformation has made on the people living in the research villages, I am inclined to argue that, in view of the rate of economic growth attained in the recent past, the numerical increase of Halpatis has not been so excessive as to preclude their absorption into the labor process. The economic acceleration and transformation that has taken place in the region of Chikhligam and Gandevigam can be seen from the fact that landowners and landless alike derive a steadily growing component of their livelihood from access to non-agricultural sources. In the case of landless households who depend almost entirely on income from their labor power, employment in agriculture decreased strongly between 1962 and 1987. The average figures that I have reported in illustration of this trend, however, conceal significant differences in this respect between Chikhligam and Gandevigam. The former shows signs of an absolute decline nowadays and only rather less than a quarter of male Halpatis can truly be called agricultural laborers. In Gandevigam their number has shown a marginal increase and they still form the majority among male members of this landless caste, although even here the decline has unmistakably set in.

I shall not elaborate in this study on the growing importance of wage labor performed by Halpatis outside agriculture and away from their village within the framework of this study. The results of my research on that subject will be published separately in a monograph. One question that immediately arises, however, is why agriculture does not absorb more labor or, in other words, why is it that this sector, which dominated the economy for so long, provides a livelihood for Halpati households in the field-work area only to a limited extent? This question cannot be answered other than in conjunction with a report on agrarian dynamics, or lack of them, in the two villages under study.

*Chapter 14*

# Continuity and Change in the Local Pattern of Agricultural Production

**How has agricultural production developed during the last 30 years? For a start, the crop system in Chikhligam has not undergone any drastic change.**

CROP SCHEDULE 1989–90 IN CHIKHLIGAM (HA'S)

| | |
|---|---|
| Main crops: | rice 104, fruit trees 34, vegetables 14, sugar cane 12, pulses 8, wheat 5, groundnuts 3, plantains 1. |
| Second/third crops: | vegetables 43, pulses 14, rice 5. |

For major landowners, the cultivation of paddy and mangoes still form the basis of their farm management. Late in the 1970s the mango harvest had deteriorated so badly that some orchard owners decided to replace the old trees by new ones. This meant a temporary drop in earnings but yields have now returned to their former level. A few large farmers have started to cultivate sugar cane which is harvested and processed by two factories at some distance from the village. Due to the transport costs involved and to the low priority given by the factories in their harvest schedule to this outlying block of land the crop is still not very profitable. Some time ago, however, an Anavil Brahmin from Chikhligam, who had been *sirpanch* at the time of my initial fieldwork, took the initiative to establish a co-operative sugar factory within the sub-district. His appointment as head of the Taluka Development Committee made it possible for him to canvass support for this plan. As chairman of the regional branch of the Congress Party he had the political contacts that were needed to gain the approval of authorities at higher levels. After a delay lasting several years due to political machinations and to opposition by the two mills that obtain cane from Chikhli, final approval of the plan came through in the Spring of 1991. The site of the factory has been decided upon, i.e. at some distance east of the *taluka's* main town, and construction of the new co-operative enterprise is scheduled to start at the end of the 1991 monsoon. Plant equipment required for the daily crushing of 2500 tons of cane per day has been ordered. Thousands of large and small farmers in the sub-district have taken out membership, including many landowners in Chikhligam who intend to grow more sugar cane in future.

An obstacle to all this is the limited acreage that is at present under irrigation, i.e. only 57 of the 372 ha. of village land. Rather more than one-third of that total (135 ha. in 1990) is in use only as grassland. The fields devoted to agriculture (slightly less than 200 ha.) have to depend mostly on rainfall. In other words, the concentration of agricultural activities at the beginning and end of the rainy season is a perpetuation of the pattern found in the early 1960s. Irrigation water reaches less than one-third of cultivable land, an expanse that has increased only marginally since my first fieldwork. This shows that agrarian forces of production in Chikhligam have undergone hardly any development at all. The irrigation canal brought to the village in 1953 is still under-utilized to the same degree as then, irrigating no more than a total of 20 ha. The landowners say this is because they are the last to receive irrigation water, so that they run the risk of being deprived at the time when their need is greatest. But being at the tail end of the irrigation scheme is not their only complaint. They object strongly to government's refusal to subsidize the construction of secondary and tertiary canals. The wide dispersion of the field and the undulating nature of the terrain undoubtedly also play a role. Officials at the district office of the Soil Conservation Department told me, however, that only in very few instances had they been able to persuade large landowners in Chikhligam to level their unproductive land with the use of cheap credit that only had to be repaid after some time and then in 15 annual instalments. The inclination to make agrarian capital more productive will in all likelihood increase in view of the imminent expansion in sugarcane cultivation, a crop which has brought prosperity to farmers elsewhere in south Gujarat (Breman 1985).

Land owned by the Dhodias is undoubtedly more intensively

cultivated. Apart from the paddy which they grow primarily to meet the needs of their own households, these small farmers have discovered the rapidly growing market for horticultural products and specialize in crops such as vegetables and *chillies*. Some have dug wells and installed oil-driven pumps to intensify the yield of their land. Compared to the Anavils, Dhodias are far more active in agriculture, but in their case the marginal size of their farms prevents a more drastic increase in production.

An indication of the low capital intensity of agriculture is that the 16 households belonging to the dominant caste, who still own a disproportionate share of cultivable land and certainly of the most fertile acreage, make very little use of modern agricultural equipment. For example, they own only two of the popular mini-tractors. The Anavils were the most rapacious lenders from the local cooperative credit society operating under firm control of their dominant caste position (Breman 1974: 108). Contrary to regulations, the funds taken up by the village elite were not used for investment in agriculture. Moreover, a large part of such loans was never repaid, with the result that at the end of the 1980s the District Cooperative Bank had to declare its Chikhligam branch bankrupt. The prominent defaulters, who included the former village head, did not suffer at all. The real losers were the Dhodias who made up the majority of members.

The use of draught cattle has sharply declined. The keeping of buffaloes and especially cows, however, has been given new impetus by the establishment of a dairy cooperative in the sub-district, with a cold store sited on the highway close to the center of the *taluka*. A local branch of this co-operative has been set up in Chikhligam in 1977 and charged with collecting milk from members, most of whom are Dhodias who have only one or two cows each; Anavils are not interested in keeping milch animals for commercial purposes. To have just a few is only profitable for those willing to look after their own cattle.

My conclusion is that the Anavil Brahmins in Chikhligam are basically not prepared to behave as agricultural entrepreneurs, managing a technologically well-equipped and efficiently organized farm. Only 5 of the 16 heads of households in this dominant caste can be identified as farm managers. A larger number leave cultivation of their land partially or even wholly to persons whom they call partners in an apparent attempt to avoid the word tenant because of the legal implications that this term involves. Kolis from neighboring villages, who are deemed trustworthy and who require more land in order to become viable farmers, are usually considered for such partnerships. When growing rice, production costs and yield are usually divided equally between owner and operator. Land intended for sugarcane is rented at a fixed tariff of Rs 1600 per acre, a sum of money that far exceeds the financial strength of most Dhodias.

Finally, the lack of interest in agriculture among the present generation of Anavils is shown by the quantity of land that has been sold during the past few years. Good agricultural land in the village has changed hands for Rs 30,000 to 50,000 per *bigha* (somewhat more than a quarter hectare), but purchasers willing to pay such an amount are difficult to find. Dhodias who have earned sufficient money outside agriculture to be able to afford a modest expansion to their smallholdings are not the only or even the most important buyers. In the 1980s five Koli farmers from a neighboring village bought land Chikhligam and subsequently settled there.

The crop system in Gandevigam has undergone more drastic change, as Table 7 shows.

TABLE 7
GANDEVIGAM: LAND-USE IN 1960–61 AND 1989–90

| | *1960–61* | | *1989–90* | |
|---|---|---|---|---|
| *Crop* | *hectare* | % | *hectare* | % |
| Fruit trees | 53 | 33 | 100 | 64.5 |
| Grassland | 27 | 17 | 0.5 | — |
| Pulses | 23 | 14.5 | — | — |
| Rice | 22 | 14 | 14 | 9.0 |
| Vegetables, elephant's foot, ginger | 15 | 9.5 | 10 | 6.5 |
| Plantains, papaya, coconut | 14 | 9.0 | 6.0 | 4.0 |
| Sugarcane | 4.5 | 3.0 | 24.5 | 16.0 |
| (of which planted more than once a year) | 17 | 10.5 | 10 | 7.5 |

Fruit tree cultivation has increased greatly in Gandevigam and its surrounding villages. Large farmers in particular have turned almost all their land into orchards. Nothing is left of the dry *jarayat* which was anyway only used for grassland. Even the rice crop, of which I had predicted on the basis of my findings in 1971 that production would further increase, has lost much of its significance. Out of the second crops that used to be planted in immature orchards in order to give shade to young trees, only plantains are still grown to any significant extent. Elephant's foot, various sorts of ginger, black pepper and other garden products for which the region used to be famed, have almost disappeared. Only the acreage planted with sugarcane has expanded.

Clearly, there is no sign of agricultural stagnation in Gandevigam. Traditionally, the majority of village land has been held by the dominant Anavil Brahmins, most of whom are still willing to invest in their property. They do not sell their land but, on the contrary, have upgraded its fertility and value in recent years by converting the remaining grassland into orchards and by irrigating dry fields. The purchase price for a *bigha* of horticultural land has risen to Rs 200,000 to 300,000, but the virtual lack of demand means that market transactions in land are rare.

D is already an old man, but he still goes out early each morning to inspect his property and to apportion the work for the day. Time is like money that you can use only once, he says. On a few plots located outside the village he grows roses. In the nearby towns there is a rapidly growing market for these flowers, which are made into garlands. Within a few years he has met all the costs needed to turn the bare and dry soil that was unsuited to agriculture into a flourishing rose garden. His business administration shows that he invests at least Rs 75,000 each year in order to make his property more productive.

S is the son of the farmer who, many years ago, was the first to plant *chiku* trees. By purchasing land outside the village with which to extend his orchard property, S has become the largest fruit producer in the region. In the middle of the 1980s, after being elected chairman of the region's agricultural co-operative, he drew up a plan for canning mango pulp. He has now stepped down from his leading position on the board, but a very successful start has been made with his agro-industrial initiative.

The Kolis of Gandevigam are also good farmers but do not own enough land to allow them to equal the agrarian success of the

Anavils. Orchard owners among them have increased in number, but the growing number of fruit trees has led to a reduction in other crops which they used to grow. Those Kolis who make the most intensive use of their land cultivate paddy, vegetables, plantains and sugarcane. They usually market their own produce because the regional cooperative shows little interest in anything other than the sale of fruit. Koli women sell the vegetables early in the morning in urban markets or peddle them, a basket on their heads, in neighboring villages. The increased cultivation of sugarcane is largely due to Koli farmers. The mill in Gandevi *taluka* accepts cane only from growers who belong to its cooperative society, but Kolis are rarely members because few of them can afford the price of a share with which they have to buy membership. The problem may be solved by delivering the cane to the mill in the name of another member-farmer, usually an Anavil Brahmin, who naturally expects a commission for such mediation. It is far more lucrative, however, to process the cane into *gur*, which is then sold to illegal distillers of alcohol, and this is done by the majority of Koli producers.

Living and working in the shadow of Anavils as they do, few Kolis are able to depend on their farming income alone. The smallness of their holdings entails that, with few exceptions, they have to combine agriculture with other activities. The necessity to do so has also increased because, now that fruit growing has become by far the most important source of income, only a very few Anavils are still prepared to allot land on a sharecropping basis to their former tenants. Several Kolis are small traders in agricultural commodities. Their business is seasonal in nature, buying windfalls or fruit that becomes damaged during harvesting, or is over-ripe. The regional co-operative accepts only fruit in perfect condition.

The cooperative organization in charge of local sales for agrarian products, which has a membership of farmers belonging to 8 villages in the immediate vicinity, is flourishing as never before. The modest accommodation of the early 1960s has been replaced by a compound with a two-storey office building and a number of sheds and stores, sited just outside Gandevigam on the territory of the adjacent village. The manager in charge has a permanent staff of 60, approximately half of whom work in local branches established in the affiliated villages.

The main purpose of the cooperative is to market the agrarian

TABLE 8
GROWTH OF THE REGIONAL COOPERATIVE, 1960–90

| *Year* | *Membership* | *Sales value of agrarian products (rupees)* |
|---|---|---|
| 1960–61 | approx. 700 | approx. 700,000 |
| 1970–71 | approx. 900 | approx. 900,000 |
| 1984–85 | 1754 | 11,815,336 |
| 1989–90 | 2474 | 17,714,211 |

*Source*: Annual Reports of the Regional Cooperative.

products of its members. However, it also has a number of stores in which farm equipment as well as daily consumer goods can be bought: fertilizers, insecticides and pesticides, agricultural implements, cattle feed, cement, wood for construction, roof tiles, clothing, rationed grains and sugar. In the mid-1980s the annual turnover of such articles had reached almost one *crore* of rupees, with a net profit of over half-a-million rupees. The co-operative also owns a rice mill, a jeep and four tractors. The vehicles together with driver are hired out to members for transport purposes and, in the case of the tractors, also for land levelling. Such wide-ranging services explain why membership has increased far more than the number of farmers who sell their produce to the cooperative. In 1989–90, 96 percent of purchases consisted of fruit; the almost negligible remainder was made up of plantains and *suran* (elephant's foot). During the latest season, *chikus* and mangoes were bought from 956 and 878 members respectively, most of whom were one and the same farmers, belonging to the caste of Anavil Brahmins, the dominant landowners in the area. Nowadays, *chikus* are grown more than mangoes because the yield of the latter is far more susceptible to climatic variations. In 1989–90, a heavy storm in the middle of the growing season caused a great deal of damage to the region's fruit trees, with the consequence that business results lagged behind those of the previous three years. The 1990–91 harvest was again excellent, however, and it seems that the next report of the co-operative will show a jump in sales to more than two *crores* rupees.

Mangoes are sold mainly in Gujarat, while the *chikus* are sent to Delhi, in 1990 a total of 250 lorryloads. The manager of the co-operative has a Bachelor's degree in Agriculture and provides members with information on new horticultural techniques. He is also interested in agro-industrial activities, understandably enough in view of the limited durability of the commodity which is the main trade. During the 1990 season a start was made with canning hundred tons of mango pulp. Experiments are also being made with drying *chikus*, the idea being to package and sell them as sweet crisps. During my latest visit I was given a large packet to advertise in my own country. "Good to eat when watching the telly", I was instructed to say.

After this brief overview of the present state of the region's agricultural economy and the strategic role of the landowners in it, I shall now examine the contribution made by the landworking proletariat. In particular, what are the conditions of work of the Halpatis and who are their employers?

*Chapter 15*

# Labor in Agriculture and Modes of Employment

The intensity with which Dhodias and Kolis in particular cultivate their land as small farmers demands a great deal of labor power, of which the owner's household is usually the principal or sole supplier. When the hiring of outsiders becomes inevitable during peak periods then the temporary and limited demand is almost always met from the immediate social circle, i.e. by less fortunate members of the same caste living close by. Since my first stay in the region, differentiation among members of these lower middle castes has become more visible but nevertheless has not led to any major discrepancy between their positions in the hierarchy of caste and class. It means, in effect, that small peasant households are by and large autonomous in their labor requirements. Simultaneously, the tendency that I noticed earlier in this milieu to withdraw from the agricultural labor market has been strengthened. In Gandevigam, the majority of Koli households have escaped into off-farm employment to seek the necessary addition to their income, and neither women nor men will work in the fields or orchards of the larger Anavil landowners. The situation among the Dhodias in Chikhligam is basically the same. Even those who are wholly landless manage to avoid a life as agricultural worker by migrating away from the village, eventually ending up at the bottom of other circuits of employment. Earlier, I have described this strategy by the lower middle castes as part of their attempt to evade the authority of the dominant caste as much as possible (Breman 1974: 246–47), and that conclusion still holds.

No substantial shift has occurred in the use of external labor by the various landowning castes, a finding which seems to indicate continuity in the social relations of production. Does this signify that the extremes of the local hierarchy, i.e. Anavils and Halpatis, are fated to continue their mutual involvement as agricultural employers and workers respectively? Broadly speaking, this is indeed the case, but there are major differences in the mode of employment in the two villages.

Notwithstanding the fact that the territory of Chikhligam is almost double that of Gandevigam, its large farmers can make do with far less labor. This is partly due to the fact that Anavils own a smaller percentage of the agricultural acreage, and also to their growing tendency to leave cultivation of their land to "partners" from middle castes. Also of significance is the poor fertility of the soil combined with the restricted availability of irrigation water. For every three hectares of dry land in Chikhligam there is less than one hectare on which other than rain-fed crops can be grown. In Gandevigam, on the other hand, irrigated land represents rather more than four-fifths of total acreage. This explains not only why more workers are employed in agriculture in Gandevigam, but also why they have year-round work.

The lack of interest in farming shown by Anavils in Chikhligam is a major factor for a relative and even absolute decline of the absorption of Halpatis in field labor. The far more dynamic nature of the agricultural economy in Gandevigam means that the scant increase in total Halpati agricultural workers must be due to other reasons. Why is it that the number of workers from local landless households employed by large landowners in Gandevigam lags far behind the continual expansion of agricultural production? It is conceivably due to increased fruit cultivation, which is less labor-intensive than earlier crops of paddy and plantains as well as the cultivation of a rich variety of horticultural products. On the other hand, even orchards need labor all year round, not only for harvesting but also for maintenance. Another factor might be the growing capital intensity of farm management in Gandevigam, causing the share of wage labor to decline in the relative sense and also in absolute terms, through replacement by motorized equipment. There has certainly been a considerable increase in the use of tractors for tillage and for transport in particular, but fruit-growing and picking has not been mechanized in any way at all.

The most important reason why more Halpatis are not absorbed into agriculture is undoubtedly their replacement by outside labor. The arrival of temporary migrants seeking employment is not a new phenomenon in either village, and farmers have made use of these transients in the past. While living in Gandevigam in the early 1960s I observed how a group of *kathiawadi* offered their services for land digging and leavelling, strenuous work for which these people from Saurashtra who migrated during the dry season, were wellknown. In the same village I also saw some *biladia*, contracted by orchard owners for the fruit-picking season (IBid.: 212–13). When I visited again in 1971 I noted a definite increase in the influx of migrant labor (ibid.: 245).

In various publications I have given detailed attention to the political economy of this mass circulation of labor in the countryside, on the basis of fieldwork carried out in nearby Surat district during the late 1970s and 1980s (Breman 1978, 1985, 1990). In addition to the deployment of an army of cane-cutters recruited for the duration of the harvest campaign, lalsting five to six months, individual farmers also made growing use of migrant labor. In Chikhligam and in Gandevigam the influx of seasonal workers has continued, though least in Chikhligam due to the declining interest in agriculture shown by large landowners. But even they hire small groups of migrants who trek through the countryside during the grass-cutting, paddy-harvesting and fruit-picking seasons. Transients who come to the village throughout the year are also given a wide range of chores to do. Such workers are few in number, however, and their stay short, rarely lasting more than a few days or a week at most.

These seasonal agricultural workers come mostly from the eastern hills bordering on Maharashtra. They are the tribal Dhangis and Kuknas, who make their main living from tilling their own fields. The combination of increasing population pressure and the felling of forests on which they depended for off-farm income, has caused growing numbers to migrate for part of the year to the plains of south Gujarat. Informants among them told me that a persistent drought in the early 1970s had given extra momentum to their exodus for part of the year. Anavil farmers in Chikhligam found these migrants driven from their homes by severe famine, prepared to work for an entire day for a pittance, quite often for nothing more than a meal. In "normal" years these seasonal workers come in small groups of 10 to 15 men and women to harvest the various crops in order as they themselves say, to meet their need for cash. All food rations are brought along for the duration of their absence. They can be met along the roadside, walking in single file. The adults carry pots and food supplies on their heads, tools in their hands and sometimes a child on the arm. Their movements should not be seen as aimless drifting through the rural landscape. They go straight to addresses where they have been before and are directed to new employers through established contacts. Neither do they migrate for an indeterminate time, but for a period lasting not more than a few weeks, until the grain supplies brought with them are finished. They then return home before making a following sortie with the same or another gang. Arrangements, made long in advance and sometimes clinched with a down payment by a jobber long before departure, contribute to the spasmodic rhythm that characterizes seasonal mobility. Conditioning factors such as these largely escape the comprehension of uninformed outsiders.

The massive army of cane-cutters, who are employed in tens of thousands by the management of sugar cooperatives in south Gujarat for the harvesting period between November and June, are impossible to overlook. In both Chikhligam and Gandevigam, cane destined for the mill is cut by workteams sent for this purpose by the agro-industrial enterprises. The forthcoming expansion of sugarcane acreage in Chikhligam will encourage landowners to level and irrigate fields that in their present shape are not suited to agriculture. Increase in land-use intensity, however important it may be in terms of production, is unlikely to cause Anavil Brahmins to become more active farmers, nor will it have a favorable impact on Halpati employment. Practice has shown that the agro-industries, which exercise tight control over every activity in sugarcane cultivation and in particular over its harvesting, import labor on a large scale from other regions. I have reported on the exclusion of the army of the local landless by the cooperative sugar mills in south Gujarat in separate publications.

Migrant workers are also put to work in fruit tree cultivation. Just as in the past, traders in Chikhligam buy the mangoes from various growers while still in flower. They station a couple of guards in the orchards and send their own mobile workteams to harvest the fruit, thus illustrating how the use of migrant labor has been linked to traditional commercial practices established a long time ago (Mukhtyar 1930: 193–94). Farmer-members of the co-operative in Gandevigam are not allowed to sell their fruit to outsiders and are, moreover, responsible for their own harvesting. Most orchard owners ensure the services of one or more seasonal workers who come there after the rainy season and return home just before the start of the next monsoon. These Dhangis—mostly young men but sometimes also entire families—work in the orchards of one employer for about seven months.

In this busy time of the year, the agrarian workforce in Gandevigam and the surrounding villages may increase by one-fifth to one-quarter. The incomers set up camp in the partly dry bed of the Ambika river. When I asked why they lived there in a collection of make-shift shelters, the Anavils used to tell me that it was by choice. According to the migrants, however, farmers trying to prevent the theft of fruit do not allow them to stay on in the orchards at the end of the working day. Still, the owners are only partially successful in restricting access to their property beyond the hours of work particularly after nightfall. Petty traders, who are always hanging around at the outskirts of the village, are assured of a ready supply of illicitly picked *chikus* and mangoes.

On closer examination, therefore, the decline of wage labor among the local landless in the agricultural economy has to be explained at least partly by the influx of migrant workers. I have discussed this replacement of local by alien labor extensively in the outcome of my field research in a neighboring district of south Gujarat (Breman 1985). Even apart from this circulatory trend, the growth of total agricultural output in Chikhligam was far slower than in Gandevigam, and has not shown any proportional increase in the absorption of agrarian labor. The labor cost component has lagged behind in the total rise of expenditure on production. The profitability of agriculture has increased, but in such a way that returns on capital are rising without any simultaneous and proportional improvement in the extremely low wages paid to agricultural labor. A breakdown of Halpati males working in agriculture according to the nature of their employment will help me to clarify this point.

The figures given in Table 9 show a drastic shift in the working pattern, namely, the decreasing importance of permanent farmhands and the replacement of this part of the workforce by predominantly casual labor. In Chikhligam, the proportion of Halpati males working on a daily basis has risen from 45 to 60 percent, and in Gandevigam from 20 to 58 percent. The number of day workers in

TABLE 9
EMPLOYMENT MODALITY OF HALPATI MEN IN AGRICULTURE, 1962–87

| | *1962–63* | | | *1986–87* | | |
|---|---|---|---|---|---|---|
| | *Total* | *Farm-hands* | *Day workers* | *Total* | *Farm-hands* | *Day workers* |
| Chikhligam | 38 | 21 | 17 | 32 | 13 | 19 |
| Gandevigam | 62 | 50 | 12 | 71 | 30 | 41 |
| Total | 100 | 71 | 29 | 103 | 43 | 60 |

the two villages combined has doubled, from 29 to 60. The actual increase is far less great, however, because in addition to work in the fields, members of this landless caste do all sorts of odd jobs in and around the village, such as wood cutting, well digging, and extracting sand from the riverbed. I have estimated that on an annual basis such diverse activities may occupy not less than one-third of effective work time for a sizable number of them. Agricultural labor still provides even these Halpatis with an important part of their total income, but this does not alter the fact that the landless existence can only be understood by stressing the occupational multiplicity of the rural proletariat. A characteristic of daily wage earners is that they are contracted by continually changing employers. They may work for one farmer for some time at a stretch, but neither side has any obligation to continue the relationship from one day to the next. Nowadays, a considerable amount of work is given to casual labor on a contract basis (*udhad*), a system to which other modes of wage payment apply, i.e. not in time but on a piece rate or a job basis. Arrangements like these are usually made with workers collectively. The landowner agrees to conditions of work with someone who acts as a gang leader rather than with every individual worker. Such work teams are not permanent, but are likely to break up once the job is finished.

Farmhands fall into two categories. The *rojio* is employed by a landowner who has the first right of call on his labor power. Although quite frequently, this type of labor is not paid on days when no work is available, the farmer has the right of first refusal, but he is not at liberty to replace these regular workers with temporary substitutes. Conversely, the farmhand may only seek other work on those days when his boss has no need of him. As permanent workers, *rojios* cannot profit from the somewhat higher wages that are paid in peak periods of the crop cycle. On the other hand, their security is greater than that of day workers.

The second type of farmhand is the true successor to the former *hali*. He is closest to his employer and is given tasks that go beyond field operations. Moreover, his working day lasts longer because he starts earlier in the morning and continues after other workers have gone home. Even on slack days he does all sorts of odd jobs in the farmer's house or on his premises, including some unpaid labor services. The farm hand is responsible for collecting, returning and maintaining equipment, and often transmits the boss's orders to the casual workers. He does not receive a higher wage for his longer working day but is given some perquisites: sometimes a meal or at least a cup of tea, a packet of *bidis*, some fruit during the picking season.

The farmhand's female equivalent is the domestic maid, a Halpati woman who works in the household of a member of the landowning caste. Maids always work in the mornings for between two and four hours, dependent on the chores they are expected to do: tidying the house and premises, cleaning pots and pans, doing the laundry, fetching water, preparing food for cooking, and in general being helpful to the mistress of the house. In my research villages most Anavil households have such domestic help for at least a few hours each day, usually the wife of a farm servant.

Between 1962 and 1987, the total number of permanent laborers among the Halpatis decreased by almost half, a reduction that is only partly countered by the introduction of migrants as seasonal farmhands, particularly in Gandevigam. Taking the dubious status as agricultural laborer of some day workers among the local landless into consideration, I am inclined to maintain my earlier conclusion that employment in the agricultural economy has not kept pace with the growth of production. But can it also be said that the inflow of migrants has caused the expulsion of local labor? Confirmation or negation of this hypothesis has to depend on the answer to whether Halpatis diverged into other sources of livelihood because they decided out of their own choice to do so, or because local landowners showed increasing preference for migrant workers. The situation in my two research villages differs in this respect.

As a result of their reduced interest in agriculture, Anavil landowners in Chikhligam need less labor. At the same time, however, they say that it is very difficult to find workers when these are needed. They complain particularly about the unwillingness of local landless to engage themselves on a permanent basis. The Halpatis in Chikhligam have been accustomed for several generations to leave the village for a considerable period because of their inability to find year-round employment. Already in the early 1960s, many of them left after harvest at the end of the monsoon, to work in the brick kilns in the neighborhood of Bombay. This pattern of seasonal migration probably started at the end of World War I. In his monograph written in the late 1920s about a village located not far from Chikhligam, Mukhtyar refers to the absence of Halpatis who had left for Bombay for this reason (Mukhtyar 1930: 160, 169). The same circulation, in which a sizable proportion of the landless in the location of my fieldwork is involved, continues until the present day and will be dealt within a forthcoming publication. Here, it will suffice to record that the Halpatis of Chikhligam have found another option to a life as agricultural laborer. The certainty that work migration will enable them to earn a living is a reasonable alternative to remaining in the village where there is a scarcity of employment. In the long run, work in the fields of Chikhligam as day laborer yields too little, and the greater majority of Halpatis refuse to work as farmhands. However unattractive the life of a seasonal migrant may be, younger people in particular prefer it to the dependency that is attached to a life which severely restricts social autonomy. It is interesting to note that the few permanent workers who are still available to Anavils almost all belong to Halpati households who have stayed behind in the village, and that too to a restricted number of families. Such households could be considered as still under the control of members of the dominant caste. With few exceptions, Halpatis who live in the new colony established just outside the village unanimously reject the idea of such attachment.

Although agriculture in Gandevigam needs far more workers the whole year through, Anavil farmers say that they have a plentiful supply from which to choose both regular and casual hands. The arrival of migrants who stay for several months at a time or all year round, other than in Chikhligam, plays a role in this but is not of decisive significance. The majority of large landowners find sufficient number of Halpatis who are prepared to offer their services, either as casual wage earners or as permanent farmhands. But don't the employers prefer migrant workers? The Anavils readily admit that Dhangis coming from the eastern hill tract are less difficult to handle. In other words, they are more obedient and more easily satisfied with what they are given. Dhangis do not complain if they are occasionally or rather frequently required to do more than the standard work load, even without extra payment. At the same time their behavior is unpredictable. Quite often they will leave without giving any notice at all and without apparent reason, according to their employers, or will not show up after the rainy season, notwithstanding an explicit prior agreement. You can't really depend on them, is an opinion frequently voiced by farmers, meaning in effect that they have a better hold over the local landless. Although they are always finding fault with the Halpatis, Anavils seem to accept that they cannot manage without the help of this bottom segment of the village.

Based on his unfavorable experiences a few seasons ago, B has decided not to use migrant workers again. After *Divali*, he had taken on four Dhodia families from one of the eastern *talukas* who did all the work in his orchards and who promised to return each year. Soon after *Holi*, however, they had run out on him. They left one night without having said anything of their intention. B assumed that they had returned to their village to prepare their land for the coming rains. He was left in great difficulty because their departure coincided with the peak of the mango harvest, when very few Halpatis were left without employment. Even those who were still free refused to work for him. Only by paying an extra high wage, to the great displeasure of other farmers, was he able to get the fruit picked.

How do Halpatis feel about the fact that their labor power is tolerated rather than valued in Gandevigam? At any rate, they do not react by opting out of agriculture and the village in large numbers. An important factor is the lack of any tradition of labor migration among the local landless. They have little notion of what work may be available outside. How would it be possible for them to come to know about employment opportunities at any great distance? They are able, after all, to sell their labor power for the greater part of the year within a radius of just a few kilometres. The steady demand in the immediate vicinity thus deters them from trying their luck elsewhere. Industrial estates situated in or near the towns pay a wage to beginners that is only a few rupees above that of agricultural workers. Halpatis coming from remote villages will need to spend most or all of that extra money on travelling. A return trip by bus from Gandevigam to the *taluka* town cost 2 rupees, and to Bilimora as much as 5 rupees in 1990. Moreover, fluctuations in industrial employment differ little from those in agriculture. Apart from seasonal peaks and drops in production that are coupled with the hiring and firing of labor respectively, it is quite common for newly-opened workshops to close down again after just a short time. Research carried out by Streefkerk in the early 1970s and repeated by him in 1990 shows that enterprises established on industrial estates in south Gujarat suffer from a high mortality rate (Streefkerk 1991). Although most Halpatis who find their way there manage to get similar employment somewhere else, such a change is likely to occur more than once.

When all is said and done, employment conditions for unskilled labor outside the village economy do not differ much from those of wage earners in agriculture. It follows, therefore, that if sufficient work is available in the fields for a wage that is not much lower than that of non-agricultural labor, the local landless will feel little urge to go elsewhere. In my opinion, this is the reason why Halpatis in Gandevigam, for whom alternative employment is within relatively easy reach, have nevertheless sought and obtained less access to it than their caste-mates in Chikhligam who live farther away. For the latter, the leap into the outside world becomes inevitable given the scant use that is made of their labor power in local agricultural production.

This difference does not explain, however, why farmers in Gandevigam still find that many members of the local proletariat are prepared to work for them as regular farmhands. Why do these Halpatis not prefer the greater autonomy enjoyed by casual wage earners? After all, the demand for labor is fairly constant, so that even members of the landless caste who are not in regular employment can earn enough throughout the year to meet at least their most elementary needs. Moreover, farmhands are not eligible for the slightly higher wages that are sometimes paid to casual labor, particularly at peak periods of the agrarian cycle. On the other hand, working as a permanent laborer provides a basic security and precludes the constant need to seek a new employer, with all the risks that this entails. All Halpatis do not weigh the pros and cons in the same way. In the locality just outside the boundaries of Gandevigam to which many Halpatis moved two decades ago, I found a sphere similar to that in the new colony adjacent to Chikhligam. Insofar as the landless living here choose to work as agricultural laborers, they do so because unskilled work away from agriculture and the village does not yield so much more money that it is worth the trouble and expense of seeking it. They combine agricultural labor with other jobs in the near vicinity of the village. But however resourceful they need to be in broadening the basis of their livelihood, they show no interest whatsoever in accepting the status of farm servant. These Halpatis try to avoid such an attachment to one farmer just as much as their fellows in Chikhligam. To their mind, the greater economic security cannot compensate for the restricted freedom of movement that is the lot of the successor to the *hali*. It would be premature, on the other hand, to conclude from this that those who do opt for that status may be accused of a traditional mentality, implying that to some degree they have internalized dependency on their employer.

*Chapter 16*

# Bondage and Dependency Re-examined

Patronage and exploitation were the terms with which I characterized the former relationship of bondage that used to exist between members of the landowning Anavil caste and the landless Halpatis in south Gujarat. The rights of the bondsmaster were the duties of the *hali* and the other way round, but this naturally did not imply that the exchange was an equal one.[1] I have actually described patronage as the social-political dimension of a pre-capitalist system of exploitation (Breman 1974: 36–37; 1985: 129). Mundle has gone one step further while discussing the historical background of contemporary relationships between the landowning and landless classes of south Bihar. As a political economist, he argues that the debt bondage that was still found practiced in the mid-1979s had originated at the end of the nineteenth century as a consequence of the penetration of capital into the backward agricultural economy of his region of research. The industrialization of the bonded labor system in the *malik-kamia* relationship, in his opinion, should be distinguished from forced *begari* labor that dated back to the pre-colonial and pre-capitalist era (Mundle 1976: 95–96). Mundle does not elaborate on the earlier existence of agrestic servitude, and does not specify, therefore, how the "modern" *kamiuti* system differs from the traditional state of subjection into which tribal people in the frontier area were forced by caste Hindus in the more remote past.

This transition is the central theme taken up by Prakash in his study of the social history of the *kamia-malik* tie in south Bihar. Following upon Mundle's contribution, which is mentioned rather haphazardly in a footnote, Prakash puts forward the proposition it was only in the late nineteenth century that *kamias* came to acquire the

[1] The positive terms in which I would have expressed, according to Brass, the labor bondage which earlier prevailed in south Gujarat is a grotesque misrepresentation of my writings on the subject (Brass 1991: 41).

meaning of bonded laborers. He argues that the redefinition of the relationship which held tribals captive to Hindu landlords was the result of the introduction of capitalism in a colonial context. The earlier servitude should not be described in terms of unfree labor, according to Prakash, but should be understood first and foremost as the manifestation of an unequal power configuration. It was only during colonial rule that *kamias*, until then dependent servants of dominant landlords, were reconstituted into unfree laborers bonded to big landowners.

> The discourse relevant for the constitution of kamia-malik relations in pre-modern south Bihar was not that of freedom and unfreedom, free labour versus bonded labour. Instead, the relevant discourse was one that distinguished kamias from maliks in terms of power; it constituted the Bhuniyas as kamias by treating them as unequal to upper-caste warriors; and considered the kamia's labour service for the malik as a natural consequence of that status.... (Prakash 1990b: 188).

The operation of the *hali* system in south Gujarat closely resembled the situation existing in south Bihar in the same period. I do not, therefore, disagree with Prakash as far as the overriding importance of social and political status in that setting is concerned. These dimensions were the organizing principle of the patronage concept which found its climax in the *raja* model. Nevertheless, the articulation of inequality was not a goal in itself, in my opinion, but went together with the appropriation of tribal labor power—Dublas in south Gujarat and Bhuinyas in south Bihar—for land clearance and cultivation. This was, after all, the economic base on which the ascendancy of caste Hindus who invaded these regions came to be founded. In other words, I still endorse Nieboer's original premise that claims that a shortage of labor produced by a high land-to-man ratio is a crucial ingredient leading people to force others to work for them. The same notion is the point of departure in Kolchin's important contribution to the debate on the origins of unfree labor (Kolchin 1987; see for India also Chakravarti and Sarkar 1985).

Prakash's argument that the mechanism of indenture became the *modus operandi* in the social construction of bonded labor towards the end of the last century is very illuminating. The concomitant change in valuation, the *mise-en-valeur* in economic jargon, which was mediated through colonial discourse, paralleled the transformation towards capitalism as expressed in the drive for higher yields and monetary profits on the basis of a very labor-intensive mode of exploitation. In line with the emerging relations of production, attachment was reduced to a mere employment modality which began when the farm servant received a "loan" with the obligation to work, until the "advance" received was repaid in labor terms. This implied reconstitution of the *kami* to the opposite pole of the free laborer, an interpretation that was codified in the legal-bureaucratic theory and practice of late-colonialism.

> Landlords and laborers ordered their relations as contracts founded on advances of loans. They executed written deeds representing their relations as creditor-debtor ties and they approached the courts as parties bound by contracts. Contributing to this liberation of the kamias from slavery and their reconstitution as bonded laborers was the objectification of agrarian relations. While this objectification, gathering increasing force by the late ninteenth century, founded on social relations in contractual exchange of things—advances of money, grain, and land for labor—it was the juridical process that helped define these exchanges as creditor-debtor transactions. In effect, therefore, the two processes acted in harmony, making the kamias manifest as innately free persons bonded by debts . . . . (Prakash 1990: 222).

Contrary to Mundle and Prakash, I am inclined to maintain that the idea of a total change in the institutionalization of agrarian relationships does not do justice to the tenacity with which pre-colonial social formations remained intact, albeit weakened, throughout colonial domination. The style of conduct between bondsmaster and bondsman, although embedded in a changing political economy, included features that had characterized the former behavior of patron and client. Apart from my own research, I find support for that contention in Ramachandran's recent study of the relationship between wage labor and unfreedom in the south India agricultural economy. A common bias in sketching the trajectory from unfree to free labor is the evolutionist notion of slow but steady progress. In firm rejection of such a trend, Washbrook has argued in a recent paper that the quality of the agricultural laboring existence in colonial south India dropped to a level significantly lower than that reached by the end of the eighteenth century (Washbrook 1990). This is an interesting point of view which, in the absence of solid empirical evidence, could easily degenerate into the construction of a "golden age" myth which seems to me mere populist fallacy.

In my reconstruction of the history of agrarian servitude in rural south Gujarat in the late-colonial period I have emphasized the gradual erosion of patronage. During the first half of the twentieth century, terms of employment became increasingly impersonal and more contractual in nature. Given the magnitude of the landless class and the slow growth of the economy in and out of agriculture, the exploitation of labor continued unabated. The change that I noted during my fieldwork in the countryside of south Gujarat between the 1960s and 1990s was caused by the further development of a capitalist mode of production for which the basis had earlier been laid.

Some schools of political economy see the creation of a free labor market as a defining feature of that transformation. I do not agree. I have referred to literature from which it appears that even middle-sized farmers in south Gujarat, who had started to become more prosperous during the first few decades of this century, thanks to expanded cotton production for the world market in the 1920s, began to substitute their own labor power and that of other household members by that of tied workers. At the same time, I recorded that these "new rich" distinguished themselves from the former landed elite by adopting a more business-like attitude towards their subordinates (Breman 1985: 131–35).

On the basis of both historical and contemporary research, I ascribe to the thesis that the neo-bonding of labor need not conflict with the capitalist mode of production. In another study focusing on the increased access to employment away from the village, I have extensively discussed the linkage between the spatial mobility of members of the landless proletariat and the immobilization of their labor power in the same or other economic sectors, i.e. inside and outside agriculture. Pre-capitalist institutions and mechanisms continue to exist but have undergone a functional change which makes them adaptable to capitalist management (see also Breman 1990: 572–81).

Reports that were highlighted in the Indian press about the continued existence of bonded labor practices in some parts of the country aroused public opinion and culminated in 1976 in the proclamation of the *Bonded Labour System* (*Abolition*) *Act*. The wide coverage was certainly influenced by the pressure brought to bear on leading politicians by social activists. Tales about the coercion exercised by employers on men, women and children were in obvious conflict with the constitutional rights of every citizen. Nevertheless, the assertiveness with which the state pronounced in favor of free labor held little conviction. To start, policy makers had pledged freedom from bondage often in the past, but without result, because effective action was not forthcoming. In addition, the 1976 Act was promulgated when the Emergency was in force throughout the land. Among the civic liberties that were then suspended was the right to strike, and labor interests were drastically curtailed in many other ways. Seen against this background, the abolition of bonded labor could be seen as a signal that the state wished to monopolize all means of coercion in its own hands while strongly curtailing high-handed behavior on the part of employers of labor. In my view, there was yet another motivation, namely, government's desire to accelerate the introduction of capitalism by depriving industrial relations of those elements that were considered to stem from "the feudal past". The doctrine that this action conceals is not so much that bonded labor is degrading but that it is above all uneconomic and inefficient. This is an opinion that was heard even in colonial times; see, for example, the way in which Mukhtyar sums up the effects of the *hali* system in the late 1920s in his study of a village nearby Chikhligam (see Breman 1974: 141). Nevertheless, even Mukhtyar considered its continuation a necessary evil since it was only by using coercive methods that farmers could restrain their Halpatis from migrating to the towns (Mukhtyar 1930: 170).

Recently, Brass has pointed out that landowners in India still adhere to this strategy. In fieldwork carried out in 1987 in five villages in Haryana, he found that farmers used debt-bondage to try to prevent landworkers from opting for alternative and better-paid employment. Brass speaks in this respect of capital's de-proletarianization of labor (Brass 1990: 37). In my opinion, his interpretation of the situation is debatable on various grounds. The way in which he characterizes the dependency of agricultural labor is, moreover, extremely unfortunate. In this case, de-proletarianization does not signify the repossession of means of production by a part of the landless proletariat. No, submission to capital engendered by means of a debt relationship causes Brass to deprive present-day farm workers in his research area of their proletarian status. In doing so, he gives the impression that the continuation or re-introduction of labor bonded through indebtedness does not accord with capitalist relations of production.

A similar suggestion, although phrased more carefully, is made by Patnaik in her introduction to a collection of essays on various forms of unfree labor in the Indian society past and present. According to Patnaik, debt bondage has not only endured but has even increased with the trend towards development which is accelerating in India's agriculture. She considers the manifestation of indentured labor relationships in areas such as Punjab-Haryana as essentially incompatible with a pure and complete hegemony of capitalism other than in a landlord-dominated format which, for that reason, is written off as a retrogressive variant (Patnaik 1985: 14, 25).

I do not share this point of view. For a more global argumentation I would refer to Miles's comparative analysis of the phenomenon of unfree labor in a wide range of capitalist regimes (Miles 1987). Based on the employment record of Halpatis outside agriculture and the village, I conclude that neo-bondage in various shapes remains an important mechanism in control over labor. A discussion of these practices, which go together with relations of production that are strongly capitalist in nature, falls ouside the scope of the present study. Here I shall confine my observations on this theme to the changing status of farm servants in the agricultural economy of south Gujarat.

The emphasis is primarily on Gandevigam, since in Chikhligam the majority of Halpatis already work away from agriculture and the village while the number of farmhands who are in the regular employ of a landowner constitute a small minority among the total work force in this village. This is not the case in Gandevigam where, for more than half the members of the landless caste, agricultural labor still forms the only or principal source of livelihood, while many are still engaged as farm servants in landowning households. The location of Gandevigam at a short distance from urban and industrial nuclei facilitates access to alternative employment, a circumstance that is comparable to Brass's research sites and which is seen by him as an important cause of labor bondage (ibid.: 55).

Is bondage through indebtedness in fact the mechanism by which Anavil farmers in Gandevigam manage to secure the labor power of the Halpatis? And is it conceivable that the latter, once they have been maneuvered into such a position of dependency, are willing to forego the chance of a better paid job away from the village if it should become available to them? To both questions my answer is negative. To prevent any misunderstanding, agrarian employers undoubtedly try to safegurd their labor requirements, particularly in the case of permanent workers. But they do so only to a certain degree by providing a cash loan, i.e. payment in advance for future labor performance. Halpatis are conditioned into dependency through the provision of facilities that do not actually entail any greater or advance expenditure on the part of the employer. Access to such facilities nevertheless increases the income of agricultural workers to more than their daily wage. Until some decades ago, and in compensation for loyal behavior, farmhands in Gandevigam were sometimes given a small plot of land on a share-cropping basis in order to grow plantains (Breman 1974: 157, 196). This was done in new orchards where the plantains provided shade for young growth. As the trees reached maturity, the farmers were less inclined to allow these supplementary crops. Some Anavils continue to give their ricefields out on a sharecropping basis. By granting this favor only to those Halpati households who provide at least one regular farmhand, the landowner forestalls any possibility of a Koli cultivator, for example, to insist on compliance with the Tenancy Act. However, the transaction is a matter of such sensitivity that neither party is prepared to talk freely to outsiders.

The former custom of granting cattle on a sort of sharecropping basis, which I found in both villages during my initial fieldwork (Ibid.: 192), has now disappeared. Nevertheless, more cattle may now be found in the landless milieu, particularly near the hutments in which farmhands live. The majority of milch animals owned by Halpatis are bought with the aid of loans from the regional milk co-operative. Repayment instalments and the cost of cattle feed which is also provided are deducted from the money paid for the daily delivery of milk. Only by selling every liter is the landless owner of a healthy and productive cow certain of a reasonable return. And even then, the Halpati cannot really manage without the farmer's consent and help, namely, for permission to cut grass on his land. Only regular workers are granted free access to their employers' fields and orchards. Day laborers find it far more difficult to maintain an animal because they have to depend on finding sufficient grass along the roadside or on the river bank. This is why a cow or a few goats are mostly kept only by regular farmhands. It is paradoxical that, while ownership of one or two animals increases the Halpati's security of existence, proper care of the cattle is dependent on the farmer's leniency. There is no question of benefiting from the cow for purposes of consumption; the use value has to be converted into cash with which to meet the costs of a wedding or any other major household event, so that the animal really functions as a savings bank.

At the present time, a wedding party for a Halpati youth may cost 5000 to 7000 rupees. Very few Anavils are still prepared to advance such a sum in exchange for promises of future repayment in the form of labor. Only very occasionally is a long-existing relationship between a landowning and a landless household carried forward into the next generation. With these few exceptions, farmers are no longer prepared to accept the obligation to contribute to the *rites de passage* in the lives of their workers. Then how do the landless finance such turning points in their lives which are bound to happen once in a while? One cheap solution is a collective wedding ceremony, for which social workers make all the necessary arrangements. In the vicinity of Gandevigam, such a ceremony and the meal that follows it, takes place twice a year in a village on the other side of the river. The costs are shared among couples participating in the arrangement and in this way can be brought down to the fairly modest amount of a few thousand rupees. There is little interest in this practice, however, which is purely a result of economic need. It is seldom that more than ten couples, coming from far and wide, are married collectively.

A still cheaper solution is to omit the wedding ceremony altogether, e.g. in cases of abduction which occur when one of the families refuses permission for the marriage. The couple then absconds in order to live together, risking a quarrel with or between the parents. Such a *fait accompli* has the considerable advantage of all festivities being dropped, so that costs are reduced to a minimum. What the outside world sees as a forced liaison may in fact be based on a quiet hint or even active encouragement by the parents to choose this most economic solution out of necessity. A pretense of quarrelsomeness may then be intended to prevent loss of face in front of neighbors.

However, such cases are the exception to the rule that a couple should have a "decent" marriage ceremony which, as we have seen, involves a considerable sum of money. Meager possessions are turned into cash for the occasion: a cow or goat, sometimes even the roofing tiles from the parents' house; but the proceeds are still not enough to meet the costs. Close kin cannot avoid making a contribution, and members of the immediate family are expected to obtain loans from their employers, in or outside agriculture. What is still lacking has to be bought on credit. Repayment is made after the feast out of the small sums, usually 5 or 11 rupees, with which the guests mark their presence. The custom that I encountered in the early 1960s for guests to take their own food to a wedding or funeral in order that the host should not be embarrassed, has fallen into abeyance. Halpatis who are unable to serve a meal to visitors at events marking the ups and downs of their life cycle are considered extremely poor even within their own caste.

Does the growing importance of other sources of finance for major expenses which the landless occasionally have to incure mean that the Anavils have renounced indebtedness as a means with which to manipulate their employees into a situation of dependency? That conclusion is questionable since it is still quite common practice to advance small amounts in return for labor to be provided at a later date. The farmers cannot always ignore occasional requests by regular workers for a cash advance. Such consumption loans rarely amount to more than a few dozen rupees at a time. Day workers sometimes make similar appeals to employers with whom they keep in close contact. If this occurs fairly regularly a Halpati might be expected to give priority in service to the farmer who provides such credit. He then gradually assumes the role of *rojio* on whom the employer has the right of first call. The worker has to report each day and may or may not be given work, in accordance with the need of the moment. On a yearly basis, loans seldom amount to more than 100 rupees. In effect, they represent a disguised addition to the wage, since landowners know quite well that their inferiors are incapable of repaying such a "debt". The pittance that they earn for a day's work prohibits the docking of even 1 or 2 rupees.

But why, I asked the Anavils who complained to me about such detbs, did they not pay a little more each day and so spare themselves the recurrent appeals for help? They replied that although it meant a lot of trouble, in the end it was the cheapest method. In Gandevigam, one informant said to have found that his workers did not begrudge being paid a lower daily wage so much as to be deprived of the small loans that they every now and then needed for sudden and unexpected expenses. The lesson that he had drawn from this was to pay the lowest wage of all Anavils in the village but, on the other hand, to be more liberal than his fellows with granting a small "advance" and never to ask for repayment. This farmer claimed that Halpatis liked to work for him, while his labor costs are no higher than those of other employers. The unwillingness of landowners to grant the requested help, particularly when the amount asked is a modest one, arouses ill-will among their inferiors which they are not at all reluctant to express.

During my stay in Gandevigam a few years ago, I experienced at close quarters the escalation of an argument that had started when an Anavil refused to give a servant the sum of 50 rupees for which he had asked. The Halpati responded by not reporting for work next morning; I found him in his hut, still grumbling. The farmer also refused to give in and had to make temporary use of a casual worker. When I left they had still not settled their dispute, and others told me that the man might eventually lose his job. But all concerned thought there was some chance of an agreement being reached because both parties gave the impression of wanting to compromise. When I returned to the village during the 1991 rainy season I asked how the conflict had ended. According to the farmer, the Halpati's mother had asked him not to give her son any credit because he wasted these petty loans on buying alcohol. The man stayed away even when his employer ordered him back to work. After working as a casual laborer for some time, he entered the service of another landowner in the village. A short while ago these two also quarreled and now, after two years, the man is working for his former boss again. Peace has been restored between them, but from what the Halpati told me it seems that a new argument is not inconceivable. The money that he had asked for was needed to pay for medicine for his wife. To whom else could he have turned? Another time he would do just the same.

Landowners repeatedly told me that the reason why they allowed only very low advances was the lack of good faith on the part of their subordinates. The latter in turn invariably used the plea of inability to repay but, according to their superiors, they never really meant to do what they promised.

To enforce his argument in this respect, B told me that during the last ten years he had given four Halpatis money with which to pay their marriage expenses. No, not the total amount that they needed, but nevertheless 1000 to 1500 rupees each. Not one of them continued to work for him and all had gone away without repaying one *paisa*, completely against what was agreed upon. All that money has gone down the drain. They own nothing on which he could recoup his claim, and to force them to work for him is equally out of the question. B made the additional comment that the disloyalty shown by the Halpatis is partly due to a distressing lack of solidarity among members of the dominant caste. Some Anavils, and he mentioned them by name, did not flinch from enticing a worker away from another employer. He had also lost a number of good workers in that way. What else could he do but follow that bad example?

The background to all this is a law that was promulgated many years ago, i.e. the Debt Relief Act, which declares illegal the requi sitioning of labor in repayment of a debt or loan. Although no worker would ever dream of appealing to this statute, politicians sometimes, particularly during elections, use language that employers find disturbing. The Anavils seem to reason that it is better not to take unnecessary risks. They can no longer count on the cooperation of government officials who in the past unfailingly supported them. The minimum wage announced by the government applies to both casual and regular workers. Calculated on a yearly basis, the amount is far higher than what farmers pay in reality. In none of my fieldwork villages was legal action taken against an employer for not complying with the minimum wage instruction. In Gandevigam, a few years ago, an official of the District Office threatened to prosecute a member of the landowning caste if he did not pay his Halpati worker at the legal rate. Moreover, he insisted that the worker should also be paid the balance of wages owed by his boss for the entire duration of his employment over the preceding year. This was done, and in the presence of his official benefactor the Halpati was compensated with a cash payment of wage arrears amounting to many hundred rupees. The following day, he returned this windfall to his employer. To refuse to do so, he was given to understand, would not only mean immediate dismissal but was almost certainly bound to finish his working life in the village.

This example shows the power of the Anavils, but it does more than that. Landowners can no longer rely on government's unconditional support in their attempts to subjugate labor, a change in atmosphere which I also mentioned in the report on my initial fieldwork. Even if one has to admit that government officials will rarely take action against employers who transgress the meager labor laws, the landless know all too well that attempts to manipulate them through indebtedness are no longer backed-up by state coercion. Consequently, farmers try to limit their risk by keeping the size of advances asked by their workers as low as possible, knowing that they will be able to do little if anything when the *quid pro quo* that they expect or even demand in the form of labor power is ignored.

It would be misleading to conclude from the above that it is not the farm servants who are prisoners of the landowners but the other way around. Nevertheless, this is what some Anavils in fact allege, saying that they have no alternative but to give credit in addition to the daily wage in the hope of thus satisfying their Halpati workers. This is an explanation which should be discarded as a caricature of social reality. The remuneration paid to agricultural workers for their exertions barely allows them to meet their daily needs. Nothing is left with which they can cope with the setbacks that regularly occur in their lives, let alone that they can build up a reserve for the major expenses that are sometimes unavoidable. Some Halpatis take up the life of a farm servant in the hope of finding protection against the continually threatening and sometimes acute hiatuses in their subsistence. At first sight this does not seem to be a wise decision since the daily wage is 2 rupees lower than that of a casual laborer. But this is complicated by the fact that the former is assured of more days of work per year than the latter, and allowed extra advantages. To be given a rice field on a sharecropping basis is fairly exceptional. But the more frequent bonuses in kind are very worthwhile for those who have trouble in scraping together a livelihood.

A Halpati woman told me laughingly how, when the landless returned at the end of the working day, the regular workers could be distinguished from the others. The men carry a bundle of grass on their heads and the women firewood, or *vice versa*. In more serious vein she added that such headloads equalled an amount of 4 to 5 rupees per day. People living in the new settlements which are mostly inhabited by day laborers, have to buy their firewood or liquid fuel in the neighborhood shop. That represents an expense of 1 to 2 rupees per day, and even more in the cold winter months.

Finally, a worker who is in daily contact with one and the same employer can bring more pressure to bear when in need of extra money. This is why some Halpatis choose to enter into a labor agreement which restricts their freedom of movement. At the same time, it is precisely this aspect that causes a greater proportion of Halpati workers to prefer the more risky but less restricted life of a day laborer. The landowners' strategy is certainly to maneuver the landless into a situation of dependency, and the granting of credit, though in as small amounts as possible, is unavoidable if they wish to augment their control over the workers. Such tactics are only partially successful since even regular workers among the Halpatis refuse to behave as though they are bonded through indebtedness. In effect, the petty loans, which farmers consider as an advance on wages, are seen by their inferiors as compensation for the shortfall in their wage which is essential if they are to survive. Employers see the cash advance as a claim on future services; the workers, on the other hand, consider that money to be their due for underpaid labor rendered in the past.

To continue to define the present relationship between farmer and worker in terms of bonded labor creates problems in my view. Firstly, because the dominated party refuses to accept the concomitant attachment, and secondly, because the dominating party is unable to make that bondage effectual. An indication of the basic freedom of movement of even a regularly employed farmhand is that he shows no scruples at all in deserting his employer whenever a more attractive alternative arises, and that the farmer is ultimately unable to prevent his subordinate from leaving him without having repaid outstanding loans.

But if that is the case, why is it that more Halpatis do not migrate out of the village to earn a higher wage as unattached worker elsewhere instead of allowing themselves to get trapped indefinitely and for minimal remuneration, in a subservient relationship with a landowner? This question, which has already been asked, is extra pertinent in the case of Halpatis living in Gandevigam. To put it more concretely, why can landowners in this village still find members of the landless caste willing to work for them for 10 rupees per day when they can earn twice that amount in nearby urban industrial estates? The Anavils are convinced that this is due to the physical and psychological incapability of the Halpatis. Their deficient body strength in combination with mental weakness, so runs the opinion staunchly held by their employers, means that the uneducated and uncivilized landless are worthy only of agricultural labor. They illustrate this denigrating sterotype by referring to the few Halpatis who have ventured into domains of employment elsewhere, only to return to the village and the land after just a short period. According to the Anavils, this was because they did not meet the needs of the factory boss: minimal discipline, a steady working capacity, and rhythm, and an elementary degree of intelligence.

Such denigration is quite without foundation. As I have already stated, the gap between agricultural and non-agricultural work is far smaller than might be suspected from the initial wage difference. Employment in the main town of the sub-district or in Bilimora along the railway, pays an untrained beginner a net wage that is little higher than that earned in the village. It is not until some time has passed that the monetary advantage becomes more substantial, and even then it applies only to the more trained and skilled workers. On the other hand, the 10 rupees paid daily to regular farmhands is a basic rate which is supplemented by bonuses in kind. The ultimate difference is not more than 5 rupees per day. This is still a worthwhile amount, which is why the young Halpatis in particular, seek alternatives to agricultural work in order to avoid being yoked in dependency. The difference is not so great, however, that Halpatis who work for an Anavil landowner on a casual or regular basis can be blamed for irrational economic behavior. Their meagre means of subsistence is not due to indebtedness but to the under-utilization of their labor power in combination with sub-standard payment. Farm servants among the Halpatis, in other words, do not represent a special category with strongly inferior employment conditions and a social consciousness which is lacking or immature. They are part of the proletariat to which all landless belong, both in or out of agriculture, and the village, temporarily or permanently, totally or partially.

Is bonded labor then a thing of the past in the two fieldwork villages? The discussion on this subject received new impetus in 1977 when the Gandhi Peace Foundation published a report claiming that forced labor continued to exist in various parts of the country. This publication in Gujarat asserted that no less than 171,000 agricultural workers in that state, almost 10 percent of the total, were bonded as *halis* (Marla 1981: 61). The southern parts of the state in particular were identified as pockets in which most of these bondsmen lived.

The government of Gujarat gave orders for a series of investigations into what had become a politically sensitive subject. The result of several studies which were immediately commissioned into agrarian conditions prevalent in Surat district was negative. Two rapporteurs concluded independently that Halpatis led an extremely miserable existence but that there was no evidence of labor bondage (Lal 1977; Shah 1978). On the other hand, the team entrusted with the investigation in Valsad district decided that there was indeed evidence of tied labor (Charan & Gupte 1978). Based on my own research in both districts, I am certain that hese contrasting opinions cannot be explained by any drastic diffe-rence in employment modalities. The widely divergent appraisals are caused simply by different definitions of the concept of labor bondage. The maximalists find that the strict control exercised by employers over their landless inferiors is sufficient reason to speak of bondage. The elements most frequently mentioned in the report on Valsad district are: permanent employment without a prior time limit, low remuneration in combination with indebtedness, occasionally also residence in a hut built on the farmer's land, employment of various members of the landless family by one and the same landowner, and illiteracy.

This profile is indeed indicative of the lives led by farm servants in the countryside of south Gujarat, but it is my opinion that these features, separately and together, have more to do with exploitation than with bondage (Breman 1985: 306–13). I continue to refute the hypothesis that the position of present-day farmhands in the villages of my fieldwork, as well as other types of labor in and out of agriculture, who are manipulated into a debt relationship by their permanent or temporary employers, is essentially the same as the former subjugation of *halis* by their masters. The suggestion of continuity emanating from such an opinion ignores the materialistic and ideological changes that have occurred between landowners and landless in the countryside of south Gujarat.

The situation of bondage which formerly marked the existence of farm servants has thus come to an end. Without basically wishing to detract from that conclusion, I would observe that social dynamics cannot be described in terms of contrasting black and white. Remnants of the former bondage can be seen above all in the attitudes of dominant landowners towards their regular workers. I have earlier pointed out that Halpatis who have stayed behind in the old neighbourhoods within the village do not evince the same militancy as their fellows who have moved to the newly built colonies. I suspect that at the time the move took place, those landless who continued to live in their old habitat already felt more empathy with the Anavils. They were inclined to keep their relationships more personalized. On the other hand, those who left the village have since increased their detachment from their former patrons, stressing the impersonal nature of labor relationships in which they are involved. Uma Kothari makes an interesting comment based on recent village research in Valsad district:

Men who work as *kayem mujoor* [i.e. permanent workers] for a landholding household tend to be those whose parents worked for the same household. Although they are not bonded and are no longer part of a relationship with farmers based on coercion, subjugation and dominance, which was characteristic of the *halipratha* system, they have some "obligation" through household ties, to work for the same household (Kothari 1990: 162–63).

Kothari warns in a footnote that terms such as "obliged", "forced", "required" are difficult to explain only in terms of economic necessity or monetary indebtedness. Her comment amounts to saying that the ingrained caste and class hierarchy at the local level, which can be understood only in its historical context, has created a climate through which Halpatis came to internalize their dependency to some extent. This makes it easier to comprehend why the landless sometimes consider it more or less inevitable that their labor power should be at the prior disposal of a particular landowning household. It is an observation that corresponds to my own research findings in the same region. On the other hand, it should not cause any surprise that Halpatis who live literally in their employer's shadow are very cautious when talking to outsiders about their life and work conditions. Verbal complaisance may well conceal a more recalcitrant behavior which tries to minimize infringement of a man's autonomy (see also Scott 1990).

The contrast between bonded and free labor is almost always reduced to the worker's individual status. Curtailed freedom of movement, however, does not necessarily refer solely or primarily to the particular man or woman with whom the farmer enters into an agreement. The transaction can involve more than one member of the household and may lack more personal specificity.

M is one of the few Halpatis in Gandevigam who has been granted a ricefield on a sharecropping basis. The owner has also given M permission to cut grass for his two cows. The milk that they yield is sent to the dairy cooperative and, together with half the rice harvest, provides a yearly income which is substantially higher than that of an average agricultural laborer. Cultivating the land and looking after the cattle takes up the greater part of M's time, and the farmer seldom calls upon him for work. Only when questioned further does it emerge that two of M's sons are employed as servants in the benefactor's household.

One can speak of accumulated dependency when the servant's

wife is also required to work in the farmer's house on a continual or rotating basis. Although my data show that this linkage is no longer so automatic as in the past, the combination of farmhand and maid servant is nevertheless a fairly common practice. The term *gharni Dubli*, formerly in usage, meant literally a Halpati maid who belonged to the household, an arrogation that arose from payment of the money needed by her husband and/or father at the time of marriage. In my research villages the provision of wedding money has become fairly exceptional, but a Halpati in regular employment sometimes obliges his wife or daughter (and in turn has the obligation placed upon him) to work as domestic help in his boss's household and, when necessary, to work on the land as well. Even today it frequently happens that members of the man's household, including young children, have to be ready to help when required. In fact, it is one of the considerations that causes a landowner to decide whether to take one or more Halpatis into his regular employment. Such elements indicate that mechanisms inherent to the operation of the *hali* system still continue.

The time when this type of bondage was common practice is not yet lost to the memory of the landless. In my opinion, it is an important reason for their resistance to attachment. The majority of casual wage earners among them associate that status with a condition of subjugation. To my mind, that opinion is shared by the present generation of workers in regular employment, although they will seldom say so openly. Nevertheless, their attitude and behavior show resentment towards members of the dominant caste, while the latter never stop talking about the faithlessness, indifference, insolence and other negative characteristics of their inferiors. During my last visit to Gandevigam during the 1991 monsoon period, an Anavil farmer described the difference with former times (an elastic concept) as follows: "When you came here for the first time, his father [pointing to a worker] spoke about my *wadi* [garden], my paddy and my buffalo as though they were his own. Nowadays his son thinks about me: he has grown rich by our sweat."

While writing up the results of my first fieldwork I observed that the life of a farm servant could not only be economically disadvantageous for a Halpati but, through the association with vulnerability and dependency, could also be experienced as dishonorable and deplorable (Breman 1974: 226). This notion was shared by both male and female servants. Nowadays, to be attached to a farmer's household is more than ever before considered to be inferior. Kothari makes yet another distinction in that denigration which differentiates the various layers of subordination.

Halpati women who work as domestic servants hold a low status position within their own household because of their minimal economic contribution, within their own caste because they are seen as being under the total control of Patidars and other upper castes, and in a village as a whole since they perform the most menial tasks (Kothari 1990: 184).

Kothari does not mention the molestation of wives and daughters of landless workers by their employers. This was undoubtedly a far more common occurrence in the days when Halpatis lived dispersed from one another in huts built on land belonging to their masters. Nowadays they find greater protection in living in their own neighborhoods. But their fear of such assaults is still very real, as shown by feminist research carried out among women in a village of south Gujarat.

Manjula Gulab, a young daughter of a Halpati, mentioned that she, her sister and sister-in-law were alone on occasions when her father and brother went out for work. Their hut is in the *wadi*, which could be easily approached by the Bhathela boys. On those days, she says, "each night brings with it new fears and we have to plan to cope with them. We usually sleep bundled together right here on this porch, it is very warm inside the house and also difficult to run if you are trapped. Huddled together we feel quite confident that no one would dare to take us on all at the same time" (Neera Desai 1990: 81–82).

I have noted similar expressions of intense anxiety and apprehension in my own diary. It is a very sensitive subject on which, understandably enough, few people are prepared to talk freely. Although a farmer can no longer indulge himself by openly having an affair with a worker's wife, there is a Halpati boy in Gandevigam who bears a striking resemblance to the Anavil for whom his mother works. Various male members of the dominant caste told me in plain *macho* language that this was the only manner by which to improve the intelligence level in the landless milieu. At the same time, I have recorded several cases of Halpati men who did not shy from taking violent action against employers who molested their wives or daughters.

Nowadays, a Halpati girl prefers not to marry an agricultural worker in regular employment because she wants to avoid having to work as domestic servant. The change in value that the landless have given to the idea of attachment, over the course of time, cannot better be expressed than by remembering the original meaning of the word *dhaniyamo*: he who gives wealth. It could naturally be argued that this was the expression of a wish that never achieved fulfillment and therefore amounted to a form of self-deception that typified the false consciousness of earlier generations of agricultural laborers. In the absence of contemporary evidence provided by the subordinate party, clarification of the earlier relationship between master and servant is largely a matter of subjective interpretation or even mere speculation. A sense of continuity, however, should not degenerate into denial that real change has taken place. It seems to me not insignificant that neither in Chikhligam nor Gandevigam can an agricultural laborer be found who is prepared to praise his employer as benefactor. On the contrary, members of the landless caste show no hesitation in using quite plain language with which to present Anavil landowners as their exploiters.

My conclusion is similar to that reached by Ramachandran on the basis of his empirical research in a village of Madurai district in Tamil Nadu between 1977 and 1986. The employment modality of only one man in the village accords with the given definition of a bonded laborer as someone who cannot work for any other than his or her current employer, except by physically escaping from the ambit of the coercive power exercised by that particular employer. When questioned in detail about their status, no less than 58 percent of the total landless workers in the village asserted that they were free to choose between alternative employers (Ramachandran 1990: 176, 251). The others fell somewhere in-between the polar ends of the bondage-freedom continuum.

> While it is possible, in the abstract, to differentiate between free wage labour and unfree labour, it is an altogether less tractable problem to separate freedom and unfreedom in the daily lives of workers and to separate the impact of different sets of influences on the freedom and unfreedom of the individual worker. The unfreedom of the workers who were neither bonded nor completely free to choose their employers took different forms, their freedom to choose employers was circumscribed in different ways and in different degrees. The most common manifestation of this kind of unfreedom was what has been called the right of first call of employers over workers (Ramachandran 1990: 252).

Prakash rightly observes that colonial theory and practice, which attributed the status of unfree labor to *kamias* in south Bihar, also instilled the notion of their resistance against domination by implicitly postulating free labor as a powerful social force.

After all, the practices that pronounced the kamias as unfree laborers also voiced disapproval of their bonded condition. Because bondage meant a *loss* of rights that the kamias were entitled to as human beings, the process that denied them those rights also treated that denial as extraordinary—a digression from the natural order of being (Prakash 1990: 233).

The ideology of inequality lost its social legitimacy and that transformation marked the beginning of the struggle for emancipation of the landless proletariat in the countryside of south Gujarat. The transitions that I have observed at the local level over a period of three decades are caused partly by internal dynamics, but are due also to widening social horizons. I shall refer to this broader context again later.

*Chapter 17*

# The Interplay of Growing Prosperity and Continued Poverty

## *17.1 Further Progress at the Top*

The prosperity of most Anavil households belonging to the dominant landowning caste has increased considerably. In Chikhligam, their material improvement is due only in small part to higher incomes from agriculture. There has been little if any change in crop patterns, and large farmers have been lukewarm in their response to government incentives to increase production and productivity. They have profited little if at all from the irrigation canal that reached their village in the 1950s, and only very few have made use of the soft loans more recently made available by the Land Development Bank for improving land quality. The Anavils also caused the bankrupcy of the local credit cooperative by taking out loans without repaying them. Rather than investing in agriculture, some households have reduced their agrarian capital by selling land. This was done to enable the younger generation—the boys at first but later also the girls—to go to secondary schools and eventually also to colleges to prepare themselves for a life outside agriculture and away from the village. The rural elite has benefited in particular from the greatly expanded supply of advanced education in the sub-district. The increase in the number of secondary schools from one to 21 during the last 30 years means that pupils now rarely have to travel more than 5 kilometres from their villages. In addition, a college that meets the rapidly growing need for further study has been opened in the main town.

The Anavils' drift away from village life has taken on considerable dimensions, with a handful even going abroad to more prosperous parts of the world. Five men from Chikhligam, i.e. two brothers and three sons of dominant caste households, together with their families, now reside in the USA or Great Britain. Their relatives, although only vaguely informed about the sort of work done by these emigrants, take it for granted that they lead a comfortable existence. The sharply negative attitude of late shown towards newcomers from Third World countries in the West has had the effect of putting a strong curb on migration to destinations overseas just when this had started to gain pace. Only very few applicants are able to pay the high price that nowadays is asked for a "green card". The restrictive practices abroad cause young Anavils to move into the towns and cities along the west coast, where they mostly join the middle class and apply for administrative and managerial employment in the public sector or private industry. The professions are highly regarded, and the best qualified prefer to set up their own trade as lawyer, doctor or accountant, while the less talented frequently remain in the village to look after the farm. Such a leap into the outside economy is financed by money earned from agriculture. Conversely, a small portion of the earnings of those who work "away" (*bahargam*) will flow back to the village, to be spent on consumption rather than used for productive investment. The annual incomes of those Anavil households that are still actively engaged in agriculture amount to about 40,000 to 50,000 rupees at most.

The waning interest shown by Anavil Brahmins in village affairs has facilitated their ousting from local positions of leadership. For some time now the village head of Chikhligam has been a Dhodia, member of the largest tribal caste in the region which is becoming increasingly successful in converting its numerical majority into actual power. At the supra-local level, however, the Anavils have mostly been able to maintain their social dominance. This can be illustrated by the activities of two cousins who, at the time of my initial fieldwork, officiated as police *patel* and village head respectively.

At the beginning of the country road running from the highway to the village I met D, standing ready to greet his eagerly awaited guest. Although I had written to announce my arrival on a motor bike today, late in July 1991, this was a welcome gesture that I had never before encountered. However, D needed only a few words to make it clear that I was certainly not the one for whom he stood there in the pouring rain. No, the Minister of Education was coming, and the garlands held by a small group of timid schoolgirls were intended for this highranking politician. D had managed to persuade the newly-appointed Minister to break his journey back to Ahmedabad from a working visit to south Gujarat for a short pause in Chikhligam. The procession of cars containing the Minister and his suite and district authorities, and the inevitable police party, was escorted triumphantly into the village. There was little time, and after tea and refreshments had been served in D's house, the VIP and his retinue prepared to depart again. But D was a satisfied man: the brief visit by this prominent party colleague had testified to the cordial relations between D and the pundits in the state capital. By making this connection so visible D had assured himself of a receptive welcome in the region's government offices which he visits almost daily. D who, early in the 1960s, had warned me to avoid the village locality in which the untouchable Dheds resided (he could tell me much better than anybody else all that I wished to know about them), now prided himself on being able to count the Minister of Education, a Harijan, as one of his "fast friends".

D leads the Janata Dal party in the sub-district and has long been a faithful supporter of Morarji Desai, experiencing his greatest hour when that aged politician returned to national politics after termination of the state of Emergency in 1977. I was told in great detail how D had waited for hours outside the prison building in which his famous fellow caste member had been detained in Ahmedabad. D was then nominated as campaign manager in south Gujarat, the region from which Desai made his comeback in national politics. Here is the birthplace of Morarji *kaka*, who became leader of the Janata opposition to Indira Gandhi. D still owns the telephone with which he was then equipped, the only one in the village, to his position as a local agent of the beloved leader. Without this facility, according to D, he would be unable to maintain the contacts essential to run his many "social work" schemes.

As a local level politician, D is involved in all sorts of development projects that are profitable to his rank-and-file and, last but not least, to himself. For instance, he has recently been elected to the board of the sub-district sugar factory that has now reached the stage of construction. Party politics, combined with caste membership, usually play a decisive role in the fight for leadership positions. On the board of the new cooperative enterprise, Dhodias hold 8 of the 14 seats, but all key posts are in the hands of *ujliparaj* such as D.

D's cousin B has always been active in the Congress movement. Their difference in political affiliation is based on life-long mutual rivalry. The local party positions with which B had been entrusted upto the district level have now been taken over by members of tribal castes forming the majority of the Congress electorate. B realizes that the rise of this cadre is a logical result of the affirmative action policy, but nevertheless has difficulty in accepting the outcome. Although he maintains that it was his own decision to retire, pressure from below was certainly brought to bear. It is a consolation that he still has a voice in consultations over the political line to be taken within the party or in important policy decisions made by the regional bureaucracy. His chairmanship of the Public Works Committee of the *taluka panchayat* during the early 1980s made it possible for him to further his business endeavors. He had not hesitated to combine official responsibilities with his personal interests by accepting building contracts. He experienced no difficulty in obtaining orders for the construction or repair of village roads and for government-subsidized housing schemes for landless workers. A few years ago he sold the lorry that he had earned from such contracts because his political activities left him with too little time and energy to run it efficiently. B is now taking life more easily and has the satisfaction of knowing that his three sons have all well-paid jobs because of him: the youngest is a technical operator in a sugar mill; the second holds a clerical position at the subdistrict office in the nearby town which yields bribes totalling more than his salary; the oldest remains in the village where he devotes the least possible attention to managing the farm and can count on becoming a member of the administration staff in the newly established sugar mill, a post which also offers the prospect of considerable illicit earnings.

In Gandevigam, the continual interest shown in agriculture among Anavil Brahmins probably explains why the position of village head remains firmly in their hands. Here, too, the younger generation in the farming households mostly work away from the village and seek training for similar professions as their fellow caste members of the same age elsewhere. However, the fertile soil makes agriculture too profitable for it to be relinquished. Landownership consequently remains concentrated in the hands of Anavils and the larger farmers among them have increased their fruit-growing acreage. The fruit crop has brought prosperity, and the very efficiently managed agricultural sales cooperative which has its headquarters in a neighboring village has contributed greatly to the economic success. This institution remains a stronghold of Anavil power. Its members, dispersed over eight villages of the *vibagh*, annually elect a new board. The chairman is habitually an Anavil, as were 11 of the remaining 14 board members in 1989–90. Gandevigam has the statutory right to appoint two members, and in that financial year both were Anavil Brahmins. In the village of my fieldwork, the profitability of the fruit crop has even caused a number of Anavils to give up well-paid government jobs to concentrate on farming activities.

When visiting Gandevigam in August 1991 I found my former host D to have grown old and weakened by illness. He was just recovering from malaria which has become rampant again in the region. It is no longer possible for him to start his day punctually at 6 a.m. with a tour of inspection of his properties. D complains that without his strict and continual supervision, his 5 Halpati servants will become even lazier than they already are by nature. He admits, grumbling, that the time has come for his son R to take over the daily running of the farm. R told me that he had indeed decided to take early retirement and to sell his house in Bilimora. Since I first met him, R has risen to be head manager of the LIC and thus has to give up a well-paid job. Reassuringly, however, he calculated that as a fruit-grower he would in no way be worse off. The orchards yield an annual amount of more than Rs 300,000, and also the rose gardens started by his father have proven a very lucrative sideline. The family has consequently had no problem at all in marrying off R's three daughters within a period of five years. With great pride I was told that on these festivities Rs 550,000 were spent, while settling the girls in their married life also requires a sizable sum of money. His youngest daughter's wedding, held just recently, had been attended by 2500 guests.

A number of orchard owners in the village have annual incomes of at least two *lakh* rupees, but even the more modest Anavil farmers usually earn not less than double the income of their caste members in Chikhligam.

Despite the high status accorded to employment outside agriculture, farming has been divested of the stigma of backwardness due to the introduction into rural areas of numerous facilities and conveniences that formerly symbolized the lifestyle of the urbanized well-to-do. There are now three telephones in Gandevigam and almost every Anavil household has a light motorbike (Hero Honda) or a scooter; one family even has the luxury of a car. With such means of transport or by using the bus, neighboring towns are easily accessible. But it is not absolutely necessary to go there even to see a movie. A color TV nowadays is standard equipment for the village elite, and a newly-opened video-renting shop in one of the Anavil streets already enjoys brisk custom. At the end of my stay in August 1991 I was present when the first washing machine was installed in the house of one of the local notables, on which occasion its owner remarked with a good deal of satisfaction that village life was no longer inferior to that in town. Residence in crowded urban neighborhoods where rents and the cost of daily living are so high, where the physical infrastructure is so appallingly bad and subject to enormous pollution, might well be losing its attraction for the Anavils.

### *17.2 The Growing Room for Maneuver in the Middle Ranks*

Although dwarved by the large landowners, small farmers of lower castes have also progressed, most noticeably the Dhodias in Chikhligam. The plots that they owned early in the 1960s, seldom amounting to more than 2 acres, were insufficient for their livelihood. The men in particular customarily left the village after harvesting, to return only at the end of the dry season. The money earned from seasonal work in brick kilns or saltpans on the outskirts of Bombay, was used partly to supplement their inadequate subsistence and partly to invest in agriculture or education. Only very few, usually jobbers in charge of the migrating gangs, were able to save enough to enable them to buy another plot; rather more were able to raise the productivity of their land by installing a pump to provide water for growing paddy, vegetables and *chillies*.

The path adopted by growing numbers among the younger generation is due particularly to their education: elementary and, increasingly, also secondary schooling. The level that they achieve qualifies them for non-agricultural work that requires some skill formation, facilitating their entry into industrial estates, workshops, other small-scale establishments, state jobs or service sector employment. As a tribal caste, the Dhodias have particularly benefited from the government's emancipatory policy, intended structurally to improve the position of deprived social segments. Many youngsters among the Dhodias have acquired jobs in the lower ranks of various government agencies, thanks to such positive discrimination. Incomes earned outside agriculture are then ploughed back into the agrarian side of the household to which they belong. A growing number of these smallholders have been able to buy cattle, kept not for home consumption but solely for their exchange value. Intensification of agricultural production in combination with diversified employment has thus been the dual strategy used to combat the situation of poverty in which landowning Dhodias found themselves.

Members of the Dhodia caste are said not only to work hard but to live economically within their sober means. I have heard numerous Halpatis sneer about their frugality, but Dhodias are also generally praised for their willingness to help one another. For example, if a wedding is at hand in a *falia*, its residents will surrender their month's ration cards in order that the family involved may buy cheap grain in the fair-price shop. Halpatis admire such expressions of solidarity, which are more muted in their own localities. In my opinion, the lack of such collective practices in the latter case is linked to the far higher degree of poverty among the landless. The income on which they are forced to exist allows little latitude for any strategy other than that of short-term self-interest.

The improved situation does not apply to all households in the middle ranks. Roughly one-third of Dhodias have progressed only marginally if at all. They are distinguished from more successful caste fellows mostly by their lack of landownership or other forms of capital. Such households depend for their livelihood entirely on earnings from their labor power, often making it necessary even for young children to work alongside their parents. As a result, the new generation of this deprived segment seldom goes to school and, later in life will thus, be equipped only for unskilled, low-paid and casual work. The vicious circle that keeps them imprisoned in a state of deprivation can only be broken with the aid of more fortunate kinsmen, neighbors or other members of their caste.

These variations in level of welfare are hidden from the view of outsiders. Emphasis is placed on unity within the caste, and the weak shelter behind the backs of the strong in the hope that an appeal to caste solidarity will help them to overcome their own backwardness.

It is not only in village politics that the Dhodias have gained in strength. The candidate for south Gujarat who helped the Congress Party to obtain one of its few victories in the state during national elections in the spring of 1991 was a Dhodia. The growing feeling of self-esteem among members of this caste increased when their

MP was subsequently invited to Delhi to become a junior Minister of Rural Development in the central government.

In Gandevigam, the comparable class to the Dhodias is that of the Kolis, although these occupy a slightly higher rank in the caste hierarchy. As small landowners, they have managed to free themselves of the economic ties that bound an earlier generation to Anavil households. Their growing autonomy is not only their own choice, but is also due to the refusal by Anavil landowners to maintain tenancy relations with Kolis. At the start of the 1960s, in addition to cultivating their own land, the Kolis had numerous agrarian side-activities such as hawking and trading, carting, etc. Their interest in off-farm employment has since accelerated by necessity. The younger generation is now convinced that schooling and technical training are a precondition for access to better-paid work outside agriculture, but even then a job in the formal sector has to be bought. This new custom depends on the family's ability to lay down a sum of money or for the persons in question to work for some length of time without pay. The Kolis belong to the category of Other Backward Classes and, contrary to the Scheduled Castes and Tribes, are only marginally covered by the government's positive discrimination policy. Much more than the Dhodias in Chikhligam, therefore, they are obliged to seek employment in the private sector of the economy. This explains why more artisans and half-skilled factory hands can be found among Kolis. Several young men, i.e. 4 out of a sample survey of 15 households in one street where I have collected data, are now at work in the Gulf States or have gone there in the recent past. Abroad they have never been more than *gastarbeiter*, however, forced to return home after their contract is over. This contrasts with the Anavil emigrants from Chikhligam and Gandevigam, more than 10 in all, who have managed to settle permanently with their families in the USA, Great Britain, New Zealand or Zimbabwe.

The position of the Kolis in the democratic framework that is the saving grace of Indian politics has become stronger because of their power of numbers. In Gandevigam, the Anavils have only managed to prevent the election of a member of the Koli caste, which is the largest in the village, to the position of headman by ensuring themselves of the votes of the Halpatis. For many years now, however, a Koli has presided over the elected council in the sub-district, illustrating the shift that has occurred in the social balance of forces in south Gujarat.

### *17.3 Continuous Deprivation at the Bottom*

The distribution of economic growth shows that the capital-owning classes have made significant progress; the large landowners the most and the small owners the least. But what about those landless people who form the social underclass in both villages? Their persistent poverty is obvious even on a cursory tour through the neighborhoods where they live. In most regions of India this massive and heterogeneous proletarianized workforce is known for its distressingly low level of living. It is noticeable, moreover, that even in such a rapidly expanding economy as that of south Gujarat there seems to have been no strong improvement to the lives of this proletariat, which here consists mainly of tribal Halpatis. And this is despite the fact that Gandevigam is situated in the middle of the central belt of industrialization where the increase of non-agrarian production has been most powerful, while Chikhligam lies close to the eastern border of this zone of rapid development. Why, then, has the landless sub-stratum in the two villages not benefited more from the new economic gains?

The demographic factor, usually the first to be mentioned, is at any rate not valid. Comparison of the censuses that I held in 1961–62 and 1986–87 in Chikhligam and Gandevigam shows that in the intervening period the landless population had increased by 50 and 30 percent respectively. That there was nevertheless no question of any demographic explosion can be seen from the drop in numbers of household members.

The slow-down is all the more surprising because the decreasing mortality rate, a general tendency during the past few decades, also extends to the rural underclass. True, the death rate is still higher among landless households: girls more than boys but, on the other hand, fewer adult women at post-fecundity age than men. Compared to the situation of the early 1960s, however, and notwithstanding many acute and chronic cases of illness, the poorer members of society now also live longer lives. If birth rates had continued to remain constantly high and if the migration factor were neutralized, the average size of a Halpati household should have increased rather than decreased. The fact that the trend is in the opposite direction gives reason to assume that fertility rates are genuinely falling. Between 1962–63 and 1986–87 the average size of a Halpati household in the two fieldwork locations combined dropped from 6.2 to 5.4. The latter figure corresponds with the results of research carried out in 1976–77 into the employment status of agricultural laborers in Valsad district. The size of the households to which they belonged, nearly all Halpatis, was 5.3 members (Gupte & Charan 1987: 15–16). According to a census taken in 1974–75, the average for this class of workers for Gujarat as a whole was even lower, i.e. 4.9 (Krishnaji 1989: 140).

In view of the incomplete registration of population dynamics—settlement and departure in particular—it is impossible to draw any far-reaching conclusions from this trend. However, no evidence can be marshalled with which to back up the popular notion that Halpati numbers are increasing more rapidly than those of higher castes living on the safe side of the poverty line.

Is the explanation perhaps to be found in a sliding-down the agrarian ladder of weaker sections among the landowning classes, causing a strong increase in the size of the landless proletariat with which the growth of rural employment has been unable to keep pace? There are certainly signs in many parts of India that the landless underclass has swelled, not only in absolute numbers but also in percentage. Where this inflow from above occurs, the lack of sufficient demand for labor power would add to the pressure at the bottom end on the already inadequate basis of existence. There has been no such *Verelendung* on any large scale in my two research villages, however. My data show that the majority of petty owners have managed by and large to consolidate their holdings. By adopting a dual strategy of intensifying agricultural production and of qualifying themselves for off-farm employment, the Dhodias in Chikhligam and the Kolis in Gandevigam have managed to prevent their own proletarization. The not inconsiderable number of Dhodias in Chikhligam, about one-third of the total, who are almost or entirely landless, have long lived in straitened circumstances, and have not come down in the past 30 years to their present state of stark poverty which they share with most Halpatis.

The landless milieu is characterized by high participation in the labor process, with minimal discrimination according to sex and age. The ratio between the economically active and non-active is highest in Halpati localities, averaging three out of the rather more than five members who constitute a household. Joining the workforce becomes obligatory very early, sometimes even before the more fortunate children start to attend school, and continues as best it may until old age. However, the abundant supply of low-skilled and cheap labor does not mean that there is a permanent demand for it. Employment is very erratic, and women more than men are victimized by its fluctuations. As mother, sister or daughter, females are given all sorts of caring tasks that bind them to the home and are difficult to combine with the inter-sectoral and inter-local flexibility required of rural proletarians. The trend towards marginalization of female labor in the village economy is expressed in the drop in total women's workdays per year to a much lower level than that for men. Days or weeks of long working hours alternate spasmodically with periods of forced idleness. Such discontinuity increases the need to call on the labor power of as many family members as possible in peak periods of employment. Households that have few if any dependents, a stage lasting only for a limited period, are strongest in the economic sense.

The conclusion that I have drawn from my research is that the continuous presence of an enormous army of reserve labor is due above all to inflow from outside. Transition to a capitalist mode of production has freed labor, not only socially but also spatially. As I have explained elsewhere (Breman 1985), the mobile labor coming from other districts and other states to the plain of south Gujarat is not the result of insufficient supply, whether temporary or permanent, in the towns and countryside. As alien workers are recruited, segments of the local landless are thrown out. This process of attraction and expulsion is discussed more extensively in another study *A Footloose Proletariat* to be published shortly. Here, I shall suffice with stating that the persistent poverty of the Halpatis in Gandevigam and Chikhligam is closely linked to the pattern of labor circulation that has taken on massive dimensions during the last few decades. Has there been no improvement at all to the quality of the landless existence? Despite the impression of immense poverty that was the most striking feature when I returned to the localities of my earlier fieldwork, further research has caused me to conclude that the overall situtation is not so hopeless as it was in the early 1960s. There are now fewer naked infants to be seen in the Halpati colonies, and it is far less common for people to go without meals for days on end because they lack any kind of food. Housing conditions have improved and inside the huts there are now a few more durable goods than there used to be. Adults have more than one set of clothes which was all they used to have. It is less common also for people to go around bare-footed.

In addition to these visual improvements, economic circumstances have picked up in the more structural sense for roughly a quarter of Halpati households. This is reflected in the better quality of their housing and their ownership of some minor non-essentials: a transistor radio, a wristwatch or a bicycle. In Chikhligam this progress has largely been restricted to the Halpatis who led a more autonomous existence even at the beginning of the 1960s; that is to say, they owned a small plot of land or found work as self-trained artisans in the vicinity of the village. A vanguard similar to that in Gandevigam consists of Halpatis who have managed to find some steady job, work that is often semi-skilled in nature and thus gives a somewhat higher wage than agricultural laborers used to get. The daily lives of these households shows a greater degree of regularity. They are able to save some money, which enables them to cope with illness and other misfortunes and to make such minimally necessary outlays as to ensure a life that has a rudimentary degree of dignity. Those who reach this level of consumption are also considered to be creditworthy and, on the basis of that status which is determined by the certainty rather than size of an income, are able to obtain goods and services without having to pay cash on the nail. Their social horizons expand, and expectations and actions become attuned to a longer time perspective. The children of such families learn to read and write, even though their school attendance is almost always limited to the primary school.

One should not expect too much of the advance made in this milieu. The quality of life is still extremely poor as can be seen from the example of two brothers who have managed to reach the powerlooms in Surat. The journey to and from this large industrial city takes more than 6 hours, a distance that is too long and too costly for casual day labor.

At 4 a.m. Bhanabhai leaves home to cycle to the main town of the subdistrict, from where he takes the bus to Bilimora which is on the railway line. The train brings him to Surat and he walks from the station to the powerloom shed. The Factory Law prescribes that laborers may not work more than eight hours per day. The owner's bookkeeping conforms to the law, but two twelve-hour shifts are standard practice in this branch of industry. For Bhanabhai, however, this extension is still insufficient. He can only make his long journey, to and fro, and the costs involved worthwhile by working two shifts in succession. He returns home only after having operated the powerlooms for 24 hours. Long before his arrival back in Chikhligam, his brother has left home to take his place. In this way the looms are kept turning day and night. The day off, which Bhanabhai and Chotubhai enjoy at home alternately, is mostly spent asleep in bed, before they have to prepare to leave again halfway through the next night. By replacing one another, the two brothers manage to meet their employer's demand, namely, that the machines should not come to a standstill.

Why do the brothers not move to Surat? Because their higher earnings would be completely eaten up by the exorbitant cost of living in this city. Only in combination with the yield from the plot of land that they own in the village is the job they have secured an attractive proposition.

Their working life is extremely harsh, but one that is far preferable to that of the casual wage earner, who not only earns far less but whose labor power can remain unutilized for days at a time if demand is lacking. At the household level, a difference in income of 5 rupees per day may be of critical significance. That small amount, together with sharp fluctuations in employment, will determine the quality of life. Access to skilled and steady work provides hope for a better future, but the advance that has been made may also be lost. In 1986–87 I was surprised to find nine diamond cutters among the Halpatis in Chikhligam, but a few years later only five were left. A recession in this sensitive industry meant that the other four had again become agricultural laborers, a setback that is difficult for them to come to terms with. The border line between a life of misery and one of bearable deficiency is very fragile. Acute destitution is now a rare occurrence, but a chronic condition of malnutrition is the rule rather than exception. For example, milk is seldom drunk, not even in households that own a cow. The animal's production has to be sold because the money is needed for other expenses that have higher priority.

A rough division, based on budget research among 40 Halpati households in various hamlets of Chikhligam and Gandevigam, gives the following picture for 1990–91: 30 percent had an annual income of 4200 rupees or less; 40 percent earned more than 4200 rupees but less than 6000 rupees; while only 30 percent earned more than 6000 rupees. According to current official guidelines, households able to expend 6400 rupees or more per year are able to fulfill their basic needs. I view this policy statement with some skepticism given the quality of that "surplus" existence described above, which relates at most to 25 percent of all Halpatis. According to my survey data, three-quarters of the members of this caste live below the poverty line. In both villages the poorest are undoubtedly the Halpatis belonging to the first category, who have to make do with about two-thirds of the amount minimally required for a life of human dignity. In the summer of 1991, an average family, consisting of rather more than 5 members, needed at least 15 rupees per day with which to buy and prepare sufficient food. Calculated on a yearly basis, the income of the poor majority is not much higher, or even less, than the total sum needed. The inevitability of various other expenses which, even at this level, require not less than 20 percent of the annual budget, brings me to conclude that there is considerable under-consumption. The life of these least skilled and lowest paid cannot be characterized as anything other than one of extreme misery if the household to which they belong includes more dependents than working members.

The Halpatis in the second category of between 4200 rupees and 6000, i.e. two out of every 5 households in this caste, also live beneath the poverty line. Fear of death through under-nourishment is no longer prevalent in this milieu because sufficient food is available most of the time to ensure survival. On the other hand, these households are unable to save the money that is necessary to meet their basic needs with any regularity. Their grasp on daily existence is therefore inadequate, and planning even for the near future is out of the question. A poor physical condition makes them vulnerable to illnesses that they can afford even less than those who have steady employment. To save anything from the little that they have is wellnigh impossible. Major expenditures that sometimes become unavoidable can only be made by selling the labor power of one or even all household members for advance payment. The price that has to be paid for the loans received is high because the debt relationship that they thus enter into leads to further underpayment and detracts from their freedom of action. In this way a type of neo-bondage evolves which, even to their own minds, in some respects resembles the earlier system of tied labor.

A certain degree of immobility and inflexibility, inherent to a state of poverty, applies even more to women than to men. Their subordination to men means that, even within the sphere of the household, inequality is a determinant feature of gender relationships. The marginalization of female labor, or at least of those tasks for which a wage is paid, has articulated that inequality. The rights derived by Halpati males from the money that they earn, i.e. roughly two-thirds of the total scraped together by an average household, frequently signify the negation of solidarity as a primary principle in their living together with a female partner. Under these circumstances the woman may experience severance of the marriage bond as a release if it means an end to humiliation and physical violence. It should be remembered, however, that in a situation of poverty, the single woman is extremely vulnerable, especially if she has to care for growing children. This can be illustrated with an example taken from a recent study on working women in a village situated close to my own fieldwork area. The case is that of a vegetable seller, the only breadwinner for her family, who daily undertakes a long journey, going and coming back again, in order to sell her goods. As a result, she is at home for no more than 6 out of the 24 hours each day.

> My work day actually begins at 8.30 p.m. I take the night shuttle to Surat every day. I carry with me chikus, mangoes, *datan* [twigs to brush teeth] and occasionally vegetables like drumsticks. I buy my goods wholesale from different producers in Palsana. The train reaches Surat at about 11 p.m. All of us vendors sleep on the platform until 4 a.m. Then we leave for the market. By about 10 a.m. we finish selling our goods, and return to Surat station to take the passenger back. We reach Udwada by 1 p.m. I reach home at about 2 p.m. If there is food I eat and rest awhile. Otherwise, I cook. Between 2 p.m. and 5 p.m. I rest, bathe, wash clothes and meet women who sell their produce. Then I pick my dinner and start on the journey again (Desai 1990: 91).

Over the last 30 years contrasts between landowners and landless in my research villages have increased and have been re-shaped into new oppositions. True, the poor have not gone further downhill. Based on my findings in the early 1960s, that would hardly have been possible. Even misery has a bottom line beneath which survival and reproduction become doubtful. The conditions suffered by the former generation of landless were very close to that critical line. The fear of a process of progressive pauperization, voiced on the basis of my impressions when I returned to Chikhligam and

Gandevigm 10 years after my initial fieldwork, has fortunately not materialized. The trend since the late 1970s and 1980s has gone in the opposite direction. Influenced particularly by the continuing low price of foodgrain, the real value of agricultural wages has shown a slight increase, but even so the threat to survival has sometimes been acute.

I have seen from nearby how individual Halpatis have succumbed to such a crisis. There was the case of a father and his daughters who, although critically ill, were sent away from a brick kiln; they somehow managed to reach their village, where they died soon after. A period of continual drought, which ravaged parts of west India in the second half of the 1970s, intensified the distress of the Halpatis in Chikhligam, but even then matters were not so bad as in a neighboring village where an agricultural laborer and his entire family swallowed pesticide in order to escape a slower death by starvation.

Overall, and apart from such incidents, the poor have at least not become poorer than they already were. At the same time, however, the gap between them and the more well-to-do classes has become greater. In other words, prosperity has increased far more strongly than poverty has decreased.

Major items of expenditure, which mark the transition to another phase of the lifecycle, weigh more heavily on the budget of landless households than on that of those who are better off. This can be illustrated by the figures in Table 10.

TABLE 10
CASTE-WISE DIFFERENCES IN MARRIAGE EXPENDITURE IN 1991

| *Caste* | *Rupees* |
|---|---|
| Halpati | 3,000 to 8,000 |
| Dhodhia | 10,000 to 20,000 |
| Koli | 15,000 to 30,000 |
| Anavil | 40,000 to 200,000 |

Somewhat less than the average of these amounts roughly corresponds to the annual household income of landless laborers, small farmers, and larger landowners respectively. However, the lack of cash reserves and any form of capital to be used as collateral forces poor Halpatis to repay the debts that they contract by providing future labor power. In this way, the freedom of movement of the new generation of landless is also restricted.

The caste-wise variations in marriage costs are further dependent on the degree of prosperity, with expenditure for a girl being almost always higher than the sum that has to be raised for a boy. The caste association of the Anavils has so far fruitlessly tried to forbid the payment of dowry, or at any rate to fix a maximum amount. As happened so frequently in the past, their richest members force expenditure upwards, while their less fortunate fellows think they have no other choice than to imitate such ostentation as best they may. Other castes in turn adjust their behavior to the example set by the village elite. In the case of the Dhodias and Halpatis, this has led during the past decade to replacing payment of a bride price, as prescribed by tribal tradition, with a dowry. Instead of being paid for giving up a daughter, parents now have to provide a large portion of the costs involved in marriage and chattels. This change in fact emphasizes the marginalization of female labor in the landless milieu. The Halpatis try in various ways to avoid the costs that are involved in setting up a new household. No festivities at all take place in the case of re-marriage of one or both partners. Such alliances are quite in accordance with the proletarian lifestyle. A low profile is also maintained when the young couple is not able or willing to celebrate their marriage. Abduction, whether or not incited by the girl's family, has already been discussed, as has the group wedding ceremony at the initiative of social workers.

The proletarian circumstances that mark the Halpati existence make it tempting to describe the lifestyle of Anavil Brahmins in terms of affluence. However, not all households of this caste live in great luxury and comfort. The most successful among them are a focal point of public attention, and stand as model for the ambitions of others in their social milieu. Conversely, it is those who lag farthest behind who determine the social image of inferiority which is the label of the under-privileged. On the basis of estimated household incomes, I calculate that during the last three decades, the gap between the two poles of the local hierarchy, i.e. the most and the least fortunate, has increased by a factor of one to one-and-half in the two research villages of my fieldwork.

*Chapter 18*

# State Action Against Poverty

In which way does government attempt to pursue a policy that counteracts such an unequal distribution of economic growth? I should mention at the start that, 30 years ago, bureaucracy and politics were not very actively engaged at the village level. This had just begun to change at the time of my first fieldwork. The development state first presented itself by propagating community development. This approach was based on the preconception that the village comprised a number of social categories—terms such as caste and class were avoided as much as possible—dependent on and compatible with one another in the challenge of economic and social development. The prevailing idea was that the government would fulfil a guiding role in the process of change and, standing as an arbiter above the parties, would reconcile any contradictory interests. The strongly populistic strategy gave priority to such concepts as consensus, harmony and co-operation in the public interest.

The effects of modernization policies, such as those introduced in the agrarian sector of the economy, were from the beginning clearly directed away from a more equal distribution of growth. Confronted with a series of critical poverty studies and the eruption of social unrest in different parts of the country towards the end of the 1960s, the national policy makers realized that they would have to allow for so-called target group-oriented programs. Their objective was to formulate and implement schemes biased specifically in favour of the under-privileged who, in the current jargon, were largely bypassed or ignored in mainstream development efforts. The populist nature of the political system made such an additional package imperative. After all, in south Gujarat as elsewhere, the multitudinous presence of marginal farmers and landless laborers had a decisive effect on electioneering. No broad-based party could afford to alienate itself from these massive vote-banks.

The democratic imperative that suggests that the state is accountable to society at large, and which is based on the enfranchisement of all members of society, inevitably means that the in-built tendency to cater for already privileged segments of the population has to make way for a more serious attempt to improve the fate of all national citizens. The least that was needed was sincerely meant or at least sincere-sounding promises that the needs and wants of the subordinate classes would be taken into account in the shaping of future policies. And this is what happened. To what extent has the state-sponsored target group approach reached and benefitted the inhabitants of the landless localities in Chikhligam and Gandevigam?

### *18.1 Introduction of a Minimum Wage*

In 1966, a committee set up by the Government of Gujarat recommended that a minimum wage be introduced for agricultural labor, but rejected the suggestion that its level should be fixed on the basis of the amount needed to meet the basic cost of living. The point of departure was then, and still is, that the proposed wage might not affect agricultural productivity and profitability (Breman 1985: 296–97). In 1972, for the first time, a statutory wage was introduced which obliged farmers to pay 3 rupees per day for casual labor. That rate has since repeatedly been raised: in 1976 to 5.50 rupees, in 1982 to 9 rupees, in 1986 to 11 rupees, and in 1990 to 15 rupees. A corresponding wage was also fixed for male farm servants; during the latest round this remuneration was increased to 5400 rupees per annum. The move to a higher scale is meant to follow the trend in the cost of living. It was no coincidence, of course, that the periodic revision of the base rate inevitably came whenever new elections were announced.

Throughout this period the market price for unskilled rural labor has consistently remained below the legally prescribed minimum wage. Over all these years employers have consistently paid about 30 percent less than the mandated amount. Whenever the actual daily wage begins to approximate the official rate of the moment, any positive effect is offset by rising prices. The next tariff round, when it finally comes, will be long overdue. During the first half of 1991, the wage for a day's labor fluctuated between 10 and 12 rupees. In Chikhligam and Gandevigam not a single farmer was prepared to pay 15 rupees in accordance with the latest government instructions.

It must be emphasized that even this latter amount is insufficient to cover the agricultural worker's daily expenditure. In its final report the *National Commission on Rural Labor* recommends the proclamation of a minimum wage of 20 rupees per day for the entire country. This amount was a compromise, calculated on the basis of the consumer price index pertaining at the end of 1990.

Cash wages are nowadays standard practice. The change occurred largely at the beginning of the 1960s, but some elements of payment in kind were continued. In the fruit-picking season in Gandevigam, for example, farm servants are still allowed to take home one or two mangoes, usually over-ripe or damaged fruits which they sell at the end of the day to petty traders for a few rupees. Such perquisites are confined to peak periods, however. A custom that has entirely disappeared is the annual handout to farmhands of clothing, footwear and a shawl, new or second hand. Moreover, even permanent workers are no longer entitled to a breakfast at the start of the day and a full meal at noontime. According to my informants in Gandevigam, this was due to a conflict that had remained unsolved in 1972 when a minimum wage law was first enacted. The farmers offered food and 1.25 rupees, while the landless laborers bargained for monetary remuneration of 1.50 rupees for each workday in addition to what they used to get in kind. Negotiations between the two parties broke down, and the Anavils unilaterally decided that in future the workers should be paid only in cash. Another factor was that employers, so they told me later, were fed up with the workers' continual complaints about the quality of the food. A transition to monetized wages also seems to be in harmony with capitalist labor practices. I have noticed, however, that when self-employed craftsmen such as carpenters, painters and masons go out on a job they still find it quite natural to be offered a meal, as do skilled workers in other non-agricultural sectors of the economy. Bonuses in kind on top of the money wage seem primarily to be withheld from those who find it hardest to do without these fringe benefits.

The progressive conversion to cash payment makes it fairly easy to trace the extent to which the market price for labor falls short of the legal minimum. Both parties are well informed about the rates that ought to be paid. The conventional assumption that the landless are ignorant of their right to a minimum wage for their work is incorrect. In general they know what is due to them, but are unable to claim it. In both my research villages, employers have little if anything to fear from the authorities, even though they are collectively in breach of the law.

The continual drought which afflicted large parts of Gujarat from 1985 to 1987 was a reason for the government to suspend the minimum wage act for agricultural labor during those years. Agricultural employers quite rightly considered that as an indication that the Rural Labor Inspectorate would not give them much trouble, even after repeal of the suspension. Their expectation was justified (Joshi 1990: 19–20). In my first conversation with the chief official charged with supervising compliance with the minimum wage law in the district, he admitted quite readily that he did not approve of the ordinance.

Various other protective regulations have had just as little effect. For example, workers cannot be forced to repay a debt with their labor. Farmers use this ruling as a pretext for their refusal to provide their inferiors with credit, in particular to give advances that the workers need to defray major costs such as those involved in a marriage or in persistent illness. What all this amounts to is that employers will not accept any responsibility for guaranteeing the survival of their workers. Not only does government exercise great restraint in imposing the minimum wage, it is also reluctant to enforce its observance. The rules laid down for engaging labor are based on payment for time, a daily wage for casual labor and an annual salary for permanent workers. It is more and more common, however, for farmers to pay labor on a contract basis, i.e. on piece rate, job work, and so on. The law covers such eventualities, but without being able to extend control over compliance with the meager labor legislation to that payment modality.

### *18.2 Social Security, a First Beginning*

Given the extremely low price of labor, the distribution of cheap grain has proved an important method by which to prevent further intensification of economic destitution due to uninhibited spiralling of the cost of living. During the last ten years or so, expenditure on basic needs has at least not risen faster than the average daily income of a landless household. This, in itself a favorable development, is undoubtedly due to state intervention in the market intended to ensure that low-quality foodstuffs remain within the financial reach of the underclasses. It should be noted, however, that this controlling trend is sometimes interrupted by sharp price increases for basic necessities, with a long-term negative effect on nutritional standards. The progress made in real wages also seems to be restricted to households living close to the poverty line. It is my impression that the lower the income, the less the people profit from price checks. For example, the regulation that restricts the validity of a ration card to the place of domicile means that the large numbers of migrant workers who enter or leave the villages are deprived of subsidies on daily necessities for which they have the greatest need.

Some years ago, the Gujarat government introduced various social insurance and welfare schemes aimed at reducing the threat to the survival of the most vulnerable categories. These were announced as an initial step towards a more comprehensive system of social protection. The need is obvious. According to the report of the National Commission on Rural Labor mentioned earlier, a mere 2.5 percent of India's GNP is expended on social security. Such a low percentage makes India quite exceptional in the world (NCRL 1991:x). Moreover, that budget is spent mainly if not exclusively on organized labor employed in the formal and largely urban sector of the economy. Extension of these schemes to the many times greater workforce in rural areas is still in a very early phase. For example, the government of Gujarat has entered on its own account into what is called a group insurance scheme, providing agricultural laborers between the ages of 18 and 60, or their next of kin, with a modest pension in the event of permanent disability or death. Of more recent date is the introduction of paid maternity leave for female agricultural laborers, equalling the minimum wage for six weeks for a first childbirth and four weeks for a second.

The greatest political capital is made out of the state pension for which old and disabled agricultural workers are eligible who have no income of their own and who live in destitution. In 1981 the then Congress government launched this scheme as an election stunt. A few years later, it declared that the state pension would be paid to all aged and handicapped workers without adult children and with absolutely no means of their own. The monthly allowance to which such people were entitled, i.e. 50 rupees per person and 100 rupees per family, was doubled in the summer of 1991. Although insufficient for survival, such amounts could alleviate the plight of those living in dire circumstances. That is, they would do so if they reached the targeted beneficiaries, but the chance of this ever happening, let alone on a regular basis, is small. In Chikhligam only one old woman has once been favored with receipt of a payment, a gesture which she and others (including the official in charge) saw as a prize in a lottery—in other words, an inexplicable and unrepeatable stroke of luck rather than a right which she could claim in future. In the present social climate, it is unthinkable that an aged Halpati in either of the villages would report to the authorities on his or her own initiative to demand that right. I would substantiate my impression, gained in the course of my fieldwork, that the promise of state support is not put into effect, on the basis of the approximate figures given in Table 11.

TABLE 11
A SURVEY OF PENSION PAYMENTS TO AGED AND HANDICAPPED PEOPLE IN GUJARAT (1987–91) AND IN VALSAD DISTRICT (1985–91)

| *Year* | *Gujarat* | | *Valsad District* | |
|---|---|---|---|---|
| | Number of Beneficiaries | Amount per case Rs | Number of Beneficiaries | Amount per case Rs |
| 1985–86 | n.a. | n.a. | 2881 | 494 |
| 1986–87 | n.a. | n.a. | 3346 | 549 |
| 1987–88 | 43939 | 754 | 2543 | 728 |
| 1988–89 | 53418 | 731 | 2932 | 802 |
| 1989–90 | 56470 | 923 | 3368 | 995 |
| 1990–91 | 58244 | 1507 | 3496 | 1649 |

To start with, the number of men and women who receive benefit is far less than the proportion of the population that has a right to it. Clients seem to be chosen arbitrarily, as suggested by the freakish trend taken by the figures. In successive years, the numbers of those receiving benefit in Valsad district rose by only one or two hundred; while the amount of the allowance showed a considerable increase. Here, however, a qualification should be made, that what is registered on paper does not necessarily materialize in practice. Quite frequently, allowances are only paid in part to the beneficiaries, or not at all. It follows from this that measures enacted during the last decade with the purpose of providing a safety net for the most vulnerable members of society have not yielded anything more than in the nature of a marginal significance. That verdict is supported by a recent study undertaken to determine the degree of rural poverty pertaining in various parts of Gujarat, and to examine the effect of government directives intended to combat that poverty. The author summarizes his disappointing findings as follows:

> The social protection of the poor which is being attempted through the schemes like the Maternity Benefit for Agricultural Laborers and the Group Life Insurance for the Landless Agricultural Laborers in the State have covered not even 10 percent of the target group population (Patel 1991: 127).

The future will have to show whether or not this scant efficacy is due to the inevitable teething troubles of new programs. However, improved target group identification, more forceful implementation, and sharper control over compliance with the law, all require much greater effort than most policy makers are willing to make. So far, preference has been given to an approach that will provide immediate political gain without too much demand on the government's budget. On the other hand, experience has shown that when the state takes the initial step towards acknowledging the principle of social security, expectations are aroused on the demand side which ultimately cannot remain unanswered.

## *18.3 Public Works*

In various parts of India attempts have been made to stimulate rural employment with the aid of public works, the trend having been set by the Employment Guarantee Scheme, enacted in neighboring Maharashtra in 1975. Programs of this sort aim at achieving a number of goals: to combat long-term unemployment; to raise the productive capacity of rural capital; to provide structural improvements for classes living in poverty; and finally, to bring upward pressure to bear on the low wage level through the implementation of labor-intensive projects. Such pilot endeavors, undertaken in a number of states over the last two decades, form the model for an approach that now extends throughout India: in 1980 the National Rural Employment Program, followed in 1983 by the Rural Landless Employment Guarantee Program. The latter is oriented particularly towards the landless rural masses, promising to provide 100 days employment per year to at least one member of each landless household. The result, verified in numerous evaluation reports, remains far below that promise. The meager results were summed up in a recent overview.

> ... these wage employment programs in India have had a limited positive impact on the situation of the poor. The programs have created only a fraction of total employment needed; the income effect is limited due to low wages and low employment; and the nutritional impact has also been small. The net result is that the programs have made only a marginal contribution towards the goal of poverty alleviation and even this only in a temporary fashion. The contribution to infrastructural capital and other asset creation in the economy has been far from satisfactory (Hirway & Terhal 1991: 25).

The same study points out that in Gujarat the considerable financial budget intended for public work has indeed been fully utilized, but without the objectives being even approximately realized. It should also be remembered that implementation according to the stated objectives of the two programs would have provided employment for 100 days to 6 percent of the target group at most. Even this figure, which should arouse considerable skepticism regarding the sincerity of the promised effort, has been far from achieved in practice. At the same time, a policy to stimulate rural employment is needed more than ever before, recent research having shown a decrease in numbers of days worked per year, particularly in agriculture (Bhalla 1991: 7). In the landless milieu in particular, this negative trend signifies an encroachment on an insecure existence which is not adequately compensated by the slow rise in real wages in the 1980s. Will this be changed drastically by the nationwide Jawahar Rojgar Yojana introduced in 1989–90, in which earlier public work programs have been merged? It is still too early to answer that question, but the initial signs do not give cause for optimism. One new element is that emphasis is shifted to execution at the local level. The village *panchayat* is given direct control over the money needed for implementing public works that meet official criteria: social forestry, road construction, small-scale irrigation, land improvement, sanitation, housing for the poor, community centers. The reason why village authorities were given a decisive vote was undoubtedly the wish to circumvent the bureaucracy as a self-interested party and thus to check the siphoning-off of funds. There is no guarantee, however, that in the set-up now decided upon improper spending will no longer occur, or at least will be at a far lower level. It has already been established that local notables use the budget to defray their election expenses. It is only admissible to speak of improper expenditure if one is certain that this was not the initial intention. To reject that assumption out-of-hand is to misjudge the political character of the relief work program, namely, to bond a clientele for whom few other favors are of such primary interest.

Finally, a recurrent characteristic of all these programs is the emphasis placed on encouraging self-employment. For example, the Training of Rural Youth for Self Employment (TRYSEM) is prominent among the state-sponsored poverty alleviation policies. In view of the shift in the rural mode of employment in the direction of waged labor, the persistence with which constant attempts are made to remould people who have no means of production whatsoever into self-employed craftsmen or service workers is all the more surprising. In my view, this pre-occupation with petty entrepreneurship has a lot to do with the capitalist character of the country's macro-economic policies. Within that framework it is far easier to make property more productive than to improve the quality of life for the indigent. To keep a cow, even if it is provided free or with a considerable subsidy, is not profitable for landless people who have to earn their living as casual labor. That is why almost no cattle are found in the new Halpati settlements that have arisen just outside my two research villages. Where would the inhabitants find provisions for these animals? I was told all the gory details of the chicken project fiasco in Chikhligam two years ago. Two Halpati families had been given ten chickens without any cost, together with a cage and chicken feed for one week. When that feed was finished, they went to the sub-district office to ask for a new ration, but their request was refused. During the following week, all their neighbors ate chicken, for the first and last time as they laughingly told me. In a milieu that lacks any sort of economic base, such income-generating activities are doomed to failure. People living in such circumstances also lack the room for maneuver and the reserves that are essential to make a success of self-employment.

Other protective measures have been just as ineffective. As we have seen, farmers have latched onto the regulation that workers cannot be forced to repay a debt with their labor power and must be refused credit, as a consequence. The implication is that employers deny any responsibility to guarantee elementary quality of life to their workers, while the state shows itself unwilling to enforce even that minimal obligation on which social reproduction of the rural proletariat depends.

## *18.4 Housing*

Government agencies have been more successful in settling the landless in their own localities. During the last 20 years or so, Halpatis, whose employers formerly allotted them a piece of land on which to build a hut and sometimes even to grow subsistence crops, have moved in large numbers to these new colonies throughout the region . As long ago as 1972 I met a district official who was responsible for housing members of this tribal caste. An important part of his work consisted of finding suitable plots on which to build houses for the local landless. More often than not, his negotiations with village authorities and local landowners ended with their setting aside wasteland of negligible market value. Such plots were usually situated far from the village center and linked only by a footpath or rough country road. I have visited several settlements of which only ruins are still standing. The residents left again soon after moving into the houses, when they realized that they were too far distant from their work in the village and that accessibility during the monsoon in particular was almost impossible.

The number of these new colonies has gradually increased over the last two or three decades. At the time of my first visit to the district, the habitat of the Halpatis was characterized by scattered huts built of mud with over-hanging thatched roofs that almost reached the ground; nowadays, the visitor's first impression is increasingly of new neighborhoods settled on the outer edge of the villages. In most cases, government has contributed to the building costs of the new houses. The special program that was earlier set up to provide Halpatis with shelter is still in operation. In 1990 a total of seventy-two houses were completed, with another 142 in process of construction. This is not a very impressive quantity when taken for the district as a whole, but it is not all. Some years ago, the Congress Party took the initiative to build houses for the rural poor, including the Halpatis. The scheme was given the meaningful name of Indira Awas and entrusted to the official agency in charge of the Jawahar Rojgar Yojana. During 1990, 854 houses were built for the target group in fifty-four villages. This dispersion over so many locations in Valsad district helps to give maximum visibility to the activity, an aspect which emphasizes its significance as political propaganda. In Chikhligam, eight Halpatis have become the happy owners of an Indira Awas. A budget of 10,000 rupees is available for each house, with a maximum contribution of not more than 250 rupees by the owner. The wage component eats up more than 50 percent of building costs; the rest of the construction budget covers material such as roofing tiles, wood and, more recently, bricks and cement. In comparison with their former huts, these Halpatis are now undeniably better accommodated. The interior of the house is usually one large room which can be partitioned off by its occupants. The floorspace is greater than their previous shelter, and has more light and air. Ventilation is also better because of a higher ceiling. The walls are mostly made of bamboo lathes strengthened with mud. The tiled roof in particular makes it obvious that the Halpati family no longer lives in a hut but in a real house.

Living together in a social space of their own has strengthened feelings of solidarity among the landless, and has given articulation to their collective caste identity. This predictable consequence was precisely the reason why landowners originally opposed the idea of their workers moving away from their land. In retrospect it has to be said that accommodation in separate colonies has branded them as the new Untouchables of class society. They are hidden away out of sight, in housing built on land that is difficult of access and which lacks such amenities as electricity, tap water and sanitation which are enjoyed by increasing numbers of villages. In the new Halpati quarter on the outskirts of Chikhligam, ten brick houses were erected in 1990 as part of a district scheme in which the former Anavil village head was involved as a contractor. For him it has been a very lucrative business, but not a single Halpati has benefited from it. The house construction was of such poor quality that it was far too dangerous to live in them. Collapsed roofs and broken walls provide witness to the sloppy supervision carried out during the construction, but also of bureaucratic indifference to what is nothing but jerry building, not only in this case but on a truly massive scale. Over time, my stay in such a milieu has brought me to the realization that the majority of slum dwellers are to be found not in urban India but in the countryside.

### *18.5 Schooling*

At the start of the 1960s, two-thirds of the population of south

Gujarat was illiterate. That proportion has gradually decreased. According to the census held in Spring 1991, fifty-five of every hundred inhabitants in Valsad district are now literate. Complacency regarding such progress must lessen when it is realized that the figures are only relative. More people than ever before are unable to read or write. The majority of these are women: although the number of females who attend some years of schooling has increased considerably in the intervening period, they are only slowly catching up with the males.

TABLE 12
DECREASE OF ILLITERACY IN SOUTH GUJARAT 1961–91
(percentages)

| | *Valsad district** | | | *Gandevi* taluka | | | *Chikhli* taluka | | |
|---|---|---|---|---|---|---|---|---|---|
| *Year* | *Total* | *Male* | *Female* | *Total* | *Male* | *Female* | *Total* | *Male* | *Female* |
| 1961 | 65.7 | 55.6 | 75.9 | 54.0 | 43.0 | 65.0 | 70.3 | 59.5 | 80.7 |
| 1971 | 62.5 | 53.7 | 71.3 | 46.8 | 38.5 | 55.3 | 65.4 | 55.7 | 74.9 |
| 1981 | 53.0 | 44.3 | 62.0 | 40.0 | 32.7 | 47.8 | 53.8 | 44.3 | 62.0 |
| 1991 | 44.7 | 36.9 | 52.8 | 33.4 | 27.2 | 39.8 | 44.2 | 36.1 | 52.4 |

*The 1961 data apply to the larger Surat district of which Valsad then still formed part.

These figures show that the sub-district in which Gandevigam is situated is in advance of the general trend, while developments in Chikli *taluka* represent the average. When access to education is divided according to social milieu, the disproportionately high degree of illiteracy among tribal castes becomes obvious. Of people registered as *adivasis* less than one in five were able to read and write in 1971. Among Halpatis in Valsad district this ratio was no more than 1:6, decreasing even further to 1:11 for women. Progress in subsequent years seems mostly to have benefited Dhodias and other landowning tribal communities. In comparison, the elimination of illiteracy among members of the landless proletariat has been much slower. The absence of compulsory education, in Gujarat as in most other states of India, seems to be of decisive significance in this respect.

What proportion do these broad figures bear to the situation in the Halpati localities of Chikhligam and Gandevigam? At the time of my first fieldwork it was very exceptional for children of this caste to go to school, and among the adults whom I came to know not one was able to read or write.

When I visited an old Halpati woman living on her own during my initial fieldwork in Chikhligam, she asked me to read a letter from her son in Bombay, where he had gone with his family after the monsoon to work in the brick kilns. It was the first time in his life that her son had had a letter written for him. The reason was the sudden death of his little daughter. He wanted to let his mother know, but she had never before received news in this manner. Not knowing what to do with it she had left the letter unopened, although she realized that its contents must be important. This inability to communicate other than orally was typical of the social isolation in which the Halpatis then lived.

In Halpati streets, such total ignorance of reading and writing is now a thing of the past. About two-thirds of the men and more than three-quarters of the women are still illiterate, but among people of less than 30 years of age the proportions are far smaller. The majority of young Halpatis have been exposed at least to some education. Far more children in the appropriate age group now attend school, even if only a relatively small number do so for more than one or two years and then too, quite infrequently. Understandably enough, the seasonal migration of whole families reduces school attendance, but even children whose one or both parents live permanently in the village often get no farther than the first few standards of primary school. What causes the high drop-out rate which means that only a minority of Halpati youth, estimated at between 20 and 25 percent, are able to obtain the school leaving certificate? Girls stay away earlier and more often than boys because they are needed to do all sorts of household chores. Some complain about ill-treatment to which Halpati children are subjected more than those of other castes. They are humiliated in a variety of ways and learn how to keep a low profile if they are serious about their schooling. Discriminatory practices are not exceptional, but in my opinion a more telling fact is that education has not yet become an important norm in the life style of the landless. Lack of social control makes it easier for Halpati children to absent themselves. That a change in mentality is about to occur is shown by the pressure brought to bear on the village authorities by residents of the new housing colony outside Chikhligam for the opening of a primary class in their neighborhood school. The village head, hunting for votes in the last election campaign, promised to co-operate. Once the election results became known, however, he reneged on his pledge and instead enlarged an existing school in another hamlet in the village which is too far away for the Halpati infants. The growing interest in education has not yet expressed itself in increased attendance of the *ashram shala*, set up on the territory of Chikhligam at the end of the 1970s. This educational-based institution run along Gandhian lines provides primary school education with board and lodging for boys and girls of scheduled tribes. Although the school was established by the *Halpati Seva Sangh*, it is attended by very few children of this primary target group, and by none at all from Chikhligam. Most boarders are Dhodias or come from other tribal communities in the region.

Even low-grade education that provides little more than some elementary knowledge of reading and writing becomes attractive only when a household's income consolidates above a critical minimum. The majority of landless households in my research villages are nowhere near such economic felicity. Nevertheless, their lack of strong motivation is not due only to factors inherent in the landless milieu, as can be illustrated by the situation in Gandevigam where, in the mid-1980s, the Anavils took the unusual decision to close the local school. I was told that the quality of the two teachers was far from satisfactory, but no successors were appointed after their forced departure. The real reason, as I discovered later, was that the steady increase of Halpati pupils had aroused indignation and scorn among the Anavils. Their own children were admitted to the far better private school in a nearby village, taken there by their fathers on a scooter or moped, or by a servant on a bicycle. It was impossible for the children of landless families to attend that school, due not so much to the distance that had to be covered as to the school fees of 40 rupees per month. The old village school, re-opened two years ago under pressure from the district administration, is now attended almost exclusively by children of the Koli and Halpati castes.

The Noon Meal Scheme, launched by the Congress Party during the 1980s in the hope of winning votes, was axed when Janata politicians came to power. The formal reason was said to be large-scale swindling with food supplies and salaries for the cooks, but the rapidly rising costs formed a far greater obstacle. Competition for the electoral votes of the massive underclass made an alternative arrangement urgently desirable. This came in 1990 with the allocation of 10 kilos of grain per school month to each pupil from a family whose income is below the poverty line. In addition, a financial bonus is given for each child who completes a school year—250 rupees for a boy and 300 rupees for a girl. My first impression is that these measures form a powerful incentive for children of landless families in Chikhligam and Gandevigam to attend school. As we have seen, however, even now only a quarter at most of these children finish primary school. A mere handful continue after that, usually also to give up in the 10th class. Only three boys and one girl, all from Gandevigam, have achieved the secondary school leaving certificate. The costs of further education are prohibitive for most Halpatis, who also have no access to employment that would make such an investment profitable.

### *18.6 Health Care*

A shortage of everything that contributes towards a healthy life means that illness is common in the proletarian milieu. Money is rarely available with which to pay for medicines, and for the same reason people wait too long before consulting a doctor, sometimes even until it is too late. Many ailments, whether temporary or chronic, result from their exhausting physical labor, from food that is inadequate both in quantity and quality, from poor housing and insufficient clothing, particularly when it is cold and wet, and finally, from the primitive sanitation and hygiene characteristic of the landless life style.

The promotion of public health is entrusted to primary health centers, of which there are a number in each sub-district, servicing the population from the main town or from some of the larger villages. People from the immediate vicinity can report to these clinics for medical help which is given either free of charge or for a small fee. When it is considered necessary, the patient is referred onwards to a hospital. In practice, however, this system is both inefficient and ineffective. The range and quality of the services on offer leave a great deal to be desired. Opening hours are extremely irregular, there is a shortage of trained personnel and the equipment is substandard. It is not suprising therefore, that clientele of the centers come mostly from the poorer segments of the population who do not have the money with which to pay for a doctor and the medicines that he is likely to prescribe. Those who are able to do so, avoid visiting the public health centers with their meager facilities. All this explains why little impression was made when, in 1986, the government of Gujarat offered free medical treatment for life to couples of whom the husband or wife allowed themselves to be sterilized.

In view of the bad reputation of the primary health centers, it might be asked whether it was in the interests of public health to route the ambitious family planning program largely through this channel. After having followed a fairly uniform approach for many years, official campaigns propagating birth control now seem to differentiate the recommended technique according to social milieu. This line of thought would consider the condom and the pill to be more suitable for members of the middle classes, a segment of society which is attributed with the rationality necessary for responsible and careful use of such methods. The irregularity and vulnerability that condition the life of the poor, on the other hand, gave rise to the assumption that people in this major target group need to be convinced of the benefit of other anti-conceptual techniques such as the IUD coil or sterilization. There are no specific data that show how this distinction has worked in practice. Figures on sterilization are available, but these do not distinguish acceptors according to social class.

TABLE 13
REGISTRATION OF BIRTH CONTROL OPERATIONS IN THE PRIMARY HEALTH CENTERS OF VALSAD DISTRICT, 1964–90

| *Period* | *Sterilization* | | | *IUD* |
|---|---|---|---|---|
| | *Men* | *Women* | *Total* | |
| 1964–70 | 20,985 | 6,071 | 27,056 | 4,799 |
| 1970–80 | 74,919 | 46,405 | 121,324 | 3,397 |
| 1980–90 | 33,398 | 128,885 | 162,283 | 51,475 |

Over time, as the registered data show, the original emphasis on male targets gradually shifted towards females. The change coincided with the end of the Emergency early in 1977. Until 1973, 82 percent of all operations were carried out on men; between 1973 and 1979 that proportion dropped to 52 percent, declining even further in the following years to less than 20 percent of late. These figures should be used with great caution, however. The reported numbers of male sterilizations are probably excessive. In the early years in particular, the payment of a cash premium to the man and to the intermediary who recommended him gave rise to fraudulent practices. Large-scale falsification of accounts also occurred in the mid-1970s during the years of Emergency, when local officials were given targets and ordered to ensure that the required number of males came forward for sterilization. That the pressure to meet the set quota still gives rise to distorted figures can be seen from the recent suspension of a doctor working for a Primary Health Center in south Gujarat who had carried out only 28 of the 119 vasectomies claimed (*Gujarat Samachar*, 20.9.1991). However dubious the quality of official data may be, it is certain that government efforts are now directed more and more towards the female population.

How does the more general image that emerges from the aggregate data compare to the situation in my two research villages? A surprising factor is that in the Halpati neighborhoods it is usually not the women who have had the operation. The majority of adult males with more than two children claim in first conversations on this theme that they were sterilized some years ago, encouraged to do so by the promised monetary award which could sometimes amount to more than a month's wage. This incentive was so strong that a number of old men, long past the age of fertility, have been sterilized twice under different names. It would be premature to conclude that landless people have submitted themselves wholesale to the birth control techniques propagated by government. The repeated campaigns disseminating the message of demographic wisdom have created a climate in which people living under the poverty line unfailingly tend to claim that they have accepted what the state wants them to do.

A number of Halpatis told me in confidence that they were not acceptors of "family planning", a euphemism for sterilization, contrary to what they had said at an earlier meeting. What had brought them to give incorrect information, sometimes without my questions providing any ground for their suggestive frankness? Because they did not want any further fuss on the subject.

In fact, resistance to the idea of being operated upon is fairly strong, largely due to the fear of sexual impotence caused by loss of physical strength after the operation. However, such sentiments are quite justifiable when inflammation and other complaints about negative after-effects are caused by lack of hygiene during the operation or, as quite commonly occurs in situations of poverty, during the next few days. Naturally enough, the tales in this respect that circulate among the Halpatis have a dissuasive effect.

What, then, is the reason for the somewhat smaller number of children in landless households in comparison to other rural classes? Accepting the arguments of Krishnaji and others, I am inclined to attribute the moderate demographic increase that has already been discussed to a constantly high mortality rate. My impression is that this concerns especially young babies and women in childbirth. Finally, it is not infeasible that the marriage fertility and birthrate of segments of the rural proletariat are adversely affected by the miserable conditions in which these people live and work (Dharmalingam 1991: 307–08).

*Chapter 19*

# The Long, Long March Towards Emancipation

I wish to underline my earlier statement that, in the absolute sense at least, poverty conditions at the bottom of the village economy have not deteriorated. It must immediately be added, however, that in my area of fieldwork, the number of people who are unable to lead a life that is socially acceptable is greater than ever before. The significance of this observation is all the more important since it does not apply to a region with a stagnating economy. On the contrary, the growth process in the direct environment of Gandevigam and Chikhligam has been quite strong during the last few decades.

How do the landless workers, who have enjoyed little if any structural improvements, experience their plight?

To answer this question I must first draw attention to the fact that people stuck in such circumstances relate their own misery to the life standard and style of the more privileged classes. Halpatis compare their meager existence with the very visible signs of increasing prosperity that they see around them, primarily among their employers. The gap that divides the landless from their betters has widened. The latter were also stationed far above their inferiors in the past, but their superiority was not necessarily expressed in material comfort and lavish consumption. Masters and servants shared the same food, for example, and ate in one another's company during the noon work-break in the fields. Large farmers assured me that they ate rice only on festive days. Even in the early 1960s I encountered older Anavil women who did not wear a blouse and who tied their sari between their legs in the typical working style. I have visited various landowning households in which only one of the brothers had married; the others remained single to avoid burdening the joint budget and property. Older Halpatis remember that time and speak appreciatively of the simplicity and also the generosity of their earlier patrons. In contrast, they are bitter about the impersonalized nature of relationships with their present and complain about their employers business-like mentality in refusing to provide any kind of financial help, even when needed for small emergencies. Poverty is thus a relative idea in yet another respect, namely, in terms of time. A fundamental change has occurred in the social context within which people continue to live in conditions of misery. This is one reason why I tend to attach only limited importance to quantitative studies which, taking an imaginary poverty line as their point of departure, try to determine whether the percentage of the population living below that line has decreased or increased. Such studies ignore the fact that the value given to indicators by which poverty is measured and weighed does not remain constant over time.

Differences between the stronger and weaker segments in society cannot be perceived only through their unequal participation in the social organization of production. Differentiation in this respect also finds expression in strongly contrasting styles and standards of living. In recent times many new consumption goods and services have become available in the rural areas of western India, and the degree to which people are able to own or to use them has become an important yardstick in the awareness created about their own situation and that of others. More than ever before, social standing is reflected in material comfort. To be unable to attain the pattern and level of consumption that is considered basic, and which implies far more than a minimal number of calories, increases the burden of inferiority that characterizes life on the underbelly of society at the end of the 20th century. Now that the ideology of social inequality is on the wane, a major long-term trend, which is emancipatory in nature, relations between high and low are couched more than ever before in terms of material prosperity and poverty.

The trend which I have observed in Gandevigam and Chikhligam mirror the change process that is occurring throughout south Gujarat: the transformation of society in a capitalist fashion. These dynamics, which put a premium on property and other concrete forms of capital, explain why the Anavils have made the greatest progress while the Kolis and Dhodhias have benefited to some extent, but the majority of Halpatis, who lack any means of production, have fallen hopelessly behind. The idea that development is a process that starts at the top of the social order and gradually trickles down to incorporate ever greater sections of the population, ignores the possibility of a more dialectic interaction between improvement and shortage. The gist of the economic strategy followed in India is that the price of labor is allowed to stagnate at an extremely low level. I have seen no evidence in the research villages of the persistent assumption that people who have much or at least something, and if only for that reason, appropriate more of the surplus that is additionally created, are prepared aftar some time to share their accumulated resources with those who start with nothing at all.

I am inclined to attach more importance to the growing social consciousness of the under-priviledged, pre-conditioned by the rise of free labor, than to the marginal improvement in their material condition. This change in mentality is particularly noticeable among the younger generation and is expressed in the ever louder demand that the quality of their life and work should be ameliorated. The limited horizon that used to be symptomatic of the Halpati milieu has been broken. People now have far more contact with one another and with the outside world, and the numbers of men and women that remain trapped in the village have dropped sharply. In contrast to the situation prevailing in the south Indian village where Ramachandran carried out his research (Ramachandran 1990: 262), the rural economy in the region of my fieldwork has become much more diversified. The strong growth in employment opportunities outside agriculture has reduced the size of the landless workforce that is dependent solely or mainly on wage income earned in their own habitat. Transport, both by bus and bicycle, and new patterns of communication, have played an important role in opening-up the countryside, including the localities of the landless. Greater mobility is only partly a question of free choice arising from the wish to satisfy new needs and wants socially or culturally determined. Although these dimensions are certainly not lacking in the case of Halpatis, their work migration is due above all to economic compulsion, i.e. the need to seek employment away from their place of domicile. However, that labor mobility simultaneously enables them to escape the dependency relationship in which many were formerly tied for long periods or even permanently to one or more local farmers. Landowners are not alone in insisting on impersonalized modes of employment, in line with their strategy of keeping the subordinated at bay. In order that their identity and dignity should not be impaired, the younger generation among landless workers in particular also and even more emphatically favor more anonymous labor relations.

Social autonomy is still restricted in many ways on the underside of society. Unfreedom in varying shapes and gradations is a recurrent feature of labor relations, not only in the village but also in the outside economy and in sectors of employment other than agriculture. However, this does not alter the fact that Halpatis now conduct themselves more freely in accepting or rejecting contacts on which they are dependent for scraping together a livelihood. Dissemination of elementary literacy, although still in an early stage, means that the rural proletariat no longer has to seek the help of members of higher classes to act as their intermediaries with the wider society. Self-respect increases as the thumbprint makes way for the written signature, just as self-confidence grows when personal or household data no longer become the property of a social superior who has the power to translate them into writing. Leaders are now emerging in this milieu equipped with some knowledge of the operations of legal and civic institutions which mark the transition to the modern nation state. All this is known as a widening of horizons and is accompanied by an expansion of the social room for maneuver.

It is all still very modest and circumspect. None of the mainstream parties in Valsad district, for example, has ever selected a Halpati candidate for a major political seat or public post. The man put forward by the Congress Party during the last elections in Spring 1991 to recruit the votes of the landless was a social worker nicknamed Dinker Dubli by his fellow Anavil Brahmins. This is in his own milieu a derisive reference to his association with the Dublas, as Halpatis are still called by members of the dominant caste. Kolis and Dhodias have a far more visible presence in the political arena, with a vanguard of leaders at various levels of public intervention who act as spokesmen for, and look after the interests of, a mobilized clientele at the grassroots. The few Halpatis, who are upwardly mobile socially, frequently tend to distance themselves from the mass of their caste mates who are less fortunate. By accepting the common name of Patel, they try to conceal their low origins from outsiders. This defensive strategy is due both to the isolation of the few who are able to make a considerable leap forward and to the enormous gap that then divides such climbers from the landless milieu from which they come.

The present lack of supra-local leadership does not alter the fact that Halpati colonies manifest a growth in collective militancy. To interpret this phenomenon as an expression of a new self-confidence on the part of people who have little to lose will not be accurate. There have been signs of insubordination in the past, but they were generally covert and more dispersed. Sabotage, avoidance, a pretense of innocence, and other tried and tested methods from the arsenal of "weapons of the weak" highlighted that sphere of resentment. Such acts of resistance, strongly individualistic in nature, still occur but are now buoyed up by less hidden sentiments of solidarity. Escapist behavior on the part of those victims who are no longer able to cope with the continual misery of their lives is a part and parcel of the social climate in the habitat of the landless. This inability to get a grip on the elementary conditions that determine one's life also explains why drunkenness is anything but exceptional in these slum-like localities. Even before noon, I have frequently found Halpati men, and sometimes even women, in a drunken stupor in or near their huts. Under such circumstances, physical violence within the household or between families becomes all too common. Not all or even most Halpatis are addicted to the illegally brewed country liquor, but it needs to be emphasized that such drinking habits are not subject to the social taboo that is nowadays placed upon them in the Dhodia milieu.

Dissatisfaction with miserable conditions seems more and more to be released in outward-oriented aggression and an open show of impudence. Great indignation is felt by the landless about their exclusion from what were once common resources. Village wasteland, for example, has been converted into private ownership, and the new owners no longer allow the Halpatis to graze their animals or to gather firewood to cook their meals. They are also vexed about their limited access to the new public facilities that have been introduced.

Electricity was brought to Chikhligam in 1969, the centenary of Mahatma Gandhi's birth, when the government of Gujarat offered to provide this sign of modernization, with a high subsidy on construction costs, to villages whose populations consisted mostly of tribals. The pylons run straight through the Halpati quarter but not a single connection has been made to the houses, or at least none for which payment has been made. Babubhai obtains electricity by tapping the cable that runs above his house. He is not the only consumer in the colony who has a free supply. A little while ago an inspector came to put an end to this illegal distribution. But where would they get the money to pay for the current, he was asked. Had the inhabitants of this neighborhood no right to some light during darkness? With the aid of a couple of bottles of country liquor, the Halpatis were able to arouse the inspector's sympathy for their point of view, and he left without taking further steps.

The Halpatis understand only too well that, notwithstanding all efforts, their meager access to education and health care, for example, is due not only to the poverty in which they live but, conversely, that it keeps them imprisoned in that situation.

Balancing on the edge of subsistence, the rural proletariat tries to retain unimpaired the dignity and self-respect that accompany a life in freedom. The majority of Halpati women deliberately refuse to become maidservants in the households of large landowners, work that yields little more than a pittance and exposes them to constant humiliation. Not only are employers unwilling to provide bonuses in kind in addition to the monetary wage, but farmworkers and maidservants no longer welcome such extras, particularly in the form of leftovers. They decline "generous" gifts such as food scraps or waste such as used tea-leaves or worn clothing, that are meant to compensate for unpaid labor services. A local-level report on the operation of dairy cooperatives quotes the statement of a Halpati laborer who compared past and present conditions.

> He told me with relish about changes he had witnessed, contrasting earlier years when laborers had to eat whatever leftovers of poor quality food were served to them by their employers' wives, with the present pushing aside of their plates if they suspected that they were fobbed off with substandard fare (George 1992: 27).

Aversion, perhaps to the extent of giving preference to underconsumption, can also be noticed among young Halpati males who refuse employment offered them by Anavil landowners, even on days when no other work is available. Their disinclination seems to be due to the lack of respectability which, according to the new generation, attaches to work as agricultural laborer. Young women have a definite fancy for a husband who earns his living in other ways, preferably as a skilled and continuously employed laborer.

> This can be illustrated by the deception to which G in Chikhligam had to take recourse in order to get married. He worked as a farm servant for an

Anavil, and with that prospect in view not a single Halpati girl in the immediate vicinity was prepared to share her life with him. Eventually, G was able to find a partner in a distant village, but only after the middleman had given the parents the impression that the candidate had just completed his training as diamond cutter and thus promised to be an ideal husband for their daughter. That pretension made the wedding feast more than usually expensive for G. The bride suffered a severe shock when she realized that she had nevertheless ended up in the milieu of agricultural laborers.

Resistance from below sometimes becomes more virulent. Great animosity was aroused in the mid 1970s when large landowners in both villages started to employ armed guards to protect their orchards and fields against theft during harvest time. The use of such mercenaries revealed their loss of control by the local elite over the landless. The *simrakha* were heavy-handed, beating up children who gathered firewood and in one case even going to the extent of raping a woman working on the land. This latter incident, which occurred in Chikhligam, was reason for the Halpatis to march in procession to the village office where they demanded that the mercenaries be instantly dismissed. They got what they asked for, but only after a fight had broken out. The conflict would have resulted in the Halpatis being taken to court were it not that an Anavil witness, none other than the former village head, withdrew the statement he had made to the police. He denied his complicity later, under threat of further aggression by the landless, describing his leniency as a gesture of goodwill intended to calm feelings that were running high. Fear that violence would escalate did indeed play a role, however. The fact that such violence was not only threatened by the underdogs but that, the notables toned down their action subsequently, seems to me to illustrate a shift in the balance of power that is taking place at the grassroot level. As a matter of fact the decision of the government of Gujarat to establish a commissariat for rural labor in 1981 was the direct outcome of a similar clash in a village at a short distance from Chikhligam. Two agricultural laborers got killed in a fight which broke out between farmers and their hired mercenaries on the one hand and local Halpatis on the other hand, because the latter had dared to ask for payment of the currently prescribed minimum wages. Belatedly, the authorities seemed to acknowledge that without a machinery for implementation the protective legislation for the rural proletariat was bound to remain meaningless.

Among the more fortunate, opinion is rife that poverty is caused by a combination of deficiencies, i.e. incapability for self-improvement, and unwillingness to progress through diligent perseverance. The discrimination experienced by the landless arouses in them feelings of bitterness and revenge. Similarly, the opposite party which has for so many generations dominated the rural arena responds with extreme aggression to the ever stronger claims coming from the subaltern class for a more equitable distribution of economic growth. Although the prosperity of the Anavil Brahmins has increased, they are nevertheless no longer able to run their affairs according to their own interests and preferences. Relative deprivation on the one hand, and loss of domination on the other, represent the ingredients of a fairly torrid social climate in which minor and even major incidents are the order of the day.

We have already seen that the state is willing to give only limited support to the landless class in its effort to bargain for a better deal Bureaucratic protection of the proletariat, hesitantly accepted as a principle on paper, has achieved even less in actual practice. Moreover, in situations of conflict, Halpatis are likely to encounter opposition rather than support from state agencies and institutions. Inadequate help is then likely to veer into overt opposition or even persecution. Victimization, or the threat of it, is the official reaction with which rural proletarians have more experience than with any benevolent effect of policy intervention.

The doctrine of social equality is laid down in the Indian Constitution. In addition, universal suffrage ensures that everyone's vote carries equal weight. Why is it, then, that during the last two or three decades only marginal improvements have been made to the pitiable life of the landless class to which most Halpatis belong? The answer is that society's institutional framework is governed by laws other than that of social equality and a political voice for all. As I have observed earlier, the traditional legitimacy of inequality has gone. This does not mean, however, that the Halpatis are strong enough to fight their legal and social discrimination collectively.

A primary problem is the lack of political visibility of the Halpatis, evinced by the dearth of organized action that extends beyond the neighborhood level. In so far as their shared interests are defended it is done by outsiders, usually without the clients being present or even aware of what is going on. The landless mass has learned to place little or no trust in promises made by outside leaders such as politicians or social activists who offer their services, and have no-one in their own midst who can act from a position of strength. In the younger generation new spokesmen are emerging but their number is still quite small and, for the time being, they lack the necessary experience to operate successfully outside their own circle.

As we have seen, the settlement of Halpatis in their own colonies has promoted feelings of solidarity amongst them. Such awareness is raised by the negative stereotypes circulated by their class opponents, including the idea that those who have nothing to contribute are redundant in society. During the last few decades, the collective identity that was shaped under adverse conditions has been further articulated by state policies which made the Halpatis into a target group. The landless mass in south Gujarat has profited far less from the benign side of state intervention than other tribal castes such as the Dhodhias, who have at least some means of production. Factual evidence shows that the most vulnerable have benefited least from the effort which government felt compelled to make to ensure that distribution of the fruits of the development process should not be left entirely to the free play of market forces. The modest and reluctant premium that was placed on social backwardness resulted from a political imperative, i.e. the need for major political parties, particularly the Congress Party, to maintain or restore their credibility among the proletariat which is of diverse composition. In the middle classes that have gained in political significance in particular, the already low tolerance for pressure from below took a more militant turn when the state held out the promise of expanding the scope for the underprivileged to benefit from affirmative action programs. In the mid-1980s vehement protests were voiced in Gujarat when the federal state government attempted to facilitate access for youths from lower castes to higher professional education and to employment in the public sector by means of a quota arrangement. Controversy around *anamat*, the reservation policy, became violent in character and were the reference point for similar actions exploding in various parts of the country in the early 1990s. The militant campaigns are caused by increasing dissatisfaction among both the upwardly mobile sections of the population. They would rather deny claims for a better life voiced ever more forcibly by segments that refuse to accept their earlier inferiority and its concomitant discrimination. In the villages of my fieldwork, Anavils do not really feel that they are under direct attack from below, but they are frustrated because the pressure that is building up signals their loss of dominance. Their annoyance extends to the Halpatis, although the members of this tribal caste have derived little advantage from the positive discrimination policy. Even that little, however, is too much in the eyes of their former masters. Scoffingly, the Anavils describe their workers as *ghar jamai*, i.e. the sons-in-law of government. The village notables voice countless complaints about the negative economic and social qualities of the Halpatis. According to the members of the landowning class, politicians should teach discipline to the landless masses rather than try to pamper people who are not fit to help themselves and do not deserve to be helped by others. Does their reproduction really serve any necessary or useful purpose? Of late the Anavil Brahmins, as well as other dominant castes in the countryside have shifted their allegiance to the Bharatiya Janata Party. This trend is also followed by sections of the upwardly mobile lower castes.

In turn, the Halpatis are increasingly dissatisfied about the non-implementation of the progress that has so frequently been promised. Congress politicians in particular have tried to appease them with pledges which hold out the promise of a different kind of life. Halpatis have since the very first elections voted for candidates belonging to this party, mostly because big farmers and other employers are vociferous in their opposition to the Congress. But the Halpatis have received little if anything at all of what was pledged to them in exchange for their political loyalty. A sign that they refuse to function any longer as a humble and docile vote bank is that during elections in Spring 1991, less than half the electorate in the landless localities of Chikhligam and Gandevigam took the trouble to cast their vote. Absenteeism was formerly far less common, and to ascribe it merely to apathy would be to under-estimate the protest that this fact silently implies. The Halpatis have gone beyond the unobtrusive realm of social struggle, for which Scott has coined the term infra-politics (Scott 1990: 183). The veiled cultural struggle and hidden acts of transgression have set the stage for more open defiance.

The politics of confrontation that have exploded in various parts of India seem to be enacted mostly through mobilization on a caste basis. Conventional wisdom has it that such articulation is based on entrenched primordial sentiments belonging to an obsolete social order. That structuring principle of society has been eroded but features from the old order are still operative, according to this assumption, to evaporate eventually. However in their workplaces and particularly in the localities where they live, landless workers in Chikhligam and Gandevigam tend to appeal primarily to members of their own caste to ensure their survival and reproduction. The reservoir of solidarity thus achieved has helped to prevent a situation of anomie in which each person is not only required to be self-dependent but is also required to operate on the assumption that every other person is a potential adversary. Correspondingly, Halpatis are addressed by outsiders as a collectivity, and usually in derogatory terms. The entire social system is permeated with caste-specific stereotypes, both positive and negative, which play a part in obtaining or, conversely, in being denied access to scarce resources such as employment. In its attempt to achieve a redistribution of the development cake and to defuse class action, official policies have been instrumental in augmenting caste consciousness. In the social struggle that is going on, privileged positions are defended with arguments of an apparently universalist character, while the subaltern masses try to convert their numerical strength in numbers as a lever with which to end their intolerable exploitation.

The fight for the public as well as private spoils has hardly begun. The landless lack the organization that is preconditional to gaining a more balanced redistribution of the development spoils. It is interesting to note that the attempt to unite Halpatis on the basis of caste has had little success. In the 1970s in particular, the *Halpati Seva Sangh*, an association set up by Gandhian activists to achieve the economic and social upliftment of members of this tribal caste, was active in south Gujarat. In a report on the origins, background and strategy of this reform movement, I voiced my skepticism about the results that had so far been achieved (Breman 1974). Nothing that has happened in the meantime gives me reason to revise that opinion. The fairly largescale subsidies donated by state agencies to high-caste leaders were spent on social work that has made little impact. The ideological premise of the association was not contestation, an attempt to give voice to a counterpoint culture of tribal origin; on the contrary, much emphasis was given to the adaptation and subordination of detribalized *adivasis* within the dominant value pattern of the Hindu hierarchy. The Halpatis of south Gujarat apparently expected little from such a strategy and completely ignored the call to enrol as members of this caste union. When they still refused to do so after campaigns lasting many years, there was no other course than to close the branch offices that had been opened in the main towns of various sub-districts. Since its inception, the *Halpati seva sangh* leadership was in the hands of an Anavil social reformer who, disappointed by the frustration of his own political aspirations, defected from the Congress Party with which he had initially thrown in his lot. He deserted to join the Janata camp, a party which used to enjoy the support of the dominant caste of landowners in his area of operation. The company which he then kept was in itself sufficient to make him suspect in the eyes of his grassroots clientele. The cadre recruited by him, i.e. young Halpatis educated at the *ashram* schools set up by the *Halpati Seva Sangh*, sabotaged their leader from within the movement, embittered by his authoritarian attitude. In effect, the caste union died a silent death during the 1980s. A more genuine organisation has sprung up more recently. Late in 1992 a few educated and town-based members of this tribal caste have founded the Samrat (All) Halpati Samaj. To their own surprise a couple of hundred of Halpatis turned up at the first meeting for which invitations had been sent out rather haphazardly. The office bearers, who are not yet formally installed, come together every saturday to discuss business and further plans. They told me emphatically, when I met them in Valsad town in January 1993, that their association is non-political. "We don't want to be used as a mere vote-bank". A high priority in their programme of activities is to facilitate access to government schemes and subsidies for members of their caste. The draft of statute is, by and large a copy of the model they were shown by Dhodhia and Koli friends. I asked them who was the first to take this initiative and how they happen to know each other. It seems that they have run into each other at marriage parties. As a matter of fact, it is for that specific reason that they are interested in keeping contact. All of them are constantly asked to find a suitable partner for an educated boy or girl. It is quite clear that those Halpatis are not eager to start a political fight for their down-trodden brothers and sisters. Still this association seems to me a step forward in comparison with the *Halpati seva sangh* which functioned under the patronage of high-caste leadership. The caste-based Samaj may well turn out to be the next stage in the long process of emancipation.

The widening horizon inherent to the extension of scale in social relationships has brought the Halpatis to the realization that, together with other landless peoples, they form a rural proletariat. However, they have not yet taken the step towards a joint strategy rooted in that consciousness. This is due to the specific way in which capitalism has unfolded in the countryside of western India in the second half of the twentieth century. The results of this transformation in terms of employment modalities have to be analysed by following the Halpatis in their labor mobility. Only then is it possible to understand the structural barriers that prevent social cohesion among the various segments of the downtrodden from becoming mobilized in their search for a common destiny.

The disjunction that has been created between the home and work arenas of the landless is the subject of a monograph devoted to the footloose proletariat in south Gujarat to be published shortly.

My conclusion at the end of this study is that a formula has not yet been found by which the groundswell of social emancipation could be converted and consolidated into organized action. The state may take a less prominent role in engineering economic and social development. At least that idea has been vociferously propagated in public debates over the last few years. The ultimate consequence will be to deliver the weak into the hands of the strong. The failure of other economic sectors to accommodate the vast mass of the semi-employed in rural areas on a regular basis implies that, without strong and effective protective action, the surrender of this proletariat to the laws of free market forces can not be halted. Unchecked competition between limited demand and abundant supply of labor will otherwise lead to further victimization of the most vulnerable segments in society. My opinion remains unchanged that basic acknowledgement of the right to a life of human dignity is the most important principle marking the piecemeal progress or temporary relapse in the painful struggle for social emancipation waged in the countryside of South Asia.

Having come to the finale of my report on the conditions of rural labor towards the end of the twentieth century, I am sure of one thing: that this song of the *hali*, the bonded laborer of south Gujarat, is a song of the past.

*I go in darkness*
*I return in darkness*
*My whole life is full of darkness*
*there is no ray of light*

(Point of View, VI, 1978: 32)

# Glossary

| | |
|---|---|
| *ādivāsī* | collective name of tribal groups and castes |
| *ādu* | dried ginger |
| *Ahīr* | caste of shepherds |
| *Anāvil Brahman* | dominant caste of landlords in south Gujarat |
| *āshram* | institution of social work on Gandhian lines |
| *āvat* | term for *jajmāni* system current in south Gujarāt |
| *bāval* | a tree, *acacia arabica* |
| *bāgāyat* | garden land |
| *Baniā* | caste of Merchants |
| *bhātā* | farm servant's daily ration in grain |
| *Bhāṭhelā* | section within the Anāvil caste |
| *bīḍī* | native cigarette |
| *chākar* | farm servant, at the same time house servant |
| *chhās* | skimmed milk, residue of the preparation of *ghī* |
| *chīku* | a fruit, *sapodilla* |
| *dāl* | lentils |
| *Darjī* | caste of Tailors |
| *Deśāī* | section within the Anāvil caste; originally, designation of local power elite and revenue farmers; derived from *désh* (land) |
| *deśāīgīrī* | government compensation for lost Deśāī rights |
| *dhaṇiamo* | traditional name for master of farm servant |
| *dharmashālā* | guest house |
| *Dheḍ* | caste of Weavers |
| *Dhodia* | a tribal caste |
| *dīvālī* | Feast of Lights |
| *Dublā* | a tribal caste; also called *halpati* |
| *gām* | village |
| *Gāmit* | a tribal caste |
| *Gāruḍā* | priest caste of Untouchables |
| *Ghānchī* | caste of Oil-pressers, at present often shopkeepers |
| *ghī* | "clarified" butter |
| *govālio* | cowherd |
| *gulām* | slave |

| | |
|---|---|
| *guṛ* | roughly pressed and evaporated cane sugar |
| *hal* | plow |
| *haladar* | species of ginger |
| *hāḷī* | designation of farm servant who was traditionally bound to a landlord |
| *hāḷīprathā* | pattern of relationships in which the traditional relationship between landlords and farm servants was institutionalized |
| *harekwāḷī* | maid |
| *Harijan* | collective name for castes of Untouchables; lit., people of God |
| *hoḷī* | spring festival |
| *ināmdār* | functionary whose office entitled him to a land grant or to part of the land rent |
| *jarāyat* | high-level, dry land |
| *juvār* | species of millet |
| *kacchā* | not pure, inferior |
| *kāḷīparaj* | description of tribal groups, non-Hindus |
| *Kaṇbī Paṭēl* | dominant caste of landowners in the northern part of the plain of south Gujarat |
| *khāvati* | advance payment in money or in kind on work to be performed |
| *kodrā* | inferior species of grain |
| *Koḷī* | caste of small farmers in the plain of south Gujarat; along the coast also fishermen |
| *kyārī* | low-lying land, surrounded by embankments, used for rice cultivation |
| *mahāl* | administrative unit, smaller than a subdistrict |
| *Mochī* | caste of Shoemakers |
| *mukādam* | labor recruiter |
| *nāgḷī* | inferior species of grain |
| *Nāyakā* | a tribal caste |
| *pālkhī* | palanquin |
| *panch* | council |
| *panchāyat* | village council, caste council |
| *Pārsī* | group, descended from the Persians, who settled in south Gujarat in the seventh century |
| *paṭlā* | labor gang |
| *paṭēl* | village functionary with police authority |
| *Pedivālā* | highest status group among the *Deśāīs* |

| | |
|---|---|
| *phalia* | village quarter or street |
| *phaṇas* | jackfruit |
| *pīpar* | long pepper (a medicinal herb) |
| *pīvānum* | drink money |
| *rājā* | lord |
| *roṭlī* | pancake made of unleavened dough |
| *riāyatvārī* | system of direct land tax collected by the government without intermediaries |
| *rupee* | national monetary unit in India; the traditional subdivision into 16 annas is still used by the population in calculating small amounts |
| *sandhal* | exchange of labor |
| *sāṛī* | women's outer garment |
| *satī-patī* | movement among tribal groups in the eastern part of south Gujarāt, which has a religious undertone |
| *seer* | unit of weight, about two pounds |
| *sūraṇ* | elephant's foot (a tuber) |
| *talāṭī* | village clerk |
| *tāluka* | subdistrict |
| *ṭopī* | men's headgear |
| *ujlīparaj* | all Hindu castes without a recent tribal past |
| *Vairāgī* | local type of *sādhu* |
| *vakil* | solicitor, lawyer |
| *vāl* | a leguminous plant |
| *Vāland* | caste of Barbers |
| *vāvlā* | plot of land which, in the past, was given to a farm servant for his own cultivation |

# Abbreviations

| | |
|---|---|
| *A.A.* | *American Anthropologist* |
| *A.I.* | *America Indigena* |
| *B.J.S.* | *The British Journal of Sociology* |
| *B.T.L.V.* | *Bijdragen tot de Taal-, Land- en Volkenkunde* |
| *C.I.S.* | *Contributions to Indian Sociology* |
| *C.S.S.H.* | *Comparative Studies in Society and History* |
| *E.A.* | *Eastern Anthropologist* |
| *ETHN.* | *Ethnology* |
| *E.W.* | *Economic Weekly* (continued as *Economic and Political Weekly, E.P.W.*) |
| *H.O.* | *Human Organization* |
| *I.E.S.H.R.* | *The Indian Economic and Social History Review* |
| *I.J.A.E.* | *Indian Journal of Agricultural Economics* |
| *I.J.S.W.* | *Indian Journal of Social Work* |
| *I.S.* | *Indian Sociologist* |
| *J.A.F.* | *Journal of American Folklore* |
| *J.G.R.S.* | *Journal of the Gujarat Research Society* |
| *J.U.B.* | *Journal of the University of Bombay* |
| *M.I.* | *Man in India* |
| *S.B.* | *Sociological Bulletin* |
| *S.G.* | *Sociologische Gids* |
| *S.R.—N.S.* | *Selections from the Records of the Bombay Government, New Series* |
| *S.W.J.A.* | *Southwestern Journal of Anthropology* |

# Bibliography I

AGRICULTURAL LABOUR ENQUIRY

1955 *Report on an Intensive Survey of Agricultural Labour, 1950–51*, vol. I (All India) and vol. V (West India), Delhi

AGRICULTURAL LABOUR IN INDIA

1960 *Report on the Second Enquiry, 1956–57*, vol. I (All India) and vol. V (West India), Delhi

AGRICULTURAL LABOUR IN INDIA

1962 Ed. by V.K.R.V. Rao, Bombay

"AGRICULTURAL WAGES AND SYSTEMS OF WAGE PAYMENTS"

1948 In *I.J.A.E.*, vol. III, no. 1, "Proceedings of the Eighth Conference"

BAILEY, F. G.

1957 *Caste and the Economic Frontier*, Manchester

BAINES, J. A.

1912 *Ethnography, Castes and Tribes*, Strassburg

BAKS, C. J., ET AL.

1965 "De betrouwbaarheid van informatie in een kastesamenleving in Gujarat, India" [The reliability of information in a caste society in Gujarat, India], in *S.G.*, vol. 12, 167–74

BAKS, C. J., J. C. BREMAN AND A.T.J. NOOIJ

1966 "Slavery as a System of Production in Tribal Society," in *B.T.L.V.*, vol. 122, 90–109

BAKS, C. J.

1969 *Afschaffing van pacht; een onderzoek naar de gevolgen van de afschaffing van pacht in twee dorpen van Zuid-Gujarat, India* [An investigation into the consequences of the abolishment of tenancy in two villages of south Gujarat, India], Ph.D. thesis, Amsterdam

BARDAN, P.

1969 "Green Revolution and Agricultural Labourers" in *E.P.W.*, vol. V, no. 29–31, July 1970.

BARODA ECONOMIC DEVELOPMENT COMMITTEE, 1918–19

1920 *Report*, Bombay

BARODA ENQUIRY COMMISSION

1873–75 Vol. I, *Correspondence*, Bombay

BARTH, F.

1960 "The System of Social Stratification in Swat, North Pakistan," in E. Leach (ed.), *Aspects of Caste in South India, Ceylon and North-West Pakistan*, Cambridge University Press

BEIDELMAN, T. O.

1959 *A Comparative Analysis of the Jajmani System*, Monographs of the Association for Asian Studies, VIII, New York

BELLASIS, A. F.

1854 "Report on the Southern Districts of the Surat Collectorate," in *S.R. — N.S.*, Bombay

BENNETT, J. W. AND I. ISHINO

1963 *Paternalism in the Japanese Economy; Anthropological Studies in Oyabun-Kobun Patterns*, University of Minnesota Press

BETEILLE, A.

1965 *Caste, Class and Power: Changing Patterns of Stratification in a Tanjore Village*, University of California Press

BOMBAY PRESIDENCY

*Miscellaneous Official Publications*; Minute 15–10–1830, Malcolm Peyt

BREMAN, J.

1969 "Veranderingen in de betrekkingen tussen landheren en landarbeiders in Zuid-Gujarat, India" [Changes in the relationships between landlords and agricultural laborers in south Gujarat, India], in *S.G.*, vol. XVI, 395–402

BROOMFIELD, R. S. AND R. M. MAXWELL

1929 *Report of the Special Enquiry into the Second Revision Settlement of the Bardoli and Chorasi Talukas*, Bombay

CENSUS, IMPERIAL, OF 1881

1882 Vol. II, Bombay

CENSUS OF INDIA, 1901

1902 Vol. XVIII, Baroda, pt. 1, Bombay

CENSUS OF INDIA, 1911

1911 Vol. XVI, Baroda, pt. 1, Bombay

CENSUS OF INDIA, 1921

1922 Vol. III, Bombay Presidency, pt. 1, Bombay; vol. XVII, Baroda, pt. 1, Bombay

CENSUS OF INDIA, 1931

1932 Vol. XIX, Baroda, pt. 1, Bombay

1933 Vol. VIII, Bombay Presidency, pt. 1, Bombay

CENSUS OF INDIA, 1941

1941 Vol. XVII, Baroda

CENSUS OF INDIA, 1951

1952 *District Census Handbook: Surat District*, Poona

CENSUS OF INDIA, 1961
1962 Paper 1 of 1962, "Final Population Totals," Delhi

CENSUS 1961, GUJARAT
1964 *District Census Handbook XVI: Surat District*, Ahmedabad

CHOKSEY, R. D.
1968 *Economic Life in the Bombay Gujarat (1800–1939)*, London

COHN, B. S.
1955 "The Changing Status of a Depressed Caste," in McKim Marriott (ed.), *Village India: Studies in the Little Community*, The American Anthropological Association Memoir no. 83, 53–77
1959–60 "Some Notes on Law and Change in North India," in *E.D.C.C.*, vol. VIII, 79–93
1960–61 "The Pasts of an Indian Village," in *C.S.S.H.*, vol. III, 241–49

CORRESPONDENCE
*Between the Directors of the East India Company and the Company's Government in India on the Subject of Slavery, Account and Papers*, vol. XVI, Session 15–11–1837 — 16–8–1838, pt. LI

DAVE, J.
1946 *Halpati-mukit; hālī prathā ane mukitadānni hilachāl*, Ahmedabad

DESAI, A. R. (ED.)
1961 *Rural Sociology in India*, The Indian Society of Agricultural Economics, Bombay

DESAI, B. A.
*The Marketing of Mangoes in the Surat District*, Ph.D. thesis, Bombay

DESAI, D.
1942 "Agrarian Serfdom in India," in *I.S.*, 1–14

DESAI, G. G.
1906 *Some Experiences of a Mamlatdar-Magistrate's Life*, Ahmedabad

DESAI, I. P.
1964 *The Patterns of Migration and Occupation in a South Gujarat Village*, Poona

DESAI, M. B.
1948 *The Rural Economy of Gujarat*, Bombay
1952 "Rural Economy of Reconstituted Gujarat," in *J.G.R.S.*, vol. XIV, 111–48

DESAI, M. B. AND C. H. SHAH

1951 "Problems of Farm Labour in Gujarat," in *I.J.A.E.*, vol. VI, 4ff.

DESAI, M. H.

1929 *The Story of Bardoli, Being a History of the Bardoli Satyagraha of 1928 and Its Sequel*, Ahmedabad

DEVOS, G.

1966 "Conflict, Dominance and Exploitation in Human Systems of Social Segregation: Some Theoretical Perspectives from the Study of Personality in Culture," in A. de Reuck and J. Knight (eds.), *Conflict in Society*, Ciba Foundation Publication, London, 60–81

DEY, S. K.

1964 *Community Development: A Bird's-Eye View*, London

DONDE, W. B.

1951 *Rural Labour in the Konkan*, Ph.D. thesis, Bombay

DOSABHAI, E.

1894 *A History of Gujarat from the Earliest Period to the Present Time*, Ahmedabad

DOVRING, F.

1965 "Bondage, Tenure, and Progress: Reflections on the Economics of Forced Labor," in *C.S.S.H.*, vol. VII, 309ff.

DUMONT, L.

1966 "The Village Community from Munro to Maine," in *C.I.S.*, vol. IX, 67–89

ELPHINSTONE, MOUNTSTUART

1884 *Selections from the Minutes and Other Official Writings of the Honourable Mountstuart Elphinstone*, by G. N. Forrest, London

ENTHOVEN, R. E.

1920–22 *The Tribes and Castes of Bombay*, 3 vols., Bombay

EPSTEIN, T. S.

1962 *Economic Development and Social Change in South India*, Oxford University Press

1967 "Productive Efficiency and Customary Systems of Rewards in Traditional South India," in M. Lipton (ed.), *Themes in Economic Anthropology*, A.S.A. Monograph Series, no. 6, London

ETHERIDGE, A. T.

1868 *Report on Past Famines in the Bombay Presidency*, Bombay

FRYKENBERG, R. E.

1965 *Guntur District: A History of Local Influence and Central Authority in South India*, Oxford

GAZETTEER OF THE BOMBAY PRESIDENCY
1877 Vol. II, Surat and Broach, Bombay
1882 Vol. XIII, pt. 1, Thana, Bombay
1883 Vol. VII, Baroda, Bombay
1901 Vol. IX, Gujarat Population: Hindus, Bombay

GAZETTEER OF THE BARODA STATE
1923 2 vols., Bombay

GAZETTEER OF INDIA
1962 Gujarat State, Surat District, Ahmedabad

GOUGH, K.
1955 "The Social Structure of a Tanjore Village," in McKim Marriott (ed.), *Village India*, 36–52
1960 "Caste in a Tanjore Village," in E. R. Leach (ed.), *Aspects of Caste in South India, Ceylon and North-West Pakistan*, Cambridge
1960 "Review of T. O. Beidelman, *A Comparative Analysis of the Jajmani System*," in *E.D.C.C.*, vol. IX, 83–91

GOULD, H. A.
1958 "The Hindu Jajmani System: A Case of Economic Particularism," in *S.W.J.A.*, vol. XIV, 428–37
1964 "A Jajmani System of North India: Its Structure, Magnitude and Meaning," in *ETHN.*, vol. III, 12–41

HABIB, I.
1963 *The Agrarian System of Moghul India (1556–1707)*, London

HARPER, E. B.
1959 "Two Systems of Economic Exchange in Village India," in *A.A.*, vol. LVI, 760–78
1968 "Social Consequences of an 'Unsuccessful' Low Caste Movement," in *C.S.S.H.*, suppl. 3, *Social Mobility in the Caste System in India*, 36–65

HISTORICAL SELECTIONS FROM BARODA RECORDS
1955 *New Series*, vol. I, 1826–35, Baroda

HOLMBERG, A. R.
1967 "Algunas Relaciones entre la privación Psico-Biologica y el Cambio Cultural en los Andes," in *A.I.*, vol. 27, 3ff.

HOMMES, E. W.
1970 *Over de evaluatie van plattelands-planning in India* [On the evaluation of rural planning in India], Ph.D. thesis, Amsterdam

HOVE
"Tours for Scientific and Economic Research Made in Gu-

zerat, Kattiawar, and the Conkuns in 1787–8," in *S.R.*, no. 16, *N.S.*

ISHWARAN, K.
1966 *Tradition and Economy in Village India*, London

JAMABANDI
1907 *Revision Settlement Report of the Gandevi Taluka of the Navsari Division, 1906–1907*, Baroda

JOSHI, V. H.
1966 *Economic Development and Social Change in a South Gujarat Village*, The Maharaja Sayajirao University of Baroda

KAPADIA, K. M.
1966 "Industrial Evolution in Navsari," in *S.B.*, vol. XV, 1–24

KEATINGE, G.
1921 *Agricultural Progress in Western India*, London

KISHORE, J.
1924 "The Village Labourer in Western India," in *Hindustan Review*, 425ff.

KLOOSTERBOER, W.
1954 *Onvrije arbeid na de afschaffing van de slavernij* [Unfree labor after the abolishment of slavery], The Hague

KOLENDA, P. M.
1963 "Toward a Model of the Hindu Jajmani System," in *H.O.*, vol. XXII, 11–32

KOSAMBI, D. D.
1956 *An Introduction to the Study of Indian History*, Bombay

KOTOVSKII, G.
1956 "Indian Agricultural Labourer; a Soviet View," in *E.W.*, April 14, 443ff.

KUMAR, D. K.
1965 *Land and Caste in South India: Agricultural Labour in the Madras Presidency during the Nineteenth Century*, Cambridge University Press

LEACH, E. R.
1967 "Caste, Class and Slavery; the Taxonomic Problem," in A. de Reuck and J. Knight (eds.), *Caste and Race*, Ciba Foundation Publication, London, 5–16

"LETTER OF COLLECTOR OF SURAT, 9–4–1823"
In *Bombay Revenue Proceedings, 1823*, Range 368, vol. XXXVIII, 4816

LEWIS, O. AND V. BARNOUW
1967 "Caste and the Jajmani System in a North Indian Village," in J. M. Potter *et al.* (eds.), *Peasant Society, A Reader*, Boston, 110–34

LORENZO, A. M.

1943 *Agricultural Labour Conditions in Northern India*, Bombay

1945 "Agrestic Serfdom in Northern India," in *I.J.S.W.*, vol. V, 133–41

MAYER, A. C.

1958 "The Dominant Caste in a Region of Central India," in *S.W.J.A.*, vol. XIV, 407–27

MEHTA, B. H.

1934 "A Summary of the Economic Life of an Aboriginal Tribe of Gujarat," in *J.U.B.*, vol. II, 311–59

MEHTA, J. M.

1930 *A Study of the Rural Economy of Gujarat*, Baroda

MEHTA, S.

1930 *Marriage and Family Life in Gujarat*, Ph.D. thesis, Bombay

MOORE JR., B.

1967 *Social Origins of Dictatorship and Democracy: Lord and Peasant in the Making of the Modern World* (paperback ed.), Bostøn

MORELAND, W. H.

1920 *India at the Death of Akbar*, London

1929 *The Agrarian System of Moslem India*, Cambridge

MORRIS, M. D.

1966 "Economic Change and Agriculture in Nineteenth Century India, a Review of Kumar's Study," in *I.E.S.H.R.*, vol II, 185–209

MORRISON'S REPORT

Letter from Collector of Surat dated 13–11–1812

MUKERJEE, R.

1933 *Land Problems of India*, London

—— *The Economic History of India: 1600–1800*, Bombay

MUKHTYAR, G. C.

1930 *Life and Labour in a South Gujarat Village*, Calcutta

NAIK, T. B.

1950 "Songs of the Anavils of Gujarat," in *M.I.*, vol. XXX, 29–56

1953 "Family in Gujarat," in *J.G.R.S.*, vol. XV, 117–38

1957 "Dubla Songs," in *Vanyajati*, 141–49

1957 "Social Status in Gujarat," in *E.A.*, vol. X, 173–82

1958 "Religion of the Anavils of Surat," in *J.A.F.*, vol. LXXI, 389–96

1958 "The Strains of a Social System," in *J.G.R.S.*, vol. XX, S1–S11

NEALE, W. C.
1962 *Economic Change in Rural India: Land Tenure and Reform in Uttar Pradesh, 1800–1955*, Yale University Press

NIEBOER, H. J.
1910 *Slavery as an Industrial System*, The Hague

OOMMEN, T. K.
—— *Agrarian Tension in a Kerala District*, I.J.I.R. Reprint Series no. 15

OPLER, M. AND R. D. SINGH
1948 "The Division of Labor in an Indian Village," in C. S. Coon (ed.), *A Reader in General Anthropology*, New York

ORENSTEIN, H.
1962 "Exploitation of Function in the Interpretation of Jajmani," in *S.W.J.A.*, vol. XVIII, 302–16
1965 *Goan: Conflict and Cohesion in an Indian Village*, Princeton University Press

PANDYA, B. V.
1959 *Striving for Economic Equality*, Bombay

PAPERS . . .
1858 *Relating to a Summary Settlement of Alienated Revenues in the Bombay Presidency*, Revenue Department, Bombay

PAPERS . . .
1895 *Relating to the Settlement of the Hereditary District Officers' Watans in the Deccan and Gujarat*, W. G. Pedder, *S.R.*, no. 174, *N.S.*, Bombay

PAPERS . . .
1897 *Relating to the Revision Survey Settlement of the Chikhli Taluka of the Surat Collectorate*, *S.R.*, no. 381, *N.S.*, Bombay

PAPERS . . .
1897 *Relating to the Revision Survey Settlement of the Bardoli Taluka*, *S.R.*, no. 359, *N.S.*

PAPERS . . .
1900 *Relating to the Revision Survey Settlement of the Bulsar Taluka of the Surat Collectorate*, Bombay

PAPERS . . .
1900 *Relating to the Revision Survey Settlement of the Jalalpor Taluka of the Surat Collectorate*, *S.R.*, no. 350, *N.S.*, Bombay

PATEL, J. M.
1964 "Agricultural Labour in a South Gujarat Village," in V. S. Vyas (ed.), *Agricultural Labour in Four Indian Villages*, Sardar Patel University, Vallabhai Vidyanagar, 92–119

PATEL, S. J.
1952 *Agricultural Labourers in Modern India and Pakistan*, Bombay

POCOCK, D. F.
1957 "The Bases of Faction in Gujarat," in *B.J.S.*, vol. VIII, 295–306
1962 "Notes on *jajmāni* Relationships," in *C.I.S.*, no. 6, 78–95

PRIVILEGES . . .
1955 *Provided by the Bombay State Government for Backward Classes*, Bombay

RAMAKRISHNAN, K. C.
1948 "Systems of Wage Payments in Agriculture with Special Reference to Madras," in *I.J.A.E.*, vol. III, no. 1

REPORT . . .
1951 *of the Congress Agrarian Reforms Committee*, Delhi

REPORT . . .
1900 *of Famine Operations in the Baroda State for the Year 1899–1900*, Baroda

REPORT . . .
1950 *of the Hali Labour Enquiry Committee*, M. L. Dantwala and M. B. Desai, Bombay (not published)

REPORT . . .
1961 *of the Scheduled Areas and Scheduled Tribes Commission, 1960–61*, Delhi

REPORT . . .
1883 *on the Census of the Baroda Territories, 1881*, Bombay

REPORT . . .
1868 *on the Revenue Settlement Introduced into the Soopa Taluka*

REVISION . . .
1907 *Settlement of the Navsari Taluka of the Navsari Division, 1906–07*, Baroda

REVISION . . .
1911 *Settlement Report Palsana Taluka*

ROGERS, A.
1892 *The Land Revenue of Bombay*, London

ROWE, W. L.
1963 "Changing Rural Class Structure and the Jajmani System," in *H.O.*, vol. XXII, 41–44

ROYAL COMMISSION . . .
1927 on Agriculture in India, vol. II, pt. 1, *Evidence Taken in the Bombay Presidency*, London

SELECTION . . .
1887 *from the Letters, Despatches, and Other State Papers Preserved in the Bombay Secretariat*, Home Series, vol. I, G. W. Forrest (ed.), Bombay

SEN, B.
1962 *Evolution of Agrarian Relations in India*, New Delhi

SHAH, C. H.
1952 *Effects of World War II on Agriculture in India (with Special Reference to Gujarat)*, Ph.D. thesis, Bombay

SHAH, P. G.
1958 *The Dublas of Gujarat*, Delhi

SHAH, S. M.
1952 *Rural Class Structure in India, with Special Reference to Gujarat*, Ph.D. thesis, Bombay

SHIRRAS, G. F.
1924 *Report on an Enquiry into Agricultural Wages in the Bombay Presidency*, Government of Bombay Labour Office, Bombay

SHUKLA, J. B.
1937 *Life and Labour in a Gujarat Taluka*, Calcutta

SINGH, B.
1960 "Plan Approach to Agricultural Workers," in *Current Problems of Labour in India*, Delhi, 60–64

SINGH, M.
1947 *The Depressed Classes: Their Economic and Social Condition*, Bombay

SIVASWAMY, K. G.
1948 "Serf Labour among the Aboriginals," in *I.J.S.W.*, vol. VIII, no. 4

SIVERTSEN, D.
1963 *When Caste Barriers Fall: A Study of Social and Economic Change in a South Indian Village*, New York

SJOBERG, G.
1965 *The Preindustrial City: Past and Present* (paperback ed.), New York

SOLANKY, A. N.
1955 *The Dhodhias*, Ph.D. thesis, Bombay

SRINIVAS, M. N. AND A. M. SHAH
1960 "The Myth of Self-Sufficiency of the Indian Village," in *E.W.*, September 10, 375ff.

SRINIVAS, M. N.
1955 "The Social System of a Mysore Village," in McKim Marriott (ed.), *Village India*, Chicago, 1–35

1965 *Religion and Society among the Coorgs of South India*, reprint, London

1966 *Social Change in Modern India*, University of California Press

SYMINGTON, D.

*Report on the Aboriginal and Hill Tribes of the Partially Excluded Areas in the Provinces of Bombay*, Bombay

THORNER, D. AND A. THORNER

1962 *Land and Labour in India*, London

VEEN, K. W. VAN DER

1972 *I Give Thee My Daughter*, Assen

VREEDE-DE STUERS, S.C.L.

1965 "De Hindoe-maatschappij in overgang" [Hindu society in transition], in P. van Emst (ed.), *Panorama der Volken*, pt. 2, 263–358, Roermond

VYAS, V. S. (ED.)

1964 *Agricultural Labour in Four Indian Villages*, Sardar Patel University, Vallabhai Vidyanagar

WALLACE, R.

1863 *The Guicowar and His Relations with the British Government*, Bombay

WATSON, J. W.

1886 *History of Gujarat*, Bombay

WEBER, M.

1892 *Die Verhältnisse der Landarbeiter im ostelbischen Deutschland*, Leipzig

1922 "Wirtschaft und Gesellschaft," in *Grundriss der Sozialökonomik*, III Abteilung, Tübingen

1924 "Die ländliche Arbeitsverfassung," in *Gesammelte Aufsätze zur Sozial- und Wirtschaftsgeschichte*, Tübingen

1967 *The Religion of India: The Sociology of Hinduism and Buddhism*, transl. by H. H. Gerth and D. Martindale (paperback ed.), New York

WISER, W. H.

1936 *The Hindu Jajmani System*, Lucknow

WITTFOGEL, K.

1963 *Oriental Despotism: A Comparative Study of Total Power* (paperback ed.), Yale University Press

WOLF, E. R.

1966 *Peasants*, Englewood Cliffs, New Jersey

# Bibliography II


*Published Books and Articles*

BARDHAN, P.

1984 *Land, Labour and Rural Poverty*. Delhi, Oxford University Press.

BARDHAN (ED.)

1989 *Conversations Between Economists and Anthropologists; Methodological Issues in Measuring Change in Rural India*. Delhi, Oxford University Press.

BOSE, S. (ED.)

1990 *South Asia and World Capitalism*. Delhi, Oxford University Press.

BRASS, T.

1990 "Class Struggle and the Deproletarianisation of Agricultural Labour in Haryana (India)". *The Journal of Peasant Studies*, vol. 18, no. 1, October, 36–67.

BREMAN, J.

1974 *Patronage and Exploitation: Changing Agrarian Relations in South Gujarat, India*. Berkeley, California University Press.

1976 "A Dualistic Labour System?: A Critique of the 'Informal Sector' Concept". *Economic and Political Weekly*, vol. 11, No. 48, November 1870–6.

1977 "Labour Relations in the 'Formal' and 'Informal' Sectors: Report of a Case Study in South Gujarat, India". *The Journal of Peasant Studies*, vol. 4, No. 3 April 171–205 and vol 4, No. 4 July p. 337–59

1978 "Seasonal Migration and Co-operative Capitalism: The Crushing of Cane and of Labour by the Sugar Factories of South Gujarat". *Economic and Political Weekly*, vol. 13, no. 31, 32, 33, 1317–60.

1985 *Of Peasants, Migrants and Paupers: Rural Labour Circulation and Capitalist Production in West India*. Oxford/Delhi Clarendon Press/Oxford University Press.

——— "Extension of Scale in Fieldwork: From Village to Region in South Gujarat". Bardhan 1989: 127–36.

1990 "'Even Dogs Are Better Off': The Ongoing Battle Between Capital and Labour in the Cane Fields of Gujarat". *The Journal of Peasant Studies*, vol. 17, no. 4, July, 546–608.

CENSUS OF INDIA

1952 District Census Handbook: Surat District. Poona.

1962 Census 1961, Gujarat. District Census Handbook 16: Surat District. Ahmedabad.

1973 Census 1971, Gujarat. District Census Handbook, Part X-C-I, A & B, Valsad District. Ahmedabad.

1983 Census 1981, Gujarat. General Population Tables, Series 5, Part II-B. Ahmedabad.

1983 Census 1981, Gujarat. District Census Handbook, Part XIII, A & B, Valsad District. Ahmedabad.

1991 Census 1991, Gujarat, Series-7, Provisional Population Totals. Ahmedabad.

CHAKRAVARTI, U.

1985 "Of Dasas and Karmakaras: Servile Labour in Ancient India". U. Patnaik (ed.), 35–75.

EPSTEIN, T. S. & D. JACKSON (EDS.)

1975 *The Paradox of Poverty*. London, Macmillan.

KOLCHIN, P.

1987 *Unfree Labor: American Slavery and Russian Serfdom*. Cambridge Mass./London, The Belknap Press of Harvard University Press.

KRISHNAJI, N.

1989 *The Size and Structure of Agricultural Labour Households in India*, in: G. Rodgers (ed.) 121–50, 1930.

MARLA, S.

1981 *Bonded Labour in India*. National Survey on the Incidence of Bonded Labour, Final Report; Sponsored by the Gandhi Peace Foundation and the Academy of Gandhian Studies. New Delhi.

MILES, R.

1987 *Capitalism and Unfree Labour: Anomaly or Necessity*?. London/New York, Tavistock Publications.

MUKHTYAR, G. C.

1930 *Life and Labour in a South Gujarat Village*. Calcutta, Longmans, Green & Co.

MUNDLE, S.

1979 *Backwardness and Bondage: Agrarian Relations in South Bihar District*. Delhi, Indian Institute of Public Administration.

PATNAIK, U. (ED.)

1985 *Chains of Servitude: Bondage and Slavery in India*. Introduction, 1–34. Madras, Sangam Books.

1985 "Reflections on the Agrarian Question and the Development of Capitalism in India". Daniel Thorner Memorial Lecture. New Delhi

PRAKASH, G.

1990 *Bonded Histories: Geneologies of Labor Servitude in Colonial India*. Cambridge/New York, Cambridge University Press, Cambridge South Asian Studies No. 44.

"Bonded Labour in South Bihar: A Contestatory History", in Bose, 178–205.

RAMACHANDRAN, V. K.

1990 *Wage Labour and Unfreedom in Indian Agriculture: An Indian Case Study*. Oxford, Clarendon Press.

RODGERS, G. (ED.)

1989 "Population Growth and Poverty in Rural South Asia". *World Employment Programme*, ILO. Delhi, Sage Publications.

RUDRA, A.

1987 "Labour Relations in Agriculture: A Study of Contrasts". *Economic and Political Weekly*, vol. 22, no. 17, 25 April, 757–60.

SARKAR, T.

1985 "Bondage in the Colonial Context", in U. Patnaik. 97–126.

SCOTT, J.

1990 *Domination and the Arts of Resistance: Hidden Transcripts*. New Haven/London. Yale University Press.

SHAH, G.

1977 *Protest Movements in Two Indian States: A Study of the Gujarat and Bihar Movements*. Delhi, Ajanta Publications.

*Other publications*

BHALLA, S.

1991 "Report on the study Group on Employment Generation". Delhi National Commission on Rural Labour.

DESAI, N.

1990 *Women's Work and Family Strategies in a Rural Community in South Gujarat*. Sponsored by Indian Council of Social Science Research.

DHARMALINGAM, A.

1991 "Social Relations of Production and Fertility in a South Indian Village". Ph. D. thesis submitted to Australian National University, Canberra.

GEORGE, S.

1992 "Generalisation and Rural Development; Eleven Villages in South Gujarat, India". Paper Read at Research Seminar on Rural Development Studies, January. Institute of Social Studies, The Hague.

GUPTE, D. B. & A. S. CHARAN

1978 Report on Bonded Labour in Valsad District. South Gujarat University, Surat.

HIRWAY, I. & P. TERHAL

1990 "Towards Right to Work in India: Indian and International Experiences in Rural Works Programmes". Report submitted to the Indo-Dutch Programme of Research on Alternatives in Development. New Delhi/The Hague.

JOSHI, V.

1990 "Role of RLG in Implementation of Minimum Wages in Gujarat". Part of Wage Study Group set up by the National Commission on Rural Labour. Gandhi Labour Institute, Ahmedabad.

KOTHARI, U.

1990 "Women's Work and Rural Transformation in India; a Study from Gujarat". Ph.D. thesis submitted to University of Edinburgh, Edinburgh.

LAL, R. B.

1977 "Bonded Labour in Gujarat (Does It Exist in Gujarat?)". Tribal Research and Training Institute. Gujarat Vidyapith. Ahmedabad.

STREEFKERK, H.

1991 "Transformation in a South Gujarat Town: Bulsar, Suthars and the Relevance of Caste Twenty Years Later", in: *Economic and Political Weekly*, vol. 26, May 25.

NATIONAL COMMISSION ON RURAL LABOUR

1991 Overview. Delhi.

PATEL, B. B.

1991 "Agrarian Structure, Rural Wages and Rural Poverty: A Study of Some Aspects in Gujarat". Sponsored by the Advisory Group of Experts on Poverty Studies set up by the Department of Economic Affairs, Ministry of Finance, Government of India. Gandhi Labour Institute, Ahmedabad.

SHAH, G.

1978 "Agricultural Labourers: Are They Bonded"? Centre for Social Studies, Surat.

WASHBROOK, D.

1990 "Land and Labour in Late 18th Century South India: The Golden Age of the Pariah"? Paper read at the conference of the British Association of South Asian Studies. School of Oriental and African Studies, London.

# Index

## Objectives of the Seminar

IT IS INDEED a great honour for me to be given an opportunity to address such a distinguished group of scientists, technologists, economists, and other specialists.

As had been mentioned by my colleague, the Director General of LPN, it is the Malaysian Government's policy and intention to increase the level of domestic rice production and at the same time to increase the income of our paddy farmers. To achieve these we are at the moment concentrating on both the small farm production as well as promoting larger scale commercial estate type of paddy production, learning perhaps from our successes on large-scale production in rubber, oil palm, and cocoa. Such a goal may be realised through sustained, well-defined efforts by all concerned with the industry. One such effort is in the area of research and development (R & D). I am glad to inform the gathering that extensive R & D efforts and linkages are already in existence in Malaysia; between the local agencies such as MARDI, LPN, UPM and other universities and agencies; between Malaysia and her ASEAN neighbours, particularly under the ASEAN-Australia and ASEAN-EEC programs; between Malaysia and other government and international agencies such as ACIAR, IDRC, FAO, and UNDP programs. Such linkages have provided an opportunity for us in increasing the skill and depth of multidisciplinary, multi-prong approaches to solving the problems faced by the rice industry.

The Malaysian National Agricultural Policy has given top priority to rice production, being the staple food of the country. In line with that we have established a wide R & D coverage ranging from agricultural production activities to grain primary processing to storage and marketing. The range of disciplines covered include rice breeding, agronomy, pest management, mechanisation, water management, socioeconomic studies, postharvest grain handling, drying, storage of paddy, milling, storage of milled rice, distribution and marketing.

The varying disciplines no doubt need to be coordinated and tuned to the general as well as specific needs of the industry. Within MARDI itself we have established a special research division, the Paddy Research Division to look into the overall needs of the industry. The specific and specialised needs are being contributed by the various R & D support divisions within MARDI such as the Central Analytical Division for the basic studies on soils, agricultural engineering, water management, seed technology, biotechnology and other basic sciences support; the Techno-economic and Social Studies Division for socioeconomic support and the Food Technology Division for the postharvest and processing components. At the inter-agency and industrial level, we have management committees to scrutinise our R & D plans and strategies. At another level, we have committees to help in the implementation of the R & D activities.

All these setup and coordination efforts are necessary because, as all of us know, R & D is an expensive tool of progress and development. Thus, R & D needs to be well planned and managed. The objectives need to be clearly defined and the implementation needs to be effectively managed to ensure that the expected results are obtained not only within the time frame needed by the industry, but also to be of relevance to the dynamic nature of the industry. In other words, any solution created through R & D needs to be appropriate to the peculiarities of the industry.

This seminar I feel is doing just that — reviewing the technology of a specific problem of the industry, i.e. in preserving grain quality through the process of aeration and in-store drying. The seminar will, I understand, review the current state of knowledge on the principles of in-store drying and aeration — in particular the technical aspects, to be followed by an assessment, in both technical and economic terms, of the relevance of these principles to the handling of wet grains under the local humid climatic conditions.

These principles and thus the technology of aeration and in-store drying have considerable relevance to the current status of the Malaysian paddy industry, especially in the preventive and corrective options available in solving the local grain handling and storage problems. In fact these principles have already been utilised to a certain extent by the local paddy and rice industry. It is hoped, however, that the discussions in this seminar, on the principles involved, will further enhance our knowledge and understanding of the mechanisms involved in bringing about and maintaining evenly distributed low moisture and temperature conditions within the bulk of grain, especially under the humid tropical conditions. I wish you a most successful and fruitful seminar.

**Mohd Yusof bin Hashim**
Director-General
MARDI

# Keynote Address

IT IS a great pleasure for me to be here on this auspicious occasion and I am deeply honoured to be asked to address the assembly and to declare open this seminar on preserving grain quality by aeration and in-store drying.

On behalf of the Government of Malaysia, I extend our warmest welcome to all the distinguished guests, learned participants, and interested observers. I understand that this seminar has been organised by several international agencies in collaboration with the local organisations, namely the National Paddy and Rice Authority (LPN) and the Malaysian Agricultural Research and Development Institute (MARDI). The Group for Assistance on Systems relating to Grain After-harvest (GASGA), the Australian Centre for International Agricultural Research (ACIAR), and the ASEAN Food Handling Bureau (AFHB) have collaborated and cooperated very effectively to organise this seminar and to bring together a pool of experts of various disciplines who are distinguished leaders in their respective fields.

The Government of Malaysia attaches great importance to matters concerning rice. This is because rice has always been our staple food. It comprises more than 20% of the food consumed by the entire population and constitutes as much as 35% of the diet of the rural population.

Paddy cultivation is the most important agricultural activity next to rubber in terms of employment and land utilisation. Presently a total of 300 000 farm families are engaged in paddy cultivation involving some 477 500 ha of paddy land, which contribute to the yearly production of about two million tonnes of paddy.

Paddy growers in Malaysia are traditionally among the poorest sectors of the rural economy. However, various programs of poverty eradication had been implemented to contain and reduce the magnitude of poverty. The incidence of poverty among paddy growers declined from 88.1% in 1970 to 55.1% in 1983. This decline in the incidence of poverty was a remarkable and profound achievement because it was attained in a period characterised by unfavourable weather conditions and worldwide inflation which led to increasing consumer prices and rising input costs. However, the incidence of poverty among paddy growers continues to be high and above national average, primarily due to the existence of a large number of uneconomic holdings, exacerbated by low yields in areas outside the major irrigation schemes.

Besides the vast investments towards new irrigation facilities, measures were also instituted to increase the net income of paddy farmers especially through a price support scheme and input subsidies. The purchase price of paddy under the Guaranteed Minimum Price Scheme (GMP) was raised from US$120 per tonne in 1970 to about US$210-225 per tonne in 1979. It was further increased to US$285-300 per tonne since 1980, giving a better return to the farmers.

When we consider the large amount of resources, both human and financial, involved in producing food, it is indefensible that substantial amounts of food are lost during and after harvests for lack of conservation facilities and techniques. As such Malaysia now attaches great importance to the problem of postharvest grain losses. Since reduction of these losses would make available thousands of tonnes of food, better incomes to the paddy growers, and at the same time reduced imports and foreign exchange savings for our country would be realised.

Under the double-cropping schemes, the off-season crop has to be harvested in the wet weather. Therefore, the major problem of harvesting this crop is the drying of paddy. If the wet paddy is not dried soon after harvest, it will deteriorate in quality and cause dry matter losses. As a result, growers obtain a poor price. It was in this context that the LPN undertook a nationwide expansion program for construction of a number of rice mill complexes aimed at providing adequate drying, milling, and storage facilities. The need for sufficient processing and storage facilities became more pressing after the extensive use of combine harvesters for harvesting operations in the major rice growing areas. This is due to the fact that harvesting time will be very much shortened thereby increasing the daily production of wet paddy. This has imposed a further burden on the existing drying facilities, both in the private as well as in the public sector mainly resulting from the fact that narrow harvest peaks cannot be efficiently handled by existing facilities. New methods of handling and preservation of wet paddy have to be devised. We therefore place very great hopes that the seminar of this nature will generate new ideas that can be adopted to solve some of our pressing problems.

The Government of Malaysia has always placed great emphasis on the role of research and development in all industries. In the agricultural sector, beside the institutes of higher learning, several research bodies and agencies have been set up to undertake various research and development works. Presently, the LPN, in collaboration with other agencies, particularly MARDI, has carried out a number of research studies which include paddy aeration for silo storage, gastight storage of milled rice, and the bulk handling of paddy. To-date, some of these research findings have been implemented and incorporated in the existing facilities.

In the area of research and development, I see a great need for closer cooperation and collaboration among the various research bodies, both local and international. This is important in order to avoid possible duplication in our works thereby enabling us to utilise our resources more efficiently. It is therefore vital that the free flow and exchange of information and findings be enhanced among the various research bodies and implementing agencies. In view of the different conditions that exist in each country, the implementation of studies from abroad should be by way of adaptation to suit the local situation rather than wholesale adoption. This is necessary to avoid any possible pitfalls and costly mistakes.

The subject of this seminar is of great relevance to the needs of the paddy and rice industry in Southeast Asia. This seminar will highlight the objectives and principles of aeration and in-store drying. It will also endeavour to discuss the systems, designs, and economics for such practices. Their appropriateness and feasibility will be further illustrated by case studies conducted in various countries. I sincerely hope that the proceedings of this seminar will be compiled and be made available as an authoritative and useful text of reference for researchers and developers of grain aeration and in-store drying technologies.

The holding of this seminar reflects the close rapport and established relations which have long existed among the various organisations of the world. This healthy relationship is held in high esteem by us and it is hoped that this chain of relationships will be extended to more countries and more agencies. Whilst you are here attending this seminar, I hope you will take the opportunity to be closer to us by sparing some of your time to get to know Malaysia and its people. I am sure you will discover many similarities between our countries. You will also notice the differences and will learn to appreciate them. We always say that to know Malaysia is to love Malaysia.

**Haji Daud bin Dato' Haji Taha**
Deputy Minister of Public Enterprises
Government of Malaysia

# Objectives of Aeration and In-store Drying

# Grain Quality Considerations in Relation to Aeration and In-store Drying

R.A. Boxall and D.J.B. Calverley*

*Abstract*

This paper considers the assessment of grain quality in relation to various intrinsic and acquired properties of grain. Intrinsic properties include the shape, size, structure, and colour of the grain, and its bulk density, thermal conductivity, equilibrium moisture content, and flowability. Important acquired properties are the presence of foreign matter, physical and pest damage, and abnormal moisture content. The parameters for deterioration in storage and the role of aeration and in-store drying in maintaining quality are discussed. Attention is drawn to the need to take account of the possibly serious effects on grain quality of the drying technique itself when developing a strategy for grain drying.

THE term 'quality' will have different meanings for the different people concerned with the handling, storage, processing, and utilisation of grain, even though all will be looking for grain of 'good quality'. For example, members of the grain trade will want dry, insect-free, undamaged grain which will store well; millers will want grain which will yield a high percentage of finished product; and consumers will be concerned with appearance, flavour, and cooking properties. Variation in grain quality begins with the variety selected by the producer, and is thereafter influenced by the climatic and soil conditions during the growing season, cultivation practices, weather conditions at harvest, and harvesting techniques. It is accepted that, whatever the condition of grain at harvest, quality cannot in general be improved during storage, handling, and processing; on the contrary, it is easily lost. So what is quality?

## Definition of Quality

The dictionary definition of the word 'quality' gives it the same meaning as 'characteristic' or 'property'. In these terms every type of grain can be said to possess many properties which in turn contribute to its overall 'quality'. A consideration of the various properties or qualities either alone or together allows the grain to be graded and valued, and enables the design and development of optimum methods of handling, storing and processing. The qualities can be grouped into two broad categories, namely intrinsic and acquired qualities.

* Storage Department, Tropical Development and Research Institute, London Road, Slough, Berkshire, SL3 7HL, U.K.

### Intrinsic Qualities

Intrinsic qualities are those possessed by the whole, unblemished, ripened grain. They include shape, size, structure, and colour, and are of importance in determining the type description of the grain, whereby it can be identified and checked for purity. The physical characteristics of grain such as bulk density, thermoconductivity, equilibrium moisture content, and flowability, are intrinsic qualities important in relation to the design of grain handling and storage facilities.

Grains are living organisms and accordingly contain, in addition to minerals and water, compounds such as carbohydrates, proteins, oils, and vitamins. The nature, proportions, and distribution of these components within the grain are characteristic for each type or variety of grain and may be said to constitute its intrinsic quality of composition, which is related to the nutritional value of the grain.

Important biological properties of grains include the rate of respiration, which is a function of the storability of grain, and the germination potential, which is of importance when grain is used as seed or for malting. These too are intrinsic qualities, as also is the inherent susceptibility to infestation. All grains are vulnerable to insect infestation or infection by microorganisms, but different types or varieties are characterised by differences in their relative susceptibility to such attack. This quality is important to those who control the movement of grain through the postharvest system. Grains which are least resistant need to be moved quickly or given greater protection against infestation or infection.

### Acquired Qualities

Acquired qualities are those which are not inherent. They are extrinsic and are thus either additional to the normal complement of intrinsic qualities or are the consequence of certain intrinsic qualities being lost or modified. The importance of the acquired qualities depends upon their nature and the extent of their occurrence, but generally they are detrimental to the overall quality of the grain concerned.

Let us now consider some examples of acquired qualities. During the harvesting and threshing operations within the postharvest system, grain is liable to acquire foreign matter such as weed seeds, straw, chaff, sand, stones, soil, etc., the presence of which can have far-reaching effects on the overall quality and the value of the commodity. Organic foreign matter often provides food for insect pests or a suitable substrate for the growth of microorganisms, which are themselves foreign matter but of an active nature. As insect pests and microorganisms multiply within the grain, by their presence they increase the amount of foreign matter and contamination of the grain. Metabolic heat is generated, the grain mass is reduced, grain breakage occurs, and dust is produced. Contamination and foreign matter are terms which are frequently used interchangeably but it is important to distinguish the two. Generally, foreign matter can be readily removed from grain whereas contaminants cannot. The latter include the soluble excretions of pests and other animals, oils, pesticides, pathogenic microorganisms spread by rodents and insects, and toxins produced by certain moulds.

Physical damage, i.e. breaking, splitting, or cracking of the grain renders it more prone to infestation by insects and microorganisms. Such damage is likely to arise at any stage in the postharvest system but particularly during threshing, drying, and processing. It may have consequences on overall grain quality that become especially evident at the marketing stage.

Pest damage will be acquired by grain if it is susceptible to infestation and not protected from such damage. A range of insect pests cause damage to grain either by chewing at it or by boring in it. Moulds and bacteria will spread into and through grains causing discoloration and lesions which may be the only manifestations of infection if grain is dried before complete destruction can occur. Grain may also become discoloured as a result of heat damage caused, for example, by the development of 'hot spots' in bulk-stored grain.

Correct moisture content is important. If the percentage moisture content of the grain is above or below that which might be expected under ambient climatic conditions, or is outside the limits prescribed by a grain grading standard, the grain may be said to have acquired an abnormal moisture content. Such a moisture content may occur through inadequate or excessive drying. An abnormally high moisture content encourages infestation by both insects and microorganisms and it allows the grain to metabolise more rapidly than is desirable, resulting in unwanted chemical changes, germination, premature aging, or increases in temperature leading to heat damage and perhaps self-destruction of the grain. An abnormally low percentage moisture content usually presents fewer problems. Grain is unlikely to become badly infested by insects when very dry, and the risk of microbiological infection is low or negligible. However, when dried too rapidly or overdried, grains may become very brittle so that they are likely to split, crack, or break easily. This of course is undesirable if a whole grain such as rice is preferred by the consumer and it also presents problems for rice millers. In the milling of grains to flour, overdry grain can also cause problems.

## Attitudes to Grain Quality

Before turning to the use of drying to preserve grain quality it is important to consider the attitudes to quality that are held by those directly involved.

The subsistence farmer will attempt to minimise the risk of loss and maintain the quality of his grain to a standard which satisfies his family's needs. An important constraint is the availability of family labour to assist with all postharvest operations. Drying of the relatively small quantities (1-2 t) of grain by natural sun-drying is accomplished by regular personal attention to turning the grain and protecting it from inclement weather. Given the opportunity to increase production and achieve a marketable surplus, the farmer is encouraged to grow high-yielding varieties and to adopt multiple cropping of his land. He is then faced with a new set of problems. His capability to harvest becomes overstretched and when his second crop matures in the wet season he is faced with the difficulty of drying a larger quantity of grain.

The technological solution to this problem is difficult. It requires financial commitment; for a dryer and a source of heat, frequently using fossil fuel. A degree of new management skill is also required for the successful operation of the drying system.

The farmer's simplest solution to this problem is to sell most of the moist grain without delay in order to minimise the risk of loss. The problem is then passed to the grain procurement organisation which is liable to be inundated with grain over a very short period of time. Unless grain is purchased against a rigorously applied quality standard, which specifies a safe, low moisture content, then the procurement organisation must provide strategically placed, adequate grain drying/conditioning facilities. The provision of facilities at primary procurement level may be more realistic but major technical and economic problems can arise. The costs of drying have to be incorporated in the quality grading and price structure. The price paid to producers bringing grain to the procurement centres has to reflect the actual drying costs and the amount of water removed from the grain during the process. Thus, the penalties that a farmer suffers through his inability to dry grain can be high.

As more grain is produced, the grading standards are likely to be raised and so unless the farmer can meet the new requirements his financial rewards for increased production are likely to be reduced.

Grain quality standards become a vital issue in the successful operation of a postharvest system. It is rare to find a system satisfactorily serving all interests, from the producers' financial expectations for his crop through to the consumers' preferences for a cheap, palatable food. The inevitable perturbations which result in seasonal variations of shortage and abundance throw a burden on the procurement and marketing organisations' ability and capacity to satisfy all demands at minimal cost.

### Maintaining Grain Quality during Storage

The aim of good storage practice is to maintain conditions in the grain that will preserve those qualities or properties considered important by the end user (in marketing, processing, etc) at an appropriate level. Basically, this means prevention of damage and deterioration caused by insects, mites, and microorganisms and protection from birds and rodents. If birds and rodents can be excluded, the rate of deterioration or the loss of quality during storage will depend upon micro-environmental and climatic conditions and will be at its lowest when grain is dry and cool, since the growth rate of microorganisms depends upon moisture and temperature and that of insects mainly on temperature. The grain will, of course, be subject to the normal process of aging, during which the physical qualities and composition of the grain will change. The rate at which aging takes place will vary with the type and variety of grain concerned and will be influenced by environmental conditions. During prolonged storage, grains are liable to suffer changes in texture, colour, flavour, or nutritive value resulting from the effects of moisture and temperature (even in the absence of insects and microorganisms). The changes which occur may not necessarily render the grain unfit for consumption, but they may make it less palatable or less acceptable to the consumer on aesthetic grounds.

### Moisture

The moisture in grain is traditionally considered in terms of the percentage moisture content by weight, usually based upon the wet weight of a sample under test. However, when discussing the preservation of grain quality, it is more meaningful to consider the moisture content of the intergranular environment, or the equilibrium relative humidity (ERH) corresponding to a particular grain moisture content. The susceptibility of stored grain to deterioration is correlated to the ERH and not to the moisture content of the grain.

Microorganisms are unable to multiply when the ERH is below 65% although it is generally accepted that to protect stored grain from mould the maximum ERH should be 70%. Insects and mites, however, are most active in the range

60–80% ERH, although some species survive and multiply at 40–60% ERH and a few, for example the khapra beetle *Trogoderma granarium,* can do so at even lower levels.

### Temperature

The effects of temperature on deterioration by insects, mites, and microorganisms are quite straightforward. The temperature range favourable to the growth of microorganisms is wide, extending from below 0°C for cold-hardy species of moulds, to over 60°C for certain heat-resistant bacteria. Most storage insects are unable to complete their life cycle at temperatures outside the range 15–45°C and at temperatures below 17°C, insect activity is generally severely reduced. Thus, cooling of grain to 17°C or lower before or during storage provides an effective means of controlling insect infestation. Under certain circumstances it may be possible to cool grain to temperatures low enough to inhibit the activity of fungi.

Grain cooling can be achieved by aeration using ambient or artificially cooled air. The term aeration is usually taken to mean the forced movement of air at low airflow rates for the purpose of cooling rather than drying, although some drying may occur. In tropical climates, storage temperatures are usually in the range 25–30°C, and although some cooling by aeration may be possible to prevent grain deterioration, it is often necessary to pay more attention to reducing grain moisture content as a means of suppressing biological, particularly microbiological, activity and use other methods to control insect infestation when this occurs.

### Drying as an Effective Method of Conditioning Grain for Storage

Cereal grains attain physiological and functional maturity at moisture contents ranging between 35 and 45% depending on the crop. Appropriate moisture levels for storage are between 10 and 14% depending upon not only the crop but also factors such as storage temperature. Consequently, timely harvesting and drying at crop maturity is important for achieving a high quality product.

The objective of drying is to reduce grain respiration by removal of excess moisture. This will prevent the qualitative deterioration which may also arise from the growth of microorganisms and the activities of insects and mites. However, the process of drying may itself adversely affect grain quality unless the drying system is selected and operated with care. Whatever method is used, when grain is dried there is likely to be some change in the characteristics which affect overall quality. Generally speaking, the adverse effects on grain quality will increase with the severity of the drying conditions. Damage to the grain may arise if drying is carried out at too high a temperature or too quickly.

The cheapest method of drying is natural drying using the heat of the sun and the evaporative effect of the wind. Grain may be held in cribs or spread in thin layers on the ground or on trays or mats to dry. Natural drying, whilst generally acceptable for traditional farming needs, does leave the grain open to dust, insect infestation, and attack by birds and other vertebrates which may result in losses in quality and quantity.

In the humid tropics natural drying may be difficult or even impossible, especially in those countries where, as a result of multiple cropping, grains must now be harvested in the wet season. Under these circumstances there is a requirement for what is popularly referred to as artificial drying. This can involve the forced movement of ambient or natural air through the grain or the use of heated air as the drying medium.

Drying with natural air is possible when its relative humidity is lower than the ERH of the grain, and even in humid areas, substantial amounts of water can be removed. Although the amount of water removed is never enough to reduce the grain moisture content to a safe level, a system of drying using first ambient air and then heated air may be desirable in preserving grain quality. We shall return to this later.

Drying procedures with heated air are basically of two types, namely low temperature/high volume and high temperature/low volume. Low temperature drying, i.e. at 5–10°C above ambient, may be applied when the grain is held in temporary or extended storage systems, whereas drying at higher temperatures is generally carried out in systems specifically designed to facilitate exposure of grain to the drying air.

The principal constraint on the acceptable length of time during which drying can be completed is grain deterioration due primarily to mould growth. However, the depth of grain and the volume of air used are also important in preventing the growth of mould. The maximum drying time using low temperature air will depend upon the grain type, its physical condition and its moisture content and temperature, all of which equate to the food supply and conditions for mould growth. Fungal deterioration is a dynamic process, involving a succession of microorganisms and resulting in the breakdown of organic matter. During this process carbon dioxide is released and a measurement of the amount of $CO_2$ produced has been used by various workers (Saul and Lind 1958; Steele et al. 1969) to determine the amount of dry matter loss due to mould growth. They subsequently calculated an allowable storage loss of less than 1%, which provided an indication of grain of acceptable quality.

When the temperature of the drying air is raised, the capacity to remove moisture is increased and so the grain drying can be completed more rapidly. High temperature drying is therefore useful where large amounts of grains have to be dried in a short time. However, there are limits to both the amount of heat that can be added and the drying speed if a high quality grain is to be maintained. These limits are based upon the degreee of physical or heat damage to the grain which results in stress cracks, brittleness, breakage, discoloration, and loss of viability, and the intended future use for the grain.

### Effect of Drying on Grain Quality

When grain is overheated during drying, the quality may be affected in a number of ways. At temperatures above 45°C the embryo of the grain will be destroyed, rendering it useless as seed. Maize for the starch industry should not be dried at temperatures above 60°C or the starch separation process will be damaged, as gluten becomes hardened or toughened and clings to the starch.

Quality changes in wheat which is overheated may be manifested in low grade flour, which is unpalatable or has poor baking properties. If the temperature is high and the RH of the air is low, there is a danger that the moisture will be removed more rapidly than water can diffuse from the inner layers of the grain, and a hardening or casing may be formed. This impervious layer will prevent free diffusion of moisture and will cause grains to become so wrinkled, scorched and discoloured that they are rejected completely because of their appearance.

The brittleness in grain which arises as a result of rapid drying at high temperatures is particularly noticeable as stress cracks or checking in the kernels. Paddy, soybeans, and maize are prone to such internal cracking if dried too rapidly and subsequent handling will cause the kernels to break up. This has serious implications for grain quality when the consumer demands whole grains, as in the case of paddy/rice.

It is interesting to note that much of the qualitative deterioration which is likely to arise during drying can be effectively reduced by measures which also conserve energy, such as the processes of 'dryeration' and combination drying. In dryeration, grain is first dried at a high temperature and this is followed by a tempering period before the grain is finally cooled. Stresses which build up in the grain during high-temperature drying are relieved during tempering before the grain is subjected to the additional stresses caused by rapid cooling and which would otherwise result in cracking or breakage of the grain. Combination drying involves an initial phase of high temperature, high-speed drying, followed by further drying at lower temperatures or using natural air. Both methods extend the period of drying and lead to a reduction in the brittleness that would result in grain breakage. Table 1, summarising the work of Foster (1975), shows the relationship between the method of drying and brittleness of maize. A similar problem is recognised with paddy where late harvesting and uncontrolled drying cause cracks in rice grains that cause milling losses (Bhashyam et al. 1985).

Many of the modifications designed to improve the fuel efficiency of grain dryers are likely to have a positive bearing on grain quality. Drying stresses will be reduced if the vapour pressure difference between the moisture in the grain being dried and in the drying air can be reduced. This depends upon the addition of moisture to the drying air and can be achieved by recirculating the drying air.

**Table 1.** Effect of drying method on britleness of dried corn.

| Drying method | Sound kernels without stress cracks (%) | Breakage (%) |
|---|---|---|
| Conventional continuous flow | 8.8 | 11.3 |
| Dryeration | 60.6 | 6.7 |
| Partial heat or combination drying | 82.2 | 3.9 |
| Unheated air | 93.3 | 1.6 |

Lower airflow rates and greater grain-holding capacity per unit of dryer output tend to reduce drying speed, a major factor in causing grain brittleness and qualitative deterioration.

### Preservation of Grain Quality by Aeration

When grain has been dried satisfactorily to a 'safe' moisture content and is kept in a store that is structurally sound, it may still be at risk from the effects of serious increases in the moisture content which may be induced by prevailing climatic conditions. Grain may take up moisture directly from the ambient air in the store, or moisture migration from warmer or to cooler parts of the grain bulk may occur.

Moisture uptake is likely to occur in those regions where humidity is consistently higher as in the humid tropics or during a tropical rainy season, and where grain is stored in bags rather than in bulk. Moisture migration, on the other hand, is more likely to occur in grain bulks, especially in metal silos or bins where the grain is in contact with the walls. Under these conditions, solar heating of the silo walls can result in large temperature differentials, i.e. high temperatures at one side or at the periphery of the structure. These differentials can cause convection currents in the grain accompanied by moisture migration from high to low temperature areas. As the air is cooled, its relative humidity rises and may reach saturation at which point excess water will be deposited on the surface of the cooler grain. Localised increases in moisture content can therefore occur and create conditions favourable for the development of microorganisms and subsequent qualitative deterioration. Aeration can prevent the occurrence of these potentially serious conditions and maintain the quality of the stored grain. The passage of air through the grain will have not only a cooling effect but will maintain a uniform, low temperature throughout the bulk.

### Developing a Strategy for Grain Drying

Much attention has been devoted to increasing production of cereal crops through the use of improved, high-yielding varieties, double or multiple cropping, and improved production technology and this has inevitably led to a number of problems in the handling of grain after harvest. Farmers now have to cope not only with increased yields but also with crops which mature and have to be harvested during the wet season. Various organisations have responded to the problems by looking at the possibilities of developing improved harvesting or threshing techniques and by introducing grain dryers to replace traditional sun drying. In the humid tropics most agriculture is small-scale and the intended beneficiaries, i.e. the small-scale farmers, find the new techniques, especially grain dryers, too expensive. They therefore have to resort to the traditional method of sun drying. However, suitable drying floors are usually inadequate and periods of sunshine too short to cope with the quantities of grain involved, and so they have no option but to sell their wet paddy early or suffer considerable grain deterioration due to mould damage, germination, and fermentation. A possible solution is to provide a centralised drying system, for example at mills, but experience has shown that these facilities may be inadequate to cope with quantities of grain involved and that millers are sometimes reluctant to invest in grain dryers because of high fuel costs. As a consequence, a severe strain may be placed on the capability of the central grain-handling agencies to dry and store the grain satisfactorily.

When the existing drying facilities cannot cope with the volume of incoming wet grain, alternative methods of handling are required. De Castro et al. (1980) investigated the use of a relatively inexpensive aeration system for paddy in bags, and showed that the rate of quality deterioration, measured by the proportion of damaged and discoloured grains, could be reduced by aeration. Thus, the holding period before drying could be extended but the need to dry could not be eliminated entirely.

There is an increasing interest among central grain-handling agencies, particularly in Southeast Asia, to move from storage of grain in bags to bulk-handling and storage. Associated with this change is the need to introduce a system of in-store drying and aeration to ensure that grain quality is maintained. Methods developed for in-store drying elsewhere in the world, particularly in temperate regions, have proven successful, being cost-efficient and resulting in commercial gains in terms of high quality grain.

The principles and requirements of aeration and in-store drying to prevent grain quality deterioration are well known in the context of temperate conditions. However, aeration and drying requirements will be quite different under humid, tropical conditions and may vary considerably from place to place. For example, it has been reported that tropical aeration requirements are four times greater than the values for temperate requirements and that local conditions on the east coast of peninsular Malaysia at Khota Bharu indicate a need for an airflow rate 50% higher than that needed on the west coast at Alor Setar (Teter 1981). Also, the requirements of drying will be greater than in the temperate regions. Overall, one can expect that energy requirements for aeration and in-store drying in the humid tropics will be greater than in temperate regions and so some care must be exercised when attempting to transfer temperate region technologies to the humid tropics.

Clearly, systems must be cost effective and so there will be a need to consider energy requirements and the effects of various drying parameters in relation to the quality of grain in order to develop an appropriate aeration or drying strategy. Various aspects of grain quality have been discussed, but in developing an appropriate strategy for the humid tropics one must guard against demanding too high a standard for overall grain quality. Ultimately, consideration must be given to providing a good wholesome product, free from contamination, which is acceptable to the local consumers. Industrialised countries' conventional standards of an acceptable commodity do not necessarily apply.

## References


Bhashyam, M.K., Mohan, S.L., and Srinivas, T. 1985. Genetic stability of crack resistance in rice grains. International Rice Research News, 10(3), 6.

De Castro, R.E., Andales, S.C., and de Padua, D.B. 1980. Field trials of predrying handling systems. Proceedings of the 3rd Annual Workshop on Grains Post-Harvest Technology. January 1980, Kuala Lumpur.

Foster, G.H. 1975. Causes and cures of physical damage to corn. In: Hill, L.D. ed., Corn quality in world markets. Danville, Illinois, Interstate Printers and Publishers, 221–229.

Saul, R.A., and Lind, E.F. 1958. Maximum time for safe drying of grain with unheated air. Transactions of the American Society of Agricultural Engineers, 10, 96–99.

Steele, J.F., Saul, R.L., and Hukill, W.V. 1969. Deterioration rate of shelled corn as measured by carbon dioxide production. Transactions of the American Society of Agricultural Engineers, 12, 685–689.

Teter, N.C. 1981. Aeration of grain. Presented at Seminar on paddy deterioration in the humid tropics. GASGA October 1981, Baguio, Philippines.

# In-store Drying of Grain: the State of the Art

**F.W. Bakker-Arkema and H.M. Salleh***

*Abstract*

The objectives of in-store drying are reviewed. Among the topics discussed are: decreasing the moisture content to safe levels; cooling the dried grain to ambient conditions; and cleaning the crop before drying. Alternative in-store drying techniques are compared with high-temperature automatic batch drying. Included in the comparison are natural air drying, low-heat drying, solar drying, in-store counterflow drying, 'dryeration', and combination drying. The use of grain stirrers and biomass energy with in-store drying installations is assessed.

The main criteria to be considered in evaluating the various methods of drying are grain quality, energy efficiency, capacity, operating costs, and farmer expertise. All in-store drying methods result in improved grain quality compared with high-temperature drying. Natural air drying produces the best quality (if properly operated), and needs the least amount of energy. However, its capacity is low and it requires the most attention and expertise of the farmer. Solar in-store drying is technologically attractive, but is of questionable economic value, at least in the United States. Grain stirrers are of value in low-heat systems in preventing the occurrence of a large moisture gradient in the bin.

Furnaces fired on biomass can supply the energy required for grain drying. The concentric vortex-cell biomass furnace, integrated into an on-farm, in-store counterflow grain dryer, is evaluated for potential use in the tropics.

THE prime objective of grain drying is to decrease the moisture content of recently harvested grain from 22–30% (wet basis, w.b.) to 12–14% (w.b.) in order for the crop not to deteriorate in quality during the subsequent storage period (Brooker et al. 1974). It depends on the end-use of the grain which quality characteristic needs to be conserved; for rice, head yield and cooking quality are essential, while for wheat the baking quality needs to be maintained. If the grain is to be used for seed, the viability has to remain high.

Grain is dried in various ways (Bakker-Arkema et al. 1978). Traditional sun drying, in which the wet grain is spread out on the ground in the open air, is still practiced in many parts of the world; the resulting grain quality is marginal at best, and the system is weather dependent. Much of the grain in the tropics is still handled and dried in bags, but the transition from bag to bulk handling and drying appears only a matter of time. Therefore, only bulk drying is considered here.

In-store drying of bulk grain requires a structure to hold the grain (e.g. a bin) along with a mechanical air-movement device (e.g. a fan), and usually an air-heater; natural air drying, low-temperature drying (also called supplemental-heat drying), ammonia drying, combination drying, batch-in-bin drying, and in-bin counterflow drying are in-store drying systems; each is able to produce excellent quality grain when properly designed and operated (Bakker-Arkema 1984). High-temperature continuous-flow dryers are found at high capacity installations such as rice mills and grain terminals; often the grain quality is deleteriously affected by the high air-temperature and drying rates (McLean 1980). Only in-store drying systems will be considered in this paper.

The initial cost, the grain quality, the energy efficiency, the operating costs, the weather patterns, and the farmer expertise required are among the major criteria for selection of in-store drying systems. Since no drying system ranks first in every category, the best choice of a dryer for a particular location will be a compromise. This is true for the humid tropics as well as for other regions where grain is dried.

* Department of Agricultural Engineering, Michigan State University, East Lansing, MI 48824-1323, USA.

### In-store Drying Systems

In-store grain drying systems can be classified into: (1) high initial moisture content/high air-temperature systems, and (2) intermediate moisture content/low air-temperature systems. The dividing line in moisture content is about 21–22% (w.b.), and in air-temperature 30–40°C. A comparison of continuous-flow and in-store drying systems for paddy has been made by Bakker-Arkema et al. (1983).

#### High Moisture Content/High Temperature Systems

Batch-drying, combination drying, and in-store counterflow drying are methods for drying bulk grain regardless of the initial moisture content. The temperature of the drying air in these systems ranges between 45° and 75°C, the airflow rate between 10 and 30 $m^3$/minute/t. The success and capacity of the systems is not dependent on the weather conditions.

*Batch drying* (also called batch-in-bin drying to distinguish it from mobile, high-temperature batch dryers) is limited to the drying of relatively shallow layers of grain (1–1.5 m) in order to prevent serious overdrying of part of the grain. Drying and cooling of the moist grain takes place in 4–8 hours depending on the initial moisture content, the air temperature and the airflow rate (Brooker et al. 1974). Batch drying to safe-storage moisture contents (11–14%) is not weather dependent, requires little operator expertise, causes overdrying of some grain with a resulting quality deterioration, and is not energy efficient (Kalchik et al. 1981). Unless high capacities are required, in-store batch drying in the humid tropics should be limited to combination drying systems.

*Combination drying* or two-stage drying is a process in which high-capacity, high-temperature, high-airflow drying to an intermediate moisture content of 17.5–22% is followed by in-store low-temperature, low-airflow drying and cooling to 12–15% (w.b.) (Morey et al. 1981). The high-temperature, high-airflow phase of combination drying can be an in-store batch system (as discussed in the previous paragraph) or a mobile batch dryer. The low-temperature, low-airflow phase of combination drying can be accomplished by ambient air or by slightly heated air (i.e. natural air drying or supplemental-heat drying), and may require from two days to two months depending on the initial moisture content of the grain placed in the bin and on the temperature and airflow rate of the drying air (Morey et al. 1981). *Dryeration* is the best known combination drying system; the intermediate moisture content is 17–18% (w.b.), ambient air is used at a rate of 0.5–1.0 $m^3$/minute/t for final drying and cooling after a 6–8 hour tempering period in a special tempering bin (McKenzie et al. 1972). Several combination drying systems are compared in the next section.

*In-store counterflow drying* is a combination drying system and consists of two bins, one of which is a heated-air counterflow dryer (Bridges et al. 1981). Wet grain, regardless of moisture content, is loaded into the first bin and dried until the bottom 8–12 cm layer has reached a moisture content of 16.5–19.0% (w.b.). The partially dried, hot grain is removed from the bottom of the store by a tapered sweep auger, and is transported to a second bin where final drying and cooling take place. The air temperature in the first bin is 40–45°C for rice and 70–95°C for maize; the airflow rate is 10–30 $m^3$/minute/t, depending on the depth of grain in the bin. Ambient air is used in the second bin; the airflow rate is a low 0.5–1.0 $m^3$/minute/t. Removal of the partially dried grain from the first to the second bin is intermittent; the time between cycles of the unloading auger depends on the airflow rate, the initial grain moisture content, and the drying air temperature, and varies from 15 to 150 minutes. If the logistics allow, final drying and cooling in the second bin should be after a 4–8 hour tempering period in order to optimise the energy efficiency and grain quality of the system. In-bin counterflow drying is a very flexible in-store drying system and is being used extensively in the United States both for maize and paddy (Bakker-Arkema et al. 1980); only the relatively high initial cost might interfere with widespread adoption of the system in the humid tropics.

#### Intermediate Moisture Content/Low Temperature Systems

Natural (ambient) air drying, supplemental-heat drying (also called low-temperature drying), and trickle-ammonia drying are limited to the drying of grains with an initial moisture content below 21–22% (w.b.). The drying air temperature in these systems ranges from ambient to 1–5°C above ambient; the airflow rate is between 0.5–4.5 $m^3$/minute/t. The systems are dependent on the weather conditions. Some newly designed units are under microprocessor control (Mittal and Otten 1983).

*Natural-air drying* and *supplemental-heat drying* systems are similar; the grain is dried and cooled in the bin in which it is stored. The distinction between the systems is that no heat (except fan energy) is added to the drying air in natural-air drying, while the drying air-temperature is increased 3–5°C in supplemental-heat systems in order to decrease the relative humidity of the ambient air to below 55%. The success of a natural air drying system depends to a large extent on the moisture conditions of the ambient air at the time of drying. If the relative humidity during the wet season is consistently over 75% grain cannot be dried properly with ambient air and the use of supplemental-heat becomes mandatory (Mühlbauer et al. 1981). The minimum required airflow rate depends on the grain moisture content and on the locally expected weather conditions. This is illustrated in Table 1, in which the minimum airflows for natural-air, low-temperature (supplemental-heat), and solar drying systems are tabulated for 12 North-Central states in the United States (Pierce and Thompson 1979). Note the much higher airflow requirements for a warm, humid climate (i.e. Indianapolis) than for a cool, dry climate (i.e. Bismarck). In evaluating the data in Table 1, it should be kept in mind that the required airflows are based on 10-year weather data, and a failure of one out of ten years. Although the natural-air and supplemental-heat in-store drying systems require considerable operator expertise and are low-capacity units, both are able to produce excellent quality grain at low operating costs and energy requirements. In order for natural-air and supplemental-heat drying to be successful in the humid tropics, the minimum airflow rates along with the maximum initial grain moistures should be determined for the different provinces in each of the grain-producing countries of the humid tropics. Good weather records are therefore required.

**Table 1.** Minimum airflow requirements ($m^3$/minute/t) for one location in each of the north central region states in the USA, assuming an initial moisture content of 24% (w.b.) and a harvest date of 15 October. These values are based upon simulated drying results for a 10-year period. Increase these values by 50% for design purposes.

| | Natural air[a] | Solar drying[b] | Low-temperature[c] |
|---|---|---|---|
| Location | Next to worst | Next to worst | Next to worst |
| Chicago, IL | 3.01 | 2.41 | 2.49 |
| Indianapolis, IN | 4.57 | 2.37 | 2.09 |
| Des Moines, IA | 2.43 | 2.51 | 2.19 |
| Dodge City, KS | 2.48 | 2.36 | 2.36 |
| Lansing, MI | 3.07 | 2.56 | 2.61 |
| St Cloud, MN | 2.13 | 1.96 | 1.94 |
| Columbia, MO | 2.77 | 3.10 | 2.44 |
| Lincoln, NE | 2.31 | 2.23 | 2.00 |
| Bismarck, ND | 0.64 | 0.60 | 0.60 |
| Mansfield, OH | 2.84 | 2.12 | 2.06 |
| Huron, SD | 1.56 | 1.35 | 1.26 |
| Madison, WI | 2.50 | 2.09 | 1.97 |

[a] A 1.1°C temperature rise due to drawing the air over the fan motor was assumed.
[b] Heat supplementation includes the 1.1°C from the fan plus a solar collector capable of providing a 24 hour average temperature rise of 1.7°C when collecting 1255 J/cm²-day.
[c] 2.8°C temperature rise from electrical heater and from the fan.
Source: Pierce and Thompson (1979).

*Trickle-ammonia drying* is a technique recently developed in the United States for in-store drying of maize, mainly for animal feed (Nofsinger et al. 1979). High moisture content grain (up to 26–28% w.b) is slowly dried over a 1-5 months period with ambient air at low airflow rates (1–2 $m^3$/minute/t) with a low level of anhydrous ammonia being injected into the airstream intermittently. The ammonia is applied initially at a 0.05% level. To suppress microbial activity permanently, weekly injections of $NH_3$ are required until the grain has reached the safe storage moisture content (Hsieh et al. 1979). Other chemicals used for this process are formaldehyde, methylene-bis-propionate, and sulphur dioxide (van Cauwenberge et al. 1983). The trickle-ammonia process is energy efficient, minimises mould development, and operates over a wide range of weather conditions; at least in the United States it is also cost effective. The potential of trickle ammonia grain drying in the humid tropics should be evaluated.

*In-store solar drying* of grains has been the topic of extensive recent research (Anon. 1980; Kranzler et al. 1980; Fraser and Muir 1981). The solar energy available at a particular location depends on the latitude, the time of year, the weather conditions, and the solar collector design. Although solar grain drying has been proven technically feasible, there is a wide diversity in opinion on the economic feasibility (Morey et al. 1979; Peterson and Morrison 1984). Only in those areas of the humid tropics where the price of the solar collector construction materials and the labour costs are low, and the fossil fuel costs are high, should in-store solar drying of grains be considered. Simulation has been used extensively to find the optimum airflow rates for in-store solar drying installations in various locations (Hasnaoui 1983).

**Table 2.** Standardised energy consumption for five alternative combination drying methods in Michigan, USA (43°N latitude).

| Drying technique | Electricity[a] (kWh/ha) | Propane (L/ha) | Energy efficiency (kJ/kg) | Total energy[b] propane equiv. (L/ha) |
|---|---|---|---|---|
| Natural Air[c] (26-23-15.5% m.c.) | 343.0 | 72.9 | 3227 | 121.6 |
| Low-temperature[c] (26-23-15.5% m.c.) | 481.9 | 79.9 | 3756 | 141.1 |
| In-bin dryeration[c] (26-20-15.5% m.c.) | 94.4 | 142.1 | 4140 | 155.2 |
| In-bin counterflow (26-18-15.5% m.c.) | 103.5 | 156.2 | 4548 | 171.0 |
| Automatic batch (26-15.5% m.c.) | 33.4 | 287.9 | 6589 | 292.6 |

[a] Based on 62.5 t, initial moisture content (m.c.) 26.0% (w.b.), final m.c. 15.5%
[b] Based on 6.9 t/ha.
[c] Energy efficiency of high-temperature drying phase is 6228 kJ/kg $H_2O$.
Source: Kalchik et al. (1981).

A three-year experimental investigation of five on-farm maize drying systems has been conducted recently in the temperate climate of the northern U.S. state of Michigan (Kalchik et al. 1981). The energy requirement of the five in-store drying systems for drying maize from 26% (w.b.) to 15.5% (w.b.) is shown in Table 2. The high-temperature batch system required more than twice as much energy (292.6 litres of propane per ha) than the combination drying system of batch/natural-air (121.6 L/ha; the other three combination drying systems (e.g. batch with low-temperature, dryeration, or counterflow) fall somewhere between these energy consumption values. In terms of grain quality, Table 3 shows that ammonia drying, natural air drying, and low-temperature drying produced good quality maize. Note that in-bin counterflow drying produced considerably less stress cracking than batch drying.

It is dangerous to transfer the in-store drying results of Tables 1 and 2 directly to the humid tropics. Although there is no question that each of the systems can be successfully operated under more humid and warmer conditions, the airflow rates have to be adjusted to the local weather conditions and should also account for dry or wet season operation.

**Table 3.** Maize quality parameters for five alternative in-store comibnation drying methods, compared with batch drying.

| Drying technique | Moisture content: Initial %, w.b. | Moisture content: Intermediate %, w.b. | Moisture content: Final %, w.b | Stress cracks % | Breakage tests % | BCFM % | Viability changes % |
|---|---|---|---|---|---|---|---|
| Natural air | 26.2 | 23.1 | 14.4 | 2.8 | 11.9 | 0.0 | 43.4 |
| Low-temperature | 27.5 | 23.0 | 13.8 | 3.4 | 13.1 | 0.0 | 41.8 |
| In-bin dryeration | 24.0 | 20.0 | 15.6 | 9.0 | 13.8 | 0.0 | 63.7 |
| In-bin counterflow | 26.4 | 18.3 | 16.3 | 64.0 | 29.0 | 0.2 | 28.5 |
| Batch | 26.0 | — | 15.5 | 87.3 | 46.3 | 0.5 | 78.0 |
| Ammonia drying | 25.6 | — | 15.6 | 0 | 9.4 | 0.0 | 65.0 |

Source: Kalchik et al. (1981).

## Auxiliary Equipment

### Pre-cleaners

Pre-cleaning of the grain before in-store drying is recommended if economic use can be made of the separated material. The small pieces of broken kernels and foreign material deteriorate more rapidly than whole kernels after in-store drying of the grain. In addition, the airflow through the grain in a bin is restricted by the small particles, or fines; the effect of the fines on the static pressure in maize has been quantified by Grama and Berne (1982).

The pre-cleaning is usually accomplished with rotating cylindrical screen machines or vibrating screen devices. Brook and Foster (1981) listed the recommended screens for the different grains.

### Grain Stirrers

Overdrying of the bottom layers and non-uniform drying of the grain mass are common faults of in-store drying systems. Grain stirrers are designed to overcome these problems, and are in common use in the United States (Bern et al. 1982). A grain stirring system typically consists of one or more 51-mm augers suspended from the store roof and sidewall, and extending to near the perforated bin floor. The augers lift the grain near the bin floor towards the top of the bin surface while moving in a predetermined pattern through the bin at a horizontal travel rate of 2 to 10 mm/s.

Stirring devices have been used successfully for both high and low-temperature in-store drying systems (Colliver et al. 1979). Their use in the subtropics might not be justified because of the high initial cost. Even in the United States stirring is economical for in-store supplemental-heat drying only for certain bin diameters and commodity price levels (Loewer 1983).

### Biomass Furnaces

Biomass in the form of rice hulls, straw, maize cobs, or woodchips appears to have promise as a replacement for petroleum-based fuels in in-store grain drying (Montalembert 1983). The economics of biomass utilisation depends on the relative cost of the available biomass fuels compared with fossil fuels. Table 4 illustrates this point for the United Kingdom in 1982 (Blakeman 1982).

Michigan research is at present centred on the improvement of the combustion efficiency of biomass furnaces, on the reduction of emissions, and on the automation of the furnace controls (Mwaura et al. 1983). Two major types of combustion furnaces are found: direct combustion (Claar et al. 1981) and gasification-combustion (Richey and Foster 1980). In a direct-combustion furnace, the energy from the burned biomass is transmitted directly to an airstream which is mixed with ambient air to give the required drying air temperature; concentric-vortex direct-combustion furnaces have proven to be the most efficient (Anderson et al. 1981). In a gasification-combustion furnace, producer gas which can subsequently be burned in a conventional gas heater is generated. Barrett et al. (1981) compared the two furnace types and concluded that there are no significant differences between them in terms of heat recovery and particulate emission. Maize dried with direct-combustion energy does not contain harmful deposits (Jacko et al. 1981; Mwaura 1983).

The technical feasibility of the biomass furnace for in-store drying has been proven. The main drawback of widespread adoption is the high capital cost. Mwaura (1983) showed that the fossil fuel costs in the United States have to rise substantially over the next decade for biomass installations to become economically viable for in-store grain drying. It appears that economic viability for the biomass furnace is closer to reality in some of the countries in the humid tropics (e.g. Brazil).

**Table 4.** Fuel costs in the United Kingdom in 1982.

| Fuel | Gross cost[a] | Gross calorific value | Overall efficiency (direct-fired)(%) | Cost/ useful MJ |
|---|---|---|---|---|
| Electricity | 3.55 p/kWh | 3.6 MJ/ kWh | 100 | 0.98 p |
| Natural gas | 27.2 p/therm | 105.5 MJ/ therm | 98 | 0.26 p |
| Bulk propane | 13.4 p/therm | 26 MJ/L | 98 | 0.53 p |
| Bottle propane | £8.65/kg | 50 MJ/kg | 98 | 0.93 p |
| Diesel oil | 17.p/L | 40 MJ/L | 95 | 0.45 p |
| Coal | £60/t | 25600 MJ/t | 75 (indirect) | 0.31 p |
| Straw | £20/t | 16800 MJ/t | 65 | 0.18 p |
| Wood | £30/t | 19800 MJ/t | 70 | 0.22 p |

Source: Blakeman (1982)

[a] £1 = 100 p = US$1.40

## Conclusions

Two-stage drying, also called combination drying, is recommended for the drying of rough rice and maize with moisture contents over 20-21% (w.b.) in the humid tropics. Relatively high drying temperatures are required in the first stage; ambient air is usually sufficient for the drying process in the second stage. During wet weather, supplemented heat drying should be considered in the second stage drying. The required airflow rates in combination drying systems should be tailored to the local weather conditions.

## References


Anderson, M.E., Claar, P.W., and Bern, C.J. 1981. Corn drying evaluation utilizing a concentric-vortex furnace system. American Society of Agricultural Engineers, Paper No. 81–3015.

Anon. 1980. Low-temperature and solar grain drying handbook. Midwest Plan Service, Publication MWPS-22. Ames, Iowa, Iowa State University.

Bakker-Arkema, F.W. 1984. Selected aspects of crop processing and storage: a review. Journal of Agricultural Engineering Research, 30(6), 1–22.

Bakker-Arkema, F.W., Brook, R.C., and Lerew, L.E. 1978. Cereal grain drying. In: Pomeranz, Y., ed., Advances in cereal science and technology, Volume II. Saint Paul, Minnesota, AACC Inc.

Bakker-Arkema, F.W., Fontana, C., and Schisler, I.P. 1983. Comparison of rice drying systems. American Society of Agricultural Engineers, Paper No. 83–3532.

Bakker-Arkema, F.W., Silva, J.S., Mwaura, E.N., Rodriguez, J.C., and Kalchik, S.J. 1980. Testing alternative on-farm grain drying systems. American Society of Agricultural Engineers, Paper No. 80–3017.

Barret, J.R., Jacko, R.B., and Summer, H.R 1981. Particulate emissions from biomass furnaces. Part 2. Characterizations. American Society of Agricultural Engineers, Paper No. 81–3562.

Bern, C.J., Anderson, M.E., Wilcke, W.F., and Hurburgh, C.R. 1982. Auger stirring wet and dry corn — airflow resistance and bulk density effects. Transactions American Society of Agricultural Engineers, 25(1), 217–220.

Blakeman, R. 1982. Fuel saving in grain drying. The Agricultural Engineer, 37(2), 60–67.

Bridges, T.C., Loewer, O.J., White, G.M., and Ross, I.J. 1981. Modeling of continuous in-bin drying systems. American Society of Agricultural Engineers, Paper No. 81–3523.

Brook, R.C., and Foster, G.H. 1981. Drying, cleaning and conditioning. In: Finney, E.E., ed., Handbook of transportation and agriculture, Volume II. Field crops. Boca Raton, Florida, CRC Press, Inc.

Brooker, D.B., Bakker-Arkema, F.W., and Hall, C.W. 1974. Drying cereal grains. Westport, Connecticut, AVI Publication Company.

Claar, P.W., Buchele, W.F., and Marley, S.J. 1981. Crop residue fired furnaces for drying grain. American Society of Agricultural Engineers, Paper No. 81–1032.

Colliver, D.G., Barrett, J.R., and Peart, R.M. 1979. Optimal usage of stirring devices in low-temperature corn drying. American Society of Agricultural Engineers, Paper No. 79–3013.

van Cauwenberge, J.E., Eckhoff, S.R., Bothast, R.J., and Anderson, R.A. 1982. Comparison of the trickle-ammonia process with the $SO_2$ process for the drying of high moisture corn. Transactions American Society of Agricultural Engineers, 25(5), 1431–1434.

Fraser, B.M., and Muir, W.E. 1981. Airflow requirements predicted for drying grain with ambient and solar-heated air in Canada. Transactions American Society of Agricultural Engineers, 24(1), 208–210.

Grama, J., and Berne, C.J. 1982. Static pressures of mixtures of corn and fines. American Society of Agricultural Engineers, Paper No. 82–3526.

Hasnaoui, M. 1983. Simulation of the behaviour of a solar powered grain drying system. International Journal of Solar Energy, 1(6), 431–440.

Hsieh, R.C., Bakker-Arkema, F.W., Cuppett, S.L., and Bagley, E.B. 1979. Ammonia-assisted grain drying in fixed-bed systems. American Society of Agricultural Engineers, Paper No. 79–3563.

Jacko, R.B., Foster, G.H., and Barrett, J.R. 1982. Contaminant residues on corn dried by combustion gases from biomass furnaces. American Society of Agricultural Engineers, Paper No. 82-3523.

Kalchik, S.J., Silva, J., Mwaura, E.N., Rodriguez, J.C., and Bakker-Arkema, F.W. 1981. On-farm corn drying comparison, American Society of Agricutural Engineers, Paper No. 81–3011.

Kranzler, G.A., Bern, C.J., Kline, G.L., and Anderson, M.E. 1980. Grain drying with supplemental solar heat. Transactions American Society of Agricultural Engineers, 23(1), 214–217.

Loewer, O.J. 1983. Economics of stirring devices in grain drying. American Society of Agricultural Engineers, Paper No. 83–3014.

McKenzie, B.A., Foster, G.H., Noyes, R.T., and Thompson, R.A. 1972. Dryeration—better corn quality with high speed drying. West Lafayette, Indiana, Purdue University Co-operative Extension Service, Bulletin AE–72.

McLean, K.A. 1980. Drying and storing combinable crops. Ipswich, Suffolk, United Kingdom, Farming Press limited.

Mittal, G.S., and Otten, L. 1983. Microprocessor controlled low-temperature corn drying systems. Agricultural Systems, 10(1), 1–19.

Montalembert, M.R. 1983. Biomass resources for energy: the critical issues. Ceres, 16(1), 40–44.

Morey, R.V., Cloud, H.A., Gustafson, R.J., and Petersen, D.W. 1979. Evaluation of the feasibility of solar energy grain drying. Transactions American Society of Agricultural Engineers, 22(2), 409–417.

Morey, R.V., Gustafson, R.J., and Cloud, H.A. 1981. Combination high-temperature, ambient air drying. Transactions American Society of Agricultural Engineers, 24(2), 509–512.

Mwaura, E.N. 1983. Biomass furnace design for grain drying installations. East Lansing, Michigan State University, unpublished Ph.D. dissertation.

Mühlbauer, W., Hofacker, W., Muller, H.M., and Thaler, M. 1981. Low-temperature in-bin wheat drying under energetic and microbiological aspects. Grundlagen der Landtechnik, 31(5), 145–154.

Nofsinger, G.W., Bothast, R.J., and Anderson, R.A. 1979. Field trials using extenders for ambient conditioning high-moisture corn. Transactions American Society of Agricultural Engineers, 22(5), 1208–1213.

Peterson, W.H., and Morrison, D.W. 1984. On-farm solar drying of crops and grain — a demonstration project. Urbana, Illinois, University of Illinois, Department of Agricultural Engineering Special Report.

Pierce, R.O., and Thompson, T.L. 1979. Solar grain drying in the North Central Region — simulation results. Transactions American Society of Agricultural Engineers, 22(1), 178–187.

Richey, C.B., and Foster, G.H. 1980. A biomass gasification furnace. American Society of Agricultural Engineers. Paper No. 80–3512.

# Some Fundamental Principles and Benefits of Aeration of Stored Grain

G.R. Thorpe*

*Abstract*

Aeration is the process of blowing air through stored grain with the principal objective of cooling the grain. It differs from in-store drying in so far as the airflow rate used in aeration is an order of magnitude lower than that used for drying, and it is therefore not an energy intensive process. This paper outlines and quantifies the advantages of aeration, which include reducing insect population growth rates, lowering the rate of pesticide decay, preserving grain quality, and lessening the incidence of moisture migration. The paper also describes, with examples, the fundamental processes of heat and mass transfer that govern the rate and amount of cooling during the aeration process.

A powerful statistical tool for estimating the climatic conditions that are likely to occur during aeration is described, and the use of the method is illustrated by means of an example that draws upon temperature data published in daily newspapers. Climatic conditions that prevail in tropical regions may not be suitable for significant benefits to accrue from aeration, and it may be necessary to lower the total heat content of ambient air before it is introduced into the grain. Two methods of modifying the air state, namely the use of a suitably modified commercial air conditioner or the isothermal drying of air in a solar cooling device, are described. It is shown for example, that the solar cooling device is capable of cooling grain to less than 18°C, as opposed to 25°C in a typical tropical environment.

FOOD grains are harvested seasonally, but consumed continuously. Hence, they have to be stored for several months. If it is planned to build up a strategic stock of food grains to act as a buffer against years of poor harvest, then they may have to be stored in good condition for many years. During storage, losses may be severe because of attack by insects, mites, moulds, rodents, and so on. For example, grain losses in Southeast Asia are reported in Champ and Highley (1981) to be in the range 5–15%. In India, Krishnamurthy (1975) reports postharvest losses as being 9%, although they sometimes exceed 50% (Khare, 1975). The single largest component of losses results from insect infestation. Insect infestations also pose a threat to grain-exporting countries, which are contractually bound to present their grains to international markets free from live insect pests.

Two key variables that affect the storability of grains are their temperature and moisture content. As a general rule, reducing either or both of these variables results in an enhanced grain storage environment. Forcing air of the appropriate thermodynamic state through a grain mass can result in the grain being either cooled or dried (or a combination of the two). If stored grain is to be dried within a month or so the airflow rate must be quite high, i.e. about 20 litres per second per tonne. Even if the air temperature is not modified by heating it, in-store drying is still energy intensive because of the power required to force air through the grain. The airflow rate required to cool grain is, however, considerably less, about 1 L/sec/t and this is a process consuming small amounts of energy. If ambient air is used, the cooling process is referred to as ambient air aeration, or sometimes natural aeration. In some geographical locations, ambient air may have too high a heat content, or enthalpy, to effect sufficient cooling, in which case it must be cooled or dehumidified before being used for aeration. Aeration is generally applied when the heat content of the air expelled from the grain bulk exceeds that of the air entering the grain.

* CSIRO Division of Chemical and Wood Technology, P.O. Box 26, Highett, Victoria 3190, Australia.

## Benefits of Aeration and Grain Cooling

Cooling grain by either ambient aeration or, where necessary, refrigerated aeration is a very powerful method of protecting stored grains. Some of the benefits pointed out by Esmay, Trott, and Thorpe[1] are:

(a) It reduces, or may even completely suppress the growth of insect pest populations.
(b) The amount of pesticide required to give long-term protection from insect pests may be significantly reduced. Furthermore, if either ambient or refrigerated aeration is used to cool the grain, the rate of decay of pesticide becomes independent of the initial grain state.
(c) Grain quality is preserved, which is particularly important when storing grains for seed or for further processing.
(d) Aeration may be used for short-term preservation of damp grains while awaiting the availability of drying facilities.
(e) Grain cooling by aeration reduces temperature gradients in stored grains, and this reduces the incidence of harmful moisture migration.
(f) Aeration consumes little energy and relies on simple technology.
(g) Grain cooling is safe, does not require rigorous storage sealing standards, and it fits easily into existing grain handling strategies.
(h) By reducing population pressure, grain cooling may be expected to slow the rate at which insects become resistant to chemical pesticides.

Some of the above benefits will be quantified in this paper.

## Components of an Aeration System

Aeration systems consist of a fan, a supply duct leading from the fan to a perforated duct placed in the store, and provision for airflows into or out of the air space over the grain surface. There may also be a return duct if recirculatory fumigation is to be carried out. The components of an aeration system for bulk grain are shown in Figure 1, and the components proposed for a system to treat bagged grain in Figure 2. Hunter these proceedings describe how fans and aeration ducts are specified for a given cooling duty.

[1] M.L. Esmay, A.R. Trott, and G.R. Thorpe, Modified environment in agricultural buildings for developing countries: to be published by FAO, Rome.

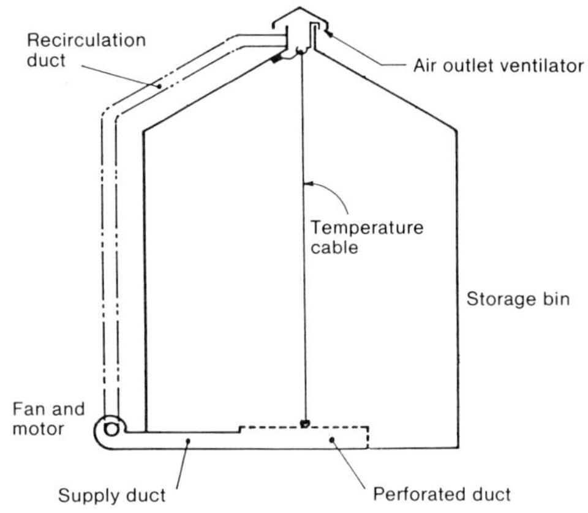


**Fig. 1.** Principal components of an aeration system for bulk grain.

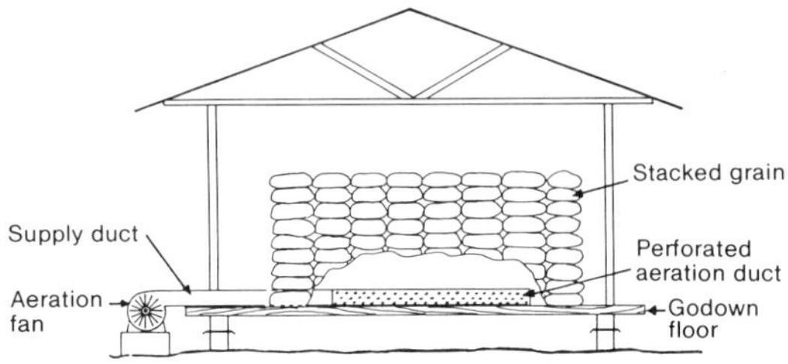


**Fig. 2.** Principal components of an aeration system for bagged grains.

## Principles of Aeration

Cereal grains are hygroscopic materials; that is, they absorb and desorb water such that their moisture contents are a function of the relative humidity and temperature of the air surrounding them. Hence, when damp grains are aerated with ambient air they may dry as well as cool. The drying of the grains has a very profound effect on the degree of grain cooling, because as the air enters the grain, latent heat of vaporisation is required to effect grain drying. Hence, it is possible for a bed of wet grain to cool to a temperature that is lower than that of the ambient air used for aeration. At normal aeration rates (1 L/sec/t), under an aeration schedule of 3 hours per day, it may take up to 10 years for the grain to be dried and cooled such that it is in thermodynamic equilibrium with the ambient air. In order to design aeration systems it is essential that these phenomena be quantified, and to do this we must consider some basic principles.

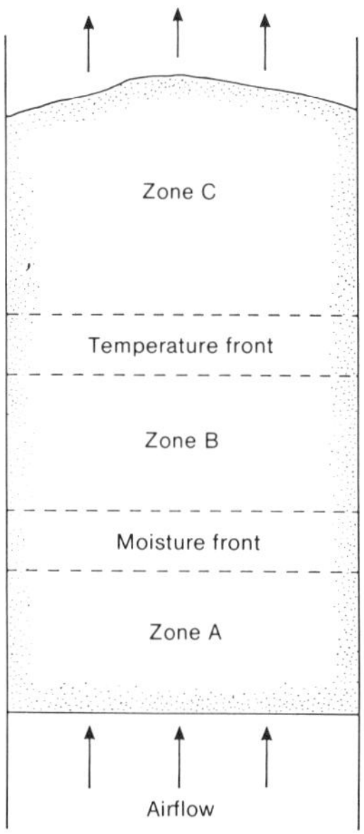


Fig. 3. The formation of temperature and moisture fronts in an aerated or dried bulk of grain.

## Fronts and Zones

It has been pointed out[1] that when air is forced through a bulk of grain, three zones are formed (A, B, and C, say), and each zone is separated by a front or change zone as shown in Figure 3. These fronts move through the grain as waves, identically in the direction of the airflow. The most rapidly moving point on a wave is called the leading edge, whilst the slowest moving point is called the trailing edge.

The grain in zone A is in moisture and temperature equilibrium with the inlet air. Its moisture content can therefore be found by reference to a pychrometric chart, such as the one shown in Figure 4. When designing drying systems, as opposed to aeration systems, the objective is usually to ensure that zone A occupies the entire bed of grain. The grain in zone C has not been affected by the cooling process and is at its original moisture content and temperature.

When designing aeration systems the airflow rate and aeration schedule are chosen so that the bed is occupied by grain mostly in zone B. But how

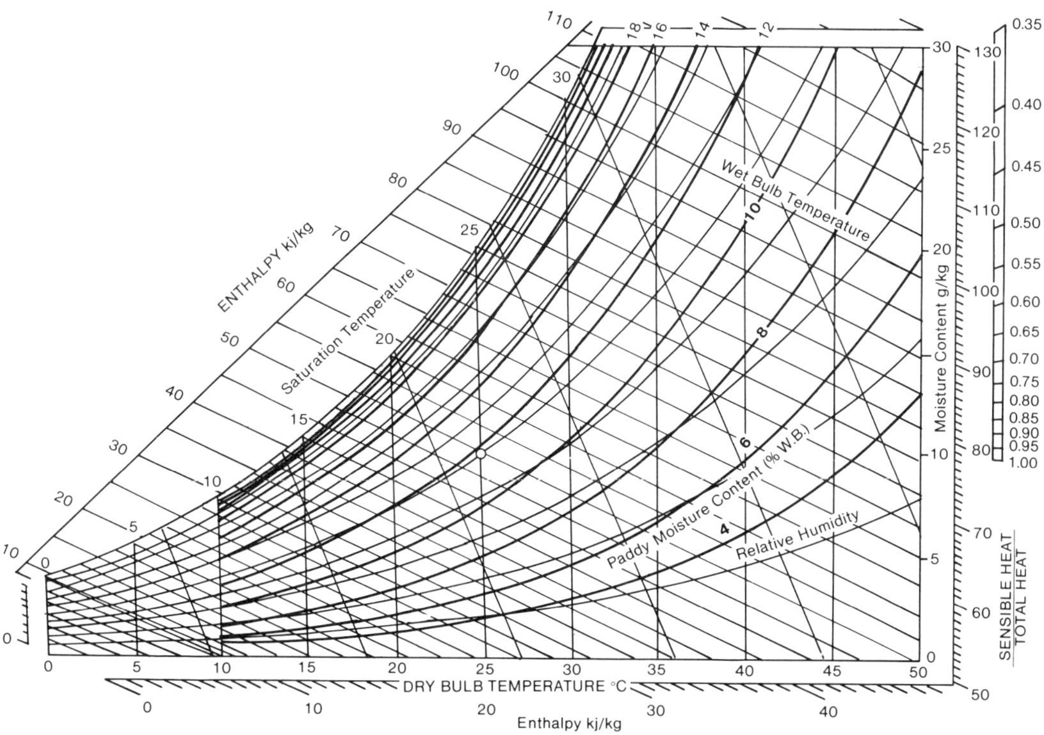


**Fig. 4.** Psychrometric chart showing equilibrium moisture content lines of Australian paddy rice, calculated from Putranon et al. (1979).

is the grain state in zone B calculated? After studying the physics of the behaviour of airflow through grains, the following design steps have been formulated:

(i) Assume that the moisture content in zone B is equal to the initial moisture content of the grain, i.e. the same as in zone C.

(ii) Estimate the temperature in zone B by finding the point of intersection of the inlet air wet-bulb temperature line and the initial grain moisture content line on a psychrometric chart. The dry-bulb temperature at the point of intersection is a first approximation to the temperature in zone B.

This initial estimate of the grain state in zone B is quite accurate, but it may be refined by carrying out the additional steps:

(iii) Estimate the difference in moisture contents between zones B and C, by using the rule that the moisture content difference is 1% for every 28°C difference in temperature between zones B and C, which is calculated in step (ii). If zone B is cooler than zone C then the moisture content in zone B is less than that in zone C, hence the difference in moisture is subtracted from that in zone C. If the temperature in zone B is found to be greater than that in zone C, the moisture difference must be added to that of zone C.

(iv) Refine the estimate of the temperature in zone B by using step (ii) with the grain moisture content of zone B set at that found from step (iii).

## Speeds of Fronts

When designing an aeration system it is essential to be able to calculate the time it takes for the grain to be cooled. Cooling usually implies that zone C has been completely expelled from the grain, and because the front separating zones A and B moves in relative terms very slowly, most of the grain is in zone B.

As indicated in Figure 3, the front separating the zones A and B is termed a moisture front, whilst the front separating zones B and C is called a temperature front. This follows from the fact that zones B and C are of similar moisture content, but different in temperature. The grain states in zones A and B are on a line of almost constant wet-bulb temperature, but their moisture contents may be very different.

The fronts can widen as they travel through the grain because the leading edges travel faster than the trailing edges. This occurs when dry grain is cooled using air with a high relative humidity. If grain is being heated and dried the fronts do not widen, as the trailing and leading edges travel with the same velocity.

Velocities of temperature fronts and moisture fronts calculated from Sutherland et al. (1971) are shown in Figures 5 and 6. These figures apply strictly to wheat, but they represent a reasonable approximation for other cereal grains. It is clear that the speeds of temperature fronts are almost independent of grain moisture content, but the velocities of moisture fronts are very dependent on

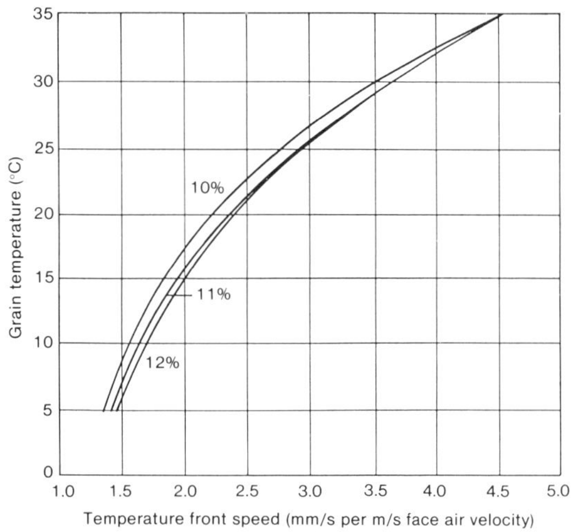


**Fig. 5.** Velocity of temperature fronts through wheat as a function of grain moisture content and temperature (calculated from Sutherland et al. 1971).

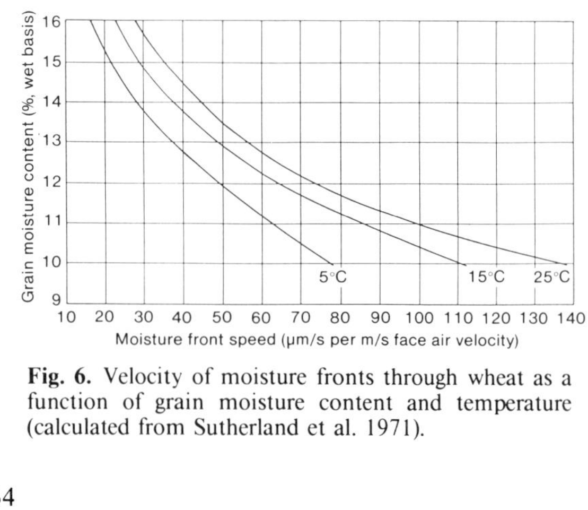


**Fig. 6.** Velocity of moisture fronts through wheat as a function of grain moisture content and temperature (calculated from Sutherland et al. 1971).

the grain moisture content. Grain temperature has an important influence on the speeds of both temperature and moisture fronts.

## Quantifying the Benefits of Aeration

It was pointed out in the introduction that the two key variables determining the storability of grains are their temperature and moisture content. We shall now examine how manipulation of these two variables quantitatively affects storage indices such as the loss of dry matter due to respiration, insect population dynamics, seed viabilility, and the rate of pesticide decay.

### Dry Matter Loss

In the hot, humid tropics grains are often harvested warm and damp. Such conditions are ideal for mould growth and unless the grain storage environment is modified by drying or cooling the grain, or both, the grain may be completely ruined within a very short time. Even under less catastrophic conditions, the loss of dry matter of the stored commodity is associated with a deterioration in quality.

Respiration may be regarded as the complete combustion of carbohydrates to carbon dioxide and water. Hexose, a typical carbohydrate, is oxidised as follows:

$$C_6H_{12}0_6 + 6O_2 = 6\ CO_2 + 6H_20 + 15778\ kJ \qquad (1)$$

This signifies that the oxidation of 1 kg of carbohydrate liberates 15778 kJ of heat, together with 1.47 kg of $CO_2$ and 0.6 kg of water. Thompson (1972) reports that for maize the rate of dry matter loss is time dependent and after a time $\theta$ seconds the fractional loss of dry matter (DM) is given by:

$$DM = 8.83 \times 10^{-4} (\exp (1.667 \times 10^{-6} \times \theta) -1) + 2.833 \times 10^{-9} \times \theta \qquad (2)$$

Equation 2 applies to shelled maize at 15.5°C and 25% moisture content (wet basis) with 30% damage. Because the rate of dry matter loss hence heat production varies with storage conditions, it is important to modify $\theta$ accordingly. An equivalent time, $\theta_{eq}$, is defined thus:

$$\theta_{eq} = \frac{\theta}{M_M M_T} \qquad (3)$$

where $M_M$ and $M_T$ are modifiers for moisture and temperature, respectively. Expressions for $M_M$ and $M_T$ are reported by Thompson (1972) to be:

(a) Temperature modifier:

For $t \leqslant 15.5°C$ or $M \leqslant 19\%$

$$M_T = 32.3 \exp (-0.1044t - 1.856) \qquad (4)$$

where t = grain temperature, °C
and M = moisture content, % wet-basis.

For $t > 15.5°C$ and $19 < M \leqslant 28\%$

$$M_T = 32.3 \exp (-0.1044t-1.856) + [(M-19)/100] \exp (0.0183t - 0.28437) \qquad (5)$$

For $t > 15.5°C$ and $M > 28\%$

$$M_T = 32.3 \exp (-0.1044t - 1.856) + 0.09 \exp (0.0183\ t - 0.2847) \qquad (6)$$

(b) Moisture modifier:
For $13 < M \leqslant 35$

$$M_M = 0.103 (\exp (455/M_{DB}\ 1.53) - 0.00845\ M_{DB} + 1.558) \qquad (7)$$

where $M_{DB}$ = moisture content, % dry basis.

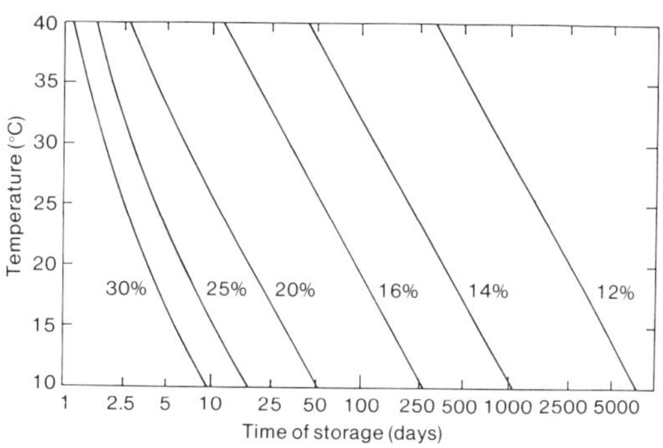


**Fig. 7.** Interaction of grain temperature, moisture content, and time for 0.5% dry matter of corn to be lost by respiration, calculated from Thompson (1972).

Equations 1 to 7 were used to calculate the storage conditions that give rise to 0.5% loss of dry matter. The results, shown in Figure 7, indicate that grain of 16% moisture content at 25°C, for example, loses 0.5% dry matter in about 50 days. Cooling the grain to 10°C extends the period to 250 days.

### Insect Population Dynamics

The rate at which grain storage pests increase is very dependent on the temperature and moisture content of the grain. Generally, a reduction in either or both of these variables reduces the rate of insect population growth. High moisture contents and temperatures are also hostile to insect pests, but these storage conditions are best avoided as they give rise to other problems associated with moisture migration and deterioration of grain quality.

Birch (1953) carried out some of the earliest work on the population dynamics of grain storage insect pests. He used the very simple concept of weekly multiplication ratio, $\lambda$, to quantify the population growth rate of insects. He represented his data by the formula:

$$N = N_0 \lambda^{\theta} \tag{8}$$

where N = number of insects after a time $\theta$ weeks

$N_0$ = initial number of insects

and $\lambda$ = weekly multiplication ratio.

The value of $\lambda$ is very dependent on grain moisture content, as can be seen from Figure 8 drawn for *Sitophilus oryzae*. Table 1 shows the increase in a population of *S. oryzae* after 20 weeks under various storage conditions.

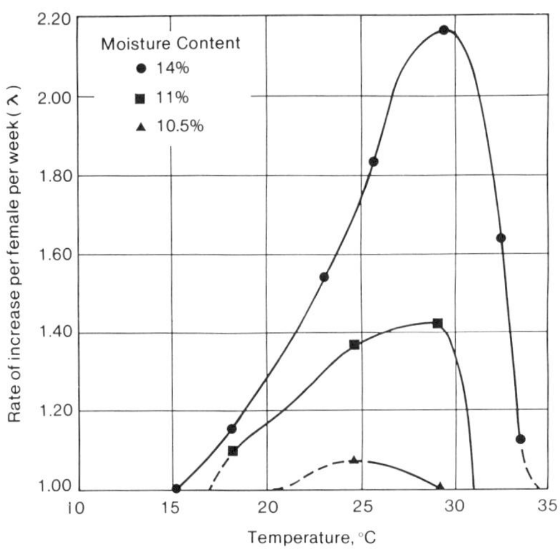


**Fig. 8.** The variation of weekly multiplication ratio of *Sitophilus oryzae* (small strain) with temperature and moisture content of wheat (after Birch 1953).

## Seed Viability

Like insect population dynamics, the viability of seeds is also very sensitive to grain moisture content and temperature. The death of seeds follows a normal distribution, and they have a mean life, $\bar{p}$, that can be represented as a function of grain temperature, t°C, and wet-basis moisture content, m, as follows:

$$\bar{p} = 10^{K_1 - K_2 m - K_3 t} \tag{9}$$

where $K_1$, $K_2$, and $K_3$ relate to the kind of seed. Values of the constants reported by Roberts (1974) are given in Table 2. These data were used to calculate the half-lives of rice at various temperatures and moisture contents, and the results are shown in Table 3.

**Table 1.** Increase in a population of *Sitophilus oryzae* after 20 weeks under various storage conditions.

| Temperature (°C) | Moisture content (%) | $\lambda$ | $N/N_0$ |
|---|---|---|---|
| 26 | 10.5 | 1.06 | 3.2 |
| 26 | 11.0 | 1.40 | 837.7 |
| 26 | 14.0 | 1.86 | 245617.3 |
| 18 | 14.0 | 1.15 | 16.4 |
| 28 | 14.0 | 2.17 | 5360368.9 |
| 33 | 14.0 | 1.13 | 11.5 |

**Table 2.** Viability constants for a variety of grains and seeds, as given by Roberts (1974).

| Species | $K_1$ | $K_2$ | $K_3$ |
|---|---|---|---|
| Wheat | | | |
| *(Triticum aestivum)* | 5.067 | 0.108 | 1.050 |
| Rice *(Oryza sativa)* | 6.531 | 0.159 | 0.069 |
| Barley | | | |
| *(Hordeum distichon)* | 6.745 | 0.172 | 0.075 |
| Broad beans *(Vicia faba)* | 5.766 | 0.139 | 0.056 |
| Peas *(Pisum sativum)* | 6.432 | 0.158 | 0.065 |

**Table 3.** Half-lives of rice under various conditions of temperature and moisture content, calculated from Roberts (1974).

| Temperature (°C) | Moisture content (%, wet basis) | Half-life (days) |
|---|---|---|
| 20 | 12 | 1750 |
| 25 | 12 | 791 |
| 30 | 12 | 357 |
| 35 | 12 | 161 |
| 40 | 12 | 73 |
| 30 | 9 | 1072 |
| 30 | 11 | 515 |
| 30 | 13 | 248 |
| 30 | 15 | 119 |
| 30 | 17 | 53 |

## Decay of Chemical Pesticides

Chemical pesticides are often admixed with grains to combat attack by insect pests. Pesticides may be effective against the target pests for many months, can be applied to grain held in poorly constructed and unsealed stores, and require little capital expenditure for their application. Although pesticides are very inexpensive compared with the value of the commodity they protect, their cost is often the largest single component incurred during storage. When applied to warm, damp grain many chemical pesticides decompose to ineffectively low levels after only a few weeks. This can be expensive because either higher dosages must be applied in the first instance, or the grain must be retreated. Drying and/or cooling grain by aeration has a beneficial effect as it slows the rate of pesticide decay. The chemical kinetics of pesticide decay have been elucidated by Desmarchelier (1978) and Desmarchelier and Bengston (1979). It can be shown that if the initial concentration of the pesticide on the grain is $C_o$ then the concentration $C_t$ after θ weeks is given by:

$$C_t = C_o \exp(-1.386 r\theta \times 10^{B(t-30)} / \theta^*_{1/2}) \quad (10)$$

where r = fractional relative humidity of the intergranular air, $\theta^*_{1/2}$ = half-life in weeks of pesticide on grain when r = 0.5 and the temperature is 30°C, B = a constant and t = grain temperature in °C.

Values of the constants $\theta^*_{1/2}$ and B are given in Table 4 for a number of commonly used grain protectants. Equation 10 was used to determine the effects of wheat temperature and moisture content on malathion residues after 6 months storage. The results are given in Table 5.

**Table 4.** Half-lives of pesticides at 30°C and 50% relative humidity, and temperature coefficient B, for a variety of pesticides (after Desmarchelier and Bengston 1979)

| Protectant | $\theta^*_{1/2}$ (weeks) | B (degrees$^{-1}$) |
|---|---|---|
| Fenitrothion | 14 | 0.036 |
| Bioresmethrin | 24 | 0.033 |
| Bioresmethrin[a] | 38 | 0.031 |
| d-Fenothrin | 38 | 0.029 |
| d-Fenothrin[a] | 40 | 0.029 |
| Pyrethrum[a] | 55 | 0.022 |
| Methacrifos | 8 | 0.055 |
| Malathion | 12 | 0.05 |
| Chlorpyrifos-methyl | 19 | 0.04 |
| Carbaryl | 21 | 0.031 |
| Pirimiphos-methyl | 70 | small |

[a]Plus piperonyl butoxide at 20 mg/kg

**Table 5.** Malathion residues on wheat held for 6 months under various conditions of temperature and moisture.

| Temperature (%) | Moisture content (%, wet basis) | Final concentration of malathion / Initial concentration of malathion |
|---|---|---|
| 30 | 10 | 0.60 |
| 30 | 12 | 0.45 |
| 30 | 14 | 0.37 |
| 30 | 16 | 0.33 |
| 20 | 14 | 0.74 |
| 40 | 14 | 0.03 |

It was stated in the 'Introduction' that a primary aim of aeration is to reduce the heat content of the grain by either cooling and/or drying it. If this objective is achieved then aeration reduces the state of pesticide decay. Aeration also has another profound effect on the rate of pesticide decay. The general rule can be stated that aeration renders the state of decay of pesticides applied to stored grains insensitive to the initial grain state (Thorpe and Elder 1982). This means that the final concentration of pesticides on aerated grain after a given storage period is about the same even whether the grain be initially warm and wet, or cool and dry, or whatever. This is illustrated in Figures 9 and 10. It is a phenomenon of great importance to pest management.

## Climatic Considerations and Aeration

The process of cooling grain using ambient air can be optimised by choosing to operate the aeration system when the total content of the air is lowest. This usually coincides with the time of the lowest ambient dry-bulb temperature. Even in tropical climates there is often a day-to-day variation in temperature, and this can be exploited: the air occurring during the coldest periods can be selected for aeration. For example, in Darwin, a tropical coastal city, the mean maximum and minimum daily dry-bulb temperatures during January, a monsoonal month, are 31.9°C and 25.2°C. Following a statistical procedure developed by Hunter (1981), it can be shown that ambient temperatures in Darwin during the coldest 15% of the time are 25.9°C or less; the mean temperature during this time is 24.7°C,

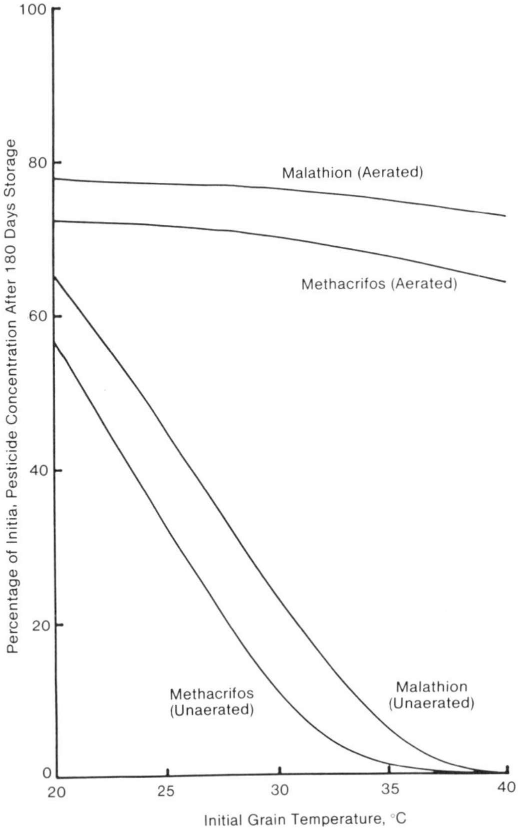


**Fig. 9.** Variation with initial temperature of pesticide concentration after 6 months storage of wheat initially at 11% moisture concentration under conditions of aeration and non-aeration.

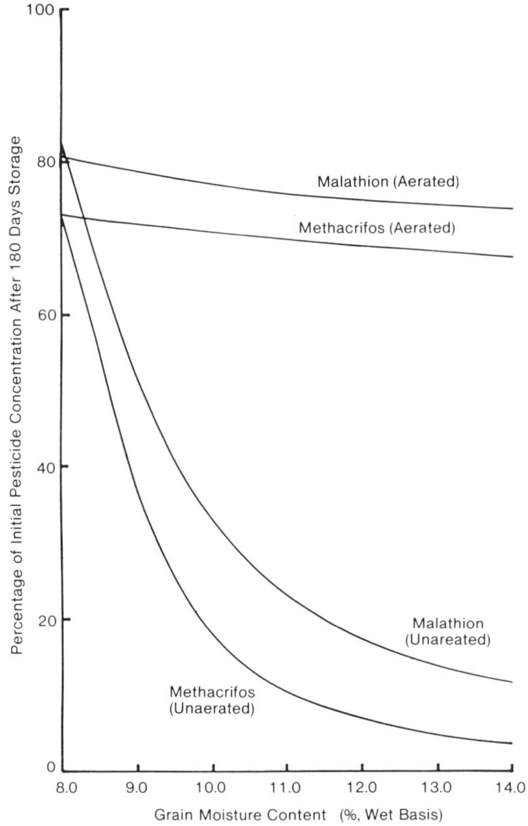


**Fig. 10.** Variation with initial moisture content of pesticide concentration after 6 months storage of wheat initially at 30°C, under conditions of aeration and non-aeration.

which is lower than the average daily mean maximum. Hunter (1981) shows that this information can be calculated from climatic data that includes daily mean minimum temperatures and some statistical information on the daily variability.

In the event of comprehensive weather data for a particular location not being collated and analysed, it is possible to make use of daily minimum and maximum temperatures published in local newspapers, or collected by research and educational institutions. Relative humidity data are not essential to the analysis but they help to ascertain the suitability of the climate for both grain cooling and drying by aeration. A simple method of analysing the data developed by A.J. Hunter (personal communication) relies on first calculating the following functions:

$$\gamma_H = (\mu_H - m)/\sigma_H \qquad (11)$$

$$\text{and } \gamma_L = (m - \mu_L)/\sigma_L \qquad (12)$$

where $\sigma_L$ and $\sigma_H$ are the standard deviations of the daily minimum and maximum temperatures and $\mu_L$, $\mu_H$ and m are the mean minimum, maximum, and median daily temperatures.

The above maximum and minimum temperatures are given by:

$$\mu_H = \sum_{n=1}^{N} v_H = \frac{v_{H1} + v_{H2} + v_{H3} + \ldots v_{Hn} \ldots + v_{HN}}{N} \qquad (13)$$

$$\mu_L = \sum_{n=1}^{N} v_L = \frac{v_{L1} + v_{L2} + v_{L3} + \ldots + v_{Ln} + \ldots v_{LN}}{N} \qquad (14)$$

where $v_{Hn}$ and $v_{Ln}$ refer to the daily maximum and minimum temperatures recorded on the nth day of the month. N is the total number of days the data are available, for example, 31 in the case of May. If one has access to archives it is worthwhile calculating the statistical data for the same month over three or four years, as there are variations in climate from year to year. If four years' data were to be collected, then ideally N = 124. However, newspapers sometimes omit some data, or may not print data on one day of the week. In this case the value of N must be adjusted accordingly.

The standard deviations may be calculated from the following standard formulae:

$$\sigma_H = [\frac{1}{N-1} \sum_{n=1}^{N} (v_{Hn} - \mu_H)^2]^{1/2} \quad (15)$$

$$\sigma_L = [\frac{1}{N-1} \sum_{n=1}^{1} (v_{Ln} - \mu_L)^2]^{1/2} \quad (16)$$

Equations 15 and 16 are easy to use. Many scientific calculators have the functions built in, or the equations are easily programmed for processing on a microcomputer.

Having calculated the standard deviations, the 14 and 86 percentiles are obtained from:

$$v_{H14} = \mu_H - 1.08\sigma_H \quad (17)$$

$$v_{H86} = \mu_H + 1.08\sigma_H \quad (18)$$

Equations of identical form apply to the percentiles around the mean minimum temperature. The daily median temperature is given by:

$$m = \frac{v_{L86} + v_{H14}}{2} \quad (19)$$

The foregoing information enables an analysis of short-term temperature variability. Its use is illustrated by the example which follows:

Table 6 shows climatic data collected at a subtropical location near the coast of a large land mass. Use the analysis to comment on the suitability of the climate for both the cooling of grain by aeration for long-term storage, and for exploiting the warmest periods of the day for in-store drying of crops.

**Table 6.** Climatic data for a sub-tropical locality.

| Day | Minimum temperature (°C) | Dew point at 0900 hours (°C) | Maximum temperature (°C) | Dew point at 1500 hours (°C) |
|---|---|---|---|---|
| 1 | 15.1 | 16 | 28.3 | 15 |
| 2 | 17.9 | 19 | 33.4 | 18 |
| 3 | 18.1 | 19 | 36.2 | 17 |
| 4 | 12.9 | 14 | 25.7 | 14 |
| 5 | 11.6 | 13 | 22.6 | 13 |
| 6 | 13.8 | 14 | 24.9 | 14 |
| 7 | 16.4 | 17 | 27.1 | 16 |
| 8 | 17.7 | 18 | 31.3 | 17 |
| 9 | 17.8 | 18 | 33.7 | 20 |
| 10 | 16.1 | 17 | 30.1 | 17 |
| 11 | 16.4 | 17 | 31.8 | 16 |
| 12 | 11.1 | 12 | 21.9 | 11 |
| 13 | 11.7 | 12 | 23.6 | 10 |
| 14 | 15.1 | 16 | 27.1 | 14 |
| 15 | 15.5 | 16 | 27.9 | 15 |
| 16 | 15.3 | 16 | 27.7 | 15 |
| 17 | 16.2 | 18 | 29.0 | 18 |
| 18 | 16.6 | 18 | 31.2 | 17 |
| 19 | 17.3 | 19 | 37.1 | 20 |
| 20 | 16.5 | 18 | 27.9 | 19 |
| 21 | 17.1 | 18 | 30.8 | 18 |
| 22 | 13.2 | 15 | 27.1 | 15 |
| 23 | 11.9 | 13 | 22.2 | 12 |
| 24 | 11.3 | 13 | 21.5 | 11 |
| 25 | 13.4 | 14 | 25.6 | 14 |
| 26 | 16.9 | 18 | 31.1 | 18 |
| 27 | 17.3 | 18 | 34.7 | 19 |
| 28 | 18.2 | 19 | 37.3 | 19 |
| 29 | 18.1 | 19 | 37.2 | 20 |
| 30 | 16.4 | 17 | 34.8 | 18 |

The daily minimum temperature is found from equation 14 thus:

$$\mu_L = \frac{15.1 + 17.9 + 18.1 + \ldots + 16.4}{30}$$

$$= 15.46°C$$

and the mean maximum thus

$$\mu_H = \frac{28.3 + 33.4 + 36.2 + \ldots + 34.8}{30}$$

$$= 29.36$$

The standard deviation of temperatures about the mean minimum temperature is found from equation 16 thus:

$$\sigma_L = \{\frac{1}{29} \cdot [(15.1 - 15.46)^2 + (17.9 - 15.46)^2 + \ldots + (16.4 - 15.46)^2]\}^{1/2}$$

$$= 2.327$$

It is found that $\sigma_H$ is 4.777 and

$$v_{L14} = 15.46 - 1.08 \times 2.327 = 12.95°C$$

and $$v_{L86} = 15.46 + 1.08 \times 2.327 = 17.97°C$$

$$v_{H14} = 29.36 - 1.08 \times 4.777 = 23.79°C$$

$$v_{H86} = 29.36 + 1.08 \times 4.777 = 34.93°C$$

Hence, the median temperature is $0.5 \times (17.97 + 23.79) = 20.88°C$.
It follows that $\gamma_L = (20.88 - 15.46)/2.327 = 2.33$ and $\gamma_H = (29.36 - 20.88)/4.77 = 1.78$.

Cooling is normally carried out during the night, typically during the coldest 15% of the time. From Figure 11 it is seen that when $\gamma = 2.33$ and the time of aeration is 15% then $z_l = -2.15$ (the negative sign arises from the fact that the aeration occurs for less than 50% of the time). The temperature, v, below which aeration occurs is given by

$$v = m + \sigma_L z_l \tag{20}$$
$$= 20.88 - 2.327 \times 2.15$$
$$= 15.88°C.$$

The aeration fan is turned on whenever the ambient temperature falls below 15.88°C.

To find the average temperature, $\bar{v}$, that occurs during aeration we read the value of $\bar{z}_l$ from Figure 12, and find it to be −2.8. This is then inserted into

$$\bar{v} = m + \sigma_L \bar{z}_l \; 20.88 - 2.327 \times 2.8 \tag{21}$$
$$= 14.36°C.$$

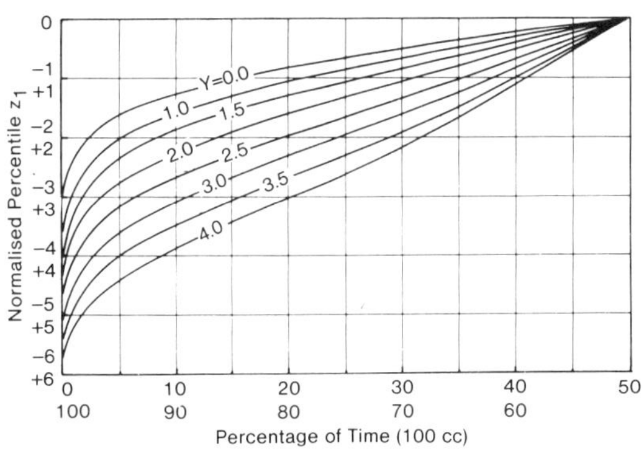


**Fig. 11.** Predicted cumulative distribution functions for ambient temperature as a function of γ (after Hunter 1981).

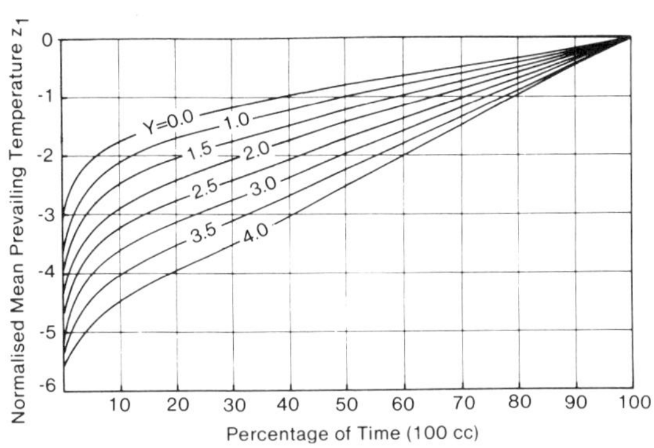


**Fig. 12.** Mean prevailing temperature versus percentage of time for which it is available (after Hunter 1981).

Hence, the average ambient temperature that occurs during aeration is about 14.4°C, which is less than the average minimum temperature. We see from Table 6 that at 0900 hours the dew point temperature is close to the minimum temperature, which indicates that the air used for aeration is close to saturation. Since the air is compressed in the fan, and the mass of air used for aeration is very small, the absorption of water by the grain is usually not serious.

Drying, in contrast to aeration for cooling, requires a high volume of air of low relative humidity. If the aeration fans are turned on during the warmest 50% of the time, the air above the median temperature, i.e. 20.88°C is chosen. The average temperature, $\bar{v}$, of this air is found from

$$\bar{v} = m + \sigma_H \bar{z}$$
$$= 20.88 + 4.777 \times 1.3$$
$$= 27.09°C$$

Examination of Table 6 shows that the dew point of the air when this temperature occurs is typically 15°C, which corresponds to a relative humidity of 50%. It is therefore clear that this climate is excellent for the in-store drying of grain, as well as for cooling by aeration.

### Refrigerated Aeration

Ambient aeration can be very beneficial in a wide range of climates, from temperate to tropical. However, in climates with a high ambient enthalpy, aeration can be considerably enhanced by reducing the enthalpy of ambient air before admitting it to the silo. The most conventional method of doing this is by mechanical refrigeration. A solar-regenerated, open-cycle, desiccant-bed grain cooling system also shows considerable promise, although it is still at the developmental stage (Thorpe and Fricke 1985).

As Esmay, Trott, and Thorpe[1] point out, the objectives and benefits of refrigerated aeration are basically those of ambient air aeration, but under similar climatic conditions the benefits of refrigeration are more marked. These benefits are achieved with the penalty of higher operating and capital costs, but the improved marketability of the grain may outweigh the costs. Particular benefits are:

(i) The storage of damp grain under tropical conditions. It can be seen from Figure 7 that 20% moisture content grain can be stored for about 30 days with only a 0.5% loss in dry matter, provided the grain can be cooled to 15°C. This temperature cannot be achieved in climates in which the ambient wet-bulb temperature always exceeds 14°C.

(ii) The protection from insect pests in pesticide-free grain. This can be achieved by ensuring that all of a grain bulk is cooled to temperatures less than 15°C (Elder et al. 1984).

(iii) The maintenance of desirable grain quality indices such as germination, milling yield, colour, taste, and so on.

## Selection of Grain Refrigeration Units

Grain refrigeration units may be built by slightly modifiying standard air conditioning equipment, or by assembling readily available air conditioning components. The most important modification to packaged air conditioning equipment is to replace the standard fan with a high-pressure fan. In a packaged air conditioning unit the standard fan normally fits into a confined space, and it is often preferable to mount an aeration fan in a separate compartment as shown in Figure 13. The main advantages of a custom-built grain refrigeration unit based on a widely used commercial air conditioner are the availability of spare parts, and the familiarity of local air conditioning mechanics with the equipment. Local refrigeration engineers may be capable of designing and assembling grain refrigeration units from standard air conditioning components such as compressors, evaporator and condenser coils, and controls. There is also the advantage of using widely available components and local expertise. Such a scheme provides local employment during both the design and construction, and maintenance phases.

It is possible to buy refrigeration units specifically designed for grain cooling, and they are fitted with suitably high pressure fans, and devices to control the relative humidity of the cooled aeration air. Such units are available in a range of capacities that are capable of cooling from 30 to 300 tonnes of grain per day. Prospective purchasers need simply specify their grain cooling needs, and the refrigeration unit suppliers will recommend complete systems. This has obvious advantages, but it must be realised that the supply of spare parts and servicing may be restricted to the particular manufacturer or its agents.

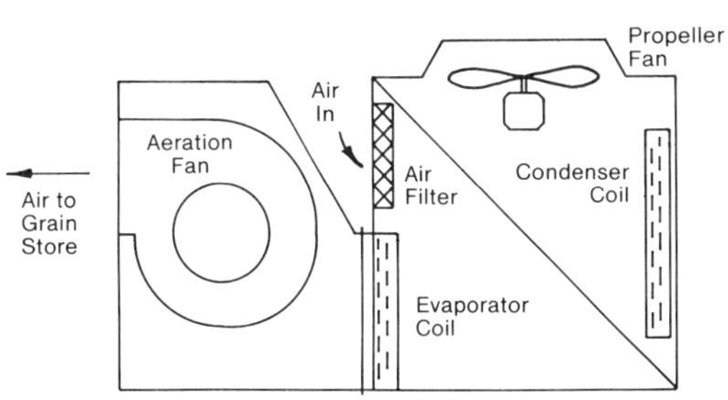


**Fig. 13.** Principal components of grain refrigeration unit constructed from a packaged commercial air conditioner and a separately mounted aeration fan.

## Open-Cycle, Solar-Regenerated, Desiccant-Bed Grain Cooling System

It has been pointed out that, particularly in humid tropical regions, the enthalpy (heat content) of ambient air may have to be reduced before it can be effectively used for aeration. We have seen that one method of reducing the enthalpy of ambient air is to cool it by means of refrigeration. A second method, now considered here, is to isothermally reduce the moisture content of the air.

Thorpe and Fricke (1985) have described a very simple-to-build and low energy grain cooling system, the principal elements of which are shown in Figure 14. During the air drying and cooling

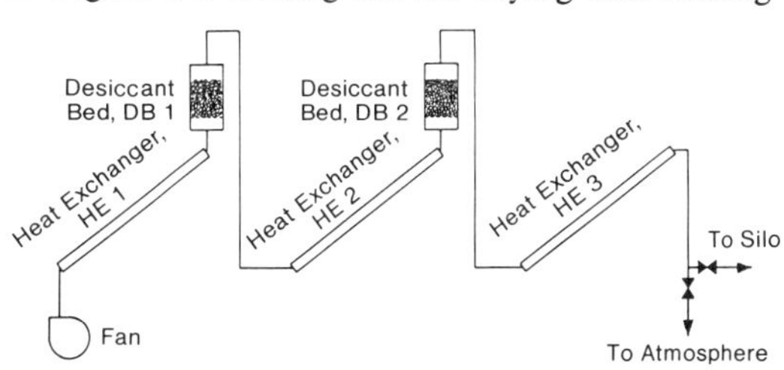


**Fig. 14.** Principal elements of a solar-regenerated, desiccant-bed, grain-cooling system.

process, night air from the aeration fan enters the first heat exchanger, HE1, and some of the heat of compression is lost to the atmosphere by convection and radiation to the night sky. The cool, humid air then enters desiccant bed, DB1, where it is dried iso-enthalpically. As a consequence of this process, the air increases in temperature and it is subsequently cooled in the second heat exchanger, HE2. The air is further dried in the second desiccant bed, DB2, before being finally cooled in the third heat exchanger, HE3.

The desiccant is regenerated by designing the heat exchangers HE1 and HE2 as effective solar collectors. Hence, during the day, ambient air leaving the fan is heated in the heat exchangers, thereby reducing its relative humidity so that the desiccant beds are regenerated for the night cycle.

The underlying principles of operation of the stage-wise system can be understood from the psychrometric chart (Figure 15). Cool ambient air

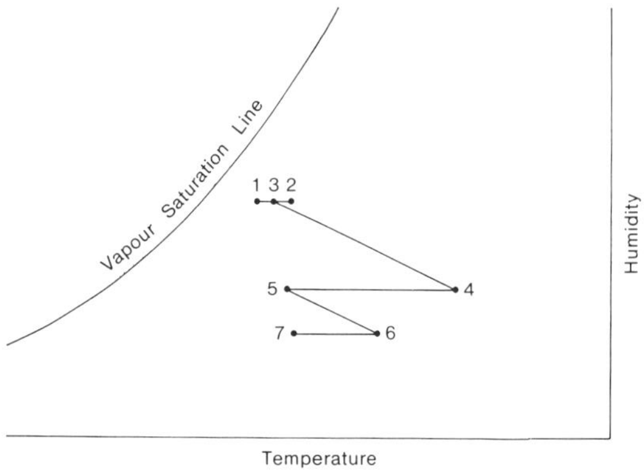


**Fig. 15.** Thermodynamic states of aeration air passing through a solar-regenerated, desiccant-bed, grain-cooling system.

with a high relative humidity, state 1, is compressed by the fan to state 2. It is this air that would normally be used for ventilating grain in ambient aeration systems, but because of its high enthalpy it has limited cooling capacity. In the desiccant-bed cooling system the air is cooled from state 2 to state 3 in the first heat exchanger, HE1. The air on entering the first desiccant bed, DB1, leaves at state 4, whence it is cooled by the atmosphere to state 5. Now this represents a reduction in enthalpy compared with natural aeration systems, but a further reduction is obtained by passing the air through the second desiccant bed, DB2, bringing it to state 6, before finally cooling it to state 7. It is this air that is used to cool the grain.

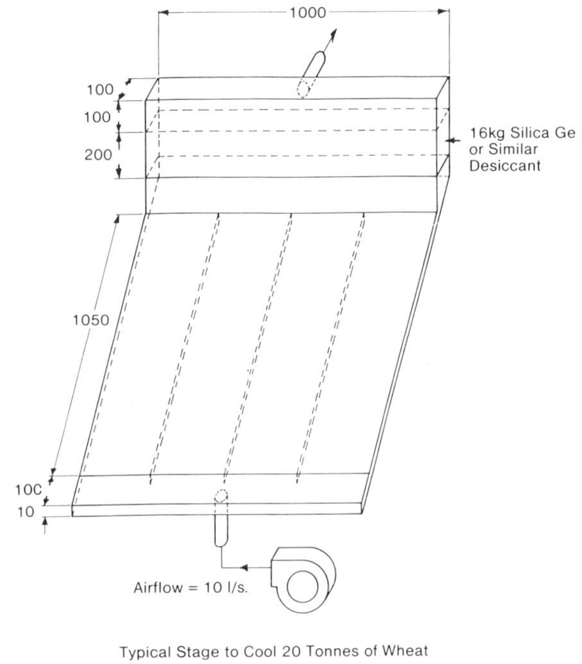


**Fig. 16.** An experimental, single stage of a solar grain-cooling system.

An experimental, single cooling stage is shown in Figure 16.

The system just described is still at the developmental stage. However, early computer modelling and experimental results indicate that under tropical conditions the enthalpy of ambient air can be reduced by about 20 kJ/kg. This means that grain cooled with ambient air to 26°C could be cooled to less than 20°C with air leaving the desiccant-bed system. Features of the solar-regenerated, open-cycle, desiccant-bed grain cooling system are:

- It is very simple to construct from readily available materials and its fabrication requires only basic sheet-metal working skills.
- It achieves a significant reduction in the heat content of ambient air, hence considerably lower grain temperatures can be achieved.
- Apart from a solenoid valve, it has no more moving parts than a conventional aeration system.
- It may prove possible to use renewable desiccants, such as cereal grains.

A program of work is under way in the CSIRO Division of Chemical and Wood Technology to develop the device to commercial realisation.

## Simulated Performance of Solar Grain Cooler

A mathematical model of the solar grain-cooling device has been formulated, and its likely performance in preserving grains in tropical climates has been predicted. In the model, the air leaving the solar cooler was used to cool maize subject to heating by respiration. The objective of using the model was to determine an effective combination of drying with ambient air, and cooling with air modified by the solar cooler. In addition, the effects of the following strategies on storage indices such as dry matter loss, pesticide residues, insect population growth, and seed viability, were investigated: (a) taking no prophylactic action, (b) in-store drying of the grain using ambient air, or (c) drying followed by cooling.

It became apparent from the model that a high flow rate of air (about 20 litres per second per tonne of grain), is necessary to prevent serious overheating and spoilage of the grain by respiration. A suitable (though not necessarily optimum) strategy is to ventilate wet grain continuously with ambient air which a drying front has passed through the bed of grain. With an airflow rate of 20 L/sec/t this takes about 40 days, and with initially 20% moisture content maize the average dry matter loss during this period is 0.36%. Drying is then followed by cooling with an airflow rate of 1 L/sec/t. Table 7 gives details of the solar cooling system, the operating logistics, and the climatic and storage conditions. After 40 days of drying, the average grain temperature and moisture content are 26.8°C and 12.7%. After a further 142 days of cooling (a total of 26 weeks), the average grain temperature is 18.2°C, and the moisture content 12.3%.

Table 8 quantifies the effects of the three storage strategies on the grain: i.e. no prophylactic action, drying only, and drying followed by solar cooling. If no action is taken, the grain is ruined by

**Table 7.** Description of solar grain-cooling system and logistics investigated.

| | |
|---|---|
| **Solar Cooler** | |
| Three unglazed absorbers, with insulation on rear of collectors. | |
| Dimensions of solar absorbers: 1.2 m long × 1 m wide × 0.01 m deep | |
| Two beds of silica gel | |
| Dimensions of beds of silica gel: 1 m long × 0.1 m high × 0.1 m wide | |
| **Operational Logistics** | |
| Drying | : 24 hours/day, 20 L/sec/t; 40 days |
| Cooling | : 6 hours/day regeneration of desiccant<br>6 hours/day aeration; 1 L/sec/t, 142 days (Total storage period 26 weeks) |
| **Climatic Data** | |
| Average conditions for drying | : Ambient dry bulb temperature: 23.6°C<br>: Ambient humidity: 1.43 g/kg<br>Solar radiation: 800 W/m$^2$<br>Temperature rise in fan: 3.23°C |
| **Air State Leaving Solar Cooler** | |
| Dry bulb temperature of air | : 21.1°C |
| Ambient humidity | : 5.9 g/kg |
| **Initial Grain State** | |
| Temperature | : 30°C |
| Moisture content | : 20% (wet basis) |

**Table 8.** Predicted performance of combined drying and cooling system for stored grain.

| | |
|---|---|
| Power requirement for drying | : 50 W/t |
| Energy consumed during drying (40 days) | : 48 kWh |
| Power requirement for aeration | : 1 W/t |
| Power requirement for regeneration | : 0.1 W/t |
| Energy consumed for aeration and regeneration (142 days) | : 1 kWh |
| **Dry matter loss** | |
| No prophylactic action | : Grain ruined by mould |
| Dry to 12.7% then store for total period of 26 weeks | : 0.6% |
| Dry to 12.7%, then cool so total storage period is 26 weeks | : 0.4% |
| **Malathion decomposition** | |
| No prophylactic action | : 94% |
| Drying then storage | : 80% |
| Drying followed by cooling | : 55% |
| **Insect population growth of** *Sitophilus oryzae* | |
| No prophylactic action | : Conditions inhospitable to insects |
| Drying then storage | : 200,000 × increase |
| Drying followed by cooling | : 20 × increase |
| **Half-life (typical cereal grains)** | |
| No prophylactic action | : 20 weeks |
| Drying then storage | : 500 weeks |
| Drying then cooling | : 2000 weeks |

overheating due to respiration, whereas drying to 12.7% m.c. followed by storage, results in a dry matter loss of 0.6%. If the grain is cooled after drying, then the dry matter loss is only 0.4%. As expected, pesticide decay is much reduced by drying and drying followed by cooling. The latter strategy would result in 7 times less malathion (an organophosphorus pesticide) having to be applied to result in a given level of pesticide residue after 6 months storage, compared with the case in which no action is taken. The effect of drying and cooling, as opposed to drying only, is particularly marked when one considers insect population growth. Because drying leaves the grain comparatively warm (about 27°C), insects are able to proliferate very quickly. Cooling to 18°C, however, has a dramatic effect on reducing the insect population growth, such that in dried and cooled grain the insect population after 6 months is 10 000 times less than in the grain that has been dried only. A quantifiable aspect of grain quality is seed viability, and here again the benefits of drying and cooling are evident from Table 7.

## Conclusions

Aeration with ambient air provides a simple, low energy method of reducing the heat content of stored grain. This has beneficial effects in so far as it can reduce the insect population pressure, preserve grain qualities such as viability, milling yield, colour, and taste, reduce the rate of dry matter loss, and reduce the rate of decay of pesticides applied to the grain. In tropical climates characterised by high ambient enthalpies, ambient aeration may be enhanced by refrigerating or desiccating the aeration air.

## Acknowledgments

Financial assistance provided for this work by the Australian Centre for International Agricultural Research and the Wheat Industry Research Council is gratefully acknowledged.

## References

Birch, L.C. 1953 Experimental background to the study of the distribution and abundance of insects. 1. The influence of temperature, moisture and food on the innate capacity for increase of three grain beetles. Ecology 34, 698–711.

Champ, B.R., and Highley, E., ed. 1981. Grain storage research and its application in Australia. Canberra, CSIRO Division of Entomology, 278 p.

Desmarchelier, J.M. 1978. Loss of fenitrothion on grains in storage. Pesticide Science, 9, 33–38.

Desmarchelier, J.M., and Bengston, M. 1979 Chemical residues of newer grain protectants. In: Evans, D.E., ed., Australian contributions to the symposium on the protection of grain against insect damage during storage, Moscow 1978. Canberra, CSIRO Division of Entomology, 108–115.

Elder, W.B., Ghaly, T.F., and Thorpe, G.R. 1984. Grain refrigeration trials in Australia. In: Ripp, B.E., et al., ed., Practical aspects of controlled atmosphere and fumigation in grain storages. Elsevier, Amsterdam, 623–643.

Hunter, A.J. 1981. A climatic model for the prediction of percentile statistics for ambient temperature. Monthly Weather Review, 109, 722–728.

Khare, B.P. 1973. Insect pests of stored grain and their control in Uttar Pradesh. Pantnagar, Nainital, G.P. Pant University of Agriculture and Technology.

Krishnamurthy, K. 1975. Post harvest losses in food grains. Bulletin of Grain Technology, 13, 33–49.

Putranon, R., Bowrey, R.G., and Eccleston, J. 1979. Sorption isotherms for two cultivars of paddy rice grown in Australia. Food Technology in Australia, 31, 510–515.

Roberts, E.H. ed. 1974. Viability of seeds, London, Chapman and Hall.

Sutherland, J.W., Banks, P.J., and Griffiths, H.J. 1971. Equilibrium heat and moisture transfer in airflow through grain. Journal of Agricultural Engineering Research, 368–386.

Thompson, L. 1972. Temporary storage of high-moisture shelled corn using continuous aeration. Transactions American Society of Agricultural Engineers, 15, 333–337.

Thorpe, G.R., and Elder, W.B. 1982. Modelling the effects of aeration on the persistence of chemical pesticides applied to stored bulk grain. Journal of Stored Products Research, 18, 103–144.

Thorpe, G.R., and Fricke, P.W., 1985. Experiments on an open-cycle solar regenerated desiccant bed grain cooling system. Presented at International conference on research and development of renewable energy technologies in Africa March-April 1985, Reduit, Mauritius.

# Requirements for Drying High Moisture Content Grain in Southeast Asia

**Dante B. de Padua***

*Abstract*

In spite of the recognised need, minimal use has been made of drying technology in ASEAN to prevent grain deterioration and toxin formation. In this paper, the handling of wet-harvested grain is redefined in the context of tropical Asian developing countries and the constraints to adoption of available drying technology are discussed.

One of the most serious constraints to investment in dryers is the incompatibility of system capacities with those actually required. Examples of establishing different drying systems and their technical and economic implications are presented.

In the Asian region, changes in the farming systems for rice and maize, brought about by the high-yielding, non-photoperiodic varieties, irrigated farms, and high fertiliser inputs, have resulted in increased yields and a double-cropping system.

One crop is harvested during the rainy season, when the ripe grain is reaped dripping wet with moisture contents of 24-28% of the wet weight for paddy, and over 35% for maize. There are many difficulties associated with this wet harvest. The fields are soggy, the mechanical harvesters have a tendency to clog, and the crop cannot be reaped and left in the field to cure and sun dry as traditionally practiced.

The development of a postharvest system to cope with the changes has been rather slow and many costly errors have been made in attempting to develop such a system. As a consequence we are still unable to handle the wet harvest. This has resulted in delays in stabilising the product, which lead to biochemical deterioration of the grain. This deterioration takes the form of general discoloration, yellowing, germinating, damage to milling quality, and aflatoxin contamination in maize. The yellowing of the kernels has been found to start on the farm and to begin within 24 hours of reaping, especially where field stacking the grain and straw is practiced. The phenomenon is associated with the activity of certain moulds and the spontaneous heating of the mass. Further development of grain yellowing in storage is also noted when the grain is improperly dried.

Immediate threshing and drying of the wet harvest is the only practical method to arrest deterioration. Research and development on dryer technology for the past 20 years in Asia have resulted in a better understanding of the theory and physics of forced convection, heated air drying. The functional relationships of drying air temperature and airflow rates to product milling quality, drying rates and uniformity of drying, the psychrometrics of the drying process, and the heat and mass transfer balance, have been established.

## Dryers in Current Use

Drying units can now be designed, developed, and manufactured locally. During the early years of the Philippines National Food Authority (NFA), some 700 units of the small 2-tonne batch dryers developed at the University of the Philippines at Los Baños were purchased to service NFA requirements. Since then, NFA has gradually replaced its small dryers with high-capacity, continuous-flow dryers. These were also locally designed and manufactured.

Thailand has developed its own version of the batch dryer and its researchers are now promoting the technology in the countryside. For its rice complexes, Malaysia has designed and constructed 30-tonne batch dryers to complement its continu-

* International Development Research Centre, P.O. Box 101, Tanglin, Singapore 9124.

ous flow dryers, and BULOG in Indonesia commissioned some 200 units of the batch dryer for their cooperatives. Local research and development has therefore led to replacement of imported units which have been found inappropriate.

One would notice, however, that the investment in heated air dryers has been made mainly by government authorities. Statistics in the Philippines indicate that only 12% of production can be handled by dryers. Most of the dryers (90%) in the country are owned by the NFA, and NFA handles only 15% of production. Obviously, the balance must be sun dried, and this is evident from the poor quality of milled rice from the wet season found in the market stalls.

### Constraints to Dryer Adoption

A study made by the Philippines National Post Harvest Institute for Research and Extension (NAPHIRE) with the assistance of the International Development Research Centre (IDRC) documented the constraints perceived by both users and non-users of dryers in the private sector. There were frequent complaints against heated air forced convection dryers (often referred to somewhat imprecisely as artificial dryers or mechanical dryers) as follows:

1. Dryer capacity too great or too little for system.
   (a) 2-tonne batch dryers introduced to farmers were observed to be too large for their requirements. For example, a farmer working 3 hectares yielding 4 t/ha and with a harvesting period of one month (25 days) would have a daily drying requirement of 0.48 t, or approximately 10 bags of 50 kg.
   (b) The 2-tonne batch dryer can easily dry two loads (4 t) per day; that is, it can service 1 ha per day. This is too small for a commercial miller. The size of dryer needed by a miller is discussed later in this paper.
2. High fixed cost required and/or limited capital available.
   (a) This problem is related by farmers to their limited farm production or by millers to their procurement.
   (b) Because of the high fixed cost and limited use, it is their opinion that a dryer cannot be justified.
   (c) Within the range of farmer or processor needs, investment in a dryer has been given a lower priority.
3. High operating (variable) costs.
   (a) These are usually mentioned in comparison to direct sun drying. The high cost of oil-based fuel for the burner and engines is cited.
   (b) Dryers are therefore thought to be unprofitable.
4. It is claimed that dryers produce poorer quality milled rice than does sun drying. Higher percentages of brokens in the milled rice, darker colour, and non-uniform drying are cited.
5. Dryers are said to be inconvenient. Heat and dust are mentioned, as is having to mix the grain in some designs of batch dryer. Limited working space, particularly where the grain is handled in bags, is also perceived as a problem.

However, some of the less obvious reasons for the unsatisfactory experiences in the user of dryers may be overlooked.

1. In almost all cases, even in plants set up by business corporations, there is no realistic analysis of drying requirements in relation to the raw material supply and the plant's downstream requirements.
2. If an economic analysis is made, in most cases the assumptions are erroneous. The viability or profitability of using a dryer is analysed as a unit in isolation. The benefit is assumed to come from increased milling recovery by proper mechanical drying, but there is no increase in milling recovery compared with sun drying. The differentials in price of wet and dry paddy cannot cover the cost of drying. In the context of integrated processing plant operations, however, there is not only a penalty cost for not having drying capability but also a benefit for having it.
3. The drying phenomenon takes place so easily: add heat to air in contact with a wet product and the product dries. Nevertheless, many of the dryers sold are not properly engineered. This may result in non-uniform drying, damage to the grain and poor milling quality, inefficient heat use, dusty operations, inconvenience and unnecessary double handling.
4. Lack of operating skill. Even the best designed dryer, when operated improperly, produces disastrous results. We have cases of operators wishing to accelerate drying by raising the operating temperature. Subsequently, the paddy owner condemns the dryer for damaging his grain.

5 The benefits of having drying capability accrue not only to the direct users but also to society in general. It should be possible to show the social profitability, or even political gain for government, from allocating resources to ensure the production of better quality grain.

### Advantages of Using Dryers

There are some positive observations made by users of dryers.

1. Owning a dryer provides more opportunities to procure stocks. There are fewer buyers on rainy days and farmers are therefore more willing to sell — even at lower prices — a situation where somebody's gain is someone else's loss.
2. A dryer provides for better management of a procurement program. Users can control volume of procurement, segregate varieties, and prevent mixing of different grades.
3. Users of dryers get better quality milled rice and better prices for their product.
4. They claim more efficient plant operations. The labour costs are lower and the labour problems fewer, there is less spillage, less contamination, and less handling of the grain.
5. By use of drying technology, farmer or marketing associations or cooperatives are able to provide better services to their members.

### Designing Dryers for Commercial Millers

The potential users of grain drying systems are:

- large commercial farmers
- farmer cooperatives or associations
- commercial millers
- government grain marketing agencies

In the Philippines, the government handles approximately 15-20% of production. There are no large commercial rice farmers, and co-operatives or associations are still the exception. Most of the grain is handled and marketed by the private commercial millers. They are therefore the logical clients for any grain-drying research and development program.

Unfortunately, a technology that suits their requirements has not been developed. For a rice businessman, his rice mill is the heart of his enterprise. Everything else, the dryer, the warehouse, the transport facility, the by-products plant, should be designed to match the production capacity of the rice mill. There are therefore some key factors for the operation of a successful drying plant.

1. The facility should be integrated with the milling process, and considered a supportive facility. The cost of drying should be analysed as part of the integrated processing system.
2. The drying plant capacity should be compatible with the market (consumer) demands and milling requirements.
3. The operators must possess technical skills. This means not only the 'how to', but also the 'why' of things, to cope with changing product characteristics and the environment.
4. Peculiar to Asia, the drying plant should be able to handle:
   - extremely wet and dirty grain;
   - two varieties, and at least two grades, of incoming paddy;
   - batches of grain delivered to the plant with varying initial moisture contents, without mixing.

### Case Study

The following example is given as a guide in establishing drying capacity needs of a commercial miller.

Example: Drying requirement for a 5 t/hour rice mill.

° For 12 hours operation/day: paddy input = 12 × 5 = 60 t/day

° For double cropping with an effective 2 months purchasing season. Total stock required = 60 × 6 month × 30 days/month.
= 10 800 t.

This assumes procuring paddy to be stored to feed the milling plant for a six-month period. Although there are only three months between harvest periods, the freshly harvested and dried paddy is cured in storage before milling. This practice is claimed to produce higher head rice yields and better cooking properties. An attempt should be made to match the volume of procurement (A) in Figure 1 to the milling requirement (B). During the rainy season harvest from August to December, procurement of volume A is possible only by using a complementary drying plant. A drying plant enables the grain businessman to keep his milling plant operating year round to meet his market requirements. It is in this context that the profitability of investment in a dryer should be analysed.

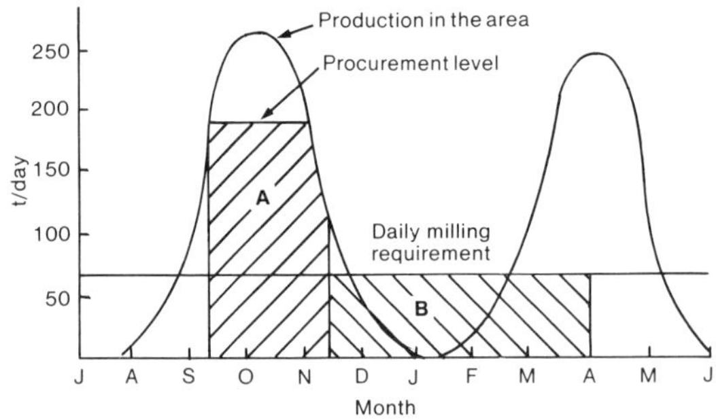


**Fig. 1.** Time course of harvesting of paddy showing procurement levels and daily milling requirements. During the wet season harvest, from August to December, procurement of volume A is possible only by using a complementary drying plant.

° Daily paddy procurement target

$$\frac{10\ 800\ \text{t}}{60\ \text{days}} = 180\ \text{t/day}$$

Options:

1. Continuous flow dryer (CFD) with tempering bins at the plant (Fig. 2). Four-stage multipass drying at 3% moisture content (m.c.) removal per pass or stage.

$$\frac{24\ \text{hours/day}}{4\ \text{passes/day}} = 6\ \text{hours to pass 180 t}$$

$$\frac{180\ \text{t}}{6\ \text{hours}} = 30\ \text{t/hour flow rate of paddy through dryer}$$

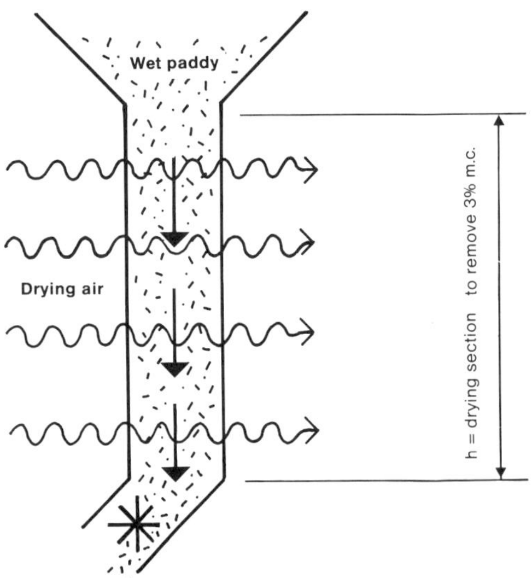


**Fig. 2.** Diagram of principles of the continuous flow grain dryer.

For a 30 t/hour flow rate, 45 minutes drying time for 3% m.c. drop
1 unit, 30 t/hour – 22.5 t batch capacity/unit
2 units, 15/t hour – 11.5 t batch capacity/unit
3 units, 10 t/hour – 7.5 t batch capacity/unit
4 units, 7.5 t/hour – 5.6 t batch capacity/unit
The height of unit will depend on dryer cross-section.
Note that a choice of more than one unit will give flexibility in handling different grades and varieties, but fixed costs will be higher, and so a cost/benefit study should be made of the various options.

2. Two-stage drying system.

   Stage 1. Drying at the farm level or collecting stations, to remove surface moisture from an initial 24–26% m.c. down to 18%. Several flat-bed dryers strategically located and sized to handle pre-drying.
   A flat-bed dryer with a capacity of 2 t/batch will take 4 hours to reduce m.c. from 26 to 18%. Therefore, it can dry 4 batches or 8 t/day. The system would need 22 units. Bigger capacity units could be designed to decrease the number required.

   Stage 2. Drying at plant, using CFD to dry 18–24% m.c. grain in two passes at 2% per pass.

$$\frac{10\ 800\ \text{t}}{90\ \text{days}} = 120\ \text{t/day}$$

   [At 18% m.c. drying period can be extended to 90 days]

$$\frac{24}{2} = 12\ \text{hours/drying time pass}$$

$$\frac{120}{12} = 10\ \text{t/hour flow rate of paddy through dryer}$$

   1 unit — 10 t/hour flow rate with a batch capacity of 5 t for a retention time of 30 minutes and 2% m.c. drop.

3. Stage 1. Same as option 2.
   Stage 2. In-store drying for 10 800 t. Size of facility will depend on number of days it will take to pull down m.c. from 18% to 14% without overdrying bottom layers. Comparative cost analysis is required since present warehouses are designed for bag storage.

While the computations are precise, farm production is variable from year to year, and competition for the raw material (paddy) has to be reckoned with. There are hills and valleys in the procurement of different grades. It would be useful to be able to generate information projecting cost and benefit of paddy as drying capacity increases, so as to be able to make an enlightened decision on what capacity to design for (Fig. 3).

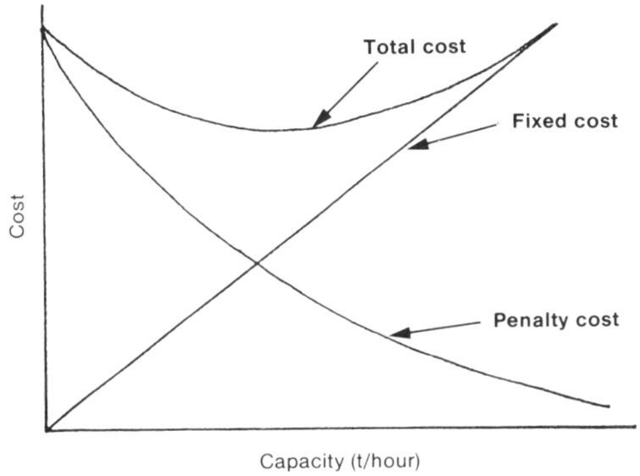


**Fig. 3.** How costs of grain drying vary with drying capacity.

4. Given the conditions where sun drying (when done properly) is still the cheapest form of drying, two-stage drying — artificial drying in combination with sun-drying — should be considered. How many hours and days of the harvest season can we depend on direct sun drying? An estimate based on records of average sunshine over a number of years should be made.

After assumptions have been validated and there is confidence in the estimates of drying capacity requirements, the investment decision on the type of technology will depend on technical performance and economic efficiency of the various options.

# Objectives of Aeration and In-store Drying Session Chairman's Summary

R.G. Bowrey*

THE four papers in this session considered the objectives of grain drying systems and discussed some procedures which can be used to achieve these objectives.

The overall objective of grain handling was considered to be the maintenance of grain quality at a level appropriate to the end use. Certain acquired aspects of quality, such as the level of foreign material, can be improved during handling but the major intrinsic quality factors, such as head yield, discoloured grain, etc. decrease during storage and the objective of the grain handling system is to slow down the rate of quality deterioration. The fact that a large decrease in quality can occur before the grain reaches the in-store dryer was mentioned, as was the detrimental effect that fast drying can have on quality.

Two papers summarised in-store dryers used in temperature climates and the danger associated with the direct transfer of this technology to tropical climates was emphasised. This is an important area that requires further fieldwork. For high moisture grain, a two-stage drying process is most appropriate. In this system the wet grain is partly dried using high temperatures and airflow rates followed by low airflow drying and cooling. However, to achieve grain cooling in tropical regions it will be necessary to use some form of refrigeration or desiccant absorber. Such systems are successful in hot dry climates, but their use in hot humid climates will cause condensation problems if the store is not airtight.

In most humid climates supplemental heating is mandatory to dry grain in storage during the wet season and much higher airflow rates will be necessary. These flow rates must be adjusted to local weather conditions as summarised in average weather records. The advantages of pre-cleaning the grain before drying is one area requiring attention as regards handling of the wet season crop.

Grain stirrers and solar heated dryers are technically effective but are unlikely to be economically feasible.

The importance of separating aeration, which is essentially a cooling process involving sensible heat transfer, and drying, which involves latent heat transfer was stressed. Cooling grain, even without drying, is useful because it retards insect population growth and increases the useful life of pesticides.

The essence of this session is contained in the following quotes from Dr de Padua's paper.

1. Utilisation of drying technology in the ASEAN region to prevent grain deterioration has been MINIMAL in spite of the recognised need.
2. The development of post harvest systems has been slow and many costly errors made. As a consequence the problem of lack of capability to handle the wet harvest persists.
3. The benefits of drying accrue to society in general.
4. Private commercial millers are the logical clients for grain drying research and a technology developed to suit their needs is required.

The solutions to these problems lie in the technologies described during this session only if they are adapted or modified to the needs of the private millers and to the local conditions.

* Allied Mills Ltd, P.O. Box 57, Mascot, N.S.W. 2020, Australia.

# Basic Principles of Aeration and In-store Drying

# Physical and Thermal Properties of Grains


## Do Sup Chung and Chong-Ho Lee*



*Abstract*

To analyse, design, and simulate grain drying and aeration systems, it is essential to have accurate information on grain physical and thermal properties such as specific gravity, bulk density, void fraction, static pressure through grains, isotherms, specific heat, thermal conductivity, heat of vaporisation, and drying rate constant.

These physical and thermal properties of grains are reviewed, particularly for rice and maize. The applicability and/or limitations of the existing data for practical or theoretical uses over a specified range of conditions are discussed. Areas in which additional research is needed are identified.


IN the past three decades considerable effort has been expended to determine basic properties of food and feed grains and other agricultural products.

Knowledge of physical and thermal properties of grain is essential in solving heat and mass transfer problems involved in grain storage, drying, aeration, refrigeration, and processing.

Some of the physical and thermal properties of grain have been determined, but only at specific levels and only for particular varieties of grain. Therefore, only limited information on physical and thermal properties of grain is available.

Simulation techniques have been used to find the optimum drying conditions and to study the feasibility of natural air drying. Recent research has emphasised the need for additional data on physical and thermal properties of grain and the variation of these properties with important factors such as moisture content.

The objective of this paper is to describe the existing data on physical and thermal parameters of grain, especially of rice and corn, and discuss their applicability and limitations for practical or theoretical purposes. Also, areas in which further data are needed are identified.

## Physical Property Parameters

There are many parameters related to physical properties of grain. In this section, several parameters which are closely related to the physical properties of grain, such as physical dimensions, volume, true and bulk densities, and porosity, are examined. Pressure drop through rice and corn is reviewed.

### Physical Dimensions and Volume of Rice and Corn

Physical dimensions of grain, such as length, width and thickness, vary according to the variety, environmental conditions, temperature, and moisture content.

As shown in Table 1, some investigators have expressed the physical dimensions as a linear function of the moisture content of rice and corn.

It seems reasonable to present the dimensions like this because physical dimensions differ with an increase in moisture content within a certain range.

The volume of grain has been measured by using an air comparison pycnometer, or the mercury or toluene displacement methods.

Wratten et al. (1969) reported that the volume of rough rice kernels ranged from 16.06 $mm^3$ to 19.17 $mm^3$ for medium grain, and from 18.35 $mm^3$ to 19.66 $mm^3$ for long grain. Also, they expressed the volume of rough rice kernels as a linear function of the moisture content of kernel within the range from 12% to 18% moisture content. Thompson and Isaacs (1967) reported the volume of hand-shelled corn kernels ranged from 269 $mm^3$ to 369 $mm^3$. Fortes and Okos (1980) expressed the kernel volume of corn as a function of temperature and moisture content of kernel.


* Department of Agricultural Engineering, Kansas State University, Manhattan, Kansas 66506, USA.

**Table 1.** Some information related to physical dimensions of rough rice and corn.

| Grain | Variety | Moisture content (%, wet basis) | Length ($10^{-3}$m) | Width ($10^{-3}$m) | Thickness ($10^{-3}$m) |
|---|---|---|---|---|---|
| Rough rice | Short grain Caloro | 10.4 - 22.6 | $7.318+1.22 \times 10^{-2}M^{a}$ | $3.358+8.90 \times 10^{-3}M$ | $2.187+8.9 \times 10^{-3}M$ |
| | Medium grain Saturn | 12 - 18 | $7.747+1.27 \times 10^{-2}M$ | $2.842+7.62 \times 10^{-3}M$ | $1.842+8.9 \times 10^{-3}M$ |
| | Long grain Bluebonnet | 12 - 18 | $8.941+5.84 \times 10^{-2}M$ | $2.388+1.65 \times 10^{-2}M$ | $1.765+1.43 \times 10^{-2}M$ |
| Corn | Pfister | 6.7 | 16.26 (0.91)[b] | 20.27 (1.07) | 12.80 (0.71) |
| | Beck | 13.96 - 21.26 | 13.30 (0.99) | 8.7 (0.76) | 4.70 (0.68) |

[a] M = moisture content (%, wet basis)
[b] Numbers in parentheses indicate the standard deviation.
Sources: Goss (1965), Wratten et al. (1969), Morita and Singh (1979), Fortes and Okos (1980).

Few reports related to the volume of rice and corn were found.

### True Density or Specific Gravity

True density is an important factor for heat and mass transfer analysis through grains. It can be measured by various instruments, including the pycnometer.

Because true density is moisture dependent, its values are sometimes presented as a function of moisture content. Measurements of the true density of rice and corn are given in Table 2. Investigators have devised a model to predict the true density of a certain range of moisture contents in which the measurement was conducted.

The true density of rice ranges from 1019 kg/m$^3$ to 1387 kg/m$^3$, depending on the variety and the variation in moisture content. The true density of corn ranges from 1190 kg/m$^3$ to about 1370 kg/m$^3$, depending on the variety and the variation in moisture content.

Some investigators have expressed the true

**Table 2.** Measurements from various sources of the true density of rice and corn.

| Grain | Moisture content (%, wet basis) | True density (kg/m$^3$) | Models[a] |
|---|---|---|---|
| **Rice** | | | |
| Caloro | 8.6 | 1358.4 | |
| Calrose | 9.2 | 1364.8 | |
| Hy Mix Early | 8.8 | 1387.2 | |
| **Rough Rice** | | | |
| Short | 14 - 22 | 1019.5 - 1037.1 | 1066.0 - 2.05 M |
| Medium | 12 - 18 | 1324.3 - 1371.8 | (1.465 - 0.0076 M)[b] |
| Long | 12 - 18 | 1362.9 - 1384.0 | (1.436 - 0.0042 M)[b] |
| Long | 14 - 22 | 1027.4 - 1054.7 | 1046.1 - 1.01 M |
| **Corn** | | | |
| Pfister | 6.7 | 1292.7 | |
| Beck 65 | 21.7 | 1236 ± 13 | |
| Seed | 16 - 44 | 1293.3 - 1190.5 | |
| Shelled | 24 - 26 | 1369.1 | $1370.0 - 0.028\left(\frac{100M}{100-M}\right)$ |
| Yellow | 9 - 27 | 1313.4 - 1284.5 | 1327.8 - 1.60 M |
| Yellow Dent | 10 - 35 | 1274.5 - 1237.0 | $1251.9+7.14M-0.597M^2+0.0019M^3$ |
| Yellow Dent | 12 - 23 | 1250.0 - 1208.1 | 1300.7 - 3.23 M |

[a]M = moisture content (%, wet basis)
[b]Models for determing specific gravity
Sources: Goss (1965), Wratten et al. (1969), Chung and Converse (1971), Gustafson and Hall (1974), Brusewitz (1975), Fortes and Okos (1980), Kim (1980), Nelson (1980), Haque et al. (1982).

**Table 3.** Bulk density of rice and rough rice.

| Variety | Moisture content (%, wet basis) | Bulk density ($kg/m^3$) | Models[a] |
|---|---|---|---|
| **Rice** | | | |
| Caloro | 8.6 | 571.1 (1.7)[b] | |
| Calrose | 9.2 | 570.7 (6.2) | |
| Hy Mix Early | 8.8 | 591.2 (9.3) | |
| **Rough Rice** | | | |
| Short | 14 - 22 | | 537.6+1.22M |
| Short | 11 - 20 | 632.0 - 664.0 | 583.6+4.27M |
| Medium | 12 - 18 | 598.3 - 648.3 | 499.7+8.33M |
| Medium | 6 - 28 | | 567.2+4.13M |
| Medium | 13.2 | 590.0 | |
| Long | 12 - 18 | 585.6 - 615.1 | 519.4+5.29M |
| Long | 9 - 11 | 561.0 - 598.0 | |
| Long | 13.5 | 710.0 - 780.0 | |
| Long | 14 - 22 | | $529.2 - 1.105M + 0.00995M^2$ |
| — | — | 576.7 | |

[a]M = moisture content (%, wet basis)
[b]Number in parentheses represents standard deviation.
Sources: Goss (1965), Wratten et al. (1969), Husain and Ojha (1969), Hosokawa and Matsumoto (1971), Agrawal and Chand (1974), Morita and Singh (1979), Vemuganti et al. (1980), Kim (1980), ASAE (1984).

density/moisture content relationship as a linear function. The linear models show that true density decreases linearly with the increase in moisture content.

In contrast to these linear models, a polynomial of the third degree with respect to moisture content was provided by Nelson (1980) to predict the true density.

### Bulk Density

Bulk density, or test weight, is generally measured by using the standard apparatus recommended by the United States Department of Agriculture. Actually, bulk density is much more important than true density in drying and storage practices. Bulk density can be changed depending on the moisture content, the amount of impurities, and the degree of filling.

The bulk density of rice and corn in most experiments was measured under the loose-fill condition. The bulk density values for rice and rough rice provided by several investigators ranged from 533 $kg/m^3$ to 780 $kg/m^3$, as shown in Table 3.

Excluding the extreme data, other data values fluctuate between 7.5% and 18.4% around the value of 576 $kg/m^3$ approved by the American Society of Agricultural Engineers (ASAE) for rough rice.

Wratten et al. (1969) and four other investigators reported the linear relationship between the bulk density and moisture content of rough rice within the range of moisture content they worked in. Kim (1980) used the polynomial of the third degree to represent the relationship between them.

The five linear models show that the bulk density of rough rice increases as the moisture content increases. This tendency is contrary to that of the true density of rough rice.

The bulk density of various varieties of corn is shown in Table 4. These values, which come from various sources, range from 448 $kg/m^3$ to 789 $kg/m^3$. Except for the bulk density values for husked ear corn and green sweet corn, the variation of the bulk density remains within the range of about 10% of 717.6 $kg/m^3$, which is the bulk density approved by the ASAE for shelled corn.

Vemuganti et al. (1980) reported that the bulk density and moisture content of flint corn and dent corn have the linear relationship. On the other hand, Brusewitz (1975) and Nelson (1980) expressed their relationship by using a polynomial of the second or higher degree. Also, Brusewitz (1975) commented that bulk density of seed corn reached a maximum value at around 30% moisture content.

However, in contrast to the trend of the values of bulk density for rough rice, the linear models for flint and dent corn show that bulk density of corn decreases with an increase in moisture content.

**Table 4.** Bulk density of corn.

| Variety | Moisture content (%, wet basis) | Bulk density ($kg/m^3$) | Models[a] |
|---|---|---|---|
| Pfister | 6.7 | 744.5 (8.6)[b] | |
| Shelled | 7 - 25 | 752.9 - 656.8 | |
| Shelled | — | 717.6 | |
| Flint | 6 - 28 | 789.1 - 644.8 | 828.5 - 6.56M |
| Dent | 6 - 28 | 779.0 - 635.5 | 818.1 - 6.52M |
| Yellow Dent | 10 - 35 | 742.2 - 638.5 | $682.9+14.22M - 0.9843M^2+0.01548M^3$ |
| Yellow Dent | 12 - 23 | 784.3 - 698.4 | |
| Seed | 16 - 44 | 734.1 - 710.3 | $1086.3 - 2.97M + 4.81M^2$ |
| Ear Husked | — | 448.5 | |
| Green Sweet | — | 448.5 | |

[a]M = moisture content (%, wet basis)
[b]Number in parentheses represents standard deviation.
Sources: Goss (1965), Brusewitz (1975), Nelson (1980), Vemuganti et al. (1980), Haque et al. (1982), ASAE (1984).

**Porosity**

Porosity changes depending on the bulk density, the degree of filling, the amount of foreign material, the amount of fine material, and the moisture content of the grain.

Porosity can be determined by using the ratio of bulk density to true density values.

Gustafson and Hall (1972) reported a linear relationship between porosity of corn and its test weight. Chung and Converse (1971) represented the porosity of yellow dent corn as a linear function of bulk density.

Wratten et al. (1969) and one other investigator expressed the porosity of rough rice as a linear function of moisture content rather than bulk density or test weight.

In this case, the degree of filling, which affects porosity, is not considered.

Therefore, it seems reasonable to represent porosity as a function of bulk density, which is

**Table 5.** Porosity values of rice and corn.

| Grain | Moisture content (%, wet basis) | Porosity (%) | Models[a] |
|---|---|---|---|
| **Rice** | | | |
| Honduras | 11.9 | 50.4 | |
| Wateribune | 12.4 | 46.5 | |
| **Rough Rice** | | | |
| Durar | 11.4 | 51.0 | |
| Taichung | 9.3 | 52.0 | |
| Kalinpong | 9.7 | 54.5 | |
| Short | 14 - 22 | 46.4 - 47.6 | 49.7 - 0.227M |
| Medium | 12 - 18 | 58.5 - 53.1 | 65.6 - 0.475M |
| Medium | 13.2 | 52.5 | |
| Long | 12 - 18 | 59.6 - 56.9 | 69.5 - 0.885M |
| Long | 14-22 | 48.4 - 50.8 | $49.4+0.064M-0.0099M^2$ |
| **Corn** | | | |
| No. 1 | 9.0 | 40.0 | |
| Yellow | 25.0 | 44.0 | |
| Yellow | 9 - 14 | 38.5 - 47.6 | |
| Yellow | 9 - 27 | | $101.0 - 0.078D_b$ |
| Shelled | 9 - 31 | 38.5 - 47.6 | |
| Yellow Shelled | 9 - 27 | | $81.4-0.056W_t$ |
| Yellow Dent | 12 - 23.4 | 37 - 42 | |
| Yellow Dent Shelled | 15.0 | 40.0 | |

[a]M = moisture content (%, wet basis), $D_b$ = bulk density ($kg/m^3$), $W_t$ = test weight ($kg/m^3$)
Sources: Thompson and Isaacs (1967), Wratten et al. (1969), Hosokawa and Matsumoto (1971), Chung and Converse (1971), Gustafson and Hall (1972), Agrawal and Chand (1974), Kim (1980), Haque et al. (1982).

moisture dependent. According to Wratten et al. (1969), the porosity of rice and rough rice ranges from 46% to 60% and decreases linearly with an increase in moisture content.

Excluding the data of Wratten et al. (1969), other data vary within a narrow range, that is, 46% through 54% (Table 5). There is little difference between these data and the value approved by the ASAE for rice.

The few data available on porosity of corn indicate that it ranges from 38.5% to 47.6%, regardless of variety and moisture variation. The porosity values of corn approved by the ASAE fall within this range. It seems desirable for the amount of fine materials to be regarded as an important factor when expressing porosity as a function of other parameters such as bulk density and moisture content.

**Static Pressure Drop**

Having exact information about the pressure drop for airflow through grain is very important in designing dryer or aeration systems as well as in choosing the correct fan. The pressure drop for airflow through grain depends on the airflow rate, the surface and shape characteristics of grain, the number, size and configuration of the voids, the variability of the particle size, and the bed depth.

Also, pressure drop is influenced by the extent of packing and the amount of foreign material. Brooker et al. (1974) reported that as the percentage fines in the mixture increases, the resistance pressure increases.

Haque et al. (1982) and Chung et al. (1985) investigated the effect of the moisture content of grain on resistance to airflow. They reported that the static pressure drop through grain increases with an increase in airflow rate as well as air velocity, and that moisture content of grain significantly affects the pressure drop.

Hukill and Shedd (1955) developed a model to predict the pressure drop over an airflow range of 0.01 through 0.20 $m^3/s.m^2$, and determined the constants involved in the equation for several grains.

Their model was revised and approved by the ASAE Committee on Technical Data.

$$\frac{P}{L} = \frac{a\,Q^2}{\ln(1+bQ)} \qquad (1)$$

where, P = static pressure (Pa)
L = bed depth (m)
Q = airflow rate ($m^3/s.m^2$)
a, b = constants

Haque et al. (1978) and Chung et al. (1984, 1985) expressed the static pressure or pressure drop as a polynomial of the second degree of airflow rate, moisture content of grain, and the percentage of fines present (Figs 1 and 2).

Models devised by several investigators to determine the static pressure drop for rice and corn are illustrated in Table 6. It can be seen that these models include only a limited number of parameters among several which may affect the static pressure drop such as airflow rate, bed depth, density, moisture content, the amount of fine materials, the degree of filling, and particle size.

In order for these models to be used for simulation work on drying and aeration systems,

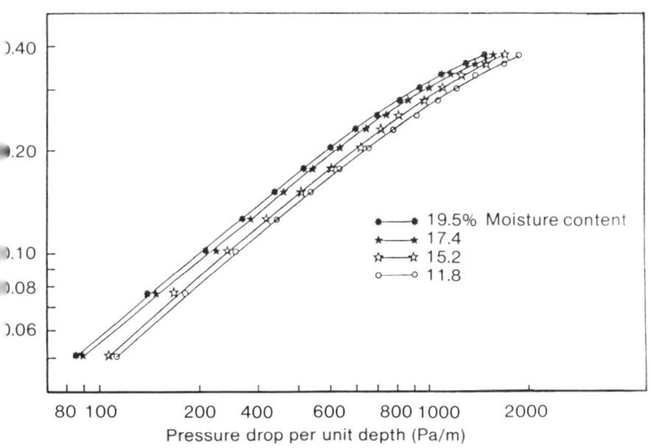


**Fig. 1.** Plots of airflow versus pressure drop for rough rice at various moisture contents (wet basis) and containing 5% foreign material.

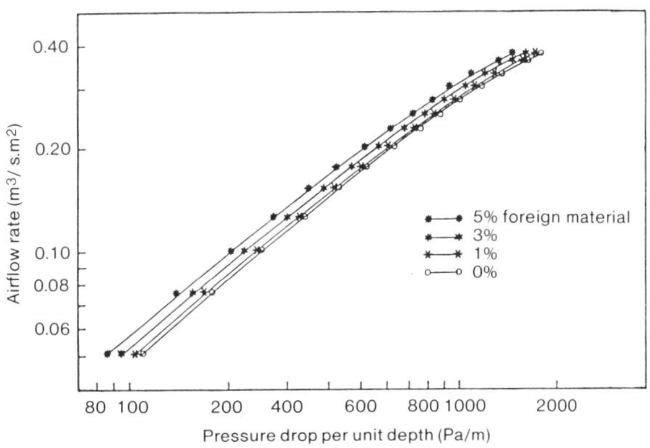


**Fig. 2.** Plots of airflow versus pressure drop for rough rice at 19.5% moisture content (wet basis) and containing various amounts of foreign material.

**Table 6.** Some models for determining pressure drop or static pressure, and airflow rate

| Model | Parameters | Value of constants |
|---|---|---|
| Gunasekaram et al. (1983)<br>$ln(Q)=a(ln\Delta p)^2+b\,ln(\Delta p)+c$ | Q = airflow rate ($m^3/s.m^2$)<br>p = pressure (Pa/m)<br>moisture content (12.8%, wet basis)<br>bed depth (5 - 30 cm) | Rough rice (long grain)<br>a = − 0.031<br>b = 0.092<br>c = −3.024 |
| Haque et al. (1978)<br>$\Delta p=C_1Q_a+C_2Q_a^2+C_3Q_a(FM)$ | $\Delta p$ = pressure drop (Pa/m<br>$Q_a$ = airflow rate ($m^3/s.m^2$)<br>(0.076 - 0.381)<br>FM = % foreign materials (0-20)<br>bed depth (45.72 cm) | Yellow dent corn<br>C1 = 436.67<br>C2 = 7363.04<br>C3 = 22525.82 |
| Haque et al. (1978)<br>$\Delta p=AV+BV^2-CMV$ | $\Delta p$ = pressure drop (Pa/m)<br>V = air velocity (m/s)<br>M = moisture content (%, wet basis)<br>(12.4 - 25.3) | Yellow dent corn<br>(clean loose-fill)<br>A = 1611.7<br>B = 4949.3<br>C = 55.1 |
| Husain and Ojha (1969)<br>$Ps=(mD+c)A^{oD+e}$ | Ps = static pressure (mm $H_2O$)<br>D = bed depth (cm)<br>Q = airflow rate ($m^3/min.m^2$)<br>(1 - 6)<br>moisture content (13.5%, wet basis) | Rough rice<br>m = 0.322<br>c = −1.32<br>D = 2.14<br>e = −0.0089 |
| Shedd (1945, 1951)<br>$Q = a\frac{p^b}{D^c}$ | Q = airflow rate ($cfm/ft^2$)<br>(10 - 70)<br>P = pressure drop (inch $H_2O$)<br>D = bed depth (ft)<br>(2 - 8) | Shelled corn with impurities<br>(20% w.b.)<br>a = 150<br>b = 0.564<br>c = 0.646<br>Cleaned shelled corn<br>a = 303<br>b = 0.422<br>c = 0.542 |
| Chung et al. (1984, 1985)<br>$SP=A(AF)+B(AF)^2$<br>$-C(AF)(MC)-D(AF)(FM)$ | SP = static pressure (Pa/m)<br>AF = airflow rate ($m^3/s.m^2$<br>(0.0508 - 0.381)<br>MC = moisture content (%, wet basis)<br>(11.8 - 19.5)<br>FM = % fine material<br>(0 - 8) | Rough rice (long grain)<br>A = 3749.15<br>B = 8289.97<br>C = 117.12<br>D = 164.23 |

further research is needed to develop models including as many as possible of the parameters which affect the pressure drop for airflow through grain.

## Thermal Property Parameters

A knowledge of the thermal properties of grain is essential when studying the problems encountered in drying and storage of grain in bulk. In this section, specific heat, thermal conductivity, isotherms, and heat of vaporisation are reviewed.

### Specific Heat and Thermal Conductivity

Various methods have been used to determine specific heat and thermal conductivity of rice and corn. The usual approach in these studies is to determine the relationship between these values and moisture content.

The ASAE (1984) has approved four linear regression equations for calculating specific heat and one for calculating thermal conductivity of rice. The ASAE (1984) also approved seven numerical values for specific heat of corn at specified moisture levels and an equation for determining the specific heat at other moisture levels.

Specific heat and thermal conductivity values of rice and corn are summarised in Table 7. These values and equations were orginated from research works conducted by Haswell (1954), Kazarian and Hall (1965), and Wratten et al. (1969).

Regardless of variety, the values of specific heat of rice range from 1.3 to about 2.0 kJ/kg.K

**Table 7.** Specific heat and thermal conductivity values for rice and corn approved by the American Society of Agricultural Engineers

| Grain | Moisure content (%, wet basis) | Mean temperature (K) | Specific heat (kJ/kg.K) | Thermal conductivity (W/m.K) |
|---|---|---|---|---|
| Rice, rough | 10.2 - 17.0 | | 1.1095+0.0448M[a] | |
| Rice, shelled | 9.8 - 17.6 | | 1.2016+0.0381M | |
| Rice, finished | 10.8 - 17.4 | | 1.1807+0.0377M | |
| Rice, rough, medium | 10.0 - 20.0 | | 0.9214+0.545M | 0.0866+0.001327M |
| Corn, yellow, dent | 0.9 | 293.7 | 1.5324 | 0.14055 |
| | 5.1 | 293.7 | 1.6915 | 0.14661 |
| | 9.8 | | 1.8338 | 0.15198 |
| | 13.2 | | | 0.17656 |
| | 14.7 | 293.7 | 2.0264 | 0.15908 |
| | 20.1 | 293.7 | 2.2232 | 0.16358 |
| | 24.7 | | 2.3739 | 0.16998 |
| | 30.2 | 293.7 | 2.4618 | 0.17240 |
| | | | 1.4654+0.0356M | |
| | 0.68 - 20.3 | 308.4 | | 0.1409+0.00118M |

[a]M = moisture content (%, wet basis)

depending on the variation of moisture content approximately 10% to 20% wet basis.

The values of specific heat of corn range from 1.5 to 2.5 kJ/kg.K with an increase in moisture content from about 10% to 30% wet basis at approximately 294K.

The values of thermal conductivity of rice range from 0.86 to 1.8 W/m.K, depending on the variety, the initial temperature, measuring methods, and moisture content of grain.

Fortes and Okos (1980) expressed the thermal conductivity of corn as a function of moisture content and temperature, whereas Wratten et al. (1969), Morita and Singh (1979), and Haswell (1954) described it as a function of moisture content.

Kim (1980) reported that the thermal conductivity of rough rice is affected by porosity.

## Adsorption and Desorption Isotherms

The accuracy of equilibrium moisture content equations is important for successful modelling and optimisation of grain drying and aeration system design. A number of theoretical, semi-theoretical, and empirical models have been proposed for predicting the moisture equilibria of cereal grains.

Some of these models have been tested for their accuracy and applicability by other investigators. Some of them are applicable within limited ranges of relative humidity, and in either the adsorption or desorption phase.

Many references have focused on the desorption phase rather than the adsorption phase. Few models are capable of accurately describing equilibrium moisture content at a relative humidity above 90%. It is known that this difficulty is caused by condensation on the grain sample, or mould development during the experimental process, especially when static methods are being used.

Zuritz et al. (1979) therefore suggested that a dynamic method is more desirable than a static one for experiment work on isotherms at levels of relative humidity above 87%. Vemuganti and Pfost (1980) also suggested that the equilibrium relative humidity method can be used in this situation.

Among the several equilibrium moisture content models, four well-known ones are the BET equation, the Smith equation, the modified Henderson equation, and the Chung-Pfost equation. The Chung-Pfost and modified Henderson models were recognised as being more accurate than other equations over a wider range of relative humidity, and for various cereal grains.

These two equations and their constants for rough rice and yellow dent corn (approved by the ASAE) are shown in Table 8.

**Table 8.** Equilibrium moisture content equations and constants approved by the American Society of Agricultural Engineers

| Equation[a] | Constants | Grain | |
|---|---|---|---|
| | | Rough rice | Yellow dent corn |
| Modified Henderson equation | | | |
| $M = \frac{1}{100}\left[\frac{\ln(1-RH)}{-K(T+C)}\right]^{1/N}$ | K | 1.9187 | 8.6541 |
| | N | 2.4451 | 1.8634 |
| $RH = 1 - Exp[-K(T+C)(100M]^{N}$ | C | 51.161 | 49.810 |
| | SEM[b] | 0.0097 | 0.0127 |
| Chung-Pfost equation | | | |
| $M = E - F \ln[-(T + C) \ln(RH)]$ | A | 594.61 | 312.40 |
| | B | 21.732 | 16.958 |
| $RH = Exp[\frac{-A}{(T+C)} Exp(-BM)]$ | C | 35.703 | 30.205 |
| | E | 0.29394 | 0.33872 |
| | F | 0.0046015 | 0.058970 |
| | SEM[b] | 0.0096 | 0.0121 |

[a]M = grain moisture (decimal, dry basis)
RH = relative humidity (decimal)
T = temperature (°C)

[b]SEM = Standard error moisture

## Heat of Vaporisation

The heat of vaporisation of water in rice and corn is defined as the energy required to vaporise moisture from the grain.

Equilibrium moisture content curves furnish the data necessary to calculate the heat of vaporisation for moisture.

Therefore, the accurate calculation of the heat of vaporisation depends entirely upon the correct choice of equilibrium moisture content data appropriate to the specific situation.

Othmer (1940) devised a convenient method to determine the latent heat based on the vapour pressure. It has been widely used because it is quite simple and useful for agricultural crops.

Gallaher (1951) described the ratio of latent heat of wheat to latent heat of free water as an exponential function with respect to moisture content.

Rodriguez et al. (1963) tested three equilibrium moisture content models such as Clausins-Clapeyron equation, Othmer equation, and BET equation after calculating the heat of vaporisation for shelled corn at 30°C through 50°C and identified the applicability of these models.

Also, they reported that the heat of vaporisation of shelled corn decreases with an increase in moisture content and approaches that of free water at high moisture contents.

Bekker and Sallans (1956) derived an equation for the differential net heat of desorption based on Othmer's method and tested the BET and Smith equations.

They concluded that the Smith equation is not reliable beyond the relative vapour pressure of 0.95, which is corresponded to about 30% moisture content at 50°C.

Johnson and Dale (1954) reported that the heat required for vaporisation of water in wheat and corn is between 1.00 and 1.06 times that for vaporisation of free water above 12.3% moisture content (wet basis), and the heat requirement is further increased below 12.3% moisture content (wet basis).

Chung and Pfost (1967) reported that the isosteric heat in the desorption and adsorption phases fell from 3508 to 2484 kJ/kg and from 3070 to 2470 kJ/kg, respectively, and the net heat of vaporisation from 1098 to 74 kJ/kg for desorption and from 658 to 60 kJ/kg for adsorption, moisture content ranging from 4 to 20% at 31°C.

They concluded that the net heat of sorption approaches zero as the moisture content increases, a conclusion also drawn by Johnson and Dale (1954).

**Table 9.** Variation of the ratio of latent heat of corn to latent heat of free water by moisture content and temperature

| Moisture content (%, wet basis) | Temperature (°C) | | | | |
|---|---|---|---|---|---|
| | 0 | 10 | 21.1 | 37.8 | 65.6 |
| 10 | 1.177 | 1.174 | 1.175 | 1.176 | 1.171 |
| 11 | 1.135 | 1.136 | 1.137 | 1.138 | 1.131 |
| 12 | 1.093 | 1.094 | 1.095 | 1.095 | 1.096 |
| 13 | 1.065 | 1.066 | 1.066 | 1.066 | 1.062 |
| 14 | 1.042 | 1.042 | 1.043 | 1.041 | 1.042 |
| 15 | 1.019 | 1.019 | 1.019 | 1.017 | 1.017 |

Source: Thompson and Shedd (1954)

Thompson and Shedd (1954) investigated the heat of vaporisation of corn and wheat by Othmer's method and concluded that for corn and wheat it decreases with an increase in moisture content as well as temperature, as shown in Table 9.

There are few numerical data on the heat of vaporisation of rice. It seems desirable to devise, for practical use, some equations for calculating heat of vaporisation directly from moisture and temperature information, such as the equation derived by Chung and Pfost (1967).

### Drying and Rewetting Rates

The accurate prediction of the drying and rewetting behaviour of deep beds of grain depends directly on the accurate description of the thin-layer process, under the assumption that the deep bed is composed of a series of thin layers.

A number of investigators have therefore tried to develop successful thin-layer drying and rewetting models for grain, especially for shelled corn. Each has developed an equation from their own data obtained under conditions of interest to them.

Thompson et al. (1968) developed a mathematical drying model to predict the performance of various types of grain dryers.

Chittenden and Hustrulid (1966) presented a method for describing and predicting the drying curves for single corn kernels under conditions of known air temperature, air relative humidity, initial corn moisture content, and kernel size.

They determined that the drying rate constant is related to the effective moisture diffusion coefficient and the effective kernel radius.

Flood et al. (1972) studied thin-layer drying and rewetting equations for a natural-air corn drying system and fixed the constants involved in their equation through some tests.

On the other hand, Troeger and Hukill (1971) developed a mathematical model for the drying rate of fully exposed corn on the basis of the diffusion law and determined their constants in terms of test variables.

Also, Sharaf-Eldeen et al. (1980) developed a model for fully exposed ear corn drying by using the general form of the solution to the diffusion equation and determined drying parameters.

They commented that their model should be beneficial in describing bulk drying of ear corn. Thompson et al. (1968), Hall (1971), Flood et al. (1972), Misra and Brooker (1980), and Li and Morey (1984) used equilibrium moisture content ($M_e$) when defining the moisture content ratio (MR). However, Hustrulid and Flikke (1959) and some other investigators introduced the dynamic equilibrium moisture content ($M_s$) rather than equilibrium moisture content ($M_e$) into models for determining drying rate of the thin-layer drying.

Misra and Brooker (1980) conducted their own experimental work for shelled yellow dent corn and compiled all useful data on thin-layer drying and rewetting in the temperature range of 2.2°C to 71.1°C from other sources.

They identified the following model as the most promising one through the preliminary analysis, then expressed the drying constants K and N as a function of air temperature (T), air humidity (H), air velocity (V), and the initial moisture content ($M_o$).

$$MR = \frac{M - M_e}{M_o - M_e} = \text{Exp}\,[-\,Kt^N] \quad (2)$$

where

MR = moisture content ratio
M = moisture content (%, d.b.)
$M_e$ = equilibrium moisture content (%, d.b.)
t = time.

Through the statistical analysis of the combined

**Table 10.** Drying constants for yellow dent corn.

| Drying constants | K | N | $R^2$ |
|---|---|---|---|
| Model I: | $K = \text{Exp}[-a+b\ln(1.8T+32) + CV]$ | $N = d\ln(H)+eM_0$ | |
| Drying | a = 7.1735<br>b = 1.2793<br>c = 0.1378 | d = 0.0811<br>e = 0.0078 | 0.967 |
| Rewetting | a = 8.5122<br>b = 1.2178<br>c = 0.0864 | d = 2.1876<br>e = −0.0167 | 0.991 |
| Model II: | $K = a + bT + cT^2$ | $N = d + eT + fT^2$ | |
| Drying (1) | $a = 2.216 \times 10^{-2}$<br>$b = 1.113 \times 10^{-4}$<br>$c = 3.435 \times 10^{-6}$ | $d = 0.5409$<br>$e = 1.498 \times 10^{-3}$<br>$f = 2.561 \times 10^{-6}$ | 0.997 |
| Drying (2) | $a = 1.026 \times 10^{-2}$<br>$b = 2.651 \times 10^{-4}$<br>$c = 2.820 \times 10^{-6}$ | $d = 0.6057$<br>$e = 1.568 \times 10^{-4}$<br>$f = 9.601 \times 10^{-6}$ | 0.981 |
| Model III: | $K = a + bT^2 + cTM_0$ | $N = d + eM_0^2 + fT^2$ | |
| Drying (3) | $a = 1.091 \times 10^{-2}$<br>$b = 2.767 \times 10^{-6}$<br>$c = 7.286 \times 10^{-6}$ | $d = 0.5375$<br>$e = 1.141 \times 10^{-5}$<br>$f = 5.183 \times 10^{-5}$ | 0.975 |

T = air temperature (K), V = air velocity (m/s), H = air humidity (%), $M_0$ = initial moisture content (%, dry basis)
Sources: Misra and Brooker (1980), Li and Morey (1984).

data obtained from their own and other investigators' experimental work, they determined the values for the drying constants K and N shown in Table 10.

They concluded that drying air temperature and the air velocity significantly affect the parameter K in the drying phase equation, and the air temperature and the initial moisture content significantly affect the parameter K in the thin-layer rewetting equation.

On the other hand, Li and Morey (1984) determined the drying constants K and N for yellow dent corn harvested in two consecutive years, by an approach similar to that used by Misra and Brooker (1980).

After testing the effect of drying air temperature, airflow rate, initial moisture content, and relative humidity on thin-layer drying rate, they concluded that drying air temperature had the greatest effect on drying rates, and airflow rate and relative humidity had smaller effects.

In addition to the above, they suggest that relative humidity can probably be neglected when developing thin-layer models for use in deep-bed drying simulation.

Few studies have focused on drying and rewetting rates in corn drying. Similarly, there are few references to drying and rewetting constants or rates for rice or rough rice.


## Acknowledgments

The authors express appreciation to the Food and Feed Grain Institute, Kansas State University, for partial support through USAID/DSAN-CA-0256.


## References and Bibliography


Agrawal, K.K., and Chand, P. 1974. Pressure drop across fixed beds of rough rice. Transactions American Society of Agricultural Engineers, 17(3), 560–563.

Agrawal, K.K., Nelson, G.L., and Clary, B.L. 1969. An investigation into the theories of desorption isotherms for rough rice and peanuts. American Society of Agricultural Engineers, Paper No. 69–890.

Agrawal, Y.C., and Singh, R.P. 1977. Thin-layer drying studies on short-grain rough rice. American Society of Agricultural Engineers, Paper No. 77–3531.

Alam, A., and Shove, G.C. 1973. Hygroscopicity and thermal properties of soybeans. Transactions American Society of Agricultural Engineers, 16(4), 707–709.

Aldis, D.F., and Foster, G.H. 1980. Moisture changes in grain from exposure to ambient air. Transactions American Society of Agricultural Engineers, 23(3), 753–760.

American Society of Agricultural Engineers, 1984. American Society of Agricultural Engineers Yearbook, 294–306.

Arora, V.K., Henderson, S.M., and Burkhardt, T.H. 1973. Rice drying cracking versus thermal and mechanical properties. Transactions American Society of Agricultural Engineers, 16(2), 320–323, 327.

Babbitt, E.A. 1945. The thermal properties of grain in bulk. Canadian Journal of Research, F23, 388–401.

Bakker-Arkema, F.W., and Patterson, R.J. 1967. Pressure drop-airflow relationship in granular materials such as grain. American Society of Agricultural Engineers, Paper No. 67–840.

Bakker-Arkema, F.W., Patterson, R.J., and Bickert, W.G. 1969. Static pressure airflow relationship in packed beds of granular biological materials such as cherry pits. Transactions American Society of Agricultural Engineers, 12(1), 134-136, 140.

Becker, H.A. 1960. On the adsorption of liquid water by the wheat kernel. Cereal Chemistry, 37(3), 309-323.

Becker, H.A., and Sallans, H.R. 1956. A study of the desorption isotherms of wheat at 25°C and 50°C. Cereal Chemistry, 33(2), 79–90.

Bern, C.J., and Charity, L.F. 1975. Airflow resistance characteristics of corn as influenced by bulk density. American Society of Agricultural Engineers, Paper No. 75–3510.

Berry, M.R. 1973. Moisture adsorption isotherms for selected foods and ingredients. Transactions American Society of Agricultural Engineers, 16(1), 137–139.

Bloome, P.D., and Shove, G.C. 1972. Simulation of low-temperature drying of shelled corn leading to optimization. Transactions American Society of Agricultural Engineers, 15(2), 310–316.

Boyce, D.S., and Davis, J.K. 1965. Pressure losses and drying rate in grain ventilated with various on-floor duct systems. Journal of Agricultural Engineering Research, 10(4), 271–276.

Brooker, D.B., Bakker-Arkema, F.W., and Hall, C.W. 1974. Drying cereal grains. Westport, CT, AVI Publishing Co Inc, 265p.

Browne, D.A. 1962. Variation of the bulk density of cereals with a moisture content. Journal of Agricultural Engineering Research, 7(4). 288–290.

Brunauer, S.P., Emmett, P.H., and Teller, E. 1938. Adsorption of gases in multimolecular layers. Journal American Chemical Society, 60, 309–319.

Brusewitz, G.H. 1975. Density of rewetted high moisture grains. Transactions American Society of Agricultural Engineers, 18(5), 935–938.

Bunn, T.M., and Hukill, W.V. 1963. Pressure pattern predictions for nonlinear airflow through porous media. Transactions American Society of Agricultural Engineers, 6(1), 32–36.

Calderwood, D.L. 1966. Use of aeration to aid rice drying. Transactions American Society of Agricultural Engineers, 16(3), 525–527, 532.

Chandra, P., Albright, L.D., and Wilson, G.E. 1977. Pressure drop characteristics of unidirectional air flow through rock beds. American Society of Agricultural Engineers, Paper No. 77–4016.

Chandra, S., and Huir, W.E. 1971. Thermal conductivity of spring wheat at low temperature. Transactions American Society of Agricultural Engineers, 14(4), 644–646.

Chang, G.S., Converse, H.H., and Martin, C.R. 1983. Bulk properties of grain as affected by self-propelled rotational type grain spreader. Transactions American Society of Agricultural Engineers, 26(5), 1543–1550.

Chen, C.S., and Clayton, J.T. 1971. The effect of temperature on sorption isotherms of biological materials. Transactions American Society of Agricultural Engineers, 14(5), 927–929.

Chittenden, D.H., and Hustrulid, A. 1966. Determining drying constants for shelled corn. Transactions American Society of Agricultural Engineers, 9(1), 52–55.

Chung, D.S., Abdelmoshin, and Kim, M.S. 1984. Resistance of grain sorghum to airflow. American Society of Agricultural Engineers, Paper No. 84–3529.

Chung, D.S., and Converse, H.H. 1971. Effect of moisture on some physical properties of grains. Transactions American Society of Agricultural Engineers, 14(4), 612–614, 620.

Chung, D.S., and Pfost, H.B. 1967. Adsorption and desorption of water vapor by cereal grains and their products, part I–III. Transactions American Society of Agricultural Engineers, 10(4), 549–557.

Chung, D.S., Gonzaga, L., Kim, M.S., and Haque, E. 1985. Resistance to airflow through rough rice mixed with foreign material. American Society of Agricultural Engineers, (In press).

Davila, S.I., Chen, L.H., and Welch, G.B. 1983. Simulation model for reversing direction airflow drying of rough rice. Transactions American Society of Agricultural Engineers, 26(6), 1842–1848.

Day, D.L., and Nelson, G.L. 1965. Desorption isotherms for wheat. Transactions American Society of Agricultural Engineers, 8(2), 293–297.

del Giudice, P.M. 1959. Exposed-layer wetting rates of shelled corn. Unpublished M.S. Thesis, Department of Agricultural Engineering, Purdue University.

Desney, R.W. 1954. The specific heat of some cereal grains. Cereal Chemistry, 31(3), 229–334.

Dunstan, E.R., Chung, D.S., and Hodges, T.O. 1973 Adsorption-desorption characteristics of grain sorghum. Transactions American Society of Agricultural Engineers, 16(4), 667–670.

Eman, A., Stroshine, R., Tuite, J., Cantone, F., Kirlies, A., Bauman, L., and Okos, M. 1981. Evaluating drying rate and grain quality parameters of corn inbreds/hybrids methodology (feed processing). American Society of Agricultural Engineers Paper No. 81–3522.

Flood, C.A., Jr, Sabbah, M.A., Meeker, D., and Peart, R.M. 1972. Simulation of a natural-air corn drying system. Transactions American Society of Agricultural Engineers, 15(1), 156–159, 162.

Fortes, M., and Okos, M.R. 1980. Changes in physical properties of corn during drying. Transactions American Society of Agricultural Engineers, 23(4), 1004–1008.

Foster, G.H. 1967. Moisture change during aeration of grain. Transactions American Society of Agricultural Engineers, 10(3), 344–346, 351.

Gallaher, G.L. 1951. A method of determining the latent heat of agricultural crops. Agricultural Engineering, 32, 34–38.

Goss, T.R. 1965. Some physical properties of forage and cereal crop seeds. American Society of Agricultural Engineers Paper No. 65–813.

Grama, S.N. Bern, C.J., and Hurburgh, C.R. 1984. Airflow resistance of mixture of shelled corn and fines.

Transactions American Society of Agricultural Engineers, 27(1), 268–272.
Grieg, D.J. 1970. The determination of the rare constant in thin layer drying of agricultural crops. Journal of Agricultural Engineering Research, 15(2), 106–110.
Gunasekaram, S., Jindal, V.K., and Shove, G.C. 1983. Resistance to airflow of paddy (rough rice) in shallow depths. Transactions American Society of Agricultural Engineers, 26(2), 601–605.
Gustafson, R.J., and Hall, G.E. 1972. Density and porosity of shelled corn during drying. Transactions American Society of Agricultural Engineers, 15(3), 523–525.
Gustafson, R.J., and Hall, G.E. 1974. Equilibrium moisture content of shelled corn from 50°C to 155°C. Transactions American Society of Agricultural Engineers, 17(1), 120–124.
Hall, C.W. 1971. Drying farm crops. Westport, CT, AVI Publishing Co. Inc., p. 245–279.
Hall, C.W. and Rodriguez-Arias, J.H. 1958. Equilibrium moisture content of shelled corn. Agricultural Engineering, 39, 466–470.
Haque, E., Ahmed, Y.N., and Deyoe, C.W. 1982. Static pressure drop in a fixed bed of grain as affected by grain moisture content. Transactions American Society of Agricultural Engineers, 25(4), 1095–1098.
Haque, E., Chung, D.S, and Foster, G.H. 1981. Pressure and velocity field in airflow through packed bed of corn mixed with fines under non-Darcy flow conditions. Transactions American Society of Agricultural Engineers, 24(6), 1595–1599, 1604.
Haque, E., Chung, D.S., and Foster, G.H. 1981. Analytical solution of the pressure and velocity field in steady-state incompressible axisymmetric Darcy flow through granular materials such as corn. Transactions American Society of Agricultural Engineers, 24(6), 1618–1620.
Haque, E., Foster, G.H., Chung, D.S., and Lai, F.S. 1978. Static pressure drop across a bed of corn mixed with fines. Transactions American Society of Agricultural Engineers, 21(5), 997–1000.
Haswell, G.A. 1954. A note on the specific heat of rice, oats and their products. Cereal Chemistry, 31(4), 341–343.
Heldman, D.R. 1975. Food process engineering, Westport, CT, AVI Publishing Co. Inc.
Henderson, S.M. 1943. Resistance of shelled corn and bin walls to air flow. Agricultural Engineering, 24(1), 367–369, 374.
Henderson, S.M. 1952. A basic concept of equilibrium moisture. Agricultural Engineering, 33, 39–32.
Henderson, S.M., Pabis, S. 1962. Grain drying theory (IV: the effect of airflow rate on the drying index). Journal of Agricultural Engineering Research, 7(2), 85–89.
Hosokawa, A., et al. 1981. Agricultural process machinery. Tokyo, Bu-Eido Publishing Co., 277 p.
Hosokawa, A., and Matsumoto, H. 1971. Bulk thermal conductivity of grains. The Journal of the Society of Agricultural Machinery, Japan, 32(4), 302–305.
Hubbard, T.E., Earle, F.R., and Senti, F.R. 1957. Moisture relations in wheat and corn. Cereal Chemistry, 34(6), 422-433.
Hukill, W.V., and Schmidt, J.L. 1960. Drying rate of fully exposed grain kernels. Transactions American Society of Agricultural Engineers, 3(2), 71–77, 80.
Hukill, W.V., and Shedd, C.K. 1955. Nonlinear airflow in grain drying. Agricultural Engineering, 36, 462–466.
Hunter, A.J. 1983. Pressure difference across an aerated seed bulk for some common duct and store cross-sections. Journal of Agricultural Engineering Research, 28, 437–450.
Husain, A., and Ojha, T.P. 1969. Resistance to the passage of air through rough rice. Journal of Agricultural Engineering Research, 14(1), 47–53.
Hustrulid, A. 1962. Comparative drying rates of naturally moist, remoistened and frozen shelled corn. Transactions American Society of Agricultural Engineers, 5(1), 64–67.
Hustrulid, A., and Flikke, A.M. 1959. Theoretical drying curve for shelled corn. Transactions American Society of Agricultural Engineers, 2(1), 112–114.
Huzayyin, A.S. 1969. Cooling and drying of high-moisture shelled corn in deep beds. M.S. Thesis, Department of Agricultural Engineering, Kansas State University.
Johnson, H.K., and Dale, A.C. 1954. Heat required to vaporize moisture in wheat and shelled corn during the process of artificial drying. Agricultural Engineering, 35(10), 705–709, 714.
Kachru, R.P., and Matthes, R.K. 1976. The behaviour of rough rice in sorption. Journal of Agricultural Engineering Research, 21(4), 405–416.
Kazarian, E.A., and Hall, C.H. 1965. Thermal properties of grain. Transactions American Society of Agricultural Engineers, 8(1), 33–37, 48.
Kim, M.S. 1980. Physical and thermal properties of grain. Ph.D. Thesis, Department of Agricultural Engineering, Seoul National University, Suweon, Korea.
Kunze, O.R., and Hall, C.R. 1967. Moisture adsorption characteristics of brown rice. Transactions American Society of Agricultural Engineers, 10(3), 448–450, 453.
Lai, F.S., and Fan, L.T. 1976. Excessive pressure drop through a bed of stored grain during aeration (research notes). Journal of Agricultural Engineering Research, 21(2), 207–209.
Lampman, W.P., and Hukill, W.V. 1968. Air resistance of perforated metal supporting grains. Transactions American Society of Agricultural Engineers, 11(6), 837–839, 843.
Li, H., and Morey, R.V. 1984. Thin-layer drying rate of yellow dent corn (as affected by drying air temperature, air-flow rate, initial moisture content, and relative humidity). Transactions American Society of Agricultural Engineers, 27(2), 581–585.
Matthes, R.K., and Kachru, R.P. 1972. Estimating the drying rate of grain using supplemental heat. American Society of Agricultural Engineers, Paper No. 72–324.
Matthies, H.J., and Petersen, H. 1974. New data for calculating the resistance to air-flow of stored materials. Transactions American Society of Agricultural Engineers, 17(6), 1144–1149.
Misra, M.K., and Brooker, D.B. 1978. Thin layer drying of shelled corn. American Society of Agricultural Engineers, Paper No. 78-3002.
Misra, M.K., and Brooker, D.B. 1980. Thin layer drying and rewetting equations for shelled yellow corn.

Transactions American Society of Agricultural Engineers, 23(5), 1254–1260.
Mohsenin, N.N. 1965. Physical properties of agricultural products. Transactions American Society of Agricultural Engineers, 8(1), 25–29.
Mohensin, N.N. 1970. Physical properties of plant and animal materials. Volume 1: Structure, physical characteristics and mechanical properties. New York, Gordon and Breach Science Publishers.
Morey, R.V., and Peart, R.M. 1971. Optimum horsepower and depth for a natural air corn drying system. Transactions American Society of Agricultural Engineers, 14(5), 930–934.
Morita, T., and Singh, R.P. 1979. Physical and thermal properties of short-grain rough rice. Transactions American Society of Agricultural Engineers, 22(3), 630–636.
Muchiri, G. 1969. Resistance to air flow through shelled corn. M.S. Thesis, Iowa State University.
Nellist, M.E., and Rees, D.V.H. 1969. The resistance to airflow of dry and soaked vegetable seeds. Journal of Agricultural Engineering Research, 14(4), 344–349.
Nelson, S.O. 1980. Moisture dependant kernel and bulk density relationships for wheat and corn. Transactions American Society of Agricultural Engineers, 23(1), 139–143.
Osborne, L.E. 1961. Resistance to airflow of grain and other seeds. Journal of Agricultural Engineering Research, 6(2), 119–122.
Othmer, D.F. 1940. Correlating vapor pressure and latent heat data. Journal of Industrial and Engineering Chemistry, 32, 841–856.
Pabis, S., and Hall, C.W. 1962. Temperature distribution and moisture movement while drying an ear corn. Transactions American Society of Agricultural Engineers, 5(2), 168–171.
Pabis, S., and Henderson, S.M. 1962. Grain drying theory (III: the air/grain temperature relationship). Journal of Agricultural Engineering Research, 7(1), 21–26.
Pfost, H.B., Maurer, S.G., and Chung, D.S. 1976. Summarizing and reporting equilibrium moisture data for grains. American Society of Agricultural Engineers, Paper No. 76–3520.
Rodriguez-Arias, J.H. 1956. Desorption isotherms and drying rate of shelled corn, 40 to 140°F. Unpublished Ph.D. Thesis, Michigan State University.
Rodriguez-Arias, J.H., Hall, C.H., and Bakker-Arkema, F.W. 1963. Heat of vaporization for shelled corn. Cereal Chemistry, 40, 676–683.
Ross, I.J. and White, G.M. 1972. Thin layer drying characteristics of white corn. Transactions American Society of Agricultural Engineers, 15(1), 175–176, 179.
Rumsey, T., and Singh, R.P. 1978. Pressure drop through a bed of food grains. American Society of Agricultural Engineers, Paper No. 78–6546.
Sabban, M.A. 1968. Prediction of batch drying performance with natural air. Unpublished M.S. Thesis, Purdue University.
Sharaf-Eldeen, Y.I., Blaisdell, J.L, and Hamdy, M.Y. 1980. A model for ear corn drying. Transactions American Society of Agrticultural Engineers, 23(5), 1261–1265, 1271.
Sharma, D.K., and Thompson, T.L. 1973. Specific heat and thermal conductivity of sorghum. Transactions American Society of Agricultural Engineers, 16(1), 114–117.
Sharma, A.D., Kunze, O.R., and Tolley, H.D. 1982. Rough rice drying as a two-compartment model. Transactions American Society of Agricultural Engineers, 25(1), 221–224.
Sharp, J.P. 1982. A review of low temperature drying simulation models. Journal of Agricultural Engineering Research, 27, 169–190.
Shedd, C.K. 1945. Resistance of ear corn to air flow. Agricultural Engineering, 26(1), 19–20, 23.
Shedd, C.K., 1951. Some new data on resistance of grains to airflow. Agricultural Engineering, 32(9), 493–495, 520.
Shedd, C.K. 1953. Resistance of seeds and grains to airflow. Agricultural Engineering, 34(9), 616–619.
Sheldon, W.H. 1960. Static pressure drop for low air flows through grain. American Society of Agricultural Engineers, Paper No. 60–302.
Sheldon, W.H., Hall, C.W., and Wang, J.K. 1960. Resistance of shelled corn and wheat to low air flows. Transactions American Society of Agricultural Engineers, 3(2), 92–94.
Shove, G.C. 1971. Low temperature drying of shelled corn. Agricultural Engineering, 52(7), 372–373.
Shove, G.C., and Hukill, W.V. 1963. Predicting pressure losses and drying rates in grain ventilation ducts. Transactions American Society of Agricultural Engineers, 6(2), 115–116, 122.
Shove, G.C., and Olver, E.F. 1967. Temperature gradients in drying grain. Transactions American Society of Agricultural Engineers, 10(2), 152–153, 156.
Smith, S.E. 1947. The sorption of water vapor by high polymers. Journal American Chemical Society, 69, 646–651.
Spencer, H.B. 1969. Pressure drop in on-floor duct drying system. Journal of Agricultural Engineering Research, 14(2), 165–172.
Strohman, R.D., and Yoerger, R.R. 1967. A new equilibrium moisture content equation. Transactions American Society of Agricultural Engineers, 10(5), 675–677.
Surniman, E.J., Bodnar, G., and Bates, G.P. 1931. Tests on resistance to the passage of air through rough rice in a deep bin. Agricultural Engineering, 12(5), 145–148.
Thompson, R.A., and Isaacs, G.A. 1967. Porosity determinations of grains and seeds with an air-comparison pycnometer. Transactions American Society of Agricultural Engineers, 10(5), 693–696.
Thompson, H.J., and Shedd, C.K. 1954. Equilibrium moisture and heat of vaporization of shelled corn and wheat. Agricultural Engineering, 35, 786–788.
Thompson, T.L., Peart, R.M., and Foster, G.H. 1968. Mathematical simulation of corn drying. Transactions American Society of Agricultural Engineers, 11(4), 582–586.
Troeger, J.M. and Hukill, W.V. 1971. Mathematical description of the drying rate of fully exposed corn. Transactions American Society of Agricultural Engineers, 14(6), 1153–1156, 1162.
van Rest, D.J., and Isaacs, G.W. 1966. Exposed-layer drying rates of small grains. Transactions American

Society of Agricultural Engineers, 9(5), 688–689.
van Rest, D.J., and Isaacs, G.W. 1968. Exposed-layer drying rates of grain. Transactions American Society of Agricultural Engineers, 11, 236–239.
Vemuganti, G.R. and Pfost, H.B. 1980. Physical properties related to drying 20 food grains. American Society of Agricultural Engineers, Paper No. 80–3539.
Vemuganti, G.R. Pfost, H.B., Chung, D.S., and Burroughs, R. 1980. Some properties of cereal grains, food legumes and oil seeds (hygroscopic, thermal, physical) affecting storage and drying. Research Report No. 18, Food and Feed Grain Institute, Kansas State University, Manhattan, KS.
Wang, C.Y., and Singh, R.P. 1978. A single layer drying equation for rough rice. American Society of Agricultural Engineers, Paper No. 78–3001.
Wang, C.Y., and Singh, R.P. 1978. Use of variable equilibrium moisture content in modelling rice drying. American Society of Agricultural Engineers, Paper No. 78–6505.
Westerman, P.W., White, G.M., and Ross, I.J. 1973. Relative humidity effect on the high-temperature drying of shelled corn. Transactions American Society of Agricultural Engineers, 16(6), 1136–1139.
White, G.M., Ross, I.J., and Klaiber, J.D. 1972. Moisture equilibrium in mixing of shelled corn. Transactions American Society of Agricultural Engineers, 15(3), 508–509, 514.
White, G.M., Ross, I.J., and Westerman, P.W. 1973. Drying rate and quality of white shelled corn as influenced by dew point temperature. Transactions American Society of Agricultural Engineers, 16(1) 118–120.
Williamson, W.F. 1965. Pressure losses and drying rates in grain ventilated with various on-floor duct systems. Journal of Agricultural Engineering Research, 10(4), 271–276.
Wratten, F.D., Poole, W.D., Chesness, J.L., Bal, S., and Ramarao, V. 1969. Physical and thermal properties of rough rice. Transactions American Society of Agricultural Engineers, 12(6), 801–803.
Zink, F.J. 1935. Specific gravity and air space of grains and seed. Agricultural Engineering, 16(11), 439–440.
Zuritz, C., Singh, R.P., Moini, S.M., and Henderson, S.M. 1979. Desorption isotherms of rough rice from 10°C to 40°C. Transactions American Society of Agricultural Engineers, 22(2) 433–436, 440.

# The Application of Psychrometrics to Grain Aeration

**Robert H. Driscoll***

*Abstract*

Current research on the dynamic relationship between air and grain properties is reviewed. Weather patterns for typical grain crop areas are considered in terms of relative humidity and temperature. There is a conflict between the economic benefits of drying grain slowly and the quality benefits of drying grain quickly. The optimum balance between the two should be sought. For example, for tropical conditions it is essential to dry high-moisture grains to a lower moisture content immediately. Great advances have recently been made in our understanding of grain/air equilibrium, yielding more accurate models of the isothermal behaviour of grain in contact with air. Such models are of great importance in developing accurate grain-drying models for use in simulation studies of grain-drying.

Daily profiles of relative humidity and temperature demonstrate that, for near-ambient drying methods, consideration of the time of day is of great importance in developing an economical drying strategy. During daylight, there is a substantial drop in relative humidity, and the ambient air is better for removing moisture. Various slow-dryer management systems are considered from the point of view of air usage, especially time control and relative humidity control. Convenient methods for representing and visualising grain drying in terms of psychrometric charts have provided simple methods of studying drying strategies. Single-grain drying, thin-layer drying, and deep-bed drying can all be represented, and charts of equilibrium front speeds allow quick estimates of drying times to be made for a range of conditions.

CONSIDERABLE research has been directed towards the problem of removing moisture from grains in tropical and subtropical climates, as evidenced by work by Renwick (1984) in Indonesia, Wahab and Dhiauddin (1984) in Malaysia, and Brooker et al. (1974) in America, and work on developing computer simulation models, as for example that of Bowden et al. (1983). The vast majority of the world's rice crops are grown in hot, humid climates where drying of the grain after harvesting is difficult. Traditionally, sun-drying is used as the solution, involving spreading the grain in thin layers on concrete or on the roadside in order to gain maximum drying benefit from the radiant and convective heat of the sun. This method is still the most commonly used method for drying crops in the world today. However, the method suffers from serious quality problems, such as suncheck-ing and discoloration, and is not a sufficiently effective drying technique for the wet season (Mendoza et al. 1981). Many areas of the tropics are now moving towards increased grain production in the dry season, for harvest during the wet season, aided by the introduction of irrigation systems. The increased land utilisation is a vital step towards grain self-sufficiency for these countries.

This trend has enormously increased the grain drying load in Southeast Asia. There is growing concern over the high proportions of grain lost due to quality problems during the wet season (Wahab and Dhiauddin 1984). The requirement of increased technology to handle the problems generated by the wet season harvest has led to new research interest in the relationship between psychrometrics and drying theory.

The following paper reviews progress in the understanding of the interactions between air and grain (Brooker et al. 1974). The areas of moisture movement in grain, and heat and mass transfer in grain drying have been avoided, being areas more relevant to forced convection modelling and grain properties. Therefore, the main emphasis is on the application of psychrometrics to the grain-drying problem, using models of grain properties from the literature without explanation or development.

* School of Food Science and Technology, University of New South Wales, P.O. Box 1, Kensington, N.S.W. 2033, Australia.

More detail can be found in reviews such as that of Sharp (1982).

The presentation has been structured as follows. Firstly, the interaction between air and grain moisture is considered for the equilibrium situation. The concepts of moisture content levels, isotherms, equilibrium relative humidities, and forms of equilibrium are prerequisites to air/grain studies, and great advances in our understanding in these areas have been made. The manner in which a grain loses its moisture is then considered, looking at the theory of two-stage drying curves, psychrometric drying plots, critical air speeds, and current models of the thin-layer drying curve. This theory is then applied and extended to a stationary deep bed, considering the drying advantages of a deep bed and the concept of moisture and enthalpy fronts. Variations in weather properties are analysed as regards their effects on an aerated grain mass. Finally, there is an overview of drying strategies, including those involving time of day and relative humidity control.

## Air/Grain Equilibrium

The moisture equilibrium characteristic for a particular grain species constitutes one of the two most vital thermodynamic properties of the grain as regards drying (Driscoll and Adamczak, unpublished report; Dung et al. 1980; Pixton 1983). The other is the rate of moisture loss of a thin layer of grain in an airstream of known temperature and humidity. The equilibrium moisture content of a food product may be expressed as a function of three variables, the atmospheric relative humidity, the product temperature, and the moisture history of the product. The last factor leads to effects such as hysteresis.

Grain is a hygroscopic material (Karon and Adams 1949), so a sample of grain will adsorb or release moisture in such a way as to come to equilibrium with its surrounding environment. Observations of the approach to equilibrium led many early researchers to the concept of dynamic moisture equilibrium, a concept since refuted by more detailed studies (Bakker-Arkema and Hall 1965). The approach to equilibrium is limited by the moisture transfer rate within the kernel (Pixton and Warburton 1968), discussed later. For a given temperature, the air relative humidity can be plotted against the moisture content of the product, by conditioning the grain gradually to different relative humidity environments and measuring the grain moisture content after equilibrium is reached. This moisture content is called the equilibrium moisture content, and the resulting plots are isotherms (Labuza 1968). Many models have been developed to describe the relationship between relative humidity and product moisture content, based on physical theories of moisture transfer or on empirical models but, because of the many different physical mechanisms involved, none is generally applicable. Examples are the BET isotherms (e.g. Day and Nelson, 1965) for low relative humidities and the Langmuir isotherms for higher relative humidities. In practice, due to the inadequacies of the physical theories, grain researchers have tended to develop their own semi-empirical models, applicable over their range of interest (e.g. Sutherland et al. 1971; Putranon et al. 1979).

If a sample of grain at a given moisture content is exposed to a small, fixed volume of air in a closed system, it takes up or releases moisture in order to change the absolute humidity of its surrounding air and so reach equilibrium. There are no substantial changes in the grain moisture content because of the large capacity of the grain to hold moisture compared with the capacity of the air. This defines the air equilibrium relative humidity for that moisture content. Expressed as a fraction, it is called the water activity of the product.

An adsorption isotherm is constructed by placing dried grain into an atmosphere of higher relative humidity than the product equilibrium relative humidity, and raising the air humidity in successive stages until saturation is reached, which takes about seven days according to Pixton and Warburton (1968), observing the product moisture content at each stage. Similarly, a desorption isotherm is obtained by successive reductions in relative humidity from saturation. The resulting curves display the characteristic sigmoid shape (Fig. 1). The adsorption isotherm gives a higher equilibrium relative humidity for a given moisture content than a desorption isotherm. The area enclosed between the two curves is related to structural changes occurring in the surface of the grain, as the moisture condensing on the surface does work on the grain structure, opening up additional polar sites for condensation in subsequent cycles (Chung and Pfost 1967; Treybal 1981). This effect is called hysteresis. Repeated cycling between wet and dry conditions reduces

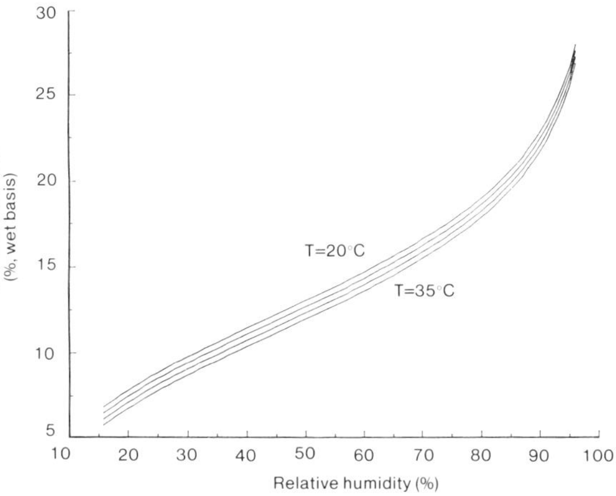


**Fig. 1.** Isothermal plots for rice variety Inga (from Putranon et al. 1979).

the hysteresis effect until, after a few cycles, it disappears completely (Chung and Pfost, 1967). However, the structural changes reduce the milling yield of the grain (Kunze and Choudhury 1972; J. Darby, pers. comm. ; Chung et al. 1984), so cycling is preferably avoided. The hysteresis effect means that the history of the grain since harvesting is significant in terms of the thermophysical and milling behaviour of the grain.

The effect of temperature on the air/grain equilibrium can be demonstrated by plotting several isotherms on the same plot. As the air temperature increases, the equilibrium relative humidity increases for a given grain moisture content. However, over a large range, the effect of temperature is small for all grains.

There are three important forms of equilibrium between a grain and its environment, and within the grain itself. The first is thermal equilibrium. It takes from about four minutes for the kernel centre to reach ambient temperature after a sudden change (Brooker et al. 1974, p. 189). The time required for moisture equilibrium varies from hours to days, or even weeks, depending on factors such as the grain type, whether the grain is husked or not, and temperature (Hsu 1984; Steffe and Singh 1979). Stress relaxation is the slowest form of equilibrium, taking over a month for grain subjected to a high vapour pressure drying gradient. A direct consequence of this factor is that, for experimental work, grains which require milling should be left for a month before milling in order to ensure repeatability of the experiments (Blakeney 1976; Chesterfield and Abbott 1974), and for commercial operations, tempering the stresses for a month ensures the maximum milling yield. For present purposes, temperature equilibrium can be assumed, since the time required for temperature equilibrium is small compared with the time required for drying. Stress equilibrium can safely be ignored, since it has no direct relationship with the grain/air interaction of stored grain. Thus, only moisture equilibrium is of direct concern in drying studies, and in many cases at sufficiently low airspeeds, even moisture equilibrium can be assumed during drying, as done by Sutherland et al. (1971), Bloome and Shove (1971), and many others.

An alternative way of plotting air/grain equilibrium data is to plot the data on psychrometric charts as lines of constant moisture content or isosteres, as in Figure 2. Such plots demonstrate that over substantial changes in temperature, lines of constant moisture content follow the lines of constant relative humidity, a reflection of the small effect temperature has on the water activity level in a grain. For example, for rice, a 10 degree change in temperature corresponds to a 2% change in relative humidity at a moisture content of about 16% wet basis. As the moisture content decreases, the effect of temperature becomes more significant.

The grain when harvested is initially not at a uniform moisture content, due to variations in position of the panicles, received sunlight, and maturity (Chau and Kunze 1982). However, except for loss of milling quality due to rewetting of the driest grains by moisture redistribution from wetter grains, this has no practical effect, the grains behaving as if the whole mass were at the

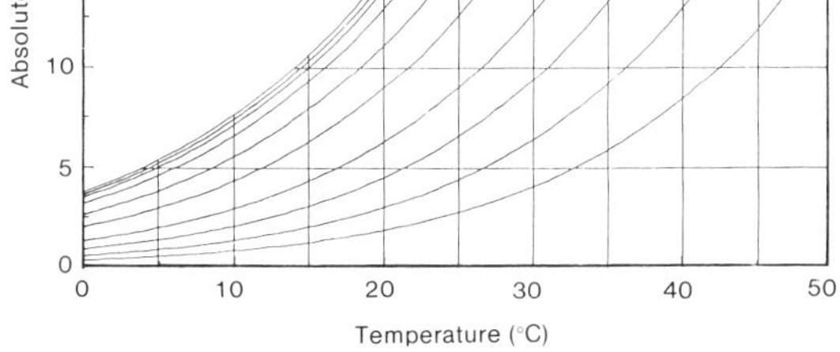


**Fig. 2.** Psychrometric plot of lines of constant moisture content for rice (from Putranon et al. 1979).

average moisture content. Thus, the isothermal data allow prediction of the drying behaviour of the grain mass. Berry and Dickerson (1973) presented isothermal data for a selection of different grains, including oats, corn, and soybeans. Over the past 20 years, large numbers of researchers have obtained detailed isothermal data for many different cereals (see, e.g., Zuritz et al, 1979; Putranon et al. 1979; Chung et al. 1984).

### Drying Zones and Air Properties

As for the isothermal data for a particular grain, there is a substantial literature on the drying rates of grain. Early literature (and occasional later papers) assumed that the application of thin-film drying equations was valid [see literature survey in Dung (1980)]. Becker and Sallans (1955) and Becker (1959) showed that diffusion theory could be used to describe moisture movement within the grain, a result confirmed by many subsequent researchers, including Hustrulid and Flikke (1959) for corn, Hukill and Schmidt (1960) for exposed sorghum kernels (they showed that the thin-film model could not be applied), Henderson and Pabis (1961), van Rest and Isaacs (1968) for shelled corn, theoretical studies by Whitaker et al. (1969), Husain et al. (1973), and Hosokawa and Motohashi (1973), tempering studies based on diffusion theory by Steffe and Singh (1979), a combination semi-empirical/diffusion/thin-film model by Dung et al. (1980), Bakshi and Singh (1980) for parboiled rice, and Aguerre et al. (1982) for rough rice. Sharma et al. (1982) showed that a two-compartment model was necessary for accurate drying rate prediction. Suarez et al. (1982) studied the application of different elemental geometries to diffusion analysis of soybeans, rough rice, wheat, and peanuts. Steffe and Singh (e.g. Steffe and Singh 1980, 1982) have done considerable research into diffusion drying behaviour for grains.

A wet, hygroscopic solid exposed to a steady-state drying regime dries in two distinct stages, separated by transitional periods (Treybal 1981). The stages are the constant drying rate period and the falling rate period. The fine detail of the mechanisms of moisture movement within the grain are not within the province of the current topic, but the two stages are briefly outlined as necessary background to the role of drying rates in air/grain interaction.

The first stage involves the removal of unbound moisture, that is, all moisture contained in the product above the critical moisture content (Treybal 1981). The critical moisture content, which is temperature dependent, is defined as the minimum moisture content at which the solid exerts a vapour pressure equal to the saturation partial pressure of the ambient air, so that the equilibrium relative humidity of the product is 100%. However, for grains, the drying rate is observed to be constant until the equilibrium relative humidity drops below about 75% (Brooker et al. 1974, chapter 8). During this stage of drying, moisture is removed at a constant rate from the solid, due to the steady state balance which arises between removal of heat from the solid by evaporation of moisture and the convective transferral of heat from the air to the solid. Provided the air flow is sufficient to remove the evaporated moisture, the product temperature drops during the initial transient period to the air wet bulb temperature and stays at that temperature throughout the constant drying rate period, an effect observed in grain by many researchers well before grain-drying theories were very advanced (Boyce 1966; Person et al. 1966; Cromarty 1964). The minimum critical air speed depends on factors such as the air humidity and temperature, but is an order of magnitude less than air speeds used in slow dryers.

As the moisture on or near the surface is removed, dry spots start appearing on the grain and the rate of evaporation starts to decrease. Successive new balance points are reached between the rate of heat loss through vaporisation of moisture and the rate of heat transfer from the air, so the product temperature rises accordingly. After a transitional time of variable length, a second stage called the falling rate period begins and continues for the remainder of the drying process. The rate of moisture removal is increasingly controlled by the rate of moisture diffusion through the grain, either by vapour or liquid diffusion. The rate of diffusion in turn is governed by the structure of the product, the internal temperature, and the moisture gradients within the product.

The whole drying process can be plotted on a psychrometric chart of temperature against absolute humidity, as in Figure 3. The condition of the grain is plotted on the chart by its dry bulb temperature and equilibrium moisture content. The initial transition stage to the constant rate

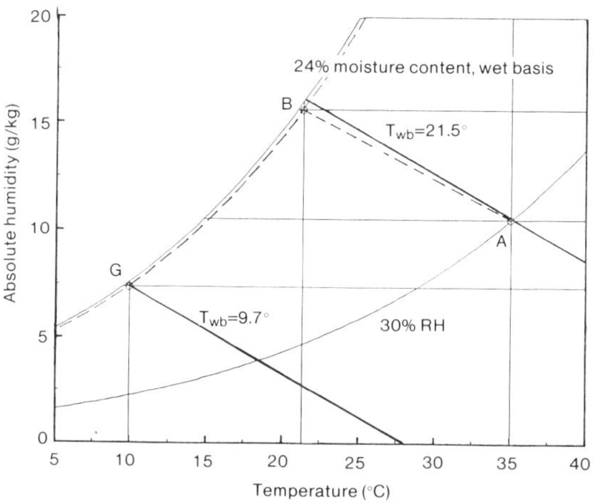


**Fig. 3.** Psychrometric representation of grain drying. A, inlet air conditions; B, intermediate grain state; G, initial grain conditions.

period consists of an enthalpy equilibrium being established between the grain and the air. On the psychrometric chart, the grain moves from its initial temperature and equilibrium relative humidity along a line of approximately constant moisture content until its temperature matches the wet bulb temperature of the air. If sufficient moisture is present, the grain moves along the 100% humidity line. The remainder of the drying process for steady state conditions is an adiabatic process and can be approximately represented on the chart by constructing the wet bulb temperature line from the inlet air point to its intercept with the 100% humidity line. The product stays at the intercept point during the constant rate period, and moves towards the air point during the subsequent transitional stage and final falling rate period. The state of the grain at any point can thus be conveniently represented by its temperature and its equilibrium relative humidity.

From the preceding discussion for a thin layer of grain, above a certain critical air speed which is the minimum sufficient to remove the evaporated moisture from the immediate vicinity of the product, no part of the drying curve is affected by the air speed, apart from minimal changes in the air/product heat transfer coefficient (Treybal 1981, p. 675–676; Henderson and Pabis 1962). For grains, field drying is generally practiced, so that the harvested moisture content is generally less than the critical moisture content, and the product possesses only bound moisture. For example, for rice the average harvest moisture content in Thailand is about 21% wet basis, well below the critical moisture content of about 30% wet basis. Acute problems occur in wet season harvests when this is not true.

## Deep Beds of Grain

For a thin layer during the falling rate period, the humidity of the exit air decreases as the product dries, so that the effective utilisation of the air passing through the grain decreases. Consequently, there are significant advantages both in handling and in air economy in drying deep beds of grain (Thompson et al. 1968), since the exit air from one layer passes into the next layer, adsorbing more moisture as it moves through the bed.

Thus, for a deep bed product, the situation is more complicated than for thin-layer drying. For an initially uniform bed subject to uniform air inlet conditions, the air will remove moisture first from the grain nearest the inlet, called the inlet layer. As the inlet layer dries and the air passing out of it drops in relative humidity, more moisture will be picked up from the second layer. This is the conceptual basis of Thompson's model of grain drying. Theoretical modelling results indicate that, for constant inlet conditions, a drying front is

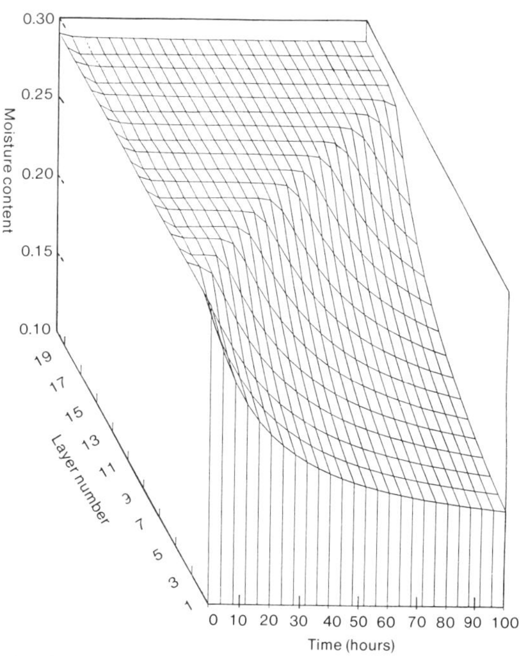


**Fig. 4.** Representation of deep-bed drying under constant inlet conditions: a plot of time versus layer number versus moisture content.

formed which moves slowly through the bed, the width of the front depending on the air speed and the difference between the initial bed equilibrium relative humidity and the inlet air relative humidity (Barre et al. 1971). Until the drying front leaves the deep bed, the exit air leaves at the humidity of the exit layer grain, so that the air passing through a deep bed is utilised in the most effective way, as shown by Figure 4. This diagram is a three-dimensional plot of time versus layer number versus moisture content, showing the complete moisture profile history of a grain mass during drying, as predicted by a near-equilibrium model. There are three main zones depicted in the diagram: the initially high moisture grain (the plateau in the top left of the diagram); the drying front (the scalloped zone); and the dried grain (seen as a new plateau area forming in the bottom right of the diagram). Slight evidence of the temperature front moving through the grain mass very quickly can be seen in the difference between the bed profiles at time zero and 4 hours after start of drying. Figure 5 shows the same drying situation, but from the point of view of the temperature profile as a function of time. The temperature front is dramatically evident in this plot.

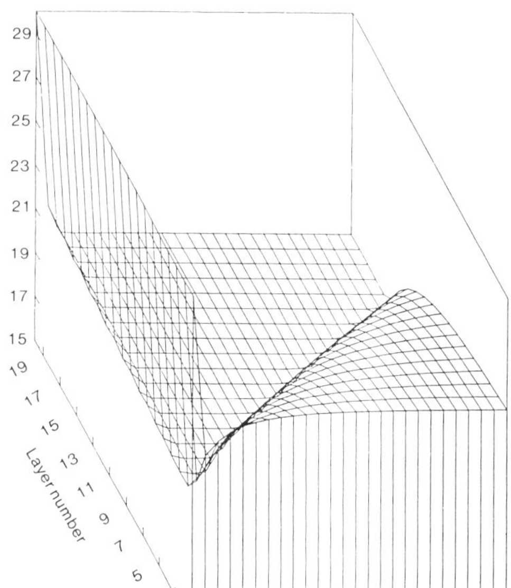


**Fig. 5.** Representation of deep-bed drying under constant inlet conditions: a plot of time versus layer versus temperature.

In contrast to thin-layer drying, air speed now becomes crucial. At slow airflow rates the air economy is highest and the pressure drop across the bed is least, the fans doing less work to force the same quantity of air through the grain, whereas at high flow rates, the drying front broadens as the rate of diffusion of moisture within the grains becomes predominant, so that the air is not given sufficient time to remove the moisture as it moves through the bed. Further, the pressure drop across the bed increases more quickly than the air speed. At low airflow rates the pressure drop is linear with speed, but at higher air speeds, a quadratic term becomes dominant (Bird et al. 1962; Teter 1982; University of New South Wales, Sydney, unpublished data).

Consequently, careful consideration must be given to designing a grain drying system, in order to locate the optimum balance between removing the moisture quickly for quality reasons, and using slow flow rates for energy conservation.

As with single layer drying, the deep bed drying process can be represented on a psychrometric chart. This first became possible around 1971 with the development of the equilibrium model of Sutherland et al. (1971), applied to wheat in Australia. It was found that, at low flow rates, the simultaneous heat and mass transfer operations occurring in a deep bed of biological products could be conveniently represented by wave theory developed by Aris and Amundsen (1973) and applied by Banks (1972). Assuming equilibrium at every point in the bed between the air and the grain temperature and moisture content, the drying process is summarised by a mass conservation and an enthalpy conservation equation, the solution of which consisted of two waves which pass through the bed, an enthalpy wave (or 'front'), and a moisture wave. The enthalpy wave normally passes through the bed at about 30 to 60 times the speed of the moisture front, the main effect on the grain being a change in its temperature. The moisture front was a change in grain moisture content which occurs at constant enthalpy. Thus, if the assumption of equilibrium is valid, the drying of a deep bed under constant conditions can be conveniently represented by a curve from the initial grain condition point along the constant moisture content line, and the wet bulb temperature line through the initial air point. The two curves intersect at a point called the B zone (the inlet air being the A zone and the initial grain state

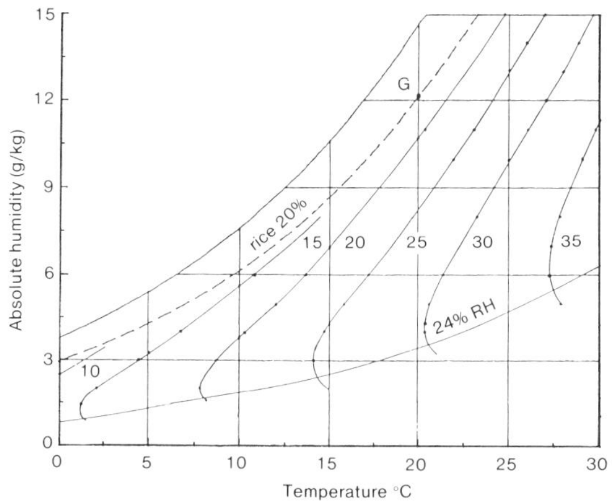


**Fig. 6** Psychrometric plot of lines of constant moisture front speeds for rice at 20% moisture content (dry basis).

being the C zone). The above method is only an approximation even if the equilibrium assumption is valid, since in practice small changes in moisture content (up to 1 or 2%) do occur over the enthalpy front and, conversely, small changes in enthalpy occur over the moisture front. However, for conceptual convenience, the above is sufficiently accurate. The above method of representation can be extended to non-equilibrium situations, provided it is clear that a single point on the drying curve no longer represents both the air and the grain state at a single point in the grain bed. In the non-equilibrium case, for a given grain point, the air point will be further along the curve towards the inlet air conditions.

Figure 6 is an example of a type of graph which can be used for estimating front speeds. On a psychrometric chart, the isostere for rice at 20% moisture content (dry basis) has been superimposed. The curves labelled 10, 15, 20, 25, 30, and 35 correspond to speeds of moisture fronts if the inlet conditions were on that curve. To obtain the actual speed of the moisture front in m/hour, multiply the number on the curve by the superficial air speed and divide by 104. For example, for rice at 20% moisture content (dry basis) being dried by air at 58% relative humidity (RH) and 20°C, with an airflow rate of 5 m/minute, the moisture front moves at speed:

$$\text{speed} = 25 \text{ (from graph)} \times 5\text{(m/minute)}/104$$

giving 0.0125 m/min or 0.75 m/hour. The graph is valid only for rice at 20% moisture content (dry basis) and under equilibrium conditions. A more general approach to estimating air speeds is described by Bowrey and Driscoll (the proceedings).

The method is also not valid when the process is not adiabatic. Non-adiabatic influences include any additional source of heat (respiration, sun, heating coils) or loss of heat (through bin walls). In the majority of situations these effects can be neglected.

Probably the single most important consequence of the growth of grain drying theory has been in the ability to predict what happens when the ambient conditions fluctuate. Figure 7 shows a drying front entering a grain bed, and conditions being held constant for 2 days. After 2 days, the dry air is replaced by air at a higher relative humidity, causing a 'wetting front' to enter the grain. The wetting front broadens as it enters the grain, so that after 10 days it has caught up with the drying front. The two fronts effectively combine, producing a single weaker drying front. Work by Sutherland et al. (1983) confirms that fronts tend to be additive in effect. This can be most clearly seen in Figure 8. A rice bed initially contains a mixture of rice at different moisture contents, as shown by the bed profile at time 0. Fluctuating ambient conditions

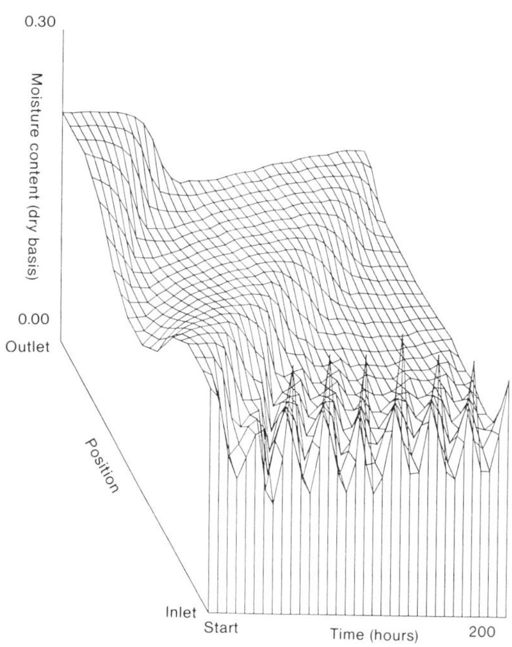


**Fig. 7.** Moisture content profile of a deep-bed of grain for a non-uniform initial bed state and varying inlet conditions.

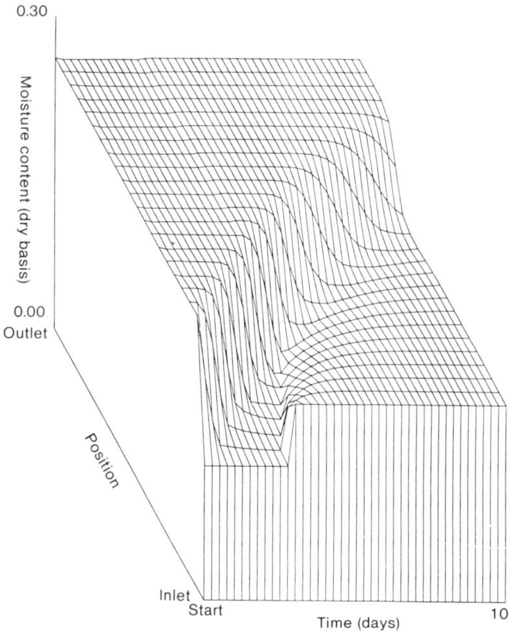


**Fig. 8.** Moisture content profile of a deep bed of grain showing an interacting drying and wetting front.

are then fed into the bed, the inlet conditions being reflected by the sharp peaks at the inlet of the bed. Despite the rapid fluctuations, the oscillations dampen down within a few layers of the inlet, producing comparatively uniform grain conditions. This point will be returned to later.

## Aeration Storage

Once the grain is dried, it must still be aerated occasionally in order to prevent the development of off odours, of hot spots caused by local insect activity, of wet spots caused by circulation currents from hot to cold areas, and of moisture migration caused by contact with atmospheric conditions of different humidity. The air must be chosen according to the climatic conditions. In temperate climates the preferred air may be early morning air. This has a low temperature, which increases the grain storage life (Hukill 1953) and high humidity, which counters the hot, dry summer days that 'suck' the moisture out of the grain, leaving it over-dried). In tropical climates the preferred air may be early evening air, when the air is cooling but still at a lower relative humidity. Ideally for very moist climates, the aeration air should be both cool and dry, and this suggests some dehumidifying system for bulk grain aeration.

Teter (1982) brought together ideas from several different sources, about designing grain storage aeration systems for humid tropical countries. He discussed how the quality parameter called dry matter loss (DML) and the air parameter called the deterioration index could be used to estimate the minimum aeration requirements for grain storage. DML has proven to be a useful quality parameter despite its known limitations and has come to be widely accepted throughout Southeast Asia and the United States. The work by Seib et al. (1980) in developing the DML parameter has resulted in a convenient equation expressing it as a function of temperature, time, and moisture content, allowing simulation studies to estimate the final quality change due to a given drying strategy. Figure 9 is a plot of the value of the DML as a function of time and position in the grain bed for the drying situation discussed in the previous section (see Figs 7 and 8). Once the grain has been dried, the DML parameter changes very slowly, so that grain near the inlet shows a much lower value for DML than grain near the exit, which takes longer to dry.

The deterioration index is a concept developed by Brooks (Mackay and Jamieson 1970), relating

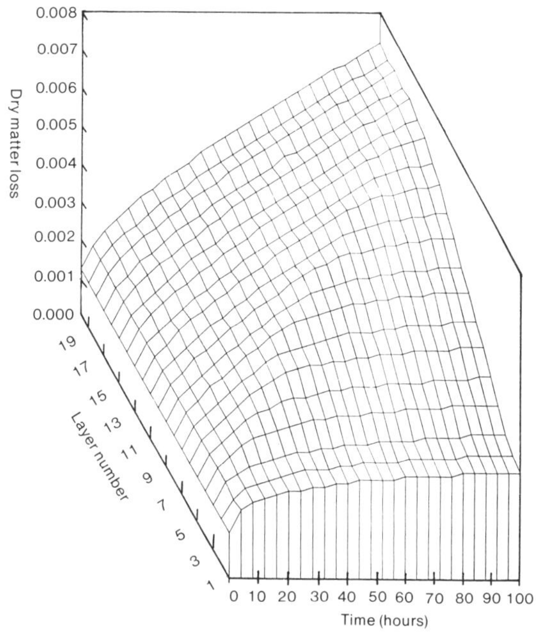


**Fig. 9.** Dry matter loss profile of a deep-bed of grain as a function of time and position during drying.

the weather conditions to the rate of growth of microorganisms and the rate of grain consumption by insects and rodents. It is defined in a similar way to the partial pressure of water vapour in the air, but only the partial pressure component above 65% RH is included in the definition:

$$DI = (\%RH - 65) \times P_{as} \times 10^{-4}$$

where DI is the deterioration index, %RH is the percent relative humidity, and $P_{as}$ is the saturation partial pressure of water vapour expressed in Pascals. For average tropical conditions, the DI varies from 0 to 10. Teter (1982) examined DI values for 13 cities in the Philippines, taking the average monthly climatic conditions. For DIs below a certain critical value that depends only on the grain type, a particular grain can be stored indefinitely, provided that it has been suitably dried. Above the critical DI value, the grain must be aerated or stored in a modified environment.

At present, a few researchers are studying circulation currents in bulk stored grain.

### Daily Weather Variation and the Wet Season

With the assistance of the Australian Department of Science and Technology and the governments of several Southeast Asian countries, extensive weather records have been collected at the University of New South Wales, Australia as part of a collaborative research project on grain drying funded by ACIAR (Mochtar et al. 1985).

Several researchers have attempted to derive mathematical formulae which would allow predictions of weather data based on knowledge of global air movements. Ward and Shapiro (1961) used a power series containing over 2000 terms, coming to the conclusion that only diurnal and annual variation was significant in the observed atmospheric oscillations of air masses. Some research has indicated that recent trends (over the past decade) are significant in predicting annual behaviour, but in general weather prediction remains uncertain and difficult to analyse. Most models are quasi-empirical (Glahn 1965). One of the most successful methods has been the Model Output Statistics method (MOS), in which a statistical relationship is determined between the predicted variable and variable determined to be weather-related, using multiple linear regression (Glahn and Lowry 1972). Most weather models are extremely complex (Balgovind et al. 1983; Yao 1983) and beyond the realms of easy application. In consequence, only diurnal and annual variations will be considered.

Figures 10 to 12 are diurnal plots of temperature and RH patterns for Kuala Lumpur, Bangkok, and Kota Bharu. The plots show the typical diurnal variation of high night RHs but strong solar dependent dips in RH during the day, even during the wet season. Figure 12 shows, in addition to temperature and RH data, diurnal profiles of absolute humidity, and indicates the general stability of absolute humidity. The direct consequence of this is that a temperature control or an RH control will both lead to a constant inlet RH, the most gentle way of drying grain. A temperature control system is an air-conditioning system which modifies the inlet air temperature in order to obtain a constant outlet temperature. An RH control system is one which modifies the inlet air temperature in order to obtain a constant RH.

Note that plots of RH and temperature versus time of day show very little evidence of the effect of the wet season, although it would be expected that higher RHs would be apparent. Furthermore, the monsoon or wet season varies greatly in character from one place to another. In one area it may involve continual rain accompanied by thunderstorms, in another intermittent rain with afternoon downpours but no storms, and so on. In many areas the wet season is not a major effect on the ambient air humidity. At Subang, Malaysia, for example, there are periods during the day when the ambient humidity drops to 60%, but as soon as the sun sets, RH climbs to 100% for the night, all year round.

Thus, for many areas, the RH is closely tied to the time of day. This is a useful factor for near-ambient drying, but the grain must be sheltered from the direct effects of the atmosphere for the wet season, so that sun drying is not possible.

### Drying Strategies

There are many factors involved in choosing what air is used in drying grain at near-ambient conditions. The air itself can be described by five main parameters, and how these five variables are chosen constitutes the selection of an air-drying strategy. The five variables are the inlet temperature, inlet RH, time of day at which air is used, air speed used (conventionally expressed in terms of cubic metres of air per tonne of grain per minute), and amount of supplementary heat added (both by

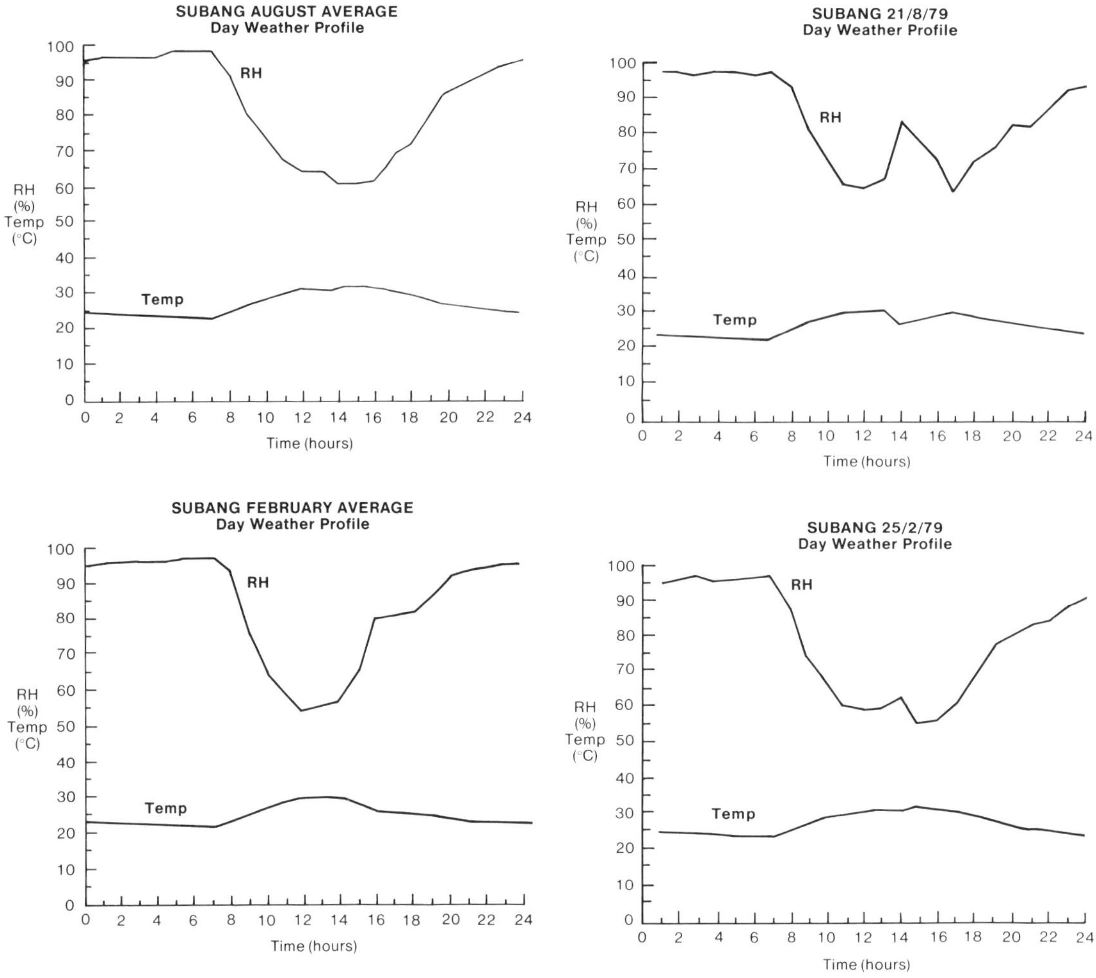


**Fig. 10.** Diurnal weather plots for Subang, Malaysia.

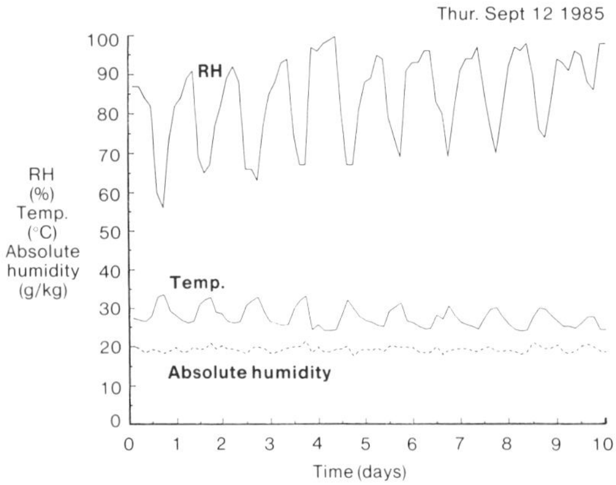


**Fig. 11.** Plots of absolute humidity, temperature, and relative humidity over a 10 day period for Bangkok, Thailand.

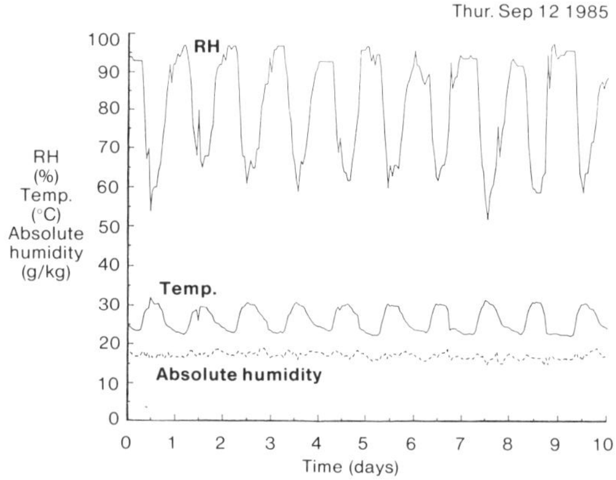


**Fig. 12.** Plots of absolute humidity, temperature, and relative humidity over a 10 day period for Kota Bahru, Malaysia.

the fan and by a mechanical heat source such as a burner). A complete drying strategy must include consideration of factors such as the grain bin dimensions and the receival moisture of the grain in addition to the air variables. In practice, there are additional constraints generated by the different costs of different heating methods, the level of trash present in the grain, the history of the grain since harvesting, the availability of different energy sources, and the urgency of keeping the mill or packaging facilities supplied with dried grain.

If ambient air is passed continuously through a deep bed, the cyclic conditions described in the previous section mean that there are times when low RH air is entering the bed (during the heat of the day), and times when the inlet air is saturated. Thus, the bottom layer will cycle in moisture content with the air conditions, drying and rewetting.

Consider firstly the period of the day when the grain is drying. A drying front will be formed, starting at the inlet and propagating into the grain mass as it removes moisture from the first few layers. After a few hours, the inlet layer will have lost sufficient moisture to be in equilibrium with the ambient conditions. As evening advances, the ambient temperature drops and the RH rises as a result, so that the air entering the grain mass is at a higher RH than the equilibrium RH of the grain. The grain will pick up moisture from the air, and a wetting front will be initiated and start propagating through the bed. There are now two fronts, a drying and a wetting front. Theoretical work by Sutherland et al. (1983) proved that fronts tend to interact (see discussion of deep-bed drying and Figs 7 and 8). The wetting front will broaden in such a way that it catches up with the drying front and interacts with it, the two fronts tending to cancel each other out, or more precisely, to add together forming one front which behaves like the average of the drying and wetting fronts.

When the many small wetting and drying fronts that are initiated during the day are added together, it becomes clear from theoretical simulation work that a grain mass possesses a high degree of inertia with respect to changes in weather conditions. The first layer of a bed will tend to cycle with the ambient weather conditions, but subsequent layers are increasingly governed by the average inlet conditions. An adequate model for most drying situations is simply to look at the average inlet air conditions, and estimate the grain bed's response on the basis of these conditions.

As a result of this fact, the most important air parameter in ambient air drying is the average RH of the inlet air. In analysing a deep bed, the grain at the outlet of the bed will be of lowest quality, because it is the last to lose its moisture. In many practical situations, it is of more benefit to remove some moisture from all of the grain quickly in order to maximise the final grain quality, by pushing a weak drying front through the bed as fast as possible. A weak drying front is achieved by using all of the available air, and is the fastest way to affect the moisture content of the top layer of the grain. Conversely, if quality is less crucial, air may be chosen which satisfies certain RH requirements. A stronger drying front will be formed, which on average moves a shorter distance per day through the grain mass than the weaker, fast front.

Examples of the applications of strategies are shown in Figure 13. Ambient 1982 air from Wagga Wagga, New South Wales, Australia was used as the input to a grain-drying simulation program, the average moisture content of the bed being monitored. Three RH based strategies were implemented. The first was to allow all of the air to pass through the grain. The first six weeks of the grain-drying history shows consistent drying. As winter approached (mid-May), the average ambient RH started to rise, so that the grain was rewetted over winter. Then in summer (September on), the grain was overdried by the hot, dry summer air. Clearly, using all ambient air is not a good strategy. The grain is rewetted, the fan costs are high, and the final product is over-dry.

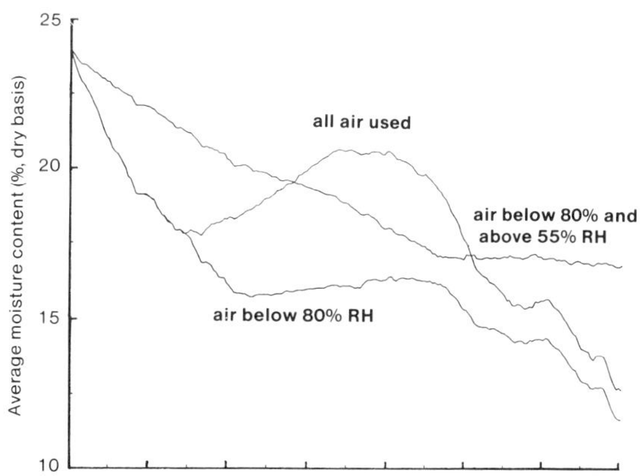


**Fig. 13.** Comparison of various relative humidity strategies for drying grain using ambient air; date versus average moisture content.

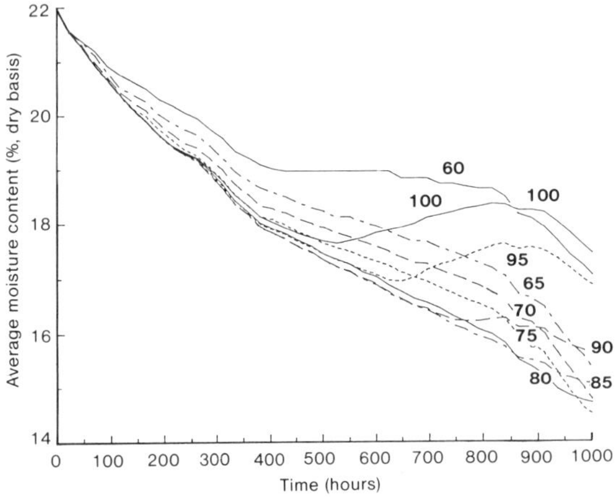


**Fig. 14.** Comparison of various relative humidity strategies for drying grain using ambient air at Bangkok; date versus average moisture content.

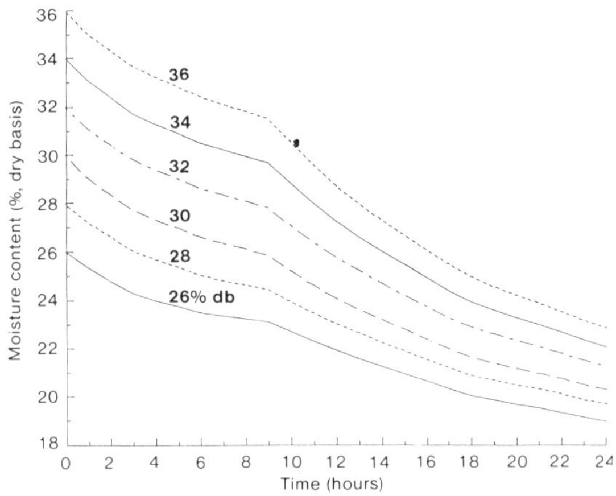


**Fig. 15.** Effect of intial moisture content for flat-bed, high-speed grain drying; time versus bed average moisture content, Bangkok.

The second strategy was to choose only air below 80% RH, in order to avoid the resetting problem of winter. The method works well until the end of May, and even over winter only slight rewetting occurs. However, the drying problem is exacerbated for summer. Again the strategy is not good.

The third strategy involved choosing air below 80% but over 55% RH, in order to avoid overdrying. Clearly this produced a final high quality grain in terms of moisture control. However, the time required to dry the grain sufficiently for milling is more than doubled, so a higher risk of grain deterioration exists. Again this is not an effective strategy. However, combinations of the above strategies suggest use of air below 80% RH until the end of May, and only occasional aeration after then.

Figure 14 relates to a two week period in Bangkok. Grain is being dried from 22% moisture content (dry basis) using ambient air. Each drying line corresponds to an RH strategy. Only air below the restriction RH was allowed to enter the grain. Clearly there is an optimum RH around 80%: above it, too much wet air enters the grain; below it, insufficient air enters the grain to dry it quickly.

Figure 15 shows the results of fast-drying simulations of grain in Thailand, where the initial grain moisture content is varied. Clearly, the receival moisture content of the grain is critical in designing a drying system which will dry the grain quickly enough to avoid quality losses.

Many other aspects of drying strategies could be investigated, including flow rates, the effects on quality of different strategies (using the dry matter loss parameter), and the implications of grain bed dimensions as, for example, were demonstrated by Bridges et al. (1980) for layer filling of bins.

## Conclusions

This paper has attempted to describe current understanding of grain drying from a viewpoint of grain properties. Research tools important to this type of study have been outlined, and methods of approach to drying situations discussed.

## References


Aguerre, R., Suarez, C., and Viollaz, P.E. 1982. Drying kinetics of rough rice grain. Journal of Food Technology, 17, 679–686.

Aris, R., and Amundson, N.R. 1973. Mathematical methods in chemical engineering Volume 2. First order PDE's with applications (2nd ed.). Engelwood Cliffs, New Jersey, Prentice-Hall.

Bakker-Arkema, F.W., and Hall, C.W. 1965. Static versus dynamic moisture equilibria in the drying of biological products. Journal of Agricultural Engineering Research, 10, 308–311.

Bakshi, A.S., and Singh, R.P. 1980. Drying characteristics of parboiled rice. In: Mujumbdar, A.S., ed., Drying '80, volume 2. Hemisphere Publishing Company, 282–288.

Balgovind, R., Dalcher, A., Ghil, M., and Kalnay, E. 1983. A stochastic-dynamic model for the spatial structure of forecast error statistics. Monthly Weather Review, 111, 701–722.

Banks, P.J. 1972. Coupled equilibrium heat and single adsorbate transfer in fluid flow through a porous medium — I: Characteristic potentials and specific capacity ratios. Chemical Engineering Science, 27, 1143–1155.

Barre, H.J., Baughmann, S., and Hamdy, M.Y. 1971. Application of the logarithmic model to cross-flow deep-bed grain drying. Transactions of the American Society of Agricultural Engineers, 14, 1061–1064.

Becker, H.A. 1959. A study of diffusion in solids of arbitrary shape, with application to the drying of the wheat kernel. Journal of Applied Polymer Science, 1, 212–226.

Becker, H.A., and Sallans, H.R. 1955. A study of internal moisture movement in the drying of the wheat kernel. Cereal Chemistry, 32, 212ff.

Berry, M.R.Jr., and Dickerson, R.W.Jr. 1973. Moisture adsorption isotherms for selected seeds and ingredients. Transactions of the American Society of Agricultural Engineers, 12, 137–139.

Bird, R.B., Stewart, W.E., and Lightfoot, E.N. 1962. Transport phenomena. New York, John Wiley and Sons.

Blakeney, A.B. 1976. Morphology of rice grain breakage. Rice Report 1976, Annual Report of World Research and Development on the Post-Harvest Utilisation of Rice.

Bloome, P.D., and Shove, G.C. 1971. Near equilibrium simulation of shelled corn drying. Transactions of the American Society of Agricultural Engineers, 14, 709–712.

Bowden, P.J., Hamond, W.J., and Smith, E.A. 1983. Simulation of near-ambient grain drying I. Comparison of simulations with experimental results. Journal of Agricultural Engineering Research, 28, 279–300.

Boyce, D.S. 1966. Heat and moisture transfer in ventilated grain. Journal of Agricultural Engineering Research, 11, 255.

Bridges, T.C., Ross, I.J., White, G.M., and Loewer, O.J.Jr. 1980. Determination of the optimum drying depth for batch-in-bin corn drying systems. Transactions of the American Society of Agricultural Engineers, 23, 228–232.

Brooker, D.B., Bakker-Arkema, F.W., and Hall, C.W. 1974. Drying cereal grains. Westport, Connecticut, The AVI Publishing Company, Inc, 185–221.

Chau, N.N., and Kunze, O.R. 1982. Moisture content variation among harvested rice grains. Transactions of the American Society of Agricultural Engineers, 25, 1037–1040.

Chesterfield, R.S., and Abbott, T.S. 1974. Effect of moisture conditions on kernel hardness of wheat and rice grain. Rydalmere, Australia, New South Wales Department of Agriculture, Chemistry Branch Report.

Chung, D.S., Garcia-Guerrero, J., and Sieb, P.A. 1984. Hygroscopic properties of rewetted corn. Presented at 1984 grains post-harvest workshop, ASEAN Crops Post-Harvest Programme, August 1984, Kuala Lumpur.

Chung, D.S. and Pfost, H.B. 1967. Adsorption and desorption of water vapor by cereal grains and their products (in three parts). Transactions of the American Society of Agricultural Engineers, 10, 549–557.

Cromarty, A.S. 1964. The chilling of moist grain. Electrical Research Association, Shinfield, Progress, 9.

Day, D.L., and Nelson, G.L. 1965. Desorption isotherms for wheat. Transactions of the American Society of Agricultural Engineers, 8, 293–297.

Dung, N.V. 1980. Aeration of paddy: the measurement and simulation of drying rates in thin layers and deep beds. Sydney, University of New South Wales, unpublished Ph.D. dissertation.

Dung, N.V., Bowrey, R.G., and Fowler, R.T. 1980. Variables affecting the drying rate of paddy rice. Food Technology in Australia, 32, 604–606.

Glahn, H.R. 1965. Objective weather forecasting by statistical methods. The Statistician, 15, 111–142.

Glahn, H.R., and Lowry, D.A. 1972. The use of model output statistics (MOS) in objective weather forecasting. Journal of Applied Meteorology, 11, 1203–1211.

Henderson, S.M., and Pabis, S. 1961. Grain drying theory I. Temperature effect on drying coefficient. Journal of Agricultural Engineering Research, 6, 169–173.

Henderson, S.M. and Pabis, S. 1962. Grain drying theory IV: The effect of airflow rate on the drying index. Journal of Agricultural Engineering Research, 7, 85–89.

Hosokawa, A., and Motohashi, K. 1973. Drying characteristics of a single grain of unhulled rice. Journal of the Society of Agricultural Machinery of Japan, 34, 372–378.

Hsu, K.H. 1984. A theoretical approach to the tempering of grains. Cereal Chemistry, 65, 466–470.

Hukill, W.V. 1953. Grain cooling by air. Journal of Agricultural Engineering, 34, 456–458.

Hukill, W.V., and Schmidt, J.L. 1960. Drying rate of fully exposed grain kernels. Transactions of the American Society of Agricultural Engineers, 2, 71–80.

Husain, A., Chen, C.S., and Clayton, J.T. 1973. Simultaneous heat and mass diffusion in biological materials. Journal of Agricultural Engineering Research, 18, 343–354.

Hustrulid, A., and Flikke, A.M. 1959. Theoretical drying curved for shelled corn. Transactions of the American Society of Agricultural Engineers, 2, 112–113.

Karon, M.L., and Adams, M.E. 1949. Hygroscopic equilibrium of rice and rice fractions. Cereal Chemistry, 26, 1–12.

Kunze, O.R., and Choudhury, M.S.U. 1972. Moisture adsorption related to the tensile strength of rice. Cereal Chemistry, 49, 684–696.

Labuza, T.P. 1968. Sorption phenomena in foods. Food Technology, 22, 15-23.

Mackay, P.J., and Jamieson, M.F.S. 1970. Food storage manual, Chapter 6. Climate in relation to food storage, 259. World Food Programme, Rome.

Mendoza, M.E.E., Rigor, A.C., and Abital, F.L. 1981. Grain quality deterioration in on-farm level of operation. Presented at regional grains post-harvest workshop, Southeast Asia Cooperative Post-Harvest Research and Development Programme. January 1981, Manila.

Mochtar, M., Driscoll, R.H., and Adamczak, T. 1985. The role of tropical weather analysis in grain drying simulation. Presented at Seminar on Grain Post-Harvest Technology, ASEAN Crops Post-Harvest Programme. August 1985, Manila.

Person, N.K. Jr., Sorenson, J.W. Jr., and McCune, W.E. 1966. Thermodynamic considerations in designing controlled storage environments for bulk grain. Transactions of the American Society of Agricultural Engineers, 9, 520.
Pixton, S.W. 1983. The importance of moisture and equilibrium relative humidity in stored products. Tropical Stored Products Information, 43, 16–29.
Pixton, S.W. and Warburton, S. 1968. The time required for conditioning grain to equilibrium with specific relative humidities. Journal of Stored Product Research, 4, 261–265.
Putranon, R., Bowrey, R.G., and Eccleston, J. 1979. Sorption isotherms for two cultivars of paddy rice grown in Australia. Food Technology in Australia, 31, 510–515.
Renwick, P.J. 1984. The role of storage drying in maintaining the quality of paddy. Presented at Grains post-harvest workshop, ASEAN Crops Post-Harvest Programme, August 1984, Kuala Lumpur.
Seib, P.A. Pfost, H.B., Sukabi, A., Rao, V.G., and Burroughs, R. 1980. Spoilage of rough rice measured by evolution of carbon dioxide. In: Proceedings of 3rd annual workshop on grains post-harvest technology. College, Laguna, Philippines, Southeast Asia Cooperative Post-Harvest Research and Development Programme.
Sharma, A.D., Kunze, O.R., and Tolley, H.D. 1982. Rough rice drying as a two-compartment model. Transactions of the American Society of Agricultural Engineers, 25, 221–224.
Sharp, J.R. 1982. A review of low temperature drying simulation models. Journal of Agricultural Engineering Research, 27, 169–190.
Steffe, J.F., and Singh, R.P. 1979. Theoretical and practical aspects of rough rice tempering. Transactions of the American Society of Agricultural Engineers, Paper No. 79-3548.
Steffe, J.F., and Singh, R.P. 1980. Parameters required in the analysis of rough rice drying. In: Mujumbdar, A.S., ed., Drying '80, volume 2. Hemisphere Publishing Company, 256–262.
Steffe, J.F., and Singh, R.P. 1982. Diffusion coefficients for predicting rice drying behaviour. Journal of Agricultural Engineering Research, 27, 489–493.
Suarez, C., Chirife, J., and Viollaz, P. 1982. Shape characterization for a simple diffusional analysis of air drying of grains. Journal of Food Science, 47, 97–100.
Sutherland, J.W., Banks, P.J., and Elder, W.B. 1983. Interaction between successive temperature or moisture fronts during aeration of deep grain beds. Journal of Agricultural Engineering Research, 28, 1–19.
Sutherland, J.W., Banks, P.J., and Griffiths, H.J. 1971. Equilibrium heat and moisture transfer in air flow through grain. Journal of Agricultural Engineering Research, 16, 368–386.
Teter, N.C. 1982. Design of grain storage aeration for the humid tropics. Presented at the 1982 winter meeting of the American Society of Agricultural Engineers, December 14–17, 1982, Palmer House, Chicago, Illinois.
Thompson, T.L., Peart, R.M., and Foster, G.H. 1968. Mathematical simulation of corn drying — a new model. Transactions of the American Society of Agricultural Engineers, 11, 583ff.
Treybal, R.E. 1981. Mass transfer operations. New York, McGraw-Hill, Inc., 784p.
Van Rest, D.J., and Isaacs, G.W. 1968. Exposed-layer drying rates of grain. Transactions of the American Society of Agricultural Engineers, 11, 236–239.
Wahab, A.R., and Dhiauddin, M.N. 1984. Bulk handling of paddy: a measure to accelerate delivery of wet grain to drying complexes. Presented at the 1984 Grains Post-Harvest Workshop, ASEAN Crops Post-Harvest Programme. August 1984, Kuala Lumpur.
Ward, F., and Shapiro, R. 1961. Meteorological periodicities. Journal of Meteorology, 18, 635–656.
Whitaker, T., Barre, H.J., and Hamdy, M.Y. 1969. Theoretical and experimental studies of diffusion in spherical bodies with a variable diffusion coefficient. Transactions of the American Society of Agricultural Engineers, 12, 668–672.
Yao, C.S. 1983. Fitting a linear autoregressive model for long-range forecasting. Monthly Weather Review, 111, 692–700.
Zuritz, C., Singh, R.P., Moini, S.M., and Henderson, S.M. 1979. Desorption isotherms for rough rice from 10°C to 40°C. Transactions of the American Society of Agricultural Engineers, 22, 433–440.

# Modelling Temperature and Moisture Changes Resulting from Natural Convection in Grain Stores

T.V. Nguyen*

*Abstract*

Moisture migration is a result of air convection currents which increase in relative humidity and transfer moisture to the grain as they pass through cooler regions in the grain store. As moisture accumulates in these regions, the grain becomes first caked and finally a moulded, decomposed mass. The accumulation of moisture also makes conditions more favourable for pest infestation.

A numerical analysis which describes the moisture migration by natural convection in stored cereal grains, under fluctuating thermal boundary conditions, was performed. The two-dimensional, transient equations, assuming local thermal and thermodynamic equilibrium, have been solved by powerful numerical methods. Results have been obtained for a rectangular bin and a bunker-type storage. They should describe the actual course of moisture migration and the temperature distributions in such structures.

COMBINED heat and mass transfer in a porous medium is a process which occurs frequently in nature as well as in various engineering endeavours. Drying and humidification in chemical processes and climate control, and movement of moisture in stored cereal grain are examples. Such processes have therefore been widely studied by scientists and engineers over many decades.

Moisture migration resulting from convection currents has been a major problem in grain storage technology, especially as regards storage in plastic enclosures, a technique which is increasing in popularity. The movement of moisture resulting from natural convection gives rise to many problems. These include accelerated development of pest infestation, mould growth, mycotoxin formation, and deterioration in quality. Under some circumstances, convection currents can be useful: for example, they promote the even distribution of fumigants through a grain bulk.

Though heat and mass transfer in porous media has received a great amount of attention recently, there has been relatively little work done on the transient character of moisture migration by natural convection in cereal grain storage. An analysis of heat and moisture transfer processes is complicated by many factors. The structure of the solid matrix varies widely in shape. It may, for instance, be composed of cells, fibres, or grains. Energy transport in such a medium occurs by conduction in both solid and fluid as well as by convection within the fluid. Mass transport occurs within the voids of the medium. The processes of heat and mass transfer are coupled, since heat is transferred as the porous medium releases or absorbs moisture. Also, the equilibrium concentration of the moisture in the medium depends on its temperature.

A detailed study of the transport processes occurring within the solid matrix and in the voids is very complicated even for a regularly shaped matrix and is impossible for the irregular void configurations which, in general, characterise porous media. The normal approach in an analysis is therefore to consider the media involved as continua. The energy and mass fluxes are then a consequence of constitutive equations which include various driving forces.

The present paper uses this approach to study the movement of moisture by natural convection in grain storages of different geometries and boundary conditions.

## Numerical Model

Consider a grain bulk of permeability $\kappa$ and porosity $\varepsilon$ which is subject to a temperature

* CSIRO Division of Chemical and Wood Technology, P.O. Box 26, Highett, Victoria 3190, Australia.

gradient $\Delta T$ and pressure P. The air has thermal expansion coefficient $\beta$, density $\rho$, and dynamic viscosity $\mu$.

The linear law of motion due to Darcy can be written as

$$\frac{\rho}{\varepsilon} \frac{\partial \underset{\sim}{V}}{\partial t} = - \text{grad } p + \rho \underset{\sim}{g} - \frac{\mu}{\kappa} \underset{\sim}{V} \quad (1)$$

where $\underset{\sim}{V}$ is the filtration velocity and $\underset{\sim}{g}$ is the gravitational acceleration.

Introducing Boussinesq's approximation which assumes that the variation of density with temperature is negligible except in the buoyancy term of the equation of motion, and a vorticity $\zeta$ defined by $\zeta = \partial v/\partial x - \partial u/\partial y$, where u and v are the x and y components, respectively, of $\underset{\sim}{V}$, equation (1) becomes

$$\frac{\rho}{\varepsilon} \frac{\partial \zeta}{\partial t} = - \beta g \rho \frac{\partial T}{\partial x} - \frac{\mu}{\kappa} \zeta \quad (2)$$

The water and energy conservation equations for continuous changes are adapted from the basic analysis given by Banks (1972), taking into account the heat of respiration

$$\varepsilon \frac{\partial w}{\partial t} + \gamma \frac{\partial W}{\partial t} + u \frac{\partial w}{\partial x} + v \frac{\partial w}{\partial y} = 0 \quad (3)$$

$$\gamma \frac{\partial H}{\partial t} + u \frac{\partial h}{\partial x} + v \frac{\partial h}{\partial y} - \frac{k^*}{\rho} \nabla^2 T - \frac{R}{\rho} = 0 \quad (4)$$

where w and W are the moisture contents of air and grain, h and H are the enthalpies of moist air and moist grain, respectively, and R is the heat of respiration.

After some manipulation and substituting the properties of grain (see Sutherland et al. 1971), equation (4) can be written as

$$\frac{\partial T}{\partial t} = \frac{1}{c_m + c_w W} \{ - \frac{h_s}{\gamma} (u \frac{\partial w}{\partial x} + v \frac{\partial w}{\partial y}) - \frac{1}{\gamma} (c_a + c_w w + w \frac{\partial h_v}{\partial T}) (u \frac{\partial T}{\partial x} + v \frac{\partial T}{\partial y}) + \frac{k^*}{\rho \gamma} \nabla^2 T + R \} \quad (5)$$

Equations (2), (3), and (5) can now be expressed in a non-dimensional form:

$$\frac{f_c}{Pr} \frac{\kappa}{\varepsilon L^2} \frac{\partial \zeta'}{\partial t'} = - Ra \frac{\partial T'}{\partial x'} - \zeta' \quad (6)$$

$$\frac{\partial w}{\partial t'} + \gamma \frac{\partial W}{\partial t'} + u' \frac{\partial w}{\partial x'} + v' \frac{\partial w}{\partial y'} = 0 \quad (7)$$

$$\frac{\partial T'}{\partial t'} = \frac{c_a}{\gamma (c_m + c_a W)} \{ - \frac{h_s}{c_a \Delta T} (u' \frac{\partial w}{\partial x'} + v' \frac{\partial w}{\partial y'}) - (1 + \frac{c_w}{c_a} w + \frac{w}{c_a} \frac{\partial h_v}{\partial T}) (u' \frac{\partial T'}{\partial x'} + v' \frac{\partial T'}{\partial y'}) + \nabla^2 T' + \frac{H^2 \rho}{\gamma \Delta T k^*} R \} \quad (8)$$

with $x' = x/L$; $y' = y/L$; $u' = uL\rho c_a/k^*$; $v' = vL\rho c_a/k^*$; and $T' = (T - T_1)/\Delta T$

The following ratios of the latent heat of vaporisation of water in grain to that of free water are used in the present study

$$\frac{h_s}{h_v} = 1 + 23 \exp(-40W) \quad \text{for wheat} \quad (9)$$

$$\frac{h_s}{h_v} = 1 + 2.566 \exp(-20.176W) \text{ for paddy} \quad (10)$$

and the relationship between $h_v$ and T,

$$h_v = 2502.39 - 2.3768T \quad (11)$$

Expressions for $h_s$ and $\frac{\partial h_v}{\partial T}$ are obtained from equations (9), (10), and (11) and substituted into equation (8) to fully specify the problem.

The governing equations of motion, mass, and energy, together with the Poisson equation, were discretised by replacing the spatial derivatives with second-order central difference approximations and the time derivatives with forward difference approximations.

The resulting finite difference equations were then solved by the Samarskii-Andreyev Alternating Direction Implicit Scheme. A combined Fourier Analysis-Fast Fourier Transform direct method was used to solve the Poisson equation (Leonardi 1984).

## Results and Discussion

The results presented here are for a rectangular storage bin of square cross-section and for a bunker-type storage. The temperature and moisture content of the grain are prescribed as initially uniform and the temperature of one or more surfaces varies with the ambient air temperature.

Figure 1 shows the results for a rectangular storage bin, one side heated by solar radiation and the top insulated. The initial grain temperature is 25°C and the temperature of the hot side is assumed to be 50°C during daytime and 25°C during nighttime. The temperature in the grain bulk changes slowly except for the layers next to the hot bin wall. From an initially uniform distribution, the temperatures at the centre and at the right wall remain the same after 12 months of

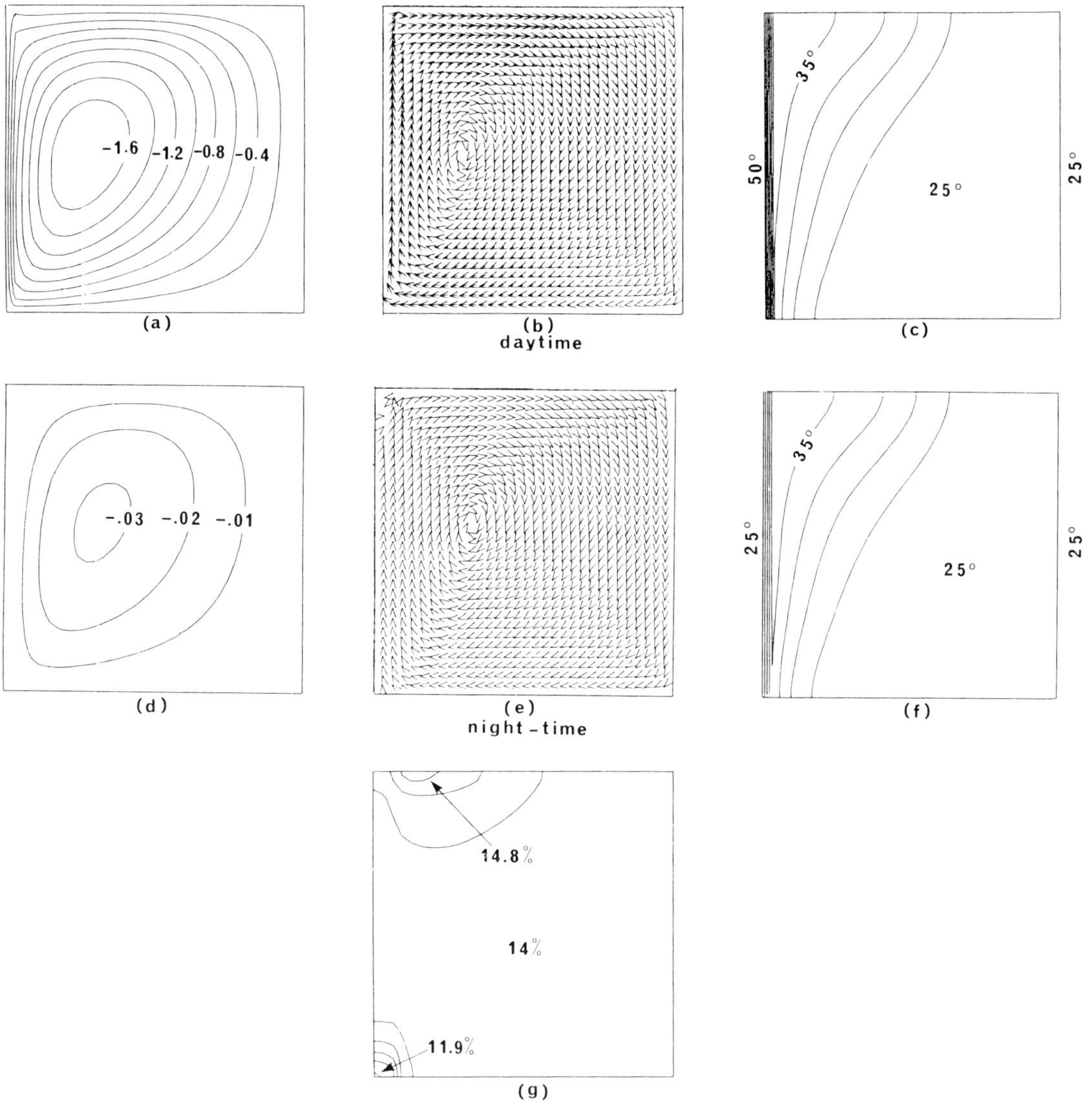


**Fig. 1.** Convective flow and temperature and moisture distribution in a rectangular grain bin with insulated top after 12 months of storage: (a,d) streamlines; (b,e) velocity vector plots, (c,f) isotherms; and (g) moisture distribution.

storage. The convection currents during daytime are quite strong, as shown in Figure 1(b) in which the lengths of the arrows represent the air velocity. The isotherms, lines of constant temperature, are distorted by these convection currents, resulting in a warmer region of grain near the top left corner. At nighttime, when large temperature gradients no longer exist, the air circulation is greatly reduced and conduction is the dominant heat transfer process near the wall. Figure 1(g) shows the distribution of moisture content in the grain store. Moisture migration is in this case a direct consequence of the convection currents. Moisture is transferred from the grain to the warm air which rises along the hot wall. It is then carried by the convection currents to the cooler region near the top left-hand corner and lost to the grain. Consequently, the moisture content increases in the region at the top of the bin and decreases in the region below.

When the top of the storage bin is not insulated but is exposed to sunshine, the airflow patterns

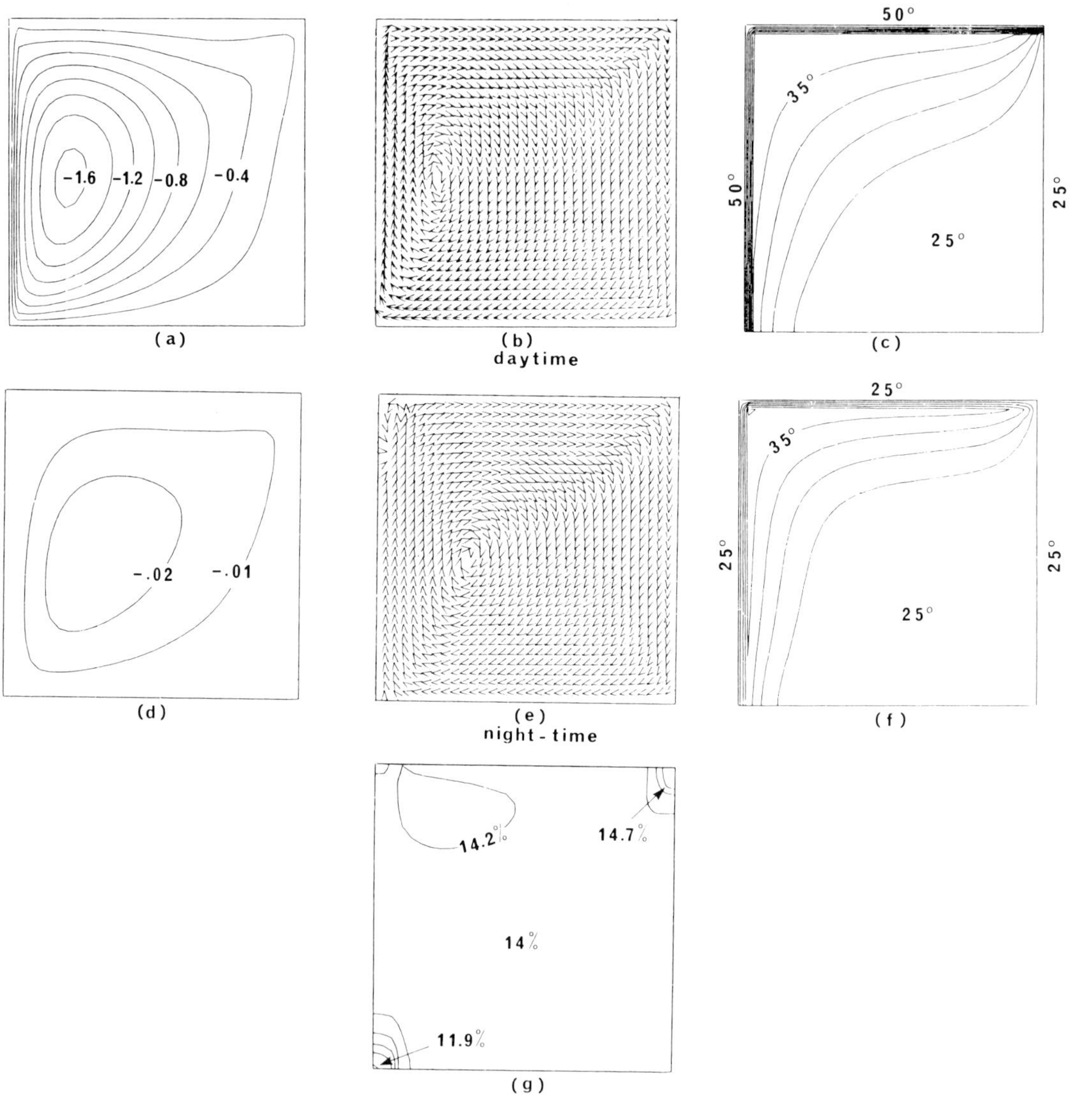


**Fig. 2.** Convective flow and temperature and moisture distribution in a rectangular grain bin with isothermal top after 12 months of storage. (a,d) streamlines; (b,e) velocity vector plots; (c,f) isotherms; and (g) moisture distribution.

and the temperature and moisture content distributions are as shown in Figure 2. The convection currents are similar to the previous case: strong during daytime and very weak during nighttime. As a result of the increased heat transfer area, a larger part of the grain store is affected by natural convection, as indicated by the isotherms. The region of increased moisture has moved to the right-hand corner and another wet spot has started to develop in the top left-hand corner.

If grain is stored at 18% moisture content (wet basis), the heat of respiration of the grain becomes quite significant (Figs 3 and 4). Convection currents now consist of one large cell occupying the whole store and two smaller inner cells, all rotating in the same direction. The strongest cell is still associated with the large temperature gradients along the hot wall. At nighttime, the circulation almost stops, except for a very weak current near the right-hand wall. The temperature of the grain bulk has risen from the initial value of 25°C to 35°C after only four weeks of storage.

In a bunker-type storage, grain is stored under plastic covers in the open. The grain is usually hot after harvesting at high temperature and the transfer of moisture by air convection currents is most serious at nights when ambient temperature is low. Figure 5 shows the simulation of a bunker with grain initially at 40°C and 10% moisture content. The temperature gradient along the sides causes cold, dense air to fall and displace hot, less dense air, which rises through the middle resulting in two distinct counter-rotating convection cells. The distortion of the isotherms by natural convection can be seen in Figure 5(c) and the redistribution of moisture in Figure 5(d).

Moisture is transferred to the region under the ridge from two lower regions along each side. It can be seen in Figure 6 that during the daytime natural convection ceases to be the dominant effect on moisture movement. Moisture may be diffused towards cooler regions as a result of the low water vapour pressure there, but there are no significant convection currents.

After another six months of storage (Fig. 5), the convection currents are more intense in the central region of the bunker and the influence of natural convection on the temperature distribution is

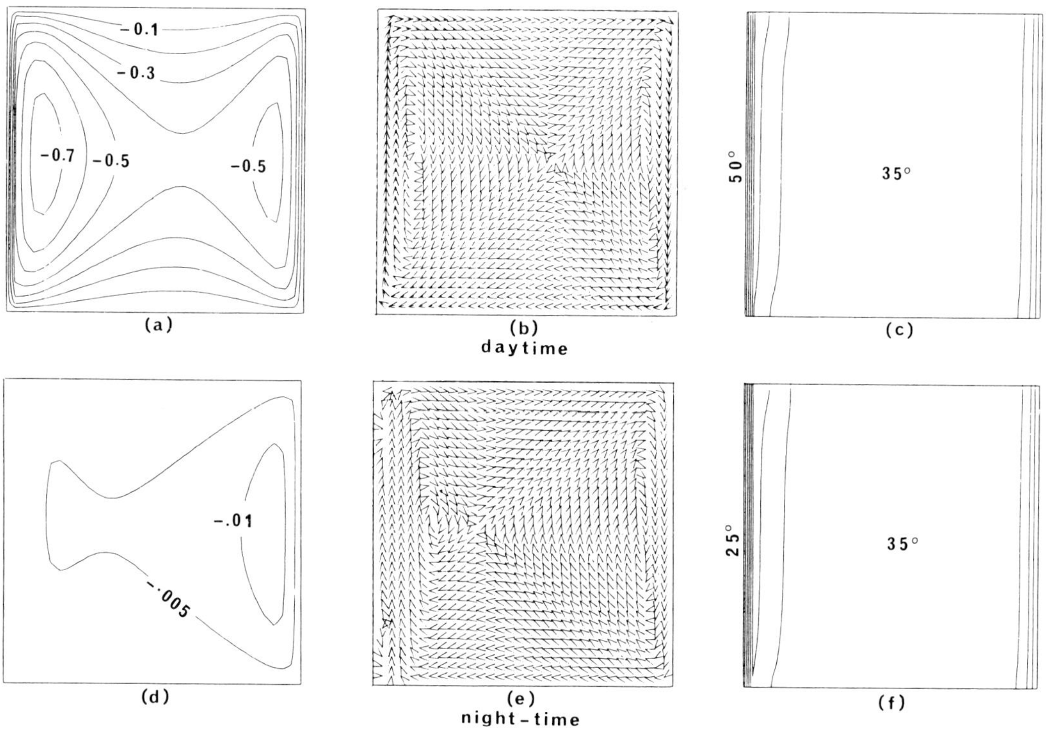


**Fig. 3.** Convective flow and temperature and moisture distribution in a rectangular grain bin with insulated top after four weeks of storage. Grain initially at 25°C and 18% moisture content wet basis: (a,d) streamlines; (b,e) velocity vector plots; and (c,f) isotherms.

more profound. Moisture redistribution now affects the whole bunker which contains an enlarged region of wet grain at the top and several dry regions below. With grain initially at 12% moisture content (w.b.), Figure 7 shows similar flow patterns, isotherms, and moisture distribution. The accumulation of moisture at the top region predicted in this study is consistent with observations in bunker storages in which caked and moulded masses of grain have been found along the ridge.

## Conclusions

A transient state, two-dimensional numerical model has been presented to simulate the temperature and moisture changes by natural convection in grain stores. Airflow patterns, and temperature and moisture distributions were obtained by solving a set of simultaneous differential equations, assuming thermal and thermodynamic equilibrium. The study, carried out on a rectangular bin and a bunker-type storage structure under fluctuating thermal boundary conditions, offers an insight into the development of the flow and temperature fields and the redistribution of moisture in the grain bulk.


## Acknowledgment

The support of the Australian Centre for International Agricultural Research is gratefully acknowledged.


## References


Banks, P.J. 1972. Coupled equilibrium heat and single adsorbate transfer in fluid flow through a porous medium. I: Characteristic potentials and specific capacity ratios. Chemical Engineering Science, 27, 1143–1155.

Leonardi, E. 1984. Sydney, Australia, University of New South Wales, Ph.D. dissertation.

Sutherland, J.W., Banks, P.J., and Griffiths, H.J. 1971. Equilibrium heat and moisture transfer in air flow through grain. Journal of Agricultural Engineering Research, 16, 368–386.


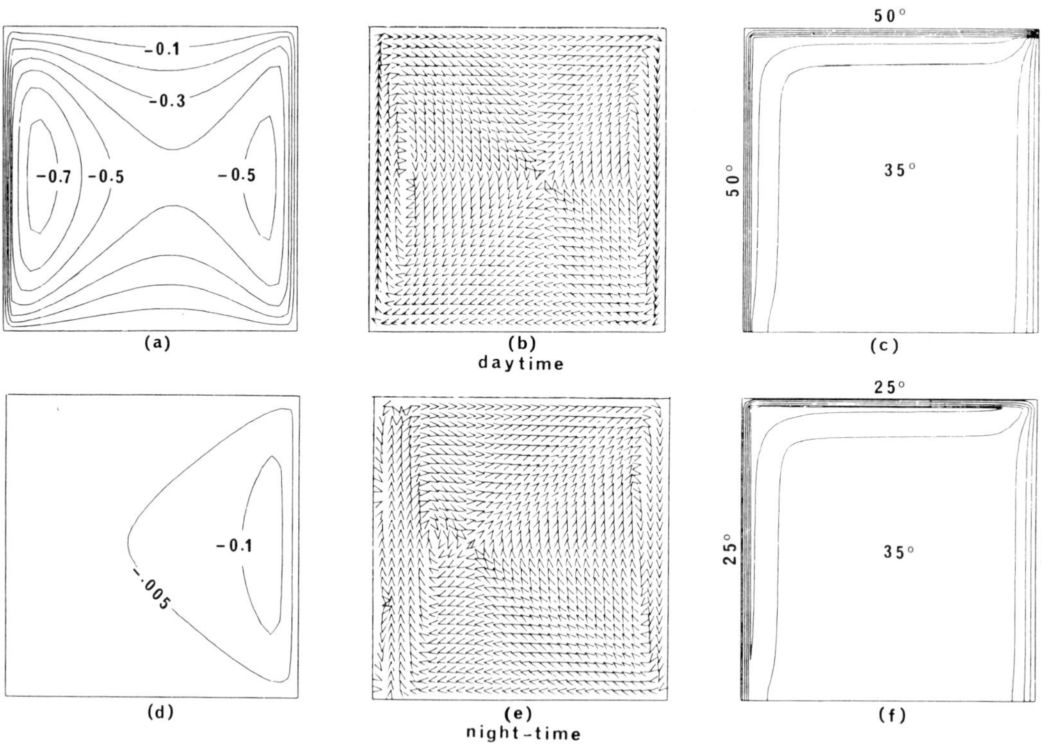


**Fig. 4.** Convective flow and temperature and moisture distribution in a rectangular grain bin with isothermal top after four weeks of storage. Grain initially at 25°C and 18% moisture content (wet basis): (a,d) streamlines; (b,e) velocity vector plots; and (c,f) isotherms.

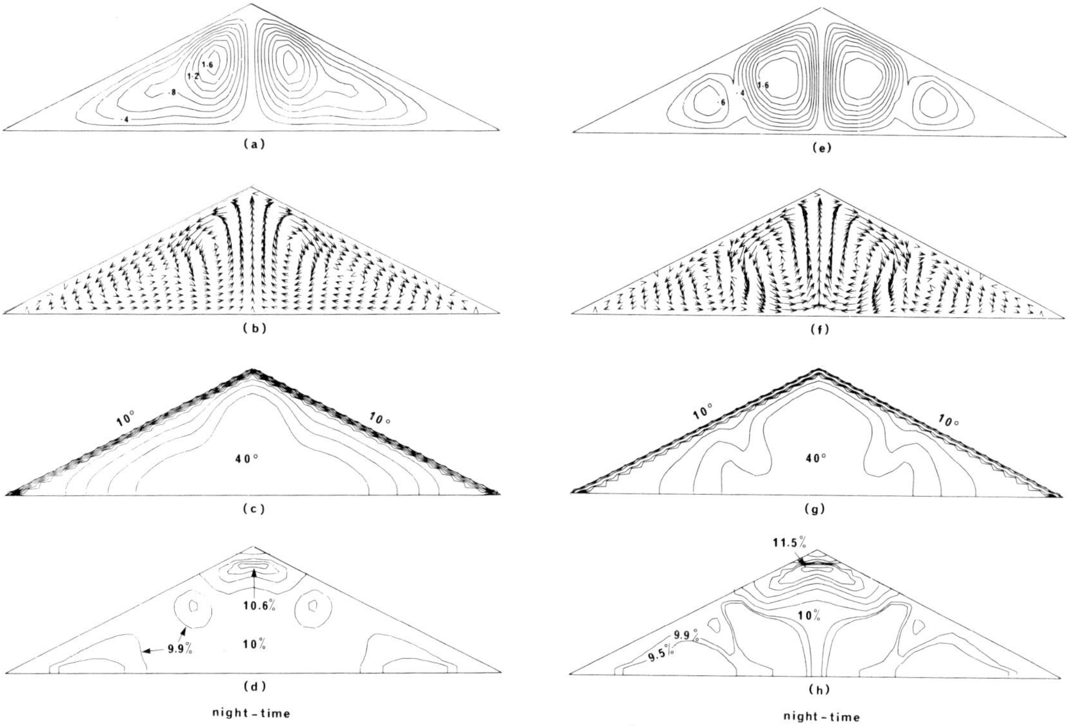


**Fig. 5.** Convective flow and temperature and moisture distribution in a bunker-type grain storage structure after 6 and 12 months storage: (a,e) streamlines; (b,f) velocity vector plots; (c,g) isotherms; and (d,h) moisture distribution.

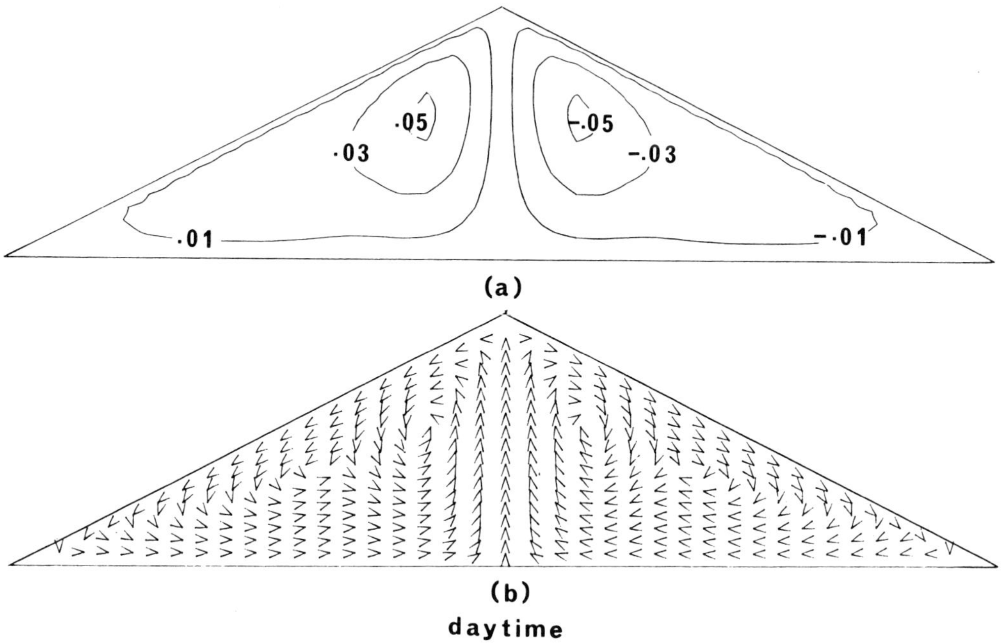


**Fig. 6.** Convective flow and temperature and moisture distribution in a bunker-type grain storage structure after six months storage: (a) streamlines; and (b) velocity vector plots.

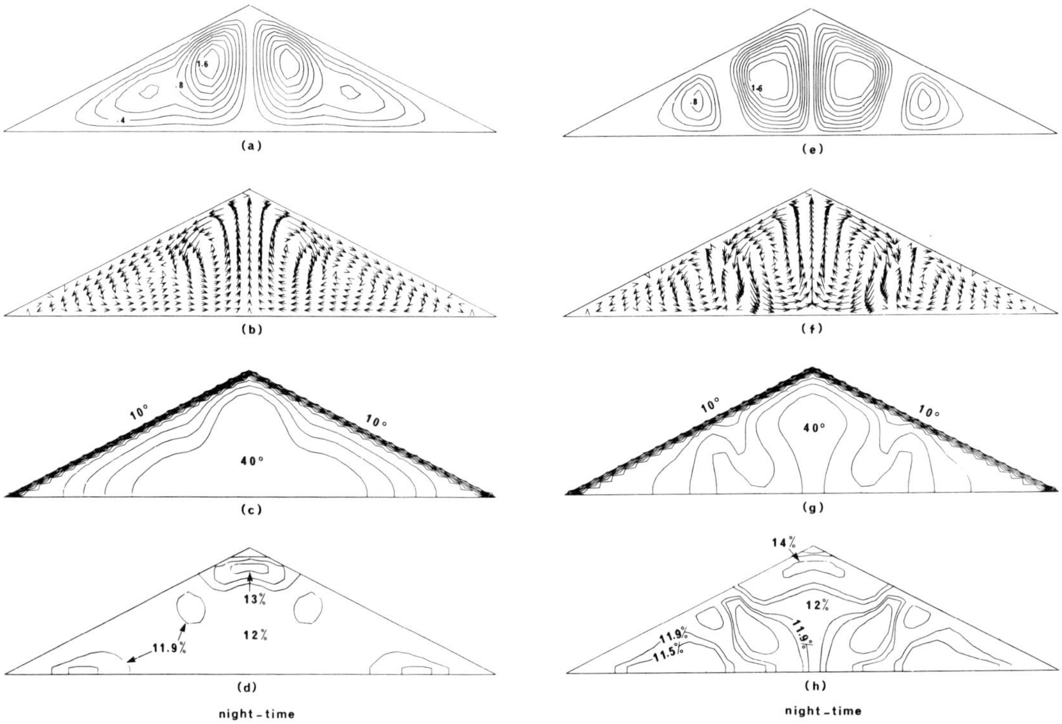


**Fig. 7.** Convective flow and temperature and moisture distribution in a bunker-type grain storage structure with grain initially at 12% moisture content (wet basis) after 6 and 12 months storage. (a,e) streamlines; (b,f) velocity vector plots; (c,g) isotherms; and (d,h) moisture distribution.

## Nomenclature

| | | |
|---|---|---|
| $c_a$ | specific heat of dry air at constant pressure | kJ/(kg dry air K) |
| $c_m$ | specific heat of dry grain | kJ/(kg dry grain K) |
| $c_w$ | specific heat of liquid water | kJ/(kg water K) |
| $f_c$ | effective conductivity factor $= k^*/k_f$ | — |
| g | acceleration due to gravity | $m/s^2$ |
| h | enthalpy of moist air | kJ/kg dry air |
| $h_s$ | latent heat of vaporisation of water in grain | kJ/kg water |
| $h_v$ | latent heat of vaporisation of free water | kJ/kg water |
| H | enthalpy of moist grain | W/m/K |
| $k^*$ | effective conductivity of packed bed | W/m/K |
| $k_f$ | thermal conductivity of air | W/m/K |
| L | height of bed | m |
| p | pressure | Pa |
| P | density of grain (kernel density) | $kg/m^3$ |
| Pr | fluid Prandtl number | — |
| R | heat of respiration | W/kg |
| Ra | modified Rayleigh number $= \frac{g\beta\rho^2 c_a}{\mu}\frac{\kappa}{k^*}\Delta T\, L$ | — |
| t | time | s |
| T | temperature | K |
| $T_1$ | temperature of cold wall | K |
| $T_2$ | temperature of hot wall | K |
| u | velocity in x direction | m/s |
| v | velocity in y direction | m/s |
| V | filtration velocity | m/s |
| w | air moisture content | kg water/kg dry air |
| x,y | coordinates | |
| W | grain moisture content (dry basis) | kg water/kg dry grain |
| $\beta$ | coefficient of volumetric expansion | $K^{-1}$ |
| $\Delta T$ | temperature difference $= T_2 - T_1$ | K |
| $\varepsilon$ | void fraction of packed bed | — |
| $\kappa$ | permeability of packed bed | $m^2$ |
| $\rho$ | density of air | $kg/m^3$ |
| $\mu$ | viscosity of air | $N\,s/m^2$ |
| $\gamma$ | $= \frac{P(1-\varepsilon)}{\rho\varepsilon}$ | — |
| $\zeta$ | vorticity | $s^{-1}$ |

# Modelling of Forced Convection in In-store Grain Drying: the State of the Art

**F. W. Bakker-Arkema***

*Abstract*

An accurate simulation model of forced convection in-store grain drying will be an important tool for the engineering design of such systems. It will contribute to faster and more efficient designs of in-store grain drying installations for the humid tropics.

In-store grain drying models fall into two categories: (1) differential equation models, and (2) heat and mass balance models. Each can be further divided into non-equilibrium and equilibrium models. The author recommends the use of non-equilibrium, differential equation models for the use of in-store grain drying simulations because of their greater flexibility and accuracy, and more fundamental character.

Successful modelling of forced convection in-store grain drying requires a knowledge of the drying rate and equilibrium moisture content of cereals. The moisture diffusion and Guggenheim-Anderson-de Boer isotherm equations are recommended for these tasks. Development of quantitative grain quality models for the humid tropics is necessary if the effect of in-store drying on quality changes is also to be modelled.

DESIGN of forced convection in-store grain drying systems has until recently been by trial and error (Hall 1957). This mode of design is time-consuming and expensive; besides, the number of experimental tests which can be conducted annually is limited due to the short-time for which moist grain is available. Simulation modelling does not have these disadvantages.

Successful modelling requires an adequately accurate simulation model. If the model is to be used for optimisation and control as well as simulation, the model should be fast (in terms of computer time) as well as accurate (Brook and Bakker-Arkema 1980).

An in-store grain drying model can be used by the engineer as a design tool for (Brooker et al. 1974): (1) sizing of the dryer dimensions, (2) predicting the dryer performance, (3) evaluating operating-parameter effects, (4) investigating fan management, (5) scaling the dryer, (6) calculating the load and unload devices, (7) preliminary costing of the total in-store drying system, and (8) exploring dryer control strategies. If a grain quality model is added to the dehydration model, the effect of operating conditions on the grain quality can also be monitored (Bakker-Arkema et al. 1977).

In-store grain drying models can be classified as non-equilibrium or equilibrium models. In the equilibrium models, it is assumed that temperature (and sometimes also humidity) equilibrium conditions exist between the drying air and the grain at each level of the drying bed, and at each time period. Although the non-equilibrium models are more accurate, use of the simpler equilibrium models still appears justified for low airflow rate conditions (Sharp 1982).

This paper reviews the in-store grain drying models with respect to accuracy, ease of application to different crops, and computer time required. The relevance to the humid tropics is emphasised.

## Differential Equation Models

Forced convection in-store grain drying models consisting of a series of differential equations are the most general of the models published in the literature. Both non-equilibrium and equilibrium type models are discussed.

* Agricultural Engineering Department, Michigan State University, East Lansing, MI 48824-1323, USA

**Non-equilibrium Models**

A simulation in-store grain drying model can be developed by writing heat and mass balances on an arbitrarily selected control volume of thickness $\Delta x$ in the dryer, along with the appropriate rate equations (Brooker et al. 1974). This results in differential equations for the air absolute humidity (H), the air temperature (T), the grain temperature ($\theta$), and the average moisture content ($\overline{M}$):

$$\frac{\partial T}{\partial x} = \frac{-ha}{G_a c_a + G_a c_v H}(T - \theta) \quad (1)$$

$$\frac{\partial \theta}{\partial t} = \frac{ha}{p_p c_p + p_p c_w \overline{M}}(T - \theta) + \frac{h_{fg} + c_v (T - \theta)}{p_p c_p + p_p c_w \overline{M}} G_a \frac{\partial H}{\partial x} \quad (2)$$

$$\frac{\partial H}{\partial x} = -\frac{Gp\partial \overline{M}}{G_a \partial t} \quad (3)$$

$$\frac{\partial \overline{M}}{\partial t} = \text{a thin-layer drying or diffusion equation} \quad (4)$$

Equation (4) can be an empirical thin-layer drying rate equation or a theoretical diffusion equation, with the drying rate constant (k) or the effective diffusion coefficient (D) as a function of temperature and moisture content (Bruin and Luyben 1980). This is essential if the tempering process is modelled in paddy (Bakker-Arkema et al. 1984).

The terms $\partial T/\partial t$ and $\partial H/\partial t$ in Equations (1) and (2) are small and have been neglected by several investigators in order to speed up the computer solutions (Bakker-Arkema et al. 1974; Spencer 1969). However, Laws and Parry (1983) included the terms in their generalised in-store drying model.

In solving Equations (1) through (4) by finite difference techniques, the grain bed is divided into layers of thickness $\Delta x$ equal to 0.01–0.1 m; the air and grain properties are assumed to be uniform in each layer during each time step. The outlet air conditions of a layer are used as the inlet conditions to the subsequent layer. The solution is obtained by stepwise numerical integration along the space (x) and time (t) co-ordinates. The time step $\Delta t$ depends on the flow rate of the air and the drying rate of the grain; it varies from 1 to 100 sec (Bakker-Arkema et al. 1974).

The accuracy of the solution is improved if the shrinkage occurring in the bed during in-store drying is accounted for (Spencer 1969).

In general, the non-equilibrium partial differential equation model is of greater accuracy and applicability over a wider range of in-store drying cases than the equilibrium models to be discussed in the next section (Sharp 1982). The greater flexibility comes at the expense of increased computer time. However, since computer technology continues to develop rapidly in terms of memory cost and computing speed, it is expected that even personal computers will soon be able to rapidly solve the non-equilibrium simulation model of in-store drying.

**Equilibrium Models**

In-store equilibrium grain drying models are a simplification of the non-equilibrium model described by Equations (1) through (4). By assuming equilibrium between the air and grain temperatures ($T = \theta$), as is the case in low airflow rate, low air-temperature drying, Equations (1) through (4) are condensed to a three-equation model (Bakker-Arkema et al. 1976):

$$G_a \frac{\partial H}{\partial x} = -\rho_g \frac{\partial \overline{M}}{\partial t} \quad (5)$$

$$G_a(c_a + c_v H)\frac{\partial t}{\partial x} = -\rho_p(c_p + \overline{M}c_w)\frac{\partial T}{\partial t} - G_a((c_v - c_w) T + h_{fg})\frac{\partial H}{\partial x} \quad (6)$$

$$\frac{d\overline{M}}{dt} = \text{thin-layer drying or diffusion equation} \quad (7)$$

The thermal equilibrium model can be further simplified by assuming instantaneous thermal *and* moisture equilibrium between air and grain at each point within the grain bed at any time during the drying process due to infinite convective heat and mass transfer coefficients. This assumption leads to a two-equation in-store grain drying model (Parry 1985).

Hukill (1954) proposed a one-equation in-store drying model obtained by combining Equations (2) and (3), and assuming the latent heat to be much larger than the sensible heat, and $\rho_p$ to be constant. This results in the so-called logarithmic in-store drying model:

$$G_a c_a \frac{\partial T}{\partial x} = \rho_p h_{fg} \frac{\partial \overline{M}}{\partial t} \qquad (8)$$

By relating the travel rate of the drying zone to airflow rate, Hukill was able to obtain an analytical solution of Equation (8) for the dimensionless moisture content in the grain bed as a function of the dimensionless depth and time variables. Solutions of Equation (8) with various initial and boundary conditions have been used by Ohio State University researchers (Barre et al. 1971; Keener et al. 1978) to analyse several in-store grain drying systems. The simplicity of the Hukill (or logarithmic) models is their only advantage.

### Heat and Mass Balance Models

Boyce (1966) and Thompson et al. (1968) independently presented an algebraic model for forced convection in-store grain drying based on heat and mass balances on thin layers of grain; the model allows layer by layer calculation of the temperatures and moisture contents of grain and air in the drying bed. The original model was later modified by assuming heat and mass transfer equilibrium between the air and the grain in each layer (Thompson 1972). Several versions of the equilibrium and quasi-equilibrium heat and mass balance models have been used in the past decade with varied success to analyse low-temperature grain drying systems (Sutherland et al. 1971; Bloom and Shove 1971; Pfost et al. 1977; Morey 1979; van Ee and Kline, 1979; Mittal and Otten 1980; Pierce and Thompson 1980). The accuracy of the predictions has been restricted by the assumptions made in the derivation of this model type (Parry 1985).

Heat and mass balance drying models have made significant contributions to drying technology mainly because of their simplicity and computation speed (Sharp 1982). The differential equation models are more accurate and theoretically sounder since they are based on the laws of heat and mass transfer and of thermodynamics. In addition, increased computer power has made the differential equation models suitable for operational research applications.

### Auxiliary Simulation Properties

The application of the forced convection in-store grain drying simulation models requires auxiliary information on grain, bed and air properties ranging from kernel drying rate to bed static pressure and air-moisture psychrometrics. If grain quality is a critical criterion, a grain-deterioration or spoilage model should be an integral part of the in-store grain drying models.

### Grain Properties

The in-store grain drying model of Equations (1) through (4) shows which thermo-physical grain property values are required for solution of the model. They include the density, the bed porosity, the heat of vaporisation, the convective packed-bed heat transfer coefficient, the specific surface area, the specific heat and the static pressure.

Data and equations of these properties for grains and legumes have been collected by several investigators (Rao and Pfost 1980; Brook and Foster 1981; Hague et al. 1982).

### Drying Rate Equations

In order to model the drying of a bed of grain, the drying rate characteristics of the individual kernels have to be known for different drying conditions. The drying rate equation fulfills this purpose.

In the past decade, hundreds of papers have been published on the drying of capillary-porous rigid particles such as grains. Few are relevant to the engineering design of drying systems. Since several exhaustive reviews on the fundamental aspects of the drying of food materials have recently been published (van Brakel 1980; Bruin and Luyben 1980; Toei 1983), this paper reviews only those equations which have proven to be effective in the actual design of in-store dryers.

The drying of grain kernels is due to various transport mechanisms. Moisture movement can be caused by liquid diffusion, vapour diffusion, thermal diffusion, capillary flow, internal evaporation and condensation, surface diffusion, shrinkage pressure, or, most frequently, a combination of these mechanisms. Since knowledge of the exact mechanisms in grain drying is still limited, the choice of one mechanism (and thus one equation) over another (and thus, another equation) for the interpretation of the drying rate data appears nonsensical to the design engineer. What he desires is an equation that represents correctly the moisture and heat transfer behaviour of the grain kernels in a certain temperature/humidity range.

The drying theory of individual particles has developed from the empirical drying rate equation of Fisher (1923) to the liquid diffusion equation of Gilliland and Sherwood (1933), the capillary movement equation of Ceaglske and Hougen (1937), the vapour diffusion/capillary liquid movement equation of Krischer (1938), the evaporation-condensation theory of Henry (1939), and finally the irreversible thermodynamic equations of Luikov 1966). Recent theoretical drying papers by, for example, Fortes and Okos (1980) and Whitaker (1980) appear to offer nothing fundamentally new on the subject. The best critique on the plethora of so-called 'fundamental' papers on the topic of mass transfer in convective drying of porous media is by van Brakel (1980).

Due to the complexity of the drying process, engineers prefer to lump the effects of the different drying transport mechanisms into a simplified moisture diffusion equation with a concentration and temperature dependent diffusion coefficient (D). For a spherical or quasi-spherical non-shrinking kernel:

$$\frac{\partial M}{\partial t} = \frac{1}{r^2} \frac{\partial}{\partial r} (r^2 D \frac{\partial M}{\partial r}) \qquad (9)$$

The D-value is the apparent or effective diffusion coefficient with an Arrhenius type temperature dependency:

$$D = D_0 \exp ( - \frac{E}{R\, T_{abs}}) \qquad (10)$$

where $\bar{E}$ is the energy of activation. Values of the diffusion coefficient for some grains at different moisture contents (e.g. paddy, maize, wheat) have been tabulated by Bakker-Arkema et al. (1983b) and Steffe and Singh (1982). Unfortunately, knowledge of the precise dependence of the moisture diffusivity on temperature and moisture content of grains is limited, and this often forces engineers to employ empirical drying rate equations for the simulation models of in-store grain dryers.

The drying rate of cereals is frequently expressed in the literature by a semi-empirical relationship, the so-called 'thin-layer' equation, which relates the standardised average moisture content to time:

$$\frac{\overline{M} - \overline{M}_e}{\overline{M}_i - \overline{M}_e} = \exp (- kt) \qquad (11)$$

where k is a product constant dependent on temperature and is called the 'drying constant' with dimension equal to $s^{-1}$. (k can thus be interpreted as the ratio of the diffusion coefficient divided by a characteristic length squared.) Values of k for different grains can be found in Brooker et al. (1974).

In comparing Equations (9) and (11) for engineering applications, the diffusion Equation (9) is preferred since it allows calculation of the moisture profiles within a grain kernel. (Equation 11 only calculates the average moisture content.) Thus, the diffusion equation can simulate the change in the moisture distribution within the paddy kernels during periods of tempering.

**Sorption Isotherms**

To solve the drying rate equation discussed in the previous section, the boundary conditions need to be specified in terms of the moisture sorption isotherm which expresses the average moisture content of a grain kernel as a function of the environmental temperature and relative humidity. The literature abounds with papers on the sorption phenomena of food products (e.g. Chirife and Iglesias 1978; Pixton and Howe 1983). In fact, publication of such papers appears to continue unabated (Chinnan and Beuchat 1985). The purpose of this activity is not evident since, for design purposes, sufficient information appears to be known on the topic.

A consensus has developed recently among scientists and engineers of the superiority of the Guggenheim-Anderson-de Boer (GAB) equation for the prediction of the water activity ($a_w$) of food products (van den Berg 1984). The GAB equation has the following form:

$$\overline{M} = \frac{\overline{M}_0 CK}{(1 - Ka_w)(1 - Ka_w + CKa_w)} \qquad (12)$$

where $\overline{M}_0$ is the moisture content at monolayer coverage; and C and K are empirical temperature-dependent product constants.

Lamauro et al. (1985) recently compared several sorption isotherm equations for different products including grains. They concluded that the GAB equation gives a good fit for over 75% of the products studied, and describes the isotherms better than the two-parameter equations. The values of $\overline{M}_0$, C and K in the GAB equation for 163 biological products are tabulated in the Lamauro papers for a limited temperature range.

For grains, the two-parameter Chung-Pfost isotherm equation has proven to be effective in the simulation of different dryer designs (Sokhansanj 1984):

$$\ln a_w = -\frac{K_1}{R\,T_{abs}} \exp\left(-K_2 \overline{M}\right) \qquad (13)$$

where $K_1$ and $K_2$ are product constants. Values for the two constants of the major grain species over the full range of temperatures encountered in the grain drying process have been tabulated by Pfost et al. (1976).

The GAB equation has a theoretical basis, and thus supplies fundamental information on the thermodynamic properties (e.g. heat of vaporisation, pore structure) of a grain (Zuritz and Singh 1985). In contrast, the Chung-Pfost relationship is empirical in nature. Which equation is better for design depends on which relationship represents the experimental data best within the range of moisture contents and temperatures encountered in the dryer under study. The importance of the equilibrium moisture content in simulation has been illustrated by van Ee and Kline (1979) who showed that a 0.5% change in the equilibrium moisture content of maize has a significant effect on the simulated values of the final grain moisture content and the rate of movement of the drying front, in in-store drying installations.

**Psychrometrics**

Modelling of the thermodynamic properties of air as it passes through the bed of grain is an integral part of the in-store grain drying simulation. The relationship between dry-bulb temperature, relative humidity, absolute humidity, density, specific volume, enthalpy, dew-point temperature and wet-bulb temperature are given in Brooker et al. (1974). A package of psychrometric subroutines called SYCHART, representing the psychrometric chart in SI units can be found in Bakker-Arkema et al. (1978).

**Grain Quality**

The objective of grain drying is to dry grain without affecting the desired grain qualities. It depends on the end-use of the product which quality characteristic needs to be conserved; for paddy, head yield is the major criterion, for wheat, baking quality. In an exhaustive and excellent review on heat damage to cereal grains, Nellist (1982) recommended a more systematic approach to the topic than has been the case in the past.

For optimum design and control of grain drying equipment, it is necessary to quantify deterioration in grain quality. In the objective function, the quality deterioration of the grain acts as a penalty function. Although many data on the subject can be found in the literature, modelling of the physical condition of a biological product is not well-developed, except for some empirical relationships for the breakage increase of maize (Brook 1977), for the moulding of maize (Morey 1979), and for the baking quality of wheat (Nellist 1978).

Different quality criteria have been used for dryer design. Ghaly and Taylor (1982) monitored the micro-baking loaf volume, the turbidity, and the viability during the thermal disinfestation of wheat. Ghaly and Sutherland (1983) measured free-fatty acid content and oil yield during the high-temperature drying of soybeans. Mühlbauer et al. (1976) considered lysine content and colour in evaluating maize dried at high temperatures. Other quality design criteria considered include head yield of rice (Fontana 1983), breakage susceptibility of maize (Kusterman and Kutzbach 1981, and mould development in maize (Mühlbauer et al. 1981). It is necessary to develop quantitative models similar to the first-order kinetic equation for viability loss by Schreiber et al. (1981) for the change of each of these quality criteria occurring during drying, and link them to the mass and heat transfer in the in-store drying models.

Simulation is an essential tool in the design of in-store grain drying systems for the tropics. Differential equation models are more accurate and flexible than other types of models. The necessary auxiliary equations are available in the literature for the major cereal crops such as rice and maize.

**List of Symbols Used**

| | |
|---|---|
| $a$ | specific product surface area, $m^{-1}$ |
| $a_w$ | water activity, decimal |
| $c$ | specific heat, kJ/kg°C |
| $d$ | diameter, m |
| $h$ | convective heat transfer coefficient, kJ/s $m^2$ °C |
| $h_{fg}$ | heat of vaporisation, kJ/kg |
| $k$ | drying constant, $sec^{-1}$ |
| $P_{at}$ | atmospheric pressure, $N/m^2$ |
| $r$ | kernel coordinate, m |

t time, s
x depth coordinate, m
C product constant
D effective diffusion coefficient, $m^2/s$
G dry weight flow rate, $kg/m^2s$
H humidity ratio, kg/kg
K product constant
M local moisture content within kernel, dry basis (decimal)
$\overline{M}$ average moisture content, dry basis (decimal)
R equivalent kernel radius, m
T air temperature, °C
$T_{abs}$ absolute temperature, K
V velocity, m/s
θ grain temperature, °C
ε void ratio, decimal
ρ density, $kg/m^3$

as a subscript:
a dry air
e equilibrium
i initial
p dry grain
r radial
t at time t
v water vapour
w water (liquid)

## References


Bakker-Arkema, F.W., Brook, R.C., and Lerew, L.E. 1978. Cereal grain drying. In: Pomeranz, Y., ed., Cereal science and technology, volume 2. St Paul, Minnesota, American Association of Cereal Chemists.

Bakker-Arkema, F.W., Brooker, D.B., and Roth, M.G. 1977. Feasibility study of in-bin drying in Missouri using solar energy. In: Shove, G.C., ed., Solar grain drying conference. Urbana, Illinois, University of Illinois.

Bakker-Arkema, F.W., Fontana, C., Brook, R.C., and Westelaken, C.M. 1983a. Concurrent flow rice drying. Drying Technology, 1(2), 171–191.

Bakker-Arkema, F.W., Fontana, C., Fedewa, C.L. and Schisler, I.P. 1983b. A comparison of drying rates of different grains. American Society of Agricultural Engineers, Paper No. 83-3009.

Bakker-Arkema, F.W., Fontana, C., Kalchik, S.J., and Westelaken, C.M. 1984. Drying in commercial rice parboiling plants. Drying Technology, 2(3), 423–433.

Bakker-Arkema, F.W., Haight, J., Roth, M.G. and Brooker, D.B. 1976. Required weather data for simulation of solar grain drying. American Society of Agricultural Engineers, Paper No. 76–3020.

Bakker-Arkema, F.W., Lerew, L.E., DeBoer, S.F., and Roth, M.G. 1974. Grain dryer simulation. East Lansing, Michigan, Michigan State University, Research Report 224.

Barre, H.J., Baugham, C.R., and Hamdy, M.Y. 1971. Application of the logarithmic model to deep bed drying of grain. Transactions American Society of Agricultural Engineers, 14(6), 1061–1064.

van den Berg, C. 1984. Description of water activity of foods for engineering purposes by means of the G.A.B. model of sorption. In: McKenna, B.M., ed., Engineering and foods, volume 1. New York, Elsevier.

Bloom, P.D., and Shove, G.C. 1971. Near equilibrium simulation of shelled corn drying. Transactions American Society of Agricultural Engineers, 14, 709–712.

Boyce, D.S. 1966. Heat and moisture transfer in ventilated grains. Journal of Agricultural Engineering Research, 11(4), 255–265.

van Brakel, J. 1978. Opinions about selection and design of dryers. In: Mujumdar, A.S., ed., Proceedings of the first international symposium on drying. Princeton, New Jersey, Science Press.

Brook, R.C. 1977. Design of multi-stage grain dryers. East Lansing, Michigan, Michigan State University, Ph.D. dissertation.

Brook, R.C., and Bakker-Arkema, F.W. 1980. Design of multi-stage corn dryers using computer optimization. Transactions American Society of Agricultural Engineers, 23(1), 200–203.

Brook, R.C., and Foster, G.H. 1981. Drying, cleaning and conditioning. In: Finney, E.E., ed., Handbook of transportation and marketing in agriculture. Boca Raton, Florida, CRC Press.

Brooker, D.B., Bakker-Arkema, F.W., and Hall, C.W. 1974. Drying cereal grains. Westport, Connecticut, AVI Publishing Corporation.

Bruin; S., and Luyben, K.Ch.A.M. 1980. Drying of food materials: a review of recent developments. In: Mujumdar, A.S., ed., Advances in drying, volume 1. New York, Hemisphere Publishing Company.

Ceaglske, N.H. and Hougen, O.H. 1937. Drying granular solids. Industrial Engineering Chemistry. 29(6), 805–813.

Chinnan, M.S. and Beuchat, L.R. 1985. Sorption isotherms of whole cowpeas and flours. Lebensm. Wiss. u. Technol. 18(2), 82–88.

Chirife, J., and Iglesias, H.A. 1978. Equations for fitting water sorption isotherms of food: Part 1 — a review. Journal of Food Technology, 13(6), 159–174.

van Ee, G.R., and Kline; G.L. 1979. Foldry — a model for low temperature corn drying systems. American Society of Agricultural Engineers, Paper No. 79–3524.

Fisher, R.A. 1923. Some moisture relations of colloids. Proceedings of the Royal Society London. A1(103), 139–161.

Fontana, C. 1983. Concurrent flow versus conventional drying of rice. East Lansing, Michigan, Michigan State University, Ph.D. dissertation.

Fortes, M., and Okos, M.R. 1980. Drying theories: their bases and limitations as applied to foods and grains. In: Mujumdar, A.S., ed., Advances in drying, volume 1. New York, Hemisphere Publishing Company.

Ghaly, T.F., and Taylor, P.A. 1982. Quality effects of heat treatment of two wheat varieties. Journal of Agricultural Engineering Research, 27(3), 227–234.

Ghaly, T.F., and Sutherland, J.W. 1983. Quality aspects of heated air drying of soybeans. Journal Stored Products Research 19(1), 31–41.

Gilliland, E.R., and Sherwood, T.K. 1933. The drying of solids. Industrial Engineering Chemistry, 25(6), 1134–1136.

Hague, E., Shamed, Y.N., and Deyoe, C.W. 1982. Static pressure drop in a fixed bed of grain affected by grain moisture content. Transactions American Society of Agricultural Engineers, 25(4), 1095–1098.

Hall, C.H. 1957. Drying farm crops. Westport. Connecticut, AVI Publishing Corporation.

Henry, P.S.A. 1939. Diffusion in absorbing media. Proceedings of the Royal Society. London A1(71), 215–241.

Hukill, W.V. 1954. Grain drying. In: Anderson, J.A., and Alcock, A.W., ed., Storage of cereal grains and their products. St Paul, Minnesota, American Association of Cereal Chemists.

Keener, H.M., Meyer, G.E., Sabbah, M.A., and Curry, R.B. 1978. Simulation of solar grain drying. Wooster, Ohio, Ohio Agricultural Research and Development Center, Agricultural Engineering Series, 102.

Krischer, O. 1938. Grundgesetze der Feuchtigkeitsbewegung in Trocknungsguter: Kapillarwasserbewegung und Wasserdampf Diffusion. VDI Z. 82, 373–378.

Kusterman, M., and Kutzbach, H.D. 1981. Influence of thermal treatment on breaking strength of corn. Grundl. Lantechn. 31(5), 171–177.

Laws, N. and Parry, J.L. 1983. Mathematical modelling of heat and mass transfer in agricultural grain drying. Proceedings of the Royal Society London. A 385, 169–187.

Lamauro, C.J., Bakshi, A.S., and Labusa, T.P. 1985. Evaluation of food moisture sorption isotherm equations. Part 1: fruit, vegetable and meat products. Lebensm. Wiss. u. Techn. 18(2), 111–117. Part 2: milk, coffee, tea nuts, oilseeds, spices, starchy foods. Ibid 18(2), 118–124.

Luikov, A.V. 1966. Heat and mass transfer in capillary porous bodies. New York, Pergamon Press.

Mittal, C.S. and Otten, L. 1980. Simulation of low-temperature drying of corn for Ontario conditions. American Society of Agricultural Engineers, Paper No. 80-3519.

Morey, R.V. 1979. Fan management for ambient air drying systems. Transactions American Society of Agricultural Engineers, 22(6), 1410–1425.

Mühlbauer, W., Hofacher, W., Muller, H.M., and Thaler, M. 1981. The energy and micro-biological aspects of low-temperature drying of wheat. Grundl. Landtechn. 31(5), 145–154.

Mühlbauer, W., Huss, W., and Kuppinger, H. 1976. Farbmesische Schnellmethode zum Erkennen einer Proteinschadigung bei der Trocknung von Maiskornern mit hohen Lufttemperaturen. Grundl. Landtechn. 26(4), 128–134.

Nellist, M.E. 1978. Safe temperatures for drying grain. Silsoe, Bedford, U.K., National Institute of Agricultural Engineering, Report No. 29.

Nellist, M.E. 1982. Developments in continuous flow grain dryers. The Agricultural Engineer, 37(3), 74–80.

Parry, J.L. 1985. Mathematical modelling and computer simulation of heat and mass transfer in agricultural grain drying: a review. Journal of Agricultural Engineering Research, 32(1), 1–29.

Pfost, H.B., Maurer, S.G., Chung, D.S., and Milliken, G.A. 1976. Summarizing and reporting equilibrium moisture data for grains. American Society of Agricultural Engineers, Paper No. 76-3520.

Pfost, H.B., Maurer, S.G., Grosh, L.E., Chung, D.S., and Foster, G.H. 1977. Fan management systems. American Society of Agricultural Engineers, Paper No. 77-3526.

Pierce, R.O., and Thompson, T.L. 1980. Management of solar and low-temperature grain drying systems. Part 1: Operational strategies with full bin. Transactions American Society of Agricultural Engineers, 23(4), 1020–1023.

Pixton, S.W., and Howe, R.W. 1983. The suitability of various linear transformations to represent the sigmoid relationship of humidity and moisture content. A review. Journal of Stored Products Research, 19(1), 1–18.

Rao, V.G., and Pfost, H.B. 1980. Physical properties related to drying twenty food grains. American Society of Agricultural Engineers, Paper No. 80-3539.

Schreiber, H., Mühlbauer, W., Wasserman, L., and Kuppinger, H. 1981. Influence of drying on the quality of wheat — study of the reaction kinetics. Zeitung Lebensmittel Untersuchung und Forschung, 73(1), 169–175.

Sharp, J.R. 1982. A review of low temperature drying simulation models. Journal of Agricultural Engineering Research, 27(2), 169–190.

Sokhansanj, S. 1984. Grain drying simulation with respect to energy conservation and grain quality. In: Mujumdar, A.S., ed., Advances in drying. New York, Hemisphere Publishing Company.

Spencer, H.B. 1969. A mathematical simulation of grain drying. Journal of Agricultural Engineering Research, 14(3), 226–235.

Steffe, J.F., and Singh, R.P. 1982. Diffusion coefficients for predicting rice drying behavior. Journal of Agricultural Engineering Research, 27(6), 489–493.

Sutherland, J.W., Banks, P.J., and Griffiths, H.J. 1971. Equilibrium heat and moisture transfer in airflow through grain. Journal of Agricultural Engineering Research, 16, 368–386.

Toei, R. 1983. Drying mechanism of capillary porous bodies. In: Mujumdar, A.S., ed., Advances in drying, volume 2. New York, Hemisphere Publishing Company.

Thompson, T.L. 1972. Temporary storage of high-moisture shelled corn using continuous aeration. Transactions American Society of Agricultural Engineers, 15(3), 333–337.

Thompson, T.L., Peart, R.M., and Foster, G.H. 1968. Mathematical simulation of corn drying — a new model. Transactions American Society of Agricultural Engineers, 11(4), 582–586.

Whitaker, S. 1980. Heat and mass transfer in granular porous media. In: Mujumdar, A.S., ed., Advances in drying, volume 1. New York, Hemisphere Publishing Company.

Zuritz, C.A., and Singh, R.J. 1985. An equation to compute the heat of evaporation of water for rough rice during drying. Drying Technology, 3(3), 421–435.

# Basic Principles of Aeration and In-store Drying Session Chairman's Summary


## C.W. Bockhop*


THE four papers presented during this session provided an excellent discussion of the basic principles of drying and storage, as well as of aeration of grain in storage. The problems were well defined.

The modelling of the temperature and moisture changes and the natural convection that occurred in grain storage structures was well illustrated in the third paper.

A good case was made for modelling of the forced convection in grain drying by Dr Bakker-Arkema who, through illustration, pointed out the value of proper modelling in the designing of drying systems.

In discussing Dr Driscoll's paper on the application of psychrometrics to grain drying, Dr Thorpe endorsed the strategy proposed for four-stage drying and cooling of wet grain, i.e. a holding stage, a drying stage, final drying, and a holding stage. He pointed out, however, that there need be little concern about the re-moisturising of grain under natural aeration, even under tropical conditions where the air moisture content is 0.02 kg of water/kg of dry air. This is because the aeration flow rate, typically 1 litre of air per second per tonne, is incapable of adding much moisture to a bulk of grain. Only that grain very close to the air inlet will be re-wetted. To prove this, simply do a mass balance over the grain mass. Moisture uptake by diffusion is also a very slow process. It could take decades for a grain bulk with linear dimensions of 10 m x 10 m x 10 m to reach an equilibrium with its environment.

Following Dr Nguyen's paper there was some discussion on the speed of natural air movement, which was reported to be in the range of 1 mm per second.

Also, the question was raised on how to prevent natural circulation and the formation of high moisture pockets. This can be alleviated by smaller storage structures, by sealed storage, and by cooling the grain.

A question was raised on the respiration rates cited in the paper. It was agreed that more information on respiration rates is needed. Loss of mass is not so important in low temperature drying. More important is development of proper theory concerning the growth of microorganisms.


* Department of Agricultural Engineering, International Rice Research Institute, P.O. Box 933, Manila.

# Design of Aeration and In-store Drying Systems

# Design Parameters for Storage and Handling Systems for Grain

**Norman C. Teter***

*Abstract*

Parameters are constants that define functional mathematical relationships. This paper explains the use of parameters of air psychrometrics, grain hygroscopicity, rates of dry matter loss and insect increase, resistance of grain to airflow, economics of grades and handling costs, and probability analysis of percentage loss data, for designing large and small storage systems in the humid tropics.

Controlled ventilation of both bulk and bagged storage shows economic benefits that should bring about the construction of tight storage buildings with forced ventilation by either ambient or conditioned air.

PARAMETERS are for measuring. In relation to grain storage, parameters are the constants that define the additive, multiplicative, and exponential and their inverses the subtractive, divisive, and logarithmic relationships between air, grain, and markets. Parameters for air are well known and vary little. Those for the physical and biological properties of grain are only partially known. Those for the economics of marketing are historically, politically, and sociologically variable according to the circumstances of the marketing system. Parameters are numbers, dimensional or non-dimensional, that make it possible to quantitatively design the properties that qualitatively enhance the value of stored grain. A.D. Gracey (personal communication) stated that the design should keep the grain dry, clean, pest-free, secure, handy, and whole. This paper is written to assist in rationalisation of design to meet these criteria.

Equations for the capacity, power, and strength characteristics of handling equipment are given by Teter (1981), who converted the equations published by Henderson and Perry (1976) into SI units. Some equations for air properties, grain properties, economic properties, and statistical properties are given in the Appendix in Basic computer language. These equations are of use to the engineer, but design of grain storage is not exclusively the milieu of the engineer. General specifications for storage should be drawn up by managers of grain handling systems whether they are farmers, millers, or middlemen. Baille (1982) presented categories of specifications which may be rationalised by engineers using parameters for the interrelated properties of air, grain, moulds, insects, and economics.

* Agricultural Engineer, Postharvest Processing, 6035 Sumner Street, Lincoln, Nebraska 68506, USA.

### Air Properties

Donald B. Brooker (1967) deserves credit for making his and others' psychrometric work available in the ASAE Yearbook of Standards (ASAE 1984). The program, AIRPSY (see Appendix), presents the psychrometric equations with their parameters and solutions. In addition to the usual psychrometric air properties. the deterioration index, DI, suggested by Brooks (1950) and defined by line 530 of AIRPSY, is included in air properties. Since this index skillfully combines temperature and relative humidity into one descriptive term that is related not only to multiplication of microorganisms and insect population dynamics, but also to equilibrium moisture content and rate of dry matter loss, it is useful in both design and operation of storages.

In using the air properties equations, care must be taken regarding the total atmospheric pressure of the air. AIRPSY line 250 defines this pressure as 101.325 Pa, a pressure that may be changed according to the altitude at which the calculations are made. Engineers make calculations at a constant pressure, a condition that seldom exists because wind and convection testify to the variability of total air pressure between two locations, even within the same structure. Neither convection currents nor moisture migration exist when total pressure is actually constant throughout a building. This point is made because texts differ on whether the ratio of vapour pressures, or the ratio of humidity ratios define the relative humidity of air. If total air pressure is constant, then the relative humidity is the ratio of the vapour pressure in the air to the vapour pressure at saturation of the air at this same total pressure. If, in the process of being saturated, the total pressure of the air changes, the ratio of the vapour pressures is the same as the ratio of the weights of vapour associated with weight of dry air, eliminating the need to define the difference between relative humidity and degree of saturation, a term sometimes used for the ratio of kg of water vapour in one kg of dry air to the kg of water vapour at saturation in one kg of dry air.

Either simulation, as that presented by Chang et al (1979), or data analysis of aeration and/or in-bin drying, may need the format of AIRBAL (Appendix), which contains no parameters since it is a 'book-keeping' computation, and not a definition of relationships. Both AIRPOW and AIRVEL use well known constants of the acceleration of gravity (9.807), the seconds in a minute (60), and the kg m per s mechanical equivalent of kW (102.2). But they may be more readily used in design when the equations are rewritten using parametric constants such as specific volume and pressure.

## Dry Matter Loss

Dry matter loss of grain is an oxidative process accompanied by the release of carbon dioxide, water, and heat. It is also accompanied by deterioration in grain quality which may result in visibly damaged kernels, a factor of grain grade and usually grain value. The oxidation may be caused by respiration of microorganisms, embryos, insects, or rodents in the grain. The constants of the oxidation are: for 1 kg of dry matter lost from a cereal grain, 1.07 kg of oxygen is used and 1.47 kg of carbon dioxide, 0.60 kg of water, and 15.7 MJ of heat are released. The process of ventilation or aeration removes the heat and the water generated in the process of oxidation or deterioration. Thus, design aeration rate depends upon the rate of dry matter loss and the properties of the air. Rate of dry matter loss (DML) is commonly expressed as %DML/day. Two different equations with the attendant parameters for paddy are compared by the programs called DMLFAO as taken from Hall (1975) and DMLKSU as taken from Seib et al. (1980). They are given in the Appendix. A check between the output of the two will show considerable percentage difference, especially for paddy below 15% moisture content, but since the rates of respiration are quite low at these moisture levels, the error is not alarming in practice. DMLFAO also shows the laboratory results of Bailey and Gurjar (1920) for brown and milled rice. Design of paddy aeration in $m^3$/tonne min, 'q', made according to DMLFAO has been satisfactory in the humid tropics; therefore, this method was used for programming DMLEMCDI, the program that relates %DML/day to percentage equilibrium moisture content (EMC) and deterioration index; an interrelation between respiration, grain hygroscopicity, and climate. DMLHRS is the same as the KSU equation, but it is solved for hours when a given dry matter loss is entered into the program. This determines the 'allowable hours' for storage if the acceptable level of %DML is known.

## Economics

Grain must be dried soon after it matures if its value as food or feed is to be retained. This value loss is reflected in the visibly damaged kernels which appear in time for moist or wet grain (15.5 to 30% moisture content); furthermore, moist and especially wet grain loses viability rapidly. Economic losses of value, damaged kernels, and germination are programmed in ECOPL. The value loss of paddy is taken from the Malaysian grade standards and prices for rice milled from paddy having different levels of % damaged kernels. The increase in damaged kernels in 10 days is taken from classroom laboratory exercises where students recorded the visibly damaged kernels in paddy at moisture contents from 14% to 25% held for 10 days, sealed and unsealed. The germinability is taken from studies of Roberts (1961). More economic studies of this nature are needed to accurately estimate the 'penalty cost' for failure to properly dry and to keep grain dry.

Harry Th. L. van Ruiten developed a technique of estimating the profit of various rice blends if the value and out-turn of head, large brokens, and small brokens are known and are compared with the value of various sale grades having given constituents of head, large, and small brokens. Since fissuring causes loss of head rice, and consequentially loss of sale value, this economic study may be used to evaluate the risks taken by possible rewetting of paddy by use of aeration. ECORBLD, or a program similar to it, may be used to optimise the profit from rice blending.

Two unique features of the program ECOTCOST are that the transport costs are expressed in terms of the percentage value of the product used in transportation, and the percentage value of the product at any geographic point is computed as the 'price surface' of this location, giving data for locating buying and processing plants to maintain a desired price level. The model is valid for any product and transportation costs as long as these are accurately known, so that the proper parameters may be inserted. The intersection of price surfaces between competing procurement centres forms the boundary of equal sale opportunity between the competing centres.

ECOWT uses the parametric difference between moisture content on a wet basis and on a dry basis to show the true weight difference occasioned by drying. A 5% change in moisture is not a 5% change in weight.

### Physical Grain Properties

GRAER gives a design rate for aeration of maize and paddy, 'q', $m^3$/tonne min. Grain may be aerated when the equilibrium moisture content (EMC) of the air is below the moisture content of the grain. To bring the air into a 'safe' storage moisture, the air should have a deterioration index (DI) of 3.8 or less for maize, and 5.0 or less for paddy. This concept is explained in greater detail by Teter (1982). If the diurnal variation of the DI is not available in the records, but an estimate can be made of the average daily relative humidity, the program estimates the hours that fans may be operated each day by assuming that the daily relative humidity resembles a sine curve. This is an assumption that is not borne out well by weather data, which show the daily variation in relative humidity to be less symmetrical than the sine wave. If the program shows that aeration is not possible, dehumidifying the air by use of supplemental heat, mechanical condensation of water, or chemical absorption is necessary. Some climates of the humid tropics are too wet for too long to make natural air aeration practical for cereal grain.

The resistance of grain to airflow, the airflow rate, and the practical static pressure limits on the aeration fan determine the design depth of grain storage. GRDEPTH computes the allowable depth of design, a practical matter in the selection between horizontal and silo storage. GRDEPTH as given applies only to paddy. An inverse of GRDEPTH, GRSPR, gives the Pascals of pressure for a given depth of grain subjected to a given air flow, 'q'. The parameters for resistance of grain to airflow may be calculated from the curve fitting program, SOL; then these parameters may be used in GRDEPTH to give a program for other grains.

GREMC contains the parameters for the EMC of the 11 grains listed for the Chung-Pfost EMC equation given in ASAE (1984). The program gives either the EMC when the °C and %RH are entered, or the %RH when °C and EMC are entered. The programs GRDK, GRGER, and GRVAL are the same as those given in ECOPL.

### Insect Population Dynamics

Estimations of population increase of insects may be used to optimise fumigation and to estimate benefits or losses wrought by environmental control. Some evidence indicates that *Sitophilus* spp. have a maximum rate of increase at a DI of 2.9, too dry for moulding to be a serious threat to grain. Thus, insect increase in the warm tropics may be highest at the climate for optimum storage moisture. Two models are presented for insect population dynamics for *Sitophilus oryzae*. The first, INSINC, assumes that the grain is in equilibrium at the input DI; the second, INSINC2, inputs the actual moisture content and is modelled after the work of Evans (1982). Both models assume that the rate of population increase is normally distributed as a function of the environment. This preliminary work for maize and wheat shows some promise for design and operation of insect control in structures. The gaseous environment factor has not been given parameters.

### Statistics

Statistical parameters have special meaning in design of storage and handling systems. The usual agricultural statistics method of referring to two data sets as being significantly or highly significantly different is not satisfactory for decision making in design. When a study is made of two systems, the difference between systems under study should be expressed as a probability of difference, which may range from 0.5 to 1.0, 0.5 being the probability that there is no difference between the two systems. The program STAT is taken from the equation for the normal curve given in line 50 of the program, and the parameters given by Abramowitz and Stegun (1964) for the area under the normal curve. The program calculates the area under the normal curve when the value of the number of standard deviations from the mean is entered. For any two normally distributed sets of data, the intersection of the two data distribution curves is the difference between the means of the two sets divided by the sum of the standard deviations of the two sets. This gives a direct reading of probability of difference which may be evaluated for design. If one system is better than another with the probability of 0.55 or more, and the systems cost the same, it is logical to select the system that is probably superior.

If the criterion for study is a percentage such as percentage loss, the data are not normally distributed, but have a markedly severe skew towards zero. The data must be transformed to a normal distribution for analysis of variance, and after the estimate is made for the variance of the transformed values, these parameters must be reconverted to the original percentage loss data. This process is tedious without the computer, but program STATP is designed to make these transformations, computing the standard deviations and reconverting them to percentage loss. Note that the standard deviation to the right of the mode is not equal to the standard deviation to the left for skewed data. The test for probability of difference is slightly different for skewed data because the input value 'x' into STAT is now the difference between the means divided by the sum of the standard deviation to the right of the lesser set plus the standard deviation to the left of the greater set.

## Solutions for Parameters

Considerable data are available that may be converted to parameters for use in the computer. For example, the ASAE Yearbook (ASAE 1984) contains a number of curves for the resistance of grains to airflow. For design, these curves may be estimated as straight line segments of log x vs. log y curves. SOL is a curve-fitting program which may be used to rapidly find the parameters of resistance for any grain published by taking points on the curve at intervals judged suitable for straight line log-log relationship.

## Future Storage Designs

Controlled ventilation of both bulk and bagged storage shows economic benefit that should bring about the construction of many more 'tight', even oxygen limiting, storage buildings in the future. Where ambient conditions preclude the feasibility of aeration or controlled ventilation, the conditioning of the air to bring the deterioration index to a level that is suitable for aeration should become a part of design considerations.

## References


Abramowitz and Stegun 1964. Handbook of mathematical functions. National Bureau of Standards, USA.

ASAE (American Society of Agricultural Engineers), 1984. Agricultural engineers yearbook of standards. St. Joseph, MI, USA, ASAE 314–315.

Bailey, C.H., and Gurjar, A.M. 1920. Respiration of cereal plants and grain. III. Respiration of rice paddy and milled rice. Biological Chemistry, 44,1.

Baille, T.S. 1982. Construction of warehouses. In: Paddy deterioration in the humid tropics. Eschborn, Fed. Rep. of Germany, GTZ, 121–141.

Brooker, D.B. 1967. Mathematical model of the psychrometric chart. Transactions of the American Society of Agricultural Engineers, 10, 558–560, 563.

Brooks, C.E.P. 1950. Climate in everyday life. London, Ernest Benn Ltd, 314 p.

Chang, Dong Il, Chung, D.S., and Pfost, H.B. 1979. Simulation of natural air drying of rough rice. In: Proceedings of workshop on grain post-harvest technology. College, Laguna, Philippines, SEARCA, 110–147.

Evans, D.E. 1982. The influence of temperature and grain mosture content on the intrinsic rate of increase of *Sitophilus oryzae (L).* Coleoptera: Curculionidae, Journal of Stored Products Research, 18, 55–66.

Hall, D.W. 1975. Handling and storage of food grains in tropical and subtropical areas. FAO Development Paper,No. 90. 40 p.

Henderson, S.M., and Perry, R.L. 1976. Agricultural process engineering (3rd ed.), Westport, CN, USA, AVI Publishing Co., 199–236.

Roberts, E.H. 1961. The viability of rice seed in relation to temperature, moisture content and gaseous environment. Annals of Botany, 25, 381–390.

Seib, P.A., Pfost, H.B., Sukabdi, A., Rao, V.G. and Burroughs. R. 1980. Spoilage of rough rice measured by evolution of carbon dioxide. In: Grain quality improvement. Proceedings of the 3rd annual workshop on grains post-harvest technology. College, Laguna, Philippines, SEARCA, 75–95.

Teter, N.C., 1981. Grain storage. College, Laguna, Philippines, SEARCA, 49–59.

Teter, N.C., 1982. Aeration of grain. In: Paddy deterioration in the humid tropics. Eschborn, Fed.Rep. of Germany, GTZ, 142–153.

## APPENDIX: DESIGN PARAMETERS FOR STORAGE AND HANDLING SYSTEMS FOR GRAIN

The following computational programs in the Basic language are examples of the use of parameters in designing grain storage and handling. The names and brief descriptions of the programs are followed by the printouts of the actual computational routines.

AIRBAL: Demonstrates a sub-routine useful in analysing work done by observed air properties in and out of bins, and also in simulation of drying and aeration where heat and water exchanges are made between grain and air.

AIRPAS: For an input of temperature in °C gives the Pascals of vapour pressure at saturation.

AIRPOW: Air power is printed when the total Pascals of pressure, the specific volume and the air delivery are entered.

AIRPSY: From an input of dry bulb temperature, °C, and wet bulb temperature, °C, gives:
Pascals of vapour pressure at saturation at dry bulb, C°
Pascals of vapour pressure at saturation at wet bulb, C°
Latent heat of vaporisation, J/kg of dry air
Pascals of vapour pressure in the air
Relative humidity, %
Specific volume of air, $m^3$/kg of dry air
Humidity ratio, kg of water/kg of dry air
Dew point temperature, °C
Deterioration index
Enthalpy, J/kg of dry air.

AIRVEL: Computes the velocity of air when the specific volume and velocity pressure of the air is known.

DMLEMCDI: Computes the equilibrium moisture content of paddy for a given deterioration index and then estimates the %DML (Dry Matter Loss)/Day for paddy in equilibrium with the ambient climate.

DMLFAO: Estimates dry mater loss, %DML/Day, for paddy, maize, brown rice, and rice using curves published in an FAO manual by Hall (1975) and data taken by Bailey and Gurjar (1920).

DMLHRS: Computes hours estimated for long and medium grain paddy to have a given %DML when stored at a given temperature, °C, and % moisture content.

DMLKSU: Computes dry matter loss of long and medium grain paddy using inputs of time, moisture content and temperature (from Kansas State University).

ECOPL: Estimates % value loss of paddy of a given moisture content in a given number of days, the % damaged kernels of a given moisture content in 10 days, and the loss of germination.

ECORBLD: For a given mill output estimates the value of rice blends of given grades when the market prices are known. This is developed from the work of Harry van Ruiten.

ECOTCOST: Computes the price surface, that is the % value of paddy or rice at a given geographic location related to the place of '100% value'. Parameters must be changed to fit the conomics of the geographic location.

ECOWT: Given the initial moisture content and wet of wet grain, computes the dried weight of the grain.

GRAER: Prints a recommended rate of aeration, 'q', ($m^3$/tonne min) for paddy or maize stored at a given moisture content when either the allowable hours per day of fan operation or the 24 hour average relative humidity is known.

GRDEPTH: Estimates the depth of paddy for a given aeration rate, 'q', and Pascals of static pressure.

GRDK: Estimates the increase in visibly damaged kernels in paddy held for 10 days at a moisture content above 15%.

GREMC: Calculates the equilibrium moisture content (EMC) of 11 grains for a given temperature, °C, and either % relative humidity, or % moisture content, using the Chung-Pfost equation.

GRSPR: Gives the Pascals of pressure required to aerate paddy placed at a given depth and subjected to a given air flow, 'q'. May be expanded for other grains.

GRSPR2: Same except for brown rice and rice.

GRVAL: Estimates the % value of paddy held for an input number of days at an input moisture content above 15%.

INSINC: Runs the population dynamics of *Sitophilus sp.* when the initial population, °C, %RH, and number of weeks are fed into the program.

INSINC2: A second estimate of population

dynamics of *Sitophilus sp.* for an 'out-of-equilibrium' situation where the % moisture content is an input. From data of Evans (1982).

STAT: Properties of the normal curve, 'y' values, and area under the curve for a given number of standard deviations, 'x'.

STATP: A suggested way to transform % loss data into a normal distribution for reliable anaylsis of variance and then to reconvert the statistical parameters to actual loss.

SOL: Solves the parameters 'A' and 'B' in log-log curves such as P = B*(Q^A), the typical grains pressure resistance curve.

Page A4

**AIRBAL**

```
5   DIM T(20)
10  PRINT "MASS BALANCE BETWEEN AIR AND PADDY"
40  CLEAR
50  INPUT "TYPE, 'W-CHANGE' OR 'H-CHANGE' OR 'EXIT'"; S$
60  IF S$="W-CHANGE" THEN GOSUB 90 : GO TO 160
70  IF S$="H-CHANGE" THEN GOSUB 90 : GO TO 220
80  IF S$="EXIT" GO TO 280
90  INPUT "Q (m^3/time unit)"; Q
100 INPUT " V, Specific Volume (m^3/kg)"; V
110 INPUT "NO. OF TIMES FOR t" ; N
120 FOR I=1 TO N
130 PRINT " t ";: PRINT I;: INPUT T : T(I)=T
140 NEXT I
150 RETURN
160 INPUT "HUMIDITY RATIO IN AIR OUT, HO"; HO
170 INPUT "HUMIDITY RATIO IN AIR IN, HI" ; HI
180 FOR I=1 TO N
190 SUM=SUM+T(I)*(Q/V)*(HO-HI)
200 NEXT I
205 PRINT "WATER REMOVED FROM PADDY = "; SUM; "kg/tonne"
210 GO TO 40
220 INPUT "ENTHALPY IN AIR OUT" ; HO1
230 INPUT "ENTHALPY IN AIR IN" ; H2
240 FOR I=1 TO N
250 SUM=SUM+T(I)*(Q/V)*(HO1-H2)
260 NEXT I
265 PRINT "HEAT REMOVED FROM PADDY = " ;SUM; "kJ/tonne"
270 GO TO 40
280 END
```

**AIRPAS**

```
10  PRINT "PASCALS OF VAPOR PRESSURE AT SATURATION"
20  INPUT "TEMPERATURE, C°" ; CEN
30  T=CEN+273.16
40  X= -27405.526+(T*97.5413)-(.146244*T^2)+(.12558*10^(-3)*T^3)-
       (.48502*10^(-7)*T^4)
50  Y= (4.34903*T)-(.3938*10^(-2)*T^2)
60  PAS=EXP(X/Y+16.91134)
70  PRINT "SATURATION VAPOR PRESSURE = "; PAS
80  INPUT "ANY MORE DATA, (Y/N)"; A$
90  IF A$="Y" GO TO 20 : END
```

Page A5

**AIRPOW**

```
10  PRINT "GIVES kW OF AIR POWER"
15  PRINT "*******************"
20  PRINT
30  INPUT "SPECIFIC VOLUME OF AIR , m^3/kg"; SVOL
40  INPUT "PASCALS OF TOTAL PRESSURE" ; P
50  INPUT "VOLUME OF AIR DELIVERY, m^3/min"; Q
60  PRINT
70  KW=(Q/SVOL/60)*(P*SVOL/9.80665)/102.2
80  PRINT "AIR POWER="; KW          "kW"
90  INPUT "ANY MORE DATA? (Y/N)"; A$
100 IF A$="Y" GOTO 30
110 END
```

**AIRVEL**

```
10  PRINT "VELOCITY OF AIR FOR A GIVEN VEL. PRESSURE AND SP. VOLUME"
15  PRINT "************************************************"
20  PRINT
30  INPUT "SPECIFIC VOLUME OF THE AIR (m^3/kg)"; SVOL
40  INPUT "PASCALS OF VELOCITY PRESSURE" ; P
50  PRINT
55  B=9.80665/SVOL
60  V=(2*9.80665*P/B)^.5
70  PRINT "VELOCITY IN m/s = " ; V
80  PRINT
90  INPUT "ANY MORE DATA? (Y/N)" ; A$
100 IF A$="Y" GOTO 30
110 END
```

Page A6

**AIRPSY**

```
5   DIM A(9)
7   FOR I=0 TO 8 : READ A(I) : NEXT I
10  PRINT "PROPERTIES OF AIR"
20  INPUT "TEMPERATURE, C°"; CEN
25  INPUT "WET BULB TEMPERATURE, C°" ; CWB
30  TWB=CWB+273.16 : T=CEN+273.16
40  X=-27405.526+(T*97.5413)-(.146244*T^2)+(.12558*10^(-3)*T^3)-
       (.48502*10^(-7)*T^4)
50  Y=(4.34903*T)-(.3938*10^(-2)*T^2)
60  PAS=EXP(X/Y+16.91134)
70  PRINT "SATURATION VAPOR PRESSURE= "; PAS; "Pascals"
170 LHT= 2502535.259-(2385.76424*CEN)
180 PRINT "LATENT HEAT OF VAPORIZATION=" ; LHT; "J/kg"
190 HX=-27405.526+(97.5413*TWB)-(.146244*TWB^2)+
       (.12558E-3*TWB^3)-(.48502E-7*TWB^4)
200 HY=(4.34903*TWB)-(.3938E-2*TWB^2)
210 PASWB=EXP(HX/HY+16.91134)
220 PRINT "VAPOR PRESSURE AT WET BULB SATURATION"; PASWB; "P"
230 LHTWB=2502535.259-(2385.76424*CWB)
240 PAV=500 : PATM=101325
260 BH=1006.9254*(PASWB-PATM)*(1+(.15577*PAV/PATM))
270 BG=.62194*LHTWB : B=BH/BG
280 PV2=PASWB-(B*(CWB-CEN))
290 IF PV2-PAV>50 THEN PAV=PAV+25 : GOTO 260
300 PRINT "VAPOR PRESSURE = " ; PAV; "P"
310 RH= PAV/PAS*100 : PRINT "% RELATIVE HUMIDITY=" ; RH
330 V= 287*T/(PATM-PAV) : PRINT "SPECIFIC VOLUME = "; V ; "m^3/kg"
350 HRAT=PAV*.6219/(PATM-PAV)
360 PRINT "HUMIDITY RATIO=" ; HRAT; "kg water/kg air"
390 DATA 19.5322,13.6626,1.17678,-.189693,.087453,-.0174053,
         .00214768,-.138343E-3,.383E-5
400 F=LOG(PAV*.00145)
410 FOR I=0 TO 8
420 S=F^I*A(I) : SUM=SUM+S : NEXT I
450 TAS=SUM-17.78
460 PRINT "DEW POINT TEMPERATURE="; TAS; "C°"
470 HAIR=1006.9254*CEN : HSW=4186.8*HRAT*TAS
```

```
490 HFG=(2502535.259-(2385.76424*TAS))*HRAT
500 HSV=1875.6864*HRAT*(CEN-TAS) : ENT=HAIR+HSW+HFG+HSV
510 PRINT "ENTHALPY = " ; ENT; "J/kg dry air"
530 DI=(RH-65)*PAS*10^(-4) : PRINT "DETERIORATION INDEX = " ; DI
550 INPUT "ANY MORE DATA? (Y/N)" ; A$
560 IF A$="Y" GOTO 20
```

Page A7

## DMLEMCDI

```
10 PRINT "% DML/DAY OF PADDY IN EQUILIBRIUM WITH DI"
20 PRINT : INPUT "DETERIORATION INDEX"; DI : PRINT "DI="; DI
30 IF DI>=12 THEN A=.82 : B=13.16 : GOTO 100
40 IF DI>=5 THEN A=.486 : B=9.149 : GOTO 100
50 IF DI>=2.78 THEN A=.369 : B=8.567 : GOTO 100
60 IF DI<0 THEN GOTO 20 ELSE A=.151 : B=7.96 : GOTO 100
100 P=EXP(A*DI-B) : PRINT "% DML/DAY =" ; P
110 INPUT "ANY MORE DATA? (Y/N); A$
120 IF A$="Y" GOTO 20 : END
```

## DMLFAO

```
10 PRINT "%DML/DAY FOR PADDY, LOONZAIN OR RICE"
20 INPUT "TYPE OF GRAIN, PD, LZ, OR R"; G$
30 INPUT "MOISTURE CONTENT, % wb" ; M
40 IF M>17 THEN PRINT "VALID ONLY FOR 17>M>10" : GOTO 30
50 IF G$="PD" GOTO 100
52 IF G$="LZ" GOTO 110
54 IF G$="R" GOTO 120 ELSE GOTO 20
60 D=.68*(10^(A*M-B))
70 PRINT "%DML/DAY = " ; D
80 INPUT "ANY MORE DATA? (Y/N)" ; A$
90 IF A$="Y" GOTO 20 : END
100 IF M>13.2 THEN A=.44 : B=9.08
102 IF M<=13.2 THEN A=.21 : B=6.04
105 GOTO 60
110 IF M>13.7 THEN A=.44 : B=9.41
112 IF M<=13.7 THEN A=.17 : B=5.67
115 GOTO 60
120 IF M>14.1 THEN A=.49 : B=10.48
122 IF M<=14.1 THEN A=.16 : B=5.83
125 GOTO 60
```

Page A8

## DMLKSU

```
10 PRINT "%DML FOR LONG OR MED. GR. PADDY, KSU EQUATION"
20 INPUT "TYPE OF GRAIN, LONG (L) OR MEDIUM (M)"; G$
30 IF G$="L" GOTO 50
40 A=.000914 : C=.654 : D=.03756 : E=33.61 : GOTO 60
50 A=.001889 : C=.7101 : D=.0274 : E=31.63
60 INPUT "TEMPERATURE, C°"; CE
70 T=CE*1.8+32
80 INPUT "MOISTURE CONTENT, %wb"; M
90 W=M/100
100 INPUT "STORAGE TIME, DAYS"; DAYS
110 ST=DAYS*24/1000
120 X=A*ST^C : Y=EXP(D*(T-60)) : Z=EXP(E*(W-.14))
130 DML=100*(1-(EXP(-X*Y*Z)))
140 PRINT "% DRY MATTER LOSS = "; DML
150 INPUT "ANY MORE DATA? (Y/N)"; A$
160 IF A$="Y" GOTO 20 : END
```

## DMLHRS

```
10 PRINT "DAYS FOR LONG OR MED. GRAIN PADDY TO LOSE A GIVEN %DM"
20 INPUT "TYPE OF GRAIN, LONG (L) OR MEDIUM (M)" ; G$
30 IF G$="L" GOTO 50
40 A=.000914 : C=.654 : D=.03756 : E=33.61 : GOTO 60
50 A=.001889 : C=.7101 : D=.0274 : E=31.63
60 INPUT "TEMPERATURE, C°"; CE : T=1.8*CE+32
70 INPUT "% MOISTURE CONTENT, w.b."; M : W=M/100
80 INPUT "% DRY MATTER LOSS"; ML : DML=ML/100
90 X=LOG(1-DML) : Y=EXP((T-60)*D) : Z=EXP((W-.14)*E)
100 ST=(-X/(A*Y*Z))^(1/C)
110 DAYS=ST*1000/24
120 PRINT "ALLOWABLE DAYS OF STORAGE ="; DAYS
130 INPUT "ANY MORE DATA? (Y/N)"; A$
140 IF A$="Y" GOTO 20
150 END
```

Page A9

## ECOPL

```
10 INPUT "SELECT THE LOSS PROG. : VALUE (V), %DAM. K. (DK), GERMIN-
     ATION (G) OR EXIT (E)" ; L$
20 IF L$="V" GOTO 100
22 IF L$="DK" GOTO 200
24 IF L$="G" GOTO 300
30 IF L$="E" GOTO 440
100 INPUT "DAYS AFTER 2000 HOURS ON REAPING DAY" ; D
110 INPUT "% MOISTURE CONTENT > 15%, w.b." ; M
120 S=(M-15)^.69 : T=D^.35 : V=100-(3.51*S*T)
130 PRINT "PERCENT VALUE =" ; V
140 INPUT "ANY MORE DATA? (Y/N)"; A$
150 IF A$="Y" GOTO 100 ELSE GOTO 10
200 INPUT "MOISTURE CONTENT, %M=> 16" ; MC
210 D=EXP((MC-16)*.29)-1
220 PRINT "VISIBLY DAMAGED PADDY KERNELS AFTER 10 DAYS =" ; D
230 INPUT "ANY MORE DATA? (Y/N)"; B$
240 IF B$="Y" GOTO 200 ELSE GOTO 10
300 INPUT "INITIAL GERMINATION, %" ; GI
310 INPUT "% MOISTURE CONTENT" ; CM
320 INPUT "TEMPERATURE, C°"; TE
330 A=.069*TE : B=.159*CM : K=10^(5.686-A-B)
340 PRINT "WEEKS TO LOSE 50% GERMINATION =" ; K
350 INPUT "WEEK AT WHICH GERMINATION DESIRED" ; W
360 R=W/K
362 IF R<1 THEN GOTO 400
370 G=GI*(1-(GW/100))
380 PRINT "GERMINATION AT 'W' WEEKS IS" ; G
385 INPUT "ANY MORE DATA? (Y/N)" ; C$
390 IF C$="Y" GOTO 300 ELSE GOTO 10
400 GW=7.8*(EXP(R*2)-1)
410 GOTO 370
420 GW=100-(7.8*((EXP((2-R)*2))-1))
430 GOTO 370
435 GW=100
437 GOTO 370
440 END
```

Page A10

## ECORBLD

```
10 PRINT "ESTIMATE OF VALUE OF RICE BLENDS"
30 INPUT "MILLED RICE CONTENT, %= H, LB, SB" ; MH, MB, MS
40 INPUT "SALE GRADE, %= H, LB, SB" ; EH,EB,ES
50 REM "PRICE VALUE CONSTANTS FOR RICE SALE ARE"
60 HP=3 : BP=2.4 : SP=1.32
70 HD=10*(MH-EH) : BR=10*(MB-EB) : SB=10*(MS-ES)
80 INPUT "MILLED RICE VALUE" ; MP
90 INPUT "BLENDED RICE VALUE"; EP
100 COST=MP-((HD*HP)+(BR*BP)+(SB*SP))
110 PROF=EP-COST
120 PRINT "RELATIVE PROFIT" ; PROF
130 INPUT "ANY MORE DATA? (Y/N)" ; R$
140 IF R$="Y" GOTO 30 : END
```

## ECOWT

```
10 PRINT "WEIGHT OF DRIED GRAIN"
30 INPUT "ORIGINAL WEIGHT IN kg" ; KG
40 INPUT "ORIGINAL % MOISTURE" ; MO
50 INPUT "FINAL % MOISTURE" ; M
60 DM=((100-MO)/100)*KG : DWT=DM/(100-M)*100
70 PRINT "DRIED WEIGHT=" ; DWT; "kg"
80 INPUT "ANY MORE DATA? (Y/N)" ; A$
90 IF A$="Y" GOTO 30 : END
```

## GRDK

```
10 PRINT "% DAMAGED PADDY KERNELS IN 10 DAYS"
20 INPUT "% MOISTURE CONTENT" ; M
30 IF M<=16 GOTO 45
40 DK=EXP((M-16)*.29)-1 : GOTO 50
45 DK=0
50 PRINT "% INCREASE IN DAMAGED KERNELS IN 10 DAYS" ; DK
60 INPUT "ANY MORE DATA? (Y/N)" ; A$
70 IF A$="Y" GOTO 20  :  END
```

## GRVAL

```
10  PRINT "ESTIMATE OF PADDY VALUE HELD SEVERAL DAYS"
20  INPUT "DAYS AFTER 2000 HOURS ON DAY OF HARVEST" ; D
30  INPUT "MOISTURE CONTENT, %" ; M
40  IF M<=15 THEN PRINT "VALUE IS 100%" : GOTO 70
50  W=100-(3.51*(D^.35)*((M-15)^.69))
60  PRINT "PADDY VALUE IS NOW" ; W
70  INPUT "ANY MORE DATA? (Y/N)" ; A$
80  IF A$="Y" GOTO 20  :  END
```

Page A11

## ECOTCOST

```
10  INPUT "TRANSPORT COSTS FOR PADDY (P) OR RICE (R)?" ; C$
20  REM "METHODS ARE 1(HAND), 2(OX CRT), 3(TR PAVING), 4(TR GRAV),
      5(TR DIRT), 6(TR BAD RD), 7(BOAT)" : SUM=0
30  IF C$="R" GOTO 300
40  INPUT "METHOD, DISTANCE, km = M,KM" ; M,KM
50  IF M=1 THEN SUM=SUM+(.839*KM) : GOTO 200
60  IF M=2 THEN SUM=SUM+(.325*KM) : GOTO 200
70  IF M=3 THEN GOTO 120
80  IF M=4 THEN GOTO  140
90  IF M=5 THEN GOTO  160
100 IF M=6 THEN GOTO  180
110 IF M=7 THEN SUM=SUM+(1.02+(KM^.45*.568)) : GOTO 200
120 IF KM>=32.2 THEN SUM=SUM+(.33+KM*.086)  : GOTO 200
130 SUM=SUM+(.33+KM*.099) : GOTO 200
140 IF KM>=32.2 THEN SUM=SUM+(.4+KM*.106) : GOTO 200
150 SUM=SUM+(.4+KM*.116) : GOTO 200
160 IF KM>=32.2 THEN SUM=SUM+(.4+KM*.115) : GOTO 200
170 SUM=SUM+(.4+KM*.132) : GOTO 200
180 IF KM>=32.2 THEN SUM=SUM+(.46+KM*.165) : GOTO 200
190 SUM=SUM+(.46+KM*.198)
200 INPUT "ANY MORE DATA? (Y/N)" ; D$
210 IF D$="Y" GOTO 40
220 PS=100-SUM
230 PRINT "PRICE SURFACE (% VALUE) OF PADDY=" ; PS : END
300 INPUT "METHOD AND DISTANCE (M,KM) FOR RICE" ; M,KM
310 IF M=1 THEN SUM=SUM+(.325*KM) : GOTO 460
320 IF M=2 THEN SUM=SUM+(.134*KM) : GOTO 460
330 IF M=3 THEN GOTO 380
340 IF M=4 THEN GOTO 400
350 IF M=5 THEN GOTO 420
360 IF M=6 THEN GOTO 440
370 IF M=7 THEN SUM=SUM+(.347+KM*.00616) : GOTO 460
380 IF KM>=32.2 THEN SUM=SUM+(.15+KM*.033) : GOTO 460
390 SUM= SUM+(.15+KM*.040) : GOTO 460
400 IF KM>=32.2 THEN SUM=SUM+(.13+KM*.040) : GOTO 460
410 SUM= SUM+(.13+KM*.046) : GOTO 460
420 IF KM>=32.2 THEN SUM=SUM+(.19+KM*.046) : GOTO 460
430 SUM=SUM+(.19+KM*.053) : GOTO 460
440 IF KM>=32.2 THEN SUM=SUM+(.21+KM*.066) : GOTO 460
450 SUM=SUM+(.21+KM*.080)
460 INPUT "ANY MORE DATA? (Y/N)" ; R$
470 IF R$="Y" GOTO 300
480 PR=100-SUM
```

Page A12

## GRAER

```
10  PRINT "AERATION RATE FOR MAIZE OR PADDY"
20  PRINT
30  INPUT "MOISTURE CONTENT, %" ; M
40  INPUT "TYPE OF GRAIN, PADDY (P) OR MAIZE (MA)" ; A$
50  IF A$="P" THEN GOSUB 230
60  IF A$="MA" THEN GOSUB 320 ELSE GOTO 30
230 PRINT "AERATION RATE (m^3/tonne minute) FOR PADDY"
240 A=.44 : B=6.08
250 GOSUB 420
260 END
320 PRINT "AERATION RATE (m^3/tonne minute) FOR MAIZE"
330 A=.27 : B=3.33
340 GOSUB 420
350 END
420 X=10^((M*A)-B))/1.721
505 INPUT "HRS/DAY OR AV. DAILY RH KNOWN? (HD/ACRH)" ; RH$
506 IF RH$="ACRH" GOSUB 600
507 INPUT "HRS/DAY OF FAN OPERATION=" ; HRS
510 Q=X/HRS
520 PRINT "q=" ; Q ; "m^3/tonne minute"
530 RETURN
600 INPUT "AVERAGE DAILY RELATIVE HUMIDITY (24 hrs.)" ; ACRH
610 A=100-ACRH : HRS=0
615 INPUT "ALLOWABLE RH OF AERATION AIR" ; GRH
620 FOR I = 1 TO 48
630 Y=A*SIN(.1309*I)
640 RH=ACRH+Y
650 IF RH<GRH THEN HRS=HRS+.5
660 NEXT I
670 RETURN
```

## GRSPR

```
10  PRINT "RESISTANCE OF PADDY TO AIR FLOW"
20  PRINT
30  INPUT "DEPTH OF PADDY, m" ; D
40  INPUT "AERATION RATE, m^3/tonne min." ; Q
50  V=.58*Q*D
60  P=V^1.32*D*53.7
70  PRINT "PASCALS OF PRESSURE=" ; P
80  INPUT "ANY MORE DATA? (Y/N)", A$
90  IF A$="Y" GOTO 30
100 END
```

Page A13

## GRSPR2

```
5   PRINT "RESISTANCE OF BROWN RICE OR MILLED RICE TO AIR FLOW"
10  INPUT "BROWN RICE (BR) OR MILLED RICE (MR)" ; A$
20  IF A$="BR" GOTO 100
30  A=1.08 : B=8.507 : C=1.161 : D=8.679
40  INPUT "AIR FLOW (m^3/s per m^2 of floor)" ; Q
50  IF Q>.04 GOTO 70
60  P=EXP(A*LOG(Q)+B) : GOTO 80
70  P=EXP(C*LOG(Q)+D)
80  PRINT "PASCALS PER m OF DEPTH" ; P
90  END
100 A=1.079 : B=8.5806 : C=1.122 : D=8.717
110 GOTO 40
```

## GRDEPTH

```
10  PRINT "DEPTH OF PADDY ALLOWABLE"
20  INPUT "PASCALS OF STATIC PRESSURE=" ; P
30  INPUT "AERATION RATE, m^3/tonne min." ; Q
40  A=Q^(-.565) : B=P^.431 : D=.245*A*B
50  PRINT "METERS OF DEPTH ALLOWABLE=" ; D
60  INPUT "ANY MORE DATA? (Y/N)" ; A$
70  IF A$="Y" GOTO 20  :  END
```

## SOL

```
10  PRINT "FOR WRITING EQUATIONS FOR GRAINS RESISTANCE"
20  PRINT "THIS SOLVES FOR A AND B IN LN(P)=A*LN(Q)+B"
30  PRINT
40  INPUT "X VALUES FOR X1,X2" ; W,X
50  INPUT "Y VALUES FOR Y1,Y2" ; Y,Z
60  X1=LOG(W) : X2=LOG(X) : Y1=LOG(Y) : Y2=LOG(Z)
70  A=(X2-X1)/(Y2-Y1)
80  B= -A*Y1+X1
90  PRINT "A AND B = " ; A  "and" ; B
100 INPUT "ANY MOR DATA? (Y/N)" ; A$
110 IF A$="Y" GOTO 40
120 END
```

## GRGER

Note: For obtaining the estimated viability of paddy in storage, use the ECOPL program.

Page A14

## GREMC

```
10 PRINT "CHUNG-PFOST EQUILIBRIUM MOISTURE FOR GRAINS"
20 PRINT "**************************************"
30 PRINT "Codes for grains are: Barley=BA, Beans=BE, Maize=MA, Ground"
40 PRINT "nuts in the pod=GN, Peanuts shelled=GS, Paddy=P, Sorghum=SOR"
50 PRINT "Soyabeans=SB, Durham wheat=DW, Hard wheat=HW, S. wh.= SW"
60 INPUT "CODE FOR THE SELECTED GRAIN" ; A$
70 IF A$="BA" GOTO 140
75 IF A$="BE" GOTO 150
80 IF A$="MA" GOTO 160
85 IF A$="GN" GOTO 180
90 IF A$="GS" GOTO 170
95 IF A$="P" GOTO 190
100 IF A$="SOR" GOTO 200
105 IF A$="SB" GOTO 210
110 IF A$="DW" GOTO 220
120 IF A$="HW" GOTO 230
130 IF A$="SW" GOTO 240
140 A=761.66 : B=19.889 : C=91.323 : E=.33363 : F=.050279 : GOTO 250
150 A=962.58 : B=15.975 : C=160.629 : E=.43001 : F=.062596 : GOTO 250
160 A=312.30 : B=16.958 : C=30.205 : E=.33872 : F=.058970 : GOTO 250
170 A=254.90 : B=29.243 : C=33.892 : E=.18948 : F=.034196 : GOTO 250
180 A=522.01 : B=37.903 : C=12.345 : E=.16510 : F=.026383 : GOTO 250
190 A=594.61 : B=21.732 : C=35.703 : E=.29394 : F=.046015 : GOTO 250
200 A=1099.67 : B=19.644 : C=102.849 : E=.35649 : F=.050907 : GOTO 250
210 A=328.3 : B=13.917 : C=100.288 : E=.41631 : F=.071853 : GOTO 250
220 A=921.65 : B=18.077 : C=112.35 : E=.37761 : F=.055318 : GOTO 250
230 A=529.43 : B=17.609 : C=50.998 : E=.35616 : F=.056788 : GOTO 250
240 A=726.49 : B=23.607 : C=35.662 : E=.27908 : F=.042360 : GOTO 250
250 INPUT "IS ENTRY IN %MC OR %RH? (MC/RH)" ; B$
260 IF B$="MC" GOTO 360
270 INPUT "RELATIVE HUMIDITY, %" ; RH
280 RH=RH/100 : Z=LOG(RH)
290 INPUT "TEMPERATURE,C°"; T
300 M=(E-(F*LOG((-1)*(T+C)*Z))) : MC=100*(M/(1+M)
310 PRINT "THE WET BASIS % MOISTURE CONTENT=" ; MC
320 INPUT "ANY MORE DATA? (Y/N)" ; C$
330 IF C$="Y" GOTO 20 : END
360 INPUT "MOISTURE CONTENT, % WET BASIS" ; MC
370 INPUT "TEMPERATURE, C°" ; T
380 M=MC/(100-MC) : WE=EXP(-1)*B*M) : R=EXP(WE*(-1)*A/(T+C))
390 RH=R*100
400 PRINT "THE % EQUILIBRIUM RH=" ; RH
410 INPUT "ANY MORE DATA? (Y/N)" ; D$
420 IF D$="Y" GOTO 20 : END
```

Page A15

## INSINC

```
10 PRINT "ESTIMATES SITOPHILUS POPULATION"
20 INPUT "INITIAL SIT. POPULATION" ; PO
30 INPUT "DETERIORATION INDEX OF CLIMATE" ; DI
40 REM "THE MAX. 28 DAY RATE OF INCREASE IS TAKEN AS 15"
50 R=15 : X=(2.9-DI)/3.3 : Y=.399*EXP(-X^2/2) : RI=R*Y/.399
60 RM=LOG(RI)/4
70 INPUT "WEEKS AFTER INITIAL POP. INPUT, W" ; W
80 IF W>22 THEN 250
90 IF W>11 THEN 200
100 PF=PO*EXP(RM*W)
110 PRINT "POPULATION AT 'W' WEEKS =" ; PF
120 INPUT "ANY MORE DATA? (Y/N)" ; A$
130 IF A$="Y" GOTO 20 : END
200 PF=(2*PO*EXP(RM*11))-(PO*EXP(RM*(22-W))) : GOTO 110
250 PF=2*PO*EXP(RM*11) : GOTO 110
```

## INSINC2

```
10 PRINT "ALTERNATE SITOPHILUS POPULATION MODEL"
20 INPUT "INITIAL POPULATION, PO" ; PO
30 INPUT "VAPOR PRESSURE OF AIR, Pascals" ; VP
40 INPUT "MOISTURE CONTENT OF GRAIN" ; MC
50 A=(0.0794*MC^2)-(2.509*MC)+21.81 : A=A/1000
60 Z=3.18-(A*(VP-400))
70 B=.195+1.485*Z
80 C=(EXP(B)+EXP(-B))/2 : T=4.8*C
85 PRINT "ESTIMATED WEEKS FOR 1 GENERATION" ; T
90 Y=.399/(EXP(Z^2/2))
100 R=15*Y/.399
110 RM=LOG(R)/4
120 INPUT "WEEKS FOR THE FINAL POPULATION, W"; W
130 IF W>4*T THEN GOTO 200
140 IF W>2*T THEN GOTO 220
150 PF=PO*EXP(RM*W)
160 PRINT "POPULATION ESTIMATE AT 'W' WEEKS" ; PF
170 INPUT "ANY MORE DATA? (Y/N)"; A$
180 IF A$="Y" GOTO 20
190 END
200 PF=2*PO*EXP(RM*2*T) : GOTO 160
220 PF=(2*PO*EXP(RM*2*T))-(PO*EXP(RM*(4*T-W))) : GOTO 160
```

Page A16

## STAT

```
10 PRINT "GIVES ORDINATE AND AREA UNDER NORMAL CURVE"
20 INPUT " STANDARD DEVIATION FROM THE MEAN, S" ; S
30 Y=EXP(-S^2/2)/SQR(6.283185) : SUM=0.0
40 PRINT "THE ORDINATE OF THE NORMAL CURVE =" ; Y
50 T=1/(1+.2316419*S) : DIM A(5)
70 A(1)=.319381530 : A(2)=-.356563782 : A(3)=1.781477937
80 A(4)=-1.821255978 : A(5)=1.330274429
100 FOR I = 1 TO 5
110 TERM=A(I)*T^I : SUM=SUM+TERM
120 NEXT I
130 P=Y*SUM : Q=1-P
140 PRINT "AREA UNDER THE CURVE TO 'S' AND BEYOND =" ; Q    ; P : END
```

## STATP

```
10 PRINT "ANOV OF PH LOSS DATA BY CONVERSION AND RECONVERSION"
30 INPUT "STUDY 1 OR STUDY 2 (1/2)" ; ST
40 DIM A(25)
50 INPUT "MODAL % LOSS"; MO
60 FOR I = 1 TO 20 : READ A(I) : NEXT I
70 IF ST=1 THEN DA=1 : N=10 ELSE DA11 : N=20
80 STL=0.0 : STR=0.0 : STLS=0.0 : STRS=0.0
90 FOR I=DA TO N
100 IF A(I)>MO GOTO 170
110 X=(A(I)/MO) : Y=ATN(X/SQR(1-X*X))
120 TLL=.5*63.662*Y : STL=STL+TLL : STLS=STLS+TLL^2 : GOTO210
170 P=1-((A(I)-MO)/(100-MO)) : Q=(100-MO)/MO : Z=(P^Q)
180 W=ATN(Z/SQR(1-Z*Z)) : TLR=100-(.5*W*63.662)
190 STR=STR+TLR : STRS=STRS+TLR^2
210 NEXT I
220 DATA 4.6,8.2,6.3,7.1,5.4,4.0,10.5,5.0,18.0,2.9
230 DATA 6.1,18.0,14.2,8.9,12.1,15.3,11.6,13.7,14.8,24.3
240 SS=STRS+STLS : S=STL+STR : M=S/10
250 SD=SQR((SS-2*S*M+10*M^2)/(10-1))
260 PRINT "STANDARD DEVIATION OF TRANSFORMED DATA ="; SD
270 A=50-SD : L=SIN(A*2/63.662)*MO
280 PRINT "STND. DEVIATION OF %PHL TO LEFT="; L
290 B=50+SD : C=(100-MO)/MO : D=(100-MO)^C
300 LR=100-((D*SIN(2*(100-B)/63.662))^(1/C))
310 PRINT "STND. DEVIATION OF %PHL TO RIGHT="; LR
320 B=2.3625/(LOG(LR)-LOG(L)) : A=-.1732/L^B
330 PRINT "CONSTANTS A & B IN P=>EXP(A*L^B) ARE"; A    "AND"; B
```

# Design of Air Distribution Systems and Fan Selection for Grain Aeration

## A.J. Hunter*

*Abstract*

Bulk seed is frequently aerated for the purpose of drying and cooling. Some simple formulae are presented to assist the designer of an aeration system with regard to: the control system, the selection of an airflow-rate, the air distribution system, and the selection of a suitable fan. Some two-dimensional airflow patterns are shown for aeration of stores having flat floors and vertical sides, and also for crossflow in circular, cylindrical stores.

ALL deterioration mechanisms in seeds depend on moisture content, temperature, and time, and since equilibrium wet bulb temperature increases with grain moisture content and grain dry bulb temperature, a useful indicator of the likely deterioration is the wet bulb temperature–time integral for the seedbulk. A suitable objective of an ambient aeration system is therefore to reduce the wet bulb temperature time integral accumulated during the storage period.

The essential characteristics of a good seedbulk aeration system are:

1. to make best use of the available air,
2. to treat the seedbulk as uniformly as possible,
3. to require as little maintenance as possible,
4. to require as little operator attention as possible,
5. to minimise uptake of moisture,
6. to have the least cost.

There are three identifiable features of seedbulk aeration systems. They are:

1. the fan,
2. the air distribution system,
3. the fan control system.

In practice, all three features vary widely over the range of possibilities in response to wide ranges of store geometry, climate, and seed type. Variations also arise from factors such as availability of electric power and construction materials. Airflow may be either upward or downward; more specifically we may refer to a blowing system or a suction system.

Because of the temperature rise experienced by the air as it passes through the fan, a suction system will achieve a lower seedbulk temperature than an equivalent blowing system. Against this, the relative humidity and hence the moisture-content of the wetting front will be less in the case of the blowing system (Sutherland et al. 1971). The suction system also involves reduced air pressure inside the store and so there may be ingress of liquid water if the store is not properly sealed. There is no significant difference in cost between suction and blowing aeration systems.

### The Hygroscopic Nature of Seeds

Because seeds are hygroscopic, the driving potential for cooling is the wet bulb temperature of the aerating air. It is reasonable to refer to the wet bulb temperature of the seedbulk, which is the wet bulb temperature of the interstitial air when equilibrium is reached.

In the following discussion, I shall refer to the wet bulb temperature of either the seedbulk or the air as simply 'temperature'.

### Control System for Aeration Fan

The time-proportioning controller developed by CSIRO Australia (Elder 1972) takes advantage of the lowest ambient temperatures. It is an electronic device which automatically runs the fan for a preset fraction of time. This fraction 'F' is at the discretion of the designer. Seed is usually harvested and introduced to the store at above average ambient temperature; it is usual therefore

* CSIRO Division of Chemical and Wood Technology, P.O. Box 26, Highett, Victoria 3190, Australia.

to set F equal to unity initially, and reduce it when the seedbulk temperature reaches the upper extreme of the aerating temperature. In fact, it may be shown (Hunter 1985a) from an approximate analysis that for a suction system, the optimum value of F is given by

$$F = 1 \qquad \{b < \bar{t}\}$$

$$F = \frac{\bar{t} - a}{b - a}, \qquad \{a < \bar{t} < b\} \tag{1}$$

$$\text{and} \quad F = O, \qquad \{a > \bar{t}\}$$

a and b correspond to $\bar{t}_a - s\sqrt{3}$ and $\bar{t}_a + s\sqrt{3}$, respectively, where $\bar{t}_a$ is the mean and s the standard deviation of ambient temperature, and $\bar{t}$ is the average seedbulk temperature.

For a blowing system the fraction is slightly less, to account for the penalty of the temperature rise in the fan.

## Airflow Rate

For a blowing system there is an optimum thermodynamic airflow rate because of the trade off between fan temperature rise and rate of cooling. For a suction system there is no optimum airflow rate; the greater the airflow rate the greater the rate of cooling.

It is reasonable to suggest that the best airflow rate to use will be determined by the initial cooling stage: this is when the seedbulk is hottest and the need for cooling is greatest.

It is instructive for the purposes of deciding what airflow rate to use to consider the simple model where for the purposes of the initial cooling, F is taken as unity, and the initial average seedbulk temperature $\bar{t}_0$ is higher than the average ambient temperature $\bar{t}_a$.

If q is the specific airflow rate in say $m^3/kg\ s$ then, neglecting heat conduction across the boundary of the seedbulk, it may be shown that (Hunter 1985a) the normalised seedbulk temperature is

$$\Xi_i = \frac{\bar{t} - \bar{t}_a}{\bar{t}_0 - \bar{t}_a} = e^{-\rho_s \Gamma q \theta}. \tag{2}$$

θ is time, $\rho_s$ the density of the seedbulk, and Γ is the ratio of the speed of a temperature or moisture front to the face velocity of the aerating air. For a temperature front, Γ may be approximated by

$$\Gamma = \zeta/1000$$

$$\text{where} \qquad \zeta = \frac{\partial t_{ad}}{\partial t_a} | w. \tag{3}$$

Integrating (2) over a storage period τ we find the normalised temperature time integral Ω to be given by

$$\Omega = \frac{1}{\rho_s \Gamma q \tau}[1 - e^{-\rho_s \Gamma q \tau}]. \tag{4}$$

Equations (2) and (4) are graphed in Figure 1. By putting values on storage periods of interest, and on the specific airflow rate q, one can see immediately how much benefit, by way of reduction in temperature–time integral will have been achieved. In Australia, we use q values of around 1 $l$/st or $10^{-6}$ $m^3/s$ kg. It turns out that the storage period in such a case is given approximately by the abscissa scale in Figure 1 in the units of weeks.

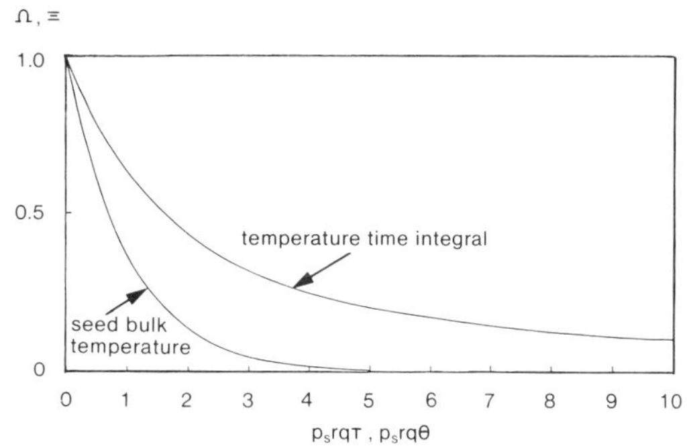


**Fig. 1.** Normalised seedbulk temperature $\Xi_i$ and normalised temperature-time-integral Ω graphed as a function of the dimensionless time parameters, $\rho_s \Gamma q \Theta$ and $\rho_s \Gamma q \tau$, respectively.

## Uniformity of Aeration between Ducts

To ensure uniformity of aeration between ducts, the criterion suggested by Holman (1960) is recommended. Holman states that the ratio of the longest path to the shortest path between the duct and the seedbulk surface should not exceed 1.5. For a level-loaded, flat-floored store, this means a spacing between ducts no greater than the depth of the seedbulk.

## Uniformity of Aeration along a Duct

The variation of static pressure along an aeration duct gives a good indication of the variation of aeration flow rate along the duct length. The static pressure variation along a duct is affected by friction losses and by static pressure regain (Hunter 1985b). (Static pressure regain is the increase in static pressure resulting from a loss of kinetic energy in the air moving along the duct.) These effects are indicated in Figures 2 and 3 for suction and blowing aeration systems, respectively.

To overcome static pressure regain problems the duct should satisfy the condition

$$\frac{A}{X} > \frac{\sqrt{\rho_s \rho_a q}}{2R\sigma} \qquad (5)$$

A is the cross-sectional area of the duct, X is the plan area of the seedbulk served by the duct, $\rho_s$ and $\rho_a$ are the densities of the seedbulk and of the aerating air, q is the specific airflow rate, σ is the fractional variation of static pressure along the duct, and R is the first Ergun coefficient for the seed defined by

$$\frac{dp}{dx} = Rv + Sv^2. \qquad (6)$$

p is static pressure of air, x the distance coordinate, v the face velocity of the air, and R and S are constant coefficients for a particular seed type.

Equation (5) may be derived by setting the maximum velocity pressure in the duct equal to the fraction σ of the static pressure difference across the seed-bulk.

To prevent static pressure problems arising from friction in the duct take

$$l < \frac{12A}{P_w f} \qquad (7)$$

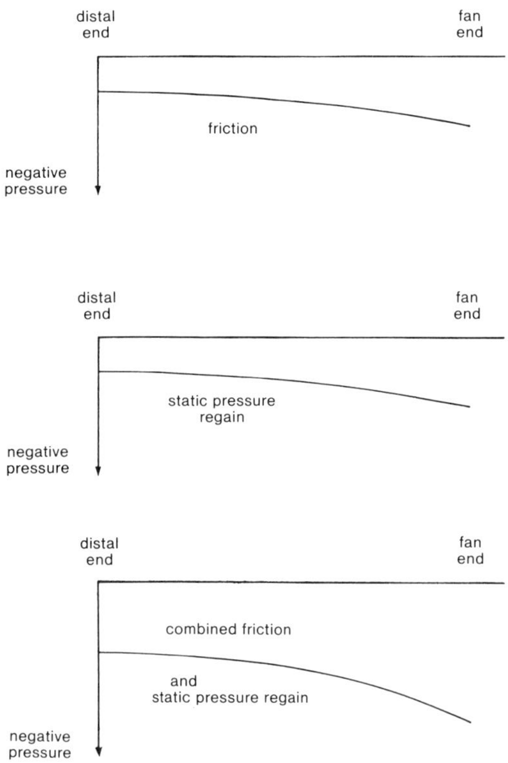


**Fig. 2.** The effect of friction and of static pressure regain on the distribution of static pressure of air along a duct for a suction aeration system.

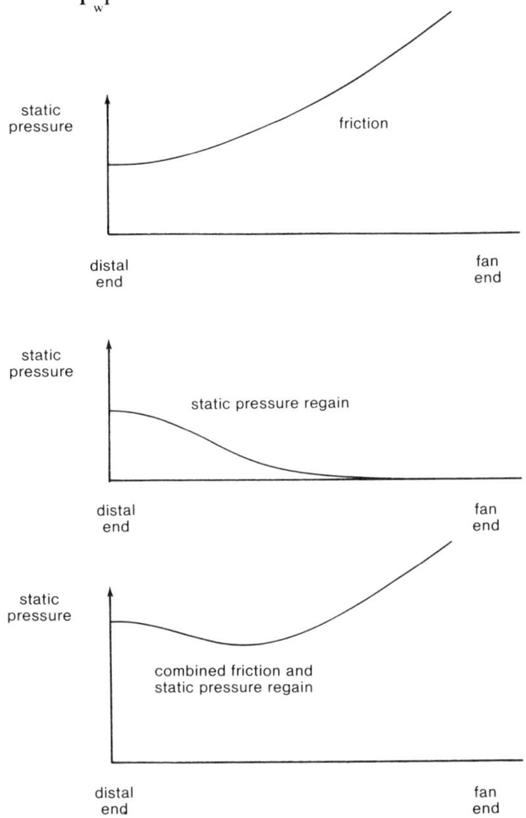


**Fig. 3.** The effect of friction and of static pressure regain on the distribution of static pressure of air along a duct for a blowing aeration system.

where $P_w$ is the wetted perimeter of the duct and f is the D'arcy-Weisbach friction factor, which may be approximated for straight aeration ducts by

$$f = \frac{1.326}{\left(ln\frac{14.88A}{P_w \varepsilon}\right)^2} \qquad (8)$$

The equivalent sand roughness height for the inner surface of the duct is ε.

Equation (7) is derived by considering friction losses alone, and then combining the result with (6).

Static pressure variation can be avoided by designing in accordance with equations (6) and (7).

## Airflow Patterns in Aerated Seedbulks

Some airflow patterns for various two-dimensional duct and store geometries were derived using conformal mapping (Hunter 1983), and are shown in Figures, 4, 5, 6, 7, 8, and 9. The flowlines shown divide the airflow into 10 equal parts.

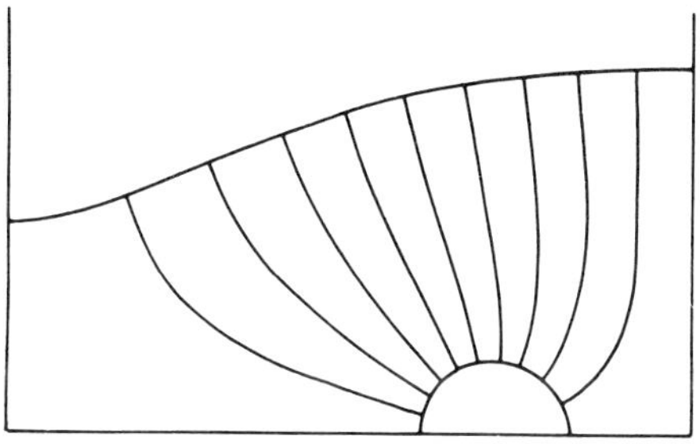

**Fig. 4.** Flow field approximating aeration of a seedbulk with a flat floor and vertical sides, using a half round duct.

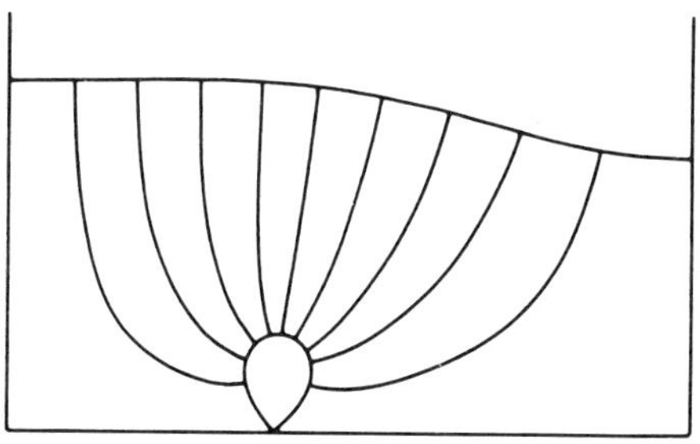

**Fig. 5.** Flow field approximating aeration of a seedbulk with a flat floor and vertical sides, using a round duct.

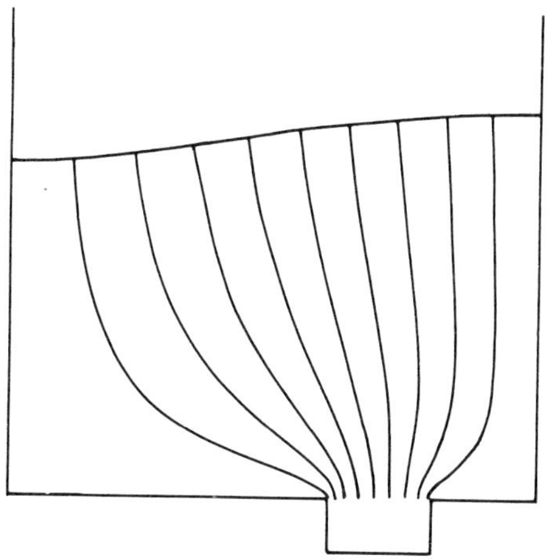

**Fig. 6.** Flow field approximating aeration of a seedbulk with a flat floor and vertical sides, using a perforated floor duct.

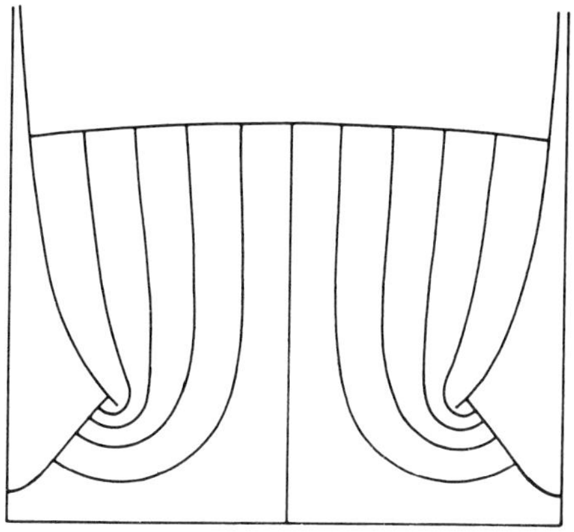

**Fig. 7.** Flow field approximating aeration of a seedbulk with a flat floor and vertical sides, using a symmetrical pair of louvre ducts.

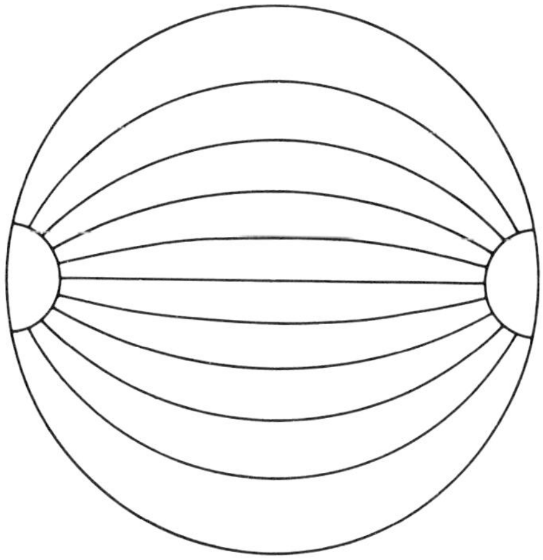

**Fig. 8.** Flow field approximating crossflow aeration of a circular cylindrical silo, using half round ducts.

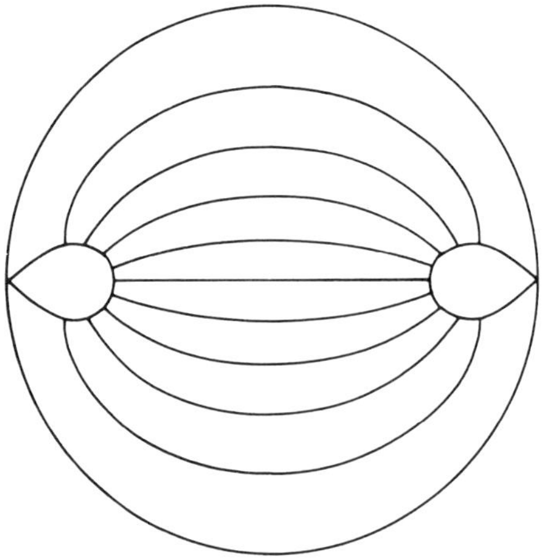

**Fig. 9.** Flow field approximating crossflow aeration of a circular cylindrical silo, using round ducts.

## Estimation of Pressure Difference across an Aerated Seedbulk

The pressure difference across an aerated seedbulk may be estimated from the laminar flow term (linear) and the turbulent flow term (square) in the Ergun equation, (6).

For a floor-mounted duct system the linear term pressure difference may be estimated from

$$\Delta p_1 = \frac{RQ}{X}(H + \Delta H) \tag{9}$$

where $\Delta H = \frac{W}{\pi} ln \frac{W}{2P_p}$

Two square law terms arise from the field close to the duct and from the field remote from the duct. They are

$$\Delta p_2 = \frac{SQ^2}{\pi l^2 P_p} \tag{10}$$

and

$$\Delta p_3 = \frac{SQ^2 H}{X^2} \tag{11}$$

respectively.

For a crossflow aeration system in a circular cylindrical store of diameter D having semicircular ducts of radius r, we have

$$\Delta p_1 = \frac{2QR}{\pi l} ln (\frac{D}{r} - 1) \tag{12}$$

$$\Delta p_2 = \frac{2SQ^2}{\pi l^2 P_p} \tag{13}$$

and

$$\Delta p_3 = \frac{SQ^2}{l^2 D}. \tag{14}$$

The total pressure difference in each case is given by

$$\Delta p = \Delta p_1 + \Delta p_2 + \Delta p_3. \tag{15}$$

Values for R and S for various seeds are given by Hunter (1983).

An allowance for packing of the seed as distinct from loose fill should be made and may be taken as increasing R and S by 20%.

## Fan Selection

Means to assist in the decision as to what airflow rate should be used have been given. Once the airflow rate has been selected, conditions have been presented which allow duct sizes to be determined. From the airflow rate and duct geometry, the pressure difference across the seedbulk can be determined as discussed above, allowance being made for supply duct losses if appropriate.

Having determined the fan duty (volume flow and total pressure) it is suggested that a direct drive fan be chosen to reduce downtime and maintenance.

When selecting a fan, the temperature rise as the air passes through the fan will be relevant if a blowing system is used. The fan temperature rise $\Delta t$ is given by

$$\Delta t = \frac{\Delta p}{\eta \rho_a C_p \zeta} \tag{16}$$

where $\eta$ is the total efficiency of the fan, and $C_p$ is the specific heat of air at constant pressure. Differences in kinetic energy of inlet and outlet flow are neglected.

The effect of the temperature rise through the fan must be considered with reference to a psychrometric chart overlaid with equilibrium isosteres for the seed type under consideration.

## Conclusions

The significant features of an aeration system are the airflow rate, the fan control system, and the air distribution system. The formulae and graphical information presented here should assist the reader in designing a practical aeration system for most types of storage and for most climatic situations. The reader will require moisture equilibrium data and Ergun coefficients for the seed type concerned.

## Acknowledgments

The author is indebted to the Australian Wheat Industry Research Council which has supported this work and to the Australian Centre for International Agricultural Research which sponsored its presentation at this seminar.

## References

Elder, W.B. 1972. New aeration techniques for the preservation of stored cereals. Proceedings of the Royal Australian Chemical Institute Cereal Chemistry Division Annual Conference, Bronte, New South Wales, September 1972, 205–212. (Aust. Pat. No. 438732.)

Holman, L.E. 1960. Aeration of grain in commercial storages. US Department of Agriculture. Marketing Research Report, No. 178, 14 p.

Hunter, A.J. 1983. Pressure difference across an aerated seed-bulk for some common duct and store cross-sections. Journal of Agricultural Engineering Research, 28 (5), 437–450.

Hunter, A.J. 1985a. Thermodynamic criteria for optimum cooling performance of aeration systems for stored seed. Journal of Agricultural Engineering Research, in press.

Hunter, A.J. 1985b. Approximate analysis of the variation of static pressure along seed bulk aeration ducts. Journal of Agricultural Engineering Research, in press.

Sutherland, J.W., Banks, P.J., and Griffiths, H.J. 1971. Equilibrium heat and moisture transfer in air flow through grain. Journal of Agricultural Engineering Research, 16 (4), 368–386.

## Notation

**Abbreviations**

dbt dry bulb temperature
wbt wet bulb temperature

**Symbols**

a average weekly minimum wet bulb temperature, °C
b average weekly maximum wet bulb temperature, °C
f D'arcy Weisbach friction factor
$l$ length of duct, m
p static pressure, Pa
$\Delta p$ difference in pressure of air, Pa
q specific airflow, $m^3/s$ kg or $l$/st
r radius of half round duct, m
s standard deviation of ambient wet bulb temperature, °C
t wet bulb temperature of seedbulk, °C
$\Delta t$ fan air wet bulb temperature rise, °C
$t_a$ dry bulb temperature of ambient air, °C
$t_o$ initial wbt of seedbulk, °C
$t_{ad}$ dbt of ambient air, °C
v face velocity of air, m/s
w moisture content of air
x distance coordinate, m
A cross-sectional area of duct, $m^2$
$C_p$ secific heat of air, J/kg K
D diameter of circular cylindrical store, m
F aeration time fraction
H height of seedbulk, m
$\Delta H$ equivalent additional height of seedbulk caused by duct constriction, m
$P_p$ perforated perimeter of duct cross-section, m
$P_w$ wetted perimeter of duct cross-section, m
Q airflow, $m^3/s$
R first Ergun coefficient, Pa s/m
S second Ergun coefficient, Pa $s^2/m$
W width of seedbulk cross-section served by the duct, m
X plan area of store, $m^2$
$\varepsilon$ equivalent sand roughness height, m
$\zeta$ differential number,

$$\zeta = \frac{\partial t_{ad}}{\partial t_a} \,|w \text{ (psychrometric chart)}$$

$\eta$ fan efficiency
$\theta$ time, s
$\rho$ density of air, $kg/m^3$
$\rho_s$ density of seedbulk, $kg/m^3$
$\sigma$ fractional variation of static pressure along a duct
$\tau$ storage period under consideration, s
$\Omega$ normalised temperature time integral
$\Gamma$ ratio of velocity of temperature front (cooling) or a moisture front (drying) to the face velocity of the air.
$\Xi_i$ normalised seedbulk temperature

The mean of any quantity is indicated by an over-bar.

# A Procedure to Estimate the Supplemental Heating Needs for In-store Drying of Paddy

R. G. Bowrey* and R. H. Driscoll†

*Abstract*

The air used to dry grain in ventilated stores provides both the energy needed to evaporate the moisture from the grain and the medium to transport this moisture out of the store. If supplemental heating is used to increase the inlet air temperature, the rate of moisture removal is increased and the airflow rate can be decreased, leading to several solutions for a particular set of circumstances. Depending on the relative costs of heating energy and electrical energy, one of these solutions is to be preferred.

This paper discusses a procedure to determine how much supplemental heating should be provided for in-store drying of paddy rice. A graph is provided to estimate the airflow rate required to dry paddy in terms of the inlet air temperature, the ambient humidity, and the initial grain moisture content. A second graph can be used to estimate the fan power required for various airflow rates. Using these graphs the total energy cost can be estimated for various amounts of heating, thereby allowing the most suitable level to be selected.

At harvest, grain is hot and moist and must be cooled, dried, or protected in other ways to prevent fungal/mould growth (Mendoza et al. 1985). In dry climates this can be achieved by aeration (Navarro and Calderon 1982) and, although this technique has applications in the tropics, the ambient relative humidity is often so high that supplemental heating is required. Increasing the temperature of the air increases both the ability of the air to carry moisture out of the store and the availability of energy to evaporate this moisture from the grain.

It has been observed by many researchers (e.g. Sutherland et al. 1971; Smith and Bailey 1983) that wet paddy under aeration cools to a temperature close to the wet bulb temperature of the inlet air (i.e. the pseudo wet bulb temperature). This occurs well before the grain dries. As the grain dries the amount of heat needed to evaporate moisture is reduced, less cooling occurs, and eventually the grain comes to equilibrium with the dry bulb temperature of the inlet air. This puts a practical limit on the maximum inlet temperature allowed for long periods of time, because holding dry grain at high temperature can affect the quality.

Both the airflow rate and the inlet temperature affect the time to dry the grain at the top of the store and if the flow rate is too low this grain will be damaged by microbial activity. The need to balance heat input and airflow rate created the need for a design procedure.

Many 'ad hoc' designs have been proposed for drying paddy at the village level. However, for larger storages in countries such as Malaysia, the Philippines, and Thailand, a more rigorous approach is needed to determine design parameters for any supplemental heating system.

## Description of Drying

Various models have been developed to predict the time/position profiles for both temperature and moisture content of grain during ventilation (Sutherland et al. 1971; Dung 1980). These models predict that, during ventilation with constant inlet air conditions, the temperature of the grain changes more rapidly than the moisture content and this sets up temperature and moisture zones which move through the grain at different speeds. A temperature (moisture) zone is a region of grain

* Allied Mills Industries Pty Ltd, P.O. Box 57, Mascot, N.S.W. 2020, Australia.

† School of Food Science and Technology, University of New South Wales, P.O. Box 1, Kensington, N.S.W. 2033, Australia.

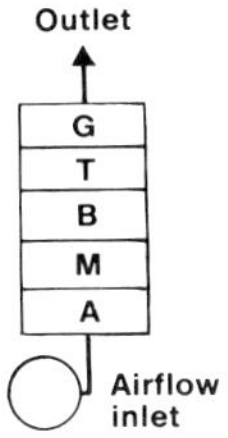


| Zone | Description | Grain Conditions |
|---|---|---|
| A | Equilibrium with inlet air | $M_A$, $T_A$ |
| M | Moisture Zone | $M_A$-->$M_B$, $T_A$--<$T_B$ |
| B | Intermediate zone | $M_B$, $T_B$ |
| T | Temperature zone | $M_B$-->$M_G$, $T_B$-->$T_G$ |
| G | Initial grain conditions | $M_G$, $T_G$ |

**Fig. 1.** Definition of zones within a bulk of drying grain.

in which the temperature (moisture) is changing. In front of the temperature zone the grain is unchanged from its initial conditions. Behind the moisture zone the grain is in equilibrium with the inlet air and in between the temperature and moisture zones there is grain which has had its temperature changed with little change in moisture content. The details of these five zones are shown in Figure 1. The nomenclature adopted in this paper is as follows:

| | | |
|---|---|---|
| H | humidity | kg/kg |
| L | grain depth | m |
| M | mass of wet grain | t |
| m | moisture content, dry basis | % |
| | fan efficiency | % |
| P | fan power per tonne of grain | kW/t |
| $Q_f$ | heat input from fan per tonne of grain | kW/t |
| $Q_h$ | heat input from heater per tonne of grain | kW/t |
| T | temperature | °C |
| v | superficial air velocity | m/minute |
| $v_f$ | drying zone velocity | m/minute |

A typical set of conditions for a ventilated paddy store in Australia is plotted on a psychrometric chart in Figure 2. The initial grain conditions are 17% moisture content (wet basis) grain at 10°C (G), dried with air at 30% relative humidity (RH) and 35°C (A). Under these conditions the grain is heated to approximately 22°C by the passage of the temperature front (B). If the initial grain temperature was 30°C the passage of the temperature front would cool the grain to 22°C. After the temperature zone has left the store the air leaves in equilibrium with grain at the pseudo wet bulb temperature and the initial moisture content (zone B). Equilibrium data are shown in Figure 3.

In both Australia and North America, grain is harvested just before the onset of winter when cold humid air is prevalent. With these low pseudo wet bulb temperatures, the wet grain near to the outlet

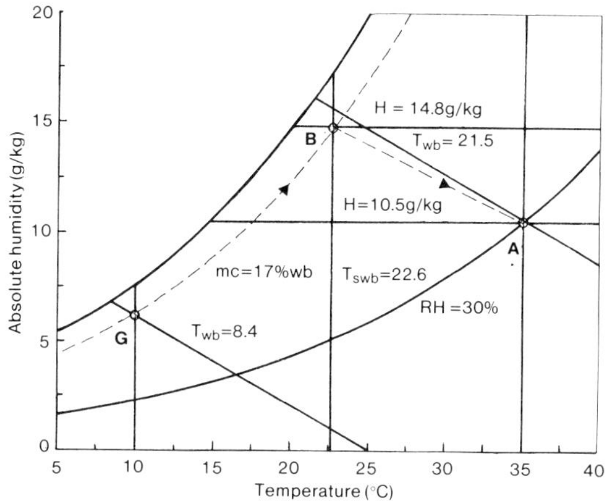


**Fig. 2.** Psychrometric representation of deep-bed drying of grain.

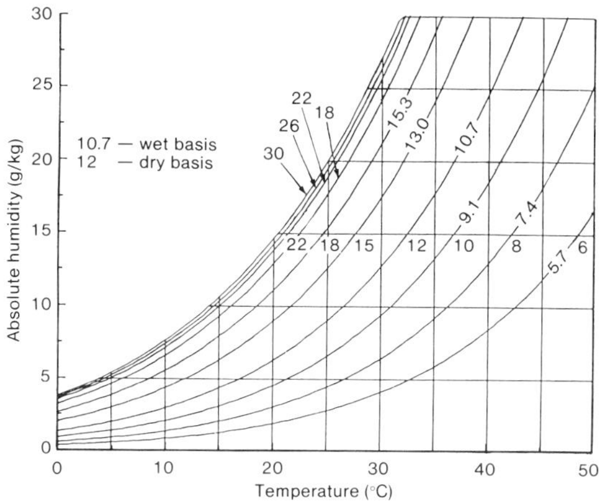


**Fig. 3.** Lines of constant moisture content for rice.

(zone B) can be kept for long periods without mould damage occurring. For this reason aeration at low airflow rates is acceptable and a holding strategy, whereby the grain is cooled with night air to about 5°C, is often employed. At this temperature, mould, bacterial and insect activity can be safely ignored for several months until the weather becomes sufficiently warm and dry for slow ambient drying. This is not true in the humid tropics and higher flow rates and supplemental heating are necessary (Teter 1982; R. H. Driscoll and T. Adamszak, pers. comm., 1984[1]).

[1] Pilot plant proposal for Malaysia. Analysis of rice drying in Bangkok.

## Working Equations[2]

For design purposes the speed of the drying zone can be estimated from the equilibrium theory which predicts that the moisture zone is a square wave, with the air leaving the store, after the passage of the temperature zone, in equilibrium with the inlet grain at the pseudo wet bulb temperature. Based on these predictions, the speed of the moisture front can be estimated from:

$$v_f = 0.185 * v * (H_B - H_A)/(M_G - M_A) \quad (1)$$

This equation is plotted in Figure 4.

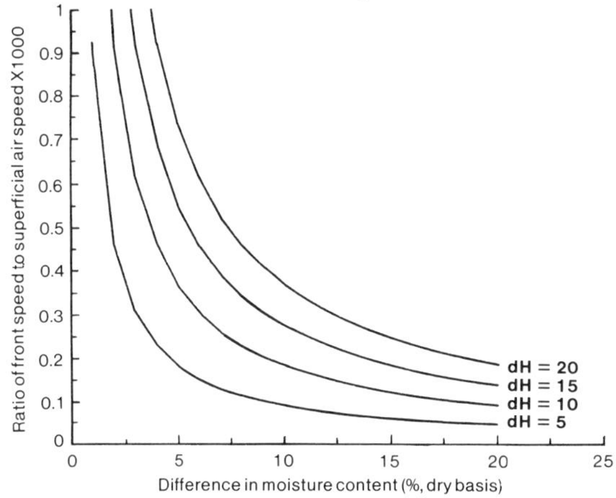


**Fig. 4.** Graphical estimation of moisture front speeds in grain.

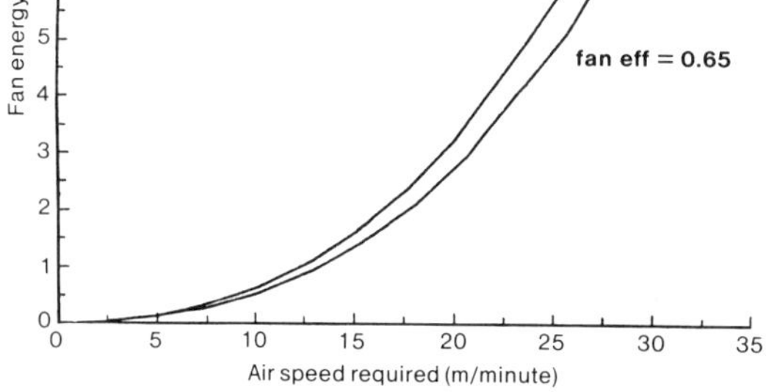


**Fig. 5.** Relationship between air speed and fan power required in an aerated grain bulk.

[2] All equations in this paper are based on the following data: grain density 650 kg/m$^3$, air density 1.2 kg/m$^3$, porosity 0.45, air heat capacity 1.0 kJ/kg K.

The fan power per tonne of grain to force the air to flow through the store can be estimated from:

$$P = (0.96 * 10^{-3}v^2 + 4.89 * 10^{-5}v^3)/kW/t \quad (2)$$

This equation is plotted in Figure 5. Because of fan inefficiencies, part of this energy heats the air and this should be allowed for when sizing the supplemental heater.

## Design Procedure

To estimate the design conditions to dry M tonnes of paddy with an initial moisture content $M_G$ in an in-store dryer using supplemental heating, the following procedure is recommended. The ambient humidity is $H_A$ and the ambient temperature $T_0$.

Select design parameters:

i. time allowed for the moisture front to pass through the store (t);
ii. grain depth (L) $\longrightarrow v_f$

Estimate design conditions:

iii. select temperature out of heater $T_A$;
iv. calculate the airflow rate using Figures 3 and 4;
v. calculate the fan power required using Figure 5;
vi. select heater to produce temperature $T_A$ at flow rate Q where $Q = M * v/(0.65L)$ m$^3$/min;
vii. estimate drying cost or other objective function;
viii. repeat until objective optimised.

### Example

10t of paddy at 30% moisture content (dry basis) are to be dried within 2 days using supplemental heating. The ambient conditions are 25°C, 20 g/kg humidity. The results of this calculation, using the procedure outlined, are shown in Table 1 for grain depths of 1 m and 2 m.

## Priorities for Further Research

The equations presented for drying speed are based on the equilibrium model. Further research will allow this equation to be replaced by more accurate equations which allow for the fact that the drying zone is not a square wave. The equilibrium data used in Figure 3 and the pressure drop data used in drawing Figure 5 were based on rice varieties and grain porosities common in Australia. In particular applications, these values should

**Table 1.** Possible design conditions to dry grain at 30% moisture content (dry basis) in two days, using air at 25°C and with 20 g/kg absolute humidity. See text for explanation.

| | $L = 1$ m $\longrightarrow V_f = 3.5 \times 10^{-4}$ m/min | | | | |
|---|---|---|---|---|---|
| $T_A$ | 30 | 35 | 40 | 45 | 50 |
| $M_A$ | 17 | 13.5 | 10.4 | 8.5 | 7.0 |
| $H_B$ | 21 | 22.5 | 24 | 25.5 | 26.5 |
| v | 24.5 | 12.4 | 9.2 | 7.3 | 6.6 |
| P | 4.0 | 0.7 | 0.4 | 0.2 | 0.2 |
| $Q_f$ | 1.4 | 0.2 | 0.1 | 0.1 | 0.1 |
| $Q_h$ | 2.4 | 3.6 | 4.2 | 4.4 | 5.0 |
| | $L = 2$ m $\longrightarrow V_f = 7.0 \times 10^{-4}$ m/min | | | | |
| $T_A$ | 30 | 35 | 40 | 45 | 50 |
| $M_A$ | 17 | 13.5 | 10.4 | 8.5 | 7.0 |
| $H_B$ | 21 | 22.5 | 24 | 25.5 | 26.5 |
| v | 49 | 24.8 | 18.4 | 14.6 | 13.2 |
| P | 24.8 | 4.1 | 1.9 | 1.1 | 0.9 |
| $Q_f$ | 8.7 | 1.4 | 0.7 | 0.4 | 0.3 |
| $Q_h$ | — | 6.3 | 7.9 | 8.7 | 9.9 |

be replaced by data more relevant to the application. Various sources of heating are available in grain drying areas, including combustion of hulls and other agricultural materials, solar absorbers, biogas, waste heat from exhaust gases, etc. The results in this paper show that temperature rises of about 10 to 20°C are very effective in reducing the fan sizes required. Consequently, relatively simple heaters can be very effective.

## Conclusions

A procedure is described to estimate the reduction in fan power permissible if supplemental heating is used to increase the temperature of the drying air. Depending on local conditions an optimum set of conditions can be selected to minimise the total drying cost while drying the grain within a nominated period of time. The procedure can be used to study the cost of changing grain depth or drying time.

## References


Dung, N. V. 1980. Aeration of paddy: the measurement and simulation of drying rates in thin layers and deep beds. Kensington, Australia, University of New South Wales, unpublished Ph.D. dissertation.

Mendoza, M. E., Rigor, A. C., Quitco, R. T., and Ilag, L. L. 1985. Grain quality deterioration on on-farm level of operation. Manila, Philippines, NAPHIRE Technical Bulletin, No. 1.

Navarro, S., and Calderon, M. 1982. Aeration of grain in subtropical climates. FAO Agricultural Bulletin, No. 52.

Smith, E. A., and Bailey, P. H. 1983. Simulation of near-ambient grain drying. Control strategies for drying barley in Northern Britain. Journal of Agricultural Engineering Research, 128, 301.

Sutherland, J. W., Banks, P. J., and Griffiths, H. J. 1971. Equilibrium heat and moisture transfer in air flow through grain. Journal of Agricultural Engineering Research, 116, 368–386.

Teter, N. C. 1982. Design of grain storage aeration for the humid tropics. Presented at the 1982 Winter Meeting of the American Society of Agricultural Engineers, December, 14–17.

# Control Systems for the Aeration and Drying of Grain

G. R. Pym* and T. Adamczak†

*Abstract*

Grain aeration and drying require a large capital investment. The effectiveness of the facilities involved can be maximised, quality can be improved, energy usage can be minimised, and labour inputs can be optimised by the application of commercially available, reliable, and economically viable controllers. Running costs can be reduced by up to 40% for a capital investment of $0.50/t.

Good intentions of management can be dwarfed by the pressure of harvest. The application of control systems to aeration and drying allows the best management option to be applied to all facilities.

This paper analyses a proposed aeration and drying system for application of possible control devices. Commercial results of the application of control systems are presented, along with the results of a simulation model of the system proposed.

Control systems allow rapid decisions to be made using past, present, and predicted conditions to achieve the objective of lowest cost, highest quality, or minimum drying time.

CONTROL systems are the means whereby a complete system can be regulated to achieve a desired result. The control can be:

1. by instruction to operators of equipment or facilities of the optimum sequence or conditions for operation — a manual system;
2. by application of simple feedback control devices dedicated to discrete operations within the system, but still regularly supervised and adjusted by a supervisor — a supervised system;
3. by application of a Programmable Logic Controller (PLC) or a computer control device capable of monitoring all conditions within the complete system, allowing rapid interaction and adjustment by the control device of all the discrete operations within the system to achieve the best management option continuously — a semi-supervised system.

Any control system requires input information. This information can range from basic details of tonnage, moisture content, and variety, through to historical information on the conditions of treatment of the grain to date and to required throughput to achieve mill requirements or to clear a wet grain backlog.

* Ricegrowers' Co-operative Mills Ltd, P.O. Box 561, Leeton, N.S.W. 2705, Australia.

† School of Food Science and Technology, University of New South Wales. P.O. Box 1, Kensington, N.S.W. 2033, Australia.

### Manual System

The most widely used form of control system is the manual control system. For a manual control system the amount of information that can be collected and analysed to make a control decision is limited by time and available manpower.

A manual control system is typified by checksheets and dials for inputs, and slides and on/off switches for outputs. Decisions are made by a person who then physically communicates the output state to the devices concerned.

### Supervised System

For a supervised system the opportunity for feedback control of some discrete operations will allow the supervisor time to review a wider range of information in order to make a controlling action. A supervised system is typified by checksheets, dials, electric or mechanical reaction inputs with adjustable setpoints, slides, and on/off switches or relays for ouputs. The 'decide' process can be performed manually but will normally include a number of inputs reacting directly with a dedicated output device to control a discrete operation in the system. The supervisor interacts with the inputs and adjusts the field of performance of an operation of the system via setpoints or by physically communicating the required output state to the output device.

## Semi-supervised System

Semi-supervised control is the realisation of a dream of industrialists. By automatically controlling a process, the role of sometimes fickle and usually expensive human resources can be oriented more towards maintenance and supervision.

For a semi-supervised system the larger amount of information it is possible to analyse in a short space of time allows the supervisor to be removed from the most routine control actions and simply respond to alarm prompts, fine tune the system, or duplicate the performance identically at other locations. He can also re-allocate priorities for the system based on the demands outside the system for rapid drying of the raw material or higher output requirements for shipping or further processing, or smoothing energy requirements from a supply authority.

The semi-supervised control system is typified by selected sensors with similar electronic outputs which can be directly wired to the PLC or computer control device. A fraction of the available input readings can be presented as a printout, graphical display or tabulated display on a Visual Display Unit (VDU), or communicated to a remote facility via a modem for inspection by a supervisor. Such a report can be expanded to include a system current status report, fault report, or audit or log of system operation for energy or raw material accounting.

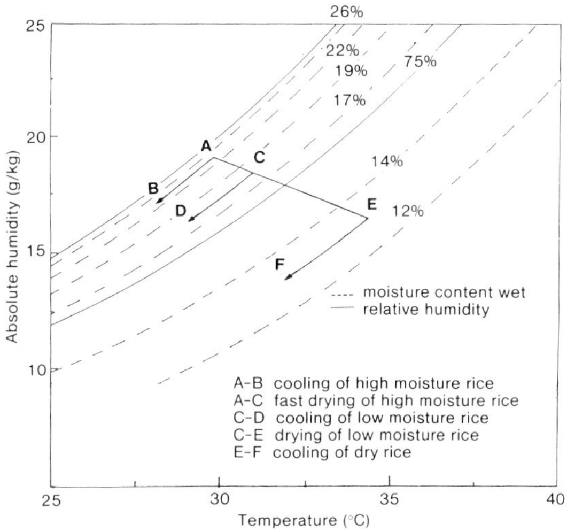


**Fig. 1.** Psychrometric chart of grain aeration and drying processes.

## Aeration

The reason for aerating grain is to achieve the final objective of the improvement of storage conditions. This may be attained by moving air of the desired properties through the grain bulk until a new microclimate is produced, which will keep the stored grain from deteriorating. In all events, aeration is aimed at improving storage conditions but not grain quality. However, it often happens that aeration prevents a decline in quality.

Since aeration may have different effects on stored grain (see Fig. 1) storage conditions may be improved in different ways. These will depend on the properties of the air and on the existing condition of the grain. Therefore, before operating the aeration system it is essential to foresee the effect aeration will have on the grain, based on the available data on the condition of the grain and the external air. Aeration control systems formalise the procedure for operation, assist in the gathering of reliable data, and can allow the application of the best management option for the economical running of an aeration system.

The objectives for operation of an aeration system are derived from the anticipated effects on the grain of the movement of air. These are:

- creating low temperatures in the grain bulk;
- equalising temperatures throughout the grain bulk;
- prevention of heating in damp grain;
- limited drying.

There are also some additional uses of the aeration system, such as introduction and recirculation of fumigants, and removal of odours and fumigant residues, though these are less frequently applied in practice.

## Cooling

Cooling of grain is the most frequently sought and advantageous objective of grain aeration. If cold air is available (during winter or on cold nights), by introducing and moving this air throughout the grain mass, its temperature will eventually be lowered (see Fig. 1.). Thus, a new environment, for all living components of the grain bulk ecosystem will be created.

Aeration is very often applied in the storage of

just-harvested damp grain, before being dried, especially where local drying capacity is insufficient. Aeration for this purpose is used mainly in temperate climates. and is intended to enable the storage of grain with excessive moisture content for short periods of time. This aeration is called in Europe 'maintenance of condition' ('ventilation de maintien') since it is aimed at maintaining the grain quality and preventing spoilage.

**Maintenance**

True maintenance of condition is applicable to the action of maintaining dry grain in a fit and saleable and consumable condition for as long as required as cost effectively as possible. An aeration control system having the ability to monitor the undesirable changes occurring in or the potential for these changes to occur in the grain bulk and summoning a remedial action is the ultimate system.

In damp grain bulks, respiration is very intensive, due to grain and microfloral metabolism. This respiration results in some loss of dry weight and produces the phenomenon called 'spontaneous heating'. Thus, heat accumulates in the grain bulk and apart from being detrimental to grain quality, the high temperatures (up to about 60°C) create temperature gradients between the heated grain and cool surrounding air. Moisture migration, brought about by the above phenomenon will seriously degrade the storage conditions and threaten the grain quality.

Aeration of paddy rice in temperate parts of Australia is carried out intermittently during the poor, ambient conditions of winter until drying can be completed in the spring. This is termed 'maintenance aeration'.

In order to prevent the deterioration processes described earlier, the aeration system, delivering high airflow rates, is put into operation almost continuously. The flow of this air through the grain bulk prevents the formation of heating foci and also eliminates the differences in grain temperatures throughout the bulk. Thus, it maintains the stored grain in a condition in which immediate damage is prevented, but it does not improve its storageability. When cool ambient is used for this aeration, some decrease of grain moisture content is obtained.

Storage of damp grain in warm climates is a problem since, at high temperatures, respiration, mould growth, and grain deterioration are accelerated. Aeration for 'maintenance of condition' would be practical in these regions during cool nights, or in the cool season only, and the aerated grain should not be too damp.

Since storage of damp grain in warm climates is very hazardous, aeration should be carried out with great care, at the highest possible airflow rates, in as shallow as possible a grain bulk, and utilising air temperatures lower than that of the grain.

It can be readily demonstrated that application of a control system will prevent overdrying. The practical reality is that application of a manual control system to overcome overdrying can be difficult, as suitable ambient conditions may only be available for short periods and at different times of the day if at all.

A significant drying effect may occur during aeration for grain cooling in large bulks. This loss of water from the grain bulk is also a loss in weight which, apart form the economic loss of saleable mass, can also indicate unnecessary running costs. A further result of overdrying can be a significant reduction in the efficiency of further processing of the grain, an example being the reduction of whole grain yield in the case of rice. In temperate climates, the loss of weight can be a significant amount before consideration of any other factors.

Equalising temperatures is a very important objective of grain aeration, especially in subtropical climates in which diurnal or seasonal fluctions in temperature occur. In this case, the purpose of aeration is not grain cooling (although the use of cool air is preferred), but the prevention of the phenomenon called 'moisture migration'.

Grain loaded into storage after harvest keeps its initial high temperature for a long time, because of its self-insulating properties. With the change of the ambient temperature during the cool season, however, the surface (and external) layers of the grain bulk become considerably cooler than the internal bulk grain. Thus, temperature gradients in the grain bulk are created, which bring about convection currents moving through the intergranular air. These currents convey the warmer air (which contains more moisture than cool air) towards the colder parts of the grain bulk. Thus, the warm air 'migrates' to the cooler surface layers. There, the cooled air deposits its excess moisture and slowly increases the grain moisture content in these parts of the grain bulk.

The moisture migration may be a slow, continuous process and, consequently, moisture will slowly accumulate in the coldest grain layers. In extreme cases, condensation of water on the grain may occur, causing rapid mould (and sometimes bacterial) spoilage. One of the typical symptoms of this phenomenon is 'crusting' of the grain surface which should be taken as a final warning that steps must be taken to prevent further damage. The most disturbing aspect of moisture migration is not the amount of damaged grain, which is usually small in proportion to the grain bulk, but the often unavoidable mixture of damaged grain with undamaged grain during unloading. This reduces the quality of the whole bulk. In addition to discoloration, mustiness, and decrease in germination, production of mycotoxins in microflora-damaged grain should also be considered. The most significant microfloral damage which has recently received worldwide attention of nutritionists is the production of mycotoxins.

Moisture migration is very marked in warm, subtropical climates in which the ambient temperature may drop considerably by night or during winter. Logically, the prevention of its occurrence will be possible by the elimination of temperature gradients throughout the grain bulk. This may be achieved by aeration with ambient air, preferably during cool nights. For this purpose, temperature should be measured throughout the aerated bulk in order to check its uniformity.

**Drying**

Proper control of grain drying operations is essential if the grain is to be marketable. Automatic control is preferred since spoiled grain due to underdrying and decreased value from overdrying very often results from manual control.

Much research has been done using various sources of energy for low temperature, grain drying systems, but only a few investigators have analysed control procedures for minimising energy consumption and even fewer have been concerned with grain deterioration and overdrying which can result from these strategies.

A dynamic programming method was used by Gunasakaran and Shove (1983) as an optimisation technique. Optimum flow rate of drying air was obtained for each day and average flow rate was compared at the same time with conventional flow rate. Savings were up to 15%. By limiting the length of drying time it was possible to eliminate the risk of grain spoilage. However, the efficient application of this technique requires a special control system and variable speed for the motor, and in many cases this may not be economical.

A critical path method was used by Colliver et al. (1979). A comparison of this technique with three other techniques — time clock control, relative humidity control, and continuous operation — was presented. The comparison was performed on five drying facilities; natural air drying, solar-assisted drying, electric heating assisted drying, off-peak power assisted drying, and stirring. Savings obtained by this method are not much higher than savings obtained by using relative humidity control or time clock control.

A microprocessor based control system was presented by Simmonton et al. (1981) for combination (high-low temperature) drying. In combination drying the grain is initially dried in a high temperature dryer to bring the moisture down to a level that the low temperature drying system can handle safely.

A 72 hour time limit is set for completion of second stage drying. Capacity of the first stage dryer is adjusted to obtain the required moisture content for the second stage of drying.

The authors suggest that a better approach might be obtained by changing 72 hour limit for second stage drying to a period variable according to weather conditions. It was also suggested that the higher the changeover moisture content the better the drying operation is.

### Control Limits or Parameters

Parameters of ambient air available for cooling of grain have been presented in Figure 2.

Figures 2a and 2b present air available for cooling when no mechanical equipment is used. The amount of this air is limited and in some circumstances there may be not enough of it to cool the grain to the required temperatures.

Figures 2c and 2d show that the required parameters of cooling air can be obtained regardless of ambient air parameters by employing equipment for mechanical cooling of the air stream.

For proper control of cooling two requirements have to be fulfilled:

- the air supplied must be of the required parameters. This can be achieved by employing wet bulb, dry bulb, and humidity sensors;

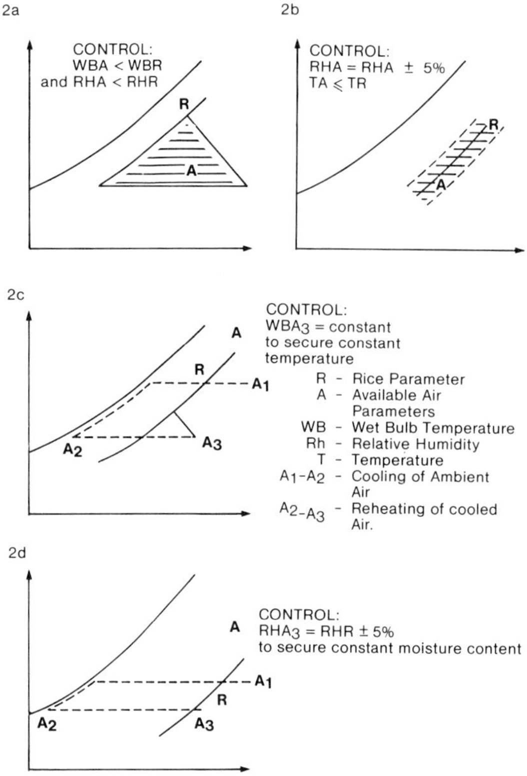


**Fig. 2.** Availability of air for cooling grain represented on psychrometric charts.
(a) Cooling of high moisture grain with ambient air.
(b) Cooling of dry grain with ambient air.
(c) Mechanical cooling of high moisture grain.
(d) Mechanical cooling of dry grain.

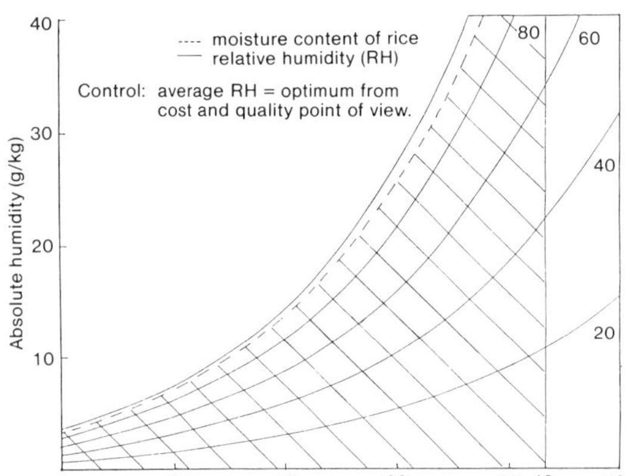


**Fig. 3.** Availability of air for drying grain represented on a psychrometric chart.

- the length of the cooling process must be controlled by means of a timer based on statistical or feedback information.

Parameters of air available for drying have been presented in Figure 3. Decision about which parameters should be chosen is dependent on availability of supplementary heating as well as on data obtained from control simulation. The results of this simulation take into account the costs of energy and quality losses.

Simulation results for in-store drying with supplementary heating for a relative humidity setpoint ranging from 45 to 75%, bin depths 3 and 6 m, velocity of supply air ranging from 1.5 m/minute to 4.5 m/minute for two locations, Alor Star and Leeton, are presented in the Appendix, Figures A1-A4. It can be seen that the optimum setpoint for most of the systems tends to be around 55% and 60% for Leeton and Alor Star, respectively. For deeper beds, the optimum tends to be in the lower range of humidity and vice versa for shallower beds.

The principle of self-tuning, constant relative humidity control is presented in the Appendix, Figures A5-A6. The main advantage of this control is that it does not require statistical analysis of the local weather data. The controller compensates for any changes in the velocity of the supply air and ambient conditions. Because the process of in-store drying takes a long time (from 100 to 2000 hours) short periods of low and high humidity can be ignored. The effects of these periods are minimised by diffusion and absorbtion phenomena, so the consequences are minimal. Long term average relative humidity is the only factor which has to be kept at an optimum level.

## Costs

### Energy

Figure 4 illustrates the increase in the cost of basic energy sources used in the aeration and drying of paddy rice in Australia. These cost increases have not been matched by increases in the price obtained for rice. In an effort to maintain returns to growers, or processor profitability, closer control of these increasingly expensive resources must be achieved. This dictates some form of control system. If the rapid rate of price increase continues then equally rapid reaction by management to control these costs is justifiable.

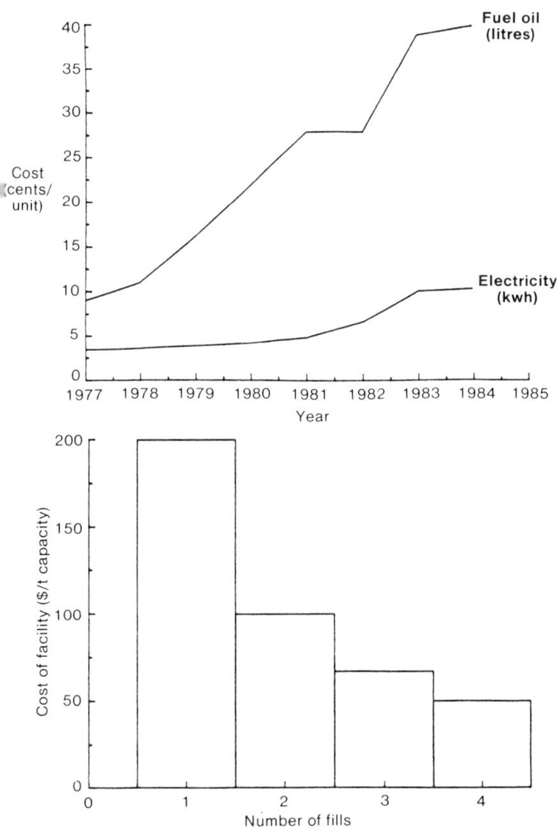


**Fig. 4.** (a) Increases in costs of electricity and fuel oil in Australia, 1977-1984.
(b) Capital cost per tonne versus utilisation of grain drying facilities.

## Quality

Costs involved in drying processes can be minimised by proper design of control systems. Examples of some of the costs which are commonly ignored are presented in Figure 5.

As can be seen losses from overdrying, underdrying (dry matter loss), or breakage can be quite high, sometimes even exceeding energy costs, and it is therefore necessary to take them into account when analysing the economics of the drying processes (see Appendix A for details).

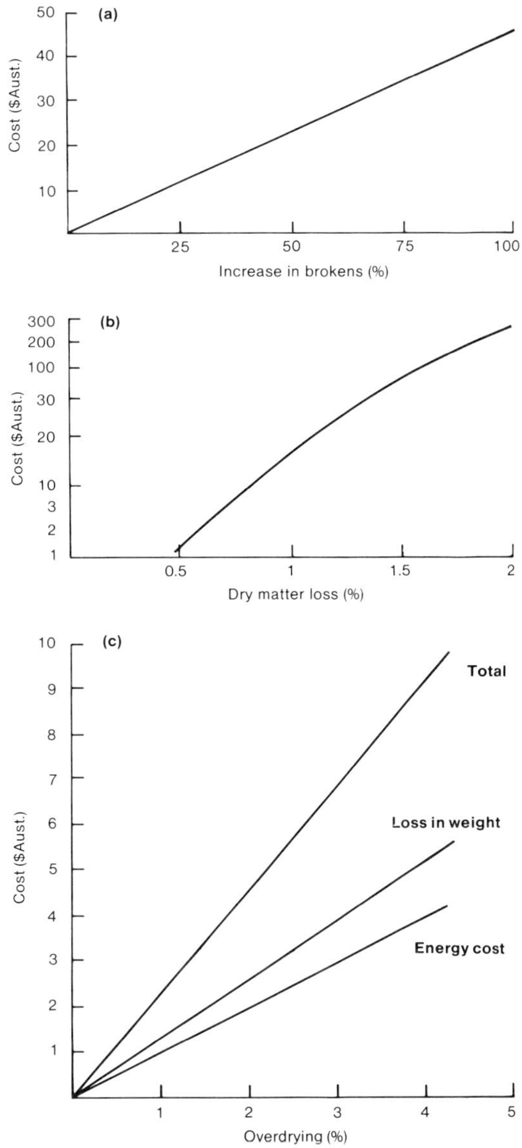


**Fig. 5.** Costs associated with drying of rice.
(a) Increase in brokens.
(b) Dry matter loss.
(c) Overdrying.

## Labour

The extent of any reduction in labour costs that may occur following implementation of a new system is hard to predict. Labour is certainly used more effectively within a well-defined system. Workers are released from more menial tasks and can apply themselves to managing the system rather than reacting to individual events.

## Facilities

The cost of energy to dry grain can exceed the cost of production (Gunasakaran and Shove 1983). The capital cost of constructing and equipping facilities for drying with services and controls is also high. The opportunity to re-use these capital intensive facilities has been graphically demonstrated in Figure 4b. This simple illustration shows the cost/tonne effect of fast turnover of grain through a drying facility. For tropical climates with two crops per year, the opportunity to process grain through a grain drying facility a number of times in each harvest reduces the marginal return required from each pass. The application of more sophisticated control systems to optimise the use of these facilities by tight control of the drying process can be demonstrated if extra usage of the facility is expected.

### System Analysis

Elder (1983) described a control system for cooling based on a time-proportioning device operated by either thermostat or humidistat. Morey et al. (1979) also used humidistat, thermostat, and time clock controls for reducing energy consumption. Simmonton et al. (1981) reported a control system for a combination dryer incorporating a logarithmic drying model and a short-term weather forecast.

Most reports in the literature identify problems of equipment inadequacies. Apart from the CSIRO controller all others cited were of one-off experimental units. Mittal and Otten (1983) comment that no automatic system is available to control the drying process under greatly varying conditions. Ricegrowers' Co-operative Mills, Australia have experience gained from operating two computer-controlled aeration and drying depots for the safe drying and storage of paddy rice in Australia for the past five years. For the 1985 harvest a further eight sheds were installed with PLC controllers to supervise fan and supplemental heater action during drying and to allow for maintenance aeration of the paddy when dry.

In the absence of equipment failures, the next most commonly reported shortcoming of supervised and semi-supervised systems is that during on-off operation wet grain at the top of bins can be put at risk. None of the devices mentioned incorporated feedback information on the grain in the bin. They relied solely on simulated predictions.

Figure 6 demonstrates the possible decision-making paths for a manual operator of a facility, and the degree to which this decision-making can be reduced by the application of from some simple discrete process controllers up to a complete system controller. This figure also highlights the time taken by a supervisor to retrieve the data

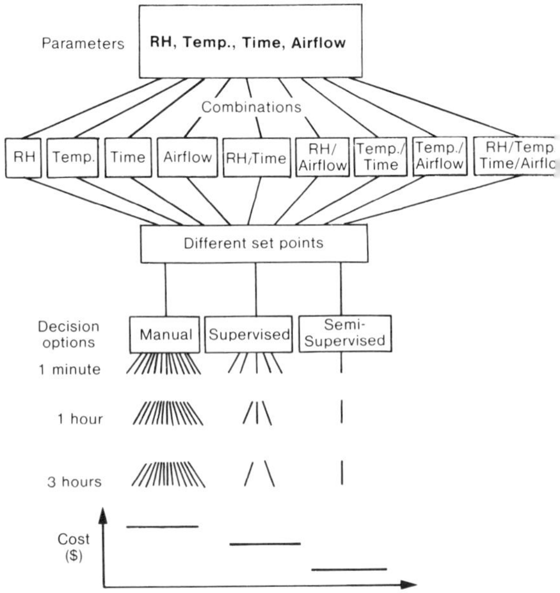


**Fig. 6.** Diagram of control parameters and decision options in grain aeration and drying.

necessary to make his decision and then act with the system control outputs. The final case is the rapid critical path selection that can be performed by a semi-supervised controller which is directly connected to inputs, has decision-making ability, and can communicate directly with all necessary outputs for tight control of the system.

Ricegrowers' Co-operative Mills Ltd, Australia operates a harvest management computer-based data information system for the manual control of the aeration and drying of 600 000 tonnes of paddy rice storage. The initial tonnage and moisture content of the paddy in the flat-bed storage bins is input to the computer. Daily moisture and temperature readings of all bins are input to the computer and together used to recommend a manual operating procedure.

The recommended manual program is manipulated by the aeration supervisor who is removed from the daily function of turning on fans and can observe and evaluate weather patterns and critically review the effect of the strategy applied to date on the objective of supply, at the lowest drying cost, dry paddy for milling.

A significant problem can arise with bulk in-store drying if adequate moisture segregation is not practiced. For a drying control system to operate effectively, the initial and final moisture contents of all the grain in the bulk must be without gross deviation. In a grain bulk the presence of significantly different grain can prove disastrous, this in no way reflecting on the efficiency or otherwise of the drying control system.

## System Comparisons

Simmonton et al. (1981) indicate the complexity of decision-making involving a combination drying system and suggest that an optimised algorithm incorporated in the software of a microprocessor based controller will allow application of proper management and control essential for successful grain drying operations (see Figure 6).

## Systems

Figure 7 indicates the increasing complexity of the available data or inputs for use in the decision making process as the system becomes more complex. It also show the increasing number of outputs that must be manipulated to achieve the level of control desired with increasingly more complex systems. The decision making process becomes more complex when the possibility of flexible use of all the facilities is considered.

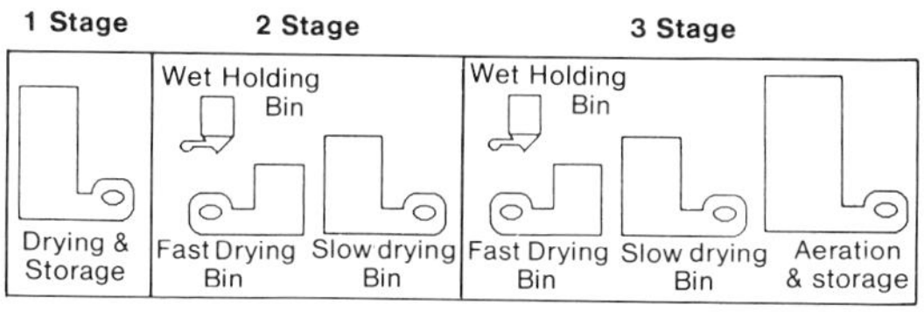


| 1 Stage | 2 Stage | 3 Stage |
|---|---|---|
| **Inputs** | | |
| Grain Moisture/Temperature<br>Ambient RH<br>Ambient Temperature<br>Time of Drying<br>Fan actually on<br>Bin level | 3<br>1<br>1<br>2<br>3<br>3 | 4<br>1<br>1<br>2<br>4<br>4 |
| **Outputs** | | |
| Fan on<br>Fan off | 3<br>3<br>Transfers H-1-2<br>Possible 2-H-1 | 4<br>4<br>H-1-2-3<br>H-2-1-3<br>H-3-1-2<br>H-1-3-2 |
| **Decision** | | |
| Fan on/off<br>Conditional on<br>-RH<br>-Temperature<br>-Time | Fan on/off<br>Conditional on<br>-RH<br>-Temperature<br>-Time<br>Bin. No 2 Available<br>Harvest Rate | Fan on/off<br>Conditional on<br>-RH<br>-Temperature<br>-Time<br>Bin. No 2 Available<br>Harvest Rate<br>Bin. No 3 Available<br>Storage needs |

**Fig. 7.** Demonstration of the increase in control functions in complex systems, such as automated grain drying and aeration.

Figure 7 presents one part of a complex situation. For a large facility with different grains being received, of different varieties and at different moisture contents and quality grades, with multiple storage bins available having multiple fans and possibly supplemental heating to supply varying milling requirements, or the maintenance of viable seed, etc., the management becomes quite complex.

Using a control system, different situations can be foreseen (modelled), the action decided upon, and that procedure used for all future occurrences.

The interaction of the individual components of the drying and aeration system can be optimised for different situations and that procedure used repeatedly.

## Results of Control

COOLING

The principles of the control system designed by CSIRO for the Australian wheat industry are shown in Figure 8.

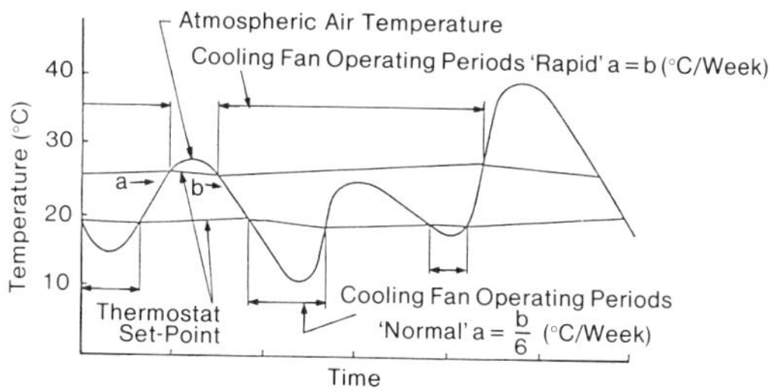


**Fig. 8.** Diagram of operations of CSIRO time-proportioning aeration fan controller.

Current models of the time-proportioning controller provide two settings: RAPID giving an average of 12 hours per day for initial cooling, and NORMAL giving 25 hours per week (on average). No initial setting is required because the controller action is self-calibrating always homing towards atmospheric temperature.

The major feature of time-proportioning control is that it ensures regular aeration through the seasons without supervision. Excessive aeration cannot occur, and the cost of operation is predictable.

However, there is no control of the inlet air humidity which in some circumstances can cause significant problems.

DRYING

An example is presented in Figure 9 of the result of management of an aeration and drying facility

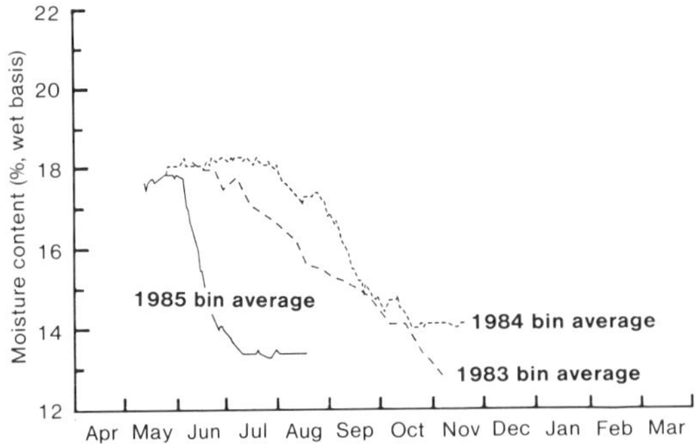


**Fig. 9.** Comparison of bin drying of paddy in Australia before and after introduction of overdrying management (1984) and supplemental heating under automatic control (1985).

by application of different control systems over a period of three years. This information is from Ricegrowers' Co-operative Mills Ltd, Australia. The facility illustrated is an aeration bin of 3000 t capacity. This bin has an airflow rate less than that typical of most Australian drying and aeration facilities: 2.56 m/min at a grain depth of 6 m. The plot for all years is the average moisture content obtained from regular manual reading of moisture/temperature sensors probed into the bin at the completion of grain inloading. There are 18 sensors located at top and middle levels in the bin.

The year 1983 shows typical historical lack of control with the priority on drying with scant regard for overdrying.

The 1984 plot shows that, even though better ambient conditions prevailed, little drying was achieved during winter due to cost management strategies applied. Also, overdrying was controlled due to application of a manual control strategy using daily weather recording thermohygrographs.

The 1985 plot illustrates the dramatic effect of the application of Liquid Petroleum Gas (LPG) burners to existing fans to achieve a 4.2°C temperature rise above ambient when turned on by a supervising PLC. The control strategy essentially relies on a relative humidity (RH) set point for burner operation with a high RH limit for system shutdown in the event of rain. The control system also monitors the ambient RH and temperature sensors and initiates alarms for gross failure of either sensor within any 16 hour period. The system also logs fan and burner operating hours and provides a maintenance aeration facility for programmable hours/week-month at programmable RH windows and at programmable time windows to take advanatage of off-peak electricity rates.

In order to optimise in-store drying, control systems simulation models for various control performances have been developed, for example a model with selftuning relative humidity setpoint.

The principle of this model is that the humidity of air supplied to the drying system is at constant average relative humidity in both systems with supplementary heating and systems based on ambient air aeration. This constant average relative humidity of supply air is obtained regardless of variable weather conditions or possible oversize of supplementary heating system. This is done by switching the fan or supplementary heating on or off.

In this way one avoids fluctuations in the bottom parts of the bin which if excessive cause additional quality losses due to rice breakage. At the same time, full control of the rice final moisture content is achieved and it can be set at an optimum level.

### Control Systems (Canada)

There are approximately 200 large commercial grain dryers and a total of more that 600 off-farm drying operations in the province of Ontario, Canada. Altogether these facilities use about 122,000 cubic metres of natural gas and propane each year to dry approximately 160 million bushels of corn. The annual cost in fuel alone is close to $22 million.

Grain drying accounts for 24% of the cost of production of grain corn, but with the new automatic systems monitoring and controlling the grain drying process, the annual costs can be cut by 10%. In addition, reduced spoilage and shrinkage, higher quality of grain resulting from the automatic controls of the process will increase the product revenue and dramatically improve the operating position of the installation.

Operating data gathered over the last year show that the advantages of the system include: reduced operator time, faster and more precise drying, reduction in unnecessary moisture pockets, and safer storage.

## Future Research Activities

Further research and development work is needed before control systems can be applied to aeration and drying facilities in the humid tropics.

The following is a brief catalogue of the information that this work must provide.

1. Determination of the safe drying conditions in the tropics: air and grain limits for RH, temperature, and time; evaluation of final moisture distribution.
2. Determination of the safe storage conditions: air limit for RH; grain limits for temperature and time.
3. Comparative evaluation of commercially available RH sensors (self-cleaning chilled mirrors, capacitive, resistive, tension) in dusty, humid environments.
4. Comparative evaluation of suitable commercially available control devices under hot, humid tropical conditions.
5. Determination of costs involved in drying of rice to lower moisture contents (the lower final moisture contents are sometimes advisable for tropical countries where long storage periods are used and no refrigeration is applied to prevent quality losses).

## Conclusions

Aeration is used to produce cooling in grains where, in the case of wet grain, self heating is possible due to the respiration of the grain and the microflora associated with it. In the case of dry grain, it is used to reduce insect activity, preserve the nutritional and aesthetic qualities of the grain, and to assist fumigant action and/or prevent overdrying.

The air parameters for drying should be optimal. The optimum is a function of drying equipment, energy costs, quality loss costs, and milling requirements. The best way to achieve it is by a semi-supervised control system, but in smaller installations this could prove too costly. However, as the cost of programmable control devices continues to drop and their power increases it becomes more realistic to plan for their inclusion in the future by the selection and application of sensing and control devices which can be readily integrated into an automated system.

The most readily applicable control system for aerated and in-store dried grain in the humid tropics would currently be a manual control system.

Developing from a manual control system to semi-supervised control has merit in allowing the operator to identify the key inputs required, the best decisions for differing situations, and the form in which the outputs should be controlled.

This then forms the strategy for the semi-supervised control system. An existing manual control system can be converted to higher levels of automation.

In guiding the operator to the key input parameters and logically developing the resultant controlling action, significant progress can be made in overcoming problems in the pressure of harvest and the stressful task of management.

## References


Baird, J. H. 1984. Recent developments in paddy drying, milling and bulk handling in the Far East. Presented at technical consultation workshop. January 1984, Manila.

Bunn, J. H., Dunlop, J. L., and Hammond, G. 1984. Energy efficient drying through computer control. In: Agricultural electronics — 1983 and beyond, volume 2. St Joseph, Michigan, American Society of Agricultural Engineers, 701–707.

Chew, T. A., and Loo, K. F. 1985. A review of the literature of the economics of drying and milling of padi in Malaysia. Presented at technical seminar on post-harvest technology, ASEAN Crops Post-Harvest Programme. August 1985, Manila.

Colliver, D. G., Brook, R. C., and Barrett, J. R. Jr. 1979. Comparison of minimal energy usage management procedures for low-temperature grain drying. Presented at the 1979 summer meeting of the American and Canadian Societies of Agricultural Engineers, June 1979, Winnipeg, Canada.

Elder, W. B. 1983. Alternative applications for the CSIRO grain aeration controller. Agricultural Engineering Australia, 12(2), 17–23.

Elder, W. B. 1984. Control of aeration. Presented at CSIRO/AGI system design seminar and workshop. July 1984, Highett, Victoria, Australia.

Gunasakaran, S., and Shove, G. C. 1983. Optimum fan operation for low-temperature grain drying — a new approach. Transactions American Society of Agricultural Engineers, June 1983, Paper No. 83-3012.

Mittal, G. S., and Otton, L. 1981. Evaluation of various fan and heater management schemes for low-temperature corn drying. Canadian Agricultural Engineering, 23, 97–100.

Mittal, G. S., and Otton, L. 1983. Microprocessor controlled low-temperature corn drying system, Agricultural Systems, 10, 1–19.

Mochtar, M., Driscoll, R. H., and Adamczak, T. P. 1985. The role of tropical weather analysis in grain drying simulation. Presented at Technical seminar on post-harvest technology, ASEAN Crops Post-Harvest Programme. August 1985, Manila.

Morey, R. V., Cloud, H. A., Gustafson, R. J., and Peterson, D. W. 1979. Evaluation of the feasibility of solar energy grain drying. Transactions American Society of Agricultural Engineers, 22, 409–417.

Navarro, S., and Calderon, H. 1982. Aeration of grain in subtropical climates. Rome, FAO Agricultural Services Bulletin, No. 52.

Persson, G. W., and Churchill, R. W. 1984. Automatic computer control of grain drying and aeration. In: Agricultural electronics — 1983 and beyond, volume 2. St Joseph, Michigan, American Society of Agricultural Engineers, 692–700.
Pierce, R. O., and Thomson, T. L. 1979. Solar grain drying in the north central region — simulation results. Transactions American Society of Agricultural Engineers, Paper No. 76-3517.
Pierce, R. O., and Thomson, T. L. 1980. Management of solar and low-temperature grain drying systems — Part 1. Operation with full bin. Transactions American Society of Agricultural Engineers, Paper No. 78-3513.
Pym, G. R. 1984. Controlled paddy rice aeration system. Presented at 6th Conference Australian technical millers association. October 1984, Canberra.
Simmonton, H. R., Allison, J. M., and McLendon, B. D. 1981. A microprocessor-based grain drying control system. Transactions American Society of Agricultural Engineers, June 1981, Paper No. 81-5027.
Teter, N. C. 1982. Design of grain storage aeration for the humid tropics. Presented at winter meeting American Society of Agricultural Engineers. December 1982, Chicago.
Van Ee, G. R., and Kline, G. L. 1979. Management strategies for corn production and low-temperature drying. Transactions American Society of Agricultural Engineers, June 1979, Paper No. 79-3032.
Webb, B. D., and Calderwood, D. L. 1978. Laboratory studies on the relations of moisture content to degree of milling and milling yield in rice. Presented at 17th Rice Technical Working Group meeting. February 1978, College Station, Texas A & M University.
Wilke, W. F., Kranzier, G. A., Bern, C. J., and Anderson, H. E. 1979. Digital electronic control system for low-temperature corn dryer. Transactions American Society of Agricultural Engineers, June 1979, Paper No. 79-3028.

## Appendix

Simulation results for in-store drying with supplementary heating, at Alor Star, Malaysia and Leeton, Australia.

The following assumptions have been made for the analysis.

1. The electricity price is 0.8¢ kWh.
2. The LPG price is 40¢ kg.
3. Dry matter loss cost has been calculated according to the formula (after Pierce and Thomson 1979):
   cost = ((dml/0.5)**4)* 0.01 *140.0
   where
   dml is the percentage of dry matter loss.
4. Brokens level has been estimated according to the formula:
   percentage of
   brokens = 0.06548*wb**2-3.837*wb + 38.82
   where
   wb is moisture content of rice, wet basis.
5. Initial rice moisture content was at 19.3% wet basis and final was 14% wet basis and the drying period was starting on 1 May 1983 and 1 May 1984 in Leeton.
6. Initial rice moisture content for Alor Star was 18% and final was 14% and drying period was starting on 5 November 1983 and 5 November 1982.
7. Paddy price is $140/t.
8. Initial percentage of brokens is 10%.

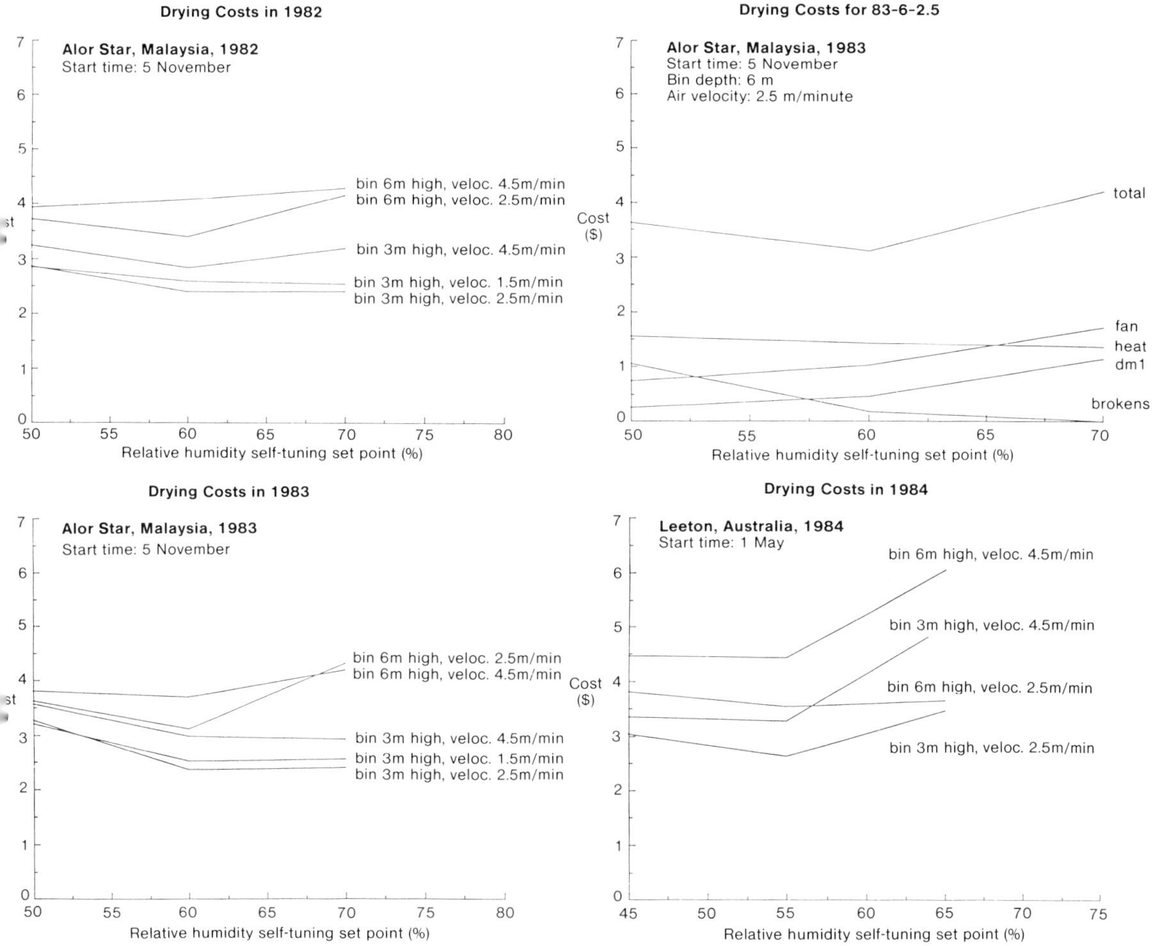


**Figs. A1-A4.** Drying costs for grain based on simulated control for different relative humidities, bed depths, and airflow velocities.

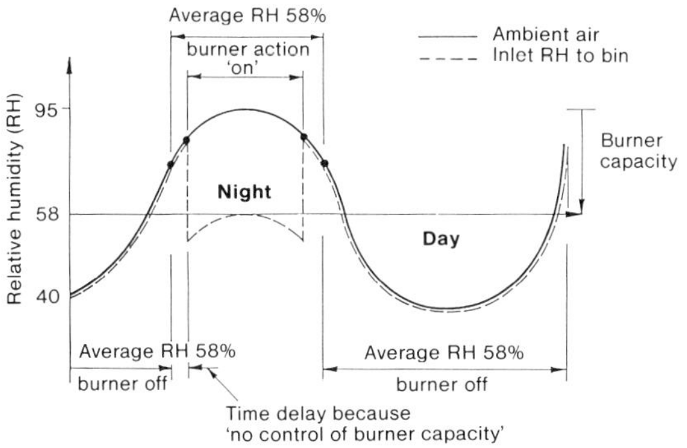


**Fig. A5.** Principle of self-tuning, constant relative humidity (RH) control, involving supplemental heating with constant heater capacity.

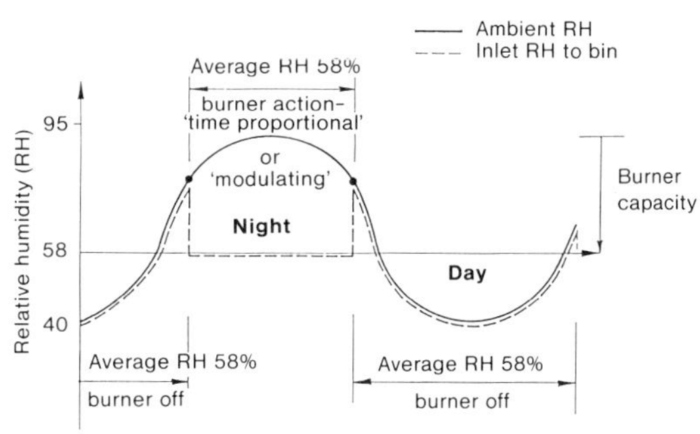


**Fig. A6.** Principle of self-tuning, constant relative humidity (RH) control, involving supplemental heating with variable heater capacity.

# Design of Aeration and In-store Drying Systems Session Chairman's Summary

## G. R. Thorpe*

THE first presentation in this interesting session was Mr Teter's paper on design parameters for storage and handling system structures. It summarised the fundamental biological and physical information that must be known before a grain handling system can be specified and designed. Many of the data are already available, e.g. we now have very sound data on the air/water vapour system. Because of variability in grain properties, equilibrium moisture data tend to be a function of variety and history. In this sense, hygroscopic data can be used as a guide. They are hard, fast, and immutable.

Likewise, parameters defining $\triangle P$ versus Q are, in economic jargon, 'rubbery'.

Mr Teter also gave data on dry matter loss for paddy, and there are, understandably, some discrepancies between two studies he referred to. I believe we must have further quantitative studies on this phenomenon because of its importance on the storage ecosystem.

Because insect population growth is much slower than the build up of heat, parameters for insect population dynamics are probably known sufficiently well.

Mr Teter places a good deal of faith in the deterioration index. I have reservations, because the index considers only the air state. The state of the grain must also be considered.

Dr Hunter's paper had four main themes:

- selection of airflow rates and aeration strategies;
- static pressure regain in aeration ducts;
- pressure drop calculations across grain bulks;
- fan selection.

The paper showed how, for a given climate, the airflow rate and aeration fraction can be chosen to minimise the temperature–time integral of the grain bulk. This assists determination of the technical optimum. The static pressure regain charts, and formulae for pressure in bulks produced by Dr Hunter, although very sophisticated are easy to use. Pressure drop formulae are also very easy to use and perhaps should be distributed in the form of computer software.

The paper by Dr Bowrey and Dr Driscoll entitled 'A procedure to estimate the supplemental heating needs for in-store paddy drying' was a paradigm of simplicity, elegance, and usefulness. It described a procedure for dryer design that uses a psychrometric chart and a simple calculator. It is not only of great practical use to the engineer, but also has considerable teaching value.

Mr Sutherland pointed out that the Bowrey-Driscoll approach is most accurate when the drying front remains within the bed. I note here that Mr Sutherland's own work is also easily programmed and potentially of great value.

I would add the caveat that these simple models, although useful, cannot account for heat of respiration. They also imply a step input of airflow rate and air state.

The last paper in the session, by Mr Pym and Mr Adamczak, dealt with control systems for aeration and drying. It highlighted the value of powerful numerical models of the drying process. These models can handle variables such as:

- air state — temperature and moisture content;
- grain state — temperature and moisture content;
- airflow rates
- stochastic influences — variations in grain state with depth and climate;
- storage design, hence capital cost;
- power requirements, hence running cost.

The integrating capacity of the models permits realistic attempts at economic optimisation to be made.

Other factors can be incorporated into other models. These include the rates of decay of pesticides, insect population growth, and dry matter loss. As yet we cannot quantify all quality indices, particularly in the humid tropics. However, I am pleased to note that this issue is being addressed in various ACIAR grain storage research projects.

* Agricultural Engineering Group, CSIRO Division of Chemical and Wood Technology, P.O. Box 26, Highett, Vic. 3190, Australia.

# Economics of Aeration and In-store Drying

# Loss Assessment — Towards a Fuller Comprehension of Postharvest Loss

R. A. Boxall, D. J. B. Calverley, and P. S. Tyler*

*Abstract*

The major developments during the past decade in connection with postharvest loss assessment and loss reduction programs in developing countries are reviewed. The relevance of the work is appraised in relation to current postharvest problems and a proposal is made for the better use of existing and future loss assessment data in food strategy planning. The prescribed methodology has been adequate for assessment of losses at farm level, but there is a lack of proven guidelines for the commercial sector. Such guidelines are needed and should embrace a wider approach that is more concerned with improving the efficiency of the system.

THIS paper briefly reviews the state of the art of loss assessment within the postharvest system. The approach that has been followed in the past, the methodology used, and the way in which measurements of loss have been utilised, are described. In the light of experience, some general conclusions are drawn and suggestions made for future development in methodology. The problems facing grain marketing organisations, especially those in Southeast Asia, in procurement and conditioning for storage of paddy, are seen as being a major area where there is concern over losses. Proposals are made for their more objective assessment.

It is concluded that the baseline against which loss has to be evaluated should be a grain quality and pricing structure that is acceptable to the consumer. The quality control objectives of marketing organisations must reflect these customer requirements and cost-effective measures should be employed to contain food quality losses within the objective limits. The implications of introducing optimal technical procedures for loss containment are therefore considered within overall policy and planning strategies.

### The Background to Loss Assessment Studies

The degree and extent of losses of grain after harvest is widely held to be a matter of concern, as also is the need to reduce loss and make improvements in the handling and storage of grain. There has been considerable activity in the field of postharvest loss assessment and loss reduction in the last 10 years. This activity was stimulated by the 1975 Resolution of the Interdepartmental Sub-Committee of the VIIth Special Session of the United Nations General Assembly, which committed member states to reduce postharvest losses by 50% by 1985. It is therefore timely to assess the achievements to date and to consider the requirements for the future.

An early review by Howe (1965) of the available information on postharvest losses, and a later one by Adams (1977), clearly showed that many of the data on the extent and types of losses were not particularly meaningful, up to those dates, and certainly did little to assist identification of targets for loss reduction programs. There was a clear need to improve and standardise the methodology for loss assessment in order to justify more firmly the development and introduction of measures designed to reduce losses in an appropriate and economical way.

Recognising this need, in mid-1978, the American Association of Cereal Chemists, with the aid of a grant from the United States Agency for International Development to the League for International Food Education, began to prepare a methodology for assessing postharvest grain losses with emphasis at the farm level. As a result of this initiative, a manual (Harris and Lindblad 1978) was published to provide the means whereby postharvest losses might be established in a

* Storage Department, Tropical Development and Research Institute, London Road, Slough, Berkshire, SL3 7HL, U.K.

standardised and meaningful way so that effective grain loss reduction efforts could be undertaken in developing countries. It was recognised that because of the enormous variability of local postharvest systems, no complete or definitive methodology for all situations would be practicable. Nevertheless, the manual provided acceptable guidelines to meet a variety of requirements and it is fortunate that many of the methodologies adopted within loss assessment and reduction projects have been based extensively on that suggested in Harris and Lindblad (1978). The manual detailed the techniques available for the measurement and interpretation of losses incurred during threshing, drying, storage, and milling for the major cereal grains as well as those arising from a range of specific causes including insects, moulds, and rodents. A major difficulty experienced in compiling the methodology was that the authors necessarily drew heavily on experiences gained from developed countries and limited specific examples from some developing countries. A detailed knowledge of the wide variety of systems in which assessments would be attempted was not available and consequently the difficulties encountered in applying the methodology could not be foreseen.

In reviewing the use to which the methodology has been put, it can be seen that most work has been devoted to farm and village level operations as was intended, and generally the prescribed techniques have proven suitable for studies in simple postharvest systems and, more specifically, in storage at the farm level. Experience has demonstrated the importance of a preliminary survey to provide an insight into the possibilities for loss reduction. Some major shortcomings of projects have arisen during the planning stages. For example, objectivity has not always been clear, particularly in relation to how, once obtained, the data on losses will influence decisions on the subsequent remedial actions to be employed. Sometimes loss assessment has been undertaken seemingly only to justify a course of remedial action which has already been decided upon. It may therefore be asked why the loss measurement was required. Also, an elaborate, time-consuming, and expensive program of gathering loss data has sometimes been embarked upon when a decision on the degree of loss reduction justified might have been made on the basis of a short preliminary survey.

The methodology of Harris and Lindblad advocated taking an early overview of the whole postharvest system in order to understand the food grain supply system before beginning an in-depth examination with a view to measuring loss. However, the methodology was designed to be applied to the on-farm sector and was not suited to detailed study of the whole system. The tendency has been to direct efforts towards one component, which is identified as in need of attention, without necessarily appreciating that this, in turn, may have far-reaching consequences to the system as a whole.

For example, the results of work at the farm level with maize, paddy, and other cereals have shown that the traditional system is efficient but problems arise when new, high-yielding varieties, or multiple-cropping techniques are introduced. Such innovations disturb the traditional capability to conserve grain, increase the risk of loss, and may place an additional strain on procurement and marketing facilities. The farmer has to face new and intractable problems with an increased quantity of produce. The traditional handling, drying, and storage systems are found to be inadequate, the characteristics of the crop are changed (often to a softer grain which is more susceptible to mould and insect damage), and the timing of the harvest and subsequent operations may be switched from a favourable dry season to an unfavourable wet season. All of these factors have made the on-farm postharvest system more vulnerable to loss. The recommendation must therefore be made that changes to traditional agricultural practices should never be introduced in isolation. Rather, they should be made only after due consideration of the full implications of the change.

### Conclusions from Experience to Date

Experience has shown that, despite the difficulties which arise when the traditional farm system is disturbed, many of the problems at farm level could be solved by integrating appropriate postharvest loss reduction techniques into the package of improvements (seed, fertilisers, improved cultivation practices, etc.) which are offered to farmers to increase production. This might be achieved by first incorporating postharvest activities into Farming Systems Research Programs which aim to link research achievements with the needs of the farmer and community by providing a realistic assessment of what changes are feasible and acceptable under existing or changing conditions. This would ensure that efforts are aimed at containing postharvest losses within the acceptable limits of the particular traditional system. The introduction of change in agriculture requires a thorough understanding of the problems it will cause, sound technical advice to meet them, and an effective training and extension program for implementation.

There are still many aspects of farm level operations needing more study and research to establish the optimum methods of harvesting, drying, handling, and storage appropriate to the total farming system. However, trained and mobile extension staff can make a significant improvement to small-scale storage by the application of known and established techniques where appropriate. One approach which is gaining favour is the use of a national team of research, training, and extension staff who identify the problems, evaluate possible solutions, make suitable recommendations, and assist the existing government extension workers to implement the recommendations. Progress is evident, but it is necessarily slow, and many farmers are still inclined to selling their produce early in an attempt to minimise their losses, thereby placing an additional seasonal burden on central procurement and storage systems already overloaded because of their need to cope with increased total production.

Early loss-assessment studies were directed towards the farm level, and particularly towards storage. On the basis of estimates that between 60 and 70% of food grain in developing countries is held at farm level, the potential benefits of loss reduction activities seemed to be greatest in this sector. However, in view of the relatively slow progress in introducing change at the farm level and the limited postharvest research and development resources relative to the crop production sector, it is perhaps wiser to concentrate what resources are available upon the commercial sector. The benefits of loss assessment/reduction activities in this sector may, after all, be greater and achieved much more quickly, at least in the short term. There is an urgent need to pay greater attention to the commercial sector now that more countries are becoming self-sufficient in food grains, with the marketed surplus produced by small farmers becoming a predominant source of a nation's food supply. The commercial sector's response to these surpluses of grain is to store for longer periods or to begin to export, and this is particularly true of countries in Southeast Asia. However, the farm sector must not be overlooked, because inevitably changes or improvements in the commercial sector will result in a reduction of quantitative and qualitative losses elsewhere, particularly if accompanied by a price incentive. Similarly, developments at farm level, although they may be slow, will have consequences for the commercial sector.

Little systematic work has been undertaken on assessing losses in commercial operations. Some attempts have been made to apply the methodology of Harris and Lindblad in this sector. The few loss assessment studies which have been done are found to be either insufficiently comprehensive or insufficiently complete in the results obtained. Many have been primarily concerned with measuring deterioration due to physical, chemical, and biological factors in storage (La Gra et al. 1982). However, it is apparent that there are overriding factors, such as deficiencies of management, which defy easy quantification.

A particular lack is of studies where losses have been evaluated in the commercial sector along with those in the farm sector in order to obtain a complete picture. A major inhibiting factor may be the absence of a prescribed methodology immediately applicable to procurement, handling, storage, and distribution operations, but there may be other reasons such as the difficulty of coordinating studies that involve both the production and commercial sectors.

One requirement might therefore be to combine a series of measurement techniques derived from the experience gained from work in the farm sector, with a description of how better to evaluate the various commercial operations. A suggested approach to solving problems in commercial storage has been outlined by Hindmarsh and McFarlane (1983) who stress the overriding importance of adapting the technical developments and improvements to fit the existing management system. Such an approach can be aimed towards reducing losses to an agreed level.

To date there does not appear to have been a quotable example of a loss assessment study being used as the tool to both identify loss and for the subsequent evaluation of the effectiveness of a loss reduction program. Until such a complete procedure has been followed and the value of the loss monitoring effort has been demonstrated, loss assessment remains hard to justify.

## Facing Up to Current Requirements

Experience with loss-assessment studies confirms that a prerequisite of national postharvest loss reduction programs is a firm government commitment together with the necessary infrastructure and finance. Sadly this commitment has sometimes been lacking even though a short-term commitment to support a project has been made. On completion of the project, national priority has tended to shift away from the postharvest sector and towards, say, crop production. Similar observations might be made about programs designed to increase production or to improve marketing. Only when such programs are brought under central control at the planning stage can an appropriate national food strategy be developed.

In addition to the central planning implications of introducing a policy to contain losses, there are obviously technical aspects which have to be considered within the operations of a marketing organisation. There is, therefore, an urgent need to adopt a broader approach to loss assessment and loss reduction which takes account of overall policies for both food production and food distribution. This is discussed under three headings.

### Food Strategy Planning

A potentially useful tool in food strategy planning is the Food Balance Sheet Equation, either for cereals and pulses generally, or for specific commodities. The equation is designed to relate national production data to actual or projected per capita consumption. It takes account of exports, imports, the use of a commodity for animal feed or as seed, and postharvest food waste. Recent experience in relation to food strategy planning in Africa has shown that substantial amounts of data of value to the preparation of the Food Balance Sheet Equation already exist or are being collected on a regular basis, but they are rarely aggregated and are therefore not fully used. Postharvest loss data may fall into this category but there is an identified need for better, objective measurements of postharvest losses of food crops.

In the past, postharvest loss data have been used largely to create a general awareness and either to draw attention to waste and inefficiency, or to focus attention upon aspects of the postharvest system in order to justify a loss reduction program and to enable priorities to be determined. However, the same data could equally be used to improve the reliability of estimates of crop yields, totals stored or handled, and the amounts available for consumption. As far as is known, there are no examples of reliable loss survey data being incorporated in national statistics for food production or utilisation. Loss figures are used, but these are usually either informed 'guesstimates' or are obtained by difference simply to explain the imbalance between two sets of data which should be in agreement.

It would seem to be a relatively simple matter to begin to incorporate postharvest loss data into national statistics which can be used for food planning. The technical resources needed to obtain such data are often well established and a standardised methodology for assessing losses is already prescribed. However, it is questionable as to how useful these data would be if the related information were incomplete. Experience has shown that production, marketing, and consumption figures are sometimes suspect. The answer to the question is that the provision of at least some credible data must be of assistance in improving the accuracy of the Food Balance Sheet Equation, but the first priority, at least in some countries, is the need to improve the overall capability to strengthen the objectivity of and capacity for data-gathering and to improve the critical assessment of data.

Ideally, the responsibility for the gathering and processing of such data must be accepted by a central Department of Statistics and Planning, which can bridge the gap which often exists between Departments of Agriculture, Food, Commerce, etc. Hitherto, the stimuli for obtaining estimates of postharvest loss have come from staff in research organisations, field services, or even aid donors specifically concerned with the postharvest sector.

In addition to achieving acceptance of loss data in overall planning, there is need to continue to monitor developmental and technical aspects of current postharvest practices. The significant changes that are taking place at the farm level and their effects on national marketing organisations have been noted earlier. Organisations find that storage operations which were satisfactory for several months storage are inadequate when the storage period is extended up to 12 months or longer.

If the currently accepted recommendations for good storage practice are put into effect, storage losses could be contained and even reduced. While evidence suggests that there exists considerable inertia and opposition to change which inhibits the application of different operations or procedures, this should not deter a continued and sustained effort to improve storage practices and encourage their adoption as far as possible.

## Technical

Insect pest infestations can always be expected and ability to control them using safe and effective pesticides cannot be assumed. Resistance of many of the important storage insects to the insecticides currently used to protect cereals or to disinfest cereal storage is now widespread. Recently, significant resistance of storage insects to phosphine has been reported (Tyler et al. 1983) and the effectiveness of phosphine disinfestation procedures requires close examination. Even when pesticides can be used, the longer periods of storage raise problems of the preservation of quality without increasing chemical residues to unacceptable levels.

In the humid tropics there are major problems in maintaining quality because of the high moisture content of grain, particularly during the wet season. Traders and millers in the private sector may procure grains during the period of the year when they can benefit most from sun drying, but during the wet season they may stop buying and farmers then have to divert their wet crops to the central grain-handling organisation.

There is no single solution to these problems and bulk storage provided with aeration and drying facilities is likely to be but one of the alternatives that can be applied as appropriate to a particular situation. The potential benefits of bulk storage and the use of aeration and in-store drying in maintaining grain quality have been described in earlier papers. The establishment of a bulk storage and handling system provided with aeration or drying facilities may show, prima facie, little advantage over seasonal bag storage systems in developing countries where in most cases absolute control of insects is not essential and where few moisture problems arise. Loss-assessment studies could be conducted to establish more precisely the benefits of the new system over the existing one by taking account of quality parameters such as percentage of discoloured grains, extent of mould damage, proportion of broken grains, and milling yields (of rice). However, we must guard against objectivity becoming confused. Loss assessment is not advisedly used to justify a change of technique under the guise of loss reduction when the real motive may be higher productivity obtained through labour or capital costs.

The loss prevention technique may lead to a significant increase in food availability (the prime aim of loss assessment/loss reduction programs), but it may achieve improvements in quality as mentioned above. Once again, care is needed in interpreting the benefit. The value of the new technique may actually be in raising the quality rather than reducing a loss. The conventional quality standards, particularly those of industrialised countries, if applied here may not be appropriate and due consideration must be given to local perception of quality. The real value of this 'improved' quality will depend upon the extent to which it is appreciated by the consumer.

Both quantitative and qualitative deteriorations occur in the humid tropics because of the high moisture content at harvest time. If wet paddy is taken into store, mould, germination, and yellowing of the rice will occur and so the need to dry the grain is obvious.

There is a widely held view that yellowing and discoloration arises only in store and that aeration and drying may well be the answer to the problem. However, studies of simulated postproduction processes made by Mendoza et al. (1981) and Mendoza and Rigor (1981) demonstrated that, at the field level, the combined effects of ambient conditions and biological factors determine how long paddy can be left unthreshed before deterioration reaches unacceptable levels. A delay of 5–6 days in threshing wet paddy will result in paddy with darkened or yellowed kernels and a loss of lustre in the rice milled from it.

The introduction of central drying facilities must be carefully considered since the economics of the operation will be dependent upon a supply of paddy which has not already deteriorated. This means that an infrastructure must be developed which allows farmers to bring wet grain to a drying facility quickly.

These examples serve to illustrate the need to

view postharvest problems as part of a total system rather than in isolation.

### Marketing Organisations

There are several reasons why marketing organisations need to store cereal grains. These include: (a) the storage of seasonal or operational stocks to meet seasonal demands and to stabilise prices; (b) the maintenance of a carry-over stock between seasons, and (c) establishment of a strategic or long-term reserve against crop failure. In every case, the marketing policy of the government will determine the scope and scale of operations. Seasonal and carry-over storage involves the collection of locally procured produce, usually into stores holding 1000 tonnes or more, and storing it for periods of up to 12 months, although longer storage of 18–24 months is not uncommon.

Deterioration during storage can be limited by ensuring that the produce received is in good condition and sufficiently dry to prevent spoilage by moisture, and through good storage management embracing a high standard of store hygiene and the use of insecticides and fumigants to control insect pests.

The efficiency of the various storage operations has not been the subject of intensive comparative study. However, it is apparent that it varies widely from good to poor, with losses ranging from small to complete. These losses will arise from the activities of, for example, insects, rodents, and birds, and the growth of microorganisms. Further losses will arise through spillage, breakage, etc., but it must be recognised that the root cause of the problem may be deficiencies of management or policy.

It can be argued that the approach to loss assessment and loss reduction has been largely symptomatic. An individual may identify the primary constraint in terms of his own discipline. An engineer, for example, may view the solution of postharvest problems in terms of a need for better grain stores or drying facilities. The entomologist, on the other hand, may see it in terms of insect pest control, while the economist may view the problem in terms of the need for credit facilities. All may be important, but so will many other views. No single approach can be considered in isolation from the system.

As a result of experiences of loss-assessment studies, a multi-disciplinary approach to removing some of the constraints has been developed. Nevertheless, in relation to the commercial sector, this approach has tended to concentrate upon technical site management with little attention to broader issues. The systematic approach implies studying the system as it exists, then devising, as a concept, the perfect system, i.e. one which is free from constraints. This may be too idealised, but one can determine the important constraints that can be removed and how best to do that, thereby arriving at a pragmatic system which, though still not perfect, is significantly better than what exists.

Marketing organisations accept some operating losses within varying limits as part of their overall running costs, but the losses are frequently poorly defined. The difficulty arises because the organisations cannot quantify grain movements in sufficiently precise terms to enable them to identify normal operating losses and losses which are preventable, e.g. caused by insects, moulds, etc. The overall level of loss accepted in marketing board operations may be arbitrarily fixed at one or two per cent of grain tonnage or value. Resources are committed to limit loss and to maintain grain quality, but often without a full analysis of cost effectiveness. It is therefore necessary to take account of not only stock losses but also the multitude of problems due to inefficiencies, overreaching of staff and management capability, inappropriate or inadequate facilities, ill-defined or inappropriate policies (for prices, stockholding, reserves, etc.), inadequate technical procedures, and inadequate crop and market intelligence. Any of these factors, alone or in combination, will give rise to low levels of cost effectiveness.

## Research Needs

The improvement of design and operating efficiency which will lead to a better understanding of acceptable levels of operating losses might be achieved by the application of systems analysis techniques. Such an approach requires the construction and operation of simulation models to test the implications of various system designs and operating procedures. It would assist in national planning of new systems for marketing organisations (i.e., to ensure optimal allocation of resources for collection, drying, storage, distribution, etc.) and it would enable the identification of constraints in existing systems and possible remedial measures.

There is a need, first of all, to develop basic

models for specific (one country) systems, and later to test and adapt the models to other systems in other countries. Experience in this field is limited; however, some progress is being made in Indonesia through a joint National Logistics Agency (BULOG)/U.K. Tropical Development and Research Institute study of minimum stock reserves. The study is concerned with food security in Indonesia, the role that BULOG plays in ensuring adequate food supplies, and the requirements (in terms of grain stocks and stock locations) necessary for BULOG to carry out its role efficiently. The objectives are to assess what decisions BULOG should make to ensure an adequate level of food security, faced as it is with changing levels of production, demand, government policies, and other factors outside its control.

Econometrics and linear programming are methodologies which have been applied to grain marketing problems elsewhere but the study in Indonesia is innovative in the way these two methodologies are used in a closely integrated manner to analyse the problems facing a food grain intervention agency. To date a set of working models has been developed and these have been shown to be capable of providing guidance on minimum stock levels at distribution points and the lowest-cost ways of moving them. However, considerable improvement and further development is needed to relate them more precisely to market conditions.

This experience is related primarily to stockholding and distribution but parameters for spoilage, methods and costs of operating various storage and handling systems, loss reduction methods, and different climatic conditions might be utilised at some stage in a basic data bank which can be drawn upon in developing future models.

This approach to 'loss assessment' is far removed from the farm level studies, the studies of specific parts or problems within the system, and those designed to evaluate a particular intervention method. All are important but we must recognise that the loss assessment activity over the past decade has not led to the widespread increases in available food that were expected. It is undeniable that action is urgently required to improve conditions in the postharvest sector and it is perhaps appropriate to look at this need in the global context. Over the decade 1970–1980, world food output per capita rose by about 0.5% [1.1% for industrial market economies; 1.4% for South Asia (Anon. 1978).)] While this rise may be taken to indicate the success of policies and technological developments aimed at increasing production, a further breakdown of the figures reveals the disturbing fact that production in the African continent actually fell by 1.1% over the period. In many African countries, physical losses of grain quality and quantity were overshadowed by various perturbations. The fall in production is in part attributable to natural disasters, e.g. the Sahelian drought, but other factors such as political instability and the insecurity caused by guerrilla activities have seriously disrupted the production, storage, and distribution of grain stocks. Added to this, a depressed world economy has hit agricultural production and distribution through shortage of currency and credit, high prices of essential inputs (fertilisers, pesticides, fuels, etc.), and weakened transport infrasture.

The real priority for governments is to face up to the need to apply integrated policies of food production, marketing, storage, and distribution. Given the present difficulties, there is a special need for assistance to enable governments to make the most effective use of the resources available to ensure that the maximum amount of produce of the highest acceptable quality reaches the consumer by the most efficient (cost-effective) means.

## References


Adams, J. M. 1977. A review of the literature concerning losses in stored cereals and pulses published since 1964. Tropical Science, 9, 1–29.

Anon. 1978. World Development Report. 1978. World Bank.

Harris, K. L., and Lindblad, C. J. (Compilers). 1978. Postharvest grain loss assessment methods. Published by the American Association of Cereal Chemists, in co-operation with the League for International Food Education, the Tropical Products Institute, the Food and Agriculture Organisation of the United Nations, and the Group for Assistance on Systems relating to Grain After-harvest, 204 p.

Hindmarsh, P. S., and McFarlane, J. A. 1983. A programmed approach to food storage improvements. Tropical Stored Products Information, 46, 3–9.

Howe, R. W. 1965. Losses caused by insects and mites in stored foods and feeding stuffs. Nutrition Abstracts and Reviews, 35, 285–293.

La Gra, J., Martinez, E., and Martinez, J. F. 1982. Identificaccion, evaluacion y reduccion de perdidas de post-cosecha. Arroz en la Republica Dominicana. Publicacion Miscelanea ICCA 358. Instituto Interamericano de Co-operacion para la Agricultura, Santo Domingo, Republica Dominicana, 165 p.

Mendoza, E., and Rigor, A. 1981. Grain quality

deterioration in on-farm level of operation. Presented at the 31st Annual Convention of the Philippine Society of Agricultural Engineers, April 1981, Quezon City.
Mendoza, E. E., Rigor, A. C., and Abital, F. L. 1981. Grain quality deterioration in on-farm level of operation. Presented at the Regional Grains Post-Harvest Workshop, January 1981, Manila.
Tyler, P. S., Taylor, R. W., and Rees, D. P. 1983. Insect resistance to phosphine fumigation in food warehouses in Bangladesh. International Pest Control, 25 (1), 10–14.

# The Economics of Grain Drying in the Humid Tropics

G. J. Ryland*

*Abstract*

A model which can be used to appraise alternative drying technologies is developed. The equilibrium or economic optimum level of drying operations is defined in terms of the economic threshold of moisture content below which point further drying reduces farmer's net returns. Economic optimum levels of drying depend on a number of critical parameters including time, quantity dried, rate of moisture content reduction, qualitative changes in the grain, as well as drying costs and anticipated returns. The paper concludes by discussing the linkages between drying operations and other postharvest operations of transport, storage, milling, and distribution, and the necessity of a systems approach to arrive at the most socially efficient pattern of drying facilities.

DRYING of grain in storage is done for a number of reasons, including preservation of grain quality, reduction of pests as part of an integrated pest management system, and to achieve a higher price for the stored product. Other indirect benefits include timely early harvests which allow better utilisation of harvesting equipment. Losses in yield in terms of both physical and qualitative loss are largely dependent upon the storage environment. If the product is not stored at a 'safe' level of moisture content then deterioration and spoilage occurs.

Drying and aeration procedures are part of the storage and inventory considerations made by all those involved in the marketing chain including farmers, grain handlers, processors, and wholesaler/retailers. Each of these agents in the marketing chain may adopt different attitudes to determine whether drying is of personal economic benefit. However, the actions of each individual agent together determine the net benefit from drying to society as a whole.

In this systems framework the main issues relate to the type, location, number, and size of drying outlets. For example, net social welfare may be maximised with a pattern of small size, decentralised drying facilities at farm level which, although operating at higher cost, are of greater benefit than investment in large-scale facilities which have lower operating cost but are offset by higher assembly costs (queuing, transport, and delays in the system).

In terms of paddy pricing policy, there may be little or no incentive to dry the product if the full subsequent resource costs of drying are not passed on to the farmer in terms of a premium[1]. Consequently, drying and aeration have important implications on policy issues relating to paddy and rice pricing systems.

In summary, there are three key issues regarding drying and aeration which I think economists have an important role in resolving. These issues are:

(1) at the level of the individual decision maker — what is the conceptual framework for determining economic threshold levels of drying?
(2) at the industry level — what is the appropriate framework for determining the type, size, number, and location of dryers? and
(3) at the national level — what is the impact of alternative pricing policies for paddy and rice on drying and aeration procedures?

In the next and subsequent sections these issues are addressed by first developing a conceptual framework in which economic analyses can be made.

* Department of Agriculture, G.P.O. Box 1671, Adelaide, South Australia 5001.

[1] It is usual to find that discounts or penalty charges for high moisture grain are less than actual drying costs because buyers of grain have the facility to blend high moisture grain with dry grain.

## The Economic Threshold Level of Drying

From the viewpoint of any individual economic agent (farmer, miller, or trader) drying is of economic benefit if the returns in terms of price premium per unit are sufficient to offset the costs of drying[2]. The economic threshold level of drying is the point below which further drying reduces profits. This concept was recently applied by Chew and Loo (1985).

The economic threshold level of drying recognises that there may be a level of moisture content above the technically defined 'safe level'. The latter, however, only results in reduced profits compared with the economic threshold level.

To express the concept of the economic threshold level of drying in mathematical and perhaps more operational form consider the following model cast in terms of the individual farmer as decision-maker.

$$\pi = P_y Y - C \qquad (1)$$

where

$\pi$ = profit per tonne of paddy to the farmer;
$P_y$ = net return per tonne of rice;
$Y$ = quantity of head rice recovered per tonne of paddy, expressed as a percentage;
$C$ = cost per tonne of drying paddy.

The head yield of rice or recovery rate Y can be expressed in terms of moisture content (m.c.), x, as follows:

$$Y = f(x). \qquad (2)$$

Similarly, the cost per tonne of drying paddy depends on the hours (H) spent drying, which can also be expressed in terms of final m.c., x.

$$H = g(x) => C = g(x). \qquad (3)$$

Substituting these relationships in (1), the optimum m.c., x, can be determined as:

$$\frac{\partial \Pi}{\partial x} = P_y \frac{\partial f(x)}{\partial x} - \frac{\partial g(x)}{\partial x} = 0. \qquad (4)$$

[2] The costs and benefits of grain drying arise throughout the entire life of the system so that an economic evaluation of the system should incorporate all cash flows over its lifetime. Further, since money has a time value, all the cash flows must be discounted to present value terms. For exposition purposes we assume that all returns and costs have been discounted.

The forms of the functions f(x) and g(x) are such that

$$\frac{\partial f}{\partial x} \leqslant 0 \text{ and } \frac{\partial^2 f(x)}{\partial x} > 0 \text{ while}$$

$$\frac{\partial g}{\partial x} \leqslant 0 \text{ and } \frac{\partial^2 f}{\partial x} > 0$$

The relationships involved are drawn in Figure 1. These relationships show that recovery rate and cost are inversely related to m.c., x.

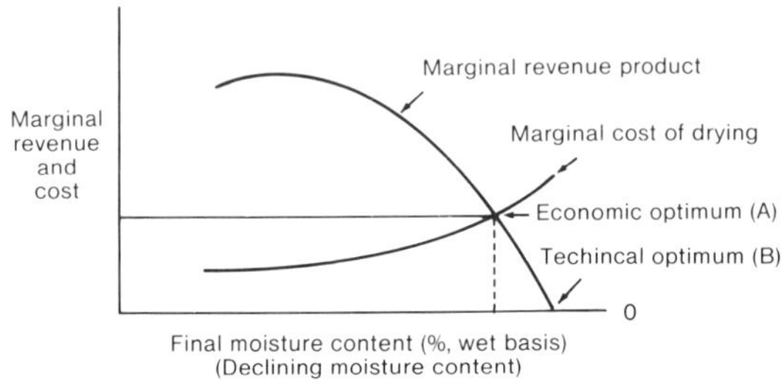


**Fig. 1** Comparison of economic and technical optimum levels of grain drying.

Using (4) it is possible to make some inferences about the economic threshold level of drying relative to the technical optimum.

The technical optimum is the point at which the first derivative of head rice yield with respect to m.c. is zero. This occurs at B in Figure 1, the level of m.c. at which head yield is maximised. The economic threshold level of drying in terms of m.c. lies to the left of the technical optimum at a higher m.c. since extra drying can never be without cost. The difference between the technical and economic threshold level of m.c. depends on the relative slopes of the drying and yield functions. Some empirical evidence suggests that marginal revenues may decline sharply while marginal costs from extra drying may rise gradually at or near the point of intersection[3].

The economic threshold level of drying will increase as marginal costs of drying increase at constant marginal revenues and the level will also increase as marginal revenues are reduced. Profit to the farmer will also be maximised if drying facilities are used in this way.

The conceptual framework described above assumes that the farmer has already installed

[3] Fredericks and Wells (1983, p. 73 et seq.) provide some evidence on drying costs and yields.

drying facilities and that these are in operation. The decision to invest in drying facilities is of a much longer-term nature, implying that the returns from drying are sufficient to pay for depreciation and interest on capital invested, as well as giving a return for management. For example, if the net return from drying is $2 after deducting operating expenses then, at an interest rate of 10%, up to $20 per tonne can be invested in drying facilities for this rate of interest. At higher interest rates, the level of investment is reduced. The implications are that determination of economic thresholds for drying, and decisions to invest in drying facilities are interrelated, in that short-run operational decisions determine longer-run, break-even levels of investment in drying facilities. This conceptual framework also provides a means of evaluating choices among alternative drying strategies.

Applied studies in the provision of drying facilities are usually conducted in the framework of conventional, cash-flow investment analysis [see, for example, Kwon (1980); Kwon and Catania (1980); Fris and Manilay (1984)]. Economic evaluation is easily made in terms of comparing the present value of benefits (B),

$$\sum_{n}\left(\frac{1}{1+i}\right)^{n} B_{n} \text{ with costs, (C), } \sum_{n}\left(\frac{1}{1+i}\right)^{n} C_{n}$$

for period, n, or determining the internal rate of return (IRR) or the interest rate, i, at which

$$\sum_{n}\frac{(B_{n}-C_{n})}{(1+i)^{n}} = 0$$

In the latter analysis, taxes, interest on capital, and depreciation are not deducted from the benefits stream as these are considered transfer payments within society. Thus, the IRR represents the average earning power of the project that benefits society as a whole.

These financial analyses, evaluate whether investment in a drying facility is economic. They do not provide any guidance on the issue of what should be the optimum level of drying and investment in drying facilities.

## Determination of the Optimal Pattern of Drying Facilities

The choice of location, size, number, and type of drying facilities depends on a number of factors including the assembly cost of the material at each proposed location, the cost of drying, the distribution cost of dried material to subsequent processes in the marketing chain, and the opportunity to increase net industry returns from drying. In terms of locational advantage a large integrated drying storage and milling plant may certainly reduce the costs of drying through economies in size and distribution costs after drying, but these cost savings have to be balanced against the increase in assembly costs (transport costs to the integrated complex and queuing) as well as the possibility of deterioration in value of the product stemming from increased delays before drying commences.

The pattern of industry organisation of drying facilities which emerges from this scenario is a tendency towards a concentration of facilities in which the cost saved from increased throughput offsets the other cost/revenue disadvantages.

On the other hand, a fragmented industry organisation of drying facilities would suggest that industry returns would be increased from a larger number of smaller size dryers. Although more costly to operate than larger facilities, they enable the product to be dried more rapidly thereby reducing spoilage as well as assembly costs.

The optimal configuration is obviously a balance of these two extreme cases. In addition, the fact that there are dryers already in existence some at farm, village, or large integrated complexes also implies that for planning purposes the investment in these facilities can be regarded as sunk costs. Here the main issue is whether facilities should be expanded or rationalised and, if so, what should be their size, number, location, and type. This is the approach we are currently taking in ACIAR Project 8344 (Ryland and Hansen 1985).

There have been a number of applied studies in this field of research. Candler et al. (1972) considered a ricemill locational problem involving the least cost allocation of paddy from 49 dryers to 16 mills in which all 16 mills had the same non-linear (concave) cost function. The optimal solution indicated that 7 mills could be closed and showed that falling average costs of milling and drying associated with large volumes more than offset the increase in transport costs associated with fewer mills.

A study by Soo Lip Tan (1971) considered the least-cost location pattern of drying facilities given 38 assembly points for paddy in the MUDA irrigated rice-growing region of Malaysia. The optimal least-cost pattern depended heavily on availabilities of paddy, with number of dryers increasing with quantity of paddy available. This would seem to indicate a tendency towards a dispersed pattern of industry organisation. Unfortunately, the available literature does not provide any analysis of situations which deal with optimal size, number, and location of mills and dryers simultaneously. One of the aims of the present ACIAR project on bulk handling in Malaysia is to develop a model which provides a simultaneous solution to this issue as well as analysing the impacts of alternative transportation and handling technologies.

### The Bearing of Drying on Paddy Pricing Policies

The choice of type, location, number, and size of drying facilities, investment in drying facilities, and operational strategies for drying may be made more complex by a grading system for paddy in which differences in paddy prices based on quality differences do not accurately reflect milling efficiency and the resource costs involved in milling and marketing of rice. An improved grading system for paddy would provide the necessary incentives for farmers to receive a price for paddy which reflects the full resource costs in improved milling efficiency. The main problem, at least with the Malaysian system, is that current discounts are insufficient to compensate millers for processing paddy of lower quality.

This means that the returns to the miller are insufficient to compensate for the extra costs involved in milling lower quality paddy. What is required is a reduced number of broader quality classes with larger price differentials sufficient to provide adequate compensation for the miller as well as providing incentive to the farmers to improve paddy quality.

While the extent of paddy price subsidy may itself be inefficient in terms of resource use, the main focus here is the efficiency of the existing grading system for paddy in providing a system of returns to the farmer which provides him with sufficient incentive to invest in drying facilities, the benefits of which are passed on to the miller in terms of increased milling efficiency. The increased returns to the miller should be sufficient to compensate him for the increased price of paddy.

In order to develop an improved grading system for paddy, there are a number of areas which require further research. These include the development of milling and drying cost functions which include quality parameters so that the relationships between cost and quality can be better understood (Fredericks and Wells 1983). There also needs to be further work on the relationship between milling efficiency and quality as there is a paucity of published information in this area, particularly in relation to observations on higher moisture contents.

### Areas for Future Research

Economists have a vital role to play in collaborating with engineers and biologists to indicate the deficiency of existing experimental results, some of which cannot be used to provide the information base necessary for economic analysis. In particular, there appears to be a lack of information relating milling efficiency in terms of head rice yield with measurable quality parameters such as moisture content, broken grains, immature grains, and yellowing. There is also a lack of technical information on the extent to which drying can be used as a substitute for pesticide application. Consequently, there needs to be additional work on the relationship of pest population levels, pesticide use, and drying.

In terms of cost functions, the economist can contribute by analysing the behaviour of drying costs with increased quality so that the extra costs associated with quality improvement can be determined. The latter will provide a basis for determining economic threshold levels of drying and for analysing alternative grading systems.

Finally, economists, engineers, and biologists should elicit the support of system scientists so that the issue of paddy transport, drying, storing, milling, and distribution of rice can be analysed as an interdependent system in which all activities in the marketing chain are examined simultaneously. There are probably considerable potential research benefits stemming from a holistic view of drying and aeration procedures rather than analysing this area in isolation.

The main benefits of each of these areas will probably be gained more from changes in paddy pricing structures than from changes in current drying and aeration procedures.

### Concluding Remarks

In this paper I have identified three issues and

avenues of research where I believe the economist can contribute either individually or in collaboration with engineers, biologists, or systems scientists. Drying and aeration procedures are operations which should be considered along with general inventory considerations of the farmer, processor, or trader. It is important that drying should be considered simultaneously with transporting the raw farm product, storing it, and transforming the raw product into a final product. The benefits from drying in terms of increased return to the economic agent can only be economic if they are sufficient to offset the extra costs involved. However, full benefits from drying may also include some indirect benefits in terms of less spoilage due to lower pest and disease incidence, while indirect costs could mean an increase in pollution through increased energy use. This means that drying itself may also have some indirect social welfare benefits as well as costs. Drying is only worthwhile if net social welfare is increased.

## References


Candler, W., Snyder, J. C., and Fraught, W. 1972. Concave programming applied to rice mill location. American Journal of Agricultural Economics, Feb. 1972, 126–130.

Chew, T. A., and Loo, K. F. 1985. A review of the literature on the economics of drying and milling of paddy in Malaysia. Presented at 8th ASEAN technical seminar on grains post-harvest technology, ASEAN Crops Post-Harvest Programme. August 1985, Manila.

Fredericks, L. J., and Wells, R. J. G. 1983. Rice processing in Peninsular Malaysia. Kuala Lumpur, Oxford University Press, 200 p.

Frio, A. S., and Manilay, A. A., ed. 1984. Coming to grips with the wet season grain handling problems in Southeast Asia. Manila, ASEAN Crops Post-Harvest Programme, 93 p.

Kwon, O. Y. 1980. An economic evaluation of solar grain drying. Canadian Journal of Agricultural Economics, Aug. 1980, 167–174.

Kwon, O. Y., and Catania, P. J. 1980. Cost-benefit analysis of solar grain drying and its policy implications. In: Proceedings of Second International Symposium on Drying '80. New York, McGraw Hill, Vol. 2, 443–446.

Ryland, G. J., and Hansen, B. R. 1985. Bulk handling of paddy and rice in Malaysia: Model development and specification. Presented at 8th ASEAN technical seminar on grains post-harvest technology, ASEAN Crops Post-Harvest Programme, May 1985, Manila.

Soo Lip Tan, 1971. A study of the optimum size, number and location of padi dryers for the MUDA area. Kuala Lumpur, University of Malaya, unpublished M.Sc. thesis.

# Some Economic Aspects of Drying of Paddy by Farmers in Selangor, Malaysia

**T. A. Chew and R. A. Ghaffar***

*Abstract*

A serious problem in Malaysia, which causes large postharvest losses, is the high moisture content of paddy delivered to National Padi and Rice Authority (LPN) complexes. A simple analysis, involving assessment of returns from versus costs of drying, is used to explain why farmers do not dry their paddy. The analysis is based on the existing paddy pricing system. Two characteristics of paddy loads resulting from the analysis are examined against a random sample of data, obtained from two government complexes, on paddy moisture deductions. Some changes to the existing paddy pricing systems, including a method for deriving 'fair' moisture deductions, are suggested.

A factor contributing to high post-harvest losses in Malaysia is the high moisture content of paddy sold to Lembaga Padi dan Beras Negara (LPN; National Padi and Rice Authority) procurement centres. These government-owned centres, which may incorporate paddy processing complexes, were set up to help rice growers market their crop. They act as a buyer of last resort; i.e. they cannot reject paddy brought in by farmers, although certain deductions are made for dirt, broken or immature grains, and excessive moisture content.

Under the present system, there is no incentive for farmers to dry their paddy. Given that the government intervenes in the paddy market through its input and product subsidy schemes, the question arises as to why incentives are not introduced into the paddy pricing system. This paper compares the costs and returns to a paddy farmer of drying his paddy with the costs and returns of not doing so. It is established that the present paddy pricing system, which includes both grading and deductions for excessive moisture, provides no incentive for paddy drying. Two characteristics of paddy loads that follow from this are examined, using a random sample of data on paddy received at two LPN complexes. Certain changes to the existing paddy pricing system, including a method for deriving 'fair' deductions for moisture content, are suggested. Some brief policy implications are also discussed.

* Faculty of Resource Economics and Agribusiness, University of Agriculture, Serdang, Selangor, Malaysia.

## The Current Paddy Pricing System at LPN Centres

LPN pays farmers the following rates for 100 kg of paddy:

| | |
|---|---|
| long-grained | $49.61[1] |
| medium-grained | $46.30 |
| subsidy | $16.54. |

The rates are for grain at 14% moisture content (m.c.). Deductions are made for moisture contents higher than 14%, as well as for dirt and broken or immature grain. Table 1 shows the actual moisture deductions made (column 2)[2] for paddy at various moisture levels (column 3) for a random sample of 232 deliveries of paddy. The data were obtained from LPN complexes in Sekinchan and Sungai Besar for their most recent paddy receivals.

Column 1 lists increasing moisture deduction for calculating the value of paddy, after deduction, shown in column 4. Column 5 shows the 'loss' suffered by farmers as a result of moisture deductions. From society's viewpoint, it is not a 'loss' at all as the column is simply a record of moisture in the paddy.

We can compare returns with and without drying as follows:

[1] All prices are in Malaysian Ringgit: $M2.4 = US$1.

[2] The subsidy and paddy prices are based on grain at 14% m.c. In the actual computation of moisture deduction based on gross weight, i.e. weight of paddy plus weight of gunny sacks, a base rate of 12% m.c. seems to be used.

**Table 1.** Returns[a] from 1000 kg paddy (gross weight)[b] at various moisture deductions

| Increasing deductions (theoretical only) (%) | Actual deduction from LPN data (%) | Moisture levels in paddy (%) | Value of paddy after deductions ($) | 'Loss' due to deductions ($) | Column 4 + Column 5 ($) |
|---|---|---|---|---|---|
| 0 | no cases | 12–13 | 628.400 | 0 | 628.400 |
| 1 | no cases | 13–14 | 622.116 | 6.284 | 628.400 |
| 2 | given 1% | 14–15 | 615.832 | 12.568 | 628.400 |
| 3 | no cases in data | 15–16 | 609.548 | 18.852 | 628.400 |
| 4 | 4 | 16–17 | 603.264 | 25.136 | 628.400 |
| 5 | 5 | 17–18 | 596.980 | 31.420 | 628.400 |
| 6 | 6 | 18–19 | 590.696 | 37.704 | 628.400 |
| 7 | 7 | 19–20 | 548.412 | 43.988 | 628.400 |
| 8 | 8 | 20–21 | 578.128 | 50.272 | 628.400 |
| 9 | 9 | 21–22 | 571.844 | 56.556 | 628.400 |
| 10 | given 10.5 | 22–23 | 565.560 | 62.840 | 628.400 |
| 11 | given 11.5 | 23–24 | 559.276 | 69.124 | 628.400 |
| 12 | given 13 | 24–25 | 552.992 | 75.408 | 628.400 |
| 13 | no cases in data | 25–26 | 546.708 | 81.692 | 628.400 |
| 14 | no cases in data | 26–27 | 540.424 | 87.976 | 628.400 |
| 15 | no cases in data | 27–28 | 524.140 | 94.260 | 628.400 |

[a] Prices are in Malaysian Ringgit: $M2.4 = US$1.
[b] Gross weight refers to weight of paddy plus weight of gunny sacks. Some minor adjustments would have to be made in columns 4, 5, and 6, if weights of gunny sacks were excluded. The actual moisture deductions, as shown in LPN records, were made in terms of the gross weight.

$$\text{with no drying, revenue} = (Y_w - D)\, P_{dp} \quad (1)$$

$$\text{with drying, revenue} = (Y_w - L)\, P_{dp} - K \quad (2)$$

where

- $Y_w$ is weight of wet paddy, say 1000 kg;
- $D$ is deduction for moisture, converted to kg;
- $L$ is weight loss in paddy from drying;
- $P_{dp}$ is price per unit weight of 'dry paddy' (14% m.c.) inclusive of paddy subsidy;
- $K$ is cost of drying, including the opportunity cost of time of the farmer or his agent.

The variables that LPN can manipulate, to induce paddy drying are D and K. L, the weight loss from drying, is a technological constant. Farmers would have an incentive to dry their paddy only if

$$(Y_w - L)\, P_{dp} - K > (Y_w - D)\, P_{dp} \quad (3)$$

or

$$(Y_w - L)\, P_{dp} - (Y_w - D)\, P_{dp} > K \quad (4)$$

or

$$DP_{dp} > (K + LP_{dp}). \quad (5)$$

In addition, to ensure that the incentive to dry is much greater at high paddy moisture levels (say, over 25%) compared with low moisture levels (say, below 20%), LPN should ensure $DP_{dp} \gg (K + LP_{dp})$ at high paddy moisture levels. We can look at our LPN data to see if equation (5) is satisfied and also if there is a difference in incentive to dry at high compared with low moisture levels.

### Is there an Incentive to Dry [$DP_{dp} > (K + LP_{dp})$]?

As an example, we look at the case of a farmer who delivers 1000 kg of paddy at 21%–22% m.c. He is given a moisture deduction of 9%, equivalent to $56.56 (Table 1). If the farmer were to dry his 1000 kg of paddy to 14% m.c., he suffers a weight reduction of about 81.4 kg[3], which is equivalent to $51.15. Thus, the farmer would only dry his paddy if K were less than $5.41 ($56.56–$51.15). Another way of looking at the situation is to say that $571.84 (Table 1) is the value the farmer gets for his paddy without drying. If the farmer were to dry his paddy from 21% m.c. to 14% m.c., his 1000 kg eventually reduces to 918.6 kg which is worth $577.25. He gains

[3] The weight reduction is derived from the definition of paddy moisture content given as

$$\%\ \text{m.c.} = \left(\frac{M}{D + M}\right) 100$$

where M is weight of moisture, and D is weight of paddy dry matter.

(577.25–571.84) dollars or $5.41[4] as before, which would hardly cover K, the cost of drying, given that the daily wage for a casual labourer is around $12.00.

Similar computations for other paddy moisture levels show that equation (5) is not satisfied.

A characteristic of paddy loads that would follow from the fact that there is no incentive for drying, would be a 'bunching' of paddy moisture levels around a certain level determined by weather and other conditions at harvest. If, on the other hand, the incentive for drying is present, there would be a greater distribution of moisture levels, including many deliveries of lower moisture content paddy. Table 2 confirms the preponderance of cases of paddy loads around 20% m.c. (i.e. for moisture deductions of about 9%) in our random sample.

**Incentive to Dry at High Versus Low Moisture Levels**

Table 1 shows some slight difference in the incentive to dry paddy at low compared with high moisture levels. At 14%–15% m.c. (gross weight of paddy), to be consistent with other deductions, a deduction of 2% should be made, but LPN data show a moisture deduction of only 1%. On the other hand, at 24%–25% m.c. an actual deduction of 13% was made, whereas it should be 12% in a consistent system. In other words, the present paddy pricing system does have higher penalties for grain at higher (over 21% m.c.) moisture levels compared with that at lower moisture levels, which is as it should be. But our contention, to be demonstrated in the next section, is that the additional penalties of 1% point and 0.5% point for 24%–25% m.c. and 22%–24% m.c. respectively, are insufficient to discourage paddy being brought in at these high moisture levels. Given our contention, it follows that the average weights of paddy deliveries at high moisture levels would not differ from those at low moisture levels. If the differential incentive factor is effective, we can argue that deliveries of high moisture paddy would be in small batches not worth the effort of drying, whereas paddy brought in at low moisture levels would be in larger loads. Table 2 shows that the mean weights of paddy delivered at high moisture levels are not significantly different from those at low moisture levels.[5]

**Changes to the Present Paddy Pricing System**

We suggest changes to the present paddy pricing system in two areas:

i) $DP_{dp}$ be made greater than $(K + LP_{dp})$ so that the incentive for drying is re-established;

**Table 2.** Different moisture deductions for various loads (LPN data)

| Moisture deductions | Number of cases | Mean weight of paddy load (kg) | Mean value of deductions for moisture ($) |
|---|---|---|---|
| **Sekinchan LPN Complex** | | | |
| 6% | 1 | 5 587.91 | 239.44 |
| | | (0.00)[a] | (0.00) |
| 7% | 12 | 6 210.82 | 313.834 |
| | | (1 438.73) | (72.73) |
| 8% | 45 | 5 522.01 | 323.02 |
| | | (1 967.14) | (113.40) |
| 9% | 84 | 4 772.78 | 316.97 |
| | | (2 101.51) | (139.56) |
| 10% | 3 | 3 937.40 | 310.096 |
| | | (2 528.77) | (199.02) |
| 11% | 25 | 4 371.96 | 362.97 |
| | | (2 491.98) | (206.85) |
| **Total** | **170** | | |
| **Sungai Besar LPN Complex** | | | |
| 1% | 1 | 1 816.20 | 11.69 |
| | | (0.00) | (0.00) |
| 4% | 1 | 1 927.20 | 51.28 |
| | | (0.00) | (0.00) |
| 5% | 19 | 1 699.64 | 58.35 |
| | | (1 181.58) | (40.52) |
| 6% | 5 | 843.36 | 35.82 |
| | | (732.38) | (31.69) |
| 7% | 7 | 1 969.63 | 94.95 |
| | | (1 749.64) | (83.88) |
| 8% | 11 | 1 501.69 | 86.08 |
| | | (1 226.34) | (70.29) |
| 9% | 9 | 2 088.73 | 132.53 |
| | | (1 044.23) | (65.70) |
| 10.5% | 4 | 2 002.82 | 145.11 |
| | | (1 844.20) | (133.97) |
| 11.5% | 4 | 1 324.48 | 110.40 |
| | | (1 038.46) | (85.59) |
| 13% | 1 | 646.92 | 61.76 |
| | | (0.00) | (0.00) |
| **Total** | **62** | | |

[a] Values in brackets are standard deviations.

[4] These calculations are not accurate because $DP_{dp}$ is based on gross weight (paddy and gunny sacks) while $LP_{dp}$ is based on net weight (paddy without gunny sacks). It is possible that over certain moisture ranges, D is less than L.

[5] One could argue that since $DP_{dp}$ was shown to be less than $(K + LP_{dp})$ previously, this section is redundant. However, with large paddy loads, $DP_{dp}$ need not be less than $(K + L) P_{dp}$ because of economies of scale in the drying process at certain moisture ranges.

ii) $DP_{dp}$ be made much greater than $(K + LP_{dp})$ at high moisture levels (over, say, 25% m.c.) so that the penalty for not drying at high moisture levels is more severe compared with that at low moisture levels (less than, say, 20% m.c.).

A 'fair' method for deriving D — 'fair' in the sense that it would be determined under free, competitive market conditions — would be as follows:

with no drying, revenue $= Y_w r_w P_r$

with drying, revenue $= (Y_w - L) r_d P_r - K$

where
- $Y_w$ is weight of wet paddy, say 1000 kg;
- $r_w$ is recovery rate of rice from wet paddy;
- $P_r$ is price of rice per kg;
- $L$ is loss in weight due to drying;
- $(Y_w - L)$ is weight of 'dry paddy' (14% m.c.);
- $r_d$ is recovery rate of rice from 'dry paddy';
- $K$ is cost of drying, all costs included.

As before, drying of paddy occurs if

$$(Y_w - L)\, r_d P_r - K > Y_w r_w P_r \qquad (6)$$

$$(Y_w - L)\, r_d P_r - Y_w r_w P_r > K. \qquad (7)$$

We can rewrite (7) as

$$(Y_w - L)\, P_{dp} - Y_w P_{wp} > K \qquad (8)$$

where
- $P_{dp}$ is unit price of 'dry paddy' (14% m.c.)
- $P_{wp}$ is unit price of wet paddy

Under free market conditions, we can expect the 'invisible hand' to exert its role and establish the following equality from (7) and (8).

$$\frac{r_d P_r}{P_{dp}} = \frac{r_w P_r}{P_{wp}} \qquad (9)$$

$$\therefore \quad \frac{r_d}{r_w} = \frac{P_{dp}}{P_{wp}}. \qquad (10)$$

From (4) earlier

$$(Y_w - L)\, P_{dp} - (Y_w - D)\, P_{dp} > K. \qquad (11)$$

Comparing (8) with (11)

$$(Y_w - D)\, P_{dp} = Y_w P_{wp} \qquad (12)$$

$$\therefore \quad \frac{P_{dp}}{P_{wp}} = \frac{Y_w}{Y_w - D}. \qquad (13)$$

Since

$$\frac{P_{dp}}{P_{wp}} = \frac{r_d}{r_w} \text{ from (10),}$$

$$\frac{Y_w}{Y_{wp} - D} = \frac{r_d}{r_w}. \qquad (14)$$

Such a moisture deduction based on the rice recovery rates, would be a 'fair' deduction[6]. As an example, let us compute the 'fair' D for paddy at 19% m.c. For paddy at 19% m.c. and 14% m.c., the recovery rates of rice from paddy are 56.62% and 61.67%, respectively (Fredericks and Wells 1983, p. 74). Using equation (14), we get:

$$\frac{Y}{Y - D} = \frac{61.67}{56.62} = 1.0892$$

If $\quad Y_w = 1000$ kg

then $\quad Y_w - D = \dfrac{1000}{1.0892} = 918$ kg

$\therefore \; D = 1000 - 918 = 82$ kg which is 8% moisture deduction.

The actual deduction, as shown in LPN data (Table 1) for paddy at 19% m.c. is only 7%.[7] This supports our suggestion that $DP_{dp}$ be made greater than $(K + LP_{dp})$, so that the incentive for drying is re-established.

Assuming that the rice recovery rate for paddy over 25% m.c. is 50%, computation of (14) yields an appropriate moisture deduction of about 18.9%. Such a deduction would be above the 14% or 15%

---

[6] This assertion is, of course, simplistic. If $r_d$, $r_w$, and K are such that $(Y_w - L)\, r_d P_r - Y_w r_w P_r \gg K$, so that there are elements of supernormal profit in the drying activity, then it becomes debatable if D as derived above would still be considered 'fair'.

[7] This is based on gross weight (paddy and gunny sacks). If net weight is used, divergence between our deduction and the LPN deduction would be greater.

currently practiced by LPN (Table 1). This supports our earlier contention that the current penalties for excessive paddy moisture (over, say, 25% m.c.) are not strong enough deterrents.

Application of (14) would result in a paddy pricing system that reflects the underlying technology in rice processing, viz. high moisture deductions at excessive paddy moisture levels to reflect the high marginal product of drying at these levels, and lower moisture deductions to reflect the lower and declining marginal product of drying as we approach 14% m.c. As shown in equation (7), the fundamental variables determining whether or not drying takes place are $r_d$, $r_w$ and K.L, the weight shrinkage from drying, is a technological constant and $P_r$, the price of rice, is irrelevant since it appears in both the drying and non-drying components. If plant breeders were to create a 'miracle' paddy where the recovery rates of rice were the same for paddy with different moisture contents, i.e. $r_d = r_w$, then the drying process would become redundant.

If storage is important and necessary and if storage of wet paddy has deleterious effects on recovery rates, then the obvious implication is that deductions for moisture content by LPN will have to be increased further.

### Cost of Drying in LPN Complexes

In Peninsular Malaysia, there are four levels which operate independently or in combination to provide paddy drying capacity. The levels are:

(i) on-farm;
(ii) farmer cooperative complexes;
(iii) private complexes;
(iv) LPN (government) complexes.

In Selangor, there is little or no on-farm drying of paddy and most cooperative complexes do not have drying facilities. Almost all drying takes place in complexes owned either by the government or the private sector. It is no accident that drying facilities are currently an integral part of milling complexes because drying is not a remunerative activity under the present LPN pricing system, which would not induce the establishment of a drying capacity separate from the milling facilities. Rather, drying would be regarded as a 'necessary' evil that cannot be avoided in the process of producing rice of acceptable quality. The LPN complexes, being buyers of last resort, set the limits on the paddy moisture deductions that can be made by private individuals. The moisture deductions given by LPN must be very attractive to rice farmers given the burgeoning paddy stocks in LPN storages. Further evidence along this line is the decline in the proportion of paddy handled by private millers, which decreased from 90% in 1974 to 67% in 1983 (Ghaffar and Hassan 1985). Also, from 1981 to 1983, 25 private millers closed down and more private processing centres may cease operation in the near future (LPN 1984). The price of milled rice has been kept constant for more than 10 years while various cost increases have eroded the marketing margin, making rice processing (which includes drying, milling, and distribution) not as remunerative as before. If the present trend continues, we can expect the role of LPN in the paddy market to increase further. A major question is: What is the cost of paddy drying in LPN complexes?

Ghaffar and Hassan (1985) estimated the average cost of drying a tonne of paddy for six LPN complexes (Table 3) as \$46.60. If adminis-

**Table 3.** Drying costs[a] per tonne for six government complexes.

| Costs | Location: Pasir Putih | Peringat | Ulu Tiram Buruk | Anak Bukit | Bukit Besar | Simpang Lima |
|---|---|---|---|---|---|---|
| **Drying Cost** | **42.51** | **30.66** | **39.45** | **59.14** | **43.37** | **64.48** |
| 1) Variable Cost | 15.31 | 16.87 | 17.94 | 43.14 | 23.06 | 33.78 |
| Unloading | 3.13 | 6.05 | 2.07 | 10.62 | 1.09 | 6.19 |
| Energy | 7.88 | 5.68 | 11.94 | 24.16 | 14.16 | 18.29 |
| Maintenance | 2.22 | 3.04 | 2.59 | 2.82 | 2.42 | 5.68 |
| Wages | 2.08 | 2.10 | 1.34 | 5.24 | 5.39 | 3.62 |
| 2) Fixed Cost | 27.20 | 13.79 | 21.51 | 16.00 | 20.31 | 30.60 |
| Emolument | 1.35 | 2.37 | 2.39 | 5.90 | 2.25 | 0.37 |
| Travels | — | — | — | — | — | — |
| Depreciation | 25.85 | 11.42 | 19.12 | 10.10 | 18.06 | 30.23 |

Source: Ghaffar and Hassan (1985, p. 18)
[a] Costs are in Malaysian Ringgit: \$M2.4 = US\$1.

**Table 4:** Average drying cost[a] per tonne of rice for mechanical dryers, Peninsular Malaysia

| Source | Date | Location | Type of dryer and number | Drying cost | Remarks |
|---|---|---|---|---|---|
| Wells, Fredericks & Gul | 1971 | Selangor | Continuous (1)<br>Batch (1) | 33.7[1]<br>26.3 | Survey of two mills |
| Runte et al. | 1973 | Kedah | Continuous (6)<br>Batch (5) | 7.6[2] | Survey of eleven mills |

[1] refers to variable drying costs
[2] includes depreciation

Source: Fredericks and Wells (1983, p. 84)
[a] Costs are in Malaysian Ringgit: $M2.4 = US$1.

trative costs are added, the average cost for drying goes up to $69.32. Compared with values quoted in the literature (Table 4), the drying cost in LPN complexes is very high, even allowing for price changes over the years. Therefore, drying in LPN complexes can by no means be considered efficient.

## Concluding Remarks

It seems clear that the present LPN paddy pricing system[8] provides no incentives for drying. This conclusion is supported by empirical evidence and theoretical calculations.

Changes in the pricing system along lines suggested, as well as re-establishing incentives for drying, could also have significant effects on existing paddy flows. Instead of paddy being dried at high cost at LPN complexes with strained facilities, it could be dried in numerous homesteads or in private dryers set up by entrepreneurs or farmer cooperatives. This is where supportive efforts in extension services and financial incentives may have to be undertaken (Cardino 1985). Overall, a diversified system of drying such as this, divorced from milling and distribution, could lead to improvements in the entire paddy production, processing, and distribution system.

[8] Why this system of paddy pricing was implemented is puzzling. Most likely it arose from the belief (right or wrong) that middlemen were 'exploiting' paddy farmers by making excessive weight deductions for moisture. LPN, in its efforts to counter this 'exploitation', probably developed the current pricing system. It is possible that LPN has other social roles to play besides ensuring the efficient marketing of paddy.

## References


Benny, J.M., ed. 1967. Proceedings of a seminar on principles and practices of artificial drying of padi. Kuala Lumpur, University of Malaya, Faculty of Agriculture Publication No. 3, 152 p.

Cardino, A. G. 1985. A socio economic study in the utilization of mechanical dryers. Presented at the ASEAN technical seminar on grains post-harvest technology, ASEAN Crops Post-Harvest Programme, August 1985, Manila.

Fredericks, L. J., and Wells, R. J. G. 1983. Rice processing in Peninsular Malaysia. Kuala Lumpur, Oxford University Press, 200 p.

Ghaffar, R. A., and Azman, H. 1985. A preliminary investigation of the costs associated with rice drying and milling in Peninsular Malaysia. Presented at the ASEAN technical seminar on grains post-harvest technology, ASEAN Crops Post-Harvest Programme, August 1985, Manila.

Len, S. C., and Kwok, C. Y. 1975. Pre and postharvesting rice technology. Kuala Lumpur, University of Malaya, Faculty of Agriculture Publication No. 7, 112 p.

LPN (Lembaga Padi dan Beras Negara) 1984. The status of the rice/padi processing industry: problems and suggestions. Presented at Seminar Peni Laian ke Atas Kecekapan Dari Segi Kos Dalam Peprosesan Padi/ Beras, LPN, Kuala Lumpur, 613 p. (In Bahasa Malaysian.)

Thet Zin, U. 1969. Survey report of rice processing in West Malaysia, parts I and II. Kuala Lumpur, Ministry of Agriculture and Fisheries, and Rome, F.A.O.

# Cost Implications of Implementing Grain Drying Systems

**L.J. Fredericks and J.P. Mercader***

*Abstract*

This paper provides an overview of grain drying in the ASEAN region, assessing its importance and giving reasons for the slow adoption of artificial drying technology. It categorises areas where drying is done into small, on-farm level, assembly points, such as cooperatives or grain traders, and government grains complexes. The economic implications of drying operations at these levels are discussed, identifying cost factors for consideration in financial analysis of drying operations.

THE introduction of high-yielding, early maturing rice varieties, coupled with irrigation and other agricultural innovations, have made possible the double cropping of paddy. With this development, significant quantities of paddy produced are being harvested in the wet season straining existing drying and other facilities. Although artificial drying technology has been more or less perfected, there are socio-economic factors affecting its adoption by the farming community. Consequently, farmers are compelled to sell undried paddy at discounted prices or wait for the sun to shine. As wet paddy deteriorates rapidly, heavy postproduction losses are incurred in the ASEAN region.

The adoption of any new technology is a long and complicated process, particularly by traditional and small-scale farmers. It is generally acknowledged that for any group to adopt a certain technology, the proposed alternative must meet the following conditions:

(a) there is a defined need for the technology;
(b) the technology is effective in solving the problem;
(c) the benefits accruing to the users are greater than the cost of the technology as well as the cost of alternative methods.

The parameters for the adoption of the technologies of paddy drying are even more numerous and complicated. Rice being a political commodity, there are policy matters related to paddy subsidies and price structures which may not satisfy purely economic criteria. Furthermore, there are socio-psychological constraints experienced by farmers who for decades have relied on sun drying. Also, the ASEAN rice production sector is generally fragmented, small-scale, labour intensive, and traditional in terms of the level of technology used. All these factors contribute to the slow adoption of artificial drying by small farmers during the wet season.

Artificial drying technologies are considered effective from an engineering point of view and their need has been established. What appears to be a major deterrent in their adoption is related to the economic aspects of artificial drying.

## Paddy Losses during Drying

Paddy postharvest losses in the ASEAN countries are estimated to range from 10 to 37%. They are made up as follows (de Padua 1979):

| | | |
|---|---|---|
| Harvesting | — | 1 to 3% |
| Handling | — | 2 to 7% |
| Threshing | — | 2 to 6% |
| Drying | — | 1 to 5% |
| Storage | — | 2 to 6% |
| Milling | — | 2 to 10% |
| TOTAL | — | 10 to 37% |

While losses attributed to drying per se range from only 1 to 5%, a significant portion of the losses incurred in storage and milling is related to drying. Improper drying, especially during the wet season, contributes to losses due to rotting and down-grading of the quality of milled rice as

* ASEAN Food Handling Bureau, 8th Floor, Syed Kechik Foundation Building, Bangsar, Kuala Lumpur, Malaysia.

characterised by a high percentage of brokens, discoloration, and mould infestation.

Of the 54.4 Mt of paddy produced in the ASEAN region, approximately 40 to 50% is harvested during the rainy season. This is of great significance, as the traditional sun drying system cannot be relied upon during the rainy season thus generating considerable strain on artificial drying facilities. It is assessed that at the current estimate of 5% loss, 2.7 Mt valued at US$326.4 million (at current prices) are wasted annually.

Losses related to drying are influenced by factors such as postharvest practices, where the drying is done, marketing operations, and other factors related to the physical environment. A JICA study on paddy losses at the farm level in Indonesia indicated 10% quantitative and 4.4% qualitative losses. In the Philippines, an ASEAN Food Handling Project study on various postharvest practices indicated that drying losses could account for reduction of monetary value by 5 to 58% due to yellowing and quality deterioration. A similar magnitude of losses was noted in Thailand where the study correlated losses in storage due to drying practices. In Malaysia, serious drying problems have been reported during the rainy season when delays in delivery to grains complexes (where drying is done) are common and inadequate drying facilities are strained by heavy deliveries of wet paddy. These studies and observations underscore the positive correlation between delays in drying and the level of deterioration of paddy.

In general, drying is undertaken to prevent the germination of seeds and to retain the maximum quality of the grain at moisture levels which do not allow the growth of bacteria and fungi and which retard the development of mites and insects. Under humid tropical conditions, the threshold level for maintaining good paddy quality is estimated at 14% moisture content (m.c.). Drying considerably affects the quality of grains and, hence, their monetary value. The world grains standard attaches value to such physical attributes of grains as colour, texture, wholeness, and purity. All these attributes are, directly or indirectly, influenced by the drying process.

## Drying Systems and Operations in ASEAN

In drying, air is used as the medium to conduct heat to the produce, causing water to vaporise; it also conveys moisture away from the drying produce. The moisture-carrying capacity of the air is dependent upon the temperature and increases with a rise in temperature. The capacity of air to remove moisture from produce depends principally upon the relationship between the moisture content of the produce and the relative humidity of the air. These are the basic principles by which natural and artificial drying operate.

There are two methods of drying: natural (sun drying and aeration) and artificial (mechanical). Sun drying is the most widely used, primarily because of its low cost. Some individuals argue that milling quality is better when paddy is sun dried. The fact, however, remains that with increased output during the wet season, sun drying becomes almost ineffective thus compelling the use of artificial dryers if the value of the crop is to be realised.

The artificial drying systems used in the region range from the simple flatbed dryers to the industrial plant-type dryers with sophisticated accessories. The drying systems vary considerably in terms of fuel used, geometric configuration, and accessories installed in conjunction with the dryer. Oil burners and liquefied petroleum gas are used although their rising cost is becoming a deterrent. Currently, efforts are being made to utilise farm by-products and other biodegradable materials as fuel. Efforts are also being made in harnessing wind and solar energies for drying purposes. Dryers with horizontal grain-holding devices can range in capacity from two tonnes in the case of flatbed dryers used on farm to several hundreds of tonnes in the case of in-store deep bins used in industrial complexes. Various sophisticated accessories have been installed in some dryers used in industrial complexes. These accessories may include mechanical loading and unloading systems such as conveyors and elevators, grain cleaning and dust control equipment, weighing devices, temperature monitoring and control panels, and safety equipment.

This paper will focus on the categorisation of dryers used at different levels of operation in paddy postproduction systems. This is primarily due to the stratification of postharvest operations in the ASEAN countries into:

(a) small on-farm operations;
(b) assembly points, such as cooperatives or grain traders; and
(c) government grains complexes.

The second rationale for this classification is the

economic implications, which vary greatly between the farm and the grains complex level.

There is a strong correlation between the point of drying and the capacities and sophistication of drying systems used. The most widely used in the region is the batch-type dryer. This has evolved from the UPLB (University of the Philippines at Los Baños) two ton flatbed dryer which had been adapted to suit various needs. At the small farm level, flatbed dryers with capacities ranging from 2 to 4 t are commonly used. Lembaga Padi dan Beras Negara (LPN) has modified this to accommodate 30 t. Similar modifications have also been made by Badan Urusan Logistik (BULOG) in Indonesia, the Thai Agricultural Engineering Division of the Department of Agriculture, and the National Food Authority (NFA) of the Philippines. The LSU continuous-flow dryer with tempering bins and drying capacities of 1 to 5 t/hour is a common fixture in government and large private grains complexes. To some extent, the Kongskilde vertical batch dryer and the Japanese batch type grain recirculating dryer have been used at assembly points and in large grains complexes. Experimental models of solar dryers and vortex wind pumps for in-store drying are being tested at the farm level and at assembly points, respectively.

The investment and operating costs of dryers vary considerably depending largely on capacity, their level of sophistication, and the fuel used.

Studies at NFA have determined the drying costs of various systems, as shown in Table 1.

The drying costs are highly variable among the types of drying systems and between models in a particular type. They range from US$2.40 per tonne using the vortex warehouse dryer, to US$37.64 using the continuous flow Cimbria dryer. Batch drying cost per tonne ranges from US$3.20 for a rice hull fed dryer, to US$15.11 for a Kongskilde dryer. Wide variation is also found with continuous flow dryers, with costs ranging from US$11.80 to US$37.64 per tonne.

**Table 1.** Comparative drying costs using various systems.

| | | | Cost/t in US$ |
|---|---|---|---|
| I. | Sun drying | | 0.14 |
| II. | Batch drying | 112 hours operation | |
| | - Rice hull fed | | 3.20 |
| | - Kerosene fed | | 4.18 |
| | - Kongskilde | | 15.11 |
| III. | Continuous flow | (600 hours/year operation) | |
| | - Satake | 2.4 t/hour | 11.80 |
| | - Woodland | 2.0 t/hour | 22.14 |
| | - Cimbria | 3.0 t/hour | 37.64 |
| IV. | Warehouse dryer | From 25 to 14% m.c. | |
| | - Vortex dryer | | 2.40 |

Source: Mangaoang (1984).

## Farm-Level Drying

The greatest resistance to artificial drying is understandably at this level of operation. Government programs have generally encouraged the use of small flat-bed dryers compatible with the volume of production in each farm whose size ranges between 0.25 and 3 hectares. Despite the need for dryers at this critical level of operation, the acceptance of farm dryers is abysmally low. The major causes of this situation are:

(a) high acquisition costs beyond the reach of small-scale farmers due to low production volume;
(b) unsatisfactory performance of some drying systems introduced;
(c) lack of technical know-how in dryer operations, resulting in inefficiency and poor milling results;
(d) capacity of dryers incompatible with farm production and other processing equipment; and
(e) the price structure of paddy, which does not seem to reflect adequately the cost of drying.

In the late 1970s, the ASEAN Food Handling Project introduced several units of flat-bed dryers to farmers in Indonesia, Thailand, and Malaysia. Their provision was supported by training and extension in their proper use and maintenance. While the farmers were generally satisfied with the performance of the dryers, there was a strong indication that they would not purchase one, primarily due to high capital and operating costs.

Several studies have indicated that, in present circumstances, artificial drying is uneconomical for the individual farmer. Studies in the Philippines have indicated that even the smallest size of locally produced, proven, and tested engine-powered equipment is quite expensive for a small farmer to acquire, maintain, and operate. A study by Lorenzana and Pablo (1983) indicated that an internal rate of return of 32.2% may be obtained if a dryer is operated for 30–60 days at 10 hours/day. However, some assumptions were considered as idealised and not to reflect actual conditions in accounting for breakdowns, percent downtime, escalating fuel costs, inadequate quantities of produce, and the possibility of sun drying between rainy days.

A 1985 study commissioned by the ASEAN Food Handling Bureau (AFHB) in West and Central Java, Indonesia, indicated that farmers only dry the paddy which they intend to keep for consumption, and sun drying is the only method used. The reason is obvious, because while sun drying is labour intensive, the opportunity cost of labour is low. Farmers, because of low production volume, reported no drying problems, even in the rainy season. Also, under current price structures and standards, the value difference between selling a tonne of paddy at 24% and at 18% m.c. is just US$4.10. When the cost of artificial drying is imputed, the farmer stands to lose.

A similar study conducted in late 1984 by AFHB and the Department of Agricultural Engineering among maize growers in Thailand indicated that artificial drying is hardly worthwhile. At any given time, the price difference between maize at 18% and 24% m.c. is marginally low and is even insufficient to compensate for the weight loss as well as the drying cost. Although there are seasonal price fluctuations, the farmers are not in a position to dry and store maize, as they have to settle their debts with the merchants soon after harvest.

**Assembly Points**

This level of operation is treated separately from government complexes for economic reasons. While in many cases drying facilities owned by cooperatives and private operators are similar to those used in government complexes, there are distinct differences in economic motivation. As profits or savings are meant to accrue to members, there is a marked consciousness in operating at the least cost or greatest benefit. There is therefore a stronger business orientation in the operations of cooperatives and private concerns. The same is true with farmers' associations whose primary objective for grouping is the minimisation of cost and the increment in benefits for individual members.

A recent study in West and Central Java, Indonesia, noted that hundreds of drying units of the Lister type and locally manufactured flatbed dryers were introduced to KUDs (Village Unit Cooperatives) over the last decade. They were provided on highly concessional terms and, in most cases, the KUDs did not have to bear the capital cost. Some KUDs have stopped using these dryers. The main reason for this is the extremely high operating costs which under the current price structure make drying an uneconomical operation for the KUDs. Table 2 shows the current price levels for paddy at various levels of moisture content. The paddy price structure does not provide any economic incentive for the KUDs to dry their paddy. The cost of drying paddy and its better quality is not reflected in a premium price.

As previous studies have indicated that drying at farm level is not economical under current conditions, attempts have been made to introduce dryers to farmers' groups. A study undertaken over

**Table 2.** Price[a] of paddy at various levels of moisture content based on price schedules used in 1984 by Village Unit Cooperatives in Indonesia.

| Moisture content (%) | Weight of paddy (kg) | Price (Rp/kg) | Amount paid (Rp) | Value of paddy at 14% m.c. | Benefits from drying (Rp) |
|---|---|---|---|---|---|
| 13.6–14.5 | 883.7 | 165.00 | 145 814 | 145 814 | — |
| 14.6–15.5 | 895.3 | 162.07 | 145 109 | 145 814 | 705 |
| 15.6–16.5 | 907.0 | 159.17 | 144 363 | 145 814 | 1451 |
| 16.6–17.5 | 918.6 | 156.24 | 143 523 | 145 814 | 2291 |
| 17.6–18.5 | 930.2 | 153.48 | 142 772 | 145 814 | 3042 |
| 18.6–19.5 | 941.9 | 150.70 | 141 938 | 145 814 | 3876 |
| 19.6–20.5 | 953.5 | 147.92 | 141 040 | 145 814 | 4774 |
| 20.6–21.5 | 965.1 | 145.52 | 140 444 | 145 814 | 5370 |
| 21.6–22.5 | 976.7 | 143.10 | 139 772 | 145 814 | 6042 |
| 22.6–23.5 | 988.4 | 140.69 | 139 054 | 145 814 | 6760 |
| 23.6–24.5 | 1,000.0 | 138.26 | 138 260 | 145 814 | 7554 |

Source: AFHB (1985).
[a] In Indonesian Rupiah: 12 Rp = US$1

a two year period in the Philippines indicated a 1:4.39 cost-benefit ratio in using the dryer. While this appears encouraging and should in fact influence price policies to reflect a premium for dried paddy, there are doubts about the assumptions used in the 'with' and 'without' project situation. The added value of approximately 30% is almost an impossibility under the normal situation. A small farmer with limited produce will not allow his crop to rot and reach a level where more than 30% of its value is reduced. Also, it is easier to aerate a small amount of paddy when it is raining (and arrest deterioration to a certain extent) and sun dry it later unless the situation is such that it rains continuously for two weeks or so; under these circumstances, the harvest may be totally lost. Also, management costs, which may need to be included in a situation where the costs and benefits are shared by many, should be included.

A study of maize traders (handling 50–50 t/day) conducted by AFHB in Thailand indicated strong interest in artificial drying by maize farmers. Most farmers sell their maize (undried) immediately after harvest due to indebtedness to traders. There may be some benefits from drying and storing, as indicated by the price fluctuations in Table 3. This illustrates that price differences over time could be more important than differences between prices for wet and dry maize at a given time. With a large volume handled at this level of operation, significant benefits could accrue to the traders by a proper timing of sale of maize. This factor could be the main reason for the strong interest in drying by the traders.

**Table 3.** Prices (THB/kg)[a] for maize at different levels of the marketing chain in Thailand.

| | Dec '83 | July '84 | Aug '84 |
|---|---|---|---|
| Farmgate price (shelling by trader) | | | |
| - m.c. 25–30% | | 2.70 | 2.00 |
| - m.c. 16% | 3.00 | 3.00 | 2.25 |
| Farmer delivers shelled maize to trader | | | |
| - m.c. 25–30% | | | 2.15 |
| - m.c. 16% | | | 2.40 |
| Wholesale price | | | |
| - local: m.c. 14.5% | | | 2.70 |
| - Bangkok: m.c. 14.5% | | | 2.90 |

Source: AFHB (1984).
[a] 25 Thailand baht (THB) = US$1.

**Grains Complex Level**

These grains complexes are operated by government agencies which have as their primary functions the maintenance of food supply and stabilisation of prices of cereals, particularly paddy and maize. They operate as terminal points for processed grains in particular. They serve as milling, storage, and distribution points for milled rice or transhipment points to other complexes or other countries. There are serious problems at this level of operation primarily influenced by factors such as marketing and price policies, quality standards, and overloading of facilities. The complexes were designed to handle only 10–20% of the national production. Currently, however, they handle much more than this. The reasons for this are many. With reduced activities by grains traders due to their lack of processing facilities, farmers tend to dump their produce on these complexes. The problem is more pronounced during the rainy season as facilities get overloaded with wet paddy. While the private sector may avoid drying problems by simply not buying, government agencies normally do not do this. This is a complicated problem which has been a subject of many discussions in the past and ways are being devised to overcome it.

In government complexes, a combination of the following systems may be used:

(a) bulk aerated bins, used for pre-drying, tempering, and final drying;
(b) aeration of plenum of bag stack;
(c) aeration of bulk paddy contained within walls of bagged paddy;
(d) flash drying;
(e) vortex wind pump machine.

As these facilities are designed to cater for a small volume of national grains production, serious problems in pre-drying and handling of wet grain occur during the wet harvest season. While efforts are being made to solve this problem by increasing drying capacities, increased paddy production and productivity have easily overtaken the improvement made. This problem has tended to be cyclical. It appears that there is a need for decentralisation of processing activities to avoid overloading of the facilities at grains complexes. Decentralisation, however, would require a review and policy re-direction to meet the objectives. Decentralisation would also imply greater participation by traders and farmers. The participation of traders in processing activities would require incentives to make their operation viable. To encourage farmers to undertake partial drying, drying facilities have to be made available and paddy price structures have to reflect the cost of drying as well as premium for better-quality grain. This implies a review of paddy quality grades and standards and the price accorded to them.

### Cost Considerations and Implications

Available economic studies on drying tend to have serious weaknesses in quantifying the economic realities of the environment. The cost implications and considerations vary according to the level of operation and, in general, these factors tend to be left out of the analysis of the economic viability of drying systems at different levels.

In general, only direct costs are reflected. Other costs related to the operation of the dryer are sometimes not considered. Often the analysis results in optimistic conclusions. These results are usually the bases for extending the drying technology to the users. But serious errors in the assumptions on costs and returns could lead to erroneous and misleading internal rates of return, benefit/cost ratios, and pay-back periods which are normally used as the bases for investments. As a result, the actual financial performance of drying operation could be much less than that in the projected analysis.

Various factors interact at varying levels due to different environmental realities. Needless to say, the concerns of small farmers differ from those of a merchant or those of a manager of a government grains complex. Assuming that the technology and other aspects have been proven satisfactory, the financial and economic aspects have to be streamlined if they are to be used as bases for recommendations to users. In general, drying cost is calculated as the ratio between the total operating and adjusted capital costs of the dryer and the total capacity of the dryer expressed in tonnes or kilograms. The items normally costed out are:

(a) acquisition cost;
(b) labour;
(c) fuel;
(d) repairs and maintenance.

While these are the obvious costs in drying, there are so-called hidden costs, the nature and magnitude of which vary according to the level of operation.

The process of investing in a dryer, whether a small 2-tonne flatbed dryer or a sophisticated continuous-flow dryer, is particularly difficult and requires careful analysis. The decision to own a dryer usually involves commitment for a considerable time to such liabilities as repayment, maintenance, and capital cost recovery. The decision rests heavily on a careful forecast and examination of detailed assumptions to justify the investment. Consideration should be given to the physical efficiency, throughput, the cost efficiency of the dryer, and the price/quality relationship for paddy.

As drying technologies improve, the emphasis has shifted from product quality and physical efficiency to cost considerations. Practically all types of dryers can now be manufactured in the region. Moreover, the high capital cost of imported drying units is a key consideration. A lower physical efficiency may be accepted for lower capital investment.

The cost of drying varies considerably due to the capital cost, the design of the dryer, fuel used, utilisation rate, drying performance and capacity, and handling methods involved.

### Drying Costs at Various Levels of Operation

In this section, an attempt is made to identify factors operating at the various levels but not normally costed.

#### Farm Level Operations

In analysing the costs and returns of dryer operation at farm level, the following factors should be considered, as they generally affect profitability adversely, thus reducing the benefit to the users:

(a) *Housing for the unit.* Currently available dryers such as the flatbed models are huge in comparison to the farmer's house. Consequently, a shed has to be built and this is an added cost. This expense may not appear much but can amount to a substantial sum within the farmer's cash flow. Also, because this shed is made of light materials, there are relatively heavy maintenance costs associated with it. The opportunity cost to the owner of maintaining the housing should also be reflected as, in general, small farmers tend to subsist on a daily income.

(b) *Cost of money.* The acquistion of drying units is usually done on a hire purchase basis. Practically all studies reflect only the related interest rates charged by financing institutions. Depending on the country and the source of funds, numerous hidden costs are incurred. There are documentation costs, loan application fees, and other bank charges which in some countries make up an additional 50% over the cost of interest on the loan. In addition, as loans are normally not granted upon application, several trips have to be made to the bank which, invariably, is distant from the farm. The cost of travel itself and the opportunity cost associated with travel could be substantial to the farmer.

(c) *Operating costs.* In most ASEAN countries, fluctuations in the prices of fuel and other materials are commonplace. As fuel constitutes a major operating expense, sensitivity analysis should be made and incorporated into the overall financial analysis to safeguard the farmers' interests. This is particularly relevant in the case of paddy whose price is generally controlled. Although the cost of fuel generally follows world market price, domestic taxes make it an expensive commodity.

(d) *Repairs and maintenance.* In most rural areas in ASEAN, the expertise for simple mechanical repairs is usually not available within the village. Mechanics, due to natural business demands, tend to be located in towns. The cost of employing the services of a repairman goes above his fees, as the farmer has to spend time and money for his own and the mechanic's transport.

(e) *Cost of money during operation.* As most small farmers tend to be subsistent from harvest to harvest, there is normally a scarcity of funds before he can sell his harvest. The cost of fuel constitutes 50% of operating costs. As the incentive for drying is to enable him to get a better price, and he has the dryer anyway, the paddy is assumed to be sold after drying. This means that he has to borrow money at normally exorbitant rates of interest to operate the dryer. This constitutes an additional expense.

**Assembly Points**

At this level of operation, the main concern is cost reduction and profitability. Generally, the volume of grain handled at this level will determine the mode of drying. In Indonesia, the village cooperatives rarely use their artificial dryers and depend heavily on sun drying. This may not be applicable for the Thai maize traders who handle 50 to 150 t/day or for the private paddy merchants handling similar volumes.

At the KUD level in Indonesia, the cost of sun drying is just Rp 2.94 compared with drying by the Lister moisture extraction unit which costs Rp 11.86/kg. Table 4 shows the comparative costs of these two drying systems. The price difference between various moisture content levels at the current pricing system (Table 2) is minimal. Under current price conditions, the maximum incremental price benefit from drying paddy (from 24% to 14% m.c.) is Rp 7.554/kg. With the current direct cost of drying using the Lister drying unit of Rp 11.86, the KUD stands to lose Rp 4.3/kg of paddy. But this does not cover the total losses incurred by the KUD due to mechanical drying. There are other costs to contend with: amortised housing cost for the unit, and repair, maintenance, and management costs.

In larger private sector operations, there are yet further costs to be considered. Before a drying unit is purchased, a feasibility study or at least a comparative financial analysis of various drying types, is made as the basis for decision making. This entails some added cost. When the unit is acquired, the staff have to be trained, and a mechanics pool may be maintained to undertake repairs and maintenance. Apart from the usual operating costs, there are replacement costs for parts that are expected to wear out during the lifetime of the machinery.

Mechanical dryers are a necessity for large-scale paddy middlemen, and the private sector has been successful in making the dryer an asset rather than a liability.

**Grains Complexes**

In grains complexes operated by the public sector, the drying facilities tend to be a mixture of sun drying, batch dryers, continuous flow dryers, and warehouse type vortex driers. The figures on drying costs in this type of operation are difficult to obtain. The drying system is an integral part of the whole processing complex and the costs of operating the complex are not broken down into the various processes involved. Storage and drying in warehouses are related operations. Furthermore, such information is normally treated as confidential and is not easily accessible.

The direct drying costs are highly variable

**Table 4.** Comparison of drying costs[a] in Village Unit Cooperatives in Indonesia.

| A. Lister Dryer | | B. Sun Drying | |
|---|---|---|---|
| *Variable Costs* | *Rp* | *Variable Costs* | |
| Diesel 200 L at Rp 245/L | 49 000 | Estimated at Rp 2.00/kg and consists mainly of labour | |
| Kerosene 100 L at Rp 175/L | 17 500 | | |
| Lubrication 2 L at Rp 800/L | 1 600 | | |
| Labour Rp 1500/t | 15 000 | | |
| Maintenance 1.2%/100 hours | 12 960 | | |
| Total variable costs per batch | 96 060 | | |
| Total variable costs per kg | 9.61 | | |
| *Fixed Costs* | | *Fixed Costs* | |
| Depreciation: | | Area required: 2 (days) × 65 $m^2$ = 130 $m^2$ | |
| Investment Rp 9.0 million; lifetime 5 years: | | Construction costs: Rp 5 000/$m^2$ | |
| 120 batches per year | 15 000 | Total costs: Rp 650 000 | |
| Interest: | | Depreciation: | |
| 20% p.a. over average value of 50% of Rp 9.0 million; | | Lifetime 20 years, 180 days per year | 180 |
| 120 batches per year | 7 500 | Interest: | |
| Total fixed costs per batch | 22 500 | 20% p.a. over average of 50% of Rp 650 000 | |
| Total fixed costs per kg | 2.25 | 180 operating days | 360 |
| Fixed plus variable cost per kg | 11.86 | Maintenance: | |
| | | 5% per year; 180 operating days | 180 |
| | | Plastic sheets: | |
| | | Rp 60 000 to cover 2 t | |
| | | 270 days useful life | 220 |
| | | Total fixed costs | 940 |

[a] In Indonesian Rupiah: 12 Rp = US$1.

depending on the fixed cost, capacity and drying rates, as well as on operating expenses. The ancillary and hidden costs are similar to those in large, privately operated assembly points, although their magnitudes may differ. Private assemblers operate as business concerns. Government complexes, on the other hand, direct their operations towards political or social goals such as the maintenance of security reserves, food supplies, and price levels. Hence, the operation of a private grains complex cannot be compared with government complexes.

## Conclusions

On the basis of currently available information, several conclusions can be drawn regarding drying operations in the ASEAN region. Firstly, it is widely recognised that artificial drying technology has been well developed in engineering terms and efforts are being made to further develop the technology. Secondly, there is the realisation that the technology has not filtered down to the users, particularly at the farm level. While problems are being encountered at all levels of operation, the problems are no longer directed at the technology per se but on matters related to marketing and price policies associated with the standards and grading system, sociological aspects, and other factors related to the adoption of artificial drying systems. These factors all relate to the economics of drying. A review of these factors appears to be necessary if continuing losses due to drying are to be arrested or minimised.

The third point relates to the financial and economic analysis of drying operations. Financial and economic analyses of drying operations are only guides for decision making. Under various conditions, they can be extremely useful as bases for investment decisions. However, under current conditions, the factors affecting drying operations should be looked at more carefully.

In the financial analyses of drying operations, it is well worth noting that the cost components at various levels vary in nature and importance, mainly because the realities in their loci of operation are different. Also, the purposes of drying vary between levels. At the farm level, where the primary concern hinges on increasing incomes, serious consideration should be given to hidden costs as well as the farmer's loss-bearing capacity, in view of his tight cash flow situation.

In the case of private grains handlers where efficiency is the rule and profitability the goal, the problems are different. Cash flow problems are more easily averted. The problem hinges on the continuing search for methods of reducing the cost of operations. This entails finding the most efficient technology, devising streamlined operations and other ways to reduce costs and enhance profitability. It is therefore not surprising that these concerns have been embodied in the goals of leading research and development institutions in developed countries.

While cost reduction and profitability are also guidelines in public sector grain complexes, it should be noted that their primary purposes are the provision of food to the people in times of adversity, protection of both producers and consumers by stockpiling to maintain prices, and the savings on foreign exchange which may be incurred by minimising imports and increasing exports. Given these mandates, the cost of grain drying may not be of primary significance at the grain complex level.

## References


AFHB (ASEAN Food Handling Bureau) 1984. Report of consultancy on maize production and marketing in Thailand: implications as regards drying methods. Kuala Lumpur, Malaysia, AFHB.

AFHB (ASEAN Food Handling Bureau) 1985. Consultancy report on drying and village level milling in Indonesia. Kuala Lumpur, Malaysia, AFHB.

de Padua, D.B. 1979. A critical review of the losses in the rice post production system in some Southeast Asian countries. Interfaces between agriculture, nutrition and food science, Philippines.

Lorenzana, J., and Pablo, R.C. 1983. Economic viability and acceptability of improved mechanical threshers and dryers in village farmers association in northeastern Philippines. In: Teter, N.C., Semple, R.L., and Frio, A.S., ed., Maintaining good grain quality. Manila, Philippines, Southeast Asia Cooperative Post-Harvest Research and Development Programme, 315–326.

Mangaoang, C.C. 1984. Cost of drying. In: Frio, A.S., and Manilay, A.A., ed., Coming to grips with the wet season grain handling problems in Southeast Asia. Manila, Philippines, ASEAN Crops Post Harvest Programme, 65–74.

# Economics of Aeration and In-store Drying Session Chairman's Summary

**Shaharuddin Hj. Haron***

FOUR papers were presented, one on loss assessment and three on the economics of drying. These four papers represented a change in focus of the seminar, from the mainly engineering aspects of the three previous sessions, to socio-economic considerations.

The first paper, presented by Mr Robin Boxall, generated a great deal of interest. He highlighted several important points on the merit of loss assessment in the developing country context and on the need to use the results obtained from such studies in future planning and development of the grain sector of the country concerned.

On the matter of the methodology of loss assessment, while this is rather well developed at the on-farm level, the same cannot be said for loss assessment in the commercial sector of the grain industry. Mr Boxall suggested that greater savings and faster corrective measures can be implemented from results obtained in surveys of the commercial sector than of the on-farm postharvest sector. This is so because on-farm activities have more complex socio-economic and political implications. The implementation of improvements resulting from findings of a loss assessment survey may well create more imbalances in the system.

Dr G.J. Ryland's paper discussed the model he has developed to determine the equilibrium or economic optimum level of drying under different situations.

He defined the economic optimum level of drying in terms of the economic threshold of moisture content below which point further drying reduces farmers' net returns. Economic optimum levels of drying depend on a number of critical parameters, including time, quantity of grain dried, rate of drying, and qualitative changes in the grain, as well as drying costs and anticipated returns. The model can be used to select the drying system appropriate to particular countries.

* LPN, P.O. Box 108, Kuala Lumpur, Malaysia.

He also discussed the linkages between drying operations and the other postharvest operations of transport, storage, milling, and distribution; and the need for a systems approach to arrive at the most socially efficient pattern of drying facilities.

The third paper, delivered by Dr Chew Teck Ann, discussed a specific case, namely some of the economic aspects of drying of paddy by farmers in Selangor, Malaysia. He stressed that the current situation, in which farmers did not attempt to dry their paddy at all is due to the fact that there is no incentive for them to do so under the current paddy pricing system.

He suggested that the following changes should be made to the present paddy pricing system:

- the penalty (deductions for high moisture grain not dried) should be made greater than the cost of drying and other associated costs borne by a farmer if he were to dry his paddy;
- the penalty should be made relatively greater for higher moisture content paddy (above 25% moisture content) than lower moisture paddy (less than 20%).

He also suggested the use of free market mechanisms to assist in determination of the penalty factor.

The final paper in the session, delivered by Mr Johnson Mercader, looked at the economics of drying in the broader perspective of the ASEAN region. He cited the importance of mechanical drying in the region, especially for the wet season harvest, and explained the reasons for the slow adoption of mechanical/artificial drying technology.

The engineering aspects of drying in ASEAN are already well researched and developed. However, more R & D studies are required on the economic and financial aspects of establishing and operating the technology, especially at the farmer and cooperative level.

This paper also categorised areas where drying is

done in the ASEAN countries and explained the characteristics of the various levels, in so far as the difficulties involved in the adoption of dryer technology are concerned. Given the socio-economic and political background, the most promising level for adoption would be the government complexes.

**Overall Findings of the Session**

1. On loss assessment
   - A multidisciplinary, broad-front approach is needed in any loss assessment exercise.
   - Data obtained need to be used in further planning and development of the grain industry concerned.
2. Economics of drying
   - The model developed on the economic optimisation of drying should be made use of in appraisals of drying systems before selection on the type, size, and location is made.
   - There is an urgent need to include drying incentives in the paddy pricing system.
   - There is also an urgent need to plan and implement more R & D studies related to the economic and financial aspects of the adoption of mechanical dryers at the appropriate level in the ASEAN region.

# Case Studies on Aeration and In-store Drying

# Silo Storage in Malaysia

K. F. Loo*

*Abstract*

A major portion of paddy in Malaysia is stored in bulk in National Padi and Rice Authority rice mill complexes. Before storage, the grain is cleaned and mechanically dried to 14% moisture content (wet basis). Bulk storage takes the form of either concrete tower silos or horizontal bulk bins. Tower silos are desirable on account of easier unloading by minimal labour. However, storage problems manifested in various degrees of grain discoloration have been encountered following use of such silos, the severity of the problems being correlated with length of storage.

Studies were carried out to determine the optimum level of aeration for avoiding grain deterioration. Three airflow rates were investigated: 0.03, 0.1, and 0.3 $m^3$ of ambient air per minute per tonne of dried grain, designated low, medium, and high flow rates. Results of a preliminary trial suggest that for storage periods up to 4 months, aeration at both medium and high flow rates reduces grain discoloration to about 2% and maintains high germination levels.

LIKE most agricultural crops, paddy is harvested during a very short period of time but is confronted with a reasonably constant demand throughout the season. In order to satisfy this demand, it is necessary to store the seasonal surplus and release it over the rest of the year (Wells and Fredericks, 1983). To ensure that the dried paddy is safely stored before milling, the National Paddy and Rice Authority (LPN) has constructed 26 rice mill complexes each with its own bulk storage facilities.

Newly harvested paddy, after cleaning and mechanical drying to 13–14% moisture content (wet basis), is stored in either flat bottom bins or concrete tower silos, depending on the type of storage facility available at a particular complex. The flat bottom bulk bins are equipped with aeration fans (Calverley et al. 1976). The tower silos do not have aeration facilities. Instead, grain turning is relied upon.

It has been observed that grains keep better in the flat bottom bulk bins than in the concrete tower silos. Very little deterioration is observed in grains stored for durations up to a year in the flat bottom bulk bins. On the other hand, grain discoloration has been observed in tower silos after only 2 months of storage. This discoloration is reflected by yellowing of rice kernels, which lowers the value of the grain.

* Division of Engineering, Lembaga Padi dan Beras Negara (LPN), 13th-17th Floors, Perkim Building, Jalan Ipoh, Kuala Lumpur, Malaysia.

## Grain Discoloration

Yellowing of rice is a common problem in Southeast Asian countries. Recent trials in Indonesia showed that it can occur both during storage and straight after harvest before drying, the latter associated with high moisture and temperature (Phillips et al. 1984). The same observation on yellow kernels was made in the Philippines, when wet paddy was not dried immediately (Mendoza and Quitco 1984). The number of yellow kernels tended to increase as drying was further delayed. Likewise, the rate of increase in yellow kernels was faster in paddy of higher moisture content.

Bramall (1984) found that there are many types of discoloured grain, ranging from the more common yellowed (stack burnt) kernels to bright red, and sometimes black grain. However, only yellowed grain is of primary concern here as other colours are mainly associated with mould spoilage occurring outside the conventional storage. The yellowing of rice is believed to be a form of non-enzymic browning displaying a heat-time relationship, i.e. it develops more rapidly at higher than lower temperatures. The availability of heat from whatever source appears to be the most critical factor.

Mohd. Nour et al. (1983) reviewed earlier studies which indicated that certain storage fungi, in particular the *Penicillium* species, can cause yellow stains on grains. It was further shown that the discoloration of rice in bulk increased with length of storage, moisture content, and activity of storage fungi.

Another possible cause of discoloration in grain during storage is a phenomenon involving enzymic activities which may, when conditions are suitable, bring about a variety of chemico-biological reactions between the chemical constituents of the grains. Linco et al. (1960) observed that, due to enzymic activity, reducing carbohydrates and free amino acids can be formed within the grain. These two elements can interact under poor storage conditions, such as when excessive heat is present, to produce brownish intermediate compounds.

A study by Teter (1979) showed that grain which had a history of grain damage due to drying delay after harvest exhibited higher rates of grain deterioration during subsequent storage.

### Typical Behaviour of an Untreated Bulk

There are several views as to the probable cause of heat build-up during storage. These include poor management of the grains and the system, inherent deficiencies in the engineering design of the storage, or a combination of both (Calverley 1977). There are many areas which need study. We need to know, for example, if the grain deterioration is due to moisture migration or seepage of water. However, it is now evident that seepage of rain water through the silo structure is at most a contributing factor, as some covered silos with full weather protection display grain deterioration as severe as that found in open silos.

### Heating Due to Infestation

A typical, newly filled silo contains paddy with a moisture content between 13 and 14% (wet basis) which is low enough to suppress microorganisms but not to reduce insect reproduction to negligible levels (Rawnsley 1976). The paddy may be attacked by insect pests harbouring within the storage structure or brought in with the grain. The insect population rapidly increases and the heat produced by the pests causes a local temperature rise, or hot spot. This can lead to migration of moisture to the cooler upper surface, resulting in mould growth there. When the temperature rises to a certain value, it forces the insects to migrate to cooler areas or results in their death. This process will continue until the insects are confined to the grain surface. There, intensive activity may continue indefinitely, the heat produced being readily lost to the atmosphere. Local observations showed that temperature usually increases fairly uniformly throughout the grain bulk.

### Moisture Migration

In an insect-free bulk, moisture migration may still occur as a result of transfer by air convection currents and by diffusion towards the cold surface (Burrel 1974). Moisture accumulates at the exposed top surface of a bulk, around the circumference of the silo, resulting in sprouting and mould growth there. At the same time, the migration of moisture by diffusion to other cooler grain layers will again lead to formation of hot spots and support the growth of microorganisms and other pests. Figure 1 illustrates diagrammatically the various transfer processes.

Heat build-up in paddy stored in concrete tower silos in Malaysia has been reported by several LPN complexes. Grain temperature has been observed to rise above 50°C at the hot spots.

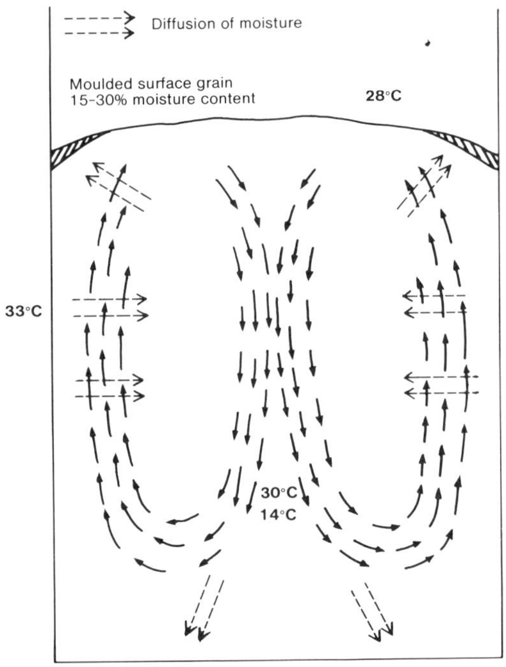


**Fig. 1.** Pattern of moisture migration to the cold surface of a hot grain bulk.

## Control Measures

From the above observations, it appears that methods for controlling deterioration can be broadly classified under two headings (Griffiths 1964).

(i) Chemical methods: the use of insecticides and fumigants to control insects.

(ii) Non-chemical methods: the use of aeration and grain turning in order to control moisture migration and biochemical deterioration.

However, the ensuing discussion focuses on the use of turning and aeration as measures to improve grain storage conditions in LPN complexes.

### Turning

Turning is the process of moving grain through air by transferring the contents of a silo into a nearby empty one. The result is a break-up of pockets of insect infestation, a slight loss of heat to the atmosphere, an averaging out of temperature and hence a reduction in maximum temperature, and the mixing of any moist grain with the rest of the bulk.

With the handling facilities available at present, grain turning in LPN silos is a tedious and time-consuming process. A complete turning of 750 t of grain in one silo may take 40 hours and often interferes with other drying and transfer operations of the plant.

### Aeration

Aeration is a more convenient strategy in that the grain need not be moved. It is achieved by blowing or drawing air through the grain mass by means of fans (forced ventilation). Normally, the air used in aeration should not only be cooler and drier than the grain to be aerated but also in sufficiently large quantity to achieve maximum effectiveness. In Malaysian conditions, like other parts of the tropics, the ambient air is not suitable for aeration purposes at all times of the day. As a compromise, aeration therefore entails the use of higher air-flow rates at times when ambient conditions are most favourable.

The cost of aeration is by no means negligible and careful design is needed to ensure that a system is as economical as possible. The horsepower and electrical energy costs for aeration rise rapidly with airflow rate and the depth of the grain (Navarro and Calderon 1982; Thorpe 1984). It was precisely for this reason that aeration fans were not installed during the construction of existing silos.

In view of the pressing need for improvement and the attainment of a safe and economical silo storage system, the Malaysian Government, through her agencies LPN and MARDI (Malaysian Agricultural Research and Development Institute), with financial assistance from the ASEAN-Australia Food Handling Programme and technical input from the SEARCA Technical Team, initiated in 1980 a research and development project to incorporate appropriate aeration facilities into the existing silo storage system and evaluate their performances in terms of the following three questions.

(i) Can aeration overcome the problem of moisture migration caused by convection currents?

(ii) What is the minimum airflow rate required and for how long and how frequently has it to be applied?

(iii) What are the limitations of aeration using ambient air during the rainy season when relative humidity is high?

The background to the storage trials, the principles involved, details of method and procedures, analysis of results and discussion of the preliminary findings have been given by Mohd. Nour et al. (1983). Here we review the problems encountered and the findings, and examine possible steps for further development of silo aeration in Malaysia.

## Silo Aeration Trials

The LPN rice mill complex at Ulu Tiram Buruk (UTB), Tanjung Karang, Selangor has been selected as the test site. It is situated within the major rice growing area in central Peninsular Malaysia.

The UTB complex has eight cylindrical silos constructed of reinforced concrete. They are arranged in a row and housed within the complex's main building. Each silo is 3.05 m in internal diameter, has a wall thickness of 0.15 m, a height of 25.91 m, and can hold 750 t of paddy at 14% moisture content (wet basis). The silo is capped with a roof slab and protected by a double-pitched asbestos-cement roof. The exposed upper half of silo walls is clad with asbestos-cement. Each silo is self-emptying by means of a conical hopper bottom at a height of 6.10 m above ground. The hopper surface is sloped at an angle of 45°.

For detection of hot spots and grain deterioration, the silo is equipped with 18 temperature sensors (thermistors). There are two sensor cables suspended from the silo top carrying 6 thermistors each. Along the wall, there are another 6 sensors, 3 on each side of the silo diametrically opposite one another. The thermistors are arranged in such a way that each one covers an effective sphere of 4 m diameter. The whole system is monitored at the plant control room.

The silo is also equipped with a vertical bucket elevator for loading and unloading with a capacity of 50 t/hour. A belt conveyor complete with discharge trolley is used for the distribution of grains at the top, while two belt conveyors at the bottom of the silo are used for removal of grain. The hopper is fitted with a remotely controlled, pneumatic shutter gate.

Five of the silos (side by side in a row) were selected for the test. Three of them were equipped with aeration facilities each with a different airflow rate. The remaining two silos acted as the controls involving grain turning from one silo to the other.

### Aeration System Selection

The design procedure of Teter (1979, 1982) was adopted to calculate the aeration requirements. Calculations showed that even under suitable conditions of low moisture content of paddy (between 13 and 14%) and normal weather, the minimum airflow rate required is 0.03 $m^3$/t/minute (Anon. 1979; Rukunudin 1980). The design day selected for normal weather is the maximum and minimum average daily humidity in the wettest month of the year on a 10 year average.

For the first trial, three airflow rates were selected namely: 0.3 (high airflow rate or HFR), 0.1 (medium airflow rate or MFR), and 0.03 (low airflow rate or LFR), the units in each case being $m^3$/t/minute.

The HFR system comprised push and suck cross-flow ventilation. It utilised two 10 horsepower (h.p.) vaneaxial fans, with two vertical perforated semicircular air ducts installed on opposite sides of the internal wall of the silo (Fig. 2a).

The MFR system was made up of a 10 h.p. centrifugal fan running on suction mode. Aeration air was drawn in at the silo top, passed through the bulk mass in vertical ventilation, and was

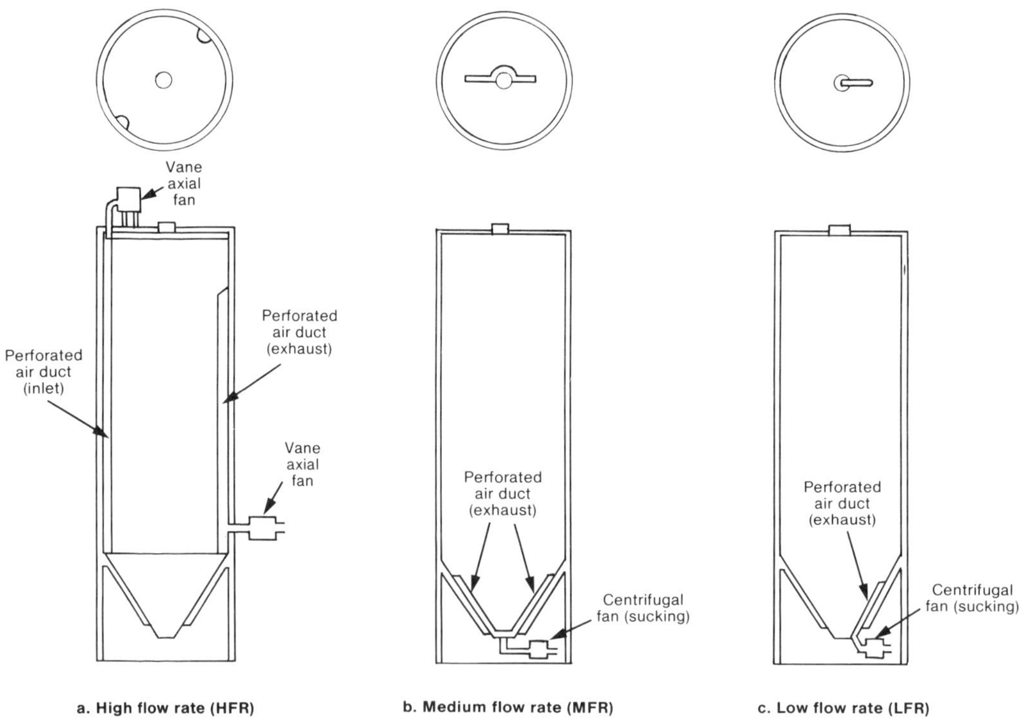


**Fig. 2.** Plan and elevation of tower silo experimental aeration systems at Ulu Tiram Buruk, Malaysia.

collected at the hopper bottom by means of two perforated semi-circular ducts installed on opposite sides of the sloping surface of the silo hopper bottom (Fig. 2b).

The LFR was similar to the MFR system except that a 0.5 h.p. centrifugal fan was used with a single, semi-circular duct placed radially on the sloping surface of the silo hopper bottom (Fig. 2c).

In all the three cases, the semi-circular air ducts were rolled from perforated steel plate with 39% opening. As noted before, two other silos were reserved for studies on the grain turning system.

**Experimental Procedures**

The effects of aerated storage at three different airflow rates compared with the control silos for grain turning, were studied for periods up to 4 months in 1981, 1982, and 1983.

Fresh, clean, and dry, good quality long-grain paddy was used in the trial. Efforts were made to ensure that the paddy used to fill all silos was of similar quality.

For complete details of experimental procedures, and an analysis of the results of the storage trials, the reader is referred to the paper Mohd. Nour et al. (1983).

**Table 1.** Reduction of grain moisture content due to aeration and grain turning.

| Treatment | Reduction in moisture content (%, wet basis) |
|---|---|
| High flow rate (0.3 $m^3$/minute/t) | 0.7 |
| Medium flow rate (0.1 $m^3$/minute/t) | 0.01 |
| Low flow rate (0.03 $m^3$/minute/t) | 1.92 |
| Control (grain turning) | 0.75 |

Source: Mohd. Nour/Jantan et al. (1983)

**Review of Findings**

The preliminary study of aeration by Mohd. Nour/Jantan et al. (1983) revealed that aerated storage at airflow rates of more than 0.1 $m^3$/t/minute can reduce yellowing in paddy bulk stored for up to 4 months in concrete tower silos to about 2% (weight ratio of yellow kernels to that of total bulk). There was also a slight drying effect. The results of the study are summarised in Tables 1–5.

The incidence of a low level of discoloured grain as well as localised high temperatures in the grain bulk indicated the presence of uneven air distribution in the aeration system. (The small incidence of yellowing could be due to uneven air distribution which in turn, was due to leakage as well as some duct perforations being closed up by paddy or broken pieces of grain. So, it is evident that fan on blowing mode would be more suitable than suction mode.) For HFR and MFR systems, there was no appreciable change or adverse effect on milling yield, brokens content, hectolitre weight, or germination test, which suggests the absence of widespread insect infestation and microorganism activity.

The study further showed that the energy cost

**Table 2.** Results of laboratory analysis of grain samples, following aeration at three airflow rates.

| Type of analysis | HFR[a] | | MFR[b] | | LFR[c] | | C[d] | |
|---|---|---|---|---|---|---|---|---|
| | Before storage | After storage | Before storage | After storage | Before storage | After storage | Before storage | After storage |
| 1. Moisture content (%, oven method) | 12.6 | 11.9 | 11.96 | 11.95 | 11.85 | 9.93 | 11.69 | 10.94 |
| 2. Percentage clean paddy | 93.9 | 92.4 | 97.3 | 96.5 | 91.3 | 93.00 | 91.5 | 87.4 |
| 3. Hectolitre wt. (kg/HL) | 54.5 | 55.6 | 55.18 | 55.91 | 55.52 | 54.95 | 55.05 | 54.95 |
| 4. Milling analysis: | | | | | | | | |
| i. Milling outturn (%) | 65.8 | 64.1 | 65.2 | 62.2 | 66.4 | 67.7 | 65.5 | 63.7 |
| ii. Head rice (%) | 67.6 | 68.6 | 67.5 | 63.0 | 68.6 | 66.9 | 69.2 | 68.5 |
| iii. Broken rice (%) | 32.4 | 31.4 | 32.5 | 35.3 | 31.5 | 28.9 | 30.8 | 28.6 |
| iv. Discoloured grains (%) | 0 | 1.55 | 0 | 1.7 | 0 | 4.2 | 0 | 2.9 |
| v. Husks (%) | 34.1 | 35.9 | 34.6 | 36.3 | 34.5 | 32.3 | 34.6 | 36.3 |
| 5. Germination level (%) | 91.0 | 85.1 | 87.4 | 89.1 | 91.0 | 48.4 | 86.67 | 53.3 |

Source: Mohd. Nour/Jantan et al. (1983)
a. High flow rate 0.3 $m^3$/minute/t.
b. Medium flow rate, 0.1 $m^3$/minute/t.
c. Low flow rate, 0.03 $m^3$/minute/t.
d. Control, grain turning.

**Table 3.** Actual (commercial level) milled rice yield (%) following aeration of paddy at three airflow rates.

| Treatments | $A_1$ | $A_2$ | $A_3$ | $A_4$ | Sample grade | Brokens | Points | Fine bran | Coarse bran | Husks |
|---|---|---|---|---|---|---|---|---|---|---|
| 1. HFR[a] (Silo 26) | 6.35 | 41.65 | 4.78 | — | — | 14.15 | 0.17 | 11.32 | 0.68 | 20.90 |
| 2. MFR[b] (Silo 27) | 2.21 | 46.71 | 2.45 | — | — | 15.14 | 0.29 | 12.40 | 1.08 | 19.72 |
| 3. LFR[c] (Silo 28) | 0.5 | 36.67 | 6.53 | 8.86 | 8.80 | 6.28 | 0.27 | 12.02 | 0.99 | 19.08 |
| 4. C[d] (Silo 29/30) | — | — | 66.98 | — | — | — | 0.35 | 10.92 | 0.89 | 20.86 |

Source: Mohd. Nour/Jantan et al. (1983)
a. High flow rate, 0.3 $m^3$/minute/t.
b. Medium flow rate, 0.1 $m^3$/minute/t.
c. Low flow rate, 0.03 $m^3$/minute/t.
d. Control, grain turning.

**Table 4.** Cost comparison for silo aeration at three airflow rates.

| Treatment | Capital cost (US$) | Operational cost (US$/t) | Quantity of paddy (t) | Storage duration (months) |
|---|---|---|---|---|
| HFR[a] | 10 840 | 1.42 | 687.47 | 3 |
| MFR[b] | 4 520 | 1.18 | 700.81 | 4 |
| LFR[c] | 3 540 | 0.45 | 318.32 | 3 |
| C[d] | — | 1.01 | 264.73 | 4 |

Source: Reproduced from Annexure 3 in the unpublished paper entitled Bulk storage padi aeration: preliminary report by LPN-MARDI.
a. High flow rate, 0.3 $m^3$/minute/t.
b. Medium flow rate, 0.1 $m^3$/minute/t.
c. Low flow rate, 0.01 $m^3$/minute/t.
d. Control, grain turning.

**Table 5.** Record of grain temperatures at various depths in a silo following aeration at three airflow rates.

| Level (below grain top surface) | Average grain temperature (°C) HFR[a] | MFR[b] | LFR[c] | C[d] |
|---|---|---|---|---|
| 1 | 36.6 | 34.4 | | |
| 2 | 36.8 | 37.3 | | |
| 3 | 33.6 | 33.0 | | |
| 4 | 36.2 | 35.7 | 34.8 | 33.0 |
| 5 | 36.1 | 36.2 | 36.2 | 34.5 |
| 6 | 28.0 | 28.7 | 28.1 | 27.5 |
| Purata | 34.6 | 34.2 | 33.0 | 31.7 |
| Ambient temperature | 32.5 | 30.0 | 30.0 | 30.0 |
| Relative humidity | 59.5% | 70.8% | 70.8% | 70.8% |

Source: Reproduced from Annexure 4 in the unpublished paper entitled Bulk storage padi aeration: preliminary report by LPN-MARDI.
a. High flow rate, 0.3 $m^3$/minute/t.
b. Medium flow rate, 0.1 $m^3$/minute/t.
c. Low flow rate, 0.01 $m^3$/minute/t.
d. Control, grain turning.

for grain turning, despite its unfavourable result, exceeded that for aeration by from 34 to 100%.

The MFR system appears to be most appropriate, as aeration can begin as soon as there is sufficient grain to cover the air ducts. In the case of the HFR system, aeration can begin only when the silo is full. The vertical ducts must be fully covered with grain to avoid short-circuiting of airflow across the head space (Mohd. Nour 1985). Another advantage of MFR is that the aeration ducts are very much easier and cheaper to install and maintain as compared with those for HFR aeration, where special equipment for accessibility is required during installation and for subsequent maintenance.

Based on the above findings, MFR aeration was extended to several LPN complexes having the same silo storage system.

## Possible Areas for Further Work

Bulk storage in concrete tower silos is superior to other forms of storage because of the ease of grain unloading. However, it is evident that there are still some technical problems to be solved.

The following areas are identified as possibilities for further research and development.

### Uniform Air Distribution

The existing air ducts for the HFR and MFR systems should be re-examined and modified to ensure even distribution of air within the grain mass. Particularly for the HFR system, care should be taken to minimise short-circuiting of air between the two vertical ducts. A more convenient and cheaper means of access to the vertical ducts should be devised for facilitating maintenance work if this system is to be adopted.

### Aeration Frequency and Duration

More research is needed on the optimisation of fan operating hours and frequency, taking into account the local weather conditions.

### Grain Cooling

The tropical climate is less suitable for aeration of bulk grain as the periods of lower air relative humidity always correspond with higher air temperature during the day, whereas higher relative humidities occur during the night when ambient temperature is low. As a result, aeration during the day risks raising the grain temperature, which will be conducive to microorganisms and insects. On the other hand, ventilation at night risks dampening the grain. A solution is perhaps to find means to cool the grain by refrigerated air. The overall economics of grain cooling needs to be investigated.

### Pest Control

As explained earlier, under tropical conditions aeration with ambient air is ineffective in controlling insect infestation. Hence, there is a need to incorporate some chemical means to arrest grain deterioration caused by insects.

In addition, suitable methods and equipment should be developed for regular cleaning and disinfestation of the internal silo wall each time the silo is emptied. This is necessary to reduce the occurrence of reinfestation of grains during storage.

### Grain Turning

Although the total energy cost for grain turning has been claimed to exceed that for aeration, a fast, independent grain turning system may be desirable for effective pest infestation control. Such a need should be further investigated.

### Reduction of the Effects of Solar Heating

Studies should also be conducted to evaluate the feasibility of having all exposed surfaces of silos painted white to minimise moisture migration due to uneven heating of silo walls.

### Up-grade Cleaning Facility

Sound grains store better. Storability of grain can be improved by ensuring its thorough cleaning before storage. There is therefore a need to examine existing cleaning facilities at the rice mill complexes with a view to up-grading them.

## Conclusion

This paper has attempted to review the findings of aeration trials carried out by a joint LPN-MARDI research team. Further development work needed to improve the feasibility of silo aeration in Malaysia is outlined.

## References


Anon. 1979. Bulk storage padi aeration project proposal. Kuala Lumpur, Lembaga Padi dan Beras Negara.

Bramall, L.D. 1984. The drying in bulk storage of high moisture grain. Project 8308 Progress Report June 1984. Canberra, Australian Centre for International Agricultural Research.

Burrel, N.J. 1974. Aeration. In: Christensen, C.M., ed., Storage of cereal grains and their products. St Paul, Minnesota, American Association of Cereal Chemists, 454.

Calverley, D.J. 1977. Problems in bulk storage of paddy: engineering aspects (structures). Presented at action oriented field workshop for prevention of post-harvest rice losses. March 1977, Alor Setar, Malaysia.

Calverley, D.J., Street, P.R., Cree, T.J., and Dendy, D.A.V. 1976. Report on a study of the technical and economic problems of off-season padi, from harvesting to post-milling storage, in the Muda area of West Malaysia. London, Tropical Products Institute.

Griffiths, H.J. 1964. Bulk storage of grain: a summary of factors governing control of deterioration. Melbourne, CSIRO.

Linco, P., Cheng, Y.Y., and Milna, M. 1960. Changes in the soluble carbohydrates during browning of wheat embryo. Cereal Chemistry, 37, 556–584.

Mendoza, M.E., and Quitco, R. 1984. Quality deterioration and value loss in grain. In: Frio, A.S., and Manilay, A.A., ed., Coming to grips with the wet season grain handling problems in Southeast Asia. Manila, ASEAN Crops Post-Harvest Programme, 15–20.

Mohd. Nour, D. 1985. Aerated bulk storage of dry padi in LPN concrete silos. ASEAN Food Journal, 1, 39–40.

Mohd. Nour/Jantan, D., Teoh, I.C., Rukunudin, I.H., Abdullah, R., and Said, S. 1983. Aerated bulk storage of paddy in vertical concrete silos. In: Teter, N.C., Semple, R.L., and Frio, A.S., ed., Maintaining good grain quality. Manila, ASEAN Crops Post-Harvest Programme, 254–265.

Navarro, S., and Calderon, M. 1982. Aeration of grain in subtropical climates. Rome, FAO.

Philips, S.I., Mitfa, R., Sintha, S., Wallbridge, A.J., Cooke, R.D., and Baker, A.A. 1984. A study of the factors involved in post-harvest yellowing of rice. Presented at grains post-harvest workshop, ASEAN Crops Post-Harvest Programme, August 1984, Kuala Lumpur.

Rawnsley, J. 1976. Storage: constraints. In: Rice post-harvest technology. Ottawa, Canada, International Development Research Centre, 133–147.

Rukunudin, I.H. 1980. UTB aeration system design. Kuala Lumpur, Malaysian Agricultural Research and Development Institute.

Teter, N.C. 1979. Grain storage. Manila, Southeast Asia Cooperative Post-Harvest Research and Development Programme.

Teter, N.C. 1982. Design of grain storage aeration for the humid tropics. Presented at the winter meeting of the American Society of Agricultural Engineers, December 1982, Chicago.

Thorpe, G.R. 1984. The benefits of aeration. Presented at CSIRO/AGI seminar on aeration system design, July 1984, Melbourne, Australia.

Wells, R.J., and Fredericks, L.J. 1983. Rice processing in Peninsular Malaysia. An economic and technical analysis. Kuala Lumpur, Oxford University Press.

# Storage and Drying Trials with Unmilled Rice in Indonesia

D.J.B. Calverley*

*Abstract*

A system for the storage/drying of unmilled rice was designed for new 3500 tonne capacity stores being built in Indonesia. The concept was to use the relatively low humidity during the latter period of the day to dry grain from 18% moisture content at intake to 15%, and later from 15% moisture content to 12% using supplementary heat. Three trials were carried out in which paddy was satisfactorily dried. In the first trial, the moisture content at intake was about 16%. Drying during limited periods of the day reduced the moisture content to about 11%. In later trials the moisture content at intake averaged 22% but at times was as high as 28%. The loading was greater than the evaporative capacity of the installation and drying was completed by continuous ventilation throughout 24 hours. The final moisture content was 14%. It was concluded that it would be possible to dry paddy using ambient air in the most adverse climatic conditions in Indonesia. There were some quality changes following extended drying times, particularly an increase in yellow grains, but it was considered that some of these changes were due to damage and deterioration which had taken place before drying commenced.

## Development of the Project

Badan Urusan Logistik (BULOG) was established in 1969 to bring order and stability to rice distribution and marketing in Indonesia by acting as an intervention and marketing agency. At this time significant changes were taking place in Indonesia's farming practices. Production of rice was increasing steadily; more area was being brought into production, more land was being irrigated, and fertilisers, pesticides, and new varieties were increasing yields and making possible more frequent cropping. These developments, however, brought their problems. New varieties mature less evenly, shed more easily, and cannot be left to dry on the stalk. Intensive cropping with more than two crops means that one harvest takes place during the rainy season. In the less populated parts of Indonesia there were harvesting problems which resulted in considerable deterioration where the crops were left unthreshed in the fields. BULOG was required to support producer prices under these conditions, often being forced to purchase lower quality unmilled rice which proved difficult to store without further deterioration and spoilage.

* Tropical Development and Research Institute, Storage Department, London Road, Slough SL37HL, Berkshire, England.

At this time BULOG was embarking on an ambitious store building program. The Advisor to the Consulting Engineers was Professor H. Pfost, late of Kansas State University. Pfost studied the historical daily weather data for Java and concluded there were periods of most days, even during the rainy season, when the ambient relative humidity fell sufficiently to dry grain to at least 15% moisture content. Pfost therefore persuaded BULOG to modify some 175 godowns under construction to permit storage of unmilled rice (paddy) in bulk and drying with high volumes of ambient air.

Also at about this time (1975) the (then) Tropical Products Institute (TPI) was initiating a collaborative project with BULOG on pest and quality control in BULOG's stores and agreed to incorporate a series of drying trials in a modified godown. TPI prepared specifications for machinery and measuring instruments which were funded by the U.K. The godown and all structural arrangements were provided by BULOG. There were substantial problems in the design and construction of godowns intended to store bulk paddy up to a height of 5.5 m. There were also serious difficulties in the commissioning and operation of the handling equipment which delayed the commencement of the trials until 1981.

## The Drying System Concept

The trials were carried out at the Food Technology and Research Centre of BULOG at Tambun, about 30 km east of Jakarta. This is in one of the most productive rice growing areas in Java, growing two, sometimes three, crops of rice each year. The major harvest starts in February during the wet season when the drying need is greatest. BULOG expected to be able to purchase paddy at about 18% moisture content (m.c.) and Pfost's concept was that this should be dried quickly to 15% m.c. using unheated ambient air, followed by a period of drying to 12% m.c. during selected periods of the day or using supplementary heat. In the event it was not possible to obtain paddy consistently at 18% m.c. and, in the second and third of three trials, paddy at moisture contents as high as 28% had to be accepted.

The capacity of the experimental godown was nominally 3500 t. The system was designed so that the crop would be loaded in layers on the drying floor at such a rate that no paddy would remain above 15% m.c. for more than 10 days. The drying capacity was rated at 3% moisture extraction (18%–15% m.c.) per 100 t intake per day. Drier ambient conditions would be selected to dry to 12% m.c. over a period of 3 months, possibly with the use of supplementary heat.

The experimental godown is 48 m long × 29 m wide (Fig. 1). The installation conforms to a conventional on-floor drying system (Fig. 2). There is a conventional twin intake arrangement at one end, each fitted with simple two-screen cleaners. The godown is filled from a single high level chain and flight conveyor and the paddy distributed with a belt grain thrower. The air supply is from a double inlet fan powered by 180 hp diesel and distributed through a central duct and above ground lateral ducts, each fitted with an air control door. The instrumentation to monitor the physical conditions of the grain was based on thermistors to indicate temperature and Reethorpe Moisture Monitors (Gough 1980) to indicate indirectly the moisture content of the grain. A comprehensive array of 600 sensors was designed for one side of the godown to measure moisture and temperature in small vertical increments in four vertical planes. A small array was designed for the other side as an aid to management to indicate the grain was being maintained in sound physical condition. It was a deliberate decision not to install automatic equipment at first, but to require each sensor to be read independently. The switching arrangements proved unsatisfactory and were later replaced by an automatic data logger, which also proved unreliable in the conditions within the godown.

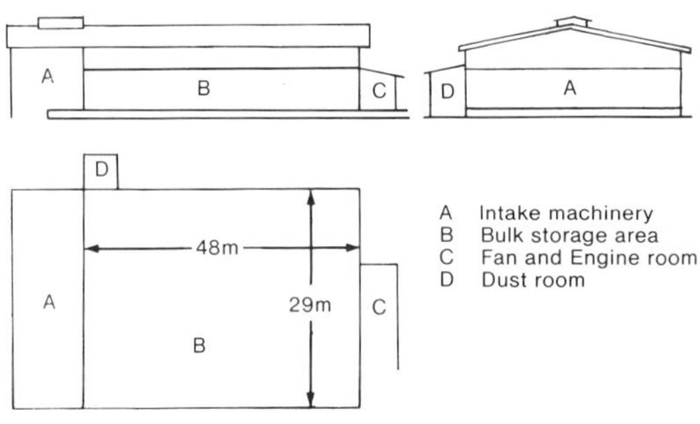


**Fig. 1.** Layout of bulk paddy godown used for storage/drying trials at Tambun, Indonesia.

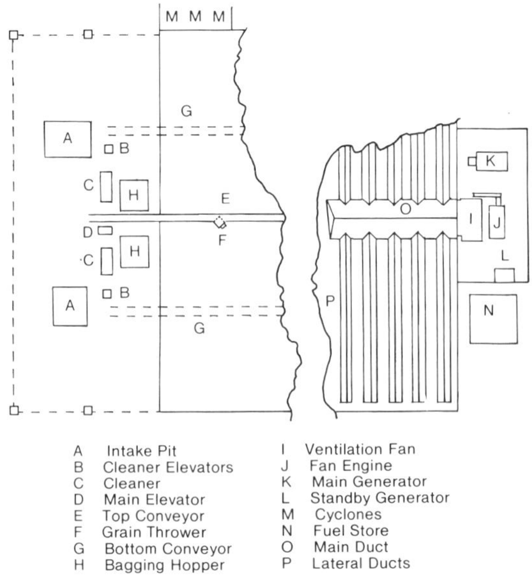


**Fig. 2.** Layout of godown equipment during storage/drying trials with bulk paddy at Tambun, Indonesia.

## Performance of the Mechanical Plant

Since the trials were intended to investigate the practicality of a drying technique, it was not considered that the suitability and performances of plant and equipment should be formally investigated. Accordingly, robust equipment of proven reliability with small grains was selected.

The machinery did not perform as well as expected. Many of the operational problems were the result of wear, which was considerably greater than expected in a number of places, and of the very high intake moisture content during the latter trials. Above about 22% m.c., the handling characteristics of paddy change markedly; it will not flow into or out of conveyors, or across cleaner screens. The problem is compounded when wet paddy is contaminated with straw, mud, and weeds, as commonly occurs with hand threshed, uncleaned paddy. This caused a particular problem with the cleaners. There was also considerable wear in the grain thrower. It was a well designed unit and problems associated with it are wholly attributable to the abrasive nature of paddy. However, the throwing action tended to separate out heavy and light particles giving an uneven distribution across the godown.

During drying, the paddy settled considerably and it then proved very difficult to initiate grain flow into the recovery conveyor. At times the grain face exhibited a negative slope. This consolidation required most of the grain to be shovelled to the discharge point, although once the grain was disturbed it flowed more or less normally. A portable auger reduced the distance that the grain had to be shovelled but the auger was quite unable to penetrate into the bulk of the undisturbed paddy.

There were breakdowns of the recovery conveyor chains fitted under the drying floor. These were of robust design and operated substantially below maximum design loading. The cause of breakdown was attributed to foreign objects falling into the conveyor, but the difficulties of repairing the conveyor during unloading operations make it important to consider alternative methods of unloading. There was also an inexplicable failure of the ventilating fan shaft during Trial II which caused delays in drying.

## Drying Trials

Loading of paddy for the first trial started on 8 April 1981. Intake was slow at first, but later increased to a maximum of 500 t in one week. However, in 6 weeks up to 27 May, only 1730 t had been loaded in to fill the northern side and, because of procurement difficulties, intake was stopped. Since it was found very difficult to procure partially dried paddy, in the two subsequent trials it was decided that paddy direct from the field would be accepted at whatever moisture content it was offered. It was calculated that extending the period of ventilation up to 24 hours per day would ensure sufficient ventilation to achieve drying at the required rate *providing the rate of intake could be regulated.* In the event, the moisture content was higher than anticipated at a mean of 22% for Trials II and III and the rate of intake could not be easily controlled. Nevertheless, paddy straight from the field was dried to a safe moisture content. This decision proved to be a significant and, in the end, an advantageous departure from the original project concept.

The rate of change of the mean moisture content of the total grain in store is illustrated in Figure 3. The lower rate of moisture content reduction in Trial I is considered to be because the fan was run for much shorter periods each day and because the exhausting ventilation air did not reach saturation level, as it did in the later trials, because of the much lower initial moisture content at intake.

Moisture content profiles measured during the drying period exhibited the classical drying front moving upwards through the grain. This is clearly shown in Figure 4 for Trial III. The plots for June also show that very little paddy had dried below 14% m.c. as a consequence of extended daily ventilation. Only when the night ventilation had to be reduced in July because of objections to the noise of the engine did the lower layers become overdried.

Ventilation during selected periods of the day during Trial I resulted in the bottom layers being dried quickly to about 11% m.c. The fan was started as soon as the ambient relative humidity (RH) fell below 80% and was stopped when it rose above about 80%. The temperature increase in the ventilating air due to the waste heat of the fan engine was 2.5°C, which led to the average RH of the ventilating air being 65% in May and 60% in

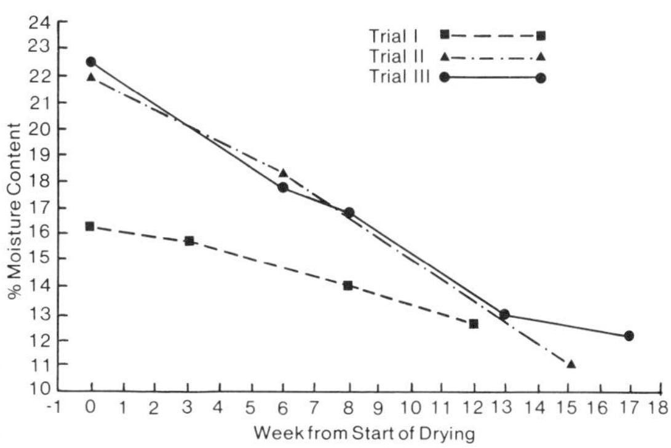


**Fig. 3.** Progress of drying of paddy during three storage/drying trials at Tambun, Indonesia.

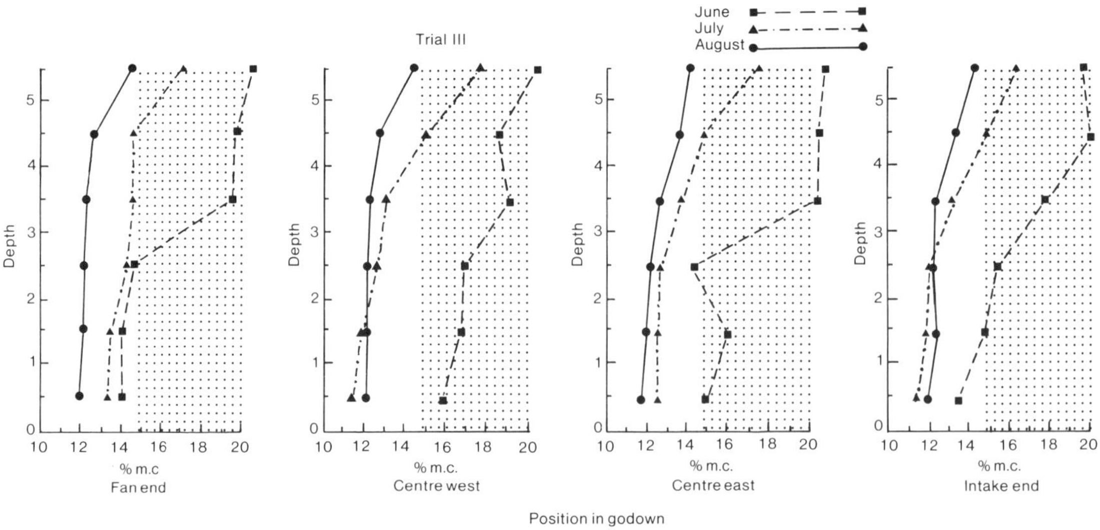


**Fig. 4.** Moisture profiles in bulk stored paddy during one of three storage/drying trials at Tambun, Indonesia.

June and July. This led to serious overdrying of the lower layers, ranging from 10.9% to 11.9% moisture content. In October 1981 (Trial I), the fan was run for 5 days continuously. The average RH of the ventilating air was 71.5%. As a result, the top layer dried from 14.1% to 12.6% m.c. and there was negligible change in the lower layers. This result was taken to show the feasibility of a management program to ventilate continuously. Later drying performances in Trials II and III confirmed that 24 hour continuous ventilation gave a higher daily rate of evaporation and less overdrying of the lower layers.

There were practical difficulties in persuading operators that it is not necessarily unwise or dangerous to leave reliable engines running unattended and there were also complaints from local residents about noise.

### Assessment of Drying Performance

The original specification prescribed an airflow rate of 0.025 $m^3$/t/sec (50 cfm/t) at 127 mm (5 in.) water gauge fan static pressure. The pressure development was based on Schedd (1953). Whilst it was expected that compaction would have some effect on fan static pressure, the very large increase in airflow resistance, more than 50% above predicted levels, was not anticipated. Consequently there was a considerable reduction in the actual evaporative capacity of the plant. During Trial I the loading rate of wet paddy was less than the actual evaporative capacity of the plant (see Table 1).

During Trials II and III, paddy was loaded into the plant faster than it was dried. Accepting that the design provided for all paddy to be dried to less than 15% m.c. within 10 days of loading, it was considered that increasing this period from 10 to 14 days would constitute a 'permissible overload'. The actual loading rates and degree of overload are given in Table 2.

The permissible overload was exceeded by 26%

**Table 1.** Storage intake rates matched to actual evaporative capacity, during storage/drying trials at Tambun, Indonesia.

| Intake moisture content (%) | Maximum daily intake (t) |
|---|---|
| 18 | 68 |
| 20 | 41 |
| 22 | 29 |
| 24 | 23 |

**Table 2.** Loading rates of paddy into plant, during storage/drying trials at Tambun, Indonesia.

| Trial no. | Intake rate (t/day) | Overload Permissible | Overload Actual | Excess over permissible |
|---|---|---|---|---|
| I | 43 | — | nil | — |
| II | 50 | 19% | 45% | 26% |
| III | 74 | 18% | 60% | 42% |

in Trial II and there was no marked increase in deterioration. In Trial III, where the overload was 42% above permissible levels, there was some marked deterioration of the upper layers due to protracted drying.

The energy consumption at 3500 kJ per kg of water evaporated was high in relation to the range 1000–2000 kJ/kg experienced in temperate climates. Apart from the greater effective drying capacity of a temperate climate, the high power requirement at Tambun was compounded by the depth of grain and its consolidation, causing the fan to operate over less efficient sectors of its performance curves. Without considering building costs, using a centrifugal fan/engine combination as at Tambun, the best overall drying rates are achieved with about 2 m depth of grain.

A computer analysis was made of the ventilation data obtained during the trials, in particular to calculate the quantities of moisture removed from grain, in relation to the quantities of ventilating air and the degree of saturation of the air after passing through the grain. A good correlation was obtained between the theoretical and actual amounts of moisture evaporated. A simple mathematical model was used to simulate the process of drying using trial data (Fig. 5). Close correlation was obtained between calculated and actual performance. It is therefore confidently expected that the model can be used to predict drying performance for similar plants under different climatic conditions and for trials at different depths of grain and rates of airflow.

A later study incorporating this model and including an examination of the climate in

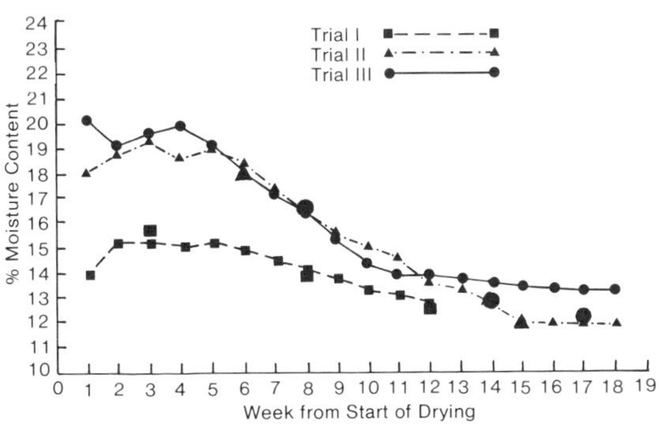


**Fig. 5.** Changes in actual and calculated moisture content during three storage/drying trials with bulk paddy at Tambun, Indonesia.

locations below 200 m in altitude, in particular the mean hourly ambient relative humidity during times of the year that drying is needed, led to the conclusion that it is possible to dry paddy using ambient air with minimal supplementary heat in the most adverse climatic conditions within Indonesia.

## Quality Changes during Storage

An attempt was made to measure dry matter losses occurring in three locations in the upper layers during a protracted period of drying, as occurred in Trials II and III. Losses ranged from 0.4–0.04% with a mean of 0.36%.

The amount of yellow and discoloured grain in the crop is considered an important indication of quality. BULOG's permissible maximum at procurement is 3%. During all three trials, the percentage of yellow and discoloured grains increased during the storage period (Table 3).

**Table 3.** Incidence of yellow and discoloured grains during storage/drying trials at Tambun, Indonesia.

| Depth of layer from surface (m) | Average rate of deterioration (%/month) | | |
|---|---|---|---|
| | Trial I 1981 | Trial II 1982 | Trial III 1983 |
| 0 | 0.62 | 0.50 | 1.53 |
| 1 | 0.56 | 0.49 | 1.30 |
| 2 | 0.54 | 0.21 | 0.91 |
| 3 | 0.52 | 0.30 | 0.96 |
| 4 | 0.48 | 0.39 | 0.70 |
| 5 | 0.38 | 0.39 | 0.92 |
| Average | 0.52 | 0.38 | 1.05 |
| Standard deviation | 0.08 | 0.11 | 0.30 |

In Trial I paddy was loaded into the store with 5.4% yellow grains. Within 4 months the quantity had increased to 9.1% in spite of the paddy being at about 12% m.c. Since it was not of particularly good quality and of unknown history, it was considered that increasing yellowing was a manifestation of damage or spoilage suffered before drying. An effort was made in Trial II to obtain better quality paddy. At intake, yellow and discoloured grains were 2.7%. After 6 months storage it was 5% and after 12 months it had increased further, but only to 12%. This was taken to indicate that once grain had manifested the predrying damage during the first period storage very little further deterioration takes place when the paddy is stored dry and under good conditions.

A linear regression analysis on the deterioration of discrete layers with respect to time showed generally that the layers which remained undried for the longer time deteriorated at a faster rate. In Trial I the surface layer deteriorated about 50% faster than layer 5. In Trial II there was no statistical difference between any of the layers. In Trial III, the top layer, being the longest undried, deteriorated at twice the rate of the bottom layer. However, it would be misleading to assume that all deterioration was the result of delayed drying. Paddy was dried almost immediately in the lower layers but it continued to deteriorate steadily in all three trials, reflecting the fact that the grain had suffered some damage and deterioration before being dried.

A linear regression analysis of monthly milling recovery tests did not give a clear picture. Trials I and III showed a reduction in recovery rate with time, but the correlation was poor. Low levels of recovery in Trial III were thought to be due to faulty laboratory techniques.

## Conclusions

All three trials demonstrated convincingly that drying grain using ambient air in the humid tropics is both possible and practical. As the mean daily RH fell during the dry season the greatest concern was overdrying. It would seem, therefore, that the original design studies underestimated the drying potential of the climate at Tambun.

Increasing intake moisture content effectively increased the evaporative capacity of the plant. The greatest evaporative capacity is achieved by continuous ventilation. It may, however, be necessary to introduce a second phase of selective ventilation to achieve the final moisture content.

Grain at higher moisture contents spoils quickly. Therefore, as intake moisture contents increase, the quantities to be added daily should be reduced. However, if the amount of wet grain loaded onto the floor in excess of drying capacity is small, the risk of deterioration is small. If the loading in excess of drying capacity is continued over a long period, deterioration will become a serious problem. It therefore has to be asked, 'what risks of deterioration are acceptable to management in the interest of least-cost operations?'

It is therefore not possible to define engineering specifications for drying installations solely on the basis of these drying trials. Where drying requirements are precisely known, least-cost designs can be prepared, but they will be less flexible in operation than designs which necessarily incorporate a capacity for contingencies, but which will cost more.

## Postscript

In relation to the present market intervention operations by BULOG, it must be concluded that BULOG's stores are too far removed from the point of harvesting, both in time and distance, to be involved in primary drying. It therefore seems evident it is in BULOG's best interests to maintain its present policy of not becoming involved directly in the procurement of wet paddy but to continue to promote and assist in every way possible the drying of paddy at cooperative and village level. However, it should be understood that whilst this project was carried out in relation to BULOG's operations in rice marketing, the conclusions and recommendations that developed from it can be applied to the drying of other durable commodities in the humid tropics without any substantial adjustment.

## References

Gough, M.C. 1980. Evaluation of a remote moisture sensor in bulk grain. Industrial and Agricultural Engineering Research, 25, 339–344.

Shedd, C.K. 1953. Resistance of grain and seeds to airflow. Agricultural Engineering, 34, 616–619.

# Drying Technologies for Maintaining Grain Quality in the Philippines

**Justin A. Tumambing***

*Abstract*

The Philippines, being a humid tropical country, is beset with the serious problem of maintaining grain quality, particularly during the wet season when grains cannot be sun dried immediately. This is brought about by the prevailing high temperature and humidity and the high moisture content (26–28%) of the harvested grains, all of which favour rapid deterioration in quality.

Mechanical dryers are seen as essential now that the government is increasing its drive for food production and self-sufficiency. Despite the perceived need for dryers, however, adoption has been very low due to some socio-economic factors and unfavourable market characteristics. These are briefly presented.

Existing drying practices in the Philippines, from traditional sun drying to mechanical drying, and their problems and costs, are discussed, as are studies on drying technologies conducted by research institutions in the Philippines.

By using in-store drying in conjunction with high-speed pre-drying, it is hoped to alleviate drying and storage problems. Initial results of a study, from laboratory experiments on rice properties to pilot-plant drying investigations, are outlined. Recommendations for improving drying strategies at farm level, in farmer associations, and at government and private processing complexes are also presented.

IN humid tropical countries such as the Philippines, grain quality deterioration is a serious problem, particularly during the wet season when most of the grains are harvested. The problem arises from the high temperatures and humidities prevailing in the grain-producing areas where grains are harvested at very high moisture content (26–28%). These conditions are conducive to rapid deterioration in quality. In the provinces of Isabela and Iloilo, two of the leading rice producing areas in the Philippines, the months of September to November are peak months of harvest during the wet season. These months have an average monthly rainfall of about 250 mm and an average monthly relative humidity of 84% (Tables 1 and 2). These conditions result in grain being harvested at high moisture contents favouring the growth of microorganisms. Quality deterioration is manifested by grain discoloration, yellowing of paddy and milled rice, decrease in head rice and milling yield, increased number of chalky kernels, high percentage of dry matter loss and, in some cases, formation of aflatoxins. Aflatoxins are carcinogenic substances produced by fungi when exposed to high moisture and temperature.

According to Teter (personal communication) an estimated decrease in quality of rough rice from premium grade to sub-standard grade will occur in 10 days if the grain is held at a moisture content of about 20.8% (wet basis). At higher moisture contents, a more rapid loss in quality will occur.

Present on-farm handling practices exacerbate the situation. Farmers often leave their fresh harvest in either conical or rectangular field stacks for several days before threshing, which is frequently delayed due to lack of threshers or manpower. Likewise, drying is delayed while waiting for sunshine or for mechanical dryers to become available. When drying is not possible, freshly threshed paddy is stored temporarily in bags.

Studies by Mendoza and Quitco (1984) of the National Post Harvest Institute for Research and Extension (NAPHIRE) simulating the delays in threshing and drying at the farm level and the delay in drying at commercial storages revealed significant occurrence of yellow kernels when threshing and drying were delayed.

A delay in threshing and drying of 1–7 days

* National Post Harvest Institute for Research and Extension, 3rd Floor, FTI Administration Building, Taguig, Metro Manila, Philippines.

**Table 1.** Paddy production, average monthly rainfall, relative humidity, temperature, and number of rainy days for Isabela, Philippines during the year 1984.

| | Production (cavans[a]) | Average monthly rainfall (mm) | Average monthly rel. humidity (%) | Average monthly temp. (°C) | Average no. of rainy days |
|---|---|---|---|---|---|
| January | 760 879 | 20.4 | 80 | 24.5 | 6 |
| February | 968 094 | 18.8 | 76 | 25.5 | 4 |
| March | 1 422 117 | 37.4 | 71 | 27.7 | 5 |
| April | 2 576 610 | 54.3 | 68 | 29.5 | 5 |
| May | 119 151 | 103.6 | 69 | 30.5 | 10 |
| June | 51 111 | 192.8 | 75 | 29.9 | 13 |
| July | 36 693 | 211.5 | 75 | 29.1 | 14 |
| August | — | 248.7 | 79 | 29.0 | 15 |
| September | 488 508 | 220.4 | 80 | 28.6 | 15 |
| October | 3 305 060 | 226.3 | 81 | 27.5 | 14 |
| November | 3 384 443 | 280.1 | 84 | 26.1 | 15 |
| December | 454 924 | 105.4 | 84 | 24.9 | 11 |

a. 1 cavan = 50 kg.

**Table 2.** Paddy production, average monthly rainfall, relative humidity, temperature, and number of rainy days for Iloilo, Philippines during the year 1984.

| | Production (cavans[a]) | Average monthly rainfall (mm) | Average monthly rel. humidity (%) | Average monthly temp. (°C) | Average no. of rainy days |
|---|---|---|---|---|---|
| January | 1 005 720 | 42.0 | 83 | 25.7 | 8 |
| February | 2 759 526 | 20.0 | 81 | 25.9 | 6 |
| March | 1 217 450 | 33.7 | 77 | 26.8 | 6 |
| April | 228 327 | 38.6 | 74 | 28.1 | 5 |
| May | 96 496 | 137.7 | 77 | 28.5 | 12 |
| June | 75 472 | 258.7 | 82 | 27.9 | 18 |
| July | 65 472 | 280.1 | 84 | 27.4 | 19 |
| August | 799 140 | 332.8 | 84 | 27.2 | 20 |
| September | 2 412 126 | 242.6 | 84 | 27.2 | 19 |
| October | 2 054 528 | 212.9 | 84 | 27.3 | 17 |
| November | 729 796 | 184.0 | 84 | 26.9 | 15 |
| December | 528 154 | 95.0 | 84 | 26.2 | 12 |

a. 1 cavan = 50 kg.

resulted in 11.7–43.3% yellow kernels for a rectangular pile of 1.0 m × 1.0 m × 1.0 m and 19.7–71% yellow kernels for a solid conical field stack with a base diameter of 3.0 m and a height of 2.5 m. Threshed paddy not immediately dried displayed 1–22% yellow kernels within 1–7 days. These high percentages of yellow kernels could amount in monetary terms to losses of US$0.46–US$1.96 per cavan (50 kg) for a rectangular pile, US$0.82–US$4.39 per cavan for a solid conical field stack, and US$0.44–US$0.82 per cavan (US$ = PHP14.00) for threshed undried paddy (Table 3). In commercial storages, the 2% maximum yellowing for premium grade set by the National Food Authority (NFA) was reached after a one week delay in drying for paddy at 23–25% moisture content and after a one month delay in drying for paddy at 20–22% moisture content.

In addition, the frequency of heavy rainstorms during the wet season means that grains are often rewetted. As a result of rewetting, grains are subjected to moisture stress which eventually results in kernel fissuring and breakage, a sure indication of loss in quality.

With the increasing drive of the government to increase food production and preserve the quality and quantity of food produced, the problem of grain quality deterioration deserves serious and immediate attention. With the increasing volume

**Table 3.** Percentage of yellow kernels, with equivalent loss in monetary value, following holding of wet paddy by various methods.

| No. of days of threshing/ drying delay | Piling method | | | | | | | | | | Delay in drying | |
|---|---|---|---|---|---|---|---|---|---|---|---|---|
| | I[a] | | II[b] | | III[c] | | IV[d] | | V[e] | | | |
| | YK(%)[f] | PHPcav[g] | YK(%) | PHPcav | YK(%) | PHPcav | YK(%) | PHPcav | YK(%) | PHPcav | YK(%) | PHPcav |
| 1 | 11.7 | 6.5 | 19.7 | 11.5 | 8.0 | 6.5 | 0.5 | 0 | 1.6 | 0 | 1.0 | 0 |
| 2 | 20.7 | 11.5 | 35.3 | 22.0 | 18.0 | 11.5 | 1.5 | 0 | 2.6 | 0 | 4.0 | 0 |
| 3 | 28.3 | 17.0 | 50.0 | 27.5 | 24.0 | 11.5 | 2.3 | 0 | 3.0 | 0 | 8.0 | 6.15 |
| 4 | 34.0 | 22.0 | 61.3 | 61.5 | 31.7 | 17.0 | 4.7 | 0 | 3.2 | 0 | 12.0 | 6.15 |
| 5 | 39.0 | 27.5 | 67.0 | 61.5 | 38.0 | 22.0 | 5.7 | 0 | 3.6 | 0 | 17.0 | 11.5 |
| 6 | 40.7 | 27.5 | 70.3 | 61.5 | 41.0 | 27.5 | 6.3 | 6.5 | 3.6 | 0 | 20.0 | 11.5 |
| 7 | 43.3 | 27.5 | 71.7 | 61.5 | 47.3 | 27.5 | 8.3 | 6.5 | 3.6 | 0 | 20.0 | 11.5 |

a. Retangular pile of 1.0 m × 1.0 m dimension.
b. Solid conical field stack with a base diameter of 3.0 m and a height of 2.5 m.
c. Hollow conical field stack with a base diameter of 3.0 m and a height of 2.5 m and an inside diameter of 0.75 m.
d. A cluster of unthreshed paddy consisting of 11 to 15 bundles built into a conical field stack after 3.0 days.
e. Bundled paddy hung on bamboo poles.
f. Yellow kernels.
g. One cavan (cav) is 50 kg.
US$1.00 = PHP14.00.

of grains harvested, the country can ill-afford the considerable penalties of allowing the grain to lose premium grade quality. The serious situation of wet grain handling therefore calls for an appropriate drying technology as the most effective method of preserving grain quality. At the same time, this drying technology must also meet the current emphasis on energy conservation, significant moves towards bulk rather than bag storage, and acceptability to end-users.

## Present Drying Practices and Problems

### Sun Drying

Sun drying is the most common method of drying high moisture paddy. In some areas, planting is timed so that paddy will mature after the rainy season to allow sun drying. Filipinos are inured to sun drying of grain and have the skills to use it to best advantage. Threshed paddy is spread on a flat, smooth surface, on concrete floors, straw mats, or hard soil, at 3 to 5 cm deep, and occasionally stirred (normally every hour) for uniform drying. Some paddy in unthreshed form is pre-dried in stacks in the field.

The temperature of the sun drying floor surface and the paddy may increase to as high as 60 to 70°C. This high, uncontrolled temperature induces severe temperature and moisture gradients within the paddy kernels resulting in a rapid drying rate and, consequently, stress checks and cracks. The resulting 'sun checking' of paddy is a common and too often, unfortunately, an accepted form of loss of rice quality. Sun drying is undoubtedly a low-cost method, but it is labour intensive and generally subjects the paddy to considerable quality deterioration.

### Mechanical Drying

The paddy harvested during the wet season, especially in October and November, is still high in moisture content (26–28%) after threshing. Threshed paddy then needs to be immediately dried. During this period, sun drying cannot be relied upon, so the use of heated air, convection type dryers seems to be necessary. Almost every conceivable type of dryer adopted from the United States, Europe, Japan and locally is available in the Philippines. These dryers range from small (2 t capacity) fixed-bed dryers, to large, expensive and sophisticated continuous-flow dryers intended for centralised operations.

The mechanical drying systems vary according to:

- Fuel used for heating the drying air (oil, liquified petroleum gas, or rice hulls).
- The geometric configuration of the grain-holding device (flat bed or vertical column).
- The type of blower or fan forcing the heated air through the grain mass (centrifugal, axial flow, or propeller type).
- Drying air temperature and time of exposure

of the grain to the drying air stream. Batch dryers use 43°C drying air temperature to dry the grain continuously while continuous-flow dryers use 60°C drying air temperature for 15–30 minutes grain exposure time.

- Provisions for mechanical loading and unloading of the grain (bucket elevators, belt conveyors, oscillating rockers, or metering rollers).
- System of controlling and monitoring drying air temperature, grain moisture, and fire safety controls.
- Provisions for grain cleaning and dust control, and automatic weighing devices (sophistication required in industrial plant dryers).

The following types of dryers are in varying degrees of use in the Philippines:

- The University of the Philippines at Los Baños (UPLB) 2 t flat-bed dryer (rice-hull fed or kerosene fed).
- The International Rice Research Institute (IRRI) 2 t vertical-bin batch dryer (rice-hull fed or kerosene fed).
- The Louisiana State University (LSU) continuous-flow dryer with tempering bins and drying capacities of 1–5 t/hour (rice-hull fed or kerosene fed).
- The Danish Kongskilde vertical batch dryer, 2 t/hour (kerosene fed).
- The Danish Cimbria continuous flow dryer, 3 t/hour (kerosene fed).
- The Woodland continuous flow dryer, 2 t/hour (kerosene fed)
- The Japanese (Satake) batch-type recirculating dryer, 2.4 t/hour (kerosene fed).
- The Kuizon reversible airflow batch dryer, 1–6 t/batch (rice-hull fed).

### Cost of Drying

Drying cost is usually computed as the ratio between the total operating cost of the dryer and its total capacity. It is used as one of the parameters in assessing the economic viability of a particular dryer and is usually expressed on a per tonne, per cavan, or per kilogram basis.

In computing the drying cost, factors to be considered are:

- drying capacity (expressed in t/hour)
- utilisation rate (hours or days/week)
- power and fuel used
- drying efficiency
- handling methods involved.

Studies made in the Philippines on costs of drying have varied in their assumptions, making it difficult to compare the costs of different drying systems.

### Sun Drying

Villaroel and Cardiño (1984) reported that in Camarines Sur, farmers and private millers incur an average sun drying cost of US$2.40 per tonne and US$3.00 per tonne, respectively (US$1 = PHP10.00).

### Batch Dryers

Mangaoang (1984) computed the drying cost of a rice-hull fed, flat-bed dryer to be US$5.80/t (US$1 = PHP7.50). This was less than the drying cost of a kerosene-fed dryer, which was estimated to be US$7.80/t. The computations were based on a 12 hour operating time, using a 6.8 hp gasoline engine.

In 1982, the National Food Authority (NFA) made a comparative study of the above dryers using actual costs. The results, which were based on a 600 hour/year drying operation, estimated the drying cost of a kerosene-fed dryer to be US$18.94/t (US$1.00 = PHP10.00). This was less than the drying cost of a rice-hull fed dryer (US$20.28/t). On the other hand, the Kongskilde, large capacity batch-type dryer (2 t/hour) had an estimated drying cost of US$21.16/t.

### Continuous-Flow Dryer

In 1982, NFA also conducted a study on the drying costs of three makes of continuous-flow dryers: Cimbria, Satake, and Woodland. The results of the study were: Satake (2.4 t/hour), US$16.52/t; Woodland (3 t/hour), US$31.00/t; Cimbria (3 t/hour), US$52.70/t.

Among these three units, the Satake dryer was found to be the least expensive to operate, due to its capacity and high drying rates. The Cimbria dryer had the highest drying cost, because of its very high fixed cost.

### Constraints to the Adoption of Mechanical Dryers

In spite of the recognition of the need for mechanical dryers to ease the drying problem, improvement of existing drying systems and the development of an energy-efficient and cost-effective drying system is still an urgent need. Tumambing (1984a) noted that the rate of adoption of, or investment in mechanical drying systems seems to have been very low, particularly in the private sector which handles about 50% of paddy production.

During the dry season, harvested paddy is commonly sun dried. During the wet season, only about 12% of the harvested paddy is dried by mechanical dryers. Of this, 10.8% is handled by the NFA and only 1.2% by the private millers and traders. At present, the total number and capacity of mechanical dryers available are not sufficient to dry the wet paddy, especially during the peak of the wet season harvest. On the other hand, mechanical dryers are underused during the dry and off-season harvests.

De Padua et al. (1984) summarised the constraints to mechanical drying as follows:

- Small-scale farming and low production volume at the farm level. Mechanical drying is rarely done by individual farmers simply because they do not have the harvest volume that would justify investment in, or use of mechanical dryers. Farmers claim that their low production volume does not permit them to make use of mechanical dryers.
- High fuel cost. While low production is a major constraint faced by small-scale farmers, mechanical drying has been practiced by farmers with large farms, by private millers, and by certain farmers' groups. However, many have stopped the practice due to their claim that rising fuel costs make mechanical drying no longer profitable.
- Lack of technical knowledge on drying. Farmers and millers who lacked understanding of the technical aspects of mechanical drying stopped the practice because, as they noted, mechanical drying gave them poor quality processed paddy. However, the poor quality may have been due to: (i) drying of paddy that was already spoiled; (ii) improper adjustment of temperature, airflow, and other drying components.
- Preference for sun drying over mechanical drying. It was observed that farmers and grain processors tend to make sun drying the basis of comparison in considering investments in mechanical dryers. Features of sun drying, such as simplicity and ease of operation, minimum operating expenses, and low capital costs, are the usual characteristics looked for by most farmers or millers intending to invest in dryers.
- Unavailability of dryers and ignorance of farmers of the existence of such equipment.

De Padua et al. (1984) also noted that, aside from the constraints in mechanical drying caused by factors in the micro-environment surrounding the farmer or grain trader/miller, other restraining factors present in the grain marketing system can be identified. These factors are enumerated as: (i) milled rice quality deficiencies, such as brokens, reduced head rice yields, and kernel discoloration, caused by the drying process but which are given little importance in pricing; (ii) existence of a market for low quality milled rice does not discourage millers from producing an inferior product; and (iii) inadequate price support by the government to cover the cost of drying.

## Drying Studies in the Philippines

### Warehouse-Type Dryer

Halos et al. (1983) describe the warehouse-type dryer using non-conventional sources of energy, developed by Dr Jeon of the International Rice Research Institute (IRRI). The heat source for drying is provided by a centre-tube furnace fuelled with agricultural wastes such as rice hulls, coconut husks, and sawdust. A vortex wind machine is to circulate the drying air inside the warehouse. The dryer (Fig. 1) has a capacity of up to 8 tonnes per batch, making it suitable for cooperative or village-level operation, or by a group of farmers or rice mill operators. It can be used for drying and storing different commodities in separate lots. Six tonnes of paddy can be dried from 20% to 14% moisture content in 6 hours using a drying air temperature of 39–42°C and air suction rate of 9.83 $m^3$/minute per cubic metre of paddy. In terms

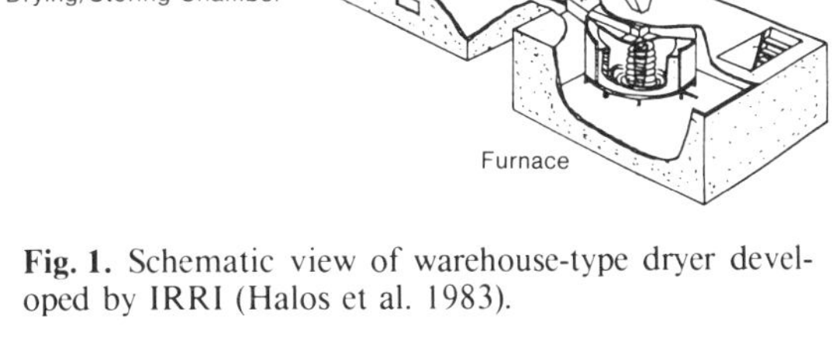


**Fig. 1.** Schematic view of warehouse-type dryer developed by IRRI (Halos et al. 1983).

of milling quality of dried paddy, milling recovery was found to be 67.5% and head rice yield 83.5%. A tonne of paddy dried from 25% to 14% moisture content entails a nominal cost of US$1.34–US$1.41 (US$1.00 = PHP10.00). This dryer will certainly be acceptable as an intermediate technology but application will differ widely due to variable weather conditions and agronomic practices between one area and another.

### African Bush Dryer

Manilay (1984) describes a natural convection, African-type dryer developed by the Silliman University College of Agriculture (SUCA), suitable for drying maize. The dryer (Fig. 2) is made up of indigenous materials and uses firewood as fuel for the oil drum surface and heat exchanger. Air above the drum surface is heated and is induced upward by a removable plywood hood placed above a 2.4 m × 2.4 m × 0.25 m tray with perforated sheet metal flooring. At 25.3 cm depth of grain, the dryer is claimed to dry about a tonne of shelled corn from 23% to 14% moisture content in 5 hours.

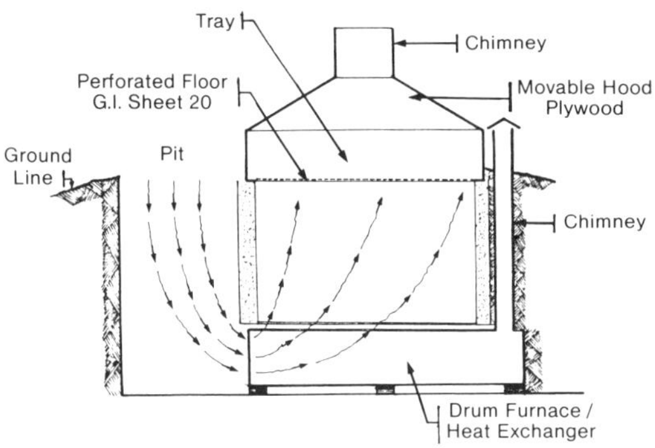


**Fig. 2.** Schematic drawing of the SUCA pit-type maize dryer.

### IRRI Batch-Type Rotary Drum Dryers

Two batch-type rotary drum dryers, one heated directly, the other indirectly, applying the heat conduction principle, were designed by IRRI for drying higher moisture content grain down to 18% (Espanto et al. 1985). Both dryers (Figs 3 and 4) are manually operated and have a capacity of 25 kg per batch. Heat is provided by burning wood or rice-hull charcoal briquettes in a movable stove positioned beneath the drum. A batch of paddy, maize, or peanuts can be partially dried from 28% to 18% moisture content in 54 minutes, 33% to 13% in 80 minutes, and 63% to 42% in 110 minutes, with a grain temperature of 56°C. Thermal efficiency values were calculated to be 18.9% for the direct-heating dryer and 10.2% for the indirect-heating dryer. Potential milling recovery and head rice recovery values of the partially dried paddy were relatively high, but grain viability was relatively low. The dryers, which are still at the pilot stage, have potential but further study is required to ensure their adoption for farm-level pre-drying of grains.

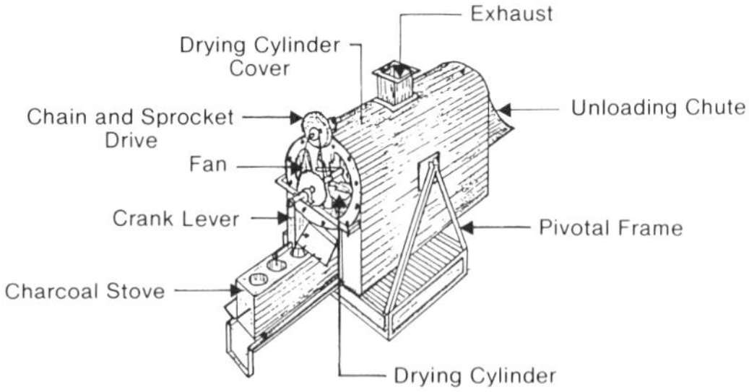


**Fig. 3.** Indirectly heated rotary drum grain dryer.

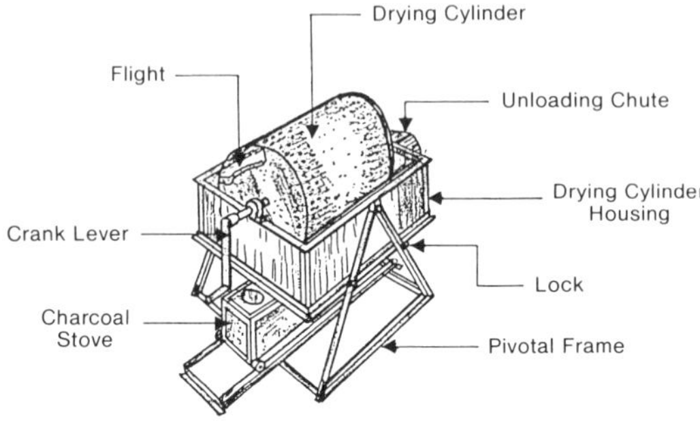


**Fig. 4.** Directly heated rotary drum grain dryer.

### Pre-drying Handling of High Moisture Paddy Using Aeration

Rapusas et al. (1978) report a laboratory investigation on the application of aeration as a means of maintaining the quality of high moisture paddy held in temporary storage when drying is delayed. The experiment was conducted during the dry season using paddy with an initial moisture content of 24–27%. The results of the study revealed that, in the dry season, non-aerated paddy held in sacks maintained its premium grade quality for up to a 3-day delay in drying. Paddy held in sacks and aerated 8 hours a day with an airflow rate of 28 $m^3$/minute/t showed no decline in grade up to 9 days, while paddy in bulk aerated with an airflow rate of 35 $m^3$/minute/t lasted for 14 days without losing its initial grade quality. In the wet season, the non-aerated paddy held in sacks showed a loss in grade within a day of delay in drying. However, paddy held in aerated sacks and in aerated bulk at airflow rates of 28 $m^3$/minute/t and 35 $m^3$/minute/t respectively, maintained its quality up to 2 days in the former case and 3 days in the latter.

Field studies of the pre-drying handling of wet paddy conducted by de Castro et al. (1980) showed that delay in drying wet paddy greatly affected grain quality and milling recovery. Wet paddy with moisture contents of 23 to 26% maintained its milled rice quality within acceptable limits for about one week by aerating at an airflow rate of 6.0 to 35.0 $m^3$/minute/t for 8 hours a day. The milling recovery of aerated wet paddy decreased by about 1% for every 5 days delay before drying. Head rice yield was also found to decline, due to the increasing percentage of damaged and discoloured kernels.

### In-Store Drying of High Moisture Grains in the Philippines

The Philippines, through its food marketing agency, the NFA, has been taking serious steps to shift from bag to bulk storage to accommodate increasing grain production, and at the same time eliminate the substantial cost of jute and plastic bags and reduce the risk of storage losses. About 17 bulk storage units integrated with drying facilities and materials handling equipment are installed throughout the country. The Agroprocessing and Marketing Directorate of NFA intends to instal additional units as part of its project to introduce commercially improved grain post harvest technologies in the Cagayan Valley region. These units vary from 250 t capacity, cylindrical Butler silos to 5000 t capacity, on-floor rectangular steel bins.

However, de Padua (1975) suggested that the successful operation of these silos is very limited, especially in long-term bulk storage. In most bulk storage units, heating, caking, fermentation, and insect infestation occur even when commodities are stored for only three to four months. These problems have been attributed mainly to mismanagement of the grain or the system itself and inherent engineering deficiencies in design or both. Many of the bulk storage systems have no provisions for turning the grain from one silo to another. The aeration systems of some silos perform poorly. Some silos have no aeration system. These variations in design and practice reflect a lack of relevant information, particularly on the physics of bulk storage as applied to humid tropical conditions.

The pressing and severe problems in drying and bulk storage call for a drying technology that will preserve the quality of the grain in an energy-efficient and cost-effective way. The drying technology should provide flexibility in meeting the different drying requirements at the level of both farmer association and government and private processing complex, for both dry and wet seasons.

### Drying Strategy

Considering the high moisture content of the harvested paddy and the unfavourable weather conditions, especially during the wet season, Tumambing (1984) suggested that it is essential that the grain moisture content be brought down quickly to a level low enough to prevent grain deterioration. This consideration, coupled with the current emphasis on grain quality and energy conservation, suggest that grain drying has to be done in several stages, each one within an allowable storage time.

A two-stage or 'combination drying' system, in which a fast-drying stage is followed by unheated or minimally heated air drying in storage, offers potential for meeting the above needs. The fast-drying phase can be done by continuous-flow dryer, a batch dryer, or a flash dryer. The grain would be dried rapidly to a moisture content safe enough (18% wet basis) for final drying to be done in storage. In-store drying would then be accomplished by moving ambient or supplementally heated air slowly through the grain bulk to bring the moisture content down to the safe storage and milling level of 13.5–14% over a period or days or weeks. An in-store drying method which uses little or no heating fuel is currently being used in Australia. It is energy efficient and produces better grain quality as compared with conventional heated-air dryers.

NAPHIRE, on behalf of the Philippine Council for Agriculture and Resources Research and Development (PCARRD) is currently undertaking a research project based on the above drying strategy. The study is being carried out in collaboration with the University of New South Wales (UNSW) and Ricegrowers' Cooperative Mills Ltd (RCM) under the auspices of the

Australian Centre for International Agricultural Research (ACIAR). The project is a pioneering study on in-store drying of high moisture grains in the humid tropics. It seeks to solve the drying and storage problem by continuing the development of this technology and apply it to the drying of grain in bulk storage in the Philippines. Similar projects are under way in Malaysia and Thailand.

The two-stage or 'combination drying' system would provide flexibility in meeting the drying requirements imposed by different seasons and variable weather conditions. When the grain matures early during favourable field-drying weather, especially in the dry season, the fast phase of the drying operation could be partially eliminated and most of the grain dried in storage. On the other hand, the high-speed dryer could be run overtime to handle grain maturing late or during the wet season when there is inclement weather and heavy procurement.

## Theoretical Considerations

Basic data have been gathered on the current situation in the postharvest industry in the Philippines, particularly as regards drying and storage, on drying requirements and energy resources for supplemental heating in drying, and on weather patterns and thermophysical properties of local rice varieties.

Weather records and data on rice properties were used as inputs to the computer simulation drying models developed at the University of New South Wales to investigate the feasibility of drying paddy in bulk storage in the Philippines. The results of the computer simulation suggest that, based on the climate conditions in Isabela, paddy could be dried using ambient air during the dry season. However, some form of supplemental heating would be necessary during the wet season. In practice, supplementary heating would be needed in both seasons to make the drying operation independent of weather conditions.

## Physical Properties of Local Rice Varieties

The basic data on grain properties that are needed for simulating the drying process were determined for six local rice varieties (IR-42, IR-50, IR-52, IR-56, IR-58, and Sinang-Domeng).

Bulk density was measured for grains with 10–22% w.b. moisture content range. Together

**Table 4.** Equilibrium moisture contents[a] of six common Philippine rice varieties at different relative humidities and temperatures.

| Relative humidity (%) | IR-42 | IR-50 | IR-52 | IR-56 | IR-58 | Sinang-Domeng | F-value | Coefficient of variation (%) |
|---|---|---|---|---|---|---|---|---|
| | | | | 21.1°C | | | | |
| 52.9 | 10.32 | 10.78 | 10.20 | 10.02 | 10.61 | 10.84 | 1.94 ns | 3.56 |
| 59.0 | 11.52 | 10.94 | 12.76 | 10.96 | 11.24 | 12.52 | 53.75** | 0.93 |
| 75.4 | 13.76 | 13.00 | 13.59 | 12.98 | 13.15 | 14.06 | 21.28** | 1.26 |
| 88.8 | 16.10 | 15.52 | 16.16 | 15.71 | 15.62 | 16.43 | 3.64* | 2.05 |
| 92.9 | 19.05 | 18.32 | 19.26 | 19.08 | 18.72 | 20.62 | 9.88** | 2.14 |
| | | | | 26.7°C | | | | |
| 27.5 | 7.36 | 7.46 | 7.40 | 7.11 | 7.03 | 7.34 | 9.05** | 1.32 |
| 58.0 | 11.38 | 10.75 | 12.09 | — | — | 11.56 | 9.04** | 3.20 |
| 75.3 | 13.48 | 12.85 | 13.47 | 12.52 | 12.64 | 13.67 | 125.80** | 0.58 |
| 86.6 | 16.06 | 15.12 | 15.59 | 14.85 | 14.93 | 15.86 | 406.55** | 0.26 |
| 91.5 | 18.77 | — | 18.45 | 17.87 | 17.68 | 18.40 | 8.56** | 1.45 |
| | | | | 32.2°C | | | | |
| 25.0 | 6.72 | 7.11 | 6.74 | 6.32 | 6.55 | 6.74 | 1.86 ns | 4.90 |
| 56.0 | 10.66 | 11.19 | 10.93 | 10.99 | 11.23 | 11.05 | 8.07** | 1.12 |
| 75.2 | 13.26 | 12.31 | 13.14 | 12.18 | 12.34 | 12.84 | 2.00** | 3.65 |
| 84.3 | 14.90 | 14.23 | 14.79 | 14.40 | 14.03 | 15.27 | 8.85** | 1.85 |
| 90.1 | 17.29 | 16.56 | 17.22 | 16.81 | 16.97 | 17.85 | 15.58** | 1.15 |

a. % wet-weight basis. Mean of three replications.
ns — not significant at 5% level of significance.
* — significant at 5% level of significance.
** — significant at 1% level of significance.
Missing data are not presented due to inconsistency of results.

with the grain density, the bulk density was used to calculate grain porosity. At lower moisture contents of 10–12%, bulk densities ranged from 539 kg/m³ (Sinang-Domeng) to 580 kg/m³ (IR-56), grain densities from 1083 kg/m³ (IR-56) to 1130 kg/m³ (Sinang-Domeng) and, hence, porosities from 0.465 (IR-56) to 0.523 (Sinang-Domeng).

Equilibrium moisture content (EMC) of the local rice varieties (equilibrated over super-saturated salt solutions) was measured at three temperatures (21.1°C, 26.7°C, 32.2°C) and five relative humidities (25, 60, 75, 85, 90% RH). The data (Table 4) revealed that at 75% RH, paddy moisture content ranged from 12.98 to 14.06% at 21°C, from 12.52 to 13.67% at 26.7°C, and from 12.18 to 13.26% at 32.2°C. These results indicate that, based on the recommended 14% final grain moisture content after drying, some rice varieties may be exposed to drying air with relative humidity slightly higher than 75%. This is particularly true in the case of ambient air drying.

**Table 5.** Summary of the results of low temperature in-store grain drying experiment conducted during January to February and April to June 1985 in the Philippines.

| | | January to Feb. 1985 | | | April to June 1985 | | | | | |
|---|---|---|---|---|---|---|---|---|---|---|
| | | RUN 1[a] | RUN 2[a] | RUN 3[a] | RUN 1[a] | RUN 2[a] | RUN 3[a] | RUN 4[a] | RUN 5[a] | RUN 6[b] |
| VARIETY | | IR-62 | IR-42 | IR-50 | IR-56 | IR-56 | IR-60 | IR-58 | IR-90 | IR-75 |
| Initial moisture content | | 21.20 | 20.95 | 22.54 | 23.00 | 16.78 | 17.64 | 17.28 | 25.20 | 29.16 |
| Final moisture content | % wet basis | | | | | | | | | |
| Average | % wet basis | 13.90 | 11.97 | 12.95 | 11.95 | 13.19 | 11.09 | 10.70 | 10.69 | 13.05 |
| Minimum | % wet basis | 13.00 | 10.95 | 11.80 | 10.80 | 12.60 | 9.80 | 9.65 | 9.88 | 14.95 |
| Maximum | % wet basis | 14.80 | 13.60 | 14.40 | 14.95 | 14.45 | 13.60 | 13.10 | 13.74 | 14.95 |
| Ambient air temperature | | | | | | | | | | |
| Average | °C | 26.1 | 26.2 | 27.4 | 29.4 | 28.1 | 28.1 | 31.5 | 30.7 | 26.6 |
| Minimum | °C | 21.8 | 21.9 | 19.4 | 25.5 | 24.4 | 25.3 | 27.8 | 26.7 | 23.6 |
| Maximum | °C | 30.2 | 31.7 | 33.3 | 33.9 | 32.2 | 34.4 | 35.0 | 34.9 | 31.7 |
| Plenum air temperature | | | | | | | | | | |
| Average | °C | 28.6 | 29.4 | 32.8 | 36.11 | 33.9 | 33.9 | 37.22 | 35.0 | 31.6 |
| Minimum | °C | 25.0 | 25.0 | 27.2 | 32.2 | 30.0 | 32.8 | 32.2 | 30.7 | 28.5 |
| Maximum | °C | 32.1 | 36.9 | 38.0 | 38.9 | 37.8 | 39.4 | 42.2 | 39.4 | 36.9 |
| Ambient relative humidity | | | | | | | | | | |
| Average | % | 72.0 | 77.3 | 71.8 | 70.1 | 79.8 | 68.2 | 67.9 | 66.0 | 84.8 |
| Minimum | % | 56.5 | 54.0 | 59.0 | 55.0 | 65.0 | 57.5 | 54.0 | 42.6 | 59.8 |
| Maximum | % | 86.5 | 96.0 | 93.0 | 97.0 | 91.0 | 81.0 | 84.0 | 86.0 | 97.0 |
| Plenum relative humidity | | | | | | | | | | |
| Average | % | 62.5 | 63.9 | 52.9 | 53.1 | 58.0 | 49.4 | 50.6 | 51.8 | 62.9 |
| Minimum | % | 49.8 | 39.0 | 35.0 | 42.0 | 50.0 | 42.5 | 36 | 33.6 | 44 |
| Maximum | % | 73.5 | 80.0 | 71.0 | 62.0 | 68.0 | 55.0 | 62.5 | 69.0 | 70.4 |
| Bed depth | m | 2.15 | 2.30 | 2.10 | 2.20 | 2.20 | 2.15 | 2.15 | 2.20 | 2.00 |
| Bulk density | kg/m³ | 599.3 | 566.6 | 606.6 | 592.1 | 578.7 | 607.1 | 592.2 | 578.8 | 605.0 |
| Airflow rate | m³/minute/m³ paddy | 9.0 | 5.4 | 6.0 | 3.6 | 3.6 | 3.7 | 1.9 | 3.6 | 4.0 |
| Air velocity | m/minute | 17.7 | 12.5 | 12.5 | 8.0 | 8.0 | 8.0 | 4.0 | 8.0 | 8.0 |
| Plenum static pressure | Pa (in $H_2O$) | 1868 (7.5) | 897 (3.6) | 897 (2.5) | 623 (2.5) | 623 (2.5) | 623 (2.5) | 311 (1.25) | 623 (2.5) | 623 (2.5) |
| Mass of wet grain | kg | 1000 | 1000 | 1000 | 1023 | 934 | 970 | 961 | 1000 | 950 |
| Mass of dried grain | kg | 880 | 869 | 899 | 888 | 891 | 904 | 898 | 850 | 836 |
| Water removed | kg | 120 | 131 | 101 | 135 | 43 | 66 | 63 | 150 | 114 |
| Drying period | | Jan. 19–25 | Jan. 27–Feb. 4 | Feb. 9–15 | April 11–17 | April 22–28 | May 8–12 | May 16–22 | June 7–17 | June 23–29 |
| Fan operating time | hours | 65 | 107 | 87 | 110.17 | 79.83 | 59.33 | 86.59 | 109.78 | 158.00 |
| Drying rate | kg $H_2O$/hour | 1.85 | 1.22 | 1.15 | 1.22 | 0.54 | 1.10 | 0.73 | 1.37 | 0.72 |
| Electric power requirement[c] | kW | 0.91 | 0.30 | 0.30 | 0.14 | 0.14 | 0.14 | 0.04 | 0.14 | 0.14 |
| Total electric energy consumption | kW/h | 59.15 | 32.10 | 26.40 | 15.42 | 11.18 | 8.31 | 3.46 | 15.37 | 22.12 |
| Specific energy requirement | kW/h/kg $H_2O$ | 0.49 | 0.24 | 0.25 | 0.11 | 0.26 | 0.13 | 0.05 | 0.10 | 0.19 |
| Specific energy costs[d] | PHP/ton | 118.30 | 64.20 | 52.8 | 30.15 | 23.94 | 17.13 | 7.20 | 30.74 | 46.57 |
| | PHP/cavan | 5.92 | 3.21 | 2.64 | 1.51 | 1.20 | 0.68 | 0.36 | 1.54 | 2.34 |

a. Intermittent drying operation (9:00 AM – 11:00 PM).
b. Continuous drying operation.
c. Efficiency of the fan n = 0.477.
d. Based on PHP2.00 per kW/h.

Sinang-Domeng, IR-42, and IR-52 varieties had relatively higher EMC values, probably because these varieties are more glutinous.

## Pilot Bin Drying Studies

The pilot plant dryer found most suitable for effective and efficient drying of grains under humid tropical conditions such as occur in the Philippines is a two-stage 'combination drying' system (Tumambing 1985). The dryer comprises two drying bins. The first bin, for high-speed drying, has a base area of 0.78 $m^2$ and is filled with grain to a depth of 0.36 m. It is designed to dry 30% moisture content grain at 26°C and 72% RH to less than 18% m.c. in no more than 20 hours. The second bin is an in-store drying bin with a base area of 0.78 $m^2$ filled to a depth of 2.2 m (1 t) designed to dry grains with 18% moisture content at 28°C and 68% RH to 14% m.c. in 21 days. Each bin is equipped with a heater, centrifugal blower, humidity and temperature sensors, and computerised controller.

**Table 6.** Summary of the results of April to June 1985 high-speed grain drying experiment in the Philippines.

| | RUN 2[a] | RUN 3[a] | RUN 4[b] |
|---|---|---|---|
| VARIETY | IR-56 | IR-50 | IR-58 |
| Initial moisture content % wet basis | 22.27 | 22.09 | 20.47 |
| Final moisture content % wet basis | | | |
| Average | 16.78 | 17.64 | 17.28 |
| Minimum | 15.60 | 16.80 | 16.70 |
| Maximum | 17.80 | 18.90 | 17.75 |
| Ambient air temperature °C | | | |
| Average | 31.2 | 30.4 | 32.8 |
| Minimum | 27.8 | 26.1 | 29.7 |
| Maximum | 33.9 | 33.3 | 35.0 |
| Plenum air temperature °C | | | |
| Average | 34.3 | 33.2 | 43.7 |
| Minimum | 28.9 | 27.2 | 35.5 |
| Maximum | 33.9 | 33.3 | 35.0 |
| Ambient relative humidity % | | | |
| Average | 68.0 | 74.1 | 66.1 |
| Minimum | 57.0 | 59.0 | 58.0 |
| Maximum | 90.0 | 88.0 | 77.5 |
| Plenum relative humidity % | | | |
| Average | 57.6 | 63.7 | 37.9 |
| Minimum | 49.0 | 52.0 | 28.0 |
| Maximum | 79.0 | 74.0 | 55.0 |
| Bed depth m | 0.36 | 0.36 | 0.36 |
| Airflow rate $m^3$/minute/$m^3$ paddy | 83.3 | 83.3 | 83.3 |
| Air velocity m/minute | 30.0 | 30.0 | 30.0 |
| Plenum static pressure Pa (in $H_2O$) | 623 (2.5) | 623 (2.5) | 623 (2.5) |
| Total weight of grain kg | 1000 | 1025 | 1000 |
| No. of batches | 7 | 7 | 7 |
| Weight per batch kg | 142.9 | 146.4 | 142.9 |
| Total weight of $H_2O$ removed | 56 | 55 | 39 |
| Drying period | April 19–20 | May 6–7 | May 14–15 |
| Total fan operating time hours | 14.50 | 13.75 | 6.00 |
| Fan hours per batch | 2.10 | 1.96 | 0.68 |
| Drying rate kg $H_2O$/hour | 4.55 | 4.00 | 6.50 |
| Electric power requirement[c] kW | 0.51 | 0.51 | 5.51 |
| Total electric energy consumption kW hour | 7.40 | 7.01 | 33.0 |
| Specific energy requirement kW hour/kg $H_2O$ | 0.11 | 0.13 | 0.85 |
| Specific energy cost[d] PHP/cavan[e] | 0.74 | 0.70 | 3.30 |

[a] Ambient air drying with supplemental heat from fan compression.
[b] With 5 kW electric heater.
[c] Efficiency of fan n = 0.477.
[d] Based on PHP2.00 per kW hour.
[e] 1 cavan = 50 kg.

A pilot plant drying experiment was conducted during the periods January to February and April to June (dry season) 1985 in Isabela, Philippines, to determine the proper combination of airflow rate and initial moisture content at a given grain depth and ambient air condition. Results of the experiment were very promising (Tables 5–7). The use of ambient air with supplemental heat from fan compression was successful as a drying strategy. One tonne of paddy 2.0–2.3 m deep was dried safely from an initial moisture content of 18–25% to about 14% in 60–158 hours, equivalent to a 5–12 day drying period. Optimum airflow rates were found to be 1.9 $m^3$/minute/$m^3$ paddy (4 m/minute equivalent air velocity) for paddy with 18% initial moisture content, and 3.6 $m^3$/minute/$m^3$ (8 m/minute equivalent air velocity) for paddy with 23–25% initial moisture content. Specific energy costs for optimum airflow rates were US$0.39–US$1.63 per tonne or US$0.02–US$0.08 per cavan (US$1.00 = PHP18.50).

Quality analysis of in-store dried paddy samples indicated no significant difference between them and naturally dried controls in terms of brown rice recovery, milled rice recovery, and germination rate, while quality was slightly improved in terms of crack ratio and head rice yield and recovery. Head rice yield was in the range of 45.7–59.4% of paddy weight, equivalent to a head rice recovery of 70.1–89.3% of milled rice weight. Lower head rice yield was a result of the high percentage of immature rice present. Crack ratio varied from 0.0025 to 0.22, depending upon the history of the grain. The highest value of yellow rice obtained was 1.95%, below the tolerable limit of 2.3% for premium grade rice. Dry matter loss varied from 0.05–0.87%, with the long varieties having higher

**Table 7.** Paddy condition after the January to February and April to June 1985 low temperature in-store grain drying tests in the Philippines, compared with a naturally dried control sample.

| TEST MONTH | Expt. run | Variety | Brown rice recovery (%) | | Milled rice recovery (%) | | Head rice recovery (%) (head rice yield %) | | Broken rice (%) | | Immature rice (%) | | Yellow[a] rice (%) | | Crack ratio | | DML (%) | Germination rate (%) | |
|---|---|---|---|---|---|---|---|---|---|---|---|---|---|---|---|---|---|---|---|
| | No. | | C | ISD | C | ISD | C | ISD | C | ISD | C | ISD | C | ISD | C | ISD | ISD | C | ISD |
| January to February 1985 | 1 | IR-62[b] | 78.6 | 79.5 | 70.7 | 67.5 | 83.2 (58.8) | 88.0 (59.4) | 16.8 | 12.0 | 3.0 | 3.15 | 1.90 | 1.95 | 0.02 | 0.02 | 0.26 | — | — |
| | 2 | IR-42[b] | 74.8 | 75.6 | 62.2 | 64.9 | 89.0 (55.4) | 89.3 (58.0) | 11.0 | 10.7 | 11.3 | 13.7 | 0.50 | 0.45 | 0.099 | 0.094 | 0.27 | — | — |
| | 3 | IR-50[c] | 76.4 | 76.8 | 66.9 | 65.4 | 86.9 (58.1) | 84.2 (55.1) | 13.1 | 15.8 | 10.4 | 11.95 | 1.10 | 1.05 | 0.103 | 0.065 | 0.87 | — | — |
| April to June 1985 | 1 | IR-56[b] | 76.9 | 77.3 | 67.2 | 66.8 | 78 (52.4) | 79.7 (53.2) | 22.0 | 20.3 | 17.9 | 19.1 | 2.40 | 1.00 | 0.082 | 0.162 | 0.26 | 95 | 97 |
| | 2 | IR-56[b] | 75.5 | 75.4 | 64.7 | 64.4 | 68.6 (44.4) | 70.9 (45.7) | 31.4 | 29.1 | 23.2 | 22.9 | 0.32 | 0.25 | 0.089 | 0.091 | 0.05 | 98 | 98 |
| | 3 | IR-60[c] | 76.9 | 76.8 | 65.7 | 68.6 | 76.2 (50.1) | 77.2 (53.0) | 23.8 | 22.8 | 31.5 | 31.7 | 0.38 | 0.32 | 0.237 | 0.221 | 0.10 | 91 | 95 |
| | 4 | IR-58[c] | 75.8 | 76.4 | 66.2 | 67.1 | 85.6 (56.7) | 86.2 (57.8) | 14.4 | 13.8 | 13.6 | 12.7 | 1.30 | 1.02 | 0.156 | 0.149 | 0.12 | 95 | 95 |
| | 5 | IR-90[c] | 72.8 | 73.4 | 63.5 | 64.2 | 79.4 (50.4) | 85.5 (54.9) | 20.6 | 14.5 | 14.2 | 16.06 | 2.14 | 2.80 | 0.016 | 0.015 | 0.63 | — | — |
| | 6 | IR-75[c] | 72.8 | 73.7 | 62.5 | 63.8 | 68.6 (42.9) | 83.1 (53.0) | 31.4 | 16.9 | 30.5 | 29.4 | 0.90 | 0.90 | 0.011 | 0.002 | 0.51 | — | — |

[a] No yellow rice was observed during April to June 1985 experiment.
[b] Medium grain.
[c] Long grain.
C — Naturally dried control sample.
ISD — In-store dried sample.

values. No visible mould growth was found in the dried samples.

The results of the experiment also showed that intermittent operation of the drying system is more energy efficient than continuous drying, provided that the drying time is within the allowable storage time of the grain. The energy needed to drive the fan and the energy costs are substantially less than for conventional heated-air dryers.

**Conclusions and Recommendations**

The need to dry high moisture paddy immediately after harvest, particularly during the wet season, necessitates the use of mechanical dryers to preserve grain quality. Many drying technologies are available but widespread adoption has yet to be realised due to some socio-economic factors and unfavourable market characteristics. In spite of the drying technologies available, it seems that there is a need for a dryer that would provide the flexibility to meet the different drying requirements for both wet and dry seasons. At the same time, the drying technology should be energy efficient, cost effective, and produce highest quality milled rice.

In-store drying, complemented by pre-drying using a high-speed dryer, has great potential to solve the drying and storage problem. In line with the government's move to shift from bag to bulk storage, however, successful operation of this drying strategy will be limited to commercial-scale drying, i.e. drying at private and government processing complexes, and will demand a well-designed structure and capable technical operators. Aside from that, a premium should be paid for high-quality milled rice.

The most recent detailed recommendations for improving grain drying in ASEAN countries were those presented at a 1984 workshop on wet grain handling sponsored by the ASEAN Crops Post-Harvest Programme (Frio and Manilay 1984). It is useful to restate them here:

1. The constraints to mechanical drying at the farm level are linked to the sociological and economic conditions prevailing in the marketing system. Activities should be geared more towards the improvement of traditional drying practices such as sun drying, natural air drying, drying using inexpensive fuel, and on-farm partial drying.
2. Farmers, by joining together in associations, could afford the use and/or purchase of mechanical dryers. However, the success of this approach depends upon the willingness of farmers to organise themselves in this way. The Irrigators' Association in Isabela, Philippines is considered as a promising group for the adoption of grain postharvest technology, particularly dryers.
3. Pre-drying and holding of wet grains, which pose the most serious problems, could be handled by both the government complexes (NFA) and the private millers. The latter could function as 'satellite drying centres', where the grain is partially dried to 18% moisture content to safely hold it before final drying and storage.
4. In the government complexes, pre-drying and holding of the grain could be done by the following systems:
   - use of bulk-aerated bins, in which may be pre-dried, as well as tempering/holding bins before the final drying process
   - aeration of plenum containing bag stacks
   - aeration of bulk paddy contained within walls of bagged paddy
   - flash drying using high airflows and temperatures with very short exposure and residence times of the grain mass
   - vortex wind pump machine as developed by IRRI, using biomass fuel or solar energy in combination with biomass.
5. Private millers should be encouraged to do more drying in their plants:
   - by making pre-drying technology available
   - by creating adequate incentives through pricing policy. Subsidies may be provided for purchases of drying units for partial drying. Cracked grain content should be penalised in the incentive pricing scheme to encourage investment in mechanical dryers
   - by making millers more aware of their social responsibility in preserving good quality grain for the country.
6. Research and development efforts should focus on:
   - development of appropriate partial drying technology both at goverment complexes and private mills
   - provision of mechanical drying incentive scheme for incorporation in the pricing and grading system
   - a socio-economic study of the marketing system of millers, traders and farmers, and their practices, relationships, etc. as information to influence grain marketing policies.

## References


De Castro, R.E., Andales, S.C., and de Padua, D.B. 1980. Field trials of pre-drying handling systems. In: Proceedings of the 3rd Annual Workshop on Grains Post Harvest Technology. College, Laguna, Philippines, Southeast Asia Cooperative Post Harvest Research and Development Programme, 126–145.

de Padua, D.B., Frio, A.S., and Manilay, A.A. 1984. Coming to grips with the wet season grain handling problems in Southeast Asia. Presented at the FAO/UNDP regional workshop on rice post harvest technology. July 1984. Cabanatuan, Nueva Ecija and Santiago, Isabela, Philippines.

de Padua, D.B. 1975. Rice post-production technology in the Philippines. Supplementary paper No. 2, Second Agricultural Policy Conference, UPLB Center for Policy and Development Studies. July 1975, Los Baños, Laguna, Philippines.

Espanto, I.H., Andales, S.C., Belonio, A.T., and Jeon, Y.W. 1985. Performance evaluation of the IRRI batch-type rotary drum dryers. Presented at the Technical Seminar on Post Harvest Technology, ASEAN Crops Post Harvest Programme. August 1985, Manila, Philippines.

Frio, A.S., and Manilay, A.A., ed. 1984. Coming to grips with the wet season grain handling problems in Southeast Asia. Manila, ASEAN Crops Post Harvest Programme, 93p.

Halos, L.S., Jeon, Y.W., and Bockhop, C.W. 1983. Design and performance of a multi-purpose dryer using non-conventional energy sources. Presented at the 33rd Annual Convention of the Philippines Society of Agricultural Engineers. April 1983, VISCA, Baybay, Leyte.

Mangaoang, C.C. 1984. Cost of drying. In: Frio, A.S., and Manilay, A.A., ed., Coming to grips with the wet season grain handling problems in Southeast Asia. Manila, ASEAN Crops Post Harvest Programme, 65–74.

Manilay, A.A. 1984. The African dryer as a dryer for maize. Post Harvest Quarterly, 8(4), 5–6.

Mendoza, E., and Quitco, R. 1984. Quality deterioration and value loss in grain. In: Frio, A.S., and Manilay, A.A., ed., Coming to grips with the wet season grain handling problems in Southeast Asia. Manila, ASEAN Crops Post Harvest Programme, 15–20.

Rapusas, R.S., de Padua, D.B., and Lozada, E.P. 1978. Pre-drying handling of high moisture paddy. In: Proceedings of the 1st Annual Workshop on Grains Post Harvest Technology. College, Laguna, Southeast Asia Cooperative Post Harvest Research and Development Programme, 243–259.

Tumambing, J.A. 1984a. Rice post production practices in the Philippines. Manila, National Post Harvest Institute for Research and Extension, 45 p.

Tumambing, J.A. 1984b. Progress report No. 2. ACIAR Project 8308 on in-store drying of high moisture grains in tropical climate. Manila, National Post Harvest Institute for Research and Extension, 13 p.

Tumambing, J.A. 1985. July 1984–June 1985 annual report for ACIAR Project 8308 on in-store drying of high moisture grains in tropical climate. Manila, National Post Harvest Institute for Research and Extension, 68 p.

Villaroel, R.F., and Cardiño, A.C. 1984. A socio-economic study on the utilization of mechanical dryers. In: Frio, A.S., and Manilay, A.A., ed., 1984. Coming to grips with the wet season grain handling problems in Southeast Asia. Manila, ASEAN Crops Post Harvest Programme, 54–64.

# Paddy Drying in Thailand

**Somchart Soponronnarit***

*Abstract*

This paper describes the postharvest practices of Thai farmers. It also reviews previous work on on-farm grain dryers and drying with unheated air in Thailand.

A study was conducted on an integrated paddy drying and storage solar hut installed at a farmer's house in the central region of Thailand. It was found that the unit was technically and socio-economically feasible. It was suitable for drying 15 t of paddy per month and for storing 10 t.

Drying in bulk storage of high moisture paddy was studied under Bangkok's climatic conditions. It was technically feasible. However, two-stage drying was recommended, i.e., fast drying in the first stage until the moisture content is less than 20% (wet-basis), followed by drying in bulk storage with unheated air at a relatively low airflow rate.

RICE is the most important crop grown in Thailand. During 1983-84, 19.55 Mt of paddy were produced at an average yield of 2.04 t/ha (Anon. 1984).

The present cultivation techniques of Thai farmers are very labour intensive. The size of farms is relatively small. In the northeast, farm size is generally between 1.6–4.8 ha, while in the north it averages less than 3.2 ha. Farm sizes in the south and central regions are approximately 1.6–3.2 and 3.2–4.8 ha, respectively. There are few farms over 4.8 ha (Anon. 1976).

Rice may be grown once or twice a year depending upon the availability of water. The year crop is harvested in the dry season, while the second crop is harvested in the rainy season. The planting of a second crop has become more common, due to the expansion of irrigation systems. However, farmers have been faced with spoilage of the second crop when moisture cannot be reduced fast enough by field drying during the rainy season.

This paper reviews current postharvest practices in Thailand, particularly as regards the need for grain drying. Previous investigations on various types of grain dryers are outlined. The paper then gives the results of studies on an integrated drying and storage hut installed at a farmer's house and on laboratory studies of in-store drying and storage of paddy. The former concerns small-scale drying and storage of paddy, while the latter concentrates on the technical feasibility of in-store drying of bulk paddy in humid, tropical climates.

* School of Energy and Materials, King Mongkut's Institute of Technology, Thonburi Campus, Sukswas 48 Road, Bangkok 10140, Thailand.

## Methods of Harvesting, Drying, Threshing and Storage

Most harvesting is done by hand, using a sickle, and it is often necessary to employ workers from other areas. For the year crop, the plants (with a grain moisture content of 20–22%, wet basis) are cut and left in the fields for a few days to reduce the grain moisture content by sun drying. The rice stems are then bundled and transferred to the threshing floor either by hand or using carts, tractors, or boats.

A different method is used for harvesting the second crop. Small bundles of cut plants are placed on the stubble with the panicles directed downward. Using this method, the grain will not hold water when there is rain. The bundles are transferred to the threshing floor when they have dried sufficiently. The harvest period is 2–14 days or more, most Thai farmers spending more than 14 days on harvesting.

Threshing the paddy by beating the cut plants on logs is quite common in the northeast of Thailand. Treading on the panicles, or driving cattle or farm tractors over the cut plants, is the usual threshing method in the central and the southern regions.

Engine-driven drum threshers are now relatively popular.

There are two methods used for cleaning the threshed paddy. In the north and northeast, winnowing baskets are usually employed. The other method of cleaning is by winnowing machine. These have been extensively used in many areas, especially in the central region. The machines may be hand or engine powered.

Paddy is stored for three main purposes: family consumption, sale, and seed for planting. Storage barns may be permanent buildings constructed from wood, with the floors elevated above the ground level to avoid flooding. Paddy is loaded and unloaded via doors. Semi-permanent barns are made from bamboo-matting walls covered with a mixture of soil and cattle dung. Each type is roofed to protect the contents from the sun and rain. There is no ventilation during storage. Some farmers do not store paddy either because they lack storage facilities or from a need to sell their crop immediately to obtain cash.

The storage period of paddy depends on its purpose. Those who grow rice for sale store paddy for a few months. For consumption and planting, paddy is stored for one year or perhaps longer. The main problems encountered during storage include invasion by birds, insects and rodents, yellowing of grain kernels, moulding, and poor germination because of rain. These causes of loss apply to both the year and second crop, but are more severe for the latter.

Some rice mills purchase wet paddy from the farmer, particularly during the rainy season. In this case, sun drying on a concrete apron is usually practiced. A few rice millers use high speed mechanical dryers.

## Need of Grain Dryers and Comparison of Heat Sources

As noted in the previous section, cut plants are left in the field to sun dry. During the dry season, paddy is always safely dried in a few days by this method. However, it will not dry quickly enough during the rainy season. This leads to a loss in paddy quality, in terms of reduced head rice yield, yellowing, and moulding. This is a serious problem facing Thai farmers. To ameliorate it other methods of drying must be introduced. An attractive possibility is a mechanical dryer. This is composed of a bin as a drying chamber, a blower driven by an internal combustion engine or an electric motor, and a heat source such as electricity, gas, kerosene, diesel oil, agricultural waste, or solar energy.

An electrical heat source is quite simple, but operating costs are high. Gas and diesel oil are less expensive than electricity but control of heating is more complicated. Agricultural wastes, such as rice hulls, may be the most practical if they are available on site so that there are no transportation costs involved. Solar energy is freely available, but the solar collector is always expensive. If a collector cheap enough for farmers could be designed, solar-powered drying may be feasible.

Successful drying can be achieved by using either a high airflow rate and high drying temperatures, or a low airflow rate and near ambient drying temperatures. The latter may be called in-store drying and is interesting because of its low drying cost and the high milling quality of the product, particularly in the case of paddy.

## Review of Previous Work on Grain Drying

Research and development studies concerned with on-farm grain drying have been conducted in Thailand for several years. Most of the dryers developed are of the batch type with drying capacities of 1–2 t and batch drying times of a few hours to a few days.

Srihavong (1978) studied hot air drying of paddy at 43, 49, and 60°C. The grain bed was 0.9 m deep and the airflow rates were 6.7, 13.3, 20, and 26.7 $m^3/min/m^3$ of grain. It was concluded that the drying temperature should not exceed 49°C to maintain high milling quality. To avoid high moisture gradients, the bed depth should not exceed 0.6 m.

A 2 t capacity farm grain dryer (fixed, flat-bed) was developed by the Ministry of Agriculture and Co-operatives, Thailand (Anon. 1977). Grain was dried by hot air at temperatures from 43–48°C. Heat was supplied by diesel oil or by burning rice hulls. When 1 t of paddy was dried with an airflow rate of 107 $m^3/min$, the drying rate was about 1.92% per hour. Fuel consumption was estimated to be 0.7 and 1.07 litres/tonne of paddy/1% (wet-basis) of moisture reduction, for a 6 kW diesel engine and a burner, respectively.

The International Rice Research Institute has developed farm grain dryers having fixed flat beds and fixed vertical beds (IRRI 1978). A 457 mm diameter tube, axial fan driven by a 1.5 kW electric motor or 2.3 kW internal combustion engine is used for a 1 t capacity fixed flat-bed dryer. For IRRI's 2 t capacity fixed vertical-bed dryer, a 533 mm diameter tube, axial fan driven by a 2.2 kW electric motor or a 3.7 kW internal combustion engine is employed. Air may be heated by burning kerosene or rice hulls. Their performance is similar to that of the dryers designed by Ministry of Agriculture and Co-operatives, Thailand (1977).

A low cost, solar rice dryer was developed by Exell and Kornsakoo (1977). The dryer is made of bamboo poles covered by a transparent plastic sheet. Air is naturally circulated above an absorber made of burnt rice hulls through the paddy layer, the optimum thickness of which is determined to be approximately 100 mm. It was able to dry about one tonne of paddy per sunny day and the same amount every three or four days during the rainy season.

All of the above dryers were implemented in rural areas but none was accepted by Thai farmers. For the case of the dryer designed by Exell and Kornsakoo (1977), Amyot and Sirisambhand (1982) pointed out that its non-acceptance was due to the following socio-economic factors.

1. Paddy is harvested as quickly as possible. As a result, the dryer can not keep up with the supply of wet grain. In order to use the dryer effectively, the farmer would have to stagger his harvest over a longer period. However, it is not possible to do this, because of unavailability of irrigation water during the planting season and the difficulty of finding agricultural workers at harvest time.
2. Rice traders and millers are usually uninterested in buying paddy in small batches.
3. The price received by the farmers for the paddy dried by solar dryer is not higher than that received for field-dried paddy.
4. The dryer is too flimsy. The plastic sheeting is constantly being torn by village dogs and children, or it simply disintegrates from exposure to the sun after a few months and has to be replaced.

Lhoste and Louis (1985) conducted a survey in Nakornpathom province. They found that some farmers stagger their planting and harvesting. It was also found that the price received for paddy depends on quality and moisture content. However, the most important problem would be lack of funds to invest in a dryer, because most farmers are in debt.

To make dryers more attractive to farmers, drying and storage should be integrated, so that paddy can be stored after drying and sold in large quantity when the price is reasonably high. The drying/storage unit needs to be durable even although this may increase its cost. In addition, it should be easy to construct and operate. Most importantly, some farmers already have storage huts, and if it were possible to modify these into drying/storage facilities this would result in minimal disturbance to existing practices.

Apart from the hot-air drying described above, drying paddy with ambient air has also been studied in Thailand. Uyeda (1984) designed and tested a simple on-farm paddy dryer. Natural air is forced through a 300 mm deep, gable-roof shaped paddy bed, by a 770 mm diameter axial flow fan driven by a gasoline engine with a continuous rated output of 2.8 kW at 1800 rpm. About 2 t of paddy can be dried each batch. From 7 runs of experiments during the wet season, it was concluded that paddy was successfully dried by natural air (final moisture content about 12–13%, wet basis). Estimated average drying rate (from 4 runs) is about 0.92% (wet-basis)/tonne of dry paddy/hour, and the average fuel consumption is 1.09 litres/tonne of dry paddy/1% (wet-basis) moisture reduction.

Drying paddy with ambient air was also studied at the Asian Institute of Technology, Bangkok by Jindal and Argarwalla (1979). Four runs of drying were conducted during the early dry season. Paddy was successfully dried with an airflow rate of 2.5–10 $m^3$/min/$m^3$ of grain at a depth of 0.6 m. The success was confirmed by a further 6 runs conducted by Pordesimo (1981) during the summer. The latter study used airflow rates between 4–7.5 $m^3$/min/$m^3$ of grain and a paddy depth of 0.6–0.9 m.

## An Integrated Paddy Drying and Storage Solar Hut

An integrated drying/storage solar hut was designed with characteristics approaching the ideal dryer discussed in the previous section. It was constructed and tested at a farmer's house in Kampaengsaen, Nakornpathom, Thailand. The aim was to study its suitability from both technical and socio-economic points of view.

### Design and Construction

The integrated paddy-drying/storage solar hut is

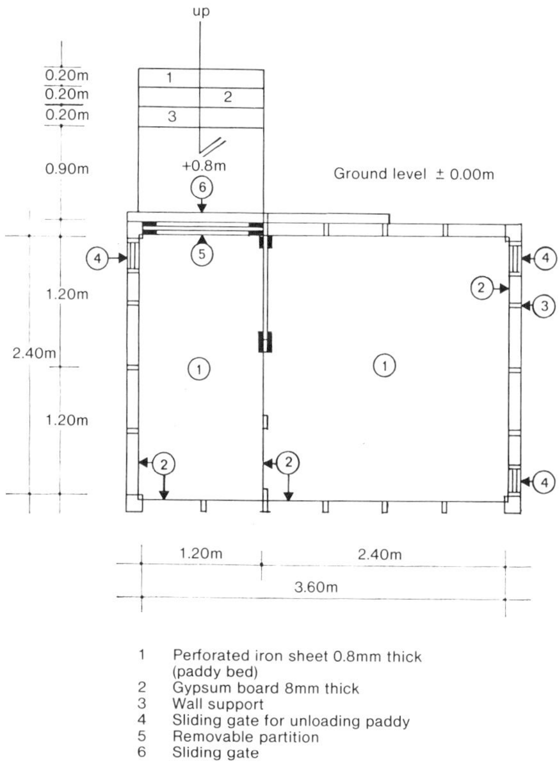


**Fig. 1.** Floor plan of integrated paddy drying and storage solar hut.

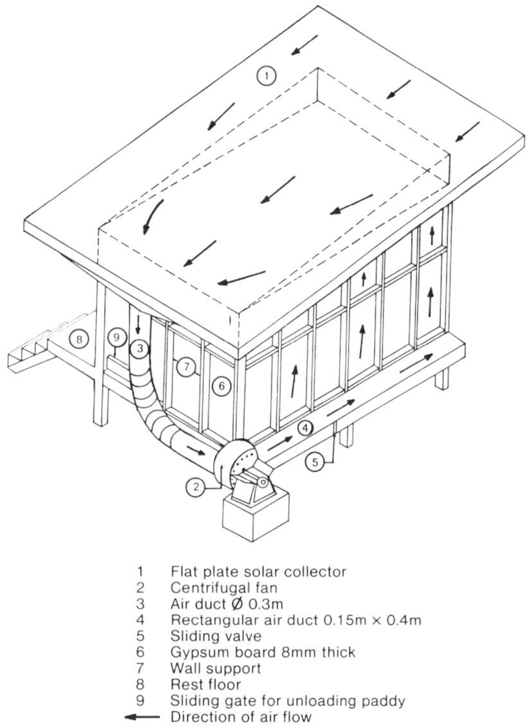


**Fig. 2.** Perspective drawing of integrated paddy drying and storage solar hut, showing direction of flow of heated air.

composed of a 18.6 m$^2$ flat plate, solar air heater, two drying/storage rooms, a fan, and an air duct system.

The solar air heater is similar to that designed by Soponronnarit (1984) and Soponronnarit and Tiansuwan (1984). It was built on the roof of the hut from corrugated, galvanised steel sheets (the absorber). In order to simplify construction, it has no transparent cover. Air flows through a 20 mm gap between the absorber and styrofoam 25 mm thick.

The space inside the hut is divided into two rooms having perforated floor areas of 2.9 and 5.8 m$^2$ (Fig. 1). About 1 t of paddy (0.6 m deep) will be dried in the smaller room and then transferred to the larger room for storage. However, both can be used for drying and/or storage. The maximum storage capacity is about 10 t at a depth of 2 m.

A 0.3 m diameter centrifugal fan, with forward-curved blades and driven by a 3.7 kW diesel engine is used to draw heated air from the solar air heater and pass it through the paddy bed in which drying takes place (see Fig. 2). The fan may be driven by tractor engine or electric motor. Normally, it will be operated during the daytime at 1500–1900 rpm, which corresponds to an engine speed of 1200–1500 rpm and a power requirement of 2.2–2.9 kW. Details of the method of calculation can be found in Soponronnarit et al. (1985b).

## Measurements

Ambient air temperature, wet bulb temperature, inlet and outlet air temperature of the solar collector, and inlet and outlet air temperature of the blower were measured using mercury thermometers and recorded at hourly intervals. Air velocity profile in the duct was measured using a pitot tube. Global solar radiation was recorded using a bimetallic radiation recorder. Grain moisture content at the bottom, middle, and top layers (15–30 samples from three places in each layer) was measured by a 'Kett' moisture meter (ricetester model D) and then calibrated with the moisture content measured by the oven method (100°C for 72–96 hours).

## Results and Discussion

Daily efficiency of the solar collector varied from 7–29% at an airflow rate of 0.027–0.041 kg/sec/m$^2$ of collector and with high variability in wind velocity. Heat obtained from the solar air heater varied from 20 to 81 MJ/day when global solar radiation was 15 MJ/m$^2$/day.

Three runs of drying experiments were conducted during the dry season and 10 during the wet season. Table 1 summarises details from 3 runs. It is estimated from the experiments that the drying rate is 0.64 and 0.3% (wet basis)/tonne of dry paddy/hour. Fuel consumption of the diesel engine driving the fan was 1.13 and 2.08 litres/tonne of dry paddy/1% (wet basis) moisture reduction, for drying paddy during the dry and wet seasons, respectively.

Head yield of paddy after milling varies from 45–50% and germination is greater than 92%.

About 7 t of dried paddy was stored in the solar hut for 5 months (January-May). It was found that the quality of the stored product remained high. There was no evidence of losses due to birds or rats.

The farmers who participated in this project were interested in drying wet paddy harvested during the wet season in order to prevent it from spoilage due to rain. They were also interested in storing paddy and selling it when the price is reasonably high. Their interest arose from the ease of operation of the hut and because it did not conflict with their traditional postharvest practices. However, they would not accept solar hut drying during the dry season because paddy can be dried easily in the field within 2 or 3 days. This is in spite of farmers knowing that losses in quantity and quality of paddy may be significant.

The solar hut is suitable for storing about 10 t of paddy and for drying 15 t of paddy per month. It costs 27 600 THB of which 21 700 THB is for materials (except engine) and the remainder is for labour (50 man-days) (during November 1985, 26 Thailand baht (THB) = US$1).

Farmers would benefit from the use of the drying/storage solar hut. For the year crop (harvested during the dry season), the benefit accrues from reduction of losses, better milling quality, and storage of paddy for selling at times of higher price. For the second crop (harvested during the wet season), it comes from reduction of losses, better milling yields, better quality (no yellow grains or moulds), and better prices for selling dry paddy. The net benefits are estimated to be 500 and 400 BHT/tonne for the year crop and second crop, respectively. The solar hut can dry and store 10 t of paddy per season. The net benefit per year is therefore 5000 + 4000 = 9000 BHT and the break-even point would be 27 600/9000 = 3 years.

More details of this study can be obtained from Soponronnarit et al. (1985b).

### Laboratory Studies of In-store Drying

In-store grain drying is popular in temperate areas. The advantages of the method are that

**Table 1.** Results of wet season drying tests in Thailand using the integrated paddy drying and storage solar hut.

| Experimental conditions | Test No. 1 | 2 | 3 |
|---|---|---|---|
| Rice variety | RD 21 | RD 7 | RD 21 |
| Mass of paddy after drying, kg | 620 | 572 | 1172 |
| Thickness of paddy bed, m | 0.14 | 0.35 | 0.71 |
| Average initial moisture content, % (wet basis) | 27.5 | 17.7 | 17.8 |
| Average final moisture content, % (wet basis) | 16.2 | 14 | 15.1 |
| Mean ambient temperature, °C | 30.5 | 31.3 | 30 |
| Mean ambient relative humidity, % | 71 | 65 | 69 |
| Mean inlet temperature, °C | 33.4 | 34.2 | 32.7 |
| Mean inlet relative humidity, % | 62 | 55 | 59 |
| Total temperature rise, °C | 2.9 | 2.9 | 2.7 |
| Temperature rise by solar energy, °C | 1.8 | 1.4 | 1.3 |
| Temperature rise by fan, °C | 1.1 | 1.5 | 1.4 |
| Flow rate, m$^3$/min/t of dry paddy | 51.1 | 52.4 | 22.8 |
| Total useful energy from solar collector, MJ | 78 | 20.5 | 27.5 |
| Total diesel oil consumption, L | 13.7 | 4.48 | 7.5 |
| Diesel oil consumption, L/t of dry paddy/1% (wet basis) moisture content reduction | 1.69 | 2.1 | 2.46 |
| Drying time, hours | 20.8 | 7.4 | 12 |

drying costs are low and grain quality after drying is significantly improved. In tropical countries such as Thailand its utility is still questionable because of significant levels of microorganism activity and grain respiration stimulated by high ambient temperatures. The purpose of the study reported here was to determine, by way of laboratory experiments in a small test bin, the technical feasibility of in-store paddy drying under tropical climates.

**Design of Drying Bin**

Simulation studies of drying paddy in bulk storage using the Hukill (1947) drying model and the Seib et al. (1980) model of dry matter loss, show that, for Bangkok's climate, airflow rates of 1.8, 5.5, and 15.2 $m^3/min/m^3$ of grain are required to dry paddy from initial moisture contents of 20, 22, and 24% (wet-basis) to the final moisture content of 14% (wet-basis) so that dry matter loss is 0.5%. The corresponding drying times are 315, 126, and 54 hours, respectively. It is calculated that air has to be heated by 5–6°C for the worst month of 30 year weather data (Soponronnarit et al. 1985a). Paddy with a dry matter loss less than 0.5% will generally be classified as 1st grade.

A drying bin was designed according to the specifications of R.H. Driscoll and T.Adamczak of the University of New South Wales, Australia (Driscoll and Adamczak, personal communication). A 0.75 kW centrifugal fan and a 1.5 kW heater were selected by using the above simulation result. Air conditions could be controlled either by constant heating or by limiting the maximum relative humidity. The bin is made of galvanised sheet steel 1.5 mm thick, and has a diameter of 0.75 m and a height of 2.75 m. It is insulated with glass wool 25 mm thick and then covered by aluminum sheet. The maximum thickness of the paddy bed is 2 m. During operation, ambient air is sucked by the fan and delivered to the plenum chamber through a perforated steel sheet, and then distributed through the paddy bed.

**Measurements**

Moisture content of the grain was determined by the oven method. Grain samples of approximately 30 g were dried in an oven at 100°C for 72–96 hours. Grain and air temperatures, and dry and wet bulb temperatures of ambient air were measured by thermocouple (Chromel-Alumel, type K) and recorded by data logger ('Minitrend 205') at hourly intervals. The accuracy was about ±0.5°C. The profile of air velocity in the duct was measured by a hot wire anemometer. Visible mould, milling quality, yellow grains, and seed germination were determined after drying.

## Results and Discussion

Eight runs of experiments on in-store paddy drying were conducted at King Mongkut's Institute of Technology, Thonburi, Bangkok during the wet season of 1984–1985. Harvested paddy was usually kept for about 2 days before drying, i.e., in rice panicles placed in the field for the first day and with threshed paddy in sacks for the following day. Each run involved approximately 420 kg of wet paddy (18–28%, wet-basis) at a bed thickness of 1.6 m (0.7 $m^3$). Airflow rates depended on the initial moisture content of paddy. They varied from 1.5–4.6 $m^3/min/m^3$ of grain. All paddy was successfully dried to 13% (wet-basis) moisture content in 71-180 hours. Head rice yield varied from 51–60% and germination was higher than 96%. No visible mould or yellow grain observed. The results are summarised in Table 2.

Drying with heated air was tested. Air was

**Table 2.** Results of laboratory tests of in-store paddy drying in Thailand.

| Experimental conditions | Test No. 1/84 | 2/84 | 3/84 | 1/85 | 2/85 | 3/85 | 4/85 | 5/85 |
|---|---|---|---|---|---|---|---|---|
| Rice variety | RD 21 | RD 21 | RD 21 | RD 21 | RD 21 | RD 21 | — | — |
| Mean ambient temperature, °C | 31.3 | 31.5 | 31.2 | 30.6 | 30.4 | 30.8 | 30.7 | 32 |
| Mean ambient relative humidity, % | 81.8 | 82 | 82.2 | 79.7 | 83.2 | 82.4 | 81.2 | 81.4 |
| Mean temperature of drying air, °C | 34.9 | 34.7 | 35.4 | 36 | 42.1 | 33.8 | 33.3 | 33.5 |
| Mean relative humidity of drying air, % | 67 | 69 | 65 | 59 | 46.3 | 70.2 | 70.7 | 75.2 |
| Mean airflow rate, $m^3/min/m^3$ of grain | 2.88 | 4.61 | 2.26 | 4.55 | 1.46 | 3.3 | 2.89 | 1.49 |
| Mean initial moisture content, %, wet basis | 19.9 | 22 | 17.6 | 28.3 | 20.7 | 24.2 | 22.4 | 18.0 |
| Mean final moisture content, %, wet basis | 13 | 13 | 13 | 13 | 13 | 13 | 13 | 13 |
| Drying time, hours | 85 | 76 | 71 | 92 | 121 | 112 | 96 | 180 |

constantly warmed by a 0.5 kW electric heater for about 9–11 hours at night. Mean temperature rises were 5.4 and 11.7°C for runs 1/85 and 2/85, respectively. Paddy in the bottom layer was significantly overdried (9.6 and 7.3%, wet-basis), while the average moisture content of the whole bin was 13% (wet-basis).

Runs 1/84, 3/84, and 3/85 involved unheated air. However, mean temperature rises were relatively high, i.e., 3.0–4.2°C. This was due to the use of butterfly valves to reduce the airflow rate. As a result, the paddy at the bottom layer was overdried (9.9–10.3%, wet-basis). Runs 2/84, 4/85, and 5/85 also used unheated air but, rather than using the butterfly valves, surplus air was bypassed. It was observed that the higher the airflow rate through the paddy bed, the higher the mean temperature rise. However, mean temperature rises were lower than those obtained in runs 1/84, 3/84, and 3/85, i.e., 3.2, 2.6, and 1.6°C for the airflow rates of 4.6, 2.9 and 1.5 $m^3/min/m^3$ of grain, respectively. This led to a reduction in overdrying of grain, i.e., 10.5, 10.8, and 11.1% (wet-basis) at the bottom layer, corresponding to the above temperature rises, respectively. The comparison is relatively reasonable because the mean relative humidity of the ambient air was around 80–82% for all runs. It is therefore suggested that low airflow rates or apparent air velocities in the bin should be used, in order to avoid overdried grain as much as possible. Operating costs are also expected to be significantly reduced. Consequently, the maximum initial moisture content must be limited. If it is 20% (wet-basis), the minimum airflow rate required to dry paddy successfully is 1.5 $m^3/min/m^3$ of grain for Bangkok's average climatic condition (Soponronnarit et al. 1985a). This gives an apparent air velocity in the bin of 2.4 m/min for a bed depth of 1.6 m, as was used during the experiments. It should be noted that the bed depth becomes an important parameter when the economics of in-store drying are analysed.

Ambient conditions during the experiments were close to average for the wet season. They may be worse in some years. It is therefore recommended that a supplemental heat source of a few degrees of temperature rise should be provided.

Since initial moisture content is often higher than 20% (wet-basis), two-stage drying is recommended; i.e., fast drying to 18–20% moisture content (wet-basis) followed by in-store drying.

## Conclusions and Recommendation

The integrated paddy drying and storage solar hut seems to be an ideal dryer for the farmer who harvests about 10 t of paddy per season and grows two crops per year. However, energy consumption for drying is still high. This is due to the use of a centrifugal fan which is needed to overcome the significant pressure loss in the system. It would be more economic if a tube axial fan which can provide high flow rate at low static pressure could be employed. In this case, unheated air should be used in order to reduce the pressure loss occurring in the solar air heater, air duct, etc. Consequently, overdried grain would not occur. It is therefore not necessary to transfer dried paddy from the drying room to the storage room because there is no need to separate them. The process then becomes in-store drying. If harvesting is staggered into 5–6 lots of 1.5–2 t each, spread over a month, it may be called layer drying i.e., the next coming layer will be loaded into the hut when the previous one is already dried. If 10 t of paddy having an initial moisture content of 20% (wet-basis) is expected to be successfully dried at a bed depth of 2 m, the minimum airflow rate and estimated power for a fan are approximately 0.4 $m^3/sec$ and 200 W, respectively. If drying is completed in 381 hours, electrical energy consumption will be 76 kW-h or 7.6 kW-h/t of paddy.

On a larger scale, as in the grain industry, in-store drying is also favoured due to low operating costs and high quality results. It is suggested that two-stage drying is necessary, i.e., fast drying in the first stage until the moisture content is less than 20% (wet-basis), followed by in-store drying using a relatively low airflow rate. Supplemental heating may be necessary for the months which are unfavourable for drying.

## Acknowledgments

The author would like to thank National Energy Administration, Bangkok, which supported the project on an integrated paddy drying and storage solar hut and the Government of Australia which provides funds through the Australian Centre for International Agricultural Research for the project of drying in bulk storage of high moisture paddy.

Also, he would like to thank his colleagues Mr B. Kunjara, Miss W. Watabutr, and Mr A. Therdyothin who contributed in the first project, and Dr R. Chirarattananon who contributed in the second. Thanks are also due to Mr C. Karnjanaboon, former graduate student, and Miss P. Preechakul, and Mr Y. Thirawanichkul, graduate students, who conducted the research during their study.

## References


Anon. 1976. A survey of post harvest information of some crops 1976. Bangkok, Thailand, Department of Agriculture, Ministry of Agriculture and Co-operatives, 180 p.

Anon. 1977. Design/development of farm dryer. Bangkok, Thailand, Ministry of Agriculture and Co-operatives, Division of Agricultural Engineering, Department of Agriculture.

Anon. 1984. Agricultural statistics of Thailand crop year 1983/84. Bangkok, Thailand, Center for Agricultural Statistics, Office of Agricultural Economics, Ministry of Agriculture and Co-operatives.

Amyot, J., and Sirisambhand, N. 1982. Social and economic factors in the process of technology transfer. Bangkok, Thailand, Chulalongkorn University Social Research Institute.

Exell, R.H.B., and Kornsakoo, S. 1977. A low cost solar rice dryer. Bangkok, Thailand, Asian Institute of Technology, Research Report.

Hukill, W.V. 1947. Basic principles in drying corn and grain sorghum. Agricultural Engineering, 28(8), 335–338, 340.

IRRI (International Rice Research Institute) 1978. Operator's manual: IRRI batch dryer and IRRI vertical bin batch dryer. Los Baños, Laguna, Philippines, Agricultural Engineering Department, IRRI.

Jindal, V.K., and Agarwalla, J.K. 1979. A model to simulate paddy drying with unheated air. Presented at grain post harvest technology workshop, Southeast Asia Co-operative Post Harvest Research and Development Programme. January 1979, Jakarta, Indonesia.

Lhoste, B., and Louis, M. 1985. Sechage et stockage du paddy a la ferme dans la plaine centrale Thailandaise: problem et perspectives. Laboratoire de Biotechnology Solaire, Ecole Nationale Superieure Agronomique de Toulouse, France, 43 p.

Pordesimo, L.O. 1981. Development and evaluation of simulation models for the drying of paddy with unheated air. Bangkok, Thailand, Asian Institute of Technology, Masters dissertation.

Seib, P.A., et al. 1980. Spoilage of rough rice as measured by carbon dioxide evolution. In: Proceedings of the 3rd Annual Workshop on Grain Post Harvest Technology, January 29–30, Kuala Lumpur, Malaysia.

Soponronnarit, S. 1984. Solar thermal processes in Thailand: a study on forced convection hut drying. Bangkok, Thailand, National Energy Administration, 110 p.

Soponronnarit, S., and Tiansuwan, J. 1984. Low cost solar air heater: application to paddy drying. Presented at regional seminar on simulation and design in solar energy applications. May 1984. Thonburi, Bangkok, Thailand, King Mongkut Institute of Technology.

Soponronnarit, S., Karnjanaboon, C., and Chirarattananon, R. 1985a. A feasibility study on in-store paddy drying. Presented at National seminar on technology for rural development, February 1985, Khonkaen, Thailand.

Soponronnarit, S., Watabutr, W., Therdyothin, A., and Kunjara, B. 1985b. A study on solar paddy drying. Bangkok, Thailand, National Energy Administration, 134 p.

Srihavong, R. 1978. Experimental data obtained from paddy drying. Journal of the Thai Society of Agricultural Engineering, 3(1), 16–24.

Uyeda, K. 1984. Simple and low cost on-farm paddy dryer. Bangkok, Thailand, Agricultural Machinery Center, KURDI, Kasetsart University.

# A System for Small-Scale Farm Drying and Storage of Paddy in Korea

Hong-Sik Cheigh*, Key-Sun Kim†, Myung-Gon Shin†, and Tai-Wan Kwon†

*Abstract*

For the purpose of loss reduction and quality retention of paddy rice at farm level in Korea, an in-bin drying and storage (IBDS) system (size: 1.8 × 2.4 × 3.2 m, capacity: 2–4 t of paddy), which consists of a fan and air ducts, was developed. Low-temperature, paddy drying with the IBDS system during the harvesting season was examined. The grain could be dried to a safe storage moisture content (14–15%) within 180–220 hours, by continuous aeration with ambient air. Even wet paddy (initial moisture content, 16–17%) could be safely stored for a year with proper fan operation. No significant changes in quality, in terms of milling yield, acid value, and germination were observed during the drying or storage. The IBDS system is considered one of the most economical and labour-saving systems for on-farm drying and storage of paddy in Korea. Information about it is being disseminated throughout the country.

RICE is the major staple food grain in Korea. Annual production is estimated at about 6 Mt. Per capita consumption during 1984 was about 135 kg. Paddy rice is harvested mainly in October. The conventional postharvest system of paddy rice in Korea is very laborious and considerable losses accompany it. The estimated quantitative loss of paddy rice is assessed in the range 8-10% depending upon the variety of rice, drying and storage conditions, and other environmental factors (Cheigh et al. 1980). In view of the labour shortage in rural areas in recent years, and the chronic food grain shortage of the nation, development of more efficient postharvest practices to reduce grain losses and the amount of labour required has therefore been an ever-present necessity.

In Korea, paddy harvesting begins from the end of September and continues for a month. Mature paddy with a moisture content of 22-24% is harvested mostly by hand, using sickles. In recent years, however, mechanical harvesters have been increasingly introduced to save labour. Traditionally, the cut plants are allowed to dry in the field to about 16-18% grain moisture content before threshing. The threshed grains are then further sun dried to a final moisture content of around 15%. Mechanically harvested paddy is often dried by fuel-fired commercial dryers. Although there are several methods of farm storage, most paddy for storage is bagged and stacked on the floor of storage house. In the course of drying and storage in such manner, paddy is exposed to birds, rodents, insects, and microorganisms, inviting considerable grain loss.

As Korea is located in the temperate zone, there are four distinct seasons. The average annual precipitation is approximately 1300 mm, most of it occurring between April and August. Only the frost-free period of seven months between April and October is suitable for cultivation. Most of harvesting activity takes place in the autumn between September and November, before the weather becomes too cold. As shown in Figure 1, from the end of September the monthly mean temperature stays below 15°C until the following April. The relative humidity remains at about 75%. These conditions suggest that the climate could be exploited in drying and storing of paddy in bulk, provided a proper system could be developed.

It would, of course, be possible to reduce such paddy losses if we were to introduce the mechanised systems which are quite popular in the industrialised nations. Regrettably, however, such systems are either technically or economically

* Department of Food Science and Nutrition, Pusan National University, 30 Jangjun-Dong, Dongrae-Ku, Pusan, Korea

† Division of Biological Science and Engineering, Korea Advanced Institute of Science and Technology, P.O. Box 131, Dongdaemun, Seoul, Korea.

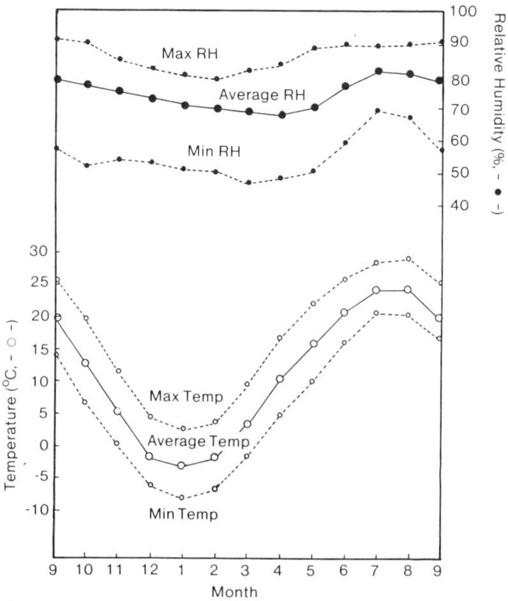


**Fig. 1.** Monthly average temperature and relative humidities at Suwon, Korea, 1971–1980.

unsuited to Korean rural conditions. We therefore focused our attention to how we might exploit our inherent climatic condition to meet postharvest requirements.

Under the joint sponsorship of the governments of Korea and West Germany, we undertook a research project on the development of drying and storage methods for paddy suited to small-scale farm use. This project was part of the research program on Integrated Rural Development Technology (Kwon 1983). From it we developed the KAIST (Korea Advanced Institute of Science and Technology) in-bin drying and storage (IBDS) system for paddy rice for the small farmer who harvests 2-4 t of paddy. Grouping by farm size showed that over 40% of all farm households lie within this range.

In this paper, we discuss the drying characteristics and storage capability of the IBDS system.

### KAIST IBDS System for Paddy Rice

In developing and designing the KAIST IBDS system at farm level, the following attributes were considered desirable: capacity according to crop size; maximum natural ventilation; possible mechanical aeration; rodent proofing; prevention of water permeation from the floor as well as the walls; and low-cost facility with easy operation and low energy consumption.

Given these, it was decided to develop a natural air drying and storage system suitable for farm household operation. The idea of natural air drying, or in-bin low temperature drying, for grains is not new. It is in common use in Western countries, but only in large-capacity installations using facilities unsuited to small, individual farms (Cheigh 1982). Also, a recent study highlighted the possibility of using natural air for similar purpose in Korea (Cheigh et al. 1982, 1985). Another factor taken into account during development of the IBDS system was the threshing of paddy while the rice plants are still wet would substantially reduce the losses arising from longer handling and exposure to the possibility of rewetting by rain. In addition, in order to be accepted by rural infrastructure in Korea, the system should be of low cost and easy to construct and operate by farmers. Figure 2 presents the suggested direction for improving the present paddy drying and storage practices in Korea using the KAIST IBDS system.

Considering those factors, we have developed an in-bin drying and storage (IBDS) store (1.8 × 2.4 × 3.2 m) which is equipped with a fan (0.5hp) to blow air, and an air duct system (main duct and

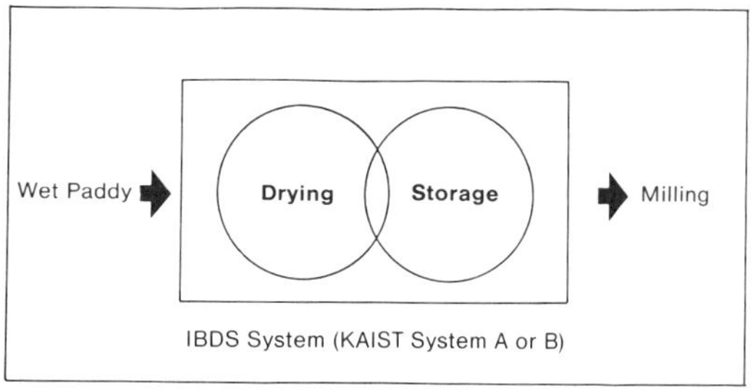


**Fig. 2.** Recommended paddy drying and storage system at farm level in Korea.

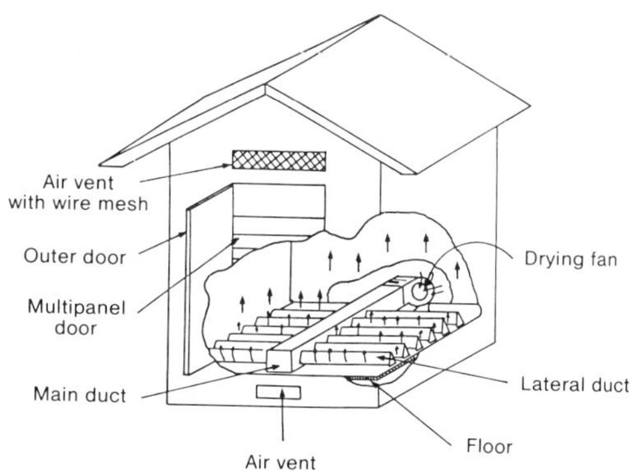


**Fig. 3.** Schematic drawing of the in-bin drying and storage (IBDS) system for paddy in Korea.

lateral) ensuring even air distribution through loaded grain bulk of 3-4 t, as shown in Figure 3. The store is constructed with cement blocks, insulation materials, wire-mesh, and waterproof paint, and therefore excludes rodents and moisture. Furthermore, the floor is above ground to prevent water and heat from that source. Finally, the loading and unloading portal has two doors, the inner one consisting of separate wooden panels to facilitate loading and unloading. Three tonnes of freshly harvested paddy can be loaded into the store for drying and year-round storage by blowing cool and dry air as needed.

## In-Bin Drying of Paddy Using the IBDS System

A series of trials (1981-1983) with the KAIST IBDS system was carried out for the in-bin paddy drying, and the following results on drying capabilities for paddy in September-October 1982 are given as an example (KAIST 1983). Milyang #30, Indica type variety of paddy, was used for in-bin drying with KAIST IBDS system, which is located in Seoul. A total of 3.0 t of paddy was manually loaded into the system. The initial moisture content of the paddy was 21.4% and an aeration rate of 5.0 m$^3$ air/m$^3$ paddy /min was used.

**Table 1.** Results of paddy drying experiments using the in-bin drying and storage (IBDS) system.

| | |
|---|---|
| Initial moisture content %, wet basis | 21.6 |
| Final moisture content %, wet basis | 13.2 |
| Drying period | 27 Sept. – 9 Oct. 1982 |
| Air temperature average, °C | 18.6 |
| Relative humidity average, % | 63.3 |
| Drying potential, kg $H_2O$/kg air | 0.0018687 |
| Airflow rate m$^3$ air/m$^3$ paddy/min | 5.0 |
| Grain depth, m | 1.1 |
| Mass of wet grain, kg | 3 000 |
| Mass of dried grain, kg | 2 710 |
| Water removed, kg | 290 |
| Drying time, hours | 211 |

Experimental conditions and results are shown in Table 1, and a schematic diagram of the experimental IBDS system in Figure 4. Figure 5 shows the changes in moisture content of paddy at various depths during the drying period of 8.8 days (211 hours). At first, the drying rate of lower layers is much higher than that of upper layers. However, at the last stage of the drying, it was decreased rapidly. It took 4-5 days for lower layer, 6 days for middle layer, and 8 days for upper layer to be dried

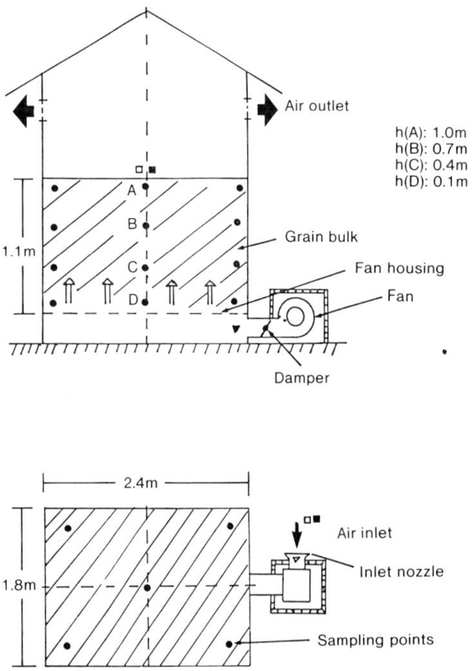


**Fig. 4.** Plan and elevation of an experimental in-bin drying and storage (IBDS) system for paddy in Korea.

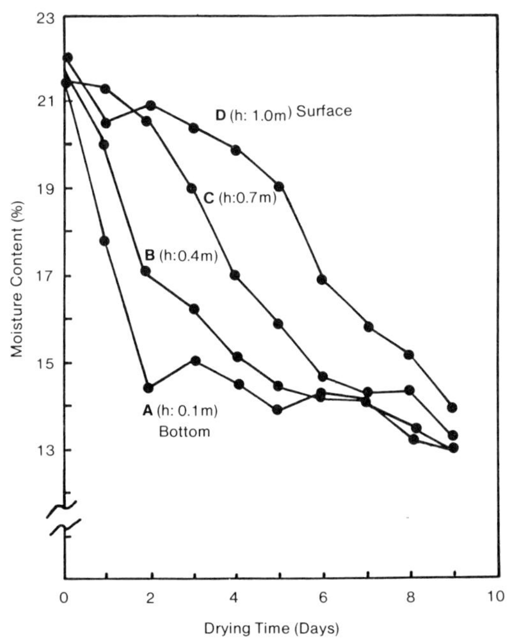


**Fig. 5.** Changes in the moisture content at various depths in paddy being dried using the in-bin drying and storage (IBDS) system.

to 14.7-15.2% moisture content, which is considered as safe level for storage.

The difference in moisture content before and after drying was about 8.4% and the mean drying rate was calculated as 0.040% per hour. The climatic data are presented in Figure 6. During the period of drying, diurnal temperatures fluctuated between 8.5 and 23.5°C and relative humidity from 24-94%. The average temperature was 18.6°C and the average relative humidity 63.3%. Generally, the air had a high drying potential (0.001869 kg $H_2O$/kg air) for in-bin drying of paddy and, as expected, this was reflected in the experimental results. Also, there were no significant changes in grain qualities such as milling yield, germination rate, amount of free fatty acids, and presence of microorganisms.

### Storage of Paddy Rice Using the IBDS System

A comprehensive comparison between the storage capabilities of the KAIST IBDS System and traditional bulk storage was made throughout the year in 1979-1980. Paddy (Milyang #15 variety) was loaded manually to each store, two IBDS and one conventional. Paddy in the IBDS store designated Store 1 had a higher moisture content than that in Store 2. The amounts and initial moisture contents of paddy are shown in Table 2. At the beginning of the experiment, the paddy used had a somewhat higher moisture content than is considered safe for storage.

**Table 2.** Initial condition of paddy used in the 1979–1980 storage experiment comparing the in-bin drying and storage (IBDS) system, and the traditional storage method.

| Store | Store condition | Paddy loaded (t) | Initial Moisture content (%) | Forced aeration |
|---|---|---|---|---|
| 1 | IBDS system (1) | 2.1 | 18.1–19.2 | When needed |
| 2 | IBDS system (2) | 2.1 | 16.5–17.2 | When needed |
| 3 | Traditional store | 0.9 | 15.7–16.1 | None |

Temperature and moisture content of grain were measured once a week during storage. Changes of paddy qualities such as free fatty acid, hardness, milling yield, and germination rate were checked periodically.

The weekly grain temperatures at the centre of each type of store are shown in Figure 7. Grain temperature decreased until February and thereafter increased continuously to reach its initial level by the end of April. No significant differences of grain temperature were noted between the two IBDS stores.

The changes in moisture content observed at various grain depths during storage of paddy in IBDS Store 1 were as follows. The paddy showed some drying at the surface. Since there was a

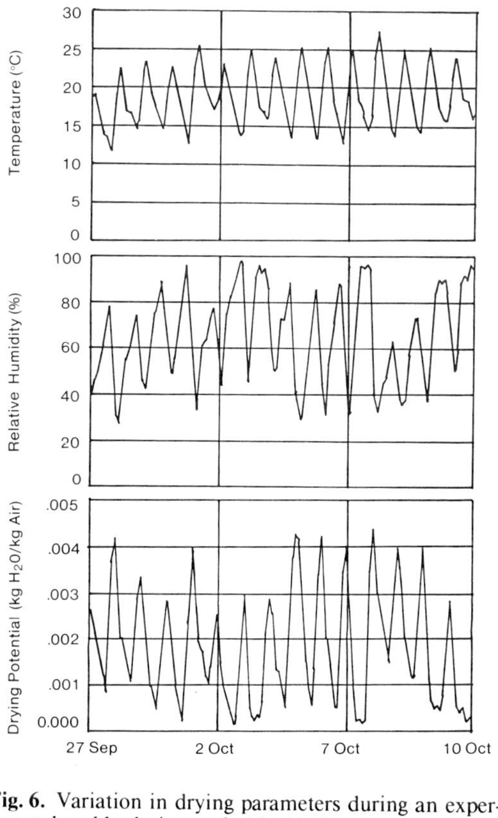


**Fig. 6.** Variation in drying parameters during an experimental paddy drying period in 1982.

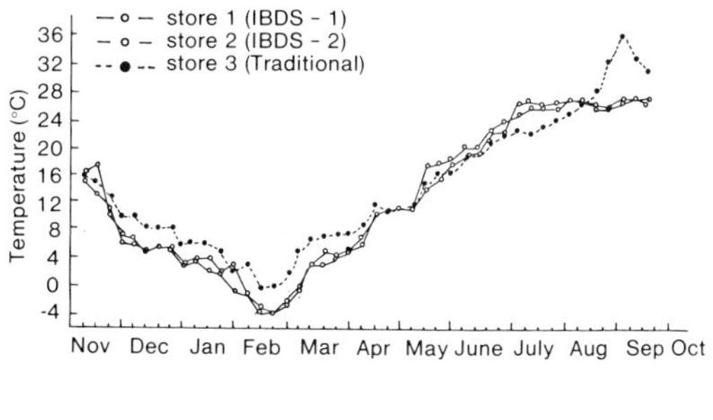


**Fig. 7.** Variation in grain temperature at the centres of three stores used in a paddy storage experiment in Korea.

tendency towards increasing moisture content from May, aeration was conducted. From this time the moisture contents of paddy stored in Store 2 (16-17% moisture content) was slightly decreased. However, the moisture content of paddy held in the traditional store increased by more than 1% at the bottom of the store, indicating absorption of moisture from the cement floor.

Some physico-chemical properties and milling data on paddy during storage were measured. There were no significant changes in free fatty acid levels, germination rate, or milling yield of paddy stored in the various stores during the term of the experiment. However, free fatty acid values rose after July, indicating that some chemical changes occur during summer.

It should be noted that even the paddy initially at higher moisture content (about 17-18%) can also be stored safely with proper aeration. However, some of the paddy rice in the bottom layer of the traditional store was contaminated by microorganisms and had deteriorated, even though its moisture content, at 15-16%, was not particularly high.

### Dissemination of Developed IBDS System

To be economically viable and socially acceptable to farmers, new facilities or equipment must be simple, durable, trouble free and, if possible, locally made rather than imported. Also, the process must be simple and use little energy. In the conventional paddy-handling system in Korea the threshed grains are finally sun dried by spreading manually on a number of mats for 2–3 days under the sunshine. In the evening, the drying grain must be gathered up and held under shelter until the next sunny day. When it rains, the spread-collection operation has to be further repeated. The dried paddy rice is then bagged and stacked in a conventional flat store. The traditional system therefore requires a number of steps to be repeated. In addition, the flat store is usually exposed to factors causing considerable losses of grain. However, IBDS system is characterised by a low investment cost of about US$450 for each system, minimal requirement for labour, less attention and a lower energy consumption of 28 kWh/t (Kwon et al. 1985). It is also simple to operate.

However, although a new technology may be economically viable and socially acceptable, farmers in general are rather conservative and reluctant to accept new methods. Therefore, it may be desirable to involve farmers in the development of such technologies at an early stage of work, rather than imposing an unfamiliar system upon them. This approach not only gains the farmers' understanding, but also generates pride in participation. We made a special effort to work with farmers at village level, to use locally available materials and equipment of a durable nature, and to develop the simple processes for this technology.

We have made every possible effort to disseminate the IBDS system throughout the country, by the following activities:

- Technical training on the KAIST IBDS system for Government officers.
- Lectures to farmers and farm leaders on the construction and utilisation of the system.
- Publication and distribution of a booklet on the system.
- Construction and demonstration of the system in rural areas.
- Technical assistance to farmers wishing to install the system.

During 1982, about 30 IBDS systems were built and used by farmers in Kyongki Do Province from funds granted by Myon Branch-National Agricultural Cooperative Federation (NACF). In addition, 90 IBDS systems throughout the country (10 in each Province) were built with the support of the Office of Rural Development (ORD) as demonstration models. During 1982–1984, a considerable number of IBDS systems were built voluntarily by farmers.

As a result, we can now boast of a very successful dissemination of the IBDS technology throughout the nation. We, of course, gave the technology wide publicity through mass media, open lectures, travelling workshops, and answering farmers' letters and telephone calls on the subject. However, there was also strong support for the dissemination activities from Government agencies and other organisations concerned, and enthusiastic acceptance and cooperation from the farmers as well. As of 1984, there are already some 10 000 units of the IBDS systems installed. The number is expected to double by the end of 1985.

We were indeed very fortunate that we could use our own climatic conditions to develop effective postharvest systems for paddy in Korea.

## References


Brooker, D.B., Bakker-Arkema, F.W., and Hall, C.W. 1974. Drying cereal grain. Westport, AVI Publishing Company.

Cheigh, H.S. 1982. Paddy drying and storage processes at farm level in Korea. In: The techno-economic feasibility study for the integrated programme for rural development technology. Seoul, Korea Advanced Institute of Science and Technology, 1–68.

Cheigh, H.S., Kim, S.K., Kim, C.J., and Rhim, J.H. 1980. Studies on the improvement method of grain storage (IV). Seoul, Korea Institute of Science and Technology, KIST Report BS G 406-147805(1), 25–30.

Cheigh, H.S., Mühlbauer, W., Rhim, J.H., and Shin, M.G. 1985. In-bin drying of paddy with ambient air: influence of drying parameters on drying time, energy requirements and quality. Korean Journal of Food Science and Technology, 17, 25–32.

Cheigh, H.S., Rhim, J.H., and Kim, S.K. 1982. In-bin drying of high-moisture paddy with continuous blowing of ambient air. Korean Journal of Food Science and Technology, 14, 271–275.

KAIST (Korea Advanced Institute of Science and Technology) 1983. Prevention of post-harvest loss of grains at village level. In: Third interim report on integrated research programme for rural development technology (extension phase). Seoul, Korea Advanced Institute of Science and Technology, 1–50.

Kwon, T.W. 1983. Introduction to integrated research programme for rural development technology. In: Proceedings of the seminar on integrated rural development technology, October-November 1983. Seoul, Korea Advanced Institute of Science and Technology, 35–42.

Kwon, T.W., Shin, H.K., and Kim, K.S. 1985. Low energy postharvest technologies for paddy rice and fruits in Korea. In: Proceedings of the workshop on energy conserving technologies for the post-harvest system. May-June 1984. Warsaw, Poland, Institute of Agricultural and Food Economics, 328–346.

# Grain Aeration in Australia

**J.W. Sutherland***

*Abstract*

Grain aeration began in Australia in 1959 and is now applied to about 30% of all permanent storage. In the southern regions of the country, aeration of unsprayed, bulk-stored grain with ambient air has been used to obtain satisfactory insect control. In northern New South Wales and in Queensland, however, refrigeration of the aeration air has been found to be necessary because of the higher ambient temperatures. Nevertheless, aeration with ambient air has proven valuable in safely holding high moisture content grain sorghum in Queensland until drying can be carried out.

Case studies of 13 aeration systems are described, comprising 7 trials of natural aeration (5 in large stores and 2 in on-farm stores), 4 trials of refrigerated aeration in large stores, and 2 trials of aeration of high moisture content grain. Many of these trials involved experiments over more than one season.

Aeration has been shown to be beneficial for controlling grain insect populations, eliminating temperature gradients, reducing moisture migration, slowing down the degradation of pesticides, and preserving grain quality. Running costs of aeration systems are comparable with costs of insecticide treatment. The application of Australian aeration technology in the humid tropics is discussed.

MOST Australian wheat and barley is harvested and placed into bulk storage during November and December, with grain temperatures ranging from 25°C to 40°C and moisture contents below 12% (wet basis). The storage period ranges generally from about 6 to 12 months. Wheat and barley used to be stored in Australia without treatment of any kind. This practice led to problems of insect activity which damages the grain and raises the temperature above the already high initial value, moisture migration which results in caking and mould growth at cold surfaces, and reduction in grain quality resulting from prolonged storage at high temperatures. The factors governing grain deterioration in bulk storage are described by Griffiths (1964), and experimental evidence of this phenomenon was obtained by Sutherland (1965, 1968).

Insects are the main cause of grain deterioration and dramatic reductions in their numbers were observed in the early 1960s when malathion, an organo-phosphorus pesticide was first introduced. However, it soon became apparent that stored products insects were becoming resistant to malathion and by 1973 resistance had been reported in 70 countries (Dyte 1974). Malathion is now virtually not used in Australia, but other chemicals have been introduced which are effective against grain insects (e.g. fenitrothion and bioresmethrin). However, these are also expected to have a finite useful life. Murray (1979) discussed the problem of insect resistance in Australia, which is contractually bound as a grain-exporting nation (approximately 70% of the annual wheat production of about 15 Mt is exported) to present its cereal grains onto the international market free from insects. In addition, customer concern about chemical residues remaining in food grains (termed 'pesticide shyness') is increasing. It is for these reasons that research has been carried out in Australia into non-chemical methods of grain insect control.

One such non-chemical method of insect control is grain aeration or ventilation, which is the process of cooling a grain bulk by passing through it air of suitable temperature and humidity. The basic principles of heat and mass transfer in air flow through fixed beds of grain have been described by Sutherland et al. (1971). They involve the formation of temperature and moisture fronts which move through the grain bed in the direction of the air flow. These fronts can interact which leads to an attenuation of grain temperature and moisture content profiles (Sutherland et al. 1983).

* CSIRO Division of Chemical and Wood Technology, P.O. Box 26, Highett, Victoria 3190, Australia.

Grain temperatures of 15–18°C prevent reproduction in insect pests such as *Rhyzopertha dominica* and *Tribolium* species and severely limit it in others such as *Sitophilus* species (Evans 1981). The dryness of Australian wheat also contributes to the lowered rates of reproduction of *Sitophilus*. Cooling to suppress insect activity is the main purpose of aeration, but there are other important benefits, such as the reduction of temperature gradients and consequently the suppression of moisture movement, removal of heat resulting from grain respiration, decreased chemical protectant breakdown rates, maintenance of grain freshness, expulsion of odours, and distribution or removal of fumigants.

## Use of Grain Aeration

An aeration system comprises a fan and ductwork to distribute the air through the grain bulk, a controller to operate the fan only when aeration will be beneficial and, for conventional systems, temperature monitoring equipment to observe and optimise the cooling performance. In areas where ambient temperatures are too high to achieve adequate insect control, the air must be chilled before it is passed into the grain. In this case, it is also necessary to insulate the walls and roof of the grain store.

Grain aeration is widely used in Australia. About 30% of the permanent storage capacity have aeration facilities. The situation in the five main grain-growing States is shown in Table 1. The use of aeration at present ranges from 50% in New South Wales to 1% in Western Australia. Aeration has been applied to vertical silos (concrete and steel) and horizontal sheds of various shapes and sizes. The largest system known is at the Geelong terminal in Victoria, where two sheds with a combined capacity of 549 000 t and 43 silos with a combined capacity of 86 000 t, giving a total of 635 000 t, are under aeration. The system uses about 200 fans with a power input of over 1MW. The smallest commercial system which has operated in Australia was a 0.4 kW CSIRO farm aeration unit (Elder 1969) connected to a 100 t farm silo of maize near Newcastle in New South Wales (Wedd 1974). A description of experiments on grain aeration carried out by CSIRO is given in the next section.

## Trials of Grain Aeration Systems

### Natural Aeration

LARGE STORES

*Melton, Victoria.* The first trial of grain aeration in Australia was in 1959 at Wallaroo in South Australia, on barley stored in vertical steel silos. The first aeration in Victoria and the first CSIRO aeration trial was with wheat at Melton near Melbourne in 1963. The trial was conducted over three years and is described by Sutherland (1968). The store was a 2700 t capacity rectangular steel shed (25.3 m long by 15.8 m wide with wall height 7.6 m). The aeration system consisted of three half-round perforated ducts (7.3 m long and 1.2 m diameter) each connected to a non-overloading centrifugal fan, direct-coupled to a 3.7 kW motor. The design airflow rate was 1.5 L/sec/t. The airflow direction was upwards (a blowing system), and the fans were controlled by a high air wet-bulb temperature limit (Griffiths 1967) and a high air relative humidity limit.

Satisfactory insect control, at an acceptable cost and without the use of insecticides, was obtained. With no treatment, 11% moisture content wheat became badly insect infested, its temperature reached 38°C, and surface moisture migration and caking occurred. Aeration cooled the wheat of 10.5–12% moisture content to 9°C in eight months resulting in almost complete absence of

**Table 1.** Practice of grain aeration in Australia (August 1985). (Information supplied by the grain handling authority in each State.)

| State | Permanent storage capacity (t) | Aerated storage (t) | Aerated storage % of capacity |
|---|---|---|---|
| Queensland | 1 688 850 | 243 000 | 14 |
| New South Wales | 6 208 100 | 3 087 200 | 50 |
| Victoria | 2 500 000 | 1 104 180 | 44 |
| South Australia | 4 200 000 | 1 670 400 | 40 |
| Western Australia | 7 500 000 | 51 000 | 1 |
| Total | 22 096 950 | 6 155 780 | 28 |

insects. Moisture contents near the ducts increased to 14–16.5% leading to slight outloading difficulty in 1963 after 2300 hours of aeration, even though a high relative humidity limit of 75% was employed. In 1964 (1300 hours of aeration) and 1965 (1400 hours of aeration) no problems arose despite no relative humidity limit being used, showing that if care is taken to avoid excessive fan operation the relative humidity can generally be dispensed with. The capital cost of the aeration equipment, including three temperature detection cables but excluding a temperature measuring instrument, was nearly $AU6000, or $AU2.20 per tonne of store capacity (during November 1985, 1.5 Australian dollars ($AU) = US$1). The average energy used was 3 kWh/t and the average running cost was 9 cents/t (3 cents/kWh). This corresponds to 0.2 kWh/t per 100 hours of aeration or 0.6 cents/t per 100 hours. The setting of the wet-bulb temperature limit was lowered when continued aeration failed to produce any further reduction in the average grain temperature, showing that a temperature front had moved through the grain bed.

*Beulah, Victoria.* A further trial of aeration of unsprayed wheat was performed at Beulah in north-western Victoria in 1965 (Sutherland 1966a). The storage complex (Fig. 1), with a capacity of 9400 t of wheat, consisted of two identical units each made up of two large concrete bins (29.6 m high and 8.2 m diameter with a capacity of 1200 t), two small concrete bins (27.7 m high and 4.7 m diameter with a capacity of 300 t) and one steel annexe bin (13.1 m wall height and 14.5 m diameter, with a capacity of 1700 t). Each large bin had a 4.1 kW centrifugal fan connected to a half-round perforated duct (3.4 m long and 1.2 m diameter). One 2.2 kW centrifugal fan supplied two small bins, each having one half-round perforated duct (1.8 m long and 0.8 m diameter). Each steel annexe bin had a 3.7 kW centrifugal fan connected to a single submerged rectangular duct (7.9 m long by 1.2 m wide). The design airflow rate for each bin was 0.8 L/sec/t. The aeration fans were controlled by a high wet-bulb temperature limit and a high relative humidity limit on the ambient air.

**Fig. 1.** Storage complex at Beulah, Victoria (9400 t of wheat).

**Fig. 2.** Steel storage at Murtoa, Victoria (22 400 t of barley).

**Fig. 3.** Concrete storage at Rainbow, Victoria (3400 t of barley).

The average wheat temperature in one large concrete bin decreased to 9°C after 1100 hours of aeration (downwards airflow — suction) in six months. In one steel annexe bin the average wheat temperature fell to 8°C after 1500 hours of aeration (upwards airflow — blowing) in eight months. In both silos, the average temperature was below 10°C by mid-July. No evidence of deterioration in grain quality was discovered, and no insects were obvious on inspecting the grain surface during the year. It was not until the last sampling that careful sieving uncovered a few live secondary insects, mainly at the peak of the bulk. No insects were found on outloading the silos even though no insecticide had been applied to the wheat. In a test on an unaerated bin, insect numbers increased, giving rise to a steep temperature gradient near the grain surface. The capital cost of the complete aeration system was $A16 000 ($A1.70 per tonne of store capacity), and the running cost was estimated to be 5 cents/t of wheat.

*Murtoa, Victoria.* The first trial of aeration combined with insecticide (malathion) treatment in Australia was carried out at Murtoa in north-western Victoria in 1965 (Sutherland 1966b). The storage (Fig. 2) holding 22 400 t of barley consisted of 14 steel silos (14.9 m wall height and 14.5 m diameter with a capacity of 1600 t of barley). Each bin had two half-round perforated ducts (4.6 m long and 0.8 m diameter) along the hopper bottom, connected to a non-overloading centrifugal fan, direct-coupled to a 3.7 kW motor. The design airflow rate was 1.0 L/sec/t of barley, and wet-bulb temperature and relative humidity control was employed.

Aeration in two bins (one with upwards airflow and the other downwards) reduced malting barley temperatures to 8°C after 1200 hours of aeration in eight months of storage. No insects were found, and cooling of the grain prolonged the life of the malathion, which has been found to degrade rapidly above 20°C and 12% moisture content (Minett et al. 1968). The capital cost of the complete aeration system was about $A24 000 ($A1.10 per tonne of store capacity). The running cost was 5 cents/t (0.4 cents/t per 100 hours of aeration) which was comparable with the cost of application of malathion spray at 12 parts per million.

In a further aeration trial at Murtoa in 1966 on two bins of unsprayed wheat, one was cooled from 30°C to 10°C after 1100 hours of aeration in eight months and the other was cooled from 27 to 10°C after 1000 hours of aeration in eight months (Elder and Sutherland 1970).

*Rainbow, Victoria.* Another trial of aeration combined with malathion treatment was carried out at Rainbow in north-western Victoria in 1967 (Sutherland and Elder 1969). For this 3400 t barley storage (Fig. 3) each of the two large concrete bins (25.9 m high, 10.7 m diameter and holding 1360 t) was fitted with a 4.1 kW two-stage axial fan, and each of the two small concrete bins (25.9 m high, 5.5 m diameter and holding 340 t) was fitted with a 2.2 kW centrifugal blower. Submerged aeration ducts (V-shaped in plan) were used being 7.8 m long and 0.8 m wide (each leg of V) in the large bins and 3.4 m long and 0.4 m wide in the small bins. The design airflow rate was 1.0 L/sec/t of barley. Aeration was in a downwards direction and controlled by a high wet-bulb temperature limit and a high relative humidity limit. The relative humidity control was achieved by measurement of the wet-bulb depression of the air and not by the use of hair elements as in the previous systems.

In one large bin, the average centreline temperature was reduced from 30 to 13°C after 1200 hours of aeration in seven months. In the other large bin, the average temperature fell from 26 to 13°C after 1070 hours in eight months. Average barley moisture contents remained unchanged, there was no drop in germination percentage, and no insects were found during inspection at the peak of the bulk, through the bulk, and on outloading.

*Kingaroy, Queensland.* A trial aeration of peanuts-in-shell was carried out at Kingaroy in southern Queensland over two seasons (1970-71 and 1971-72). Three concrete bins (26 m high, 6 m diameter, and holding 170 t of peanuts-in-shell) were each fitted with a 0.6 kW axial fan giving an airflow rate of 2.5 L/sec/t.

All the fans were connected to a CSIRO time-proportioning aeration controller (Elder 1972).

Aeration combined with malathion application kept the peanuts-in-shell in good condition for a storage period of 10 months (Ghaly 1978). Analysis of bulk samples showed the average infestation level of the aerated bins to be significantly less that that of the unaerated bins. Free fatty acid content was also less in the aerated bins. There were no significant differences in viability and appearance of the kernels between aerated and unaerated peanuts.

FARM STORES

*Albury, New South Wales.* Two steel farm silos holding 28 and 44 t of oats were aerated for a period of two years (1969 and 1970) using two outlets of a 0.4 kW CSIRO farm aeration unit (Fig. 4) at Albury in southern New South Wales (Elder 1969). The silos had a wall height of 4.5 m, with the larger one having a diameter of 4.9 m and the smaller one 4.0 m (Fig. 5). Two full-round aeration ducts (each 2.4 m long and 0.15 m diameter) were

**Fig. 4.** CSIRO unit for farm aeration of grain.

**Fig. 5.** Steel farms silos at Albury, New South Wales (28 and 44 t of oats).

placed along the hopper bottom of the larger silo, and one duct (2.0 m long and 0.15 m diameter) supplied the smaller silo. The design airflow rate was 1.6 L/sec/t of oats.

The oats in both silos were kept in good condition even though no chemical treatment was applied (Elder 1971). The average centreline temperatures were kept between 8 and 18°C, apart apart from the first three months after harvest. Grain temperatures near the north wall remained high enough to permit some insect reproduction for much of the storage period. This was partly offset by some drying of the grain near the walls, as a result of repeated conduction heating and aeration cooling. The level of insect infestation was kept under control, although there was a large seasonal variation. Insects generally congregate at the peak of the grain bulk where they are accessible for chemical treatment if required. The moisture content of the oats reached 13% (wet basis) near the aeration ducts in both bins, and the average increased from 10 to 11%. Germination tests showed no deterioration in quality after two years storage.

*Dookie, Victoria,* Aeration of two steel farm silos, each having a capacity of 54 t of wheat, was performed at a flow rate of 2 L/sec/t using a CSIRO farm aeration unit and controller at Dookie in north-eastern Victoria. The flat-bottomed bins were 4.5 m diameter and had a wall height of 4.5 m. The full-round aeration ducts were 3.7 m long and 0.15 m in diameter.

In the first trial (1969-70), unsprayed wheat of 11% moisture content was stored successfully under aeration for 18 months, whereas an unaerated bin had to be fumigated. Wheat temperatures were reduced from 27 to 18°C in three months in the aerated bins (Williams and Elder 1979). The number of insects found in the control silo at outloading was 70 times the number (live plus dead) in one of the aerated silos. The other aerated silo was exposed to the heat of the sun and became infested at the north-western wall.

Greater cooling at the north-western wall of the exposed silo was achieved in a second trial (1971-72), by placing the aeration duct in the north-western sector of the bin floor (Ghaly 1984). This suppressed insect numbers to the same level as for the partly shaded silo. After four months of aeration, the average wheat temperatures at the centre of the silos ranged from 11 to 21°C for the next 14 months. The average wheat moisture contents increased by 2% as a result of aeration.

### Refrigerated Aeration in Large Stores

*Dalby, Queensland.* The first refrigerated aeration system for grain in Australia was installed at Dalby, southern Queensland in 1967 (Sutherland et al. 1970). The aim was to extend aeration into areas of Australia where the climate is too warm for the use of ambient air to be effective. The concrete storage (with a capacity of 12 300 t of wheat) consisted of two identical units with each comprising one large central bin (24.4 m high, 10.3 m diameter and holding 1500 t) surrounded by eight smaller bins (each 24.4 m high, 6.4 m diameter and holding 580 t). A packaged refrigeration unit of about 10 kW cooling capacity (designed and built by CSIRO) was connected to two 580 t bins as shown in Figure 6. Ambient air was drawn through louvres and a wire screen into the evaporator and the condenser (Fig. 7). The evaporator (0.84 m by 0.48 m) consisted of 6 rows of 16 mm diameter copper tubes with aluminium

Fig. 6. Refrigerated aeration unit at Dalby, Queensland (connected to two 580 t silos).

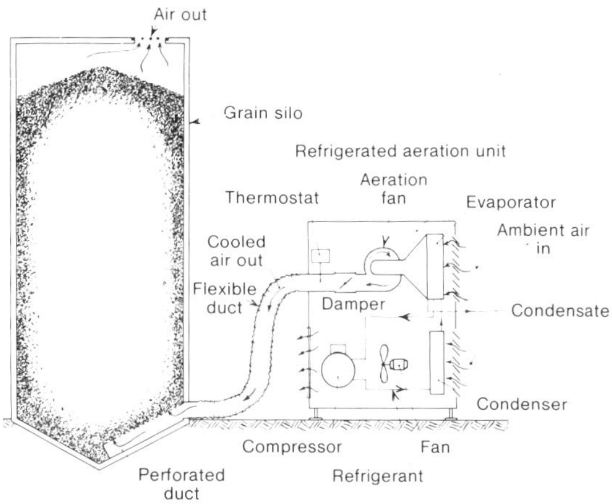


Fig. 7. Layout of refrigerated grain aeration system at Dalby, Queensland.

fins every 3 mm. Air from the evaporator passed through a 3.7 kW centrifugal fan and was discharged through a butterfly damper into the insulated aeration duct. The damper was controlled by a motor and thermostat so that the airflow rate was varied to keep the air temperature leaving the unit at about 10°C. Because of the heat added by the fan motor, the air relative humidity was reduced to below 80% for all rates of airflow. The refrigeration compressor was rated at 3.0 kW and used R22 refrigerant.

Results of three trials with wheat carried out over two seasons (Table 2) showed that insecticide-free wheat could be safely stored for 10 months, and that insect populations could be held at relatively low levels (Sutherland et al. 1970). The effect of low airflow rate and consequent inadequate cooling at the uninsulated walls of the silo was shown by the much higher infestation level in Trial No. 2. Although insect control was good in the other two trials, it was concluded that insulation or direct cooling of the silo walls would be necessary to render all parts of the grain bulk inhospitable to insects. The high electrical energy consumption of 34 kWh/t for Trial No. 1 was reduced to 18 kWh/t for Trial No. 3 by more

Table 2. Summary of results of three grain refrigerated aeration trials at Dalby, Queensland.

| Trial No.<br>Season | 1<br>1967-68 | 2<br>1967-68 | 3<br>1968-69 |
|---|---|---|---|
| Average airflow rate (L/sec/t of silo capacity) | 0.45 | 0.15 | 0.6 |
| Quantity of wheat (t) | 594 | 594 | 570 |
| Storage period (months) | 10 | 10 | 10 |
| Mean grain temperature (°C) | | | |
| Initial | 26.4 | 33.0 | 28.3 |
| Final | 10.4 | 12.1 | 15.5 |
| Time to cool to 15°C (months) | 0.6 | 2.1 | 0.6 |
| Mean grain moisture content (%, wet basis) | | | |
| Initial | 10.8 | 10.7 | 11.1 |
| Final | 12.5 | 10.9 | 10.9 |
| Total operating time (hours) | 5600 | 5600 | 1900 |
| Energy used (kWh/t of silo capacity) | 34 | 11 | 18 |
| Number of live insects | | | |
| Inloading | 0 | 0 | 0 |
| Outloading | 0 | 35 | 3 |
| Quantity of grain sieved (kg) | | | |
| Inloading | 76 | 72 | 82 |
| Outloading | 214 | 94 | 91 |

Source: Elder et al. 1984.

selective operation of the refrigeration unit (1900 hours as against 5600 hours) by employing a thermostat to sense grain temperature. Grain moisture contents near the aeration duct were also reduced in Trial No. 3 as a result of reduced aeration.

*Brookstead, Queensland.* In the next trial of refrigerated aeration, at Brookstead in southern Queensland, a 5400 t capacity squat concrete-walled corrugated-iron roofed silo (9.6 m wall height and 28 m diameter) was insulated externally with a 50 mm thickness of spray-applied polyurethane foam covered with white paint (Fig. 8), with the object of maintaining all parts of the grain bulk to be refrigerated below 15°C (Elder et al. 1975, 1976). In the first experiment the existing aeration system was used to blow air through the grain. Return ducts were connected from the headspace to the two aeration fans to recirculate the air. The headspace air was cooled via a separate recirculation system comprising a belt-driven, double-inlet centrifugal fan (designed to deliver a flow rate of 5 $m^3$/sec against a pressure of 370 Pa) and a chilled brine 8-row cooling coil (2.1 m by 1.1 m with fins at 3 mm pitch) connected to a commercial chiller set of 70 kW nominal capacity (Fig. 9). The chiller incorporated two 11 kW compressors and two air-cooled condensers fitted with three 0.6 kW propellor fans. Each system was charged with 10 kg of Freon 22 and the brine passed through two shell-in-tube heat exchangers in series. The brine system comprised a 0.75 kW centrifugal pump circulating 390 kg of 22% by weight ethylene glycol solution through the chiller, cooling coil, mixing tank, orifice flow meter, and flow switch, via 50 mm diameter insulated P.V.C. pipes.

**Fig. 8.** Grain refrigerated aeration system on an insulated 5400 t squat silo at Brookstead, Queensland.

In the second and third seasons, the cooled air was again discharged into the roof space but was then drawn downwards through the grain, thereby filtering out the dust created during filling of the silo (Fig. 9). The existing linear aeration ducts were replaced by two concentric ducts (one at the perimeter of the silo and the other at ¼ radius) connected directly to the cooling system. Each duct could be operated independently so that cooling could start as soon as the central duct was covered with grain. In the system now in use the inner duct has been removed and the chilled air is passed directly to the perimeter duct. The air flows upwards through the grain and then back to the cooling circuit.

The main results obtained at Brookstead over three seasons are given in Table 3. Cooling to less than 15°C was achieved within two months, insect control was excellent, and grain quality was unaffected (Ghaly 1976). The perimeter duct on the floor of the silo installed after the first season enabled grain temperatures below 15°C to be achieved at the wall much earlier and through the hottest months. The average grain moisture content was increased by 0.4% in each season. The average monthly energy consumptions were 3.7, 3.3, and 2.8 kWh/t over respective storage periods of 6, 9, and 10 months. The total capital cost of the system, in 1973, was $A7.40/t ($A2.60/t for insulation), and the electricity cost over six months was about 50 cents/t. The third trial was conducted using malting barley instead of wheat, and the enhanced germination and malt yield resulting from storage in the refrigerated silo has led to continued commercial use of the system for barley.

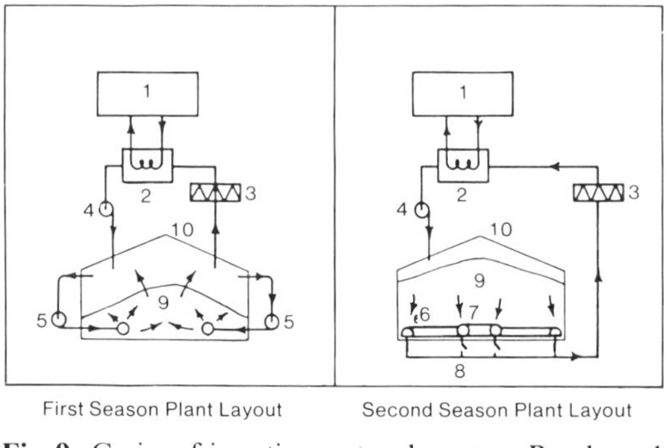


**Fig. 9.** Grain refrigeration system layouts at Brookstead, Queensland: 1. chiller; 2. cooling coil; 3. air filter; 4. fan; 5. original aeration system with recirculation duct added; 6. perimeter duct; 7. central duct; 8. duct gates; 9. grain; 10. silo.

**Table 3.** Summary of results of three grain refrigerated aeration trials at Brookstead, Queensland.

| Trial No.<br>Season | 1<br>1973-74 | 2<br>1974-75 | 3<br>1975-76 |
|---|---|---|---|
| Average airflow rate (L/sec/t of silo capacity) | 0.8 | 0.8 | 0.9 |
| Quantity of wheat (t) | 2390 | 5230 | 2230 |
| Storage period (months) | 6 | 9 | 10 |
| Mean grain temperature (°C) | | | |
| Initial | 29 | 27 | 26 |
| Final | 6 | 5 | 8 |
| Time to cool to 15°C (months) | 1.7 | 1.2 | 1.4 |
| Mean grain moisture content (%, wet basis) | | | |
| Initial | 11.0 | 11.2 | 11.3 |
| Final | 11.4 | 11.6 | 11.7 |
| Energy used (kWh/t of silo capacity) | 22 | 30 | 28 |
| Number of live insects | | | |
| Inloading | 17 | 1 | 6 |
| Outloading | 1 | 0 | 0 |
| Quantity of grain sieved (kg) | | | |
| Inloading | 384 | 543 | 255 |
| Outloading | 490 | 863 | 429 |

Source: Elder et al. 1984.

*Lah, Victoria.* The refrigerated aeration system applied to a 1700 t wheat silo at Lah in north-western Victoria (Fig. 10) was designed on the basis of computer simulation (Thorpe 1976) and the practical experience gained in Queensland (Hunter and Taylor 1980). The theoretical work showed that recirculation systems gave much lower annual costs than systems such as that at Dalby in which the air from the silo is exhausted to atmosphere. The steel silo (13.1 m wall height and 14.5 m diameter) was thermally insulated with polyurethane foam sprayed externally to a nominal thickness of 25 mm on the roof and 50 mm on the wall. The foam was protected against ultraviolet degradation by an expansive elastomeric hypalon coating. The system comprised a self-

**Fig. 10.** Refrigerated aeration system on a 1700 t silo at Lah, Victoria.

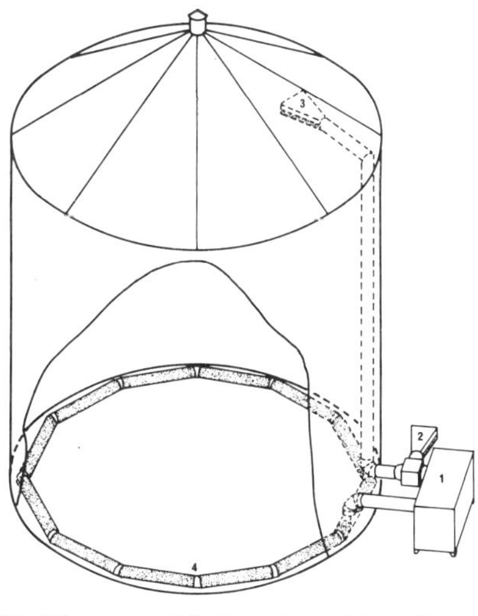


**Fig. 11.** Diagram of Lah grain refrigeration system: 1. refrigeration unit; 2. atmospheric air intake; 3. return air intake; 4. perforated air distribution duct.

contained refrigeration unit with a direct-expansion air-cooling coil connected to distribution ducting around the periphery of the site floor and a recirculation duct from the headspace (Fig. 11).

The refrigeration unit had a cooling capacity of 34 kW (i.e. 20 W per tonne of silo nominal capacity) and used a 7.5 kW compressor. The cooled air was supplied via a 7.5 kW fan through perforated half-round ducting (0.57 m diameter) arranged in a 12-sided figure on the floor of the silo as close as possible to the silo wall (Fig. 11). Return air was drawn from one side of the headspace into a circular duct (0.57 m diameter) located inside the silo. A damper and T-piece were installed in the return duct just upstream of the refrigeration unit so that atmospheric air could be drawn into the unit at inloading to overcome dust problems, and to allow the initially hot return air to be expelled. The loading hatch at the apex of the silo was left open to the atmosphere. The control circuit had thermostats to cope with defrosting the coil, to switch off the compressor when the temperature of the return air fell to 8°C, and to switch off both the compressor and the fan when the return air and ambient air temperatures were both below 11°C.

Results of two trials at Lah with unsprayed wheat (summarised in Table 4) showed that adequate insect control temperatures were achieved, energy cost was comparable to that of new protectant insecticides, and that the peripheral duct satisfactorily cooled the central grain and peak of the bulk, as well as the grain near the silo wall (Hunter and Taylor 1980). No live insects were found at either inloading or outloading of the wheat. The capital cost of the refrigerated aeration system in 1976 ws $A17.05/t of which 43% was for the thermal insulation.

**Table 4.** Summary of results of two grain refrigerated aeration trials at Lah, Victoria.

| Trial No.<br>Season | 1<br>1976-77 | 2<br>1977-78 |
|---|---|---|
| Average airflow rate (L/sec/t of silo capacity) | 1.5 | 1.5 |
| Quantity of wheat (t) | 1650 | 1750 |
| Storage period (months) | 9 | 5 |
| Mean grain temperature (°C) | | |
| Initial | 34 | 30 |
| Final | 9 | 9 |
| Time to cool to 15°C (months) | 1.2 | 1.4 |
| Mean grain moisture content (%, wet basis) | | |
| Initial | 8.8 | 8.3 |
| Final | 9.2 | 8.8 |
| Energy used (kWh/t of silo capacity) | 24 | 22 |
| Number of live insects | | |
| Inloading | 0 | 0 |
| Outloading | 0 | 0 |
| Quantity of grain sieved (kg) | | |
| Inloading | 276 | 271 |
| Outloading | 204 | 210 |

Source: Elder et al. 1984.

*Gravesend, New South Wales.* The ability of a direct-expansion refrigeration system and continuous airflow through the grain at the silo wall to provide adequate cooling performance (as at Lah) led to a concept of modular cooling units of a standard design to suit all types of grain stores (Elder 1980). Six such modified commercial air-conditioning units were installed along the south-western wall of a horizontal shed store (with a capacity of 15 000 t of wheat) at Gravesend in the northern wheat belt of New South Wales (Fig. 12). The store (96 m long by 24 m wide and 6 m wall height) comprised a concrete floor, steel grain retaining walls, and corrugated-iron gable roof. The openings at the eaves were closed off by fitting sheets of iron vertically from the top of the 6 m high grain retaining wall to the roof along the side walls, and at an angle so as to shed grain from the gable ends. The whole structure was insulated externally with 25 mm of polyurethane foam on the roof and 50 mm on the retaining walls, the insulation being covered by a white hypalon mastic coating.

Each refrigeration unit (servicing 2500 t of wheat) recirculated air drawn from above the grain surface and delivered it cooled to half-round perforated ducts (0.8 m diameter) around the periphery of the shed floor (Fig. 13). The coil of

**Fig. 12.** Refrigerated aeration system on a 15 000 t grain shed at Gravesend, New South Wales.

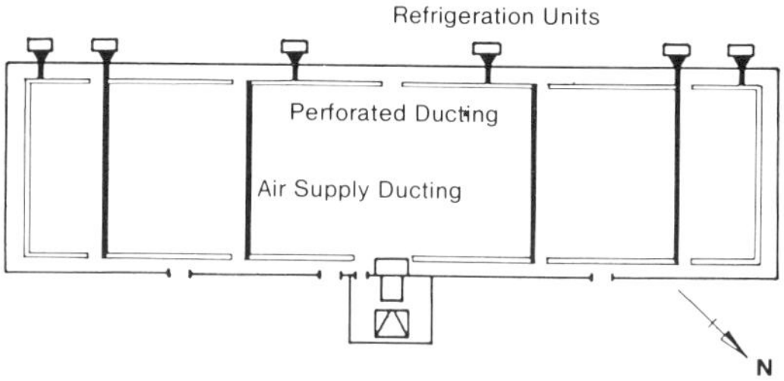


**Fig. 13.** Layout of refrigeration units and ducting in the Gravesend grain shed.

each unit had a cooling capacity of 30 kW. The power absorbed by the refrigeration condensing system and the refrigerated air recirculation fan in each unit averaged 13 kW at near full capacity in summer and 6 kW in winter.

The trials carried out at Gravesend on unsprayed wheat are described in detail by Elder and Ghaly (1984) and Elder et al. (1984) and some typical results are given in Table 5. In the 1977–78 season, the airflow rate was substantially lower than the design value, and it took three months for the average grain temperature to fall from 32 to 15°C. Nevertheless, good insect control was achieved although low insect populations could be found at the peaks of the grain surface. The average wheat moisture content increased by 1.1% (wet basis), and electricity consumption averaged 2.6 kWh/t/month. The capital cost of the system in 1977 was $A14.50/t, of which insulation accounted for 40%. In 1980–81, although the airflow rate was at the design value, the shed was only 30% full and the wheat was in three separate heaps. Insect control was reasonable, even though the wheat took 4.5 months to cool from 28 to 15°C.

**Table 5.** Summary of results of two grain refrigerated aeration trials at Gravesend, New South Wales.

| Trial No. | 1 | 2 |
|---|---|---|
| Season | 1977-78 | 1980-81 |
| Average airflow rate (L/sec/t of silo capacity) | 0.7 | 1.0 |
| Quantity of wheat (t) | 12 160 | 4335 |
| Storage period (months) | 10 | 10 |
| Mean grain temperature (°C) | | |
| Initial | 32 | 28 |
| Final | 9 | 12 |
| Time to cool to 15°C (months) | 2.9 | 4.5 |
| Mean grain moisture content (%, wet basis) | | |
| Initial | 9.9 | 10.9 |
| Final | 11.0 | 11.8 |
| Energy used (kWh/t of silo capacity) | 26 | 29 |
| Number of live insects | | |
| Inloading | 0 | 5 |
| Outloading | 5 | 19 |
| Quantity of grain sieved (kg) | | |
| Inloading | 968 | 404 |
| Outloading | 1016 | 340 |

Source: Elder et al. 1984.

## Combined Grain Cooling and Pesticide Treatment

A theoretical study by Thorpe and Elder (1982) showed that cooling of grain slowed down the rate of degradation of chemical pesticides applied to the grain. This study was confirmed by some experiments at Murtoa (north-western Victoria) in 1980 using fenitrothion (Hunter 1981). In the temperate and sub-tropical wheat-growing regions of Australia, aeration can reduce amounts of the pesticide methacrifos applied by factors of 7 and 4, respectively (Thorpe and Elder 1982). Further confirmation of this phenomenon was obtained in the experiments using malathion at Murtoa and Rainbow which were described earlier.

The combined use of cooling (with either ambient air or chilled air) and insecticides can be achieved in a number of ways. Refrigerated aeration could be used in uninsulated grain stores in addition to a low dosage of insecticide that would kill pests which could otherwise breed in warm grain near the walls and floor of the store. The insecticide would provide protection from reinfestation when the grain was outloaded from the store. Capital costs would be lower because of the absence of thermal insulation which accounts for about 40% of the total cost of the cooling system. As an alternative, grain could be cooled rapidly to 20°C, at which temperature the rate of reproduction of most insects is retarded, and a smaller dose of insecticide applied. Because of the decreased degradation of the insecticide at the lower temperature, the grain could then be held without further cooling.

## Aeration of High Moisture Content Grain

*Ardrossan, South Australia.* It was shown in a trial at Ardrossan, South Australia in 1966 that aeration enabled barley to be stored safely at a higher moisture content than usual (Doolette and Sanders 1966). The trial was conducted in two vertical concrete silos (28.3 m high, 10.7 m diameter, and having a capacity of 1700 t of barley). The silos were filled with barley having average moisture contents of 12 and 14% (wet basis), and aeration was in a downwards direction at a flow rate of 1.0 L/sec/t. It was found that after storage for nine months there were no measurable differences in quality in the barley in both silos. After 1150 hours of aeration the mean grain temperatures were 8°C (in the silo containing 14% moisture content barley) and 10°C (in the silo containing 12% moisture content barley).

*Biloela, Queensland.* A trial of aeration of grain sorghum of 16% moisture content was carried out at Biloela, near Gladstone in Queensland in 1977 (W.B. Elder, pers. comm.). Two farm silos each of 100 t capacity (3.4 m wall height and 5.8 m diameter) were used. A CSIRO farm aeration unit was connected to a full-round perforated duct (3.0 m long and 0.15 m diameter) positioned along the sloping floor of one silo, and the other silo was left unaerated. The sorghum temperature in the centre of the aerated silo decreased from 31 to 23°C in one month (after about 400 hours of aeration), whereas the unaerated silo centre temperature increased from 31 to 50°C in two weeks as a result of respiration heating. The average sorghum moisture content changed from 15.8 to 15.2% in the aerated silo and from 15.7 to 15.9% in the unaerated silo. The germination percentage changed from 92 to 89% in the aerated bin and to 80% in the unaerated bin. No mould was detected, but the unaerated sorghum had an unpleasant odour on outloading. Aflatoxins were detected in the unaerated bin but not in the aerated bin. This trial showed that grain sorghum can be stored safely for limited periods under aeration while waiting to be dried. This practice is now in commercial use in large aerated silos at Gladstone, before the sorghum is dried in a large tower dryer.

## Solar Grain Cooling for the Humid Tropics

In temperate climates, ambient air is often of sufficiently low temperature and enthalpy to provide effective cooling of a grain bulk. However, in tropical climates, which are characterised by hot and humid ambient air conditions, the amount of grain cooling achievable may be inadequate to prevent loss of grain quality. In this case, lower grain temperatures can be achieved by reducing the enthalpy of the ambient air by isothermally lowering its absolute humidity before passing it into the grain bed. This can be carried out in an open-cycle, solar-regenerated, desiccant-bed grain cooling system (Thorpe 1981; Thorpe and Fricke, 1985).

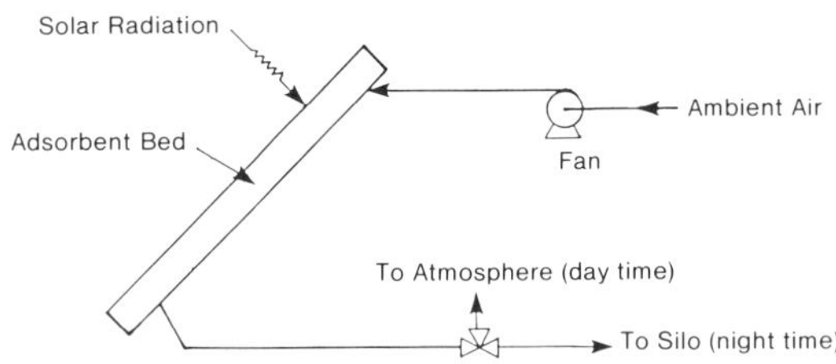

**Fig. 14.** Schematic arrangement of components in the adsorbent-bed grain cooling system.

The principle of operation of this device, which can be employed in single or multiple stages, is shown in Figure 14. During the night, ambient air is blown through a narrow bed of dry desiccant material so that the heat of sorption can be readily dissipated to atmosphere, and the air of reduced enthalpy is passed into the grain. The desiccant becomes wet, but it is dried during the day by solar radiation falling on the desiccant bed and the simultaneous passage of air through it. During the regeneration process (daytime) the air leaving the adsorbent bed is expelled to atmosphere.

Experiments have shown that by using the solar cooling system, grain temperatures up to 10°C lower than that produced by normal ambient aeration can be achieved (Thorpe 1981). The system is easily built from sheet metal, has no more moving parts than a conventional grain aeration system, can use seed grains as the adsorbent, and ensures drier grain.

## Discussion and Conclusions

Trials of a variety of aeration systems using both ambient and refrigerated air have been described. The trials used a mixture of upwards and downwards airflow, and manual and automatic control. In general, upwards flow aeration is now preferred, mainly because the heat of compression added by the aeration fan reduces the relative humidity of the air entering the grain bed. This can remove the need for a humidistat control and restrict the moisture content of the grain near the aeration duct.

When aeration began in Australia, control of fan operation was performed by manual setting of temperature and humidity limits on the ambient air. Also, it was shown by Griffiths (1967) that wet-bulb temperature control was preferable to dry-bulb temperature control, as the temperature reached by the grain is dependent on the air wet-

**Fig. 15.** Mobile unit for natural aeration of grain.

**Fig. 16.** Mobile unit for refrigerated aeration of grain.

bulb temperature and the grain moisture content. However, this practice of manual setting of limits made supervision of operation at remote country sites difficult. A number of other controller types, e.g. temperature difference and time switch have been tried, but almost all new aeration systems now installed in Australia use the CSIRO time-proportioning controller (Elder 1972). This controller requires no operator intervention, and automatically selects the coldest periods for aeration fan operation by means of a thermostat which winds down when the fan is running and winds up when the fan is not running. The controller has a rapid setting which ensures operation for 12 hours per day on average and a normal setting which allows 15 hours per week on average. The time-proportioning controller is now commercially available in Australia in both electro-mechanical and electronic forms.

The aeration systems described in this paper have all been fixed. However, mobile grain cooling units are quite possible and a few, using both ambient and refrigerated air, are in use in Australia (Figs 15 and 16). Both the units shown are mounted on trailers and are designed to cool 2000 t of wheat at airflow rates from 1.0 to 1.5 L/sec/t. They use 10 kW centrifugal fans and the mobile refrigerated aeration unit (Fig. 16) has a cooling capacity of 44 kW. The capital costs, in 1980, were about $A2500 for the mobile natural aeration unit and $A15 000 for the mobile refrigerated aeration unit (Hunter 1981).

In general, any degree of grain cooling achieved by aeration is beneficial to the grain, and thus even in the humid tropics aeration can be recommended as a worthwhile method of reducing grain storage problems. In these climates, a technique which would reduce the enthalpy of the ambient air, such as the solar grain cooler described in this paper or some form of dehumidifier, would of course be beneficial.


### Acknowledgments

The financial support of the Australian Wheat Industry Research Council for this work is gratefully acknowledged. The author thanks the Australian Centre for International Agricultural Research for the invitation to prepare this paper, and for the provision of funds to permit its presentation at the seminar.


### References


Doolette, J.B., and Sanders, P.T. 1966. Study of barley stored in bulk at Ardrossan in 1966. Adelaide, South Australia, South Australian Co-operative Bulk Handling Limited.

Dyte, C.E. 1974. Problems arising from insecticide resistance in storage pests. European Plant Protection Organisation Bulletin, 4(3), 275–289.

Elder, W.B. 1969. CSIRO develops aeration system for farm stored grain. Power Farming and Better Farming Digest, 78(10), 10–13.

Elder W.B., and Sutherland, J.W. 1970. Aeration of wheat and barley in Victorian country storages in 1966. Melbourne, Australia, CSIRO Division of Mechanical Engineering, Internal Report No. 85.

Elder W.B. 1971. Farm aeration trial continues to give good results. Power Farming and Better Farming Digest, 80(5), 12–13.

Elder W.B. 1972. New aeration techniques for the preservation of stored cereals. In: Proceedings of the 22nd Annual Conference of the Cereal Chemistry Division of the Royal Australian Chemical Institute, Bronte, New South Wales, September 1972, 205–212.

Elder W.B., Hunter, A.J., and Griffiths, H.J. 1975. Refrigeration of bulk wheat in a thermally insulated silo for control of insect infestation. Australian Refrigeration, Air Conditioning and Heating, 29(12), 38–46.

Elder W.B., Hunter, A.J., and Griffiths, H.J. 1976. Refrigeration of an existing grain storage for insect infestation control. International Institute of Refrigeration, Joint Meeting of Commissions, Melbourne, Victoria, September 1976, Paper No. 46.

Elder W.B. 1980. Grain refrigeration systems incorporating modular cooling units. Australian Refrigeration, Air Conditioning and Heating, 34(7), 8–13.

Elder W.B., and Ghaly, T.F. 1984. Insect population dynamics at the surface of refrigerated wheat bulks of up to 15,000 tonnes. Proceedings of the 3rd International Working Conference on Stored Entomology, Manhattan, Kansas, U.S.A., October 1983, 78–98.

Elder W.B., Ghaly, T.F., and Thorpe, G.R. 1984. Grain refrigeration trials in Australia. In: Ripp, B.E. ed., Controlled atmosphere and fumigation in grain storages. Developments in agricultural engineering 5. Amsterdam, Elsevier, 623–643.

Evans, D.E. 1981. Temperature manipulation for control of storage pests. In: Champ, B.R., and Highley, E. ed., Grain storage research and its application in Australia. Canberra, CSIRO Division of Entomology, 125–139.

Ghaly, T.F. 1976. Refrigerated aeration trials of stored wheat: insect control and quality changes. In: Proceedings of the 26th Annual Conference of the Cereal Chemistry Division of the Royal Australian Chemical Institute, Broadbeach, Queensland, October 1976, 24–28.

Ghaly, T.F. 1978. An aeration trial of bulk stored peanuts for the control of insect infestation and quality loss. Journal of Stored Products Research, 14(1), 45–51.

Ghaly, T.F. 1984. Aeration trial of farm-stored wheat for the control of insect infestation and quality loss. Journal of Stored Products Research, 20(3), 125–131.

Griffiths, H.J. 1964. Bulk storage of grain. A summary of factors governing control of deterioration. Melbourne, Australia, CSIRO Division of Mechanical Engineering, Report E.D.8.

Griffiths, H.J. 1967. Wet-bulb control of grain aeration systems. Melbourne, Australia, CSIRO Division of Mechanical Engineering, Circular No. 3.

Hunter, A.J., and Taylor, P.A. 1980. Refrigerated aeration for the preservation of bulk grain. Journal of Stored Products Research, 16(3/4), 123–131.

Hunter, A.J. 1981. Economics of combined cooling and pesticide treatment. In: Proceedings of the 1st Australian Stored Grain Pest Control Conference, Melbourne, Victoria, May 1981, Section 4, 20–26.

Minett, W., Belcher, R.S., and O'Brien, E.J. 1968. A critical moisture level for malathion breakdown in stored wheat. Journal of Stored Products Research, 4(2), 179–181.

Murray, W.J. 1979. Infestations in the Australian wheat industry during the past two decades — immediate short term solutions and a prospective for the future. In: Evans, D.E. ed., Australian contributions to the symposium on the protection of grain against insect damage during storage, Moscow, July 1978. Canberra, CSIRO Division of Entomology, 7–14.

Sutherland, J.W. 1965. Bulk barley storage experiment at Beulah in 1964. Melbourne, Australia, CSIRO Division of Mechanical Engineering, Internal Report No. 31.

Sutherland, J.W. 1966a. Aeration of unsprayed bulk stored wheat at Beulah in 1965. Melbourne, Australia, CSIRO Division of Mechanical Engineering, Internal Report No. 41.

Sutherland, J.W. 1966b. Aeration of bulk stored barley at Murtoa in 1965. Melbourne, Australia, CSIRO Division of Mechanical Engineering, Internal Report No. 32.

Sutherland, J.W. 1968. Control of insects in a wheat store with an experimental aeration system. Journal of Agricultural Engineering Research, 13(3), 210–219.

Sutherland, J.W., and Elder, W.B. 1969. Aeration of wheat and barley in Victorian country storages in 1967. Melbourne, Australia, CSIRO Division of Mechanical Engineering, Internal Report No. 68.

Sutherland, J.W., Pescod, D., and Griffiths, H.J. 1970. Refrigeration of bulk stored wheat. Australian Refrigeration, Air Conditioning and Heating, 24(8), 30–34 and 43–45.

Sutherland, J.W., Banks, P.J. and Griffiths, H.J. 1971. Equilibrium heat and moisture transfer in air flow through grain. Journal of Agricultural Engineering Research, 16(4), 368–386.

Sutherland, J.W., Banks, P.J., and Elder, W.B. 1983. Interaction between successive temperature or moisture fronts during aeration of deep grain beds. Journal of Agricultural Engineering Research, 28(1), 1–19.

Thorpe, G.R. 1976. The design of refrigerated grain storages. International Institute of Refrigeration, Joint Meeting of Commissions, Melbourne, Victoria, September 1976, Paper No. 68.

Thorpe, G.R. 1981. A solar cooling system for bulk stored food grains. Agricultural Engineering Australia, 10(1), 14–19.

Thorpe, G.R., and Elder, W.B. 1982. Modelling the effects of aeration on the persistence of chemical pesticides applied to stored bulk grain. Journal of Stored Products Research, 18(3), 103–114.

Thorpe, G.R., and Fricke, P.W. 1985. Experiments on an open-cycle solar-regenerated desiccant-bed grain cooling system. Presented at the international conference on renewable energy, Commonwealth Science Council, March 1985, Reduit, Mauritius.

Wedd, S. 1974. Grain aeration under semi-coastal conditions. Glenfield, Australia, New South Wales Department of Agriculture, Technical Report No. 74/6.

Williams, P., and Elder, W.B. 1979. Cooling farm stored grain to limit insect reproduction. Agricultural Engineering Australia, 8(1), 3–11.

# Paddy Drying in Australia


## Lindsay D. Bramall*



*Abstract*

The rice industry of New South Wales, Australia began in the 1920s as a pioneering venture in the newly established Murrumbidgee River Irrigation Scheme. Early days saw mechanical harvesters stripping the rice at low moisture for bagging, before further drying, storage, and milling. For bag storage, a harvesting moisture limit of 16.5% was imposed. However, because of unfavourable weather, this was difficult to achieve. At the same time, crop size was increasing and 'suncracking' was correlated with rapid drying and wetting in the field. These circumstances eventually led to a change from bag to bulk handling. Since the late 1950s, paddy has been harvested at up to 23% moisture content and dried in deep-bed storage using high-velocity airflows. This strategy dries the paddy to a safe, millable moisture level (13.5%) under controlled conditions and maximises the final grain quality and whole grain yield. Deep-bed drying was well established by the 1970s and research has continued since then on refining the drying process with regard to aeration strategies, energy conservation, grain quality, electronic monitoring and control, pre-heating of air, and automation using minicomputers.


THE growing of rice in New South Wales (NSW), Australia began in the 1920s, thousands of years after neighbouring Asian countries. The impetus to grow rice came from the search for crops that could be grown in the newly created Murrumbidgee River Irrigation Area (MIA) in south-western NSW. However, inexperience, lack of consultant expertise, and inadequate implements resulted in a slow establishment of the industry. It was not until the 1930s that rice growing began to flourish, with annual crops jumping from hundred to tens of thousands of tonnes of paddy. Crop sizes have continued to increase since then (Table 1).

From the beginnings of rice growing in NSW in 1924 until 1956, all paddy was bagged in the field and stored in bags in the rice marketing authority's storage sheds. For satisfactory storage in bags, it was found that the moisture content of the grain should not exceed 16.5%, otherwise there was a danger of mould growth during adverse weather conditions.

In those days, the rice harvest was spread over the months of May, June, and July, when rainy weather is not uncommon, and it was frequently difficult for farmers to reduce the moisture content of the crop to 16.5% in the field. When autoheaders were introduced, the rate of stripping increased and it became more and more difficult to cope economically with the crop in bags.

Bulk handling of wheat had been practiced in Australia for a considerable time, and in 1951 the Rice Marketing Board set about investigating ways and means of handling the rice crop in bulk, as a way of achieving greater efficiency and reducing costs. At about the same time, it became evident that there were other benefits to be obtained if the paddy could be stripped and put into storage at moisture contents up to 23%.

Paddy dried to 16.5% or less in the field, was

**Table 1.** Growth of the New South Wales rice industry.

| Harvest year | Area cropped (ha) | Production (t) | Yield (t/ha) |
|---|---|---|---|
| 1925 | 64 | 232 | 3.63 |
| 1930 | 8 093 | 34 405 | 4.25 |
| 1935 | 8 847 | 36 553 | 4.13 |
| 1940 | 9 819 | 35 859 | 3.65 |
| 1945 | 9 953 | 32 809 | 3.30 |
| 1950 | 15 006 | 70 365 | 4.69 |
| 1955 | 15 651 | 99 723 | 6.37 |
| 1960 | 19 783 | 132 154 | 6.68 |
| 1965 | 24 894 | 157 622 | 6.33 |
| 1970 | 39 033 | 243 485 | 6.24 |
| 1975 | 72 316 | 374 577 | 5.18 |
| 1980 | 110 257 | 585 980 | 5.31 |
| 1985 | 118 300 | 844 000 | 6.76 |


* Ricegrowers' Co-operative Mills Ltd, P.O. Box 561, Leeton, New South Wales 2705, Australia.

subject to cracking. Faults appeared in the kernels of the grain and, during milling, these kernels tended to break more readily, thus reducing the 'mill-out'. This phenomenon is called 'suncracking.'

It was shown that if paddy could be stripped and safely stored at moisture contents up to 23% the incidence of suncracking was greatly reduced, thereby increasing the value of the crop.

In 1956, the Rice Marketing Board converted a bag shed at Leeton to bulk storage with an aeration system consisting of a series of mesh ducts and tubes connected to exhaust fans which drew air through the grain from the top and exhausted the moist air to atmosphere through the axial-flow exhaust fans.

This system demonstrated that, in the climatic conditions of the region, drying with unheated air was effective.

Another experiment was simultaneously carried out at Yenda, near Griffith, in conjunction with the Commonwealth Scientific and Industrial Research Organization (CSIRO) to obtain fundamental data on paddy drying with ambient unheated air. A pilot bin consisting of a radially aerated, cylindrical bin with perforated walls was used.

Both of these experimental grain drying/storage systems were developed to a commercial scale. The radial aeration design was the basis for construction of 60 × 100 t bins (6000 tonnes total) in the Murray Valley rice growing region, to the south of the MIA. Although this was a one-off exercise, these bins are still a functional part of the paddy storage system.

The Leeton 'mesh-duct' pilot plant was a very successful experiment and formed the basis for continuing development. From the experience gained, a rather different system was developed, based on research work carried out at the University of California. This involves blowing air through the grain from the bottom of the bulk, a system which, although simpler from an engineering point of view, proved just as effective. This principle has been maintained, the only changes being in the design of ductwork at floor level. It has proven desirable to have a flat floor. The air is therefore introduced under the floor and passes into the grain through perforated steel sheets flush with the concrete floor of the storage bin.

An alternative to this 'below floor' manifold system is the semi-circular cross section, mesh-covered 'above-floor ducting'. Both air manifolding systems are equally acceptable for aeration, but if the bin or shed is to be used for other purposes, such as storing bagged, milled rice, the semi-circular ducts may be inconvenient and interfere with access by forklift trucks.

It was determined from this early work, that paddy with an initial moisture content of 20–22% could be satisfactorily dried in bulk storage in beds of 5.5–6 m deep by properly controlled, unheated aeration, using an aeration rate of 1 cubic foot per minute per cubic foot of grain.

Considerable supervision of this operation was necessary because the rate of drying depended upon weather conditions.

### Growth of the Rice Industry

As the crop size increased through to the 1960s and 1970s, storage capacity also rose, to cope with the drying and storage requirements. The basic design followed the early trial recommendation and is reflected in the 66 horizontal storage sheds dispersed throughout the growing area today.

Paddy was received into storage sheds with little concentration on moisture segregation. Varietal, dockage, and quality segregation took precedence, and the only attempts to segregate on the basis of moisture were to keep aside the very wet paddy and very dry paddy.

As soon as bins were filled, fans were switched on and the aeration program begun. Aeration was controlled by monitoring paddy moisture by sampling at a depth of 1.5 m from the bin surface. Each bin was sampled once per week with 20 samples being taken per 1000 t bin. The results of the tests were then forwarded to Aeration Officers who determined an aeration program for each bin depending on the moisture results obtained.

This program was followed irrespective of the time of year or climatic conditions, except that no aeration was performed during periods of rain or when rain was expected. Paddy delivered for milling at higher than milling moisture content, was dried in the mill hot-air dryers, with single or double pass dependent on moisture. This generally meant that hot-air dryers were used almost continuously for the first few months of each crop year, particularly through the harvest period.

In 1979, based on experiences over immediately preceding harvests, the aeration program was critically examined by technical personnel. This led to the implementation of the following strategy for the 1979 harvest:

- segregation of paddy at receival based on moisture content;
- strict control of moisture testing to ensure at least one series of moisture measurements was completed for each bin each week;
- weekly recording of bin temperatures to enable the overall aeration program to be changed as necessary;
- changes to the aeration program to ensure use of low-speed fans in place of high-speed wherever possible;
- an overriding control of individual bins to enable account to be taken of climatic conditions, paddy needs, and the most cost-efficient drying practices available.

The relatively large size of the 1979 harvest (resulting in 674 000 tonnes) meant that a further complication arose and had to be dealt with. Mills had to be kept running at maximum throughput during harvest to mill as much paddy as possible. This meant minimising double-pass drying at mills which dramatically reduced mill throughput. During the harvest, sufficient paddy had to be dried to below 14.0% to permit safe storage in hired structures with little or no aeration.

These objectives were met during the harvest by performing a large series of half-bin filling and drying operations. By filling the bins to a depth of only 3 m, the resistance to airflow was greatly reduced, resulting in drying of the paddy in approximately one quarter of the time.

Analysis of the performance of the control strategies adopted in the drying of this crop resulted in further changes which were implemented to develop a more efficient aeration program for the 1980 harvest. These were: additional managerial support to control the entire system; upgrading moisture and temperature monitoring to enable more accurate control of drying and storage; and additional managerial aids to store and analyse drying data, so that more accurate aeration programs could be developed for each bin.

The first of these changes was achieved with the appointment of Aeration Supervisors, to liaise with the Co-operative's Technical Services Department in management of drying operations. These supervisors greatly improved the control and efficiency of all drying operations.

After exhaustive evaluation of various devices for remote sensing of moisture, it was decided that no suitable unit was available to monitor moisture in deep beds of grain. The Technical Services Department set about the design and construction of some 7000 sensors based on principles used in other devices. The sensors were completed before the 1980 harvest and were used successfully throughout that and the subsequent harvest.

Apart from construction and installation of a sensor network, a minicomputer was purchased to store and analyse all data collected, and provide an aeration program for each bin, based on contemporary moisture and temperature information.

The minicomputer was programmed to provide an updated aeration program for each bin from daily measurements of moisture and temperature. Receival and transfer data were also fed into the computer to give up-to-date details of bin status. The computer was also used to appraise each bin of paddy in terms of expected whole grain yield. Fans were then switched 'on' or 'off' manually by Aeration Officers according to the aeration program generated by the computer. The computer was programmed with various aeration strategies, the basic objective being to provide dry rice with minimum consumption of electricity.

The following general principles were used in developing this strategy: maximising of aeration in the early period of harvest, making use of the best available climatic conditions; maximising use of half-bin filling; once dry, use of minimum air and, where possible, low-speed fans.

A direct result of the implementation of these strategies was the virtual total abandonment of oil-fired, hot-air column drying and an overall reduction in electricity consumption. These economies were effected despite the fact that rice which was previously dried with column dryers required in-store bin drying and in the face of a 41% increase in the size of the crop. They are illustrated in Figures 1 and 2.

Crop tonnage increases continued to put pressure on the ability to store and dry the paddy. In 1981, a 3000 t rice bin was fitted with LPG gas burners at the aeration fan inlets. Temperature rises of 10°C resulted in marked improvements in drying rates without significantly affecting the whole grain yield, thus greatly increasing the production of dry millable paddy. From the costings it was shown that the additional expense for gas was easily offset by the reduced electricity consumption due to shorter fan running hours. So

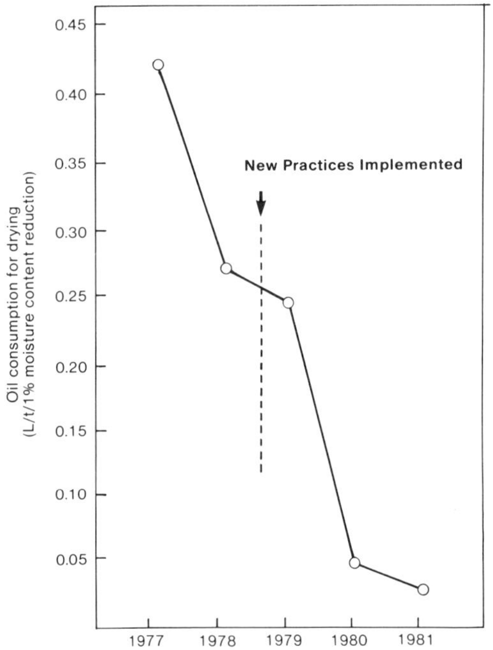


**Fig. 1.** Reduction in oil consumption for grain drying, following adoption of new practices for drying paddy in Australia.

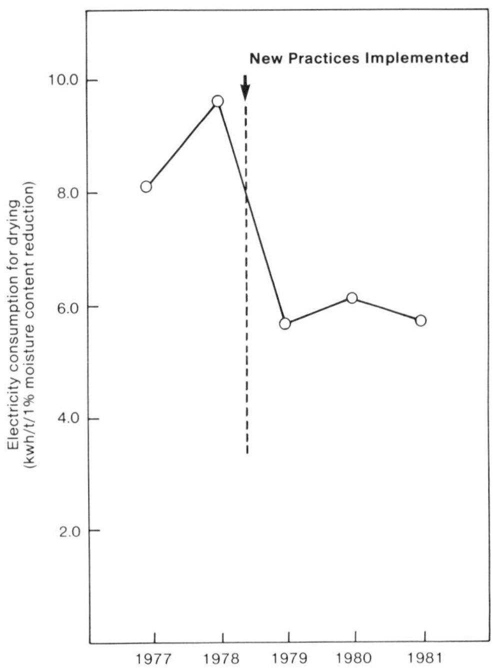


**Fig. 2.** Reduction in electricity consumption for grain drying, following adoption of new practices for drying paddy in Australia.

successful was this installation that the remaining 3 × 5000 t bins were fitted with burners. An indication of the efficiency of this 18 000 t capacity shed to act as a fast-drying system is shown by the fact that up to 100 000 t of wet paddy now passes through it each year.

In anticipation of a one million tonne crop in 1985, nine additional 12 000 t drying sheds were fitted with LPG gas burners to increase drying capacity early in the harvest and to dry moist paddy in the normally unsuitable wet weather winter.

### Fine Tuning of the Drying System

It is our belief that, in terms of horizontal bulk drying and storage of paddy, the past 60 years have seen the development of a highly efficient aeration, drying, and storage system with all areas of major impact having been investigated and refined. Such being the case we have now reached the stage where increases in efficiency will come from fine tuning of the existing strategy by implementing computer-based control and management systems.

In 1981, three 12 000 t paddy storage sheds at the Gogeldrie depot were set up as a first fully automated aeration and drying system.

Individual bins of paddy were continuously monitored for moisture content and temperature, by use of an on-site computer.

In conjunction with the measurement of paddy moisture content, the system also monitored ambient weather conditions and automatically operated the aeration fans on an individual bin basis.

This automatic drying system chose ambient air which gradually dried the grain to an acceptable moisture, as compared with the manual method which can often result in either wetting or overdrying.

The centre of the automatic drying system, the depot control computer receiving information from moisture and temperature sensors in individual bins through the depot, determined the quality of the ambient air for drying purposes and relayed 'on' or 'off' commands to individual aeration fans, as depicted in Figure 3.

Because there were more than 400 separate sensors at the Gogeldrie storage site, it was impossible to individually wire each sensor back to a central location. An advanced communi-

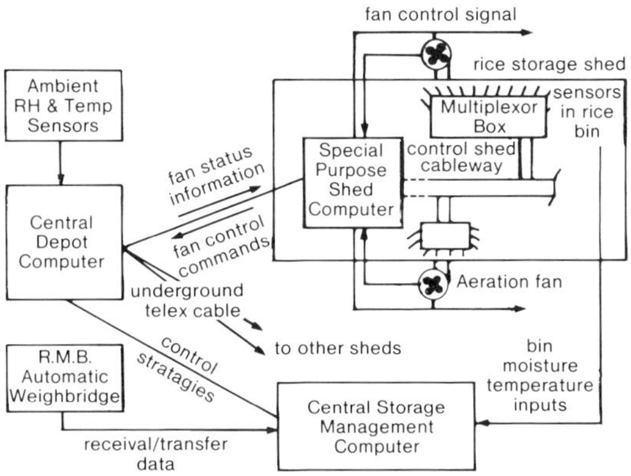


**Fig. 3.** Ricegrowers' Co-operative Mills' automated aeration system for bulk-stored paddy.

cations scheme was developed to relay sensor readings to the computer.

Each storage shed had been equipped with its own special-purpose shed computer, which acted as a relaying point by gathering bin moisture information in the shed and transmitting this to the central depot control computer via an underground telex line. The special-purpose shed computer also relayed the 'on' or 'off' commands to specific aeration fans in the shed.

The Aeration Supervisor in Leeton was kept informed by daily reports from the depot control computer on the status of grain in individual bins, the hours of fan usage, and other information needed to monitor the quality of stored grain.

As well as controlling grain drying, the depot control computer was linked to the rice marketing authority's automatic weighbridge, and was thereby able to record receivals and despatches of bulk grain at the depot.

This depot was a prototype. After it had operated for two harvests, another depot of similar size was set up using 'off the shelf' electronic hardware. This second automated depot has run very successfully for three harvests and has been used to test many of the aeration strategies from which the current system has evolved.

## The State of the Art

The immense amount of work that has gone into the development of the Australian in-store drying system has been aimed at achieving:

- a higher whole grain yield by accepting high moisture paddy into store for controlled drying;
- optimisation of the paddy moisture to a consistent 13.5% by aeration with air selected on a relative humidity/equilibrium relative humidity basis;
- a reduction in fan operating costs by air selection, fan design, and power load shedding;
- a reduction in manpower costs by implementing remote grain and air sensing systems linked to microprocessor controllers and computers.

The most recent step towards achieving even finer control has been in the area of a microprocessor-controlled aeration system with supplemental heating through modulated gas burners. The operation of fans and burners is dependent on the program held within the microprocessor. With inputs from ambient relative humidity and temperature sensors the controller operates fans to achieve the required number of hours at a preset relative humidity. If such air is not available the burners are automatically ignited to reduce relative humidity by raising air temperature by a maximum of 5°C.

The system is new and is still being assessed. However, it is expected to achieve new levels of drying efficiency, while reducing the problems of manual fan switching at remote locations and eliminating concern over the effects of overnight changes in climatic conditions.

In summary, Ricegrowers' Co-operative Mills Limited has been very active in the aeration and drying of paddy rice. Great changes have occurred in the industry as a result of its research and development work, and it is now recognised as a leader in the field of paddy drying technology.

# Grain Aeration and In-store Drying in the USA

**Do Sup Chung, Boma Kanuyoso, Larry Erickson, and Chong-Ho Lee***

*Abstract*

Grain aeration is a well accepted practice for maintaining the quality of stored grain. Successful experiences in using aeration systems have been obtained from commercial to farm level in the USA. Of the various drying methods, in-store drying has in recent years been generally accepted at the farm level in the USA.

This paper reviews several studies made of aeration and in-store drying, especially for shelled maize and rough rice, at various locations in the USA. The recommended operating conditions and management methods for grain aeration and in-store drying are discussed, as are the applicability and limitations of aeration and in-store drying methods developed in the USA. In addition, a research project on aeration of rough rice under humid air conditions currently being conducted at Kansas State University is briefly described.

GRAIN aeration is the practice of forcing air through stored grain at low airflow rates to prevent or reduce deterioration of grain quality. The practice of turning grain to equalise temperature and reduce hot spot formation can usually be successfully replaced by aeration. Preventing moisture migration by maintaining a relatively uniform temperature throughout the grain mass, and limiting biological activity by cooling the grain, are probably the most common objectives of grain aeration. Aeration may also reduce the hazard of spontaneous heating when it is necessary to hold moist grain in storage for brief periods. Some storage odours may be removed by air changes during aeration. Grain aeration systems have been used successfully in all principal grain production areas in the United States.

In-store drying can be defined as a process of slow drying with natural air or air heated by only a few degrees (1–6°C). Grain is dried and stored in the same bin. This is also known as deep bed drying and is one method of in-bin drying (Anon. 1980).

In-store drying systems are attractive to grain farmers because of relatively low initial equipment costs, minimum energy input costs, reduced fire hazards, and reduced operational management (Bunn et al. 1981).

* Food and Feed Grain Institute, Kansas State University, Manhattan, Kansas 66506, USA.

Extensive studies carried out in a number of locations in the USA have provided most of the design criteria utilised in present in-store drying systems. These criteria, however, are based upon local weather and crop conditions.

The objective of this paper is to provide information by reviewing the results obtained from several studies of grain aeration and in-store drying systems, mostly for shelled corn and rough rice at a number of locations in the USA. Recommended operating conditions and management methods for grain aeration and in-store drying are emphasised. The applicability and limitations of the two processes are discussed.

## Grain Aeration System: Some Technical Considerations

Aeration may be used to improve the grain storage conditions at either on-farm storage or commercial storage facilities (Calderwood 1975). Although the two types of grain storage are different, the principles of aeration in one type are, in most instances, applicable to the other.

Studies on aeration on flat and deep vertical bins have been discussed by Holman (1960). These studies were originally conducted in cooperation with the agricultural experiment stations of Georgia, Indiana, Iowa, Kansas, Michigan, and Texas.

A study on aeration systems, especially for corn

and wheat, in deep bins and flat storages was carried out in Iowa and Indiana by Foster and Stahl (1959). Kline and Converse (1961) carried out an extensive study on grain aeration systems in the hard winter wheat area, with the objective of describing effective methods of using aeration systems to maintain the market quality of grain. Schroeder and Calderwood (1972) discussed the design and operation of aeration systems for rough rice for short-term storage. This study was elaborated for long-term storage by Calderwood et al. (1983).

Proper management of fan operation in grain aeration systems was discussed by Brooker et al. (1974). Foster and Tuite (1982) have focused their discussion on the design and operation of the system.

**Types of Aeration Systems**

The principal parts of a grain aeration system are: (1) one or more fans to supply the required air; (2) one or more ducts to move air into or out of stored grain; (3) supply pipes to connect the fans and ducts; (4) a motor to operate each fan; (5) a control system to regulate fan operation; and (6) a storage bin in which the grains are stored and aerated.

Holman (1960), and Kline and Converse (1961) pointed out that there are three major types of aeration systems based on combinations of fans and ducts used in the system. These are: (1) a fixed fan system, with a permanently placed fan for each storage bin or for each duct in a storage; (2) a manifold system, with a fan connected to two or more storage bins or ducts through the same manifold; and (3) a portable system, with a portable fan movable from bin to bin or from one part of a storage to another part. Fans and ducts are the two major pieces of equipment needed for the proper delivery of aerating air through the grain mass in terms of the required rate and uniformity of the airflow.

**Airflow Requirements**

The airflow rate used in grain aeration systems must be adequate to cool all the grains before undesirable changes take place. The airflow rate required depends on the purpose of aeration, the kind of grain, the size and type of storage structure, and climatic conditions (Foster and Tuite 1982).

Compromises are often needed in determining the airflow rate because of limitations on fan power, especially when the resistance to airflow in the system is high.

**Table 1.** Recommended airflow rates for aerating grains in the United States.

| Grain moisture content (%) | Storage type | Airflow rates $m^3/sec/m^3 \times 10^{-3}$ | |
|---|---|---|---|
| | | Northern States | Southern States |
| 12–15 | Farm storage | 0.67–1.344 | 0.67–3.35 |
| | Flat | 0.672–1.344 | 0.896–3.35 |
| | Upright | 0.336–0.672 | 0.448–1.344 |
| 15–25 | Flat | 6.72–10.08 | |
| | Upright | 3.36–6.72 | |

Sources: Holman 1960; Hall 1980; Foster and Tuite 1982.

Table 1 shows the recommended airflow rates for the aeration of stored grain in the USA, as given by Holman (1960), Hall (1980), and Foster and Tuite (1982). The most commonly used airflow rates range from $0.672 \times 10^{-3}$ to $1.344 \times 10^{-3}$ $m^3/sec/m^3$ grain. These rates are generally adequate for reducing insect and mould activity and for keeping moisture migration and accumulation within acceptable limits.

Many researchers agree that in most farm-type storages, aeration rates in the range $1.344 \times 10^{-3}$ to $0.672 \times 10^{-3}$ $m^3/sec/m^3$ are normally used. An airflow rate of $1.344 \times 10^{-3}$ $m^3/sec/m^3$ is widely used to aerate shelled corn and soybeans, and for wheat and other smaller seed crops, $0.672 \times 10^{-3}$ $m^3/sec/m^3$ is more common (Frus 1979; Halderson and Sandvol 1980; Foster and Tuite 1982). Referring to the aeration of dry rough rice in Southern U.S. rice growing region, Calderwood (1975) pointed out that an airflow rate of $1.344 \times 10^{-3}$ $m^3/sec/m^3$ is generally considered to be adequate.

**Static Pressure and Fan Horsepower Requirements**

A fan should be selected on the basis of manufacturer's performance ratings. The static pressure, against which the fan must deliver the required airflow, is dependent upon total airflow rate, kind of grain, and grain depth. Fan horsepower requirements are determined on the basis of the total amount of airflow needed and the resistance encountered in forcing the air through the grain. The fan power required increases as the grain depth and the velocity of the air moving through the grain increases. This accounts for the limitation of air delivery through a deep bin.

A convenient method to estimate the static pressure and fan horsepower requirements has been given by Holman (1960) for aerating shelled corn, wheat, rough rice, grain sorghum, oats, and soybeans at different rates of airflow and at depths ranging from 3.05 to 45.72 m. Figures 1 and 2 show the relationship for aeration of shelled corn and rough rice, respectively.

Grain resistance to airflow is obviously essential in designing a grain aeration system. Although many research studies have been made on the resistance to airflow of many kinds of grains and seeds, less attention has been given to providing data on airflow resistance at the low range of airflow rates, which is commonly found in aeration condition. The data reported by Shedd (1953) were based on the minimum airflow rate of 0.0051 $m^3/sec/m^3$. Also, they were based on a relative clean and dry grain. There are only a few data available on the effects of moisture content and fines and broken contents on airflow resistance (Chung et al. 1984; Chung et al. 1985).

## Time Required for Cooling Grain

As aeration progresses, a cooling zone, which moves through the grain in the direction of airflow, is established. The airflow must be adequate to move the cooling zone through all the grain within an allowable time limit. This factor is the most important consideration in establishing time limits for completing a cooling stage.

Holman (1960) pointed out that the time required for cooling a bin of grain to near the outside air temperature depends on the airflow rate, the amount of cooling resulting from evaporation of moisture from the grain during aeration, and the uniformity of the airflow.

A general guide established by Holman (1960) for grain cooling time required in the central part of the United States indicates that cooling time at an airflow rate of $1.344 \times 10^{-3}$ $m^3/sec/m^3$ and under favourable conditions is about 80 hours in the summer, 120 hours in the fall, and 160 hours in the winter. The effect of seasonal differences in the amount of evaporative cooling on cooling time has been demonstrated by Foster and Tuite (1982) as seen in Figure 3.

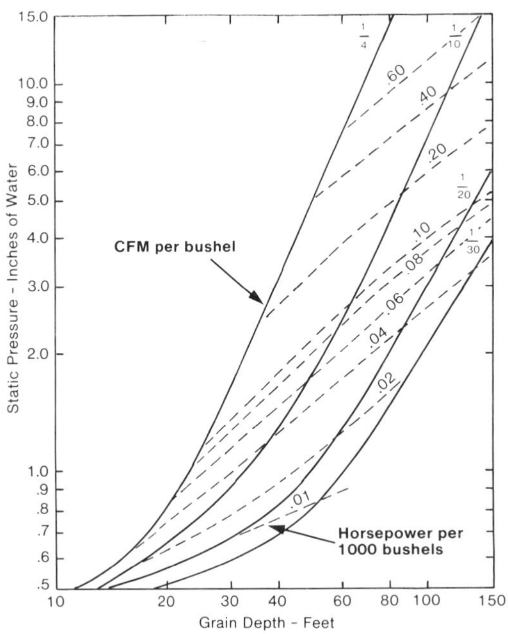


**Fig. 1.** Fan horsepower and static pressure requirements for aerating shelled maize at various airflow rates and at grain depths ranging from 10–150 feet (Holman 1960). Conversion factors: 1 cfm/bushel = $13.4 \times 10^{-3}$ $m^3/sec/m^3$; 1 hp = 745.7 J/sec; 1 inch = 2.54 cm; 1 foot = 0.3048 m; 1 bushel = 0.03524 $m^3$.

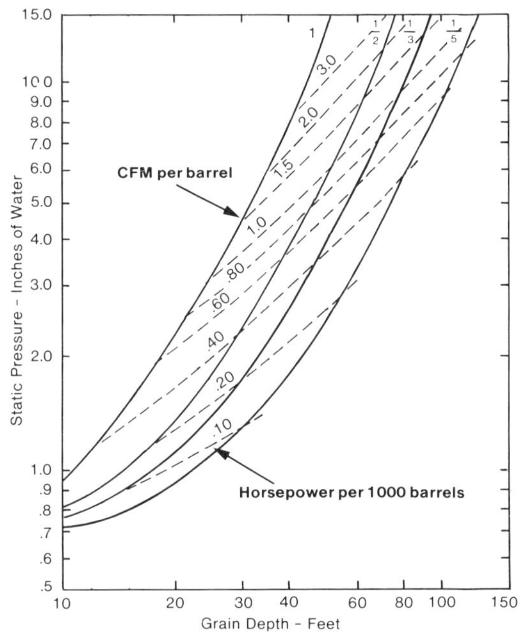


**Fig. 2.** Fan horsepower and static pressure requirements for aerating rough rice at different airflow rates and at grain depths ranging from 10–150 feet (Holman 1960). Conversion factors: 1 barrel (U.S. dry) = 3.281 bushels = 0.1156 $m^3$; 1 cfm/bushel = $13.4 \times 10^{-3}$ $m^3/sec/m^3$; 1 hp = 745.7 J/sec; 1 inch = 2.54 cm; 1 foot = 0.3048 m.

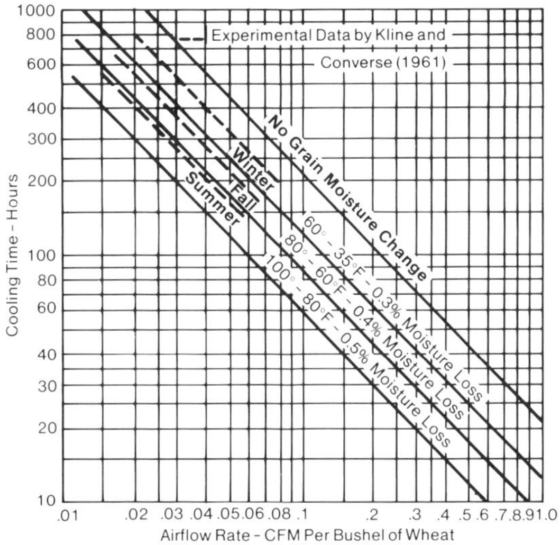


**Fig. 3.** Effect of airflow rate (cfm = cubic feet/minute) and grain moisture loss on cooling times required (Foster and Tuite 1982). Coversion factors: 1 cfm/bushel = 13.4 $\times$ $10^{-3}$ $m^3/sec/m^3$; °C = (°F − 32)/1.8.

**Table 2.** Time required for cooling rough rice, as affected by airflow rates.

| Airflow rates ($m^3/sec/m^3$ x $10^{-3}$) | Time required for cooling (hours) |
|---|---|
| 3.160 | 17 |
| 2.528 | 21 |
| 1.264 | 42 |
| 0.632 | 84 |
| 0.316 | 168 |

Source: Schroeder and Calderwood 1972.

Additional data have been produced by Schroeder and Calderwood (1972) showing the time required to cool dry rough rice under Texas weather conditions, as shown in Table 2. An airflow rate as low as 0.376 $\times$ $10^{-3}$ $m^3/sec/m^3$ was found suitable for cooling dry, rough rice during winter or for equalising temperature during other months.

## Recommended Operating Conditions in Grain Aeration

Proper operation of the grain aeration system is essential for maintaining the quality of stored grain. Various conditions and factors must be considered in establishing an effective operation of grain aeration systems.

### Grain Moisture Content and Temperature

The condition of grain to be aerated is very important and can determine the success or failure of aerating. Grain can absorb or give up moisture depending on the vapour pressure differential of the air and moisture in grain. When the grain neither loses nor gains moisture, it is in moisture equilibrium with the cooling air.

Holman (1960) determined the maximum moisture content and the corresponding relative humidity for the safe storage of grains, as given in Table 3.

**Table 3.** Maximum mixture contents for safe storage and corresponding relative humidities for different grains and seeds.

| Grain or seed type | Maximum moisture content for safe storage (%, wet basis) | Relative humidity of air at which grain is in equilibrium with air at 25°C (%) |
|---|---|---|
| Shelled corn, oats | 13.0[a] | 61 |
| Wheat (hard red winter) | 13.0[a]–13.5 | 64–68 |
| Wheat (soft red winter) | 13.0[a]–13.5 | 67–73 |
| Wheat (hard red spring) | 14.0[b]–14.5 | 7 |
| Soybeans | 11.0 | 68 |
| Rice | 13.0 | 71 |
| Pea beans | 16.0–18.0 | 76–84 |
| Grain sorghum | 13.0 | 65 |

Source: Holman 1960.
[a] Southern areas 12%.
[b] Higher moisture limits applicable because of lower average air temperatures in producing areas.

Especially during summer, grain often goes into storage at temperatures of 30°C or higher. Accordingly, grain harvested during summer and fall should be cooled as soon as the atmospheric conditions permit, to avoid deterioration. However, there is no single optimum storage temperature. Storage practices have shown that a grain temperature of 10°C is usually satisfactory, especially if the grain is to be moved during hot weather. If it is not moved, a temperature of 1.5 to 4.5°C is desirable (Holman 1960; Frus 1979).

### Atmospheric Conditions

Based on past weather records and results of grain aeration tests that have been carried out, the suggested maximum monthly operating air

**Table 4.** Suggested maximum monthly operating air temperature for cooling grain in specified areas of the USA.

| Month | South East[a] Operating temperature (°F) | South East[a] Operating temperature (°C) | South West[b] Operating temperature (°F) | South West[b] Operating temperature (°C) |
|---|---|---|---|---|
| July | 85 | 29.44 | 90 | 32.22 |
| August | 85 | 29.44 | 90 | 32.22 |
| September | 85 | 29.44 | 80 | 26.67 |
| October | 70 | 21.11 | 70 | 21.11 |
| November | 60 | 15.56 | 60 | 15.56 |
| December | 45 | 7.22 | 60 | 15.56 |
| January | 45 | 7.22 | 50 | 10.00 |
| February | 50 | 10.00 | 50 | 10.00 |
| March | 55 | 12.78 | 50 | 10.00 |
| April | 60 | 15.56 | 60 | 15.56 |
| May | 75 | 23.89 | 70 | 21.11 |
| June | 80 | 26.67 | 85 | 29.44 |

Source: Holman 1960.

[a] Allows aproximately 6 hours of operation each day at relative humidities below 80%.

[b] Allows about 12 or more hours of operation each day under normal conditions.

temperatures for cooling grains were determined by Holman (1960). Data for several areas are presented in Table 4.

It is also a common practice to aerate grain with air having an average relative humidity approximating those values listed in Table 3. However, aeration with air having a relative humidity as high as 80% has proven satisfactory when the air temperature is at least 6°C lower than the grain temperature.

Kline and Converse (1961) suggested that in summer and fall, fans may be operated at all times when the relative humidity is 90% or below and air temperatures are satisfactory. This high limit eliminates fan operation during extended periods of rain and fog. During winter, a maximum relative humidity of 80% is suggested to restrict increase in grain moisture content.

It is interesting to note that if the grain to be stored has a temperature of 35 to 38°C or higher and excess moisture, it may be desirable to run the fan when air humidities are above 80% (Holman 1960). According to McKenzie et al. (1980), the cooling front moves through the grain about 50 times faster than a drying or wetting front. Therefore, only a small fraction of the grain is rewetted during an aeration stage, even with high humidities. Kline and Converse (1961) and Foster (1976) have pointed out that the relative humidity of the incoming air is not critical for the bulk of the grain, since the air is warmed as it passes through the grain and the humidity is reduced.

### Direction of Airflow

Aeration can be accomplished by moving the air up or down through the grain. Both downward and upward aeration systems have advantages and disadvantages. In some situations, one system works better, but whether the airflow is upward or downward makes little difference in fan requirements or air delivery (Foster and Tuite 1982).

Two major reasons of moving air down through the grain are as follows: (1) downward movement counteracts the natural tendency of convection currents to move upward to cold surface grain where condensation may occur, and (2) the warm and moist exhaust air is expelled through warm grain in the lower part of the bin. The main disadvantage of moving air downward is the uncertainty of knowing when aeration is complete, since grain at the bottom is the hardest to check.

In the warmer southern area, the use of moving air upwards through the grain has more advantages. The main advantage is that heat trapped under the roof can be forced out at the top, rather than being pulled downward. A more uniform airflow can be obtained by moving the air upward, and aeration progress can be easily determined by checking the grain temperature at the top center. However, especially during very cold and freezing weather, the air moving up through the grain may be cold enough for moisture to condense on the cold grain and on the underside of the bin roof (Holman 1960).

## Grain Aeration System Management

System management is a major factor in proper operation of grain aeration systems. From it, comes the decision which will lead to determination of fan operation schedules (Foster and Tuite 1982).

### Problems in Operation

Experience has shown that the methods of operating grain aeration systems often vary from one area to another, due to the different climatic conditions. Further, the purpose of aeration is not always the same.

In northern states, where there is a wide difference in winter and summer temperatures, the prevention of moisture change during storage is probably the most important use of aeration. In the more humid southern areas of the United States, the problem of maintaining grain with moisture content near the upper limit for safe storage is the major concern, besides the problem of insect control. In these areas, aeration is used to keep the grain as cool as practical in order to stop moulding, heating, and other deterioration processes.

Larger capacity aeration systems are used to speed the cooling during the limited amount of favourable cooling weather in the south (Foster and Stahl 1959).

Another precaution should be taken when dealing with summer-harvested grains, particularly wheat and oats, which require a different operational schedule to corn and other grains harvested in the fall. It is often necessary to remove the harvest heat from wheat and oats quickly to avoid damage from heating and insects.

### Operation Schedules

To successfully aerate stored grain for long-term storage, it is necessary to calculate the number of days required to cool a bin of grain and total aeration time per year for a lot of stored grain. The number of days required depends on the percentage of each day that is suitable for aeration, and in estimating total aeration time per year, the number of cooling stages involved should be considered.

Ordinarily, grain is cooled by aerating it by successive stages during the summer, fall, and winter. There is a possibility of cooling grain in either a single stage or two stages, by postponing a certain part of the aeration process until later in the season when the air temperature is low enough to cool the grain to the desired temperature. One objection to this last method, especially cooling in a single stage, is that aeration may be so late that some grain has deteriorated or moisture migration has occurred. Typical cooling of grain in three stages is illustrated in Figure 4.

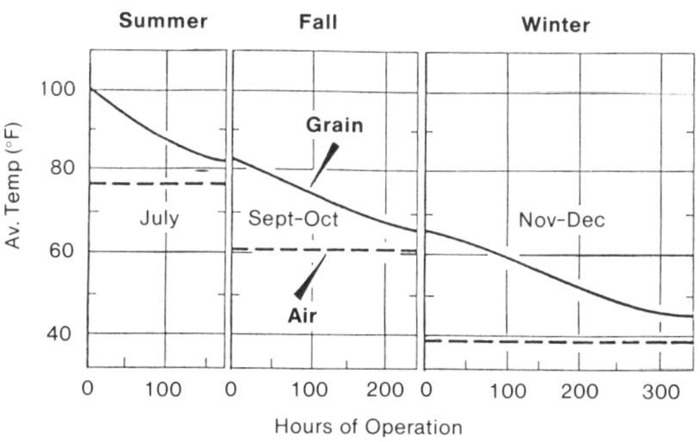


**Fig. 4.** Aeration of wheat by stages at an airflow rate of $0.672 \times 10^{-3}$ $m^3/sec/m^3$ (Holman 1960). Conversion factor: °C = (°F − 32)/1.8.

Foster and Tuite (1982) described the aeration schedule for the corn belt area in the United States as follows: (1) the grain should be aerated as soon as possible to reduce the temperature to 15–21°C for summer-harvested grain and to 10–15°C for fall-harvested grain. Summer-harvested grain normally needs to be aerated again in the early fall to reduce the temperature to 10–15°C; (2) the grain is then aerated in the late fall to cool it to between -1 and 4°C; no further aeration is recommended until spring, unless storage problems develop; (3) in March or April, grain may be aerated to warm it to 10–15°C.

The schedule outlined above was based on daily 24 hours fan operation where appropriate conditions exist. If the average relative humidity remains high for several days, the fan should not be operated except when the relative humidity is below 70–80%. Intermittent aeration can be scheduled according to forecast weather information. A control system, which is usually a combination of thermostats and a humidistat, is required for such an aeration system.

Experiments conducted by Calderwood et al. (1983) indicate that in the Texas rice area, the weather conditions are such that 100 hours or more of aeration system operating time can be expected each month from November through March using a thermostat set for fan operation at temperatures of 15°C or lower in series with a humidistat set for fan operation at relative humidites of 70% or lower.

### Recommendations for Operating Grain Aeration System

Some specific recommendations for successful aeration in the midwestern USA as described by Brooker et al. (1974) and summarised by Bakker-Arkema et al. (1978) are as follows: (1) operate the aeration fan for a few days immediately following the initial storage of the grain; (2) operate the fan only when the ambient temperature is 6–8°C below the grain temperature and the relative humidity of the air is below 70%; (3) cool the grain to between 2 and 10°C depending on the length of storage; (4) operate the fan for several hours a week; and (5) warm the grain to 21°C in the spring by reversing the process used in the fall.

## Applicability of Grain Aeration in the Humid Tropics

In warm, humid, tropical areas, where the conditions are mostly favourable for the growth and activities of microorganisms and insects, a major portion of the grain postharvest losses is still due to the inadequate facilities and improper ways of handling, drying, and storing grain materials. The prevailing atmospheric conditions in these areas are characterised by relatively high temperatures ranging from 21 to 35°C, and humidities averaging 75–80% or higher.

It is obvious that cooling by aeration to reduce the temperature to less than 18–21°C is unlikely to be achieved, because cold air is not available.

Storage structures are normally not insulated, and seasonal and diurnal climatic changes of air temperature and solar radiation therefore cause the outer portions of the grain mass to gain or lose heat and change temperature. Moisture migration may occur as a result of non-uniform temperatures within the stored mass of grain. The practice of aeration is needed in particular to equalise the grain temperature during storage. It should be carried out in conjunction with other preservation treatments, such as fumigation.

As the relative humidity increases to higher than 75%, the use of moving air through the grain should be very carefully managed, as some moisture is likely to be adsorbed from the air. Depending on local weather conditions, there may be possibilities for aerating stored grain on occasional days or for a few hours on each of a number of days when aeration will be effective. Information on this topic is still very limited and more research is needed.

## Aerating Rough Rice under Warm, Humid Conditions

A research project on aeration of rough rice under warm, humid conditions is being conducted at Kansas State University. The primary objective is to investigate the extent of heat and mass transfer occurring during aeration of dry rough rice in a storage column, under conditions which may represent the typical atmospheric conditions in humid tropical areas. Experiments are conducted with clean, dry, long grain rough rice and with the conditions as follows:

| | |
|---|---|
| Initial moisture content | 13.0–14.5% |
| Initial grain temperature | 29.44–35°C |
| Aerating air temperature | 21.11–29.44°C |
| Aerating air relative humidity | 70–90% |
| Aerating airflow rates | $1.344 \times 10^{-3}$–$5.376 \times 10^{-3}$ m³/sec/m³ |

The tests are made with the insulated grain column inside a controlled environmental chamber where both temperature and relative humidity are maintained at the desired level. The conditioned air is blown upward through the grain by a fan equipped with a variable speed motor. The temperatures and moisture contents of the grains during aeration are measured at certain positions along the bed column and at certain time intervals, until a cooling stage is completed.

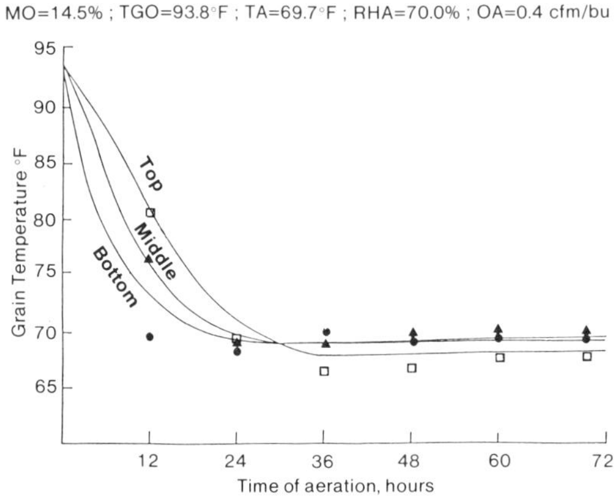


**Fig. 5.** Grain temperature versus time of aeration through rough rice. Conversion factors: 1 inch = 2.54 cm; °C = (°F − 32)/1.8; 1 cfm/bushel = $13.4 \times 10^{-3}$ m³/sec/m³.

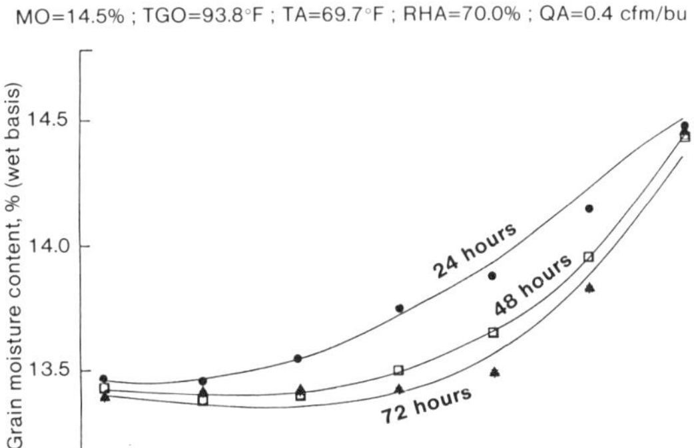


**Fig. 6.** Grain moisture content versus distance of air travel through rough rice during aeration. Conversion factors: 1 inch = 2.54 cm; °C = (°F − 32)/1.8; 1 cfm/bushel = $13.4 \times 10^{-3}$ m³/sec/m³.

Experimental work is still underway. Some preliminary experiments have been conducted and the data obtained are presented in Figures 5 to 8. Figures 5 and 6 show the changes in grain temperatures and moisture content, respectively, under the following aeration conditions: (1) initial grain moisture content, 14.5%; (2) initial grain temperature, 93.8°F (34.3°C); (3) aerating air temperature, 69.7°F (20.9°C); (4) aerating air relative humidity 70%; and (5) aerating airflow rate 0.4 cfm/bu (5.376 × $10^{-3}$ $m^3/sec/m^3$). From Figure 5, it can be found that rough rice temperature decreases to temperatures lower than the average aerating temperature of 69.7°F

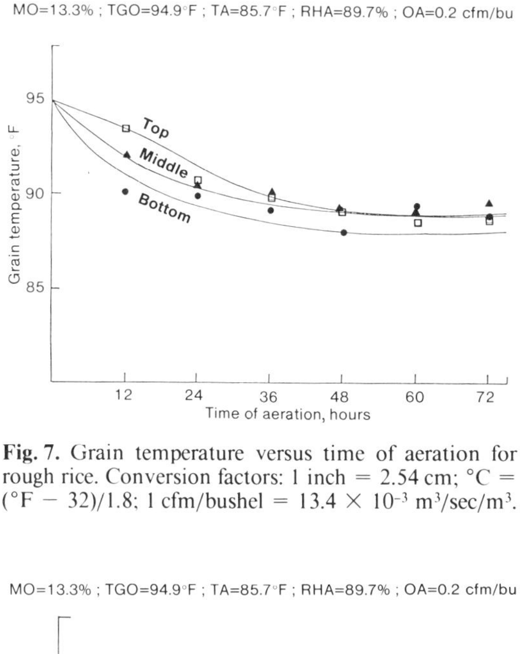


**Fig. 7.** Grain temperature versus time of aeration for rough rice. Conversion factors: 1 inch = 2.54 cm; °C = (°F − 32)/1.8; 1 cfm/bushel = 13.4 × $10^{-3}$ $m^3/sec/m^3$.

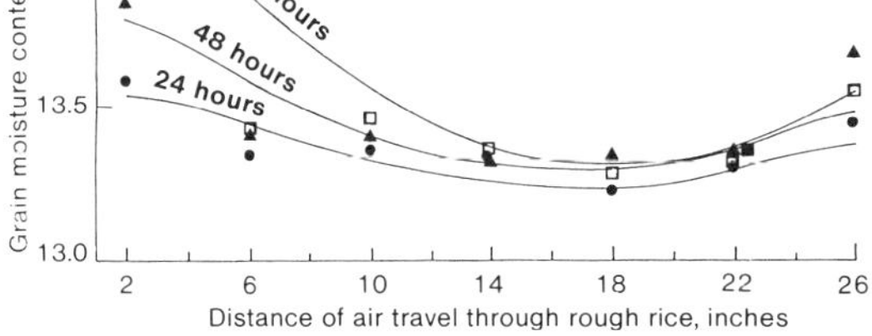


**Fig. 8.** Grain moisture content versus distance of air travel through rough rice. Conversion factors: 1 inch = 2.54 cm; °C = (°F − 32)/1.8; 1 cfm/bushel = 13.4 × $10^{-3}$ $m^3/sec/m^3$.

(20.9°C), indicating that evaporative cooling of the grain has taken place during the aeration. This might be explained by considering that the aerating air temperature is 24.1°F (13.4°C) lower than initial grain temperature, so that the vapour pressure differential between moisture in grain and vapour in air is strong enough to cause moisture transfer from the grain to the air. This is also in agreement with the decrease in moisture content, especially in the lower part of the grain bed as shown in Figure 6. Figure 5 also shows that cooling time for this aeration is about 27 hours, which is higher than the values given by Holman (1960) and Schroeder and Calderwood (1972). A better comparison could be made if aeration conditions for these other studies were known in detail.

Figures 7 and 8 show the grain temperature and moisture content changes as affected by aeration under the following conditions: (1) initial grain moisture content, 13.32%; (2) initial grain temperatures, 94.9°F (34.9°C); (3) aerating air temperature, 85.7°F (29.8°C); (4) aerating air relative humidity, 89.7%; and (5) aerating airflow rate 0.2 cfm/bu (2.688 × $10^{-3}$ $m^3/sec/m^3$). From Figure 7, it can be observed that grain temperature decreases to around 88.5°F (29.8°C). The relative humidity of the aerating air is relatively high, and aerating air temperature is less than 10°F (5.6°C) lower. Adsorption is therefore more likely to occur than desorption. This is in agreement with the increase of moisture content of the grain during aeration particularly in the lower part of the rough rice column. From Figure 7, cooling time is found to be about 48 hours, which is generally in agreement with cooling time reported previously. After 48 hours cooling, the increase in average grain moisture content is about 0.16%, which can be considered small.

## In-store Drying Systems: General Technical Background

Although in-store drying has been known and used for a considerable time since the 1950s, experience shows that the technique has not always been well adapted. As already pointed out, these systems are attractive to grain farmers; however, by using ambient air or slightly heated ambient air for the drying medium, slower rates of drying will generally be associated with the systems. This will increase the risk for grain spoilage. The potential for spoilage, in many instances, outweighs all other economic considerations (Saul and Lind 1958; Steele et al. 1969; Thompson 1972). Ambient air conditions play a major role in determining the effectiveness of in-store drying systems.

The design, management, and performance of in-store drying systems have been treated in general in a number of publications from the United States Department of Agriculture (USDA), state agricultural experiment stations, and industry. Research and experience by these groups have yielded reliable design data for most of the features of such systems. Among the most important guidelines are the recommendations given by USDA (1965) on maximum grain depth and minimum airflows for natural air drying in the USA, as shown in Table 5.

Important information on the principles of in-store drying, which is also known as low temperature drying, and recommended practices for the North Central region have been presented, giving emphasis on the system for shelled corn (Anon. 1980). Foster (1953) conducted tests to determine the airflow rates required for drying wheat and shelled corn with unheated air under Indiana weather conditions. Tests to determine the practicability of drying rough rice in storage bins at Beaumont, Texas, were made by Morrison (1954). The results obtained indicate that rough rice can be dried using unheated air under weather conditions in the Texas rice belt.

Henderson (1955) conducted studies to determine the optimum air rate for deep bed, unheated-air rice dryers for California installations, to determine the effect of various performance features upon final rice quality, and to attempt to apply the findings to other rice producing areas. He concluded that deep bed, unheated air rice dryers would produce good quality rice in California if the proper airflow rate through the mass was applied.

Sorenson and Crane (1960) made tests at Beaumont, Texas, during seven crop years (1952–53 through 1958–59) to determine the practicability of drying rough rice in-store in Texas. The results obtained emphasised the importance of time-temperature-initial moisture relationship in reducing the moisture content of rice to below 15%.

**Table 5.** USDA recommended maximum grain depths and minimum air flows for natural air drying[a].

| Type of grain | Grain moisture content (%) | Recommended depth of grain (feet) | (m) | Recommended minimum airflow (cfm/bu) | ($m^3/sec/m^3 \times 10^{-3}$) |
|---|---|---|---|---|---|
| Wheat | 20 | 8 | 2.44 | 3 | 40.3 |
| | 18 | 10 | 3.05 | 2 | 26.9 |
| | 16 | 12 | 3.66 | 1 | 13.4 |
| Oats | 25 | 8 | 2.44 | 3 | 40.3 |
| | 20 | 11 | 3.35 | 2 | 26.9 |
| | 18 | 12 | 3.66 | 1.5 | 20.2 |
| | 16 | 16 | 4.88 | 1 | 13.4 |
| Shelled corn | 25 | 6.5 | 1.98 | 5 | 67.2 |
| | 20 | 10 | 3.05 | 3 | 40.3 |
| | 18 | 12 | 3.66 | 2 | 26.9 |
| | 16 | 16 | 4.88 | 1 | 13.4 |
| Grain sorghum | 20 | 8 | 2.44 | 3 | 40.3 |
| | 18 | 10.5 | 3.20 | 2 | 26.9 |
| | 16 | 16 | 4.88 | 1 | 13.4 |
| Rice | 22 | 6 | 1.83 | 4 | 53.8 |
| | 20 | 8 | 2.44 | 3 | 40.3 |
| | 18 | 8 | 2.44 | 2 | 26.9 |
| Barley | 20 | 8 | 2.44 | 3 | 40.3 |
| | 18 | 10 | 3.05 | 2 | 26.9 |
| | 16 | 14 | 3.14 | 1 | 13.4 |
| Soybeans | 20 | 10 | 3.05 | 3 | 40.3 |
| | 18 | 12 | 3.66 | 2 | 26.9 |
| | 16 | 16 | 4.88 | 1 | 13.4 |

Source: USDA 1965.
[a] These are somewhat conservative. No location specified.

Since the advent of computer technology, a number of researchers have attempted to simulate the in-store drying of grain. Several models have been proposed to describe the heat and mass transfer processes in the in-store grain drying systems. It appears that computer simulation offers the greatest potential for predicting performance of the system.

An in-store drying system generally consists of a bin for both drying and storing grain, a fan and motor to force the air through the product, and an appropriate air distribution system. The air distribution systems commonly used are a perforated floor or various arrangements of ducts and laterals. Hall (1980) has pointed out that the success of in-store drying systems depends on proper selection and design of component parts, proper management of the systems, and desirable atmospheric conditions.

**Allowable Storage Time**

In-store drying is an effort to get the grain dried before it spoils. The key is to move the drying zone up and through the top of the grain within the allowable storage time. To select an airflow rate for a natural air drying system, information on allowable storage time is necessary.

Mould is the major cause of spoilage in grain. By controlling moisture content and temperature, mould growth is restricted and grain can be dried without significant spoilage. Grain temperature and moisture determine the allowable storage time

**Table 6.** Allowable storage time for shelled corn[a]. (Developed from Thompson, transaction of ASAE 333–337, 1972.)

| Grain temp. deg. F | Corn moisture, percent | | | | | | |
|---|---|---|---|---|---|---|---|
| | 18 | 20 | 22 | 24 | 26 | 28 | 30 |
| | | | | days | | | |
| 30 | 648 | 321 | 190 | 127 | 94 | 74 | 61 |
| 35 | 432 | 214 | 126 | 85 | 62 | 49 | 40 |
| 40 | 288 | 142 | 84 | 56 | 41 | 32 | 27 |
| 45 | 192 | 95 | 56 | 37 | 27 | 21 | 18 |
| 50 | 128 | 63 | 37 | 25 | 18 | 14 | 12 |
| 55 | 85 | 42 | 25 | 16 | 12 | 9 | 8 |
| 60 | 56 | 28 | 17 | 11 | 8 | 7 | 5 |
| 65 | 42 | 21 | 13 | 8 | 6 | 5 | 4 |
| 70 | 31 | 16 | 9 | 6 | 5 | 4 | 3 |
| 75 | 23 | 12 | 7 | 5 | 4 | 3 | 2 |
| 80 | 17 | 9 | 5 | 4 | 3 | 2 | 2 |

Source: Anon. 1980; Hellevang 1983. Developed from Thompson 1972.

[a] Conversion factor: °C = (°F − 32))/1.8

(AST) for corn which shows how long grain can be kept before it spoils, as shown in Table 6. This table shows the allowable storage time for corn, which is the time until 0.5% dry matter decomposition is reached. The data show that as moisture content or temperature increases, the allowable time for drying and storing decreases. This means that wetter or warmer grain requires higher airflow rates in order for drying to be accomplished within the allowable time (Hellevang 1983).

**Airflow through Grain**

Airflow is the major factor affecting the performance of in-store drying systems. Airflow rates largely determine drying time. In the USA, these airflow rates are measured in cubic feet per minute per bushel (cfm/bu), where 1 cfm/bu is equal to 0.01344 $m^3/sec/m^3$. From previous experiments conducted by comparing the progress of the drying front in the bins, which differ only in the way they are equipped and their energy requirements, it was found that the more air is delivered, the faster the drying front moves through the grain, the greater the amount of water removed, and the more reliable the system is. However, there are practical limits to how much airflow can be delivered to a full bin because the grain acts as a barrier. This airflow resistance is largely influenced by airflow rates, grain depth, and kind of grain. Grain cleanliness and the amount of packing are other influencing factors (Anon. 1980).

Sorenson and Crane (1960) recommended an airflow rate $12.9 \times 10^{-3}$ $m^3/sec/m^3$ for rough rice having an initial moisture content of 25% wet basis. With this airflow rate, the rice bed usually should be no deeper than 2.44 to 3.01 m (8 to 10 ft) for economical use of power. In the California rice-growing area where the climate is drier and cooler than in the Texas rice-growing area, Henderson (1958) recommended airflow rates of $5.7 \times 10^{-3}$ $m^3/sec/m^3$ and $9.2 \times 10^{-3}$ $m^3/sec/m^3$, respectively, for rough rice having initial moisture contents of 20 and 25%.

Resistance to airflow is obviously fundamental in the design of in-store drying systems. Many research studies of resistance to airflow have been done on the measurement of pressure drop in a deep bin at various depths of grain and airflow rates. The work of Shedd (1953) resulted in the availability of important data relating pressure drop with airflow rate and grain depth for many kinds of grains and other seeds, as shown in Figure

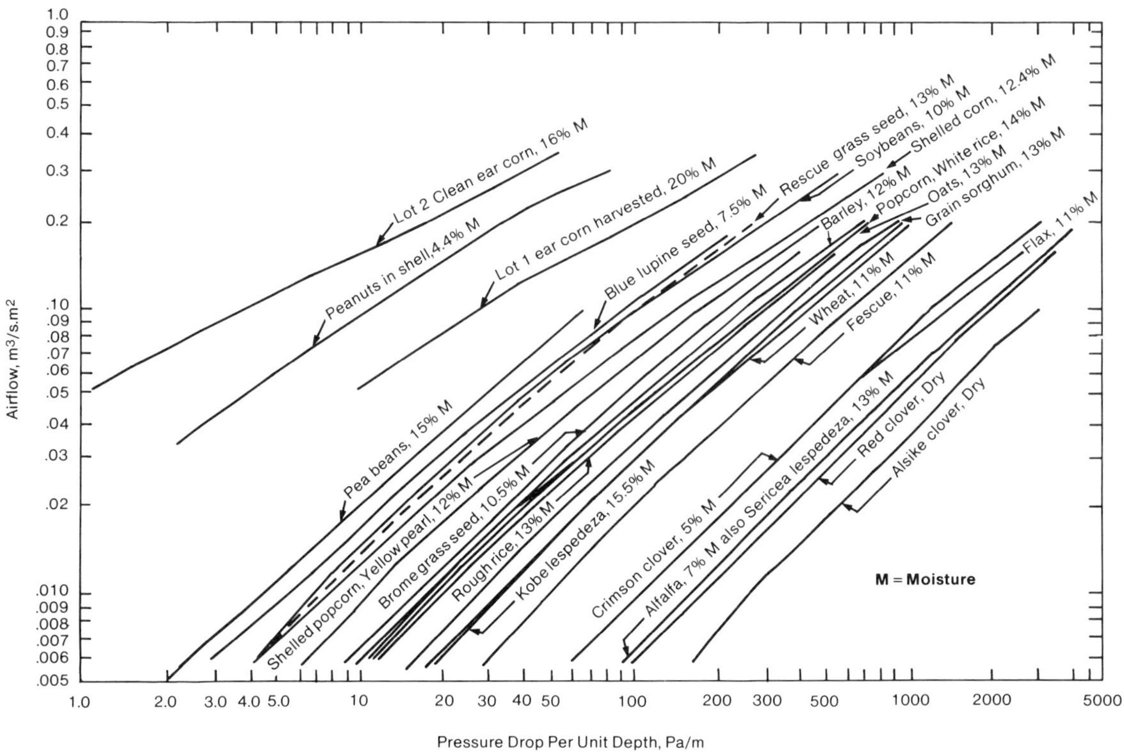


**Fig. 9.** Resistance of grains and seeds to airflow (ASAE 1982).

9. However, direct application of Shedd's data is limited to loose-filled, clean, relatively dry grain and seeds. Several attempts have been made to improve their applicability by investigating the effects of other important variables, such as moisture content, amount of broken and fine materials, and amount of packing. Chung et al. (1984) developed an empirical equation relating grain moisture and fines to static pressure of grain sorghum at flow rates ranging from 0.075 to 0.19 $m^3/sec/m^3$. They concluded that static pressure drop in grain sorghum increased with a decrease in moisture content and increased with an increase in fines and broken material. Chung et al. (1985) investigated the effects of moisture content and foreign material in rough rice on airflow resistance. The model developed, within a airflow rate range of 0.0508 to 0.304 $m^3/sec/m^3$, has the same form as that for grain sorghum developed by Chung et al. (1984). They concluded that increasing moisture content decreased the pressure drop, and increasing the foreign material also decreased the pressure drop.

In selecting a fan for a new system, Figure 10 shows the recommended full-bin airflow rate applicable in the North Central region of the United States. The eastern and southern portions of the region need higher airflows because of higher humidities and somewhat warmer temperatures at harvest. A higher airflow than the minimum in Figure 10 increases drying speed, gives an added

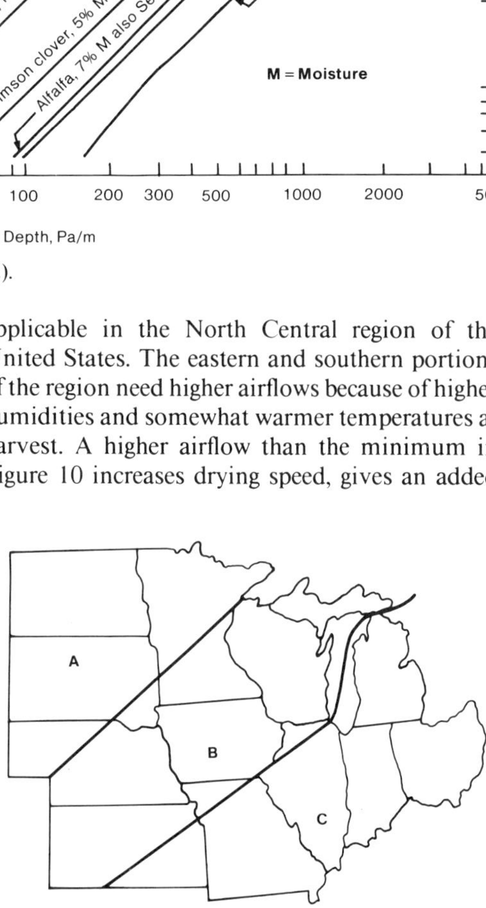


A Provide at least 13.4 x $10^{-3}$ $m^3/sec/m^3$ (1 cfm/bu)
B Provide at least 16.8 x $10^{-3}$ $m^3/sec/m^3$ (1.25 cfm/bu)
C Provide at least 20.2 x $10^{-3}$ $m^3/sec/m^3$ (1.5 cfm/bu)

**Fig. 10.** Recommendations of full-bin airflow rates for fan selection (Anon. 1980).

**Table 7.** Static pressures required for shelled corn[a].

| Grain depth feet | 0.5 | 0.75 | 1.0 | 1.25 | 1.5 | 1.75 | 2.0 | 2.25 | 2.5 | 2.75 | 3.0 |
|---|---|---|---|---|---|---|---|---|---|---|---|
| | Airflow rate cubic feet per minute per bushel | | | | | | | | | | |
| | static pressure, inches of water | | | | | | | | | | |
| 2 | | | | | | | 0.1 | 0.1 | 0.1 | 0.1 | 0.1 |
| 3 | | | 0.1 | 0.1 | 0.1 | 0.1 | 0.1 | 0.1 | 0.2 | 0.2 | 0.2 |
| 4 | | 0.1 | 0.1 | | 0.2 | 0.2 | 0.2 | 0.3 | 0.3 | 0.4 | 0.4 |
| 5 | 0.1 | 0.1 | 0.2 | 0.2 | 0.3 | 0.3 | 0.4 | 0.5 | 0.5 | 0.6 | 0.7 |
| 6 | 0.1 | 0.2 | 0.2 | 0.3 | 0.4 | 0.5 | 0.6 | 0.7 | 0.8 | 0.9 | 1.0 |
| 7 | 0.2 | 0.2 | 0.3 | 0.5 | 0.6 | 0.7 | 0.9 | 1.0 | 1.2 | 1.3 | 1.5 |
| 8 | 0.2 | 0.3 | 0.5 | 0.6 | 0.8 | 1.0 | 1.2 | 1.4 | 1.6 | 1.8 | 2.1 |
| 9 | 0.3 | 0.4 | 0.6 | 0.8 | 1.0 | 1.3 | 1.6 | 1.8 | 2.1 | 2.4 | 2.8 |
| 10 | 0.3 | 0.5 | 0.8 | 1.0 | 1.3 | 1.7 | 2.0 | 2.4 | 2.8 | 3.2 | 3.6 |
| 11 | 0.4 | 0.7 | 1.0 | 1.3 | 1.7 | 2.1 | 2.5 | 3.0 | 3.5 | 4.0 | 4.6 |
| 12 | 0.5 | 0.8 | 1.2 | 1.6 | 2.1 | 2.6 | 3.1 | 3.7 | 4.3 | 5.0 | 5.7 |
| 13 | 0.6 | 1.0 | 1.4 | 1.9 | 2.5 | 3.1 | 3.8 | 4.5 | 5.3 | 6.1 | 7.0 |
| 14 | 0.7 | 1.2 | 1.7 | 2.3 | 3.0 | 3.7 | 4.6 | 5.4 | 6.4 | 7.4 | 8.4 |
| 15 | 0.8 | 1.4 | 2.0 | 2.7 | 3.5 | 4.4 | 5.4 | 6.5 | 7.6 | 8.8 | |
| 16 | 0.9 | 1.6 | 2.3 | 3.2 | 4.2 | 5.2 | 6.4 | 7.6 | 8.9 | | |
| 17 | 1.1 | 1.8 | 2.7 | 3.7 | 4.8 | 6.1 | 7.4 | 8.9 | | | |
| 18 | 1.2 | 2.1 | 3.1 | 4.3 | 5.6 | 7.0 | 8.5 | | | | |
| 19 | 1.4 | 2.4 | 3.5 | 4.9 | 6.4 | 8.0 | 9.8 | | | | |
| 20 | 1.5 | 2.7 | 4.0 | 5.5 | 7.2 | 9.1 | | | | | |

Axial fans: above solid line. Centrifugal fans: below dashed line.

Source: Anon. 1980.
[a] Avoid static pressure above 5.0 inches of $H_2O$.
Conversion factors: 1 inch = 25.4 mm; 1 foot = 0.3048 m; 1 cfm/bu = 13.4 × $10^{-3}$ $m^3/sec/m^3$.

margin of safety, and reduces the need to finish drying in spring.

Table 7 shows the static pressure required for shelled corn for various airflows and grain depths. This table can be used as a guide to determine the maximum practical airflows. A system for less than 11.4 cm $H_2O$ static pressure is preferable to reduce drying cost (Anon. 1980). Recommendations for designing an in-store drying system were given as follows (Anon. 1980); (1) bin is equipped with full perforated floor; (2) a grain distributor is installed; (3) a minimum of airflow rate at 1 cfm/bu (0.0134 $m^3/sec/m^3$) is provided; (4) roof opening is of 1 sq ft/1000 cfm (0.055 $m^2$/1000 $m^3$/hr); (5) static pressure is designed at 11.4 cm $H_2O$ or less; and (6) bin depth is limited to 20 feet (6.1 m), 14 feet (4.3 m) is preferred.

## Managing an In-store Dryer

To handle harvest moisture conditions for a given year, three filling strategies have been suggested: (1) Single filling, when the bin is filled in 1 to 3 days; (2) layer filling, when the grain is added in layers over a period of time; and (3) controlled filling, when layer filling is managed in a specific way (Anon. 1980).

| Zone | Full-bin airflow $m^3/sec/m^3$ x $10^{-3}$ | 9-1 | 9-15 | 10-1 | 10-15 | 11-1 | 11-15 | 12-1 |
|---|---|---|---|---|---|---|---|---|
| | | Harvest date | | | | | | |
| | | Initial moisture content, percent | | | | | | |
| A | 13.4 | 18 | 19.5 | 21 | 22 | 24 | 20 | 18 |
| | 40.3 | 22 | 22.5 | 24 | 25.5 | 27 | 22 | 18 |
| B | 13.4 | 19 | 20 | 20 | 21 | 23 | 20 | 18 |
| | 40.3 | 21 | 22.5 | 23.5 | 24.5 | 26 | 22 | 18 |
| C | 13.4 | 19 | 19.5 | 20 | 21 | 22 | 20 | 18 |
| | 40.3 | 21 | 22 | 23.5 | 24.5 | 25.5 | 22 | 18 |
| D | 13.4 | 19 | 19.5 | 20 | 21 | 22 | 20 | 18 |
| | 40.3 | 20.5 | 21.5 | 23 | 24 | 25 | 22 | 18 |

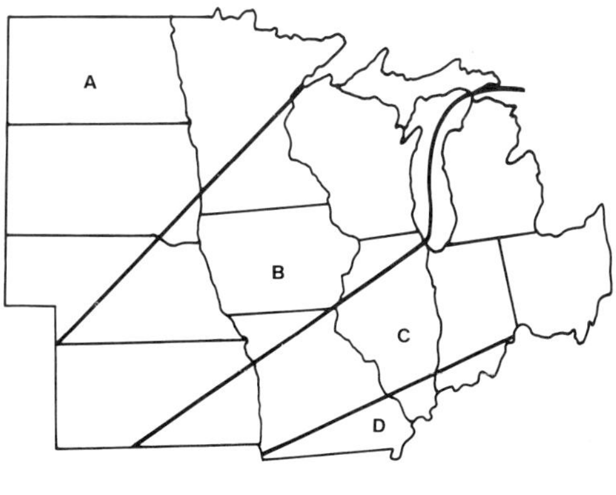


**Fig. 11.** Maximum maize moisture contents for single-fill drying. Developed by T.L. Thompson. (Anon. 1980).

For single filling systems, harvest moisture content is the limiting management factor. It is more suitable for varieties known for early field dry down. Figure 11 gives the maximum initial moisture content for single filling systems, depending on harvest date and airflow rate.

Layer fillings usually involve the first grain, which is also the wettest. Limiting grain depth to get a higher airflow rate is one way to handle this higher moisture grain, which in this case is corn.

Controlled filling is similar to layer filling except the drying front does not come through the top before another layer is added.

Sorenson and Crane (1960) recommended the following operating schedule for unheated air drying in Texas: (1) start the fan as soon as the air distribution system is completely covered; (2) operate the fan continuously as the bin is filled and until the moisture content of rough rice at 0.3 m (1 ft) below the top surface is reduced to 15%; (3) operate the fan then only when the relative humidity of ambient air is 75% or less; and (4) continue drying until the moisture content of rough rice 0.3 m (1 ft) below the top surface is reduced to 12.5%.

Regardless of filling strategy, common management practices for in-store drying systems can also be described as follows: (1) the best drying procedure is to operate fans continuously until the drying front has moved through the top of the grain. The fan is not shut off in the fall during periods of rainy weather or high relative humidity. The fan can be shut off when the drying front has moved through the top of the grain or when all the grain has been cooled to about −3.9°C; (2) the crop can be held over winter with above normal storage moisture content by cooling it with aeration and with weekly observation. Check any changes in moisture content or increase in grain temperature. Run the drying fan a total of six to eight hours about every two weeks when the outdoor temperature is between −6.7 and −1.1°C. Cover the fan after it has been shut off to avoid flowing air currents; (3) drying can normally be safely completed in the spring if needed. About a month is required, so that allowable storage time should be checked for spring conditions (Hellevang 1983).

A study made by Morey et al. (1979) leads to the conclusions that, based on the management procedures studied, continuous fan operation under appropriate fall shut-off and spring start-up procedure is preferable for management strategies.

### Supplemental Heat

Addition of supplemental heat to the air reduces grain moisture content in the lower layers of the bin, but does not reduce moisture content in the top of the bin. By controlling the addition of supplemental heat, the desired final moisture content may be more easily achieved and rewetting during an extended period of damp, cold weather, can also be prevented.

If average moisture content is to be lowered by the end of fall, increasing airflow is preferable to addition of supplemental heat, since additional airflow has more advantages in terms of reliability and drying capacity.

For an unstirred, low temperature dryer, a solar collector can provide adequate supplemental heat because: (1) the longer drying period provides an extended period for collecting solar energy, (2) night and short periods of cloudy weather seldom cause problems because grain stores energy; and (3) higher airflow requirements along with low temperature rises mean that collectors can be used (Anon. 1980). Calderwood (1979) made a study of rice drying with solar heat by collecting the heat in the daytime, storing it in a suitable medium such as a rock-bed or solar pond, then using it at night or at other times when relative humidity is high. It was concluded from this study that time for drying and hours for fan operation were lower with solar heat than with unheated air, and milling yields of rice samples dried with either solar heated air or unheated air were equivalent.

### Simulation Studies

A simulation study of natural air drying of rough rice has been conducted at Kansas State University by Chang et al. (1978). A series of simulated drying tests using official weather data for 15 years from Beaumont, Texas were taken and fan models were developed to determine minimum airflow rate and maximum bed depth of rough rice drying by natural air.

During the actual experiments conducted, a fan was operated continuously until the moisture content of the top layer of grain reached 15%. It was then operated only when relative humidity was 65% and lower. Since only Steele's data for corn storage were available as far as the dry matter loss equation is concerned, the success or failure of a given simulation was determined using two criteria: (1) allowable storage time is based on a criterion of 0.5% dry matter loss (Steele et al. 1969); and (2) in unheated air and supplemental heat drying application under Texas conditions, the moisture in the wettest layer of rice was to be reduced below 15% in 15 days or less to prevent loss in grade from discoloured kernel (Sorenson and Crane 1960). Using the fan model developed, economic airflow rate and bed depths can be recommended. For initial moisture contents 24, 22, 20, and 18% wet basis, minimum airflow rates are 0.067, 0.040, 0.027, and 0.013 $m^3/sec/m^3$, respectively, and maximum bed depths are 0.91, 1.52, 2.13, and 2.44 m. These results were similar to the USDA recommendations, as described by Sorenson and Crane (1960). From this study it is obvious that well defined and acceptable grain deterioration criteria for each kind of grain should be determined to ensure success of future simulation studies.

In a preliminary study of rough rice drying in a tropical country, Chang et al. (1978) simulated grain drying for two harvesting seasons using Costa Rica weather data (Fig. 12). In this simulation, airflow rate was 0.040 $m^3/sec/m^3$, bed depth was 1.524 m, initial moisture content was 27.0%, and initial air temperature was 26.67°C. Rough rice was dried from 20 May–28 June 1976. Final dry matter loss of top layer was 0.437%. This preliminary finding may show the possible application of rough rice drying by natural air under tropical weather conditions.

A mathematical model has also been used by Chang et al. (1979) in developing a simple dryer selection model for on-farm drying facilities to optimise costs for the drying systems.

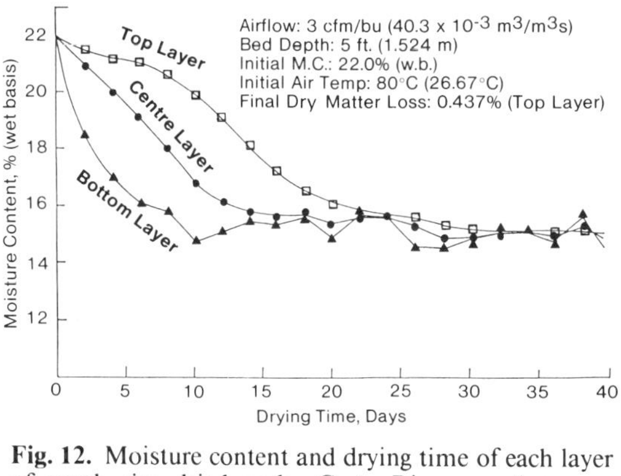


**Fig. 12.** Moisture content and drying time of each layer of rough rice dried under Costa Rican weather conditions, starting on 20 May 1978 (Chang et al. 1978).

## Acknowledgment

The authors express appreciation to the Food and Feed Grain Institute, Kansas State University for partial support through USAID/DSAN-CA-0256.

## References

Anon. 1980. Low temperature and solar grain drying handbook. Ames, Iowa, Iowa State University, Midwest Plan Service 22, 86p.

ASAE (American Society of Agricultural Engineers) 1982. Agricultural engineers handbook. St Joseph, Michigan, ASAE.

Bakker-Arkema, F.W., Brook, R.C., and Levew, L.E. 1978. Cereal grain drying. In: Pomeranz, Y., ed., Advances in cereal science and technology. St Paul, Minnesota, American Association of Cereal Chemists, 1–90.

Bunn, J.M., Lea, W.R., Jr., and Hammond, G. 1981. Potential for ambient air drying in South Carolina. Transactions American Society of Agricultural Engineers, 24, 227–232.

Calderwood, D.L. 1975. Rough rice (paddy) drying methods in the United States. Tropical Stored Products Information, 30, 13–22.

Calderwood, D.L. 1979. Rice drying with solar heat. Transactions American Society of Agricultural Engineers, 22, 1435–1438.

Calderwood, D.L., Cogburn, R.R., Webb, B.D., and Marchetti, M.A. 1983. Aeration of rough rice in long-term storage. Presented at winter meeting ASAE. December 1983, Chicago.

Chang, D.I. Chung, D.S., and Pfost, H.B. 1978. Simulation of natural air drying of rough rice. Kansas State University, Food and Grain Institute, AID/ta-C-1162.

Chang, D.I. Chung, D.S., and Hodges, T.O. 1979. Grain dryer selection model. Presented at winter meeting ASAE. December 1979, New Orleans.

Chung, D.S., Abdelmoshin, G.A., and Kim, M.S. 1984. Resistance of grain sorghum to airflow. Presented at winter meeting ASAE. December 1984, New Orleans.

Chung, D.S., Gonzaga, L., Kim, M.S., and Hague, E. 1985. Resistance to airflow through rough rice mixed with foreign material, in press.

Foster, G.H. 1953. Minimum airflow requirements for drying grain with unheated air. Agricultural Engineering, 34, 681–684.

Foster, G.H. 1976. Drying and aeration of grain and feed ingredients. In: Feed manufacturing technology. American Feed Manufacturers Association, 471–478.

Foster, G.H. 1982. Drying cereal grains. In: Christensen, C.N. ed., Storage of cereal grains and their products (3rd ed.). St Paul, Minnesota, American Association of Cereal Chemists, 78–115.

Foster, G.H., and Stahl, B.N. 1959. Operating grain aeration systems in the corn belt. Washington, D.C., USDA Marketing Research Report 337.

Foster, G.H., and Tuite, J. 1982. Aeration and stored

grain management. In: Christensen, C.N., ed., Storage of cereal grains and their products (3rd. ed.). St Paul, Minnesota, American Association of Cereal Chemists, 117–143.

Frus, J.D. 1979. Aeration of stored grain. Ames, Iowa, Iowa State University, Cooperative Extension Service, 7 p.

Halderson, J.L, and Sandvol, L.E. 1980. Aeration of grain storage. Moscow, Idaho, University of Idaho, Cooperative Extension Service Agricultural Experiment Station, 6 p.

Hall, C.W. 1980. Drying and storage of agricultural crops. Westport, Connecticut, AVI Publishing Company, 381 p.

Hellevang, K.J. 1983. Natural air/low temperature crop drying. North Dakota State University, Cooperative Extension Service, 29 p.

Henderson, S.M. 1955. Deep-bed rice drier performance. Agricultural Engineering, 36, 817–820.

Henderson, S.M. 1958. Deep-bed grain drying on the ranch with unheated air. California Agricultural Experiment Station, Leaflet 103.

Holman, L.E. 1960. Aeration of grain in commercial storage. Washington, D.C., USDA Marketing Research Report 178.

Kline, G.L., and Converse, H.H. 1961. Operating grain-aeration systems in the hard winter wheat area. Washington, D.C., USDA Marketing Research Report 480.

McKenzie, B.A., Fossen, L.V., and Stockdale, H.J. 1980. Managing dry grain in storage. In: Proceedings 1980 grain conditioning conference, Urbana-Champaign, Illinois.

Morey, R.V., Cloud, H.A., Gustafson, R.J., and Peterson, D.W. 1979. Management of ambient air drying systems. Transactions American Society of Agricultural Engineers, 22, 1418–1425.

Morrison, S. 1954. Drying rice with unheated air. Agricultural Engineering, 35, 735–736.

Saul, R.A., and Lind, E.F. 1958. Maximum time for safe drying of grain with unheated air. Transactions American Society of Agricultural Engineers, 1, 29–33.

Schroeder, H.W., and Calderwood, D.L. 1972. Rough rice storage. In: Houston, D.H., ed., Rice: chemistry and technology. St Paul, Minnesota, American Association of Cereal Chemists, 166–187.

Shedd, C.K. 1953. Resistance of grains and seeds to airflow. Agricultural Engineering, 34, 616–619.

Sorenson, J.W. Jr., and Crane, L.E. 1960. Drying rough rice in storage. Texas Agricultural Experiment Station, Bulletin B-952, 19 p.

Steele, J.F., Saul, R.L., and Hukill, W.V. 1969. Deterioration rate of shelled corn as measured by carbon dioxide production. Transactions American Society of Agricultural Engineers, 12, 685–689.

Thompson, T.L. 1972. Temporary storage of high moisture shelled corn using continuous aeration. Transactions American Society of Agricultural Engineers, 16, 333–337.

USDA (United States Department of Agriculture) 1965. American Farmers Bulletin 2214. Washigton, D.C., USDA.

# Case Studies on Aeration and In-store Drying
# Session Chairman's Summary

**Dante B. de Padua***

THE papers in this session described the small and large grain drying and storage systems existing in the Asian region; their success as well as their problems in operation.

Mr Loo Kau Fa described the different types of facilities found in the LPN complexes in Malaysia. They consist basically of continuous flow bulk handling, cleaning and drying facilities, and bulk storage of either the horizontal bin or vertical silo type. The operational problems described include poor performance of cleaners and scalpers with extremely wet and dirty grain, yellowing of grain in vertical silos, and insect infestation. Control measures undertaken include turning of grain from one silo to another, aeration, and chemical fumigation. An airflow rate of 0.1 $m^3$ per tonne of grain per minute, in a system where air is introduced at the bottom of silo, was found to be the most cost effective in controlling yellowing of the grain.

The paper of Mr Calverley, presented by Mr Boxall, dwelt on the concept of an in-store drying facility installed with BULOG in Indonesia and the results obtained after a long struggle. While the results do indicate feasibility of in-store drying, TDRI admits that such a system would have serious limitations and difficulties in solving BULOG's problems.

Mr Justin Tumambing provided the background of the drying problem in the Philippines, the state of the art of existing technologies, some basic grain moisture equilibrium studies, and initial results of pilot plant trials in two-stage drying. It was noted that the EMC of some varieties tested had a higher ERH at 14% m.c. than published values of other varieties.

Dr Somchart Soponronrarit reviewed the post-harvest practices in Thailand, and discussed an integrated drying and storage system using solar huts at the farm level.

* International Development Research Centre, P.O. Box 101, Tanglin, Singapore.

Dr Hong-Sik Cheigh discussed a system for small farm drying and storage of paddy in South Korea, which was designed to match the capacity according to cultivation, and used natural ventilation or mechanical aeration.

The system with a capacity of 2-4 tons of paddy has found wide acceptance in Korea, with over 10 000 units installed.

Mr Sutherland discussed the case studies involving 13 trials of aerating grains in Australia. Seven of these trials were on natural aeration in large stores and two on aeration of high moisture content grain. The trials indicated that aeration is beneficial for controlling grain insect populations, eliminating temperature gradients, reducing moisture migration, slowing down pesticide degradation, and preserving grain quality. The cost of aeration system was also found comparable with the costs of insecticide treatment. Aeration was recommended as a worthwhile method of reducing grain storage problems even in the humid tropics. Under humid tropical conditions, the use of a dehumidifier or an open cycle solar-regenerated desiccant-bed grain cooling system was recommended to reduce the enthalpy of ambient air.

Mr Bramall described circumstances leading to the development of a drying system for rice in Australia. The shift from bag to bulk handling prompted developmental works on drying at high velocity airflows in deep-bed storage. This method is continuously being refined with emphasis on aeration, energy conservation, grain quality maintenance, electronic monitoring and control, pre-heating of air, and automation using mini-computer.

Dr Chung discussed grain aeration and in-store drying in the USA. In-store drying is generally accepted and practiced at the farm level in the USA. He cited various studies on aeration and in-store drying. He recommended management methods under various operating conditions for grain aeration and in-store drying. Also, the limitations and application of the method were discussed. He further noted that a research project on aeration of rough rice under humid air conditions is being conducted at Kansas State University.

For the humid tropics, the consensus was to evolve a two-stage drying system, where the high moisture harvest is rapidly dried down to more manageable levels. The final drying may be done in storage, but it was recognised that the latter required a greater degree of supervision compared with the high temperature, high airflow, continuous-flow dryers favoured in the US.

There was a high degree of cost consciousness associated with the different options, and that cost/benefit analysis should be an integral part of the decision-making process.

# Concluding Session

# Summary of Seminar

B. R. Champ*

In my summary of the seminar proceedings, I do not propose to reiterate what our chairmen have so capably said. Rather, I would like to highlight one or two especially relevant issues that I feel have emerged.

It is evident to me from listening to the various papers that the seminar has been opportune in coinciding with times of major change in approaches to preserving grain quality by physical methods, such as aeration and in-store drying.

Very importantly, the *systems approach* to aeration and drying is now being fully recognised. The technologies are not being considered in isolation as in the past when facilities were developed and their performance dictated quality and the handling and storage systems that could be used. As our understanding of the basic technology has increased, we are now more able to purpose design our aeration and drying capacity to meet the needs of the industry and to interpose it into a total system that ensures that the quality of the grain as produced is preserved through to final consumption.

This leads of course to the intrusion of *economics* into the drying and aeration fields. Engineers and biologists have come to realise that they also can no longer work in isolation, and they are now working side by side with economists. The benefits are immense. The technology so developed is viable and justifiable, and hence has a better chance of introduction into the industry to the benefit of all. I trust we all agree that the cooperative approach between technologists and economists is now mandatory in all our activities.

A problem that has to be faced increasingly today is that resources for loss reduction and related research activities are limited — and by limited, I mean that the total resources available are less than those in aggregate that are required to fully address the problem. Economic assessments can be made of proposed interventions at different points where losses of commodity occur in the chain from production to final disposal. These assessments will provide an objective comparison of the potential benefits to be derived, so that priorities can be established to maximise the return from the resources available.

A critical element in all these developments has been the advances in *computer technology*. The fast, high-capacity units that are now readily available in compact form at affordable prices have revolutionised the whole area of aeration and drying technology and the integral economic considerations. Simulation modelling can now reach its full potential in taking into account all the parameters that are relevant, whether it be in determining the location and size of dryers on a regional basis, or the detail of movement of moisture through grain during drying. It is no longer necessary to make assumptions and all variables can be taken into account as, for example, came out in the discussions on the different theories of drying where equilibrium need not be assumed.

The other exciting development is of course in process control whereby manual operations can be replaced by microprocessor control with its attendant precision. The level of competence in supervision of an operation is thus that of the designer of the software not of an unskilled field worker.

Let me now move to some specific points. We have heard considerable discussion on theoretical aspects of aeration and drying. But we have also heard reference to problems in the quality of the

*ACIAR, G.P.O. Box 1571, Canberra, A.C.T. 2601, Australia.

grain that is to be preserved, and how quickly quality is lost after treatment.

The inclusion of trash with grain presented for drying is a major constraint to effective implementation of the technology. Obviously, pricing and procurement policies must be revised to help overcome this problem. Similarly, the point has been made repeatedly that the present pricing arrangements encourage farmers *not* to dry grain, and this is happening in that critical period immediately after harvest when the grain is very prone to quality deterioration. We must accept that proper drying commences from the moment of harvesting or even before then, and that this must be taken into account in design of the procurement systems.

In this context, the point was made that there is need for models to be developed to quantitatively define quality. These are needed to realise the full potential of optimisation models used in assessment of the various processes integrated into postharvest handling of grain. I believe the need for these has been recognised and that appropriate descriptors are on the way.

Reference was also made to the influence of varietal characteristics of grain on loss of moisture during drying. We therefore need to approach plant breeders to investigate the possibility of breeding new rice varieties, varieties, for example, with different water activity characteristics that permit easier drying in the humid tropics.

The problems of using aeration and in-store drying in the humid tropics are soluble. Attention has been, and will continue to be, given to providing an adequate research base for the technology, and developing from this, handling and drying systems that are compatible with the requirements of the region. It seems most probable that use of aeration and in-store drying will happen and will expand in the region. We must ensure that the benefits that will accrue from their introduction are not delayed and are maximised by ensuring that the difficult problems of transposing research to commercial practice are addressed and overcome expeditiously.

On behalf of ACIAR and GASGA, I would like to express our appreciation of the efforts of our local colleagues in LPN, MARDI, and AFHB whose hard work, carried out in a thoroughly professional manner, has made this seminar possible and successful.

I would also like to express the thanks of the organisers to the speakers, the chairpersons, and rapporteurs and to all participants for invaluable contributions.

# Closing Remarks

J.P. Mercader*

I am honoured to deliver these closing remarks for the seminar on Preserving Grain Quality by Aeration and In-store Drying. Dr Fredericks, the Director of the Bureau, tenders his apologies for not being able to be here in view of urgent official commitments.

In closing this seminar, I am obliged to look back at the discussions over the past three days. To a very large extent, I believe that the seminar has achieved its objectives, namely reviewing the current state of the art of in-store drying and aeration, assessing in both technical and economic terms the relevance of these technologies to the handling of wet paddy in the humid tropics. The 24 papers presented and the discussions they stimulated were sufficient to attain these objectives and bring about a greater awareness of the problems and opportunities in overcoming problems related to drying and aeration.

In retrospect, the seminar has affirmed some critical points and insights on the subject. These are:

(1) That drying is an integral part of the postharvest system and is interrelated with other processes in the system. In attempting to solve problems related to drying and aeration, there appears a need to consider and integrate them with other aspects such as harvesting and handling, storage, milling, and marketing and distribution.

(2) The seminar also confirmed to a large extent that the engineering aspects of drying technology have reached a stage of near-perfection and the problems related to drying are peripheral to the technology, but are nevertheless critical factors for consideration if losses in drying are to be minimised.
These are problems which hinge on paddy quality standards and price structures, policies related to marketing, and the provision of incentives for viable participation by the private sector in the postharvest handling chain. In this regard, the seminar directly and indirectly calls for policy review and redirection.

(3) The seminar is also a vivid demonstration of interest and support in solving the problems related to drying of grains under humid tropical conditions. I note with gratitude that studies on drying under humid tropical conditions are being conducted even in temperate countries. These efforts are comforting and should be encouraged. This is the very nucleus for meaningful development cooperation. As these efforts appear to be directed at solving problems in the humid tropics, they would be more meaningful if they go beyond the technological sphere of the problem and consider ancillary factors that influence the adoption of the technology. These factors appear to be the greater bulk of the problem, at least in this region.

It is encouraging to note that the seminar was actively participated in by delegates from notable research, education and development institutions from all over the world and most especially from the private sector.

In closing, I would like to think that the seminar has stirred greater interest for further research works, development efforts, and review and reformulation of policies related to maintaining grains quality. Thank you.

* ASEAN Food Handling Bureau, 8th Floor, Syed Kechik Foundation Building, Bangsar, Kuala Lumpur.

# Participants

**Australia**

**Mr T. Adamczak,** Engineer, School of Food Science and Technology, University of New South Wales, P.O. Box 1, Kensington, N.S.W. 2033

**Mr A. Blewitt,** Centre Secretary, ACIAR, G.P.O. Box 1571, Canberra, A.C.T. 2601

**Dr R.G. Bowrey,** Divisional Research and Development Manager, Allied Mills Ltd, P.O. Box 57, Mascot, N.S.W. 2020

**Mr L. Bramall,** Technical Services Manager, Ricegrowers' Co-operative Mills Ltd, P.O. Box 561, Leeton, N.S.W. 2705

**Dr B.R. Champ,** Research Program Coordinator, ACIAR, G.P.O. Box 1571, Canberra, A.C.T. 2601

**Mr James Darby,** Storage Research Officer, Ricegrowers' Co-operative Mills Ltd, P.O. Box 561, Leeton, N.S.W. 2705

**Dr R.H. Driscoll,** Researcher, School of Food Science and Technology, University of N.S.W., P.O. Box 1, Kensington, N.S.W. 2033

**Mr Jan van S. Graver,** Experimental Scientist, CSIRO Division of Entomology, Stored Grain Research Laboratory, G.P.O. Box 1700, Canberra, A.C.T. 2601

**Mr E. Highley,** Scientific Editor, ACIAR Grain Storage Research Program, G.P.O. Box 1571, Canberra, A.C.T. 2601.

**Dr A. Hunter,** Principal Research Scientist, Agricultural Engineering Group, CSIRO Division of Chemical and Wood Technology, P.O. Box 26, Highett, Vic. 3190

**Dr Tuan Van Nguyen,** Research Scientist, Agricultural Engineering Group, CSIRO Division of Chemical and Wood Technology, P.O. Box 26, Highett, Vic. 3190

**Mr Geoffrey Pym,** Development Manager (Technical), Ricegrowers' Co-operative Mills Ltd, P.O. Box 561, Leeton, N.S.W. 2705

**Dr J. Remenyi,** Research Program Co-ordinator, ACIAR, G.P.O. Box 1571, Canberra, A.C.T. 2601

**Mr B. Eric Ripp,** Consultant, Co-operative Bulk Handling Limited (W.A.), 22 Delhi Street, West Perth, W.A.

**Dr G.J. Ryland,** Chief Economist, South Australian Department of Agriculture, C/- SAGRIC International, G.P.O. Box 1671, Adelaide, S.A. 5001

**Mr J. Sutherland,** Engineering Scientist, Agricultural Engineering Group, CSIRO Division of Chemical and Wood Technology, P.O. Box 26, Highett, Vic. 3190

**Dr G.R. Thorpe,** Principal Research Scientist, Agricultural Engineering Group, CSIRO Division of Chemical and Wood Technology, P.O. Box 26, Highett, Vic. 3190

**Bangladesh**

**Dr George Srzednicki,** Technical Adviser, Bangladesh-German Seed Development Project, P.O. Box 790, Dhaka

**Burma**

**U Win Myint Thein,** Project Director, Agriculture and Farm Produce Trade Corporation, 70 Pansodan Street, Rangoon

**U Aung Than,** Project Engineer, Agriculture and Farm Produce Trade Corporation, 70 Pansodan Street, Rangoon

**England**

**Mr R. Boxall,** Entomologist, Storage Department, Tropical Development and Research Institute, London Road, Slough, Berkshire, SL3 7HL

**Mr D.J.B. Calverley,** Officer-in-Charge, Storage Department, Tropical Development and Research Institute, London Road, Slough, Berkshire, SL3 7HL

**Mr H.M. Cutler,** Agricultural Engineer, Post Harvest Technology, Econ Group Ltd, Boroughbridge Road, Ripon, N. Yorks, HG4 IUE

**Indonesia**

**Mr Rafael Pranata,** Storage Entomologist, BIOTROP, P.O. Box 17, Bogor

**Dr Mulyo Sidik,** BULOG, Jalan Gatot Subroto 49, Jakarta

**Mr Chrisman Silitonga,** Head, Research and Development, BULOG, Jalan Gatot Subroto 49, Jakarta

**Mrs Sintha Sistyanto,** BULOG, Jalan Gatot Subroto 49, Jakarta

**Mr M. Soemangat,** Department of Agricultural Engineering, Faculty of Agricultural Technology, Gadjah-Mada University, Bulaksumur, Yogyakarta

**Mr Suhargo,** Department of Agricultural Engineering, Faculty of Agricultural Technology, Gadjah Mada University, Bulaksumur, Yogyakarta

**Mr Slamet Zubaidy,** Research on Agricultural Engineering, BULOG, Jalan Gatot Subroto 49, Jakarta

**Italy**

**Mr Fernando Robayo,** Agricultural Services Division, FAO, Via Delle Terme Di Caracalla, Rome 00100

**Kenya**

**Thitai G.N.W.,** National Council for Science and Technology, P.O. Box 30623, Nairobi

**Korea**

**Dr Hong-Sik Cheigh,** Research on Food Chemistry and Processing, Pusan National University, Jangjun-Dong, Dongrae-Ku, Pusan

**Malaysia**

**Y.B.En. Daud bin Dato'Taha,** Deputy Minister of Public Enterprises

**Mr Abas Bin Abd. Wahab,** Lecturer, Fak. Kej. Jentera, University Technology Malaysia, Jalan Gurney, Kuala Lumpur

**Mr Idris Abas,** Mechanical Engineer, LPN, Tingkat 17, Bangunan Perkim, Jalan Ipoh, Kuala Lumpur

**Mr Abdullah Ali,** Senior Research Officer, MARDI, Jalan Kuala Kedah Peti Surat 105, Alor Setar, Kedah

**Associate Professor Mohd Afifi bin Abd. Mukti,** Lecturer, Fak. Kej. Jentera, University Technology Malaysia, Jalan Gurney, Kuala Lumpur 54100

**Mr Ahmad Hajazi Man,** Senior Research Officer, MARDI, Bukit Raya Pendang, Kedah

**Mr Ahmad Robin Wahab,** Senior Research Officer, MARDI, P.O. Box 12301, Kuala Lumpur 01-02

**Mr Ang Chu Suan,** Lecturer, Department of Electrical Engineering, University of Malaya, Kuala Lumpur

**Mr Arifin Bin Roslan,** Engineer, LPN, Anak Bukit, Alor Setar, Kedah

**Mr Awang Sham B. Amit,** Complex Manager, LPN, Alor Setar, Kedah

**Mr Azam Bin Ahmad,** Senior Engineer, FELCRA, Projek Seberang Perak 549, Jalan Mohkamah, Teluk Intan, Perak

**Mr Aziz Endut,** Complex Manager, LPN, Simpang 4, Kerpan, Kedah

**Mr Azizs Bin Safri,** Engineer, LPN, Tingkat 8, Bang. Pkink, Jln Tengku Maharani Puteri, Kota Bharu, Kelantan

**Mr Azizul Ghazali,** Senior Research Officer, MARDI, P.O. Box 105, Alor Setar, Kedah

**Mr Zainol Abdul Aziz,** Senior Research Officer, MARDI, P.O. Box 12301, Kuala Lumpur 01-02

**Mr Azman Hassan,** Lecturer, Faculty of Resource Economics and Agri-Business, Universiti Pertanian Malaysia, Serdang, Selangor

**Mr Berhanuddin Yusof,** Engineer, LPN, Tingkat 17, Bangunan Perkim, Jalan Ipoh, Kuala Lumpur

**Mr Boonler Somchit,** ASEAN Food Handling Bureau, 8th Floor, Syed Kechik Foundation Building, Bangsar, Kuala Lumpur

**Dr Chew Tek-Ann,** Lecturer, Faculty of Resource Economics and Agri-Business, Universiti Pertanian Malaysia, Serdang, Selangor

**Mr P K Chong,** Divisional Sales Manager, J. Whyte (Malaysia) Sdn Bhd, 9 Jalan 222, Petaling Jaya

**Mr Dhiauddin Mohd Nour/Jantan,** Food Technology Division, MARDI, G.P.O. Box 12301, Kuala Lumpur 01-02

**Ms Faridah Mat Elah,** Assistant Research Officer, MARDI, G.P.O. Box 12301, Kuala Lumpur 01-02

**Ms Fatimah M. Arshad,** Lecturer, Faculty of Resource Economics and Agri-Business, Universiti Pertanian Malaysia, Serdang, Selangor

**Dr L.J. Fredericks,** Director, ASEAN Food Handling Bureau, 8th Floor, Syed Kechik Foundation Building, Bangsar, Kuala Lumpur

**Dr Mohd Ghazali Mohayidin,** Faculty of Resource Economics and Agri-Business, Universiti Pertanian Malaysia, Serdang, Selangor

**Mr Shariful Azman Bin Haji Othman,** State Chief, Selangor/Wilayah Persekutuan/Pahang, C/-LPN, P.O. Box 108, Kuala Lumpur

**Mr Ivan B. Ho,** Managing Director, Hisco (Malaysia) SDN BHD, 1 Lorong SS13/6A, Subang Jaya Industrial Estate, Subang Jaya, Selangor

**Mr Ibni Hajar bin Haji Rukunudin,** Research Officer, Food Processing Station, MARDI, P.O. Box 1, Pendang, Kedah

**Mr Johari Bin Endan,** Faculty of Engineering, Universiti Pertanian Malaysia, Serdang, Selangor

**Mr Kamarozaman Bujang,** Pest Control Supervisor, LPN, Tingkat, Wisma Persekutuan, Jalan Hangtuah, Melaka

**Mr Kok Kang Sang,** Senior Agricultural Officer, Office of Project Director, FELCRA, Seberang Perak, 548 Jalan Mahkamah, Teluk Intan, Perak

**Mr Lee Chin Hui,** LPN, P.O. Box 108, Kuala Lumpur

**Dr Lee Nai Pin,** Wellcome Malaysia Sdn. Bhd., 7 Lorong SS13/3F, Subang Jaya, Selangor

**Mr Lim Boon Ka,** Director, LPN, P.O. Box 108, Kuala Lumpur

**Mr Lim Kean Teik,** Secretary, Kedah and Perlis Rice Millers Association, Jln Kodiang, Kubang Pasu, Kedah

**Mr Lim Ken Huat,** Engineer, Ban Eng Huat Rice Mill, 159 Jln Maarof, 59000 Bangsar, Kuala Lumpur

**Ms Kamariah Bte Long,** Research Officer, MARDI, Peti Surat 25, 36307, Sg. Sumun Hilir, Perak

**Mr Loo Kau Fa,** Director, LPN, 13-17th Floors, Bangunan Perkim, Jalan Ipoh, Kuala Lumpur

**Mr Low Kee Wah,** FELCRA, Lot 145, Jalan Bunus, Peti Surat 12254, Kuala Lumpur

**Mr Modh Bookeri Mohd Ludin,** Research Officer, MARDI, P.O. Box 25, 36307 Sg. Sumun, Perak

**Mr Hamdan Manaf,** Research Officer, MARDI, G.P.O. Box 12301, Kuala Lumpur 01-02

**Ms Faridah Mat Elah,** Assistant Research Officer, Food Technology Division, MARDI, G.P.O. Box 12301, Kuala Lumpur 01-02

**Mr J.P. Mercader,** ASEAN Food Handling Bureau, 8th Floor, Syed Kechik Foundation Building, Bangsar, Kuala Lumpur

**Mr Mohd. Nor Bin Hasan,** Senior Engineer, LPN, Anak Bukit, Alor Star, Kedah

**Mr Nujib Ismail,** Lecturer, Fak. Kej. Jentera, University Technology Malaysia, Jalan Gurney, Kuala Lumpur

**Mr Ong Seng Hock,** Research Officer, MARDI, P.O. Box 12301, Kuala Lumpur 01-02

**Mr Raffi Abdul Aziz,** Technical Assistant, LPN, Anak Bukit, Alor Setar, Kedah

**Mr Abdul Rahim Muda,** Senior Research Officer, MARDI, P.O. Box 12301, Kuala Lumpur 01-02

**Mr M Ridzuan Mat Daud,** Manager, Seed Preparation Centre, KADA, P.O. Box 127, Kota Bharu, Kelantan

**Ms Rohani Md. Yon,** Senior Research Officer, MARDI, Bukit Raya, Peti Surat 1, Pejabat Pospendang, 06707, Pendang

**Dr Roslan Abdul Ghaffer,** Lecturer, Faculty of Resource Economics and Agri-Business, University Pertanian Malaysia, Serdang, Selangor

**Mr Rosli Hassan,** Complex Manager, LPN, Utan Aji, Perlis

**Ms Ruslima Bt. Abdullah,** Assistant Research Officer, Food Technology Division, MARDI, Serdang, Selangor

**Ms Jean Sambhi,** Administrative Assistant to ACIAR, Australian High Commission, P.O. Box 921, Kuala Lumpur 01-02

**Mr Shafie Muda,** Complex Manager, LPN, Jerlun, Kedah

**Mr Shaharuddin Hj. Haron,** Director-General, LPN, P.O. Box 108, Kuala Lumpur

**Ms Shariffah Norin,** Assistant Research Officer, MARDI, Bukit Raya, P.O. Box 1, Pendang Kedah

**Mr Sulong Bin Yusof,** Complex Manager, LPN, Jitra, Kedah

**Mr Syed Hashim Waffa bin Syed Mohd. Ghazi,** Fakulti Kejuruteraan Jentera, University Technology Malaysia, Jalan Gurney, Kuala Lumpur

**Mr Tahir Daud,** Complex Manager, LPN, Bukit Raya, Pendan, Kedah

**Mr Tahir Miri,** Complex Manager, LPN, Sungei Baru, Kedah

**Professor Tan Hong Siang,** Department of Electrical Engineering, University of Malaya, Kuala Lumpur

**Mr Teoh Inn Chek,** Senior Engineer, LPN, Tingkat 17, Bangunan Perkim, Jalan Ipoh, Kuala Lumpur

**Mr Wahid bin Said,** Research Officer, MARDI, Bukit Raya, P.O. Box 1, Pendang, Kedah

**Mr Wang Yung Shyan,** Quality Control Officer, LPN, Tingkat 17, Bangunan Perkim, Jalan Ipoh, Kuala Lumpur

**Mr Ng Yuen Sang,** Chairman, Kedah and Perlis Rice Millers Association, Alor Star, C/- Jln Kodian Kubang Pasu, Kedah

**Dr Mohd Yusof bin Hashim,** Director-General, MARDI, G.P.O. Box 12301, Kuala Lumpur 01-02

**Mr Mohd. Zainal Ismail,** Research Officer, MARDI, G.P.O. Box 12301, Kuala Lumpur 01-02

**Mr Zainal Rashid Suratman,** Complex Manager, LPN, Jitra, Kedah

**Mr Zulkifli Ibrahim,** Complex Manager, LPN, Pering, Kedah

**Nepal**

**Mr Joost Gwinner,** Seed Technologist, German Agency for Technical Cooperation (GTZ), P.O. Box 1457, Kathmandu

**Philippines**

**Dr Silvestre C. Andales,** College of Engineering and Agro-Industrial Technology, University of the Philippines at Los Baños, Laguna

**Mr Leo R. Bascos,** Research Assistant, National Food Authority, Matimyas Building, 101 E. Rodriguez Sr. Avenue, Quezon City

**Dr C.W. Bockhop,** Head, Department of Agricultural Engineering, International Rice Research Institute, P.O. Box 933, Manila

**Mrs Cecilia V. De Dios,** Section Chief, Product Development Section, National Food Authority, Matimyas Building, 101 E. Rodriquez Sr. Avenue, Quezon City

**Mr Alfredo T. Gonzalez,** Pest Control Services Incorporated, 745 Aurora Boulevarde, Cubao, Quezon City

**Edilberto Ramiro Lagunda,** National Post Harvest Institute for Research and Extension, 3rd Floor, Administration Building, Food Terminal Incorporated, Taguig, Metro Manila

**Jose J. Lorenzana,** Project Leader, Isabela Post-Harvest System Project, School of Engineering, Isabela State University, Echague, Isabela

**Ponciano S. Madamba,** College of Engineering and Agro-Industrial Technology, University of the Philippines at Los Baños, Laguna 3720

**Evelyn C. Malaguit,** Research Chemist, National Food Authority, Matimyas Building, 101 E. Rodriquez Sr. Avenue, Quezon City

**Mr Ruben E. Manalabe,** Chief, Facilities Design and Development Division, National Post Harvest Institute for Research and Extension, 3rd Floor, Administration Building, Food Terminal Incorporated, Taguig, Metro Manila

**Edgar Manebog,** National Post Harvest Institute for Research and Extension, 3rd Floor, Administration Building, Food Terminal Incorporated, Taguig, Metro Manila

**Mrs Leonila Parrilla,** C/- M Villanueva, 8 Lourdes Street, Pasay City

**Roque C. De Pedro,** Visayas State College of Agriculture, Baybay, Leyte 7127-A

**Miss Maria Teresa M. Pescador,** BILEG Mannalon Incorporated, Poblacion, Asingan, Pangasinan

**Mr Maximo G. Ramos,** Deputy Executive Director, National Post Harvest Institute for Research and Extension, 3rd Floor, Administration Building, Food Terminal Incorporated, Taguig, Metro Manila

**Rodolfo P. Roa,** Regional Engineer, National Food Authority, Region 10, Cagayan, Deoko City

**Lilia G. Rosete,** National Food Authority, Matimyas Building, 101 E. Rodriquez Sr. Avenue, Quezon City

**Mrs G. Sabio,** Research Assistant, National Post Harvest Institute for Research and Extension, 3rd Floor, Administration Building, Food Terminal Incorporated, Taguig, Metro Manila

**Mr R.L. Semple,** Post-Harvest Entomologist, ASEAN Crops Post-Harvest Programme, C/- National Post Harvest Institute for Research and Extension, 3rd Floor, Administration Building, Food Terminal Incorporated, Taguig, Metro Manila

**Hilarion P. Susarno,** Provincial Manager, National Food Authority, San Francisco Agusan del Sur

**Mr Justin Tumambing,** Project Leader, National Post Harvest Institute for Research and Extension, 3rd Floor, Administration Building, Food Terminal Incorporated, Taguig, Metro Manila

**Mr H.T.L. van Ruiten,** ASEAN Crops Post-Harvest Programme, C/- SEARCA, Los Baños College, Laguna 3720

**Mauricio R. Valdez,** Division Chief, National Food Authority, Matimyas Building, 101 E. Rodriguez Sr. Avenue, Quezon City

**Dr Yong Woon Jeon,** Associate Engineer, International Rice Research Institute, P.O. Box 933, Manila

**Singapore**

**Dr Shuit Hung Ho,** Department of Zoology, National University of Singapore, Lower Kent Ridge Road, Singapore 0511

**Dr Dante de Padua,** International Development Research Centre, P.O. Box 101, Tanglin 9124

**Mr Boon Kok San,** Department of Zoology National University of Singapore, Lower Kent Ridge Road, Singapore 0511

**Sri Lanka**

**Mr K.B. Palipane,** Rice Processing Research and Development Centre, Jayanthi Mawatha, Anuradhapura

**Thailand**

**Dr S. Ilangantileke,** Assistant Professor, Division of Agricultural and Food Engineering, Asian Institute of Technology, P.O. Box 2754, Bangkok 10501

**Mr Kriang-krai Mekvanich,** Head, Farm Mechanization Promotion Centre, Room 105, Farm Mechanization Subdivision, Department of Agricultural Extension, Bangkhen, Bangkok 10900

**Mr Pradit Kongkatong,** General Manager, Mah Boon Krong Rice Mill, 88 Mooz Tivanont Road, Bangkader, Pathumthani

**Dr Maitri Naewbanij,** Researcher, Storage and Processing Section, Department of Agriculture, Bangkhen, Bangkok 10900

**Mr Woatthichai Narkrugsa,** Agricultural Industry Department, Faculty of Agricultural Technology, King Mongkut's Institute of Technology, Lardkrabang Campus, Bangkok 10520

**Dr Somchart Soponronnarit,** Assistant Professor, King Mongkut's Institute of Technology, Thonburi Campus, Bangmod, Rasburana, Bangkok 10140

**Ms Sriwai Singhagajen,** Storage and Processing Section, Department of Agriculture, Bangkhen, Bangkok 10900

**Ms Pismai Srisukphrasert,** Chief, Research Administration, Field Crops Research Institute, Department of Agriculture, Bangkhen, Bangkok 10900

**Mr Peter Thomson,** General Manager, Pacific Seed (Thai) Limited, P.O. Box 15, Phraphuttabat, Saraburi 18120

**United States of America**

**Professor F. Bakker-Arkema,** Department of Agricultural Engineering, Michigan State University, East Lansing, Michigan 48824

**Professor Do Sup Chung,** Department of Agricultural Engineering, Kansas State University, Shellenberger Hall, Manhattan, Kansas 66506

**Mr N. Teter,** Consulting Agricultural Engineer, 6035 Sumner Street, Lincoln, NE 68506

**West Germany**

**Mr. B. Bohnstedt,** Deutsche Gesellschaft fur technische Zusammenarbeit (GTZ), FB171 6236, Eschborn P.O.B. 5180

**Mr R. Harnisch,** Deutsche Gesellschaft fur technische Zusammenarbeit (GTZ), GTZ Projekt fur Nacherntefragen, Pickhuben 4, 2000 Hamburg 11

**Dr W. Mühlbauer,** Principal Investigator, Agricultural Engineering Department, University of Stuttgart-Hohenheim, 7000 Stuttgart 70, Garkenstr. 9, Hohenheim